Springer Tracts in Advanced Robotics 100

Editors

Prof. Bruno Siciliano
Dipartimento di Ingegneria Elettrica
e Tecnologie dell'Informazione
Università degli Studi di Napoli
Federico II
Via Claudio 21, 80125 Napoli
Italy
E-mail: siciliano@unina.it

Prof. Oussama Khatib
Artificial Intelligence Laboratory
Department of Computer Science
Stanford University
Stanford, CA 94305-9010
USA
E-mail: khatib@cs.stanford.edu

More information about this series at http://www.springer.com/series/5208

Henrik I. Christensen · Oussama Khatib
Editors

Robotics Research

The 15th International Symposium ISRR

Springer

Editors
Henrik I. Christensen
Robotics and Intelligent Machines
Georgia Institute of Technology College of Computing
Atlanta, GA
USA

Oussama Khatib
Department of Computer Science
Stanford University
Stanford, CA
USA

ISSN 1610-7438 ISSN 1610-742X (electronic)
Springer Tracts in Advanced Robotics
ISBN 978-3-319-29362-2 ISBN 978-3-319-29363-9 (eBook)
DOI 10.1007/978-3-319-29363-9

Library of Congress Control Number: 2016930274

Printed on acid-free paper

This Springer imprint is published by Springer Nature
The registered company is Springer International Publishing AG Switzerland

Series Foreword

Robotics is undergoing a major transformation in scope and dimension. From a largely dominant industrial focus, robotics is rapidly expanding into human environments and vigorously engaged in its new challenges. Interacting with, assisting, serving, and exploring with humans, the emerging robots will increasingly touch people and their lives.

Beyond its impact on physical robots, the body of knowledge robotics has produced is revealing a much wider range of applications reaching across diverse research areas and scientific disciplines, such as follows: biomechanics, haptics, neurosciences, virtual simulation, animation, surgery, and sensor networks among others. In return, the challenges of the new emerging areas are proving an abundant source of stimulation and insights for the field of robotics. It is indeed at the intersection of disciplines that the most striking advances happen.

The *Springer Tracts in Advanced Robotics* (*STAR*) is devoted to bringing to the research community the latest advances in the robotics field on the basis of their significance and quality. Through a wide and timely dissemination of critical research developments in robotics, our objective with this series is to promote more exchanges and collaborations among the researchers in the community and contribute to further advancements in this rapidly growing field.

As one of robotics pioneering symposia, the International Symposium on Robotics Research (ISRR) has established over the past two decades some of the field's most fundamental and lasting contributions. Since the launching of STAR, ISRR and several other thematic symposia in robotics find an important platform for closer links and extended reach within the robotics community.

This fifteenth edition of "Robotics Research," edited by Henrik Christensen and Oussama Khatib, offers in its thirty-seventh-chapter volume a collection of a broad range of topics in robotics including aerial vehicles, perception and mapping, planning, control, systems, and integration. The content of these contributions provides a wide coverage of the current state of robotics research: the advances and challenges in its theoretical foundation and technology basis, and the developments in its traditional and novel areas of applications.

ISRR culminates with this important reference on the current developments and new directions in the field of robotics, which also marks an important milestone in the history of STAR since the first volume published in 2003: 100 volumes—a collective note of thanks from the series' editors to all volume authors and editors for having contributed to this record!

Naples, Italy
November 2015

Bruno Siciliano
STAR Editor

Preface

The International Symposium on Robotics Research (ISRR) is a series of biennial symposia, which began in 1989, and is sponsored by the International Foundation of Robotics Research (IFRR). ISRR is the longest running series of robotics research meetings and dates back to the very earliest days of robotics as a research discipline. The first meeting was organized by Mike Brady and Richard Paul and took place in Bretton Woods (New Hampshire, USA) in August 1983. In the following years, the ISRR symposia were held successively in Kyoto (Japan) 1984, Gouvieux (France) 1985, Santa Cruz CA (USA) 1987, Tokyo (Japan) 1989, Hidden Valley PA (USA) 1993, Herrsching (Germany) 1995, Shonan Village (Japan) 1997, Snowbird UT (USA) 1999, Lorne (Australia) 2001, Siena (Italy) 2003, San Francisco CA (USA) 2005, Hiroshima (Japan) 2007, and Lucerne (Switzerland) 2009. The ISRR symposia are conceived to bring together in a small group setting researchers from academia, government, and industry to assess and share their views and ideas about the state of the art of robotics, and to discuss promising new avenues for future research.

The Fifteenth International Symposium of Robotics Research was held in Flagstaff, Arizona on December 9–12, 2011. Nearly 80 participants from the major institutions of robotics research around the world joined the meeting. The technical program featured 37 contributions, selected from open submissions and invited contributions by the program committee and the members of IFRR. The program was organized around oral presentation in a single-track format and included for the first time a small number of interactive presentations.

The symposium contributions contained in this volume report on a variety of new robotics research results. The technical program was organized in 10 sessions covering a broad spectrum of robotics research. The session topics included perception, manipulation, grasping, vehicles and design, navigation, control and integration, estimation and SLAM. In addition to the technical sessions, the program included two forums: (i) the *Frontier Forum* was chaired by Prof. Hirochika Inoue (JSPS), with the participation of Robert Ambrose (NASA), Thomas Bongrath (KUKA), Herman Bruynincks (KU Leveun), Steve Cousin (Willow Garage),

Sadao Kawamura (RSJ), Kazuhiro Kosuge (IEEE-RAS), Yoshi Nakamura (Tokyo University), Gill Pratt (DARPA), Chuck Thorpe (OSTP), and Richard Voyles (NSF); (ii) the *Pioneer Forum* was chaired by Ruzena Bajcsy (Berkeley) with the participation of Bob Bolles (SRI), Rodney Brooks (Rethink Robotics), Raja Chatila (LAAS), Paolo Dario (SSSA), Shigeo Hirose (TITECH), John Hollerbach (Utah), Hirochika Inoue (JSPS), and Yoshiaki Shirai (Ritsumei University). These forums brought a global view of the field and generated much discussion on the challenges in robotic research and its future perspective. The technical program was complemented by a rich social program and a unique technical field visit to NASA test site in Arizona with spectacular demonstrations of NASA's robotic platforms.

We are grateful to Robert Ambrose and his team for organizing the exceptional field visit to NASA test site. We would like also to express our special thanks to Marie Johnson for all the efforts she devoted to the management and local organization of the symposium.

The greatest words of thanks go of course to the authors and participants who have all contributed to the success of this symposium by bringing an outstanding program, excellent technical presentations, and stimulating and insightful discussions.

Atlanta, GA, USA Henrik I. Christensen
Stanford, CA, USA Oussama Khatib

ISRR 2011 Committee

Symposium Chairs
Henrik I. Christensen, Georgia Tech
Oussama Khatib, Stanford University

Program Committee
Peter Corke, Queensland Institute of Technology
Makoto Kaneko, Osaka University
Rudiger Dillmann, Karlsruhe Institute of Technology
Raja Chatila, LAAS/CNRS
Hirochika Inoue, AIST
John Hollerbach, Utah University
Roland Siegwart, ETH-Zurich

The International Foundation of Robotics Research

President
Oussama Khatib

Executive Officers
Henrik Christensen
Gerd Hirzinger
Yoshiko Nakamura

Honorary Officers
Ruzena Bajcsy
Robert Bolles
Rod Brooks
Shigeo Hirose
Hirochika Inoue
Takeo Kanade

Dan Koditschek
Bernie Roth
Tomomasa Sato
Yoshiaki Shirai

Members
Fumihito Arai
Herman Bruyninckx
Raja Chatila
Wan-Kyum Chung
Peter Corke
Paulo Dario
Rudiger Dillman
Gregory Hager
John Hollerbach
Masayuki Inaba
Makoto Kaneko
Vijay Kumar
Matthew Mason
Roland Siegwart

Contents

Part I
Aerial Vehicles

Progress on "Pico" Air Vehicles

Robert J. Wood, Benjamin Finio, Michael Karpelson, Kevin Ma, Néstor O. Pérez-Arancibia, Pratheev S. Sreetharan, Hiro Tanaka and John P. Whitney

Abstract As the characteristic size of a flying robot decreases, the challenges for successful flight revert to basic questions of fabrication, actuation, fluid mechanics, stabilization, and power—whereas such questions have in general been answered for larger aircraft. When developing a flying robot on the scale of a common housefly, all hardware must be developed from scratch as there is nothing "off-the-shelf" which can be used for mechanisms, sensors, or computation that would satisfy the extreme mass and power limitations. This technology void also applies to techniques available for fabrication and assembly of the aeromechanical components: the scale and complexity of the mechanical features requires new ways to design and prototype at scales between macro and MEMS, but with rich topologies and material choices one would expect in designing human-scale vehicles. With these challenges in mind, we present progress in the essential technologies for insect-scale robots, or "pico" air vehicles.

1 Introduction

Over the past two-plus decades there have been multiple research projects aimed at the development of a flapping-wing robotic insect. These include a butterfly-like ornithopter from the University of Tokyo [1], the "Micromechanical Flying Insect" project at Berkeley [2, 3], and the Harvard "RoboBee" project [4]. These efforts are motivated by tasks such as hazardous environment exploration, search and rescue, and assisted agriculture—similar to the tasks cited for many autonomous robots regardless of scale or morphology. Using swarms of small, agile, and potentially disposable robots for these applications could have benefits over larger, more complex individual robots with respect to coverage and robustness to robot failure.

R.J. Wood (✉) · B. Finio · M. Karpelson · K. Ma · N.O. Pérez-Arancibia · P.S. Sreetharan · H. Tanaka · J.P. Whitney
Harvard School of Engineering and Applied Sciences, 33 Oxford St, Cambridge, MA 02138, USA
e-mail: rjwood@seas.harvard.edu

H.I. Christensen and O. Khatib (eds.), *Robotics Research*,
Springer Tracts in Advanced Robotics 100, DOI 10.1007/978-3-319-29363-9_1

But the interest in these types of robots goes well beyond the expected tasks; use as tools to study open scientific questions (e.g. fluid mechanics of flapping flight, control strategies for computation and sensor-limited systems) and the necessary technological achievements that must accompany these projects are the real motivations.

Work in unmanned aerial vehicles has a rich history that spans from scientific inquiry to congressional policy.[1] In 1997, the United States Defense Advanced Research Projects Agency (DARPA) announced its “Micro Air Vehicle” program which defined an MAV as being 15 cm or less in largest linear dimension, have a range of 10 km, peak velocities over 13 m/s, and operate for more than 20 min.[2] Performers in this program developed multiple successful MAV prototypes including the Black Widow and Microbat [6] as well as some of the first examples of piezoelectrically actuated MAVs [2, 7]. In 2005, DARPA again pushed the limits of aerial robotics by announcing the “Nano Air Vehicle” program[3] which had the requirements of 10 g or less vehicles with 7.5 cm maximum dimension, able to fly 1 km or more. Results include the 16 cm, 19 g “Nano Hummingbird” from Aerovironment,[4] the maple seed-inspired Lockheed “Samarai”,[5] and a coaxial helicopter from a Draper Labs led team.[6] There are also a number of recent commercially-available flapping-wing toy ornithopters and RC helicopters on the scale of micro air vehicles such as the Silverlit ‘iBird’ and the Wowwee Flytech toys.[7]

Using these trends, we define a “pico” air vehicle as having a maximum takeoff mass of 500 milligrams or less and maximum dimension of 5 cm or less. This is in the range of most flying insects [8], and thus for pico air vehicles we look primarily to insects for inspiration. An example prototype pico air vehicle, a prototype from the Harvard RoboBee project[8] is shown in Fig. 1.

Regardless of the classification, the challenges of creating effective flying robots span many disciplines. For example, fluid mechanics changes as a function of characteristic length and velocity: micro air vehicles on the scale of large birds (Re > 10,000) exist in a regime of turbulent flow and steady lift to drag ratios greater than 10 [8]. Nano air vehicles may exist in the transition region (1000 < Re < 10,000) and thus the impact of boundary layer separation (and potential reattachment) becomes particularly relevant. Whereas for pico air vehicles

---

[1]Section 220 of the National Defense Authorization Act for Fiscal Year 2001 states that, “It shall be the goal of the Armed Forces to achieve the fielding of unmanned, remotely controlled technology such that… by 2010, one-third of the aircraft in the operational deep strike force aircraft fleet are unmanned” [5].

[2]http://www.defense.gov/releases/release.aspx?releaseid=1538.

[3]DARPA BAA-06-06.

[4]http://www.avinc.com/nano/.

[5]http://www.atl.lmco.com/papers/1448.pdf.

[6]http://www.draper.com/Documents/explorations_summer2010.pdf.

[7]http://www.wowwee.com/en/products/toys/flight/flytech.

[8]http://robobees.seas.harvard.edu.

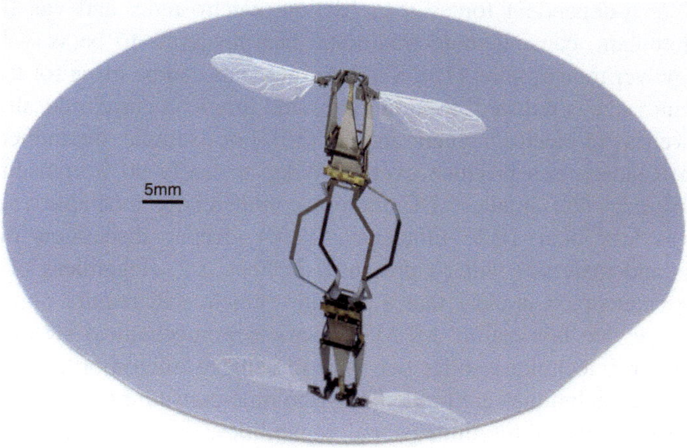

Fig. 1 Example of a recent prototype of a "RoboBee". These two-wing, 100 mg robots are capable of controlled thrust and body moments

(Re < 3000), the flow is almost entirely laminar and thus so-called *unsteady* mechanisms can be employed to enhance lift beyond what would be achievable from constant velocity alone. Nonetheless, it appears that the energetic cost for flight—when considering a metric similar to cost of transport—increases with decreasing characteristic length. Where we could expect a larger-scale aircraft (tens of meters in characteristic dimension) to stay aloft for many hours or even days, flight times for micro, nano, and pico air vehicles are expected to be on the order of an hour, a few tens of minutes, and less than ten minutes respectively [9].

Similar scaling trends also exist for device manufacturing. It is useful to make a distinction between *feature size* and *characteristic size* as pertaining to a vehicle: the former refers to the smallest dimension of the mechanical components of the system—the pitch of gear teeth, thickness of a constituent material, and length of a flexure are examples—while the latter is more descriptive of the overall scale of the vehicle and can refer to the wingspan, chord length, or some similar quantity. We make the argument that as the characteristic size of a vehicle is reduced, feasible approaches to fabrication and assembly of the various propulsion and control mechanisms makes a distinct transition between more standard approaches using "off-the-shelf" components and machining and assembly tools to requiring entirely novel methods. This is one of the fundamental challenges for creating a pico air vehicle. As the feature sizes of the mechanical components are decreased below around 10–100 μm, the designer can no longer rely on standard macro-scale machining techniques. Even high resolution CNC mills,[9] with positioning accuracy down to one micron, require end mills that are rare or non-existent below 50–100 μm. Furthermore, the physics of scaling dictates that as the feature size is

[9]For example, Microlution 5100: http://microlution-inc.com/products/5100.php.

decreased, area-dependent forces (e.g. friction, electrostatic, and van der Waals) become dominant, causing more traditional bearing joints to be very lossy with respect to power transmission [10]. Similar arguments can be made for transducers. As the feature size is reduced, friction losses and limits on current density decrease the effectiveness of electromagnetic motors [10]. For example, the induction motor in a Tesla Roadster can produce over 7 kW/kg at nearly 90 % transduction efficiency[10] whereas the smallest DC motors commercially available can produce 39 W/kg at less than 18 % efficiency.[11,12] A deeper discussion of actuator geometries and materials will be presented in Sect. 2.2. Regardless of the transduction mechanism, it is clear that a pico air vehicle will require non-traditional solutions to device fabrication. MEMS (microelectromechanical systems) surface micro machining techniques offer one path to achieve micron-order feature sizes. However, these techniques are hindered by the time-consuming serial nature of the process steps, limited three dimensional capabilities, and the high cost of prototyping using MEMS foundries. A solution for fabrication and assembly of a pico air vehicle will be described in Sect. 2.4 and examples of both ends of the fabrication spectrum are shown in Fig. 2.

Challenges for control are also scale-dependent. Larger-scale vehicles can take advantage of passive stability mechanisms (e.g. positive wing dihedral) and generally have larger mass and power capacity for various sensors and computer architectures. An insect-scale device will have significantly reduced payload capacity as compared to a micro or even nano air vehicle. Therefore, the control challenges for a pico air vehicle are currently centered around flight stabilization using limited sensing and computation capabilities. This is in contrast to "higher-level" control problems of autonomous navigation [12] and coordination of multiple unmanned air vehicles [13].

Beyond aeromechanics, actuation, fabrication, and control, there are numerous additional issues including power, system-level design, integration, and mass production. Thus the challenges for a pico air vehicle are daunting, yet form a set of exciting and well-posed engineering questions and scientific opportunities. The remainder of this article will discuss recent progress in a number of these areas.

2 Overview of a Pico Air Vehicle

This article will focus on some of the key components of a flapping-wing pico air vehicle, as shown in Fig. 3, based on the design of the Harvard RoboBee. These components make up the majority of the power and mass budget for the pico air

---

[10]http://www.teslamotors.com/roadster/specs.

[11]SBL02-06H1 from Namiki: http://www.namiki.net/product/dcmotor/pdf/sbl02-06ssd04_01_e.pdf.

[12]Note that this does not include drive circuitry, which is also exacerbated at small scales.

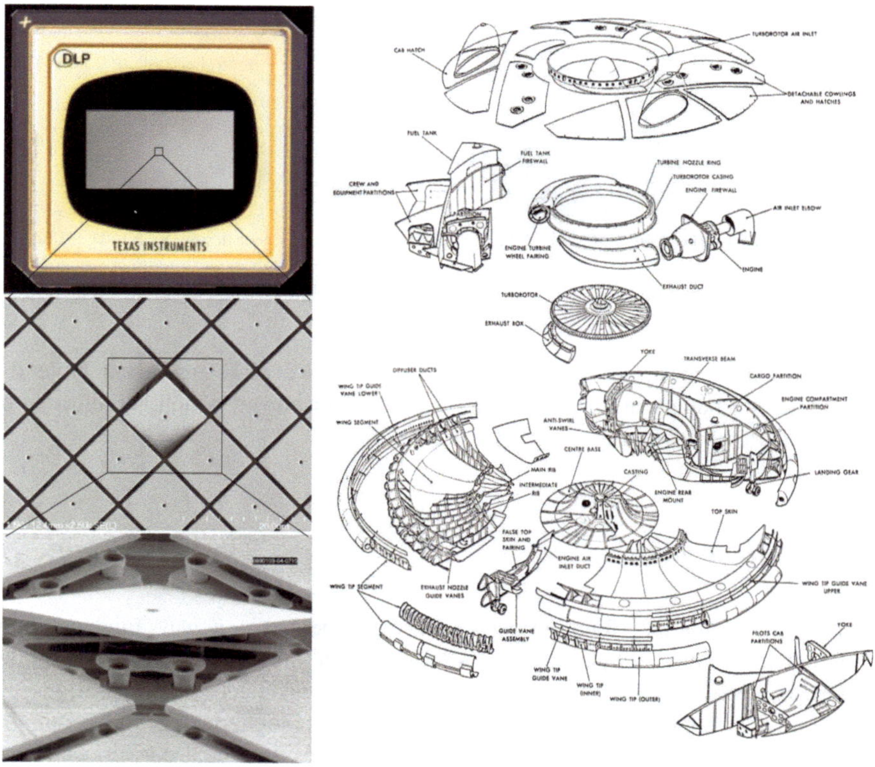

Fig. 2 At two ends of the fabrication and assembly spectrum: MEMS surface micromachined mirrors from a Texas Instruments DLP display (*left*, images courtesy of Jack Grimmet and Martin Izzard, Texas Instruments) and a "nuts-and-bolts" approach to assembly of a complex macro-scale device: an experimental "human-scale" hover-capable aircraft, the "Avrocar" [11] (*right*)

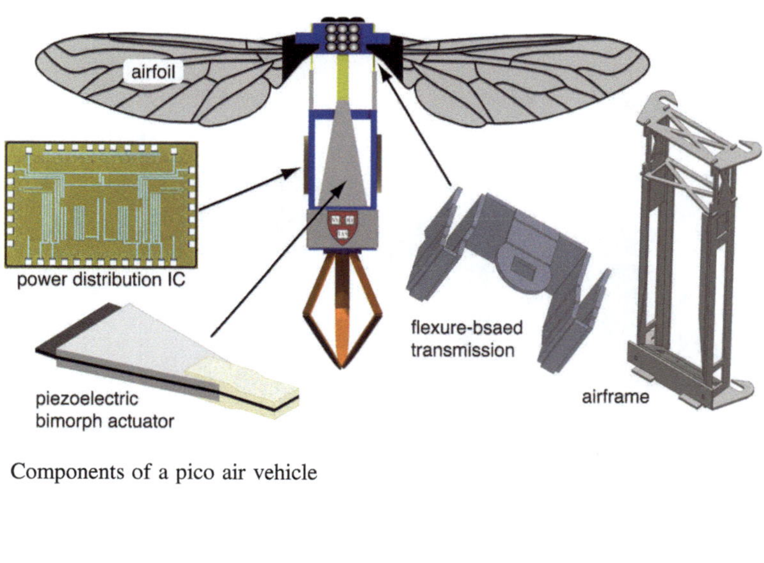

Fig. 3 Components of a pico air vehicle

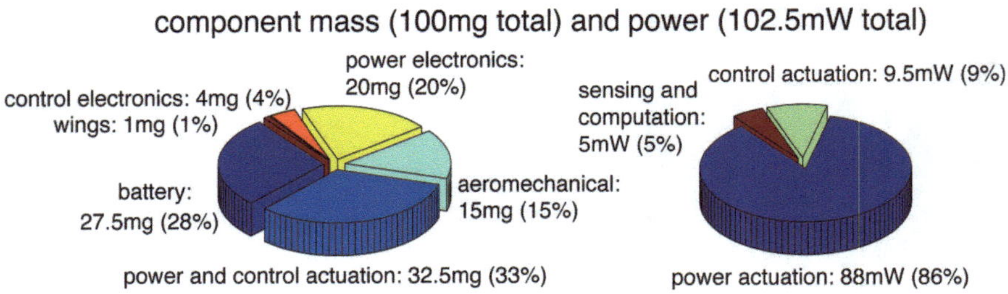

Fig. 4 Mass (*left*) and power (*right*) budgets for a 100 mg robot, derived using the method in [9]

vehicle, which is shown in Fig. 4 for a hypothetical 100 mg robot. Note the dominance of battery and actuator mass and actuator power, which is indicative of the energetic cost of flight at these scales.

2.1 Aeromechanics

Due to the scaling of fluid properties, insects operate in a way fundamentally different from conventional aircraft. Although there are many, sometimes subtle, differences between the flight apparatuses of individual species, in general, insects have one or two pairs of wings, driven in multiple rotational degrees of freedom by flight musculature, and powered by metabolic processes which convert chemical energy for flight. For a flapping-wing pico air vehicle, we derive some design principles from Dipteran (two-winged) insects. We assume that each wing has two rotational degrees of freedom (DOFs): flapping and rotation about an axis approximately parallel to the leading edge (i.e. pronation and supination). During flapping, the up-stroke and downstroke are assumed to be nominally symmetric with no stroke plane deviation. The wing motion is illustrated in Fig. 5. Thinking

Fig. 5 Illustration of one-half cycle of wing motion (i.e. the down-stroke) for a Diptera assuming negligible stroke plane deviation. *Top row* lateral perspective. *Bottom row* dorsal perspective

about the propulsion mechanism as a lumped-parameter 2nd order system, the dominant components are the inertia of the wing, potential energy storage in the muscles, plates, and joints of the thorax, and the damping due to interaction between the wing and the air. As with Diptera and other insects which use *indirect* flight muscles, we assume that the wing drive for a flapping-wing pico air vehicle will also operate at resonance to amplify the wing stroke [8].

In order to control motion in these two DOFs, the actuators are coupled to the wings using a flexure-based articulated transmission mechanism (see Fig. 3). Previous designs utilized a spherical five-bar mechanism to map two linear actuator inputs to the two wing DOFs [14]. Kinematically, this allows direct control over the phasing of wing rotation and any asymmetries in between the upstroke and down-stroke. However, dynamic coupling between the two degrees of freedom creates challenges for controlled trajectories at the flapping resonant frequency. Instead, an under-actuated version of the transmission includes a passive flexure hinge at the wing base such that flapping is commanded by a single power actuator and rotation is passive [15]. Tuning the dynamics of the system at design time places the rotational resonance well above the flapping resonance such that we can assume quasi-static wing rotation while driving the actuator at the first flapping resonant frequency. There is evidence that wing rotation in some insects is driven by inertial and aerodynamic forces, as opposed to directly activated by thoracic musculature [16–18].

The presence of unsteady flow features arising from wing-wake and wing-wing interactions, aeroelastic coupling between fluid pressure and wing bending [19, 20], and the significance of vortex formation and shedding [21] result in challenges for a succinct description of the relationship between wing properties (geometric, inertial, and elastic), wing motions and deformation, and resulting flow and force generation. To study the aeromechanics of flapping-wing flight, researchers have employed a variety of methods including dynamic scaling [21], computational fluid dynamics (CFD) methods [22], and direct biological observation [23]. Each of these has led to significant insights into the details of flow structure, performance of many flying insects, and the function of various morphological features. A combination of these methods, the *blade-element* method [24], merges quasi-steady flow analysis with empirically-fit force and torque coefficients which hide all the unsteady terms behind these coefficients. This has allowed designers to study a variety of wing morphologies and trajectories. However, in some cases, the aeroelastic interaction between strain energy in the airfoil and fluid pressure may have a non-negligible effect on the dynamics and energetics of the vehicle. In such cases, it is valuable to study the fluid mechanics using either a moving-mesh CFD code or at-scale experiments which do not make any scaling assumptions on wing compliance.

Given the ability to manufacture insect-scale airfoils, such as the *Eristalis* wing in Fig. 6 [25], and actuate with insect-like trajectories and wingbeat frequencies, we have begun multiple experiments which are aimed at a deeper understanding of the fluid mechanics for a pico air vehicle with emphasis on learning design rules to enhance aerodynamic efficiency—and thus overall performance of the robot. Recent experiments include:

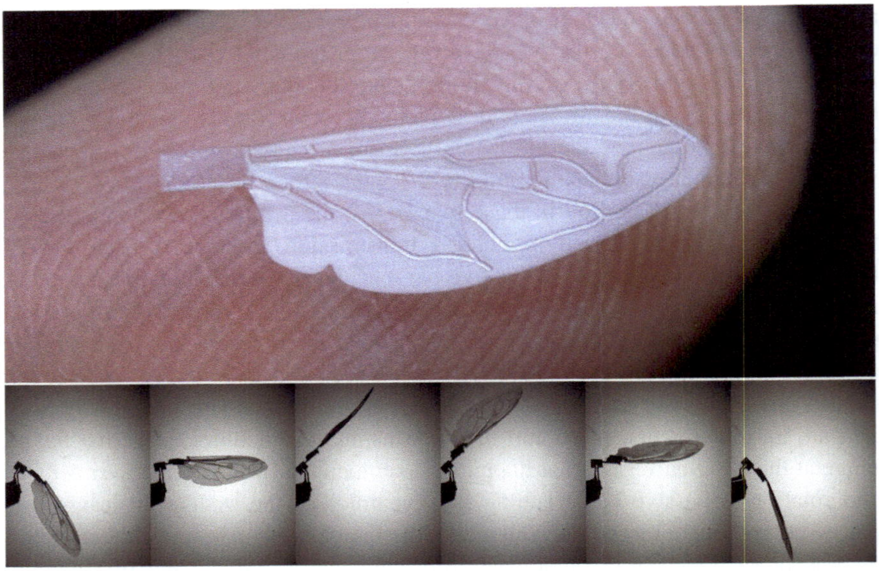

Fig. 6 Photo of a micromolded polymer wing mimicking the features of a *Eristalis* wing (*top*). This wing was created in a single molding step and includes veins ranging from 50 to 125 μm thick, 100 μm corrugation, and a 10–20 μm membrane. A sample of the wing motion (dorsal perspective) at 150 Hz flapping frequency taken from a high speed video (*bottom*)

- Multiple methods to create biomimetic airfoils and verification that the static characteristics are consistent with biological wings [25, 26].
- Established a blade-element based model of under-actuated flapping dynamics (i.e. passive rotation) and validated using a custom-made single flapping-wing, high resolution force sensing [27], and high speed motion reconstruction [24].
- Explored the function of distributed versus localized wing compliance on lift force generation [28].

2.2 Actuation

As previously discussed, the physics of scaling requires us to seek an alternative to electromagnetic actuation for a pico air vehicle. But there are more subtle reasons for this as well. Even if the power densities and efficiencies were comparable, the unloaded RPM of a rotary electromagnetic motor will typically increase with decreasing size, thus requiring substantial gearing to produce useful work and increasing the overall complexity of the transmission system. Furthermore, as we are assuming a reciprocating flapping motion, a rotary motor would require additional transmission components (and rotary bearings) to convert the rotation to wing flapping, again increasing part count and complexity. Instead we look to oscillatory

Table 1 Qualitative comparison of actuation technologies

Type	Example	Efficiency	Toughness	Bandwidth[a]	Max. ε	Max. σ	Density
Bulk piezo.	PZT-5H[b]	◔	◔	●	○	◕	●
Single crystal	PZN-PT[c]	◔	○	●	◔	◑	●
SMA	Nitinol[d]	○	◕	◔	◑	●	●
IPMC	Nafion[e]	◔	◑	◔	●	◔	◑
EAP	DE[f]	◔	◑	◑	●	◔	◔
Electromag.	Brushless[g]	◕	NA	◑	NA	NA	●

● highest, ◕ high, ◑ moderate, ◔ low, ○ lowest
[a]Depends upon structure geometry
[b]From Piezo Systems: http://www.piezo.com
[c]Single crystal piezoelectric ceramics, see [33]
[d]Shape memory alloy: http://www.dynalloy.com
[e]From DuPont, see [34]
[f]Dielectric elastomers, see [31]
[g]For example, 0308 DC micro-motor from Smoovy: http://www.faulhaber-group.com

actuators based on induced-strain materials. Induced-strain materials respond to an applied stimulus with a simple change in geometry. There are multiple options including piezoelectric, electroactive polymers, solid-state phase transitions, electrostriction, and thermal expansion. There have also been many demonstrations of relatively simple geometries for producing linear actuation from electrostatic forces [29], clever piezoelectric linear motors,[13] piezoelectric stacks and "moonie" type actuators [30], and many dielectric elastomer configurations [31]. Each material and actuator morphology can be evaluated based on the standard metrics of blocked force, free displacement (and thus energy), density (and thus energy density), bandwidth (and thus power density), and efficiency. However, the focus is not only on performance, but also practicality. Therefore, additional considerations include fabrication complexity, cost, robustness, the drive method, and linearity of the input–output response and any related control issues. Table 1 qualifies actuation options relative to some of these metrics. A more comprehensive study of actuation choices for a pico air vehicle is presented in [32] with reference to multiple flapping-wing design break points.

Given the needs of a pico air vehicle, we chose clamped-free bending bimorph polycrystalline piezoelectric actuators as a local minimum in complexity while meeting the key specifications for bandwidth, power density, and efficiency. Furthermore, we can rapidly prototype many geometries and obtain all necessary materials commercially. Note that the use of these piezoelectric actuators also

[13]"Squiggle" motors: http://www.newscaletech.com/squiggle_overview.html.

carries some important scaling decisions since we are assuming a resonant primary drive. The resonant frequency will monotonically increase with decreasing size (this trend can be seen clearly in insects [8]). For quasi-static operation of piezoelectric actuators, power density will increase roughly linearly with operating frequency. Thus for smaller devices, this type of actuator is attractive and can out perform insect flight muscle by a factor of two or more [35]. The opposite trend is true as well: it is clear that, for direct-drive transmissions, above a certain size these actuators will not be able to deliver sufficient power due to a fixed (either fracture or breakdown-limited) energy density and reduced operating frequencies. The specific cutoff is highly dependent on the details of the vehicle design and will not be discussed here. Finally, we do not assume that piezoelectric actuation is the best choice for all functions of a pico air vehicle. As discussed in Sect. 2.3, we divide actuation between power delivery and control. The previous discussions have focused on maximizing resonant power delivery in order to generate thrust to maintain flight, however the requirements for a control actuator could be rather different than a power actuator, thus a hybrid solution is a potentially viable option.

2.3 Control

The challenges for control for a pico air vehicle are not in planning and navigation, but rather more fundamental topics of stabilization, sensing, and electromechanical design. Flapping-wing robots similar to the one in Fig. 1 are designed such that the mean lift vector passes through the center of mass and the periodic drag forces are symmetric on the upstroke and downstroke, thus there are nominally zero body torques during flight. However, fabrication errors and external disturbances can easily excite instabilities in the roll, pitch, or yaw angles which need to be actively suppressed. Figure 7 displays a typical behavior in the absence of any controller or constraints on the body degrees of freedom for a flapping-wing pico air vehicle. It is worth noting that the robot in Fig. 7 survived over ten such crashes without any damage, which demonstrates the robustness of the materials and components that constitute the robot.

Our control efforts to date have concentrated on (a) development of the thoracic mechanics to enable modulation of wing trajectories and hence body torques,

Fig. 7 When driven open-loop, the RoboBee prototypes are very unstable in body rotations and crash shortly after takeoff

(b) exploration of appropriate sensor technologies, and (c) methodologies for controller synthesis and related demonstrations. Recent progress in these areas include:

- We have demonstrated the ability to generate lift greater than body mass and perform uncontrolled takeoff experiment such as shown in Fig. 7 [4]. This provides the baseline aeromechanical design and allows us to quantify the thrust the robot can achieve to help bound payload for sensing and power.
- The original designs presented in [4] only had the ability to control thrust and one body torque (i.e. pitch torques). We have demonstrated the ability to generate bilateral asymmetry in stroke amplitude using multiple thoracic mechanics configurations [36, 37]. This involves a morphological separation of power and control actuation similar to the role of the *indirect* and *direct* flight muscles in the thoracic mechanics of Dipteran insects [38].
- Similarly, we have performed experiments with stroke plane deviation as an alternative method for torque generation in [39].
- Beyond modulating the wing trajectory, we have performed torque measurement experiments which verify that there is a one-to-one relationship between dorsoventral mean stroke angle bias and the resulting pitch torque [40].
- Through collaborations with Centeye, Inc,[14] insect-inspired optical flow sensors have been integrated on-board a gliding micro air vehicle [41].
- Work at U.C. Berkeley has prototyped a number of insect-inspired inertial and horizon-detection sensors such as a biomimetic haltere (similar to the Coriolis force sensing structures in Diptera [42]) and photoreceptive ocelli similar to the horizon detection sensors in insects [43].
- Finally, we have implemented an adaptive control scheme to control the mean lift force during flapping [44].

These efforts are primarily focused on the standard feedback control strategy in which a disturbance is detected by a proprioceptive sensor, a computer chooses a compensatory action according to some control law, and the action is then implemented by a system of amplifiers and electromechanical structures. We refer to devices which perform such complex tasks without the intervention of electrical circuits (i.e. analog or digital computers) as examples of *mechanical intelligence*. There are many everyday examples including windshield wipers, whippletrees, and automobile differentials. In these examples, feedback control is performed as a consequence of the mechanical design. For example, automobile differentials automatically distribute equal torques to the wheels regardless of differences in wheel velocities. We have applied this concept to the passive regulation of wing motions by a modified version of the flexure-based transmission called PARITy: "Passive Aeromechanical Regulation of Unbalanced Torques" [45]. The PARITy design equally distributes torques to the wings in response to perturbations, due to either external disturbances or fabrication errors, without the need for sensors or

[14]http://www.centeye.com.

computation. This allows an active controller to operate on a much longer time scale since short time scale perturbations are eliminated, thereby reducing the required sensor bandwidth and computation power.

2.4 Fabrication

The integrated circuit revolution of the 1950s and 1960s now enables the majority of the consumer electronics that are enjoyed daily. As these techniques evolved in the 1980s to include electromechanical components, an even greater space of applications emerged including sensors, optics, and even actuation [46]. Microrobots have been made using MEMS surface and bulk micromachining techniques [47, 48]. However, there are many drawbacks to using integrated circuit (IC) and MEMS technologies to create a pico air vehicle. First is the dramatic difference between the material properties of silicon and insect tissue: the former being rigid and brittle while the latter exhibits a large range of material properties, is generally quite resilient, and is approximately the density of water. Second, although the suite of techniques for high resolution machining is an appealing aspect of MEMS processes, the resulting structures are typically "2.5D", with high aspect ratio components being extremely challenging in terms of machining or requiring hinged structures [49]. Finally, although MEMS foundries exist (e.g. the Multi-User MEMS Process, MUMPS[15] and Sandia's SUMMiT[16]), cost and turn-around time are generally prohibitive to rapid prototyping. With the advent of *mesoscopic* manufacturing methods [50], we have demonstrated key components of the flight apparatus of robotic insects [15, 51] and recently the first demonstration of a 60 mg flapping-wing device which can produce thrust greater than its body weight [4] has proven the feasibility of creating insect-scale flying robots using these techniques.

Mesoscopic manufacturing based on lamination and folding is depicted in Fig. 8. Here a spherical five-bar mechanism is created in three steps. First, the constituent materials—typically thin sheets of polymers, metals, ceramics, or composites—are laser micromachined to the desired planform geometries. These layers are then aligned and laminated using thermoset sheet adhesives and a heated press. Second, the quasi-planar devices are released using a final laser machining step. Lastly, the devices are folded into their final configuration. In the case in Fig. 8, tabs and slots are integrated to assist with alignment during folding, although there are other methods to ensure precision in this final step including fixturing, surface tension, differential thermal expansion, and even embedded actuation [52]. This process enables the development of articulated components with any number of DOFs, layered actuators such as the piezoelectric bending

[15]http://www.memscapinc.com.

[16]http://www.mems.sandia.gov/tech-info/summit-v.html.

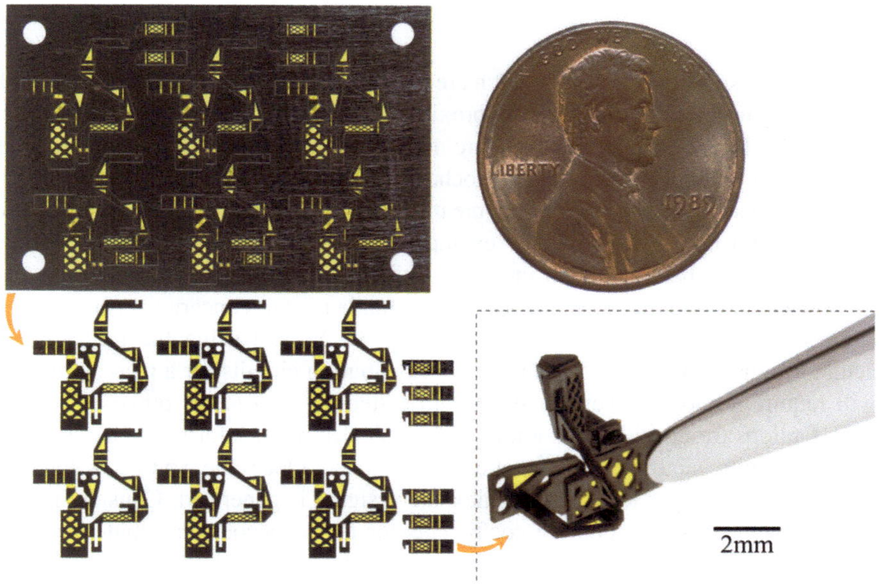

Fig. 8 Example of the process flow for articulated microstructures. In this example, six spherical five-bar linkages are created by a sequential micromachining and lamination process, then folded into the final configuration (*inset*)

actuators described in Sect. 2.2, and integrated electronics, all with feature sizes ranging from micron to centimeter. The concept of folding as an assembly process has been further developed into the a larger space of applications for "Programmable Matter" using robotic origami to produce arbitrary shapes and functional structures [53].

2.5 Power

The power source for a pico air vehicle is the most significant delimiter to flight time [9]. Options for power storage include electrochemical (i.e. batteries and fuel cells [54]), electrostatic (i.e. capacitors and supercapacitors), and mechanical (i.e. elastic strain energy).[17] As with all components, practicality is a fundamental consideration. Existing batteries have poor energy storage (approximately 500 J/g based on existing small-scale lithium batteries from Fullriver[18]) compared to fuels such as gasoline which can be two orders of magnitude greater. But energy density alone is not sufficient to describe the effectiveness of a candidate power source.

[17]Note that this only refers to storage, not transduction or harvesting.

[18]http://www.fullriver.com/.

Conversion efficiency, storage/packaging, and operating conditions should also be considered.

There are sub-gram batteries which are commercially available (see footnote 18). While the lower end of this range (approximately 200 mg) could be acceptable for a pico air vehicle, smaller batteries are feasible, though rare or non-existent as commercial products. Since the electrochemical reactions are scale-independent (at least for the scales considered here), creating smaller batteries becomes an exercise in fabrication and packaging. For example, it is possible to dice and repackage lithium-polymer batteries in an inert atmosphere.

Power distribution efficiency is also a fundamental concern. Assuming the source will have a voltage of approximately 3.7 V, and using the piezoelectric actuator dimensions from [51], the power distribution circuits for a pico air vehicle will require a boost conversion stage with a step-up ratio in the range of 50–100 [55]. Options for boost conversion include piezoelectric transformers, charge pump ladder circuits, and electromagnetic transformers. Once the source voltage is boosted to the proper level, the actuator drive signal is generated. Considering the low electromechanical coupling coefficients for many piezoelectric materials, it is essential to recover remaining charge from one half cycle of the harmonic oscillation of the thorax and use for the next half cycle. Charge recovery circuits for bimorph actuators have been developed [56] and a custom integrated circuit which generates the periodic drive signal and coordinates charge recovery has been created and demonstrated for a flapping-wing robot [57]. Therefore, the power source is the key remaining technology required to bring the pico air vehicle in Fig. 1 to power autonomy.

3 Next Steps

The progress on pico air vehicles reported in this article is the tip of the iceberg. The next steps include:

- **Power source**: Characterization of batteries and other viable power sources (including supercapacitors and micro fuel cells) under appropriate loading conditions.
- **Integration**: The best demonstration for any core technology involves integration onto a flight-worthy robot.
  - *On-board sensors*: Continued collaboration with manufacturers of optical flow sensors (Centeye, Inc.), aiming to demonstrate a flight-worthy sensor and use in altitude control experiments.
  - *On-board power electronics:* Integrating the components from Fig. 9 into the airframe utilizing the layered manufacturing technique described in Sect. 2.4.

Fig. 9 Components (*left*) and a complete tapped-inductor-based 20 mg boost conversion circuit (*right*)

- **Accelerator-based computation**: The RoboBees project is exploring compute architectures which employ highly specialized integrated circuits to perform a single task (such as control or sensor processing) extremely efficiently.
- **System-level design and optimization**: Finally, while much attention has been paid to each component, there has been few efforts for system-level optimization for vehicles of this scale. The work in [9] suggests the most promising areas to focus design efforts and how improvements to the performance of any subsystem will contribute to increased flight time.

Acknowledgements This work was partially supported by the National Science Foundation (award number CCF-0926148), the Army Research Laboratory (award number W911NF-08-2-0004), the Office of Naval Research (award number N00014-08-1-0919-DOD35CAP), and the Air Force Office of Scientific Research (award number FA9550-09-1-0156-DOD35CAP). Any opinions, findings, and conclusions or recommendations expressed in this material are those of the authors and do not necessarily reflect the views of the National Science Foundation.

References

1. H. Tanaka, K. Hoshino, K. Matsumoto, I. Shimoyama, in *IEEE/RSJ International Conference on Intelligent Robots and Systems* (Edmonton, Alberta, Canada, 2005)
2. R. Fearing, K. Chang, M. Dickinson, D. Pick, M. Sitti, J. Yan, in *IEEE International Conference on Robotics and Automation* (2000)
3. R. Fearing, S. Avadhanula, D. Campolo, M. Sitti, J. Yan, R. Wood, in *Neurotechnology for Biomimetic Robots* (The MIT Press, 2002), pp. 469–480
4. R. Wood, IEEE Trans. Robot. **24**(2) (2008)
5. National Defense Authorization Act, Fiscal Year 2001 (2000) Public Law 106-398, 106th Congress

6. M. Keennon, J. Grasmeyer, in *AIAA/ICAS International Air and Space Symposium and Exposition: The Next 100 Years* (Dayton, OH, 2003)
7. A. Cox, D. Monopoli, D. Cveticanin, M. Goldfarb, E. Garcia, J. Intell. Mater. Syst. Struct. **13**, 611–615 (2002)
8. R. Dudley, *The Biomechanics of Insect Flight: Form, Function and Evolution* (Princeton University Press, 1999)
9. M. Karpelson, J. Whitney, G.Y. Wei, R. Wood, in *IEEE/RSJ International Conference on Intelligent Robots and Systems* (Taipei, Taiwan, 2010)
10. W. Trimmer, J. Sens. Actuators **19**, 267 (1989)
11. Avro Canada VZ-9AV Avrocar (2010). http://www.nationalmuseum.af.mil/factsheets/factsheet.asp?id=10856
12. S. Shen, N. Michael, V. Kumar, in *IEEE International Conference on Robotics and Automation* (Shanghai, China, 2011)
13. T. Stirling, D. Floreano, in *Proceedings of the 10th International Symposium on Distributed Autonomous Robotics Systems* (2010)
14. S. Avadhanula, R. Wood, D. Campolo, R. Fearing, in *IEEE International Conference on Robotics and Automation* (Washington, DC, 2002)
15. R. Wood, in *IEEE/RSJ International Conference on Intelligent Robots and Systems* (San Diego, CA, 2007)
16. A. Ennos, J. Exp. Biol. **140**, 161 (1988)
17. A. Ennos, J. Exp. Biol. **140**, 137 (1988)
18. A. Bergou, S. Xu, Z. Wang, J. Fluid Mech. **591**, 321 (2007)
19. S. Combes, T. Daniel, J. Exp. Biol. **206**(17), 2989 (2003)
20. S. Combes, T. Daniel, J. Exp. Biol. **206**(17), 2999 (2003)
21. M. Dickinson, F.O. Lehmann, S. Sane, Science **284**, 1954 (1999)
22. R. Mittal, G. Iaccarino, Ann. Rev. Fluid Mech. **37**, 239 (2005)
23. C. Ellington, C. van der Berg, A. Willmott, A. Thomas, Nature **384**, 626 (1996)
24. J. Whitney, R. Wood, J. Fluid Mech. **660**, 197 (2010)
25. H. Tanaka, R. Wood, J. Micromech. Microeng. **20**(7) (2010)
26. J. Shang, S. Combes, B. Finio, R. Wood, Bioinspir. Biomim. **4**(036002) (2009)
27. R. Wood, K.J. Cho, K. Hoffman, J. Smart Mater. Struct. **18**(125002) (2009)
28. H. Tanaka, J. Whitney, R. Wood, J. Integr. Compar. Biol. **51**(1), 142 (2011)
29. W. Tang, T.C. Nguyen, M. Judy, R. Howe, J. Sens. Actuators A: Phys. **21**(1–3), 328 (1990)
30. R. Newnham, A. Dogan, Q. Xu, K. Onitsuka, J. Tressler, S. Yoshikawa, in *Proceedings of IEEE Ultrasonics Symposium*, vol. 1 (Baltimore, MD, 1993), pp. 509–513
31. R. Pelrine, P. Sommer-Larsen, R. Kornbluh, R. Heydt, G. Kofod, Q. Pei, P. Gravesen, in *Proceedings of International Society for Optical Engineering*, vol. 4329 (2001), pp. 335–349
32. M. Karpelson, G.Y. Wei, R. Wood, in *IEEE International Conference on Robotics and Automation* (Pasadena, CA, 2008)
33. J. Yin, B. Jiang, W. Cao, I.E.E.E. Trans, IEEE Trans. Ultrason. Ferroelectr. Freq. Control **47** (1), 285 (2000)
34. S. Lee, H. Park, K. Kim, Smart Mater. Struct. **14**(6), 1363 (2005)
35. E. Steltz, R. Fearing, in *IEEE/RSJ International Conference on Intelligent Robots and Systems* (San Diego, CA, 2007)
36. B. Finio, J. Shang, R. Wood, in *IEEE International Conference on Robotics and Automation* (Kobe, Japan, 2009), pp. 3449–3456
37. B. Finio, B. Eum, C. Oland, R. Wood, in *IEEE International Conference on Robotics and Automation* (St. Louis, MO, 2009)
38. B. Finio, R. Wood, Bioinspir. Biomim. **5**, 045006 (2010)
39. B. Finio, J. Whitney, R. Wood, in *IEEE/RSJ International Conference on Intelligent Robots and Systems* (Taipei, Taiwan, 2010)
40. B. Finio, K. Galloway, R. Wood, in *IEEE/RSJ International Conference on Intelligent Robots and Systems* (San Francisco, CA, 2011)

41. R. Wood, S. Avadhanula, E. Steltz, M. Seeman, J. Entwistle, A. Bachrach, G. Barrows, S. Sanders, R. Fearing, Robot. Autom. Mag. **14**(2), 82 (2007)
42. G. Nalbach, J. Comp. Physiol. A **173**, 293 (1993)
43. W. Wu, L. Schenato, R. Wood, R. Fearing, in *IEEE International Conference on Robotics and Automation* (Taipei, Taiwan, 2003)
44. N. Pérez-Arancibia, J. Whitney, R. Wood, in *American Controls Conference* (San Francisco, CA, 2011)
45. P. Sreetharan, R. Wood, J. Mech. Des. **132**(5), 051006 (2010)
46. K. Petersen, Proc. IEEE **70**(5), 420 (1982)
47. R. Yeh, E. Kruglick, K. Pister, J. Microelectr. Mech. Syst. **5**(1), 10 (1996)
48. B. Donald, C. Levey, C. McGray, I. Paprotny, D. Rus, J. Microelectr. Mech. Syst. **15**(1), 1 (2006)
49. K. Pister, M. Judy, S. Burgett, R. Fearing, J. Sens. Actuators A: Phys. **33**, 249 (1992)
50. R. Wood, S. Avadhanula, R. Sahai, E. Steltz, R. Fearing, J. Mech. Des. **130**(5) (2008)
51. R. Wood, E. Steltz, R. Fearing, J. Sens. Actuators A: Phys. **119**(2), 476 (2005)
52. J. Paik, E. Hawkes, R. Wood, J. Smart Mater. Struct. **19**(12), 125014 (2010)
53. E. Hawkes, B. An, N. Benbernou, H. Tanaka, S. Kim, E. Demaine, D. Rus, R. Wood, Proc. Natl. Acad. Sci. **107**(28), 12441 (2010)
54. M. Tsuchiya, B. Lai, S. Ramanathan, Nat. Nanotechnol. **6**(282) (2011)
55. E. Steltz, M. Seeman, S. Avadhanula, R. Fearing, in *IEEE/RSJ International Conference on Intelligent Robots and Systems* (Beijing, China, 2006)
56. D. Campolo, M. Sitti, R. Fearing, IEEE Trans. Ultrason. Ferroelectr. Freq. Control **50**(3), 237 (2003)
57. M. Karpelson, R. Wood, G.Y. Wei, in *Symposium on VLSI Circuits* (Kyoto, Japan, 2011)

Aerial Locomotion in Cluttered Environments

Dario Floreano, Jean-Christophe Zufferey, Adam Klaptocz, Jürg Germann and Mirko Kovac

Abstract Many environments where robots are expected to operate are cluttered with objects, walls, debris, and different horizontal and vertical structures. In this chapter, we present four design features that allow small robots to rapidly and safely move in 3 dimensions through cluttered environments: a perceptual system capable of detecting obstacles in the robot's surroundings, including the ground, with minimal computation, mass, and energy requirements; a flexible and protective framework capable of withstanding collisions and even using collisions to learn about the properties of the surroundings when light is not available; a mechanism for temporarily perching to vertical structures in order to monitor the environment or communicate with other robots before taking off again; and a self-deployment mechanism for getting in the air and perform repetitive jumps or glided flight. We conclude the chapter by suggesting future avenues for integration of multiple features within the same robotic platform.

D. Floreano (✉) · J.-C. Zufferey · A. Klaptocz · J. Germann
Laboratory of Intelligent Systems, EPFL Lausanne, Lausanne, Switzerland
e-mail: dario.floreano@epfl.ch

J.-C. Zufferey
e-mail: jean-christophe.zufferey@epfl.ch

A. Klaptocz
e-mail: adam.klaptocz@epfl.ch

J. Germann
e-mail: jurg.germann@epfl.ch

M. Kovac
Wyss Institute, Harvard University, Cambridge, USA
e-mail: mirko.kovac@wyss.harvard.edu

H.I. Christensen and O. Khatib (eds.), *Robotics Research*,
Springer Tracts in Advanced Robotics 100, DOI 10.1007/978-3-319-29363-9_2

1 Introduction

Many environments where robots are expected to move present complex structure, such as walls, furniture, ceilings, trees, bushes, rocks, and so forth, that we generically refer to as clutter. For example, search-for-rescue robots may be deployed in semi-collapsed buildings with debris on the ground or in forests with trees and vegetation; monitoring robots may be asked to explore buildings and houses; and environmental robots may need to disperse within urban environments to collect pollution information.

Several articulated wheeled and legged robots have been developed for locomotion over irregular and cluttered terrain, but these robots tend to be rather slow in heavily cluttered environment and may get stuck or flip over unstable objects. The significantly longer time required to explore and find the required information may compromise the entire mission.

In this chapter, we propose to use small robots capable of moving in 3 dimensions in order to quickly and safely locomote through cluttered environments. Flight is an example of 3D locomotion that would allow robots to rapidly explore cluttered environments as long as there is an aperture sufficiently large to allow them to fly through. Here we show that although cluttered environments present several challenges for robots moving in the air, such as small size and good perception, they also present several opportunities for the robots to learn about their surroundings and pause to communicate, monitor, and save energy.

In the following sections, we present a few design features that allow small robots to rapidly and safely move in 3 dimensions through cluttered environments: (a) a perceptual system capable of detecting obstacles in the robot's surroundings, including the ground, with minimal computation, mass, and energy requirements; (b) a flexible and protective framework capable of withstanding collisions and even using collisions to learn about the properties of the surroundings when light is not available; (c) a mechanism for temporarily perching to vertical structures in order to monitor the environment or communicate with other robots before taking off again; (d) a selfdeployment mechanism for getting in the air and perform repetitive jumps or glided flight.

Each section introduces a specific feature and validates it with experimental results obtained with a custom-made robot. In the closing section, we discuss ways in which these features could be brought together within a single robotic platform in order to obtain an agile and resilient robot for locomotion in cluttered environments.

2 Vision-Based Flight

Flight in cluttered and indoor environments brings enormous constraints in terms of size and energy because the flying platform must be lightweight to be maneuverable and small to pass through doorways or between obstacles such as buildings, posts and trees. Therefore most perceptual systems such as scanning laser range finders typically used by terrestrial robots [1] are too heavy and bulky to fit on small flying robots [2]. An alternative consists in taking inspiration from the visual system of the insect compound eye for its ability to extract visual information from an almost omnidirectional field of view with small computational and energetic requirements.

Approximately two-thirds of the neurons in the insect brain are dedicated to visual information processing [3, 4]. Biologists have unraveled a significant part of their functioning. They discovered for instance that optic flow plays a predominant role in flight control by providing information on distance to surrounding objects [5–7]. Interestingly optic flow can be estimated using relatively low-resolution vision sensors, which translates to small packages and limited processing needs [8].

Based on this consideration, researchers have explored what can be classified as 2D optic-flow-based control strategies. They developed autonomous systems moving on flat surfaces [9–13], or constrained flying robots in the form of tethered helicopters [14] or horizontally flying systems [15–17]. Here instead, we aim at controlling aircraft moving in 3D and relying on roll and pitch movements in order to steer. Airplanes and helicopters in translational flight indeed use rolling and pitching movements to alter their trajectory [18].

Optic flow is the perceived visual motion of surrounding objects as projected onto the retina of an observer [19]. Assuming a mobile observer moving in an otherwise stationary environment, the motion field describing the projection of the object velocities onto its retina depends on its self-motion (amplitude and direction of rotation and translation), the distance to the surrounding objects, and the viewing directions [20]. This intricate combination of effects makes it generally difficult to extract useful information out of optic flow, especially with 3D moving systems. However, in translating aircraft, one can estimate self-rotation using rate gyroscopes, whereas translation can be assumed to be aligned with the longitudinal axis of the plane with an amplitude that can be measured by an onboard airspeed sensor (anemometer or Pitot tube). In these conditions, optic-flow can be derotated using the gyroscopic signals in order to produce an output that is proportional to the proximity of objects in the considered viewing direction [2, 18].

Aiming at simple 3D control strategies that can fit any small translating flying robots with limited processing power, we propose to follow a reactive paradigm where perception is directly linked to action without intermediary cognitive layers [19, 21–23]. Since optic flow can be turned into proximity information as seen previously, the simplest way of achieving reactive behaviours such as obstacle avoidance, ground following or lateral stabilization is to linearly combine a set of derotated optic-flow sensors into rolling and pitching commands [24].

Such a control scheme is implemented on our 10 g microflyer (MC2) to demonstrate fully autonomous flight in an enclosed environment. The robot is equipped with two linear camera extracting optic-flow in 3 viewing directions as shown in Fig. 1. Two rate gyroscopes provide information to derotate the optic-flow estimates in order to map them into proximity signals. A small anemometer is used to regulate the airspeed.

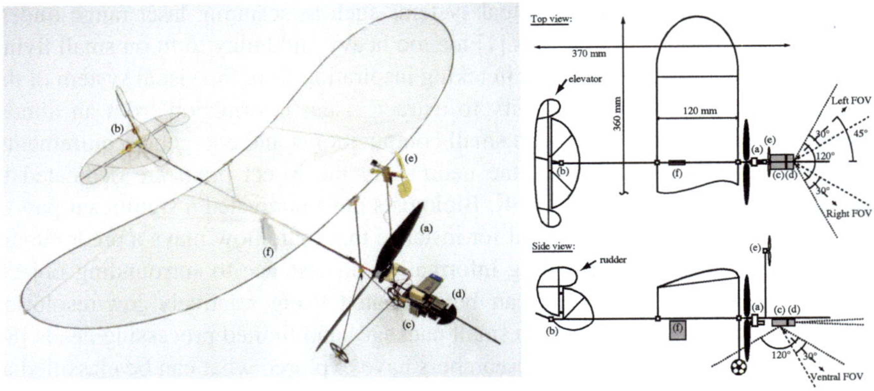

Fig. 1 The 10 g MC2 microflyer. The on-board actuators and electronics consists of **a** a 4 mm geared motor with a lightweight carbon-fiber propeller, **b** two magnet-in-a-coil actuators controlling the rudder and the elevator, **c** a microcontroller board with a bluetooth wireless communication module and a ventral camera with its pitch rate gyro, **d** a front camera with its yaw rate gyro, **e** an anemometer, and **f** a 65 mAh lithium-polymer battery. Reprinted from [25]

Figure 2 shows the coverage of the frontal field of view by the MC2 onboard cameras (left) as well as the mapping of the optic-flow values into command signals (right) for its rudder and elevator, which will directly affect the rolling and pitching rates, respectively. The gains Ω allow to tune how the plane reacts to the proximity of objects perceived in the 3 viewing directions. What is typically desired is that an object perceived in the ventral area pitches up the aircraft, which will make it climb and therefore get away from the perceived object. Similarly, the left and right weights are set so that laterally detected object will roll the aircraft away from them.

In-flight tests are carried out in a randomly textured arena of 7 by 6 m. Once switched on, the microflyer swiftly takes off due to its airspeed controller applying full thrust when reading zero airspeed. Once in flight, the robot will get repelled by the ground under the effect of the ventral optic-flow detector sensing the proximity of it. As soon as a wall is perceived in one of the two lateral viewing directions, the microflyer will roll in the opposite direction. Once the aircraft is tilted, the ventral detector will not be oriented towards the ground anymore, but towards the closeby wall. It will therefore produce a pitching up reaction, which will in turn help the aircraft to steer away from the corresponding wall. As soon as the perceived proximities decrease close to zero, the microflyer will naturally get back to a level and almost straight flight as it is naturally stable by design. A video showing this autonomous behaviour is available at http://lis.epfl.ch/microflyers. More detailed description of the results can be found in [2, 24].

This Braitenberg-inspired control strategy can easily be generalized to more than three viewing directions [18] in order to increase robustness by limiting the regions that are not covered by an optic-flow detector. To demonstrate how this can be done, a larger outdoor flying robot is fitted with up to 7 optic-flow detectors

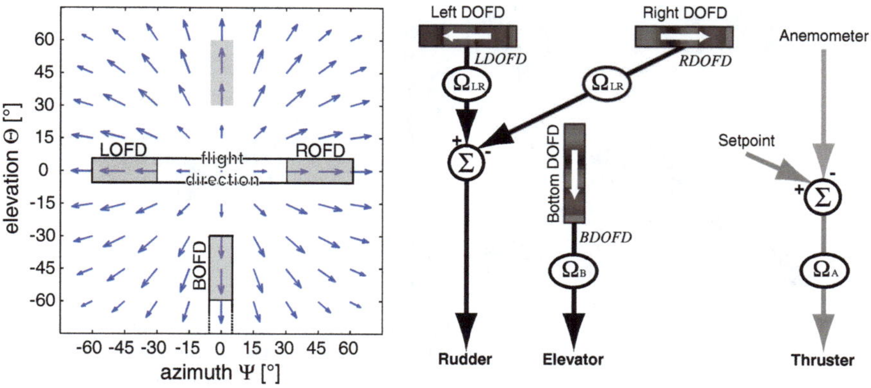

Fig. 2 *Left* An azimuth-elevation graph displaying the typical optic-flow field experienced in the frontal area during straight motion. The zones covered by the cameras mounted on the MC2 are represented by the two thick *rectangles*. The *gray* zones within the thick *rectangles* define the three viewing directions in which optic flow is extracted. The corresponding optic-flow detectors (OFD) are prefixed with L, B, and R for *left*, *bottom* and *right*, respectively. A fourth OFD could have been located in the *top* region, but since this microflyer never flies inverted (due to its passive stability) and gravity attracts it towards the ground there is no need for sensing obstacles in the dorsal region. *Right* The control strategy allowing for autonomous operation of the MC2. The three OFDs are prefixed with D to indicate that they are derotated (as explained in the text). The signals produced by the *left* and *right* DOFDs, i.e. LDOFD and RDOFD, are subtracted to control the rudder, whereas the signal from the *bottom* DOFD, i.e. BDOFD, directly drives the elevator. The anemometer is compared to a given setpoint to output a signal that is used to proportionally drive the engine in order to maintain airspeed reasonably constant. The *Ω ellipses* indicate that three gain factors are used to tune the resulting behaviour. Reprinted from [24]

covering the viewing directions ranging from left to bottom to right at 45° with respect to the flight direction (Fig. 3).

Here again, the basic idea of the underlying control strategy is to use weighted sums of all proximity signals coming from the various viewing directions as commands for pitching and rolling rates (Fig. 4). We name this generalized control strategy "optiPilot".[1] The set of gains (or weight distribution) is chosen in order to achieve repulsion from all obstacles that could be sensed by any of the optic-flow detectors.

Equipped with this set of divergent optic-flow detectors, the robot is capable of taking-off automatically as it get repelled by the ground and laterally stabilized, follow the underlying terrain at a preset height depending on the strengths of the set of gains, reject lateral and longitudinal perturbations, and avoid collisions with obstacles such as trees (Fig. 5), ground or water [18]. In addition, this optiPilot control strategy can be used as a low-level control layer to ensure flight stabilization

[1]Patent # PCT/IB2008/051497 & US 2011/0029161.

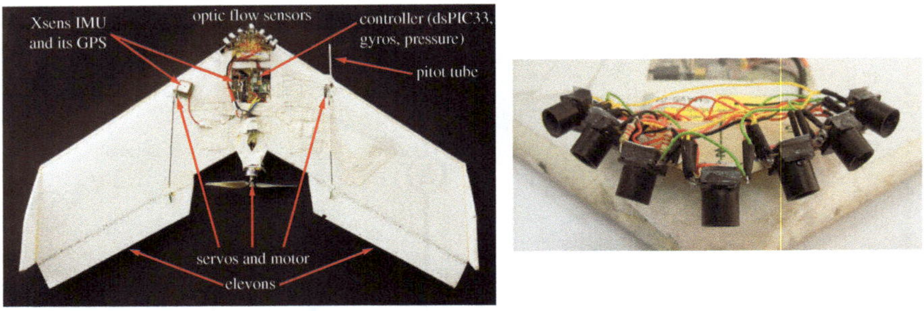

Fig. 3 *Left* The swinglet flying wing used for the experiments. It has a wing span of 80 cm and a total weight of 400 g including about 50 g of additional sensor payload. No particular efforts have been made at this stage to reduce the weight of the sensor. *Right* Visual front-end composed of seven optic computer mouse sensors pointed at 45° eccentricity with respect to the aircraft roll axis. Reprinted from [18]

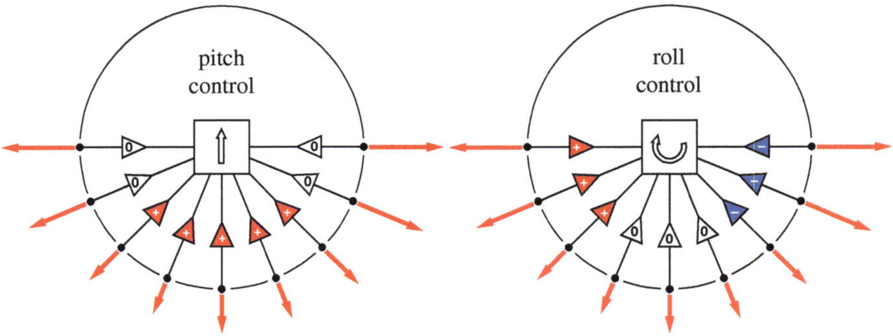

Fig. 4 Mapping translation-induced frontal optic flow (represented by the *arrows* in the periphery) into roll and pitch control signals. The *left* (resp. *right*) diagram represents a possible weight distribution that will make the aircraft pitch (resp. roll) away from any seen objects. The *arrows* in the center indicate pitch (resp. roll) direction for a positive pitch (resp. roll) signal. Reprinted from [18]

Fig. 5 Reactive tree avoidance maneuver. From *left* to *right* swinglet approaching a tree, climbing in reaction to the perceived ventral optic flow generated by the tree, and passing above the tree

and collision avoidance while a GPS-driven higher layer takes care of the trajectory control [26]. Videos of these behaviours are available for download at http://lis.epfl.ch/microflyers.

3 Surviving and Exploiting Collisions

As the environment increases in clutter, space for maneuvering decreases and the risk of collisions increases. Even the best perceptual systems can fail due to lack of light or contrast for optic flow, symmetric or ambiguous information, or the small size of obstacles. Collisions are thus inevitable, but most human-made systems are not designed to withstand them. Exposed blades are especially sensitive and result in catastrophic failures after contact. Though some recent platforms consider protecting propellers and moving parts [27–29], it is often included as an afterthought and cannot protect from large collisions. Some helicopters that can land autonomously and take off again exist [30], though only if they land on their feet. No provisions are made for landing upside-down, or for collisions with obstacles that cause loss of flight control.

Insects however have evolved resilient, lightweight and flexible bodies that allow them to frequently collide with windows or low contrast walls and continue flying. Even when falling upside-down, insects can right themselves using their legs and wings and quickly return to the air [31].

We have taken the protective and flexible bodies of insects as a design principle for a new class of flying robots that can withstand collisions, resume flight, and even fly against obstacles. Flying indoors requires small size to fit through doors and windows, and thus a maximum dimension of 40 cm was chosen for this design. Forward flight is useful for optic flow-based avoidance algorithms but the ability to hover is also required for maneuvering close to obstacles. To best fulfill the size and flight-mode requirements a hybrid airplane-rotorcraft design was selected that features two counter-rotating propellers for ample thrust in hover mode, an elevator and rudder for steering and a wing for forward flight and stability.

Besides the typical aerodynamic considerations that apply to all flying platforms, two additional requirements were included specific to cluttered environments:

- **Collision robustness**: The platform must be able to withstand collisions at full speed with hard objects such as walls. The ability to remain airborne after light contact with objects is also beneficial.
- **Autonomous self-recovery**: The platform can take off again after contact that results in a fall to the ground from any possible falling position without any human intervention.

These two capabilities were included through intelligent design of the robot's morphology (Fig. 6a). The teardrop-shaped wing built using a single flexible carbon fibre rod absorbs the force of frontal collisions. A second carbon fibre rod

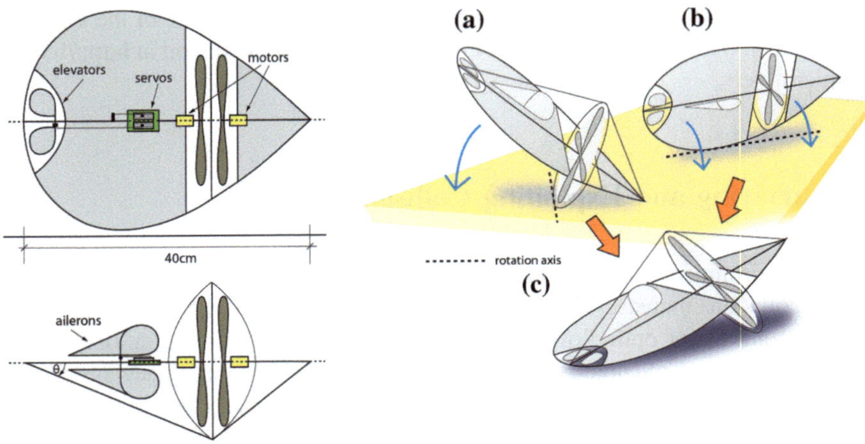

Fig. 6 *Left* Initial design of platform morphology to resist collisions and to upright autonomously after a fall. *Right* Passive uprighting based on platform morphology. Whether the platform falls on its front (**a**) or its side (**b**) gravity will act on its COG and subsequently place it into stable takeoff position (**c**). Reprinted from [32]

surrounds the propellers, protecting them from side impacts. Sensitive electronics, control surfaces and propellers are all housed on the main fuselage within these two protective rods, which is decoupled from the wing (and thus the force of a collision) through a spring. Using lightweight carbon fibre, mylar wings and miniature 3D-printed plastic components keeps the weight of the platform at a mere 20.5 g, thus minimizing the kinetic energy that must be dissipated in a collision.

The shape and position of the carbon rods along with the intelligent placement of the centre of gravity (COG) are also central to the autonomous self-recovery capability. Whether the platform lands on its front (Fig. 6a) or on its side (Fig. 6b) gravity acting on the COG will always rotate the platform about its protection ring or wing to return it to takeoff position (Fig. 6c). The dimensioning of the different platform components and the placement of the COG is optimized to find the best balance between aerodynamic stability and self-recovery abilities (more information on this process can be found in [32]).

During remote-controlled flight tests the prototype flying platform proves to be an agile flyer in both hover and forward flight modes. The two 14 mm propellers are each powered by a 6 mm DC motor, control surfaces are actuated by two miniature servo-motors, and energy is provided by a 110 mAh battery (enough for around 10 min of flight). Transition between the two modes is smooth and easily controllable, partially due to the low placement of the COG. During flight tests the platform was intentionally flown against walls and the ceiling during both hover and forward-flight modes to qualitatively assess its robustness to collisions and self-recovery capabilities. Several observations were made during these tests:

- Light contact with walls do not always cause the platform to fall to the ground. It can in fact fly along the wall, its front tip grazing the surface. This behavior resembles insects flying against a window pane looking for an exit.
- After collisions with an object that cause a fall to the ground, the prototype always settles to one of two stable positions on the ground, and in most cases can take off again without human intervention.

To further test the platform's resilience to collisions, the platform was systematically dropped from a height of 1 m from a variety of different starting positions. High-speed video was taken of each collision to try to analyze the deformation of the structure during a collision. As the platform hits the ground, the shock is partially absorbed by the spring at the nose of the platform (Fig. 7d), and partially by the deformation of the wing. Figure 8 shows frames from a typical collision and subsequent righting of the platform after a head-on collision with the ground.

The prototype presented here is a first step towards developing flying robots capable of surviving and recovering from collisions that are inevitable in cluttered environments. As with the robust exoskeletons of insects, airframe design and platform morphology must take into account not only aerodynamic constraints but also the ability to cope with this contact. The gravity-based self-recovery strategy presented above is only a first step and imposes severe limitations on the flight capabilities of the robot and the collisions it can recover from. As the environment gets more complicated, active recovery systems (mimicking the legs used by insects to recover when they fall on their backs [31]) will be required to push away from obstacles before taking flight.

As flying platforms become capable of surviving collisions, they can start using this contact to their advantage. Equipping the robot with a combination of sensors such as strain gages or artificial skin can allow it to detect the force of contact and

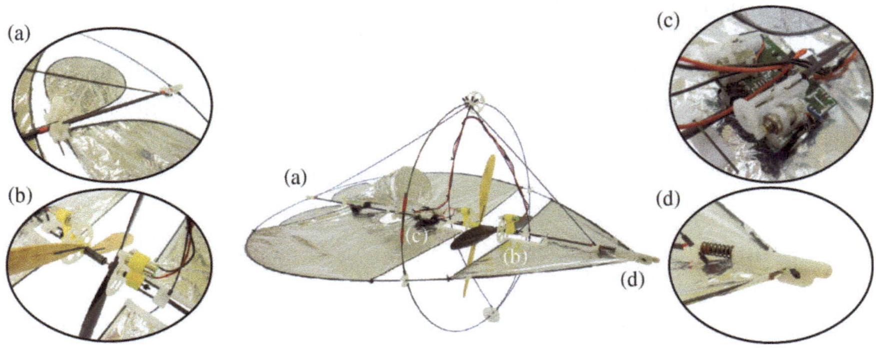

Fig. 7 The prototype flying platform with details of various subsystems. **a** Depicts the linked dual elevator assembly and the connector between the wing and the back of the main bar, both printed using a 3D printer. **b** Shows the coaxial motor assembly, linked using miniature ball bearings. **c** Is the off-the-shelf motor-control board that features two on-board linear servos. **d** Details the spring mechanism for absorbing frontal impacts and separating the wing from the main axis of the prototype. Reprinted from [32]

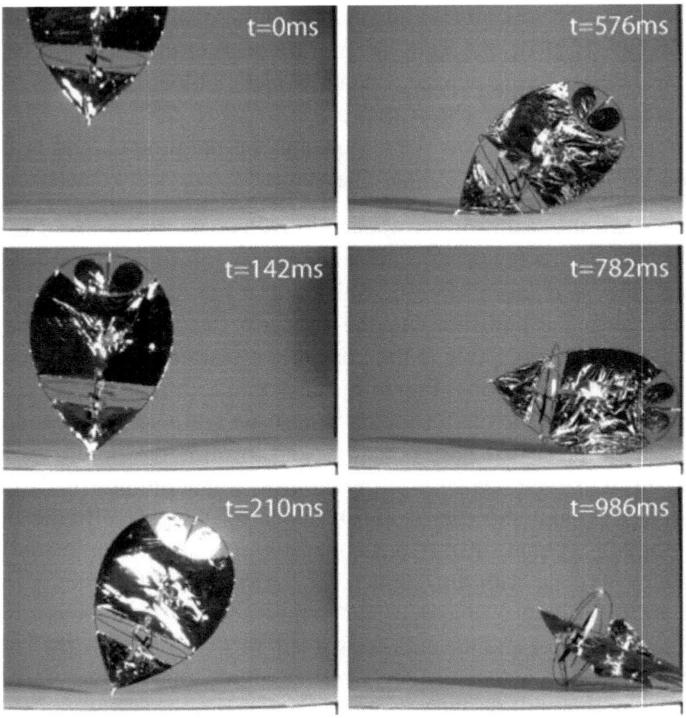

Fig. 8 Time-sequence of a typical head-on collision with the ground and subsequent self-recovery, taken with a high-speed camera. The platform rolls onto its side before rolling into takeoff position. Reprinted from [32]

gather information on its environment. This haptic information can be used to navigate in dark or low-contrast situations such as caves when other sensory modalities fail. These behaviours can be directly inspired from insects that bump into windows looking for an exit or even humans following walls with their hands in the absence of light. Attachment mechanisms can also be integrated into the platform, allowing it to perch on walls and save energy.

4 Perching

Power management is very important for small flying robots where the typical autonomy is in the order of 10–20 min and the motors consume most of the available energy [33]. Cluttered environments with vertical surfaces and ceilings offer the opportunity of temporarily perching to power off the propellers and monitor the environment or communicate with other robots from an elevated position. To date, only a few solutions exist to successfully perch for MAVs and most of them either require complex aerial maneuvers or expose the platform to high impact forces when attaching [33–36]. In this section, we present a perching (i.e. attachment and detachment) mechanism for MAVs that does not rely on complex control strategies but flies head-first into the substrate and dampens the impact forces to avoid potential damage to the robot when colliding (for a detailed description of the mechanism the reader may refer to [37]).

In our mechanism design we assume that the mechanism will be mounted on the tip of a flying robot, which is flying at a constant forward velocity towards a surface. The principle of the mechanism consists of two arms that are charged using a torsion spring (Fig. 9). Once the MAV impacts the surface, the spring is released by a mechanical trigger and the two arms snap forward to stick needles into the surface. In order to detach a motor pulls back two strings that are attached to the arms. Once the arms are pulled back, a small magnet fixes them in their charged position. In case that the detachment would not succeed immediately, this mechanism could discharge and recharge again several times to pull the needles out of the wall.

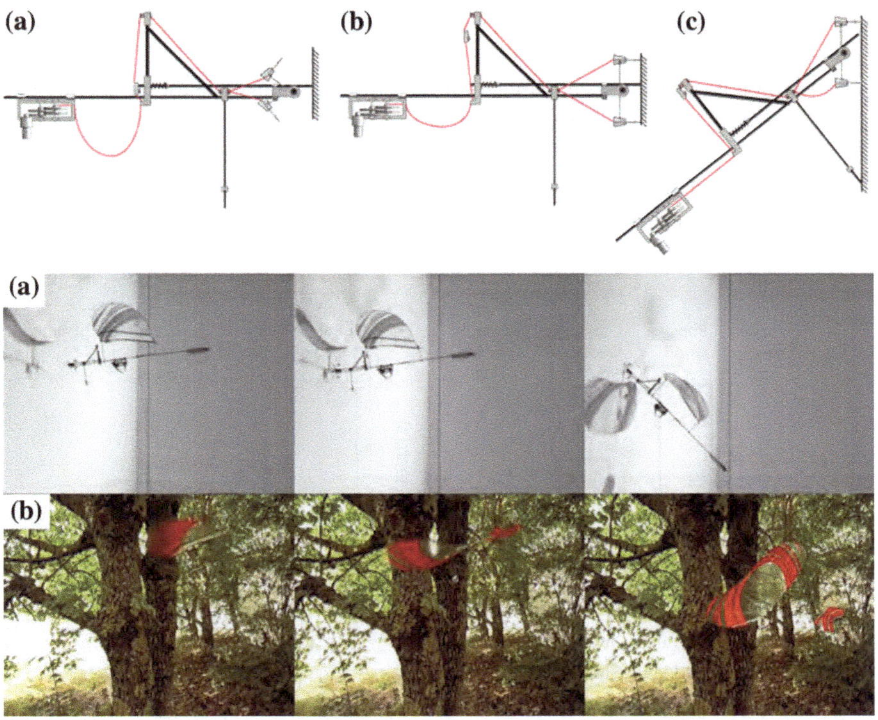

Fig. 9 *Top* Attachment sequence in CAD: in flight, the perching mechanism is in a charged state; **a** once it touches the surface, the trigger separates the magnets and the arms snap forward and **b** stick the needles into the substrate; **c** finally, the mechanism settles in its stable position on the surface. *Bottom* Perching sequence of the microflyer testbed to **a** a wallpaper wall, and **b** a maple tree. Reprinted from [37]

We dimension the torsion spring and the mass of the arms in a way that the robot is decelerated while the arms are snapping forward and has zero velocity in the moment when the needles penetrate into the surface. This is a necessary condition to avoid that the MAV crashes into the surface or that the snapping would bounce it off the surface.

The fabricated prototype has a total weight of 4.6 g. We evaluate attachment to different substrates by launching the mechanism at painted concrete, composite hardboard wood, poplar wood and poplar bark. In order to obtain a security margin of how well the perching mechanism can support the flying robot when perched to the wall, we measure the weight that the mechanism can hold until it detaches and define the security factor (SF) to be the maximal weight divided by the weight of the mechanism. In Fig. 10 we can observe that the security factor varies from 12 to 91 and in general is lower for harder than for softer substrates.

In addition, we evaluate the reliability of the perching mechanism on the four substrates. The results in Fig. 10 show that the attachment is successful on all substrates. Also the detachment is successful in all cases, but we observe that the effort to detach is different depending on the substrate.

To demonstrate that the perching mechanism can successfully be integrated on a MAV, we illustrate a complete perching sequence to a wallpaper wall and a maple tree in Fig. 10.

A limitation of this design is that, although the mechanism enables attachment to a wide range of surfaces, it cannot perch to very hard surfaces such as glass or metal surfaces. To address this, one possible solution would be the combination of different attachment mechanisms for different situations (e.g. magnetic [33] or synthetic gecko-skin [38]).

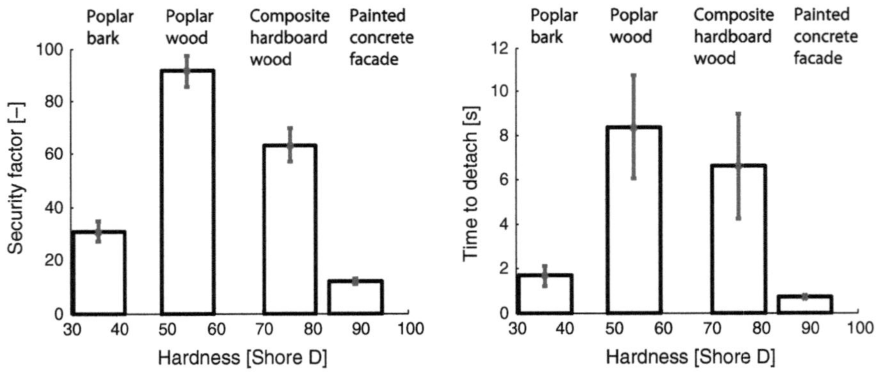

Fig. 10 *Left* Security factor for attachment to four different substrates. *Right* Time to detach from four different substrates. Reprinted from [37]

5 Jumping and Gliding

As shown in the previous sections, temporarily pausing between movements can be beneficial for power management and for the ability to sense the environment for extended periods of time at a specific location. For example, the surface a small flying robot could be covered with flexible solar cells which would allow the system to recharge its batteries while on the ground and move again when it has acquired sufficient energy. Another way to move with high energetic efficiency is to adopt jumping locomotion to move on the ground. Jumping is especially adapted for small robots because the environment appears bigger when the robot decreases in size. In nature, many small animals, such as locusts, springtails, click beetles and fleas use jumping as their main means of locomotion, as it allows them to overcome relatively large obstacles despite their small body size. In robotics, a variety of jumping robots have been presented so far. For an overview on the locomotion capabilities of these robots, we summarize their jumping performance in Table 1.

Table 1 Performance of existing miniature jumping robots

Name	Mass (g)	Size (cm)	Jump height (cm)	Jump distance (cm)	Jump height per mass (cm/g)	Jump height per size[a] (–)	Jump height per mass and size[a] (cm/(10^2· cm · g))
Class 1: Able to perform standing jumps							
Closed elastica jumper [45]	30*	30.5	20	70	1.18*	1.16	3.86
Voice coil jumper [46]	42*	3	5	0	0.12*	1.67	3.97
Spherical crawling/rolling robot [47]	5*	9	20	5	4.02*	2.23	44.62
Class 2: Able to perform standing jumps with on-board energy							
Grillo [43]	8	5	5	20[b]	1.25	2	25
EPFL jumping robot v1 [44]	7	5	138	79	20.12	28.17	402.36
Class 3: Able to perform repetitive standing jumps with on-board energy							
Microbot [48]	11	46	38	0	3.45	0.83	7.51
Michigan jumper [49]	42	11	15	11	0.37	1.4	3.36
EPFL jumping robot v2 [50]	9.8	12	76	81	8.31	6.79	69.21
Class 4: Able to perform repetitive steered standing jumps with on-board energy							
Jollbot [51]	465	29.4	18.4	0	0.04	0.63	0.13
Scout [52]	200	11	30	20	0.15	2.8	1.4
Mini-Whegs [53]	190	10.4	22	22	0.12	2.25	1.18
EPFL jumping robot v3 [39]	14.3	18	62	46	4.49	3.56	24.92

*Weight without batteries or control unit
[a]Jumping height at 90°
[b]Value N/A, here calculated assuming a take-off angle of 45°

In this section, we describe the EPFL jumper v3 [39] which is a miniature jumping robot with a mass of 14.3 g that uses the same principles for repetitive jumping and uprighting as locusts or fleas. The main requirement in the development of the jumping mechanism is to build a lightweight propulsion unit for jumping robots, where the jumping height and take-off angle can be adjusted. For small jumping systems it is most beneficial to first slowly charge an elastic element and then use the legs as catapult to jump [40–44]. The working principle in our design is to charge a torsion spring using a low power motor and then release its energy to quickly extend a four bar leg linkage to perform the jumping movement.

The implemented jumping mechanism (Fig. 11a) uses a 4 mm DC motor to turn a cam by way of a four stage gear box. The jumping height and take-off angle can be adjusted by adjusting the geometry of the legs. A jump can be executed every 3 s with a power consumption of 350 mW. The reader may refer to [39, 44] for a more detailed explanation and characterization of the jumping principles used.

The ability to jump repetitively and to steer its jump is implemented using a carbon cage (Fig. 11b) around the jumping mechanism. After landing, the jumping mechanism charges for the next jump and the cage passively uprights until the only contact with the ground is the base of the cage. Once upright, the entire jumping mechanism is inside the cage and can rotate around its vertical axis using a second DC motor around the main rod (Fig. 11c).

In order to reduce the risk of damaging the legs on landing, the charging of the jumping mechanism starts already during the aerial phase to better protect the legs inside the cage. As the center of gravity is in the lower part of the structure, the robot settles in a stable upright position and is ready to steer and jump again. The motor to steer and the motor of the jumping mechanism are remotely controlled using a miniature infra red controller and a 10 mAh Lithium Polymer battery. The 10 mAh provided by this battery theoretically allow for 6.3 min of continuous

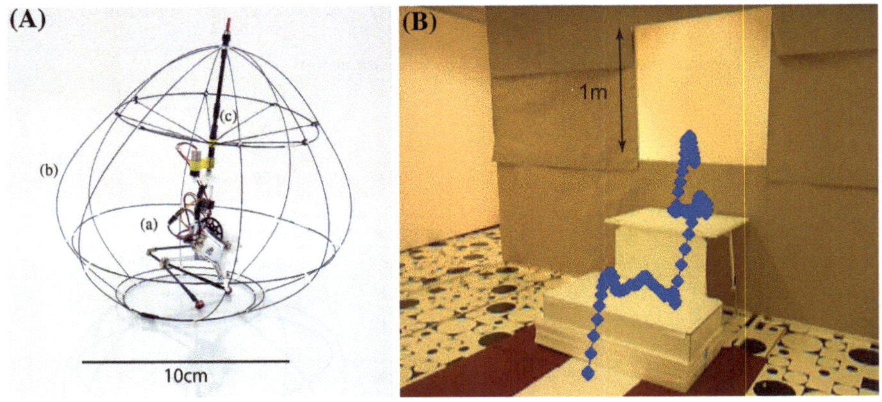

Fig. 11 **A** EPFL jumper v3. (*a*) Jumping mechanism, (*b*) cage, (*c*) main rod. **B** Trajectory of the jumping robot successfully climbing two stairs of each 50 cm height and jumping into a window. Reprinted from [39]

recharging of the jumping mechanism or approximately 108 jumps. The completely functional remote controlled prototype has a maximal dimension of 18 cm and a mass of 14.33 g including batteries and electronics.

As a demonstration of the ability of our prototype to successfully perform steered jumps in cluttered environments, we build an obstacle course which consists of two stairs with a height of 45 cm each and a window of 1 m $\times$ 1 m (Fig. 11). We place the robot on the ground at 10 cm distance to the first stair and aim at locomoting with sequential steered jumps up the stairs and through the window, all without human intervention on the scene. Depending on the operating skill of the human operator the window can be entered in approximately four jumps (see [39] for three successful passages of this obstacle course).

It has been suggested [43, 51, 55] that wings could be used to prolong the flight phase of a jumping system. For lack of an existing term for this hybrid jumping and gliding locomotion, we introduce the term 'jumpgliding'. As the first miniature jumpgliding robot that is capable of successive jumpgliding locomotion without human intervention, we present the EPFL jumpglider [54] (Fig. 12a), a 16.5 g system that can jump and transition to a steered gliding phase. Figure 12b illustrates the locomotion capabilities of the EPFL jumpglider. It shows the trajectory of a jump from an elevated position of 2.53 m height, a stable gliding phase, three sequential jumps to progress on level terrain and finally a jump off the table to glide down to the floor.

In [56] we evaluate under which conditions the addition of wings to a jumping robot gives added benefits compared to jumping without wings. The potential benefits which are considered are (i) the ability to prolong jumps using wings and (ii) the reduction of potentially destructive impact forces that have to be absorbed by the robot structure on landing. The results indicate that wings can prolong jumps originating from elevated starting positions, but not those occurring on level ground. A jumping robot without wings, such as the EPFL jumper v3 is a better solution for locomotion on level terrain. However, wings can both reduce the impact forces and help maintain an upright orientation on landing, allowing the robot to reliably perform repetitive jumps and safely descend elevated positions and stairs.

6 Towards Adaptive Morphologies

This chapter has described four features, or design principles, for small robots to move in the air in cluttered environments, such as a visual system with large field of view and integrated gyroscopes for perception and stabilization, a flexible cage for surviving and exploiting collisions, a perching mechanism to save power and increase information and communication tasks, and a self-deploying mechanism for repeated jumps or glided flight. Locomotion in cluttered environments may require adaptive forms of locomotion that can perform multiple types of locomotion, such as walking and jumping, flying and rolling, climbing and gliding, etc.

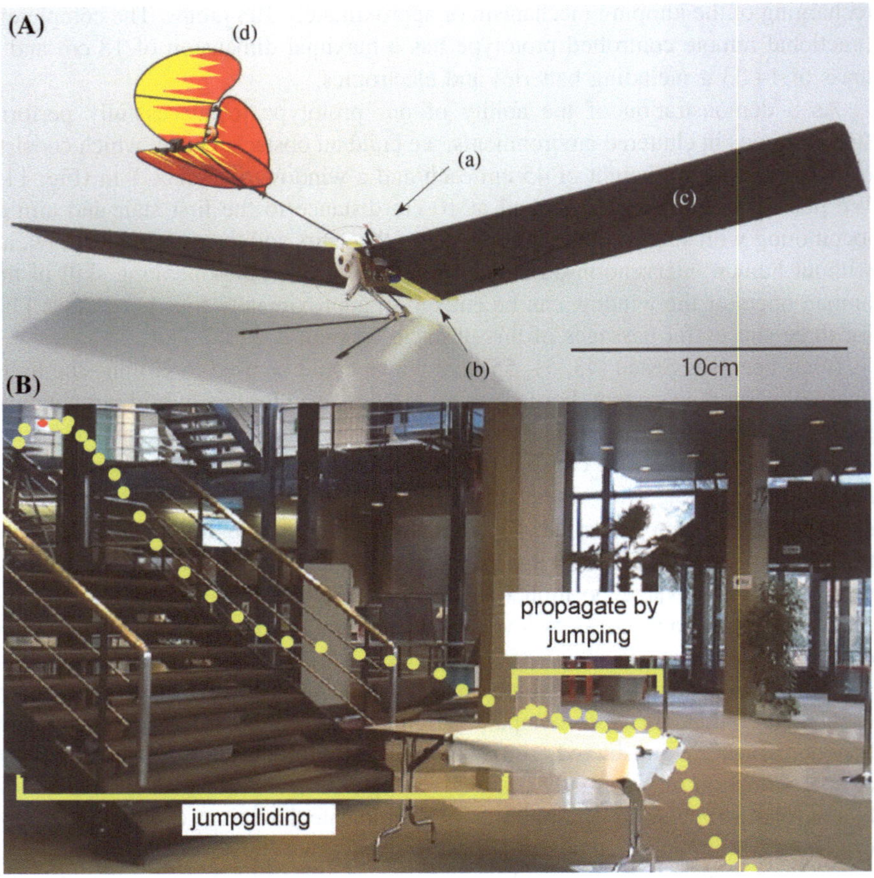

Fig. 12 **A** EPFL jumpglider. (*a*) Jumping mechanism, (*b*) CNC cut Polyimide frame, (*c*) wings, (*d*) tail with rudder. **B** Illustration of the locomotion capabilities of the EPFL jumpglider. Reprinted from [54]

Most existing robots are designed to exploit only a single locomotion strategy, such as rolling, walking, flying, swimming, or jumping. This greatly limits their flexibility and adaptability to different environments where specific and different locomotion capabilities could be more efficient. Multi-modal locomotion capabilities could be implemented by incorporating different actuation mechanisms within the same robot. For example, the jumping and gliding robot presented here is an example of a robot with two actuation systems (one of them passive) for two locomotion modes. Another more efficient strategy would consist in achieving multiple locomotion modes with less actuation systems and motors. In order to do that, the robot could be endowed with an adaptive morphology that enables the transition between multiple states and reuse of the same actuation system for different purposes. For example, we can modify our posture and four appendices for walking, climbing, swimming, and crawling. Flexible robots, with highly integrated perceptual systems, and adaptive morphologies represent a promising solution for highly resilient and efficient robots capable of moving through cluttered, unknown, and dynamic environments.

Acknowledgements These works have been sponsored by several grants of the Swiss National Science Foundation, including the NCCR Robotics, by EPFL, and by the Science and Technology Division of Armasuisse.

References

1. R. Siegwart, D. Nourbakhsh, I. Scaramuzza, *Introduction to Autonomous Mobile Robotics*, 2nd edn. (MIT Press, 2011)
2. J.-C. Zufferey, *Bio-inspired Flying Robots: Experimental Synthesis of Autonomous Indoor Flyers* (EPFL/CRC Press, 2008)
3. N. Strausfeld, *Atlas of an Insect Brain* (Springer, 1976)
4. G. Taylor, H. Krapp, Sensory systems and flight stability: what do insects measure and why. Adv. Insect Physiol. **34**, 231–316 (2008)
5. M. Srinivasan, M. Lehrer, W. Kirchner, S. Zhang, Range perception through apparent image speed in freely-flying honeybees. Vis. Neurosci. **6**, 519–535 (1991)
6. L. Tammero, M. Dickinson, The influence of visual landscape on the free flight behavior of the fruit fly drosophila melanogaster. J. Exp. Biol. **205**, 327–343 (2002)
7. M. Egelhaaf, R. Kern, H. Krapp, J. Kretzberg, R. Kurtz, A. Warzechna, Neural encoding of behaviourally relevant visual-motion information in the fly. Trends Neurosci. **25**(2), 96–102 (2002)
8. M. Land, Visual acuity in insects. Annu. Rev. Entomol. **42**, 147–177 (1997)
9. P. Sobey, Active navigation with a monocular robot. Biol. Cybern. **71**, 433–440 (1994)
10. D. Coombs, M. Herman, T. Hong, M. Nashman, Real-time obstacle avoidance using central flow divergence and peripheral flow, in *International Conference on Computer Vision* (1995), pp. 276–283
11. J. Santos-Victor, G. Sandini, F. Curotto, S. Garibaldi, Divergent stereo for robot navigation: a step forward to a robotic bee. Int. J. Comput. Vis. **14**, 159–177 (1995)
12. M. Srinivasan, J. Chahl, K. Weber, S. Venkatesh, H. Zhang, in *Robot Navigation Inspired by Principles of Insect Vision*, ed. by A. Zelinsky. Field and Service Robotics (Springer, 1998), pp. 12–16
13. J. Serres, F. Ruffier, S. Viollet, N. Franceschini, Toward optic flow regulation for wallfollowing and centring behaviours. Int. J. Adv. Rob. Syst. **3**(27), 147–154 (2006)
14. F. Ruffier, N. Franceschini, Optic flow regulation: the key to aircraft automatic guidance. Robot. Auton. Syst. **50**(4), 177–194 (2005)
15. T. Neumann, H. Blthoff, Behavior-oriented vision for biomimetic flight control, in *Proceedings of the EPSRC/BBSRC International Workshop on Biologically Inspired Robotics* (2002), pp. 196–203
16. L. Muratet, S. Doncieux, Y. Brire, J. Meyer, A contribution to vision-based autonomous helicopter flight in urban environments. Robot. Auton. Syst. **50**(4), 195–209 (2005)
17. J. Humbert, J.K. Conroy, C. Neely, G. Barrows, *Wide-Field Integration Methods for Visuomotor Control* (Springer, 2009)
18. A. Beyeler, J. Zufferey, D. Floreano, Vision-based control of near-obstacle flight. Auton. Rob. **27**(3), 201–219 (2009)
19. J. Gibson, *The Perception of the Visual World* (Houghton Mifflin, Boston, 1950)
20. J. Koenderink, A. van Doorn, Facts on optic flow. Biol. Cybern. **56**, 247–254 (1987)

21. J. Gibson, *The Ecological Approach to Visual Perception* (Houghton Mifflin, Boston, 1979)
22. V. Braitenberg, *Vehicles—Experiments In Synthetic Psychology* (The MIT Press, Cambridge, MA, 1984)
23. A. Duchon, W.H. Warren, L. Kaelbling, Ecological robotics. Adapt. Behav. **6**, 473–507 (1998)
24. J. Zufferey, A. Beyeler, D. Floreano, *Optic Flow to Steer and Avoid Collisions in 3D* (Springer, 2009), pp. 73–86
25. J.-C. Zufferey, A. Klaptocz, A. Beyeler, J.-D. Nicoud, D. Floreano, A 10-gram vision-based flying robot. Adv. Robot. J. Robot. Soc. Jpn. **21**(14), 1671–1684 (2007)
26. J.-C. Zufferey, A. Beyeler, D. Floreano, Autonomous flight at low altitude using light sensors and little computational power. Int. J. Micro Air Veh. **2**(2), 107–117 (2010)
27. J.F. Roberts, T. Stirling, J.-C. Zufferey, D. Floreano, Quadrotor using minimal sensing for autonomous indoor flight, in *European Micro Air Vehicle Conference and Flight Competition (EMAV2007)* (2007)
28. P. Oh, M. Joyce, J. Gallagher, Designing an aerial robot for hover-and-stare surveillance, in *12th International Conference on Advanced Robotics, 2005. ICAR'05. Proceedings* (IEEE, 2005), pp. 303–308
29. D. Schafroth, S. Bouabdallah, C. Bermes, R. Siegwart, From the test benches to the first prototype of the mufly micro helicopter. J. Intell. Rob. Syst (2008)
30. S. Saripalli, J. Montgomery, G. Sukhatme, Vision-based autonomous landing of an unmanned aerial vehicle, in *IEEE International Conference on Robotics and Automation*, vol. 3 (2002)
31. L. Frantsevich, Righting kinematics in beetles (insecta: Coleoptera). Arthropod Struct. Dev. **33**(3), 221–235 (2004)
32. A. Klaptocz, G. Boutinard-Rouelle, A. Briod, J.-C. Zufferey, D. Floreano, An indoor flying platform with collision robustness and self-recovery, in *2010 IEEE International Conference on Robotics and Automation (ICRA)* (IEEE, 2010), pp. 3349–3354
33. J. Roberts, J. Zufferey, D. Floreano, Energy management for indoor hovering robots, in *IEEE/RSJ International Conference on Intelligent Robots and Systems (IROS2008)* (2008), pp. 1242–1247
34. M.L. Anderson, C.J. Perry, B.M. Hua, D.S. Olsen, J.R. Parcus, K.M. Pederson, D.D. Jensen, The sticky-pad plane and other innovative concepts for perching uavs, in *Proceedings of the 47th AIAA Aerospace Sciences Meeting* (2009)
35. R. Cory, R. Tedrake, Experiments in fixed-wing uav perching, in *AIAA Conference on Guidance, Navigation, and Control* (2008)
36. A. Lussier-Desbiens, M. Cutkosky, Landing and perching on vertical surfaces with microspines for small unmanned air vehicles. J. Intell. Rob. Syst. **57**(1), 313–327 (2010)
37. M. Kovac, J. Germann, C. Hurzeler, R. Siegwart, D. Floreano, A perching mechanism for micro aerial vehicles. J. Micro-Nano Mechatron (2010)
38. D. Santos, B. Heyneman, S. Kim, N. Esparza, M.R. Cutkosky, Gecko-inspired climbing behaviors on vertical and overhanging surfaces, in *IEEE International Conference on Robotics and Automation*, 2008, pp. 1125–1131
39. M. Kovac, M. Schlegel, J.-C. Zufferey, D. Floreano, Steerable miniature jumping robot. Auton. Rob. **28**(3), 295–306 (2010)
40. T.J. Roberts, R.L. Marsh, Probing the limits to muscle-powered accelerations: lessons from jumping bullfrogs. J. Exp. Biol. **206**(15), 2567–2580 (2003)
41. R. M. Alexander, *Elastic Mechanisms in Animal Movement* (Cambridge University Press, 1988)
42. M. Burrows, Biomechanics: Froghopper insects leap to new heights. Nature **424**(6948), 509 (2003)
43. U. Scarfogliero, C. Stefanini, P. Dario, Design and development of the long-jumping "grillo" mini robot, in *IEEE International Conference on Robotics and Automation* (2007), pp. 467–472
44. M. Kovac, M. Fuchs, A. Guignard, J. Zufferey, D. Floreano, A miniature 7 g jumping robot, in *IEEE International Conference on Robotics and Automation (ICRA2008)* (2008), pp. 373–378

45. A. Yamada, M. Watari, H. Mochiyama, H. Fujimoto, A robotic catapult based on the closed elastica with a high stiffness endpoint and its application to swimming tasks, in *IEEE/RSJ International Conference on Intelligent Robots and Systems* (2008), pp. 1477–1482
46. P. Zhang, Q. Zhou, Voice coil based hopping mechanism for microrobot, in *IEEE international conference on Robotics and Automation* (2009), pp. 1783–1788
47. Y. Sugiyama, M. Yamanaka, S. Hirai, Circular/spherical robots for crawling and jumping, in *IEEE International Conference on Robotics and Automation* (2005), pp. 3595–3600
48. S. Dubowsky, S. Kesner, J.S. Plante, P. Boston, Hopping mobility concept for search and rescue robots. Ind. Rob. Int. J. **35**(3), 238–245 (2008)
49. J. Zhao, R. Yang, N. Xi, B. Gao, X. Fan, M. W. Mutka, L. Xiao, Development of a miniature self-stabilization jumping robot, in *IEEE/RSJ International Conference on Intelligent Robots and Systems* (2009), pp. 2217–2222
50. M. Kovac, M. Schlegel, J.-C. Zufferey, D. Floreano, A miniature jumping robot with selfrecovery capabilities, in *IEEE/RSJ International Conference on Robotics and Automation* (2009), pp. 583–588
51. R. Armour, K. Paskins, A. Bowyer, J.F.V. Vincent, W. Megill, Jumping robots: a biomimetic solution to locomotion across rough terrain. Bioinspiratoin Biomimetics J. **2**, 65–82 (2007)
52. S.A. Stoeter, P.E. Rybski, N. Papanikolopoulos, Autonomous stair-hopping with scout robots, in *IEEE/RSJ International Conference on Intelligent Robots and Systems*, vol. 1 (2002), pp. 721–726
53. J.M. Morrey, B. Lambrecht, D. Horchler, R.E. Ritzmann, R.D. Quinn, Highly mobile and robust small quadruped robots, in *International Conference on Intelligent Robots and Systems*, vol. 1 (2003), pp. 82–87
54. M. Kovac, W. Hraiz, O. Fauria, J.-C. Zufferey, D. Floreano, The EPFL jumpglider: a hybrid jumping and gliding robot. Review (2011)
55. M. Kovac, J. Zufferey, D. Floreano, *Towards a Self-deploying and Gliding Robot* (Springer, 2009)
56. M. Kovac, J.-C. Zufferey, D. Floreano, Hybrid jumping and gliding locomotion for miniature robotics. Review (2011)

Opportunities and Challenges with Autonomous Micro Aerial Vehicles

Vijay Kumar and Nathan Michael

Abstract We survey the recent work on micro-UAVs, a fast-growing field in robotics, outlining the opportunities for research and applications, along with the scientific and technological challenges. Micro-UAVs can operate in three-dimensional environments, explore and map multi-story buildings, manipulate and transport objects, and even perform such tasks as assembly. While fixed-base industrial robots were the main focus in the first two decades of robotics, and mobile robots enabled most of the significant advances during the next two decades, it is likely that UAVs, and particularly micro-UAVs will provide a major impetus for the third phase of development.

1 Introduction

The last decade has seen many exciting developments in the area of micro Unmanned Aerial Vehicles (UAVs) that are between 0.1–0.5 m in length and 0.1–0.5 kg in mass. Just as the incorporation of 2-D mobility reinvigorated robotics research in the 1990s, the ability to operate in truly three-dimensional environments is bringing in new research challenges along with new technologies and applications. Indeed by some estimates [51], the UAV market is estimated to exceed $60 B in the next 3 years, and this forecast is conservative since it does not account for the thousands of micro-UAVs that are likely to be fielded in the near future.

Our focus in this work is on UAVs that have gross weights of the order of 1 kg and below; although as described in [5, 8, 30, 40] the platform development represents a challenge in its own right. While commercial products ranging from 5 to 350 g are available, most of these products do not carry the sensors and processors required for autonomous flight. Many of these small aircrafts do not have the endurance required for missions of longer than 5 min. Longer endurance requires

V. Kumar (✉) · N. Michael
Department of Mechanical Engineering and Applied Mechanics, GRASP Laboratory, University of Pennsylvania, Philadelphia, PA, USA
e-mail: kumar@seas.upenn.edu

H.I. Christensen and O. Khatib (eds.), *Robotics Research*,
Springer Tracts in Advanced Robotics 100, DOI 10.1007/978-3-319-29363-9_3

bigger batteries, and with the current energy densities of Li-polymer batteries (of the order of several hundred Watt-hr/kg), the mass fraction used by batteries is significant, often between 25–50 % of the gross weight.

There are many types of micro-UAVs that are in various phases of research, development and practice. Fixed-wing aircrafts are less adept than rotor crafts at maneuvering in constrained, 3-D environments. While avian-style flapping wing aircrafts provide more agility, our limited understanding of the aerodynamics and the fluid-structure coupling in such aircrafts presents a formidable challenge [10]. Insect-style flapping wing vehicles provide the ability to hover in place while also enabling forward flight [2]. However, it is unclear that they represent a significant advantage over rotor crafts or ducted fans in terms of efficiency, endurance, or maneuverability, and they do incur a significant increase in complexity [43].

There are two configurations of rotor crafts that have gained acceptance in the research community. Co-axial rotor crafts, exemplified by the Skybotix Coax [8], are equipped with two counter-rotating, co-axial rotors and with a stabilizer bar [6]. Prototypes of less than 300 g (without sensors or processors) with a hover time of nearly 20 min make them attractive for robotics applications. In addition, the stabilizer bar confers passive mechanical stability making them easy to control.

However, we argue (see next section) that multi-rotor aircrafts exemplified by quadrotors currently represent the best bet in terms of maneuverability and their ability to carry small payloads. Hence the rest of this paper will address the mechanics and control of quadrotors, and approaches to state estimation, mapping, planning, exploration and manipulation.

2 Rotor Craft Designs and Scaling Laws

In this section, we explore the effect of choosing length scales on the inertia, payload and ultimately angular and linear acceleration. In particular, we can analyze maneuverability in terms of the robot's ability to produce linear and angular accelerations from a hover state. If the characteristic length is L, the rotor radius R scales linearly with L. The mass scales as L^3 and the moments of inertia as L^5. On the other hand the lift or thrust, F, and drag, D, from the rotors scales with the cross-sectional area and the square of the blade-tip velocity, v. If the angular speed of the blades is defined by $\omega = \frac{v}{L}$, $F \sim \omega^2 L^4$ and $D \sim \omega^2 L^4$. The linear acceleration a scales as $a \sim \frac{\omega^2 L^4}{L^3} = \omega^2 L$.

For multi-rotor aircrafts like the quadrotor, thrusts from the rotors produces a moment with a moment arm L. Thus the angular acceleration $\alpha \sim \frac{\omega^2 L^5}{L^5} = \omega^2$. However, the rotor speed also scales with length since smaller motors produce less torque which limits their peak speed because of the drag resistance that also scales the same way as lift.

There are two commonly accepted approaches to scaling: Froude scaling and Mach scaling [55]. Mach scaling is used for compressible flows and essential assumes that the tip velocities are constant leading to $\omega \sim \frac{1}{R}$. In other words, the rotor speed scales inversely with length. Froude scaling is used for incompressible flows and assumes that for similar aircraft configurations, the Froude number, $\frac{v^2}{Lg}$ is constant. Here g is the acceleration due to gravity. This yields $\omega \sim \frac{1}{\sqrt{R}}$. Neither Froude or Mach number similitudes take motor characteristics into account. It is clear that the motor torque (τ) scales with length. The surface area, which goes as $R^2 \sim L^2$, and the volume of the core which scales as $R^3 \sim L^3$, are both important variables governing motor performance. It turns out Froude scaling ($\omega \sim \frac{1}{R}$) is consistent with $\tau \sim L^2$ while Mach scaling is consistent with $\tau \sim L^3$. While the reality might be somewhere in between, these two limiting cases are meaningful for our analysis. Froude scaling suggests that the acceleration is independent of length while the angular acceleration $\alpha \sim L^{-1}$. On the other hand Mach scaling leads to the conclusion that $a \sim L$ while $\alpha \sim L^{-2}$. In other words, smaller aircrafts are much more agile. Note that this conclusion is based on the assumption that the propeller blades are rigid, the efficiency of the blade is independent of the length scale and the inertia associated with the blades can be neglected. These factors can be important but considering the inertia of the blade further emphasizes the benefits of scaling down—longer blades require larger cross-sections to minimize stresses and the inertia grows faster than L^5.

For other types of rotor crafts, including co-axial rotor crafts, the linear acceleration scales the same way but the angular acceleration does not. This is because the moment arm associated with the rotors is exactly L. This moment arm does not scale the same way with coaxial helicopters. Similarly the scaling law for conventional helicopters and ducted fans appears to be different. Thus if our objective is to build small, highly maneuverable aircrafts, multi-rotor helicopters like the quadrotor appear to be the best configuration. While rotorcrafts with six and eight rotors have been developed and are commercially available [3], the main benefits appear to be redundancy due to the number of rotors and increased safety because of the compactness of a six-rotor design over a four-rotor design.

There are three design points that are illustrative of the quadrotor configuration. We use the Pelican quadrotor from Ascending Technologies [3] equipped with sensors (approx. 2 kg gross weight, 0.75 m diameter, and 4000 rpm nominal rotor speed at hover), consuming approximately 400 W of power. The Hummingbird quadrotor from Ascending Technologies (500 g gross weight, approximately 0.5 m diameter, and 5000 rpm nominal rotor speed at hover) consumes about 75 W. Attempts to develop a smaller quadrotor at the University of Maryland [34] suggest that a quad rotor without sensors of mass 62 g, 0.075 m diameter and 9000 rpm rotor speed consumes a little over 10 W of power.

3 Control

3.1 Dynamics

The dynamics of quadrotors can be simplified to rigid body dynamic models with approximations to the aerodynamic forces [32]. In Fig. 1, the inertial frame, $\mathscr{A}$, is defined by the triad $\mathbf{a}_1$, $\mathbf{a}_2$, and $\mathbf{a}_3$ with $\mathbf{a}_3$ pointing upward. The body frame, $\mathscr{B}$, is attached to the center of mass of the quadrotor with $\mathbf{b}_1$ coinciding with the preferred forward direction and $\mathbf{b}_3$ perpendicular to the plane of the rotors pointing vertically up during perfect hover (see Fig. 1). Let $\mathbf{r}$ denote the position vector of the center of mass C in $\mathscr{A}$. The vehicle has mass m and the components of the inertia tensor is given by the $3 \times \ 3$ matrix J along the principal axes $\mathbf{b}_i$. The rotation matrix describing $\mathscr{B}$ in $\mathscr{A}$ is given by $R \in SO\,(3)$, while the angular velocity of the vehicle, $\Omega \in \mathbb{R}^3$, is defined as

$$\dot{R} = R\hat{\Omega}$$

where the operator $\hat{\cdot}$ is defined such that $\hat{x}y = x \times y$ for all $x, y \in \mathbb{R}^3$.

The forces on the system are gravity, in the $-\mathbf{a}_3$ direction, the lift forces from each of the rotors, F_i, and the drag moments from the rotors M_i, all in the $\mathbf{b}_3$ direction. Each rotor has an angular speed ω_i and produces a lift force $F_i = k_F\ \omega_i^2$ and drag moment $M_i = k_M \omega_i^2$. The constants, k_F and k_M, are related to the drag and lift coefficients, the cross sectional area and the rotor speed as discussed in Sect. 2. However, for a specific rotor, it is quite easy to determine these empirically. The thrust input is given by:

$$u_1 \sum_{i=1}^{4} F_i$$

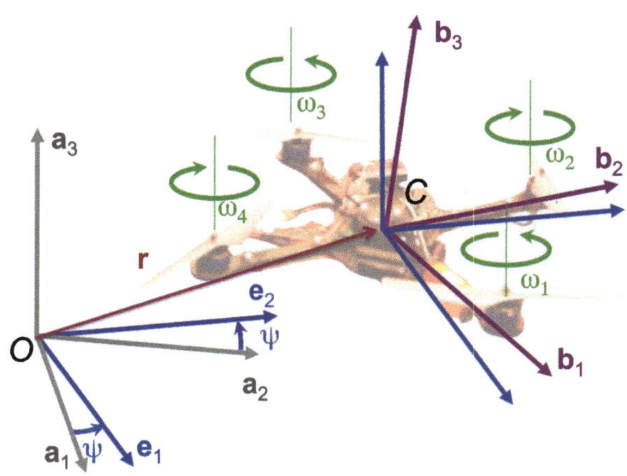

Fig. 1 The vehicle model. The position and orientation of the robot in the global frame are denoted by **r** and R, respectively

while the moment input vector is

$$\mathbf{u}_2 = L\begin{bmatrix} 0 & 1 & 0 & -1 \\ -1 & 0 & 1 & 0 \\ \mu & -\mu & \mu & -\mu \end{bmatrix}\begin{bmatrix} F_1 \\ F_2 \\ F_3 \\ F_4 \end{bmatrix}$$

where L is the distance of the rotor axis from C, and $\mu = \frac{k_M}{Lk_F}$ is a non dimensional F coefficient that relates the drag (moment) to the lift (force) produced by the propellor blades.

The dynamic model is given by:

$$m\ddot{\mathbf{r}} - mg\mathbf{e}_3 = u_1 R\mathbf{e}_3 \tag{1}$$

$$J\dot{\Omega} + \Omega + J\Omega = \mathbf{u}_2 \tag{2}$$

where $\mathbf{e}_3 = [0, 0, 1]^{\mathrm{T}}$.

3.2 Control

The control problem, to track smooth trajectories ($R^{\mathrm{des}}(t)$, $\mathbf{r}^{\mathrm{des}}(t)$) $\in SE(3)$, is challenging for several reasons. First, the system is underactuated—there are four inputs (u_1, $\mathbf{u}_2$) while $SE(3)$ is six dimensional. Second, the aerodynamic model described above is only approximate. Finally, the inputs are themselves idealized. In practice, the motor controllers must generate the required speeds to realize these inputs. The dynamics of the motors and their interactions with the drag forces on the propellers can be quite difficult to model, although first order linear models are a useful approximation.

The first challenge, the underactuation, can be overcome by recognizing that the quadrotor is differentially flat. See [36, 38] for a discussion of differential flatness. To see this, we consider the outputs $\mathbf{r}$ and ψ as shown in Fig. 1, and show that we can write all state variables and inputs as functions of the outputs and their derivatives. Derivatives of $\mathbf{r}$ yield the velocity $\mathbf{v}$, and the acceleration,

$$\mathbf{a} = \frac{1}{m}u_1\mathbf{b}_3 + \mathbf{g}$$

By writing the unit vector:

$$\mathbf{e}_1 = [\cos\psi, \sin\psi, 0]^{\mathrm{T}}$$

we can define the body frame from ψ and $\mathbf{a}$ as follows:

$$\mathbf{b}_3 = \frac{\mathbf{a}-\mathbf{g}}{\|\mathbf{a}-\mathbf{g}\|}, \mathbf{b}_2 = \frac{\mathbf{b}_3 \times \mathbf{e}_1}{\|\mathbf{b}_3 \times \mathbf{e}_1\|}, \mathbf{b}_1 = \mathbf{b}_2 \times \mathbf{b}_3$$

provided $\mathbf{e}_1 \times \mathbf{b}_3 \neq 0$. This defines the rotation matrix R as a function of $\mathbf{a}$ and ψ . To write the angular velocity and the inputs as a function of the outputs and their derivatives, we write the derivative of acceleration or jerk,

$$\mathbf{j} = \frac{1}{m}\dot{u}_1\mathbf{b}_3 + \frac{1}{m}u_1\Omega \times \mathbf{b}_3$$

and finally, the snap or the derivative of jerk:

$$\mathbf{s} = \frac{1}{m}\ddot{u}_1\mathbf{b}_3 + \frac{2}{m}\dot{u}_1\Omega \times \mathbf{b}_3 + \frac{1}{m}u_1\dot{\Omega} \times \mathbf{b}_3 + \frac{1}{m}u_1\Omega \times (\Omega \times \mathbf{b}_3)$$

where

$$\dot{\Omega} = J^{-1}(\mathbf{u}_2 - \Omega \times J\Omega)$$

From the equations above it is possible to verify that there is a diffeomorphism between the 18×1 vector:

$$[\mathbf{r}^{\mathrm{T}}, \mathbf{v}^{\mathrm{T}}, \mathbf{a}^{\mathrm{T}}, \mathbf{j}^{\mathrm{T}}, \mathbf{s}^{\mathrm{T}}, \psi^{\mathrm{T}}, \dot{\psi}^{\mathrm{T}}, \ddot{\psi}^{\mathrm{T}}]$$

and

$$R \times \left[\mathbf{r}^{\mathrm{T}}, \dot{\mathbf{r}}^{\mathrm{T}}, \Omega^{\mathrm{T}}, u_1, \dot{u}_1, \ddot{u}_1, \mathbf{u}_2^{\mathrm{T}}\right]^{\mathrm{T}}$$

Accordingly define the vector of flat outputs to be:

$$\mathbf{z} = [\mathbf{r}, \mathbf{v}, \mathbf{a}, \mathbf{j}, \psi, \dot{\psi}]^{\mathrm{T}} = [\mathbf{z}_1, \mathbf{z}_2, \mathbf{z}_3, \mathbf{z}_4, z_5, z_6]^{\mathrm{T}}$$

We can also define a vector of fictitious inputs

$$\mathbf{v} = \left[\mathbf{v}_1^{\mathrm{T}}, v_2\right]^{\mathrm{T}}$$

related to the original inputs by a nonlinear transformation of the form:

$$\begin{bmatrix} \mathbf{v}_1 \\ v_2 \end{bmatrix} = \mathbf{g}(\mathbf{z}) \begin{bmatrix} \ddot{u}_1 \\ \mathbf{u}_2 \end{bmatrix} + \mathbf{h}(\mathbf{z}) \quad (3)$$

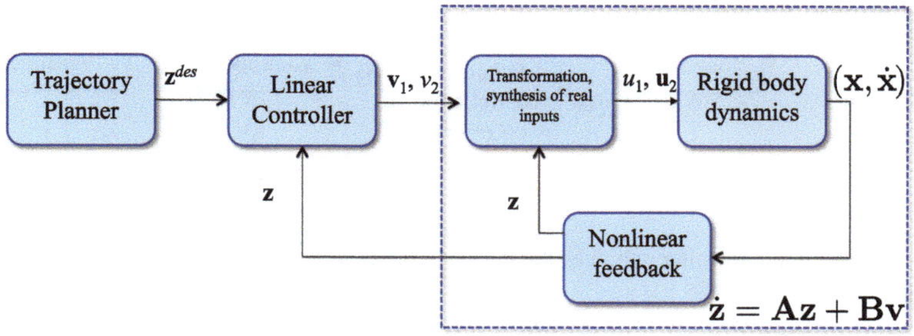

Fig. 2 Nonlinear feedback allows us to reduce the nonlinear system to a linear system (4)

so the state equations are linear:

$$\dot{\mathbf{z}} = \mathbf{Az} + \mathbf{Bv} \tag{4}$$

with

$$\mathbf{A} = \begin{bmatrix} \mathbf{0}_{3\times3} & \mathbf{I}_{3\times3} & \mathbf{0}_{3\times3} & \mathbf{0}_{3\times3} & \mathbf{0}_{3\times1} & \mathbf{0}_{3\times1} \\ \mathbf{0}_{3\times3} & \mathbf{0}_{3\times3} & \mathbf{I}_{3\times3} & \mathbf{0}_{3\times3} & \mathbf{0}_{3\times1} & \mathbf{0}_{3\times1} \\ \mathbf{0}_{3\times3} & \mathbf{0}_{3\times3} & \mathbf{0}_{3\times3} & \mathbf{I}_{3\times3} & \mathbf{0}_{3\times1} & \mathbf{0}_{3\times1} \\ \mathbf{0}_{3\times3} & \mathbf{0}_{3\times3} & \mathbf{0}_{3\times3} & \mathbf{0}_{3\times3} & \mathbf{0}_{3\times1} & \mathbf{0}_{3\times1} \\ \mathbf{0}_{1\times3} & \mathbf{0}_{1\times3} & \mathbf{0}_{1\times3} & \mathbf{0}_{1\times3} & 0 & 1 \\ \mathbf{0}_{1\times3} & \mathbf{0}_{1\times3} & \mathbf{0}_{1\times3} & \mathbf{0}_{1\times3} & 0 & 0 \end{bmatrix}, \quad \mathbf{B} = \begin{bmatrix} \mathbf{0}_{3\times3} & \mathbf{0}_{3\times1} \\ \mathbf{0}_{3\times3} & \mathbf{0}_{3\times1} \\ \mathbf{0}_{3\times3} & \mathbf{0}_{3\times1} \\ \mathbf{I}_{3\times3} & \mathbf{0}_{3\times1} \\ \mathbf{0}_{1\times3} & 0 \\ \mathbf{0}_{1\times3} & 1 \end{bmatrix}$$

This obviously makes the control problem trivial. See Fig. 2 for a graphical description of the controller design.

There are several difficulties following this naive approach. First, the linear controller based on (4) works only if the dynamics can be effectively linearized. This in turn depends on the cancelation of the dynamics in (3) which is difficult because the dynamic model only represents an approximation of the aerodynamic forces and our knowledge of the parameters in the model is not perfect. While parameter estimation and adaptive control techniques (e.g., [39]) can be used to learn and adapt to these parameters, it is often not possible to get access to the low level signals involving higher order derivatives of the state and the inputs.

Indeed, the second challenge is to derive estimators that yield the extended state, $\mathbf{z}$, which includes not only the position and velocity, but also the acceleration and jerk. Knowledge of the thrust (u_1) and attitude ($\mathbf{b}_3$) allows us to estimate acceleration. Similarly, measuring the derivative of the thrust ($\dot{u}_1$), which is related to the rate of change of motor speeds, and the angular rates ($\mathbf{\Omega}$) allows us to estimate the jerk. However, this information is not usually available from motor drivers.

However, this model of exact linearization is useful since it allows us to design trajectories in the 18-dimensional space of flat outputs and their derivatives which are guaranteed to respect the dynamics and constraints we might want to impose on the state variables.

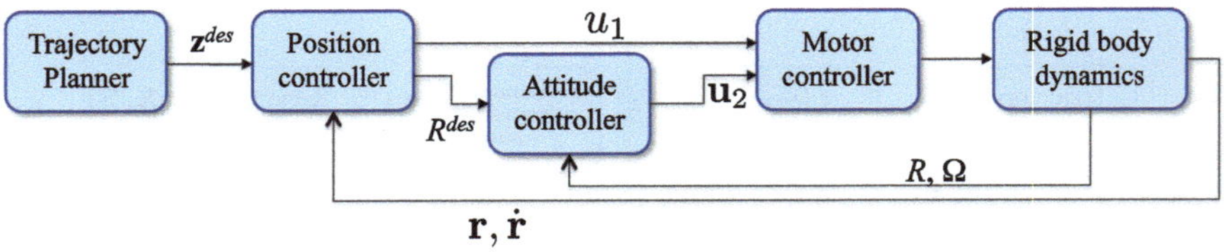

Fig. 3 The attitude controller achieves the desired orientation, which is in turn computed from the errors in position

In most previous work [8, 14, 42], the control problem is addressed by decoupling the position control and attitude control subproblems as illustrated in Fig. 3. The position controller is obtained by projecting the position error (and its derivatives) along $\mathbf{b}_3$ and applying the input u_1 that cancels the gravitational force and provides the appropriate proportional plus derivative feedback:

$$u_1 = m\mathbf{b}_3^{\mathrm{T}}\left(\ddot{\mathbf{r}}^{\mathrm{des}} + K_d\left(\dot{\mathbf{r}}^{\mathrm{des}} - \dot{\mathbf{r}}^{\mathrm{des}}\right) + K_p\left(\mathbf{r}^{\mathrm{des}} - \mathbf{r}^{\mathrm{des}}\right) - \mathbf{g}\right). \quad (5)$$

The attitude controller varies based on the representation which is either using Euler angles, quaternions or rotation matrices. Euler angle representations have singularities and are suitable only for small excursions from the hover position. In most cases, it is sufficient to use linear controllers that are based on the linearization of the plant dynamics around the hover position [8, 15, 28, 32, 42]. The use of quaternions permits the exact cancellation of dynamics and a nonlinear controller that is exponentially stable almost everywhere in $SO(3)$ [53]. A similar result with rotation matrices is available in [22]. In both these papers, the error is defined on the rotation group and does not require the error to be small.

In [22], the two controllers are shown to result in a nonlinear controller that explicitly track trajectories in $SE(3)$. The key idea is to design exponentially converging controllers in $SO(3)$ using an accurate measure of the error in rotations instead of taking linear approximations:

$$\hat{e}_R = \frac{1}{2}\left(\left(R^{\mathrm{des}}\right)^{\mathrm{T}} R - R^{\mathrm{T}} R^{\mathrm{des}}\right) \quad (6)$$

which yields a skew-symmetric matrix representing the axis of rotation required to go from R to R^{des} and with the magnitude that is equal to the sine of the angle of rotation. Computing the proportional plus derivative of the error on $SO(3)$ and compensating for the nonlinear inertial terms gives us:

$$\mathbf{u}_2 = J(-k_R e_R - k_\Omega e_\Omega) + \Omega \times J\Omega, \quad (7)$$

If we do not consider constraints on the state or the inputs, (6–7) achieve asymptotic convergence to specified trajectories in $SE(3)$ [22]. From a practical standpoint it is possible to neglect the nonlinear $\boldsymbol{\Omega} \times J\boldsymbol{\Omega}$ term in the controller and achieve satisfactory performance [26]. Finally, as shown in [28], it is possible to combine this controller with attitude only controllers to fly through vertical windows or land on inclined perches with close to zero normal velocity.

Trajectory controllers allows the robot to build up momentum and reorient itself while coasting with the generated momentum.

3.3 Adaptation and Learning

The dynamic models suffer from two types of limitations. First, such parameters as the location of the center of mass, the moments of inertia and the motor time constants are not precisely known. Second, the aerodynamic models are only approximate.

The first difficulty is overcome using parameter estimation algorithms. Because the unknown parameters appear linearly in the equations of motion (as in the case for robot manipulators [9, 37, 54]), we can write the state equations in discrete time as follows,

$$\mathbf{y}_{k+1} = \theta^T \Phi_k$$

θ is the *parameter vector*, Φ_k and $\mathbf{y}_k$ are the regressor and the measurement at the *k*th time step. A simple linear least-squares method can be used to estimate the unknown parameters as shown in [27] either in a batch or in a recursive algorithm provided the dynamics are persistently excited. These methods can also be used to determine the offsets in IMU readings and for online calibration [47].

Adapting to varying aerodynamic conditions such as those encountered in narrow passages or perturbations due to wind gusts is harder because of the interaction between the time scales of estimation and control. Model Reference Adaptive Control techniques can be used in such settings, although it is necessary to get good measurements of the inputs (motor currents or speeds) and state variables for effective adaptation.

Iterative learning has been used effectively in [25, 28] for acrobatic maneuvers. Such techniques allow the robot to learn trajectories and inputs without knowing a precise aerodynamic model.

Regardless of the specific platform, it is unlikely that a conventional model-based approach to control can work without a robust adaptation mechanism. The small length scales and inertias lead to variations in dynamics that are very difficult to model and impossible to reason about in real time. However it is also unlikely if purely data-driven approaches can be used for control of micro-UAVs. While apprenticeship methods and variants of reinforcement learning algorithms (see, for example, [1]) have achieved remarkable results, they require an expert human operator to generate data for model and control identification. Further, it is unclear if these methods can generalize the results to cases not a priori encountered, where training data is not available. Indeed, in much of the work considered in our own group [28, 50], it is very challenging if not impossible for a trained human operator to fly the robot in the specified manner.

4 Planning

Incremental search [23] and sampling based techniques [21], which are excellent for planning in configuration spaces, are not particularly well-suited for planning motions for underactuated systems. RRT methods and their variants can solve problems with dynamic constraints. For example, in [47], a RRT planner is used to generate trajectories online through a cluttered environment with models acquired by a laser and a camera, but for dynamic models obtained by linearization around the hover operating point. However, the complexity of a 12-dimensional state space with four inputs makes such techniques impractical for planning fast motions through constrained environments. Smaller problems, for example planning motions in the plane, can be solved using reachability algorithms [12], but it is difficult to explore using the full state space using such approaches.

An alternative approach is to use a combination of planning algorithms for configuration spaces along with controller synthesis techniques to ensure the UAVs can execute the planned trajectory. For example, RRT-like techniques have been used with LQR-like control synthesis techniques to find trajectories and sufficing (and even optimal) control policies [49]. Similarly, uncertainty in dynamics and estimation can be addressed using LQG techniques with motion planners [52]. However, techniques like this have yet to be applied to 3-D motion planning of UAVs.

Model predictive control (MPC) techniques represent a third approach that can be used to solve planning and control problems for underactuated systems [19, 56]. These techniques are promising since they combine open loop (optimal) motion planning with feedback control—by generating open loop trajectories based on environmental models periodically with a time interval that is much smaller than the horizon of planning, corrective motions can be generated to accommodate changes in the environment. However, with such approaches, convergence guarantees are difficult to prove. It is possible to prove stability of the MPC algorithm when the linearized model is fully controllable about the goal position [56] (which is generally possible when the goal corresponds to a static hover position), or if a control Lyapunov function can be synthesized for goal positions [16]. Guarantees aside, the synthesis of optimal control solutions even with a finite horizon and a terminal cost function can be difficult with limiting on-board processing resources. Thus it appears to be difficult to directly apply such techniques to the trajectory generation of a quadrotor with guarantees.

It appears that a hierarchical approach that combines incremental search or sampling based techniques in configuration space with optimal control techniques that refines configuration space trajectories in state space is the best framework to solve such problems. If a configuration space planner can be used first to establish waypoints and constraints, optimal trajectories that respect these constraints and the dynamics of the UAV can be generated as a second step. In [26], the property of differential flatness is used to develop an algorithm that enables the generation of optimal trajectories through a series of keyframes or waypoints in the set of positions and orientations, while ensuring safe passage through specified corridors and satisfying constraints on achievable velocities, accelerations and inputs. Since the cost function and all the constraints can be written as algebraic functions of the flat output vector, $\mathbf{z}$, the general setting reduces to solving the problem:

$$\min_{\mathbf{v}(t)} \int_0^T L(\mathbf{z})dt, \quad s.t.\, g(\mathbf{z}) \leq 0 \tag{8}$$

A simple choice for $L(\mathbf{z})$ is the square of the norm of the input vector, which turns out to be the equivalent of finding the trajectory that minimizes the snap and the yaw acceleration along the trajectory. It also has the added benefit of yielding a convex cost function. Recall that trajectories in this flat space automatically satisfy the dynamic equations of motion. Thus the only constraints in $g(\mathbf{z}) \leq 0$ are those on the position (obstacles), velocity (maximum angular rates because of gyro saturation), accelerations (saturation of the IMU), and inputs (propellers can only exert positive lift). All except the position constraints are linear. By linearizing the position constraints the optimization in (8) becomes a convex program. The unconstrained problem, the minimum snap trajectory optimization, yields an analytical solution—a seventh degree polynomial function of time for which we can introduce a polynomial basis for the trajectories. We can similarly use polynomial functions (if necessary of higher order) to satisfy all the constraints in (8). The resulting trajectories have interesting time scaling properties [26] and can be refined efficiently for different values of T to obtain the fastest trajectory to satisfy all the constraints. Finally the quadratic program can be solved in real time quite efficiently, and even in a distributed MPC-like setting for multiple quadrotors at speeds approaching 20 Hz [50].

5 State Estimation and Perception

State estimation is a fundamental necessity for any application involving autonomous UAVs. However, platform design, mobility and payload constraints place considerable restrictions on available computational resources and sensing. The agility and three-dimensional mobility of the vehicle require sensors that provide low-latency information about the three-dimensional environment surrounding the vehicle. Although in open outdoor domains, this problem is seemingly solved with onboard GPS, IMU and camera sensors [44, 46], indoor domains and cluttered outdoor environments still pose a considerable challenge. In such complex environments, the vehicle must be able to localize, detect or perceive obstacles, generate trajectories to navigate around the obstacles and track the trajectories with reasonable accuracy. Any failure to successfully achieve any of these requirements may in fact lead to a complete failure of the vehicle. Further, outdoor environmental effects (e.g. obscurants [45], wind, direct sunlight, GPS-shadowing) and indoor structural considerations (e.g. obstacles, tight corridors, vehicle-induced wind [32]) can challenge the consistency and accuracy of estimation algorithms that are not designed to directly consider these issues.

The fusion of information from multiple onboard sensors such as IMU, laser and cameras (monocular, stereo and RGB-D) do much to address these issues but come with a cost on processing demands, payload and power. Thus there is a real need to find a balance between sensor availability, onboard and offboard processing and operating conditions (which in turn lead to restrictions on the kind of environments in which the UAV can operate).

Initial developments in the area focused on systems capable of navigating indoor environments with algorithms leveraging laser and IMU information to generate a map and localize within the map [4, 13]. Processing for estimation and mapping is shared between local and external computational resources. Unlike the previous work, a monocular camera approach is employed for full pose estimation and localization of ground terrain in GPS-denied environments in [7] but with offboard processing. However, a major concern with offboard processing is the need to maintain uninterrupted, low-latency communication. While this is possible in some indoor and outdoor environments, it inhibits the ability of the system to operate autonomously throughout more general and complex environments. Additionally, the added time cost of external information exchange reduces the performance of the onboard feedback control due to the communication incurred time-delays.

To address these issues, in [47] we considered a similar problem but required the development of an implementation that permitted all processing to occur on the vehicle in real-time. The advancements made in this work were in the form of system design and algorithm optimization to permit autonomous navigation using an IMU, camera and laser to generate three-dimensional maps throughout large and multi-story environments using only limited onboard processing. Further, as all processing occurred in real-time and on the vehicle, we were able to leverage the feedback from the state estimation to drive model-based adaption to account for external disturbances due to gusting wind and ground effects.

Thus far, the discussion focuses on autonomous navigation, where the vehicle plans and controls to goals provided by an external entity. A remaining question is the introduction of perception, planning and control to permit autonomous exploration, where perception algorithms must also allow the UAV to reason about the environment to determine control policies that will yield maximal information for mapping. However, a major challenge in moving toward this direction is the lack of three-dimensional sensors that can be mounted on UAVs, which are required for 3-D exploration. Unfortunately, rich sensor sources such as three-dimensional laser range finders and omni-directional cameras either do not fit the vehicle payload constraints or are prohibitive given the limited computational resources. As such, it is necessary to focus on new algorithmic methods to explore an environment given limited sensing and computational resources. A current strategy we are pursuing in ongoing research [48] is the application of stochastic-differential equations to establish information frontiers between spatial regions that represent the known, explored environment and regions that represent the unexplored environment. The approach strives to find a balance between the computationally complexity of analyzing a full three-dimensional map and the limited field-of-view of onboard sensing. The area of autonomous exploration and perception is clearly an area with rich research possibilities that will become increasingly viable as computing and sensing options improve in time.

6 Other Challenges

6.1 Scaling and SWaP Constraints

One of the key challenges in creating small autonomous UAVs are the so-called size, weight and power constraints. Packaging constraints are severe. Sensors and processors have to be smaller due to the limitations on payload. Because of this, it is difficult to create autonomous quadrotors (with onboard computation and sensing) at small length scales. The smallest autonomous quadrotors capable of exploring, mapping and scouting an unknown three-dimensional building-like environment have a characteristic length of approximately 0.75 m, mass of a little less than 2 kg., and power consumptions over 400 W leading to a mission life of around 10–20 min [47]. The main reason for the size is the need to carry three-dimensional sensors like Hokuyo laser range finders or Microsoft Kinect cameras. This in turn leads to high power consumption. Many impressive advances have been made in mapping and estimation for autonomous navigation using just an IMU and a camera [7]. Recent results point to algorithms that yield estimates of 3-D metric information from just monocular vision combined with a good IMU [20, 35]. This suggests that the sensor payload challenges associated with scaling can be overcome in the near future.

However, the net payload constraints are still significant if the UAV needs to be able to transport or manipulate a payload. Since the linear acceleration scales with L (Sect. 2), it is impossible to design small UAVs that are able to overcome this fundamental constraint. Current UAVs with $L \sim 1$ m have a maximum payload of around 1 kg. One way to overcome this constraint is by using multiple UAVs to cooperatively transport or manipulate payloads. Recent work suggests that the challenges in coordinating multiple UAVs and adapting individual vehicles to constraints imposed by other vehicles is possible in different settings ranging from payloads suspended from UAVs [11, 17, 18, 31] to payloads rigidly grasped by UAVs [29].

6.2 Grasping and Manipulation

There are many challenges in aerial grasping for micro-UAVs. The biggest challenge arises from their limited payload. While multiple UAVs can be coordinated to carry payloads with grippers [29], the end effector or gripper has to be light weight and capable of grasping complex shapes. Second, the dynamics of the robot are significantly altered by the addition of payloads. While this can be beneficial to tasks when aerial robots need to sense the payload that has been grasped, it is important to also be able to compensate for and adapt to changes in the dynamics caused by the grasped payload. It is clear that the design of claws for grasping represents a challenging mechanism design problem where the compliance and damping must be finely tuned to grasping. Finally, all the challenges associated with grasping objects (approaching, contacting, and securing the grasp) make this a significant challenge.

Preliminary work in this direction has appeared in conferences over the last 2 years. The difficulties associated with the analysis of the flight dynamics and stability are explained with the help of an approximate model in [41]. The mechanics and design for aerial grasping are addressed in [27, 29]. Parameter estimation methods for estimating the grasped payload and the ability to adapt to the payloads are investigated in [27]. The application to construction of structures is discussed in [24] in which the sensed disturbance forces are used to verify successful grasping and assembly. Micro-UAVs afford opportunities for truly three-dimensional grasping since they can, in principle, approach objects or assemblies from any direction, and because they can sense disturbance forces without additional sensors. This is a fertile area of future research.

6.3 Adaptation to Complex Environments with Changing Dynamics

As discussed earlier, it is very difficult to model micro-UAVs with a high degree of precision because of the complexity of modeling air drag, the interactions between the motor, rotor and the fluid through which the propellor blades must move, the dynamics of the flexible propellor blade and the different nonlinearities and saturation effects in the sensors and actuators. And such difficulties get compounded when the rigid body dynamics interact with the aero dynamics and the fluid-structure coupling effects become significant, as is the case in flapping-wing vehicles or rotor crafts with long blades. As discussed earlier in Sect. 3.3, adaptive control and iterative learning techniques can be used to handle some of these challenges. However, parameterizing the set of uncertainties and ensuring the appropriate level of sensing and actuation to identify these parameters may not always be possible. Methods such as the ones described in [25, 27, 28] are good starting points for such studies.

The effects of changes in the aerodynamics in three-dimensional environments are much harder to study. A study of wind gusts in [57] illustrates the challenges in modeling and experimentation. For small aircrafts, small, local variations in wind conditions can be significant. Transitions between indoor and outdoor environments can induce large perturbations. Even without wind gusts, changes in elevation can dramatically alter the lift generated by individual propellers resulting in significant disturbances to the vehicle. Some of these phenomena are studied for modestly changing environments in [47] where the inputs required to compensate for the changes can be parameterized by a small set of trim parameters. In these studies the sensed information was limited to gross position and velocity information which in turn limits the level of adaptation that is possible. If aerial vehicles are to become as reliable and easy-to-use as ground vehicles, it is necessary to develop techniques that will enable safe and robust low-level navigation behaviors in complex environments.

7 Conclusion

Micro UAVs are potentially game changers in robotics. They can operate in constrained three-dimensional environments, explore and map multi-story buildings, manipulate and transport objects, and even perform such tasks as assembly. Our recent experiments with quadrotors in collapsed buildings in Sendai, Japan in July 2011 [33] demonstrated many benefits of using autonomous quadrotors for mapping unknown environments, searching in collapsed buildings and exploration in settings that are too dangerous for human rescue workers. Just as the advent of mobile robots led to a flurry of activity with new research problem areas, micro-UAVs will inevitably lead robotics research in new and exciting directions.

References

1. P. Abbeel, Apprenticeship learning and reinforcement learning with application to robotic control. Ph.D. thesis, Stanford University, Stanford, CA, 2008
2. Aeroenvironment nano hummingbird (2011). http://www.avinc.com/nano
3. Ascending Technologies, GmbH. http://www.asctec.de
4. A.G. Bachrach, Autonomous flight in unstructured and unknown indoor environments. Master's thesis, MIT, Cambridge, MA, 2009
5. C. Bermes, Design and dynamic modeling of autonomous coaxial micro helicopters. Ph.D. thesis, ETH Zurich, Switzerland, 2010
6. C. Bermes, D. Schafroth, S. Bouabdallah, R. Siegwart, Modular simulation model for coaxial rotary wing mavs, in *Proceedings of The 2nd International Symposium on Unmanned Aerial Vehicles* (2009)
7. M. Blosch, S. Weiss, D. Scaramuzza, R. Siegwart, Vision based MAV navigation in unknown and unstructured environments, in *Proceedings of the IEEE International. Conference on Robotics and Automation* (Anchorage, AK, 2010), pp. 21–28

8. S. Bouabdallah, Design and control of quadrotors with applications to autonomous flying. Ph. D. thesis, Ecole Polytechnique Federale de Lausanne, Lausanne, Switzerland, 2007
9. J. Craig, P. Hsu, S. Sastry, Adaptive control of mechanical manipulators, in *Proceedings of the IEEE International. Conference on Robotics and Automation*, vol. 3 (1986), pp. 190–195. doi:10.1109/ROBOT.1986.1087661
10. L. Faruque, J.S. Humbert, Dipteran insect flight dynamics. part 2: lateral-directional motion about hover. J. Theoret. Biol. **265**(3), 306–313 (2010). doi:10.1016/j.jtbi.2010.05.003
11. J. Fink, N. Michael, S. Kim, V. Kumar, Planning and control for cooperative manipulation and transportation with aerial robots. Int. J. Robot. Res. **30**(3) (2011)
12. J.H. Gillula, H. Huang, M.P. Vitus, C.J. Tomlin, Design of guaranteed safe maneuvers using reachable sets: autonomous quadrotor aerobatics in theory and practice, in *Proceedings of the IEEE International. Conference on Robotics and Automation* (Anchorage, AK, 2010), pp. 1649–1654
13. S. Grzonka, G. Grisetti, W. Burgard, Towards a navigation system for autonomous indoor flying, in *Proceedings of the IEEE International. Conference on Robotics and Automation* (Kobe, Japan, 2009), pp. 2878–2883
14. D. Gurdan, J. Stumpf, M. Achtelik, K. Doth, G. Hirzinger, D. Rus, Energy-efficient autonomous four-rotor flying robot controlled at 1 kHz, in *Proceedings of the IEEE International. Conference on Robotics and Automation* (Roma, Italy, 2007)
15. H. Huang, G.M. Hoffman, S.L. Waslander, C.J. Tomlin, Aerodynamics and control of autonomous quadrotor helicopters in aggressive maneuvering, in *Proceedings of the IEEE International. Conference on Robotics and Automation* (Kobe, Japan, 2009), pp. 3277–3282
16. A. Jadbabaie, J. Hauser, On the stability of receding horizon control with a general terminal cost. IEEE Trans. Autom. Control **50**(5), 674–678 (2005)
17. Q. Jiang, V. Kumar, The direct kinematics of objects suspended from cables, in *ASME International Design Engineering Technical Conference and Computer and Information in Engineering Conference* (2010)
18. Q. Jiang, V. Kumar, in *The inverse kinematics of 3-d towing*, ed. by J. Lenarcic, M.M. Stanisic. Advances in Robot Kinematics (2010), pp. 321–328
19. H. Kim, D. Shim, S. Sastry, Nonlinear model predictive tracking control for rotorcraft-based unmanned aerial vehicles, in *Proceedings of the American Control Conference*, vol. 5 (Anchorage, AK, 2002), pp. 3576–3581
20. L. Kneip, A. Martinelli, S. Weiss, D. Scaramuzza, R. Siegwart, Closed-form solution for absolute scale velocity determination combining inertial measurements and a single feature correspondence, in *Proceedings of the IEEE International Conference on Robotics and Automation* (2011), pp. 4546–4553
21. S.M. Lavalle, *Planning Algorithms* (Cambridge University Press, 2006)
22. T. Lee, M. Leok, N. McClamroch, Geometric tracking control of a quadrotor uav on SE(3), in *Proceedings of the IEEE Conference on Decision and Control* (2010)
23. M. Likhachev, G. Gordon, S. Thrun, ARA*: anytime A* with provable bounds on sub-optimality. Adv. Neural Inf. Process. Syst. **16** (2003)
24. Q. Lindsey, D. Mellinger, V. Kumar, Construction of cubic structures with quadrotor teams, in *Proceedings of Robotics: Science and Systems* (Los Angeles, CA, 2011)
25. S. Lupashin, A. Schollig, M. Sherback, R. D'Andrea, A simple learning strategy for high-speed quadrocopter multi-flips, in *Proceedings of the IEEE International Conference on Robotics and Automation* (Anchorage, AK, 2010), pp. 1642–1648
26. D. Mellinger, V. Kumar, Minimum snap trajectory generation and control for quadrotors, in *Proceedings of the IEEE International Conference on Robotics and Automation* (Shanghai, China, 2011)
27. D. Mellinger, Q. Lindsey, M. Shomin, V. Kumar, Design, modeling, estimation and control for aerial grasping and manipulation, in *Proceedings of the IEEE/RSJ International Conference on Intelligent Robots and Systems* (San Francisco, CA, 2011)

28. D. Mellinger, N. Michael, V. Kumar, Trajectory generation and control for precise aggressive maneuvers with quadrotors, in *Proceedings of the International Symposium on Experimental Robotics* (Delhi, India, 2010)
29. D. Mellinger, M. Shomin, N. Michael, V. Kumar, Cooperative grasping and transport using multiple quadrotors, in *International Symposium on Distributed Autonomous System* (Lausanne, Switzerland, 2010)
30. B. Mettler, Modeling small-scale unmanned rotorcraft for advanced flight control design. Ph. D. thesis, Carnegie Mellon University, Pittsburgh, PA, 2001
31. N. Michael, J. Fink, V. Kumar, Cooperative manipulation and transportation with aerial robots. Auton. Rob. **30**(1), 73–86 (2011)
32. N. Michael, D. Mellinger, Q. Lindsey, V. Kumar, he grasp multiple micro uav testbed. IEEE Robot. Autom. Mag. (2010)
33. N. Michael, S. Tadokoro, K. Nagatani, K. Ohno, Experiments with air ground coordination for search and rescue in collapsed buildings (2011). Working Paper
34. D.S. Miller, G. Gremillion, B. Ranganathan, P.D. Samuel, S. Zarovy, M. Costello, A. Mehta, J.S. Humbert, Challenges present in the development and stabilization of a micro quadrotor helicopter, in *Autonomous Weapons Summit and GNC Challenges for Miniature Autonomous Systems Workshop* (2010)
35. A.I. Mourikis, N. Trawny, S.I. Roumeliotis, A.E. Johnson, A. Ansar, L. Matthies, Vision-aided inertial navigation for spacecraft entry, descent, and landing. IEEE Trans. Robot. **25**(2), 264–280 (2009)
36. R.M. Murray, M. Rathinam, W. Sluis, Differential flatness of mechanical control systems: a catalog of prototype systems, in *Proceedings of the 1995 ASME International Congress and Exposition* (1995)
37. G. Niemeyer, J.J. Slotine, Performance in adaptive manipulator control, in *Proceedings of the IEEE Conference on Decision and Control*, vol. 2 (1988), pp. 1585–1591. doi:10.1109/CDC.1988.194595
38. M.J.V. Nieuwstadt, R.M. Murray, Real-time trajectory generation for differentially flat systems. Int. J. Robust Nonlinear Control **8**, 995–1020 (1998)
39. R. Ortega, M.W. Spong, Adaptive motion control of rigid robots: a tutorial, in *Proceedings of the IEEE Conference on Decision and Control*, vol. 2 (1988), pp. 1575–1584. doi:10.1109/CDC.1988.194594
40. D. Pines, F. Bohorquez, Challenges facing future micro air vehicle development. AIAA J. Aircr. **43**(2), 290–305 (2006)
41. P. Pounds, A. Dollar, Hovering stability of helicopters with elastic constraints, in *ASME Dynamic Systems and Control Conference* (2010)
42. O. Purwin, R. D'Andrea, Performing aggressive maneuvers using iterative learning control, in *Proceedings of the IEEE International Conference on Robotics and Automation* (Kobe, Japan, 2009), pp. 1731–1736
43. J. Ratti, G. Vachtsevanos, Towards energy efficiency in micro hovering air vehicles, in *IEEE Aerospace Conference* (Big Sky, MT, 2011), pp. 1–8
44. S. Saripalli, J.F. Montgomery, G.S. Sukhatme, Vision-based autonomous landing of an unmanned aerial vehicle, in *Proceedings of the IEEE International Conference on Robotics and Automation* (Washington, DC, 2002), pp. 2799–2804
45. K.W. Sevcik, N. Kuntz, P.Y. Oh, Exploring the effect of obscurants on safe landing zone indentification. J. Intell. Robot. Syst. **57**(1–4), 281–295 (2010)
46. C.S. Sharp, O. Shakernia, S.S. Sastry, A vision system for landing an unmanned aerial vehicle, in *Proceedings of the IEEE International Conference on Robotics and Automation*, vol. 2 (Seoul, Korea, 2001), pp. 1720–1727
47. S. Shen, N. Michael, V. Kumar, 3D estimation and control for autonomous flight with constrained computation, in *Proceedings of the IEEE International Conference on Robotics and Automation* (Shanghai, China, 2011)
48. S. Shen, N. Michael, V. Kumar, Exploration and control for autonomous mapping with aerial robots. Technical report, University of Pennsylvania, 2011

49. R. Tedrake, LQR-Trees: feedback motion planning on sparse randomized trees, in *Proceedings of Robotics: Science and Systems* (Seattle, WA, 2009)
50. M. Turpin, N. Michael, V. Kumar, Trajectory design and control for aggressive formation flight with quadrotors, in *Proceedings of the International Symposium of Robotics Research* (Flagstaff, AZ, 2011)
51. U.S. Military Unmanned Aerial Vehicles (UAV) Market Forecast 2010–2015 (2011). http://www.uavmarketresearch.com/
52. J. van der Berg, P. Abbeel, K. Goldberg, LQG-MP: Optimized path planning for robots with motion uncertainty and imperfect state information. Int. J. Robot. Res. **30**(7), 895–913 (2011). doi:10.1177/0278364911406562
53. J. Wen, K. Kreutz-Delgado, The attitude control problem. IEEE Trans. Autom. Control **36** (10), 1148–1162 (1991)
54. L. Whitcomb, A. Rizzi, D. Koditschek, Comparative experiments with a new adaptive controller for robot arms. IEEE Trans. Robot. Autom. **9**(1), 59–70 (1993). doi:10.1109/70.210795
55. C.H. Wolowicz, J.S. Bowman, W.P. Gilbert, Similitude requirements and scaling relationships as applied to model testing. Technical report, NASA, 1979
56. J. Yu, A. Jadbabaie, J. Primbs, Y. Huang, Comparison of nonlinear control design techniques on a model of the caltech ducted fan, in *IFAC World Congress, IFAC-2c-112* (1999), pp. 53–58
57. S. Zarovy, M. Costello, A. Mehta, A. Flynn, G. Gremillion, D. Miller, B. Ranganathan, J.S. Humbert, P. Samuel, Experimental study of gust effects on micro air vehicles, in *AIAA Conference on Atmospheric Flight Mechanics, AIAA-2010–7818* (American Institute of Aeronautics and Astronautics, 2010)

Part II
Perception and Mapping

Unsupervised 3D Object Discovery and Categorization for Mobile Robots

Jiwon Shin, Rudolph Triebel and Roland Siegwart

Abstract We present a method for mobile robots to learn the concept of objects and categorize them without supervision using 3D point clouds from a laser scanner as input. In particular, we address the challenges of categorizing objects discovered in different scans without knowing the number of categories. The underlying object discovery algorithm finds objects per scan and gives them locally-consistent labels. To associate these object labels across all scans, we introduce *class graph* which encodes the relationship among local object class labels. Our algorithm finds the mapping from local class labels to global category labels by inferring on this graph and uses this mapping to assign the final category label to the discovered objects. We demonstrate on real data our algorithm's ability to discover and categorize objects without supervision.

1 Introduction

A mobile robot that is capable of discovering and categorizing objects without human supervision has two major benefits. First, it can operate without a hand-labeled training data set, eliminating the laborious labeling process. Second, if human-understandable object labels are needed, automatic discovery and categorization leave the user with a far less tedious task of labeling categories rather than raw data points. Unsupervised discovery and categorization, however, require the robot to understand what an object constitutes. In this work, we address the challenges of unsupervised object discovery and categorization using 3D scans

J. Shin (✉) · R. Siegwart
Autonomous Systems Lab, ETH Zurich, Zurich, Switzerland
e-mail: jiwon.shin@mavt.ethz.ch

R. Siegwart
e-mail: rsiegwart@ethz.ch

R. Triebel
The Oxford Mobile Robotics Group, University of Oxford, Oxford, England
e-mail: rudi@robots.ox.ac.uk

H.I. Christensen and O. Khatib (eds.), *Robotics Research*,
Springer Tracts in Advanced Robotics 100, DOI 10.1007/978-3-319-29363-9_4

from a laser as input. Unlike other object discovery algorithms, our approach does not require presegmentation of background, one-to-one mapping between input scan and label, nor a particular object symmetry. Instead, we simply assume that an entity is an object if it is composed of two or more parts and occurs more than once.

We propose a method for robots to discover and categorize objects without supervision. This work especially focuses on categorization of the discovered objects. The proposed algorithm is composed of three steps: detection of potential object parts, object discovery, and object categorization. After segmenting the input 3D point cloud, we extract salient segments to detect regions which are likely to belong to objects. After detecting these potential object parts, we cluster them in feature and geometric space to acquire parts labels and object labels. Reasoning on the relationship between object parts and object labels provides a locally-consistent object class label for each discovered object. Processing a series of scans results in a set of discovered objects, all labeled according to their local class labels. To associate these local class labels, we build a *class graph*. Class graph encodes the dependency among local class labels of similar appearance, and smoothing the graph results in a distribution of the global category labels for each local class label. Marginalizing out the local class labels gives the most likely final category label for each discovered object. We demonstrate on real data the feasibility of unsupervised discovery and categorization of objects.

Contributions of this work are two-folds. First, we improve the object discovery process by extracting potential foreground objects using saliency. Instead of relying entirely on perfect foreground extraction, our algorithm takes the foreground segments only as potential object parts and performs further processing on them before accepting them as object parts. It can thus handle imperfect foreground extraction by removing those potential object parts deemed less fit to be actual object parts. Second, we propose a novel categorization method to associate the locally-consistent object class labels to the global category labels without knowing the number of categories. Our algorithm improves the results of categorization over pure clustering and provides a basis for on-line learning. To our knowledge, no other work has addressed the problem of unsupervised object categorization from discovered objects.

The organization of the paper is as follows. After discussing related work in Sect. 2, we introduce a saliency-based foreground extraction algorithm and explain the single-scan object discovery algorithm in Sect. 3. In Sect. 4, we propose a method for associating the discovered objects for object categorization. After the experimental results in Sect. 5, the paper concludes with Sect. 6.

2 Related Work

Most previous work on unsupervised object discovery assume either a presegmentation of the objects, one object class per image, or a known number of objects and their classes [1, 4, 13]. In contrast, [16] proposed an unsupervised discovery algorithm that does not require such assumptions but instead utilizes regularity of patterns in which the objects appear. This is very useful for man-made structures such as facades of buildings. [2] developed a method to detect and segment similar objects from a single image by growing and merging feature matches.

Our work builds on our previous work [17], which gives nice results for single scenes but does not address the data association problem across different scenes. Thus, the above algorithm cannot identify instances of the same object class that appear in different scenes. In contrast, this approach solves the data association problem and introduces a reasoning on the object level, instead of only assigning class labels to object parts.

An important step in our algorithm is the clustering of feature vectors extracted from image segments. Many different kinds of clustering algorithms have been proposed and their use strongly depends on the application. Some classic methods such as the Expectation-Maximization (EM) algorithm and *k*-means clustering assume that data can be modeled by a simple distribution, while other methods such as agglomerative clustering are sensitive to noise and outliers. To overcome these problems, alternative approaches have been proposed. [11] presented a *spectral clustering* algorithm, which uses the eigenvectors of the data matrix to group points together, with impressive results even for challenging data. Another recent clustering approach is named *affinity propagation*, proposed by [5]. It clusters data by finding a set of exemplar points, which serve as cluster centers and explain the data points assigned to it. This method avoids the pitfalls of a bad initialization and does not require the number of clusters to be prespecified. In this work, we use affinity propagation to cluster image segments in feature space.

Our object categorization method is inspired by the *bag of words* approach [3]. Outside of document analysis, the bag of words method has been applied in computer vision, e.g., for texture analysis or object categorization [10, 15]. Our work uses it to bridge the gap between reasoning on object parts and object instances.

3 Object Discovery

This section describes the algorithm for discovering objects from a single scan. Figure 1 depicts the overall process of the object discovery. Our single-scan object discovery algorithm is based on our previous work [17], which treats every segment as a potential object part and accepts them as objects if after inference, any nearby segment has the same class label as itself. This algorithm, however, has several disadvantages. First, because the original algorithm considers all segments as potential object parts, it makes many false neighborhood connections between foreground and background segments. This results in object candidates composed of real object parts and background parts. Second, it has relatively high false-positive rate because it cannot differentiate clutter objects from real objects. Third, it wastes computation by extracting feature descriptors on background

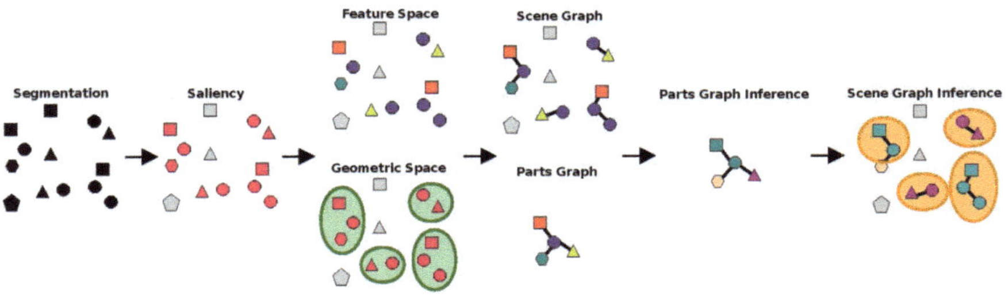

Fig. 1 Overview of the discovery process (best seen in *color*). After performing segmentation on input data and extracting salient segments, the algorithm clusters the salient segments in feature and geometric space. The clusters are then used to create scene graph and parts graph, which encode the relationship between object parts and objects. Running inference on the graphs result in the discovery of four objects as shown on the *right*

segments. In this paper, we introduce saliency-based foreground extraction algorithm to overcome these problems.

3.1 Extraction of Potential Object Parts

A simple way to seperate foreground from background is to fit planes into the data and remove all points that correspond to the planes. This removes all wall, ceiling, and floor parts as in, e.g., [4], but can cause at least two problems. First, it may also remove planar segments close to a wall or floor that are actually object parts and thus should not be removed. Second, it is often insufficient to define background as planar because background may be truly curved or non-planar due to sensor noise.

Inspired by computer vision [7], we suggest a different approach for foreground extraction using *saliency*. The idea is to classify certain parts of an image as visually more interesting or *salient* than others. This classification determines saliency based on difference in entropy of a region to its nearby regions. Most work on saliency has been on 2D images, but [6] uses saliency for object recognition in 3D range scans. Their technique, however, remaps depth and reflectance images as greyscale images and applies 2D saliency techniques to find salient points. This work detects salient segments in true 3D by processing depth values of range data directly.

Our saliency algorithm computes saliency at point level and segment level. Point saliency provides saliency of a point while segment saliency represents saliency of a segment. A *point saliency* s_p is composed of a *local* saliency s_l and a *global* saliency s_g. Local saliency s_l is defined as

$$s_l(\mathbf{p}) = \frac{1}{s_l^{max}} \sum_{\mathbf{p}' \in \mathcal{N}(\mathbf{p})} \mathbf{n} \cdot (\mathbf{p} - \mathbf{p}'), \tag{1}$$

where $\mathbf{n}$ is the normal vector at a point $\mathbf{p}$, and $N(\mathbf{p})$ defines the set of all points in the neighborhood of $\mathbf{p}$. To obtain a value between 0 and 1, the local saliency is normalized by the maximum local saliency value s_l^{max}. Intuitively, local saliency measures how much the point $\mathbf{p}$ sticks out of a plane that best fits into the local surrounding $N(\mathbf{p})$. This resembles the plane extraction technique mentioned earlier.

Points that are closer to the sensor are more likely to belong to foreground and thus globally more salient than points that are far away from the sensor. We capture this property in global saliency. Global saliency s_g is defined as

$$s_g(\mathbf{p}) = \frac{1}{s_g^{max}} ||\mathbf{p}^{max} - \mathbf{p}||, \tag{2}$$

where $\mathbf{p}^{max}$ denotes the point that is farthest away from the sensor origin. As in local saliency, global saliency is normalized to range between 0 and 1.

We define segment saliency s_s for a segment $\mathbf{s}$ as a weighted average of the local and global saliency for all points which belong to the segment and multiply it by a size penalty α, i.e.,

$$s_s(\mathbf{s}) = \alpha \left(\frac{1}{|\mathbf{s}|} \sum_{\mathbf{p} \in s} w s_l(\mathbf{p}) + (1 - w) s_g(\mathbf{p}) \right), \tag{3}$$

where $\alpha = \exp(-(|\mathbf{s}| - |\mathbf{s}_{mean}|)^2)$ penalizes segments that are too big or too small as they are likely to originate from a wall or sensor noise; $|\mathbf{s}|$ denotes the size (number of points) of the segment $\mathbf{s}$; and w weighs between local and global saliency. The weight w depends on the amount of information contained in local and global saliency, measured by entropy of the corresponding distributions. Interpreting s_l and s_g as probability distributions, we can determine entropy h_l and h_g for local and global saliency by

$$h_l = - \sum_{i=1}^{N} s_l(\mathbf{p}_i) \log s_l(\mathbf{p}_i) \tag{4}$$

$$h_g = - \sum_{i=1}^{N} s_g(\mathbf{p}_i) \log s_g(\mathbf{p}_i), \tag{5}$$

where $N = 20$ in this work. As a saliency distribution with lower entropy is more informative, we set the weight w as $w = \frac{h_g}{h_g + h_l}$, which is high when local saliency has low entropy and low when it has high entropy. The weight ensures that more informative entropy distribution contributes more to the final saliency.

Segment saliency $s_s(\mathbf{s})$ ranges between 0 and 1. We consider a segment salient if its saliency is higher than 0.5 and accept it as a potential object part. Only these potential object parts $\mathcal{S}$ are further processed for object discovery. Figure 2 shows a scene after salient segments are extracted.

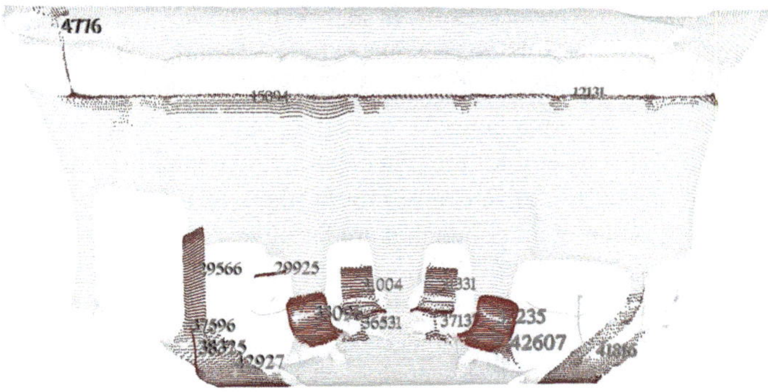

Fig. 2 An example image after saliency computation. *Colored* segments are considered salient and thus treated as potential object parts. Numbers indicate segment IDs

3.2 Object Discovery for a Single Scan

Once we extract potential object parts $\mathcal{S}$, next step is to reason on them to discover objects. The object discovery step on single scan is based on our previous work [17]. The underlying idea behind our object discovery algorithm is that object parts which belong to the same object are frequently observed together, and hence by observing which parts occur together frequently, we can deduce the object class label for these parts. Using this idea, a brief summary of the algorithm is as follows. Given the potential object parts $\mathcal{S}$, we extract a feature vector $\mathbf{f}_i$ for each potential object part $\mathbf{s}_i$. The feature vector $\mathbf{f}_i$ is composed of spin images [8], shape distributions [12], and shape factors [18]. To determine which set of potential object parts originate from the same *parts type* $\mathcal{F}_i$, we cluster these parts in feature space using affinity propagation [5]. Affinity propagation implicitly estimates the number of clusters C, resulting in clusters $\mathcal{F}_1, \ldots, \mathcal{F}_C$. These clusters define the discovered object parts types.

Clustering in feature space provides parts types, but it does not define which parts belong to the same object *instance*. To obtain the object instances, we perform another clustering on the potential object parts $\mathcal{S}$ but this time in geometric space. As object parts for the same object instance are physically close, clustering in geometric space enables us to group together potential object parts that belong to the same object instance. The geometric clustering algorithm connects every pair of potential objects whose centers are closer than a threshold ϑ_g, and this results in a collection of connected components. The number of connected components K defines the maximum number of object classes present in the scene, and each cluster $\mathcal{G}_i$ of the resulting clusters $\mathcal{G}_1, \ldots, \mathcal{G}_K$ corresponds to an object instance.

Given parts types $\mathcal{F}_1, \ldots, \mathcal{F}_C$ and object classes $\mathcal{G}_1, \ldots, \mathcal{G}_K$, we can then assign a class label $\mathcal{G}_i$ to each potential object part $\mathbf{s}_i$. We determine the assignments by reasoning on the labels at two levels. First, on a more abstract level, the statistical

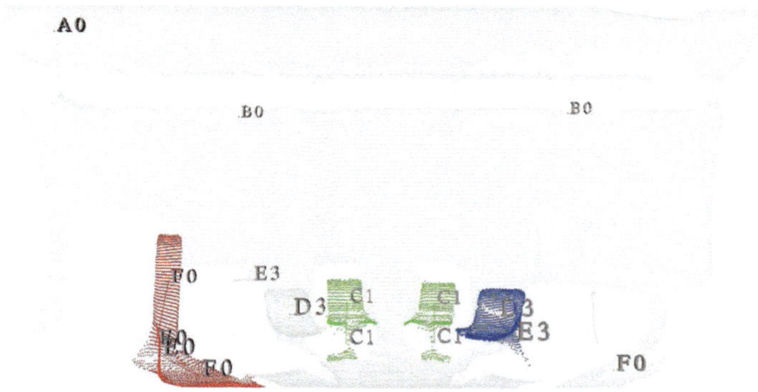

Fig. 3 Result of object discovery of the scene shown in Fig. 2. Discovered objects are *colored* according to their class labels. *Letters* indicate the parts types and *numbers* indicate object classes. Notice that not all potential object parts are accepted as object parts

dependency of class labels $\mathcal{G}_1, \ldots, \mathcal{G}_K$ across different parts types $\mathcal{F}_1, \ldots, \mathcal{F}_C$ is encoded in a Conditional Random Field (CRF) [9] named *parts graph*. Parts graph exploits the fact that object parts that co-occur frequently in the same object instance are more likely to belong to the same object class. For example, back rest and seat, both of which belong to a chair, are frequently found together while seat and shelf, which belong to different objects, are not. The second level of reasoning propagates parts types to object class relationship onto a finer level by combining the class labels obtained from the parts graph with the local contextual information from actual scenes. This is encoded using another CRF called *scene graph*. Performing inference on the parts graph provides the most likely object class label $\mathcal{G}_i$ per parts type $\mathcal{F}_i$ while inference on the scene graph leads to the object class label $\mathcal{G}_i$ per object part $\mathbf{s}_i$. After for all object instances, all their parts are labeled with the most likely object class label, we accept those object instances that contain at least two parts with the same class label as discovered objects $\mathcal{O}_1, \ldots, \mathcal{O}_N$. Figure 3 shows an example of the outcome of the discovery algorithm.

4 Object Categorization

Object discovery algorithm of the previous section is able to find object classes for which at least two instances occur in a given scene. It uses appearance and geometry, i.e., similarity of features and structures, to find several instances of objects that are most likely to define a class in one given scene. In this paper, we go one step further and try to find object *categories*, i.e., object classes that are consistent across a sequence of input scenes. This, however, is not straightforward. As the object discovery process is entirely unsupervised, the resulting local class labels

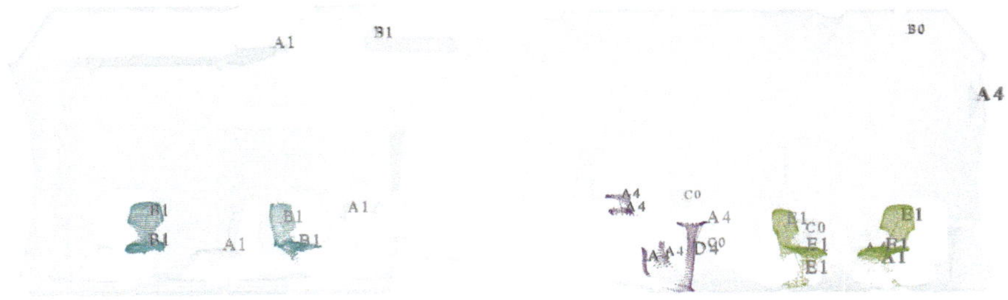

Fig. 4 Objects found in two different scenes. Segments of the same local object label have the same *color* locally

are not unique over a given number of input scans. This means that an object class might be associated with a class label $\mathcal{G}_1$ when one scene is observed, but the same object class might have a different class label $\mathcal{G}_2$ if observed in a different scene. An example of this is shown in Fig. 4. To identify object instances of the same class from different scenes, we need to solve the *data association problem*. Unfortunately, this problem is intractable in general as it involves a correspondence check between every pair of object classes that are found in different scenes. One simple way to address this correspondence problem is to join all scenes into one big scene and run the discovery algorithm on the big scene. Unfortunately, this approach has two major drawbacks: first, the number of connected components K in this big scene would be very large. This heavily increases the computation time of the algorithm and decreases its detection performance because it fails to sufficiently restrict the number of potential object classes. And second, it limits the possibility of running the object discovery in an online framework, which is one major goal of this work. The reason here is that the parts graph would need to be re-built every time a new scene is observed, which decreases the efficiency of the algorithm.

This work addresses the data association problem by introducing a third level of reasoning named *class graph*. The key idea behind the class graph is to find a mapping from local class labels to global category labels. Unlike the parts graph and the scene graph, the class graph models the statistical dependencies between labels of object class instances rather than object parts. Details of the class graph is explained in Sect. 4.2. Next section describes object feature vector for representation of object instances, which are the building blocks of class graph.

4.1 Object Representation

Object feature vector enables a compact representation of object instances. Object feature vector $\mathbf{o}$ is composed of a histogram $\mathbf{h}$ of visual word occurrences and a shape vector $\mathbf{v}$. The histogram $\mathbf{h}$ captures object appearance while the shape vector

$\mathbf{v}$ captures object volume. To compute the histograms, we take the *bag of words* approach and represent an object as a collection of visual words. Bag of words requires visual vocabulary to be defined, and we determine the visual vocabulary by clustering the object parts feature vector $\mathbf{f}$ of all discovered objects. Each cluster $\mathcal{F}_i^*$ is a word in the visual vocabulary $\mathcal{F}_1^*, \ldots, \mathcal{F}_{C^*}^*$, and the total number of words in the vocabulary C^* is equal to the number of clusters C^*. With the visual vocabulary, representing an object as a histogram is simplified to counting the number of occurrences of each visual word in the object. In traditional bag of words approaches, every feature makes a contribution to the bin corresponding to the visual word that best represents the feature. Such approaches, however, do not take into account the uncertainty inherent in the assignment process. Hence, in our work, each object part feature vector $\mathbf{f}$ contributes to all bins of the corresponding histogram $\mathbf{h}$, where the contribution to a bin is determined by the probability $p(\mathbf{w}_i|\mathbf{f})$ of the feature vector $\mathbf{f}$ belonging to the visual word $\mathbf{w}_i$. We compute this probability by nearest-neighbor.

In addition to the histogram $\mathbf{h}$, object feature vector $\mathbf{o}$ contains a shape vector $\mathbf{v}$, which represents object's physical properties. The shape vector $\mathbf{v}$ is composed of three elements—size in horizontal direction, size in vertical direction, and object's location in vertical direction. The horizontal and vertical spans provide the bounding volume in which the object resides. The vertical location gives an estimate on where the object is likely to be found.

4.2 Class Graph

Once the object feature vectors $\mathbf{o}_1, \ldots, \mathbf{o}_{N^*}$ are computed for all discovered objects $\mathcal{O}_1, \ldots, \mathcal{O}_{N^*}$, we determine the mapping from local class labels $\mathcal{G}_1, \ldots, \mathcal{G}_M$ to global category labels $\mathcal{G}_1^*, \ldots, \mathcal{G}_{K^*}^*$ using *class graph* $\mathfrak{C}$. Class graph $\mathfrak{C}$ consists of the node set $\mathcal{V}_{\bar{\mathbf{o}}} = \{\bar{\mathbf{o}}_1, \ldots, \bar{\mathbf{o}}_M\}$ and the edge set $\mathcal{E}_{\bar{\mathbf{o}}} = \{(\bar{\mathbf{o}}_i, \bar{\mathbf{o}}_j)|D(\bar{\mathbf{o}}_i, \bar{\mathbf{o}}_j) < \vartheta_{\bar{\mathbf{o}}}\}$. The nodes are the local class labels $\mathcal{G}_1, \ldots, \mathcal{G}_M$ represented as mean object feature vectors $\bar{\mathbf{o}}_1, \ldots, \bar{\mathbf{o}}_M$, and the edges connect similar local class labels, where the similarity between two local labels is the distance between their mean object feature vectors. The threshold for object similarity $\vartheta_{\bar{\mathbf{o}}}$ is set to 0.5.

To assign global category labels $\mathcal{G}_1^*, \ldots, \mathcal{G}_{K^*}^*$ to local class labels $\mathcal{G}_1, \ldots, \mathcal{G}_M$, we need to find the number of global categories K^*. As mentioned earlier, Affinity Propagation (AP) implicitly determines the number of clusters, and therefore, we cluster the mean object feature vectors $\bar{\mathbf{o}}_1, \ldots, \bar{\mathbf{o}}_M$ by AP clustering. The number of clusters K^* resulting from AP clustering is the maximum number of global categories, and the clusters $\mathcal{G}_1^*, \ldots, \mathcal{G}_{K^*}^*$ are the initial global category labels for the local class labels $\mathcal{G}_1, \ldots, \mathcal{G}_M$. Smoothing this initial mapping determines the final mapping from local class labels to global category labels. Figure 5 shows the overall steps of categorization by class graph.

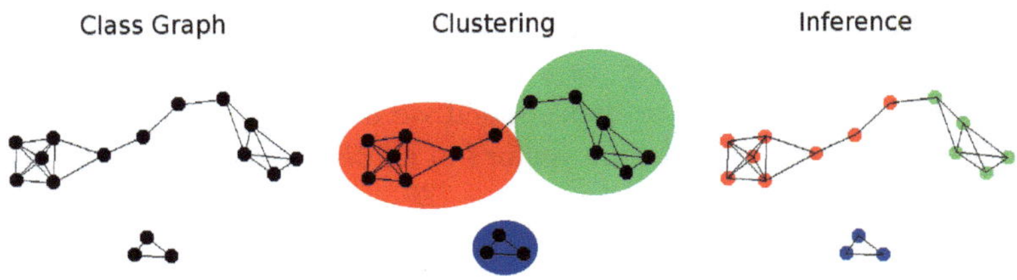

Fig. 5 Categorization by class graph. Local class labels, represented as mean histograms, are the nodes of the graph, and the links between two similar nodes form the edges. Clustering the local class labels provides the initial mapping from local class labels to global category labels. Running inference on the class graph provides a distribution of category labels for each local label. These distributions are then used to determine the category label for each discovered object

4.3 Smoothing

Class graph $\mathfrak{C}$ captures the dependency among the local class labels $\mathcal{G}_1, \ldots, \mathcal{G}_M$ but it does not assign a category label $\mathcal{G}_i^*$ to each local label $\mathcal{G}_i$. To determine the category labels, we apply probabilistic reasoning. We treat the nodes of the graph as random variables and the edges between adjacent nodes as conditionally dependent. That is, the global category label $\mathcal{G}_i^*$ of a local class label $\mathcal{G}_i$ depends not only on the local evidence $\bar{\mathbf{o}}_i$ but also on the class labels $\mathcal{G}_j^*$ of all neighboring labels $\mathcal{G}_j$. For example, if the local class label $\mathcal{G}_i$ is strongly of category $\mathcal{G}_i^*$,, based on its evidence $\bar{\mathbf{o}}_i$, then it can propagate its category label $\mathcal{G}_i^*$ to its neighbors $\mathcal{G}_j$. On the other hand, if its category label is weak, then its category label $\mathcal{G}_i^*$ can be flipped to the category label $\mathcal{G}_j^*$ of its neighbors. This process penalizes sudden changes of category labels, producing a smoothed graph. We perform the smoothing again using a Conditional Random Field (CRF).

Our CRF models the conditional distribution

$$p(g \mid \bar{\mathbf{o}}) = \frac{1}{Z(\bar{\mathbf{o}})} \prod_{i \in \mathcal{V}_{\bar{o}}} \varphi(\bar{\mathbf{o}}_i, g_i) \prod_{(i,j) \in \mathcal{E}_{\bar{o}}} \psi(\bar{\mathbf{o}}_i, \bar{\mathbf{o}}_j, g_i, g_j), \tag{6}$$

where $Z(\bar{\mathbf{o}}) = \sum_{g'} \prod_{i \in \mathcal{V}_{\bar{o}}} \varphi(\bar{\mathbf{o}}_i, g_i') \prod_{(i,j) \in \mathcal{E}_{\bar{o}}} \psi(\bar{\mathbf{o}}_i, \bar{\mathbf{o}}_j, g_i', g_j')$ is the *partition function;* $V_{\bar{o}}$ are the local classes; and $\mathcal{E}_{\bar{o}}$ are the edges between the local classes. Our formulation of the CRF is slightly different from the conventional approaches in that our feature similarity function f_n of the node potential $\log \varphi(\bar{\mathbf{o}}_i, g_i) = w_n \cdot f_n(\bar{\mathbf{o}}_i, g_i)$ is the conditional probability $p(g_i \mid \bar{\mathbf{o}}_i)$. Likewise, the feature similarity function f_e of the edge potential $\log \psi(\bar{\mathbf{o}}_i, g_i) = w_e \cdot f_e(\bar{\mathbf{o}}_i, \bar{\mathbf{o}}_j, g_i, g_j)$ is also defined as a conditional probability $p(g_i, g_j \mid \bar{\mathbf{o}}_i, \bar{\mathbf{o}}_j)$. The feature functions f_n and f_e hence range between 0 and 1, simplifying the weighting between node and edge potentials to scalars. In supervised learning with CRFs, node weight w_n and edge weight w_e are learned from training data. In this unsupervised work, however, we cannot learn these values as there is no training data available. We therefore determine node weight w_n and edge weight w_e manually using an appropriate evaluation measure on a validation set. Figure 8 in Sect. 5 shows the effect of setting different combinations of node weight w_n and edge weight w_e.

As mentioned in Sect. 4.2, the object feature vector clustering provides the total number of global object categories C^* and the initial mapping from local class labels $\mathcal{G}_1, \ldots, \mathcal{G}_M$ to global category labels $\mathcal{G}_1^*, \ldots, \mathcal{G}_{K^*}^*$. Using the clusters, we can model the feature similarity function $f_n = p(g_i \,|\, \bar{\mathbf{o}}_i)$ of node potential $\varphi(\bar{\mathbf{o}}_i, g_i)$ as

$$p(g_i \,|\, \bar{\mathbf{o}}_i) = \frac{p(\bar{\mathbf{o}}_i \,|\, g_i) p(g_i)}{\sum_{g'} p(\bar{\mathbf{o}}_i \,|\, g') p(g')} \tag{7}$$

where $p(\bar{\mathbf{o}}_i \,|\, g_i) = p(\bar{\mathbf{h}}_i \,|\, g_i^{\bar{\mathbf{h}}}) p(\bar{\mathbf{v}}_i \,|\, g_i^{\bar{\mathbf{v}}}) = \exp(-||\bar{\mathbf{h}}_i - \bar{\mathbf{h}}^{g_i}||) \exp(-||\bar{\mathbf{v}}_i - \bar{\mathbf{v}}^{g_i}||)$ and $p(g_i) = 1 - \frac{1}{|g_i|+1} \cdot p(\bar{\mathbf{o}}_i \,|\, g_i)$ measures how well $\bar{\mathbf{o}}_i$ fits to the cluster center g_i, and the global category prior $p(g_i)$ reflects how likely the category exists. A cluster with more members are more likely to be a true object category than a cluster with fewer members, and hence $p(g_i)$ is proportional to the size $|\, g_i \,|$ of the category.

We define the edge feature as

$$p(g_i, g_j \,|\, \bar{\mathbf{o}}_i, \bar{\mathbf{o}}_j) = p(g_i \,|\, \bar{\mathbf{o}}_i, \bar{\mathbf{o}}_i) p(g_j \,|\, \bar{\mathbf{o}}_i, \bar{\mathbf{o}}_j), \tag{8}$$

where $p(g_i \,|\, \bar{\mathbf{o}}_i, \bar{\mathbf{o}}_j) = p(g_i \,|\, \bar{\mathbf{o}}_{ij})$ and $p(g_j \,|\, \bar{\mathbf{o}}_i, \bar{\mathbf{o}}_j) = p(g_j \,|\, \bar{\mathbf{o}}_{ij})$ are estimated by a mean object feature vector $\bar{\mathbf{o}}_{ij}$. The probabilities $p(g_i \,|\, \bar{\mathbf{o}}_{ij})$ and $p(g_j \,|\, \bar{\mathbf{o}}_{ij})$ are computed by the nearest-neighbor.

To infer the most likely labels for the nodes of the class graph $\mathfrak{C}$, we use max product loopy belief propagation. This approximate algorithm returns the labels $\mathcal{G}_i^*$ which maximizes the conditional probability of Eq. 6. For the message passing, we take the generalized Potts model approach as commonly done and incorporate the edges in the inference only when g_i and g_j are equal. This results in the propagation of the belief only between equally-labeled nodes. The inference step continues until convergence and provides the distribution of global category labels $\mathcal{G}_1^*, \ldots, \mathcal{G}_{K^*}^*$ for every local class label $\mathcal{G}_i$.

To find the category label $\mathcal{G}^*$ for each discovered object $\mathcal{O}$, we compute the category which maximizes the assignment probability

$$P(g \,|\, \mathbf{o}) = \sum_{\bar{\mathbf{o}}'} p(g \,|\, \bar{\mathbf{o}}') p(\bar{\mathbf{o}}' \,|\, \mathbf{o}). \tag{9}$$

The probability of the category for a given local label $p(g \,|\, \bar{\mathbf{o}}')$ can be read directly from the class graph $\mathfrak{C}$, and the probability of the local object class given an object $p(\bar{\mathbf{o}}' \,|\, \mathbf{o}) = \exp(-||\bar{\mathbf{o}} - \mathbf{o}||)$ is computed as the object's similarity to the class mean. Discovered objects are accepted as objects when the probability of its most likely category label is greater than 0.5. Figure 6 shows the results of categorization of the two scenes shown in Fig. 4.

Fig. 6 Objects found in two different scenes. Segments of the same object label have the same *color*

5 Results

This section presents the results of running the algorithm on scans from real world scenes. The data set was collected using a nodding SICK laser with a width of 100° and a height of 60°. Each set was captured at the horizontal resolution of 0.25° and the vertical resolution of 15° a second. All scenes were static. The test set was a set of 60 scans from four offices. In total, these data sets contained 208 objects, including chairs, couches, poster boards, trash bins, and room dividers.

We first tested the effect of including saliency in the discovery step. Figure 7 qualitatively shows the difference in object discovery with and without saliency computation. Including saliency improves the precision[1] of discovery from 44 to 84 % while decreasing recall from 83 to 74 %. That is, while including the saliency step does eliminate some true objects, it is much more effective at eliminating none objects than the same algorithm without the saliency step.

Quantitatively, we computed V-measure [14] of our algorithm. V-Measure is a conditional entropy-based external cluster evaluation measure which captures the cluster quality by homogeneity and completeness of clusters. It is defined as

$$V_\beta = \frac{(1+\beta) * h * c}{(\beta * h) + c}, \tag{10}$$

where h captures homogeneity, c completeness, and β the weighting between homogeneity and completeness. A perfectly homogeneous solution has $h = 1$, and a perfectly complete solution has $c = 1$. Figure 8 shows the quality of clustering with varying node and edge weights and the effect of object distance threshold on the quality of clustering. Left graph indicates that the results of our algorithm is robust to the change of node and edge weights, but smoothing improves the overall results

[1]A discovered object is considered true positive if it originates from a real object and false positive if it is not a real object. False negative count is when a real object is not discovered.

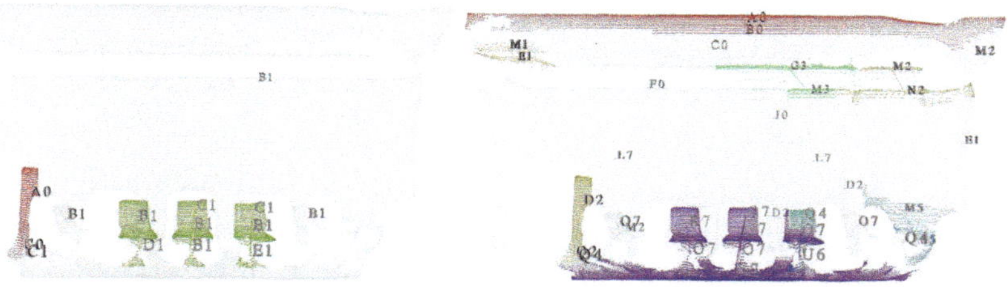

Fig. 7 The results of object discovery with (*left*) and without (*right*) saliency computation. All connected segments are considered objects for categorization. Objects are *colored* by their local class label

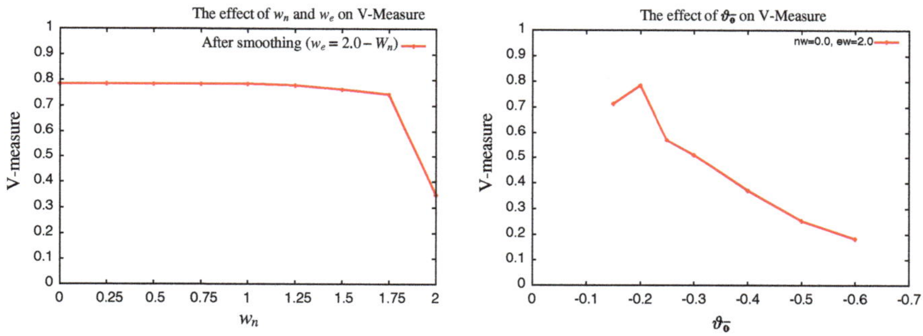

Fig. 8 Evaluation of our categorization step using V-measure. *Left* graph shows the effect of node and edge weights on v-measure. *Right* graph shows the effect of the object distance threshold on v-measure

Fig. 9 Effect of the object distance threshold on precision and recall

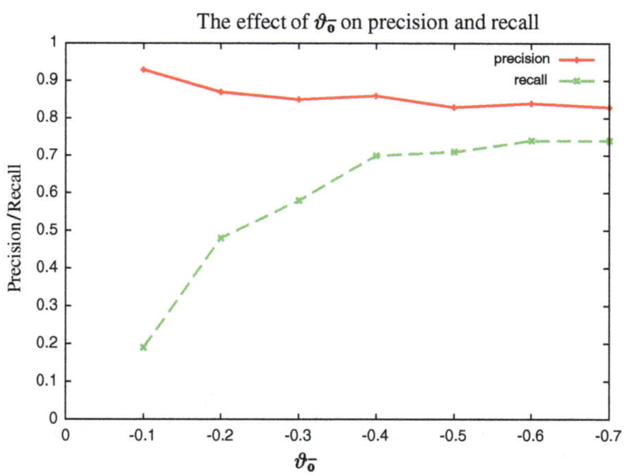

over pure clustering. Right graph shows that the quality of clusters depends on the object distance threshold $\vartheta_{\bar{o}}$, which indicates that the initial clustering result influences the final categorization quality.

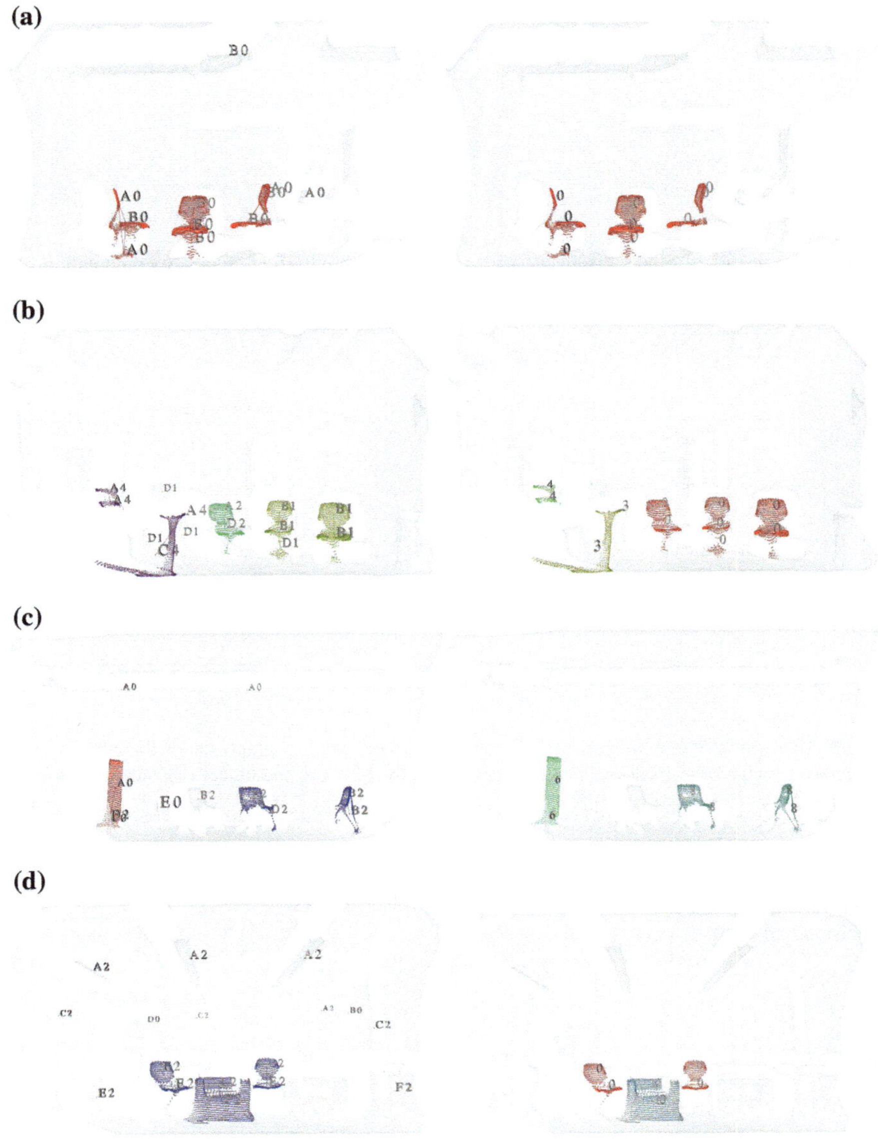

Fig. 10 Results of category discovery. *Left* images contain objects discovered through the object discovery process, and *right* images are the same objects after categorization. Objects in the *left* images are *colored* according to their local class labels while objects in the *right* images are *colored* by their global category labels. Notice that the categorization step can correct incorrect classifications of the discovery step. **a** Room 1. **b** Room 2. **c** Room 3. **d** Room 4

Figure 9 shows precision and recall[2] of the algorithm for varying object distance threshold $\vartheta_{\bar{o}}$. Not surprisingly, precision drops and recall increases as the threshold increases. This is because higher threshold results in fewer categories, which in turn means more of the discovered objects are accepted as categorized objects.

Figure 10 shows qualitative results. Left images are the results of performing object discovery per each scan, and right images are the corresponding images after categorization. Discovered objects are colored according to their local class label, i.e., with respect to other objects within a single scan, while categorized objects are colored according to their global category label, i.e., with respect to all other objects of the data set. The categorization step is able to assign the same global category labels to objects with different local class labels as shown in Fig. 10b while assigning different global category labels to objects with the same local label as shown in Fig. 10d. In addition, the chairs found in different scene are correctly labeled to be the same type as shown in Fig. 10a, b, d.

6 Conclusion and Outlook

We presented a seamless approach to discover and categorize objects in 3D environment without supervision. The key idea is to categorize the objects discovered in various scenes without requiring a presegmented image or the number of classes. Our approach considers objects to be composed of parts and reasons on each part's membership to an object class. After objects are discovered in each scan, we associate these local object labels by building a class graph and inferring on it. We demonstrated our capability of discovering and categorizing objects on real data and performance improvement class graph smoothing brings over pure clustering.

Our approach has several avenues for future work. First, we can use the results of categorization for object recognition. Once the robot has discovered enough instances of an object category, it can use the knowledge to detect and recognize objects, much the same way many supervised algorithms work. Our algorithm simplifies creating training data to converting robotic class representation to human representation. Another direction for future work is on-line learning. While the proposed approach allows the robot to reason on knowledge gained over time, the knowledge is updated in batch. This limits the availability of new information until enough data is collected for the batch processing. A robot that can process incoming data and update its knowledge on-line can utilize the new information immediately and adapt to changing environment. Extending our work to handle categorization on-line will thus make unsupervised discovery and categorization more useful for robotics.

[2]In computing precision and recall, we did not take into consideration the correctness of the category labels. Any real object that got categorized was considered true regardless of its label.

References

1. S. Bagon, O. Brostovski, M. Galun, M. Irani, Detecting and sketching the common. IEEE Comput. Vis. Pattern Recognit. (2010)
2. M. Cho, Y. Shin, K. Lee, unsupervised detection and segmentation of identical objects. IEEE Comput. Vis. Pattern Recognit. (2010)
3. G. Csurka, C. Bray, C. Dance, L. Fan, Visual categorization with bags of keypoints. in *Workshop on Statistical Learning in Computer Vision*, ECCV (2004)
4. F. Endres, C. Plagemann, C. Stachniss, W. Burgard, Unsupervised discovery of object classes from range data using latent Dirichlet allocation. in *Proceedings of Robotics: Science and Systems* (2009)
5. B.J. Frey, D. Dueck, Clustering by passing messages between data points. Science **315**(5814), 972–976 (2007)
6. S. Frintrop, A. Nuechter, H. Surmann, J. Hertzberg, Saliency-based object recognition in 3D data. in IEEE/RSJ *International Conference on Intelligent Robots and Systems* (2004)
7. L. Itti, C. Kock, and E. Niebur, A model of saliency-based visual attention for rapid scene analysis. IEEE Trans. Pattern Anal. Mach. Learn. **20**(11), 1254–1259 (1998)
8. A.E. Johnson, M. Hebert, Using spin images for efficient object recognition in cluttered 3D scenes.IEEE Trans. Pattern Anal. Mach. Learn. **21**(5), 433–449 (1999)
9. J. Lafferty, A. McCallum, F. Pereira, Conditional random fields: probabilistic models for segmenting and labeling sequence data. in *Proceedings of International Conference on Machine Learning* (2001)
10. T. Leung, J. Malik, Representing and recognizing the visual appearance of materials using three-dimensional textons. Int. Conf. Comput. Vis. (1999)
11. A. Ng, M. Jordan, Y. Weiss, On spectral clustering: analysis and an algorithm. Adv. Neural Inform. Process. Syst. (2002)
12. R. Osada, T. Funkhouser, B. Chazelle, D. Dobkin, Shape distributions. ACM Trans. Graphics **21**(4), 807–832 (2002)
13. M. Ruhnke, B. Steder, G. Grisetti, W. Burgard, Unsupervised learning of 3D object models from partial views. in *IEEE International Conference on Robotics and Automation.* Kobe, Japan (2009)
14. A. Rosenberg, J. Hirschberg, V-measure: a conditional entropy-based external cluster evaluation measure. in *Joint Conference on Empirical Methods in Natural Language Processing and Computational Natural Language Learning* (2007)
15. J. Sivic, B. Russell, A. Efros, A. Zisserman, W. Freeman, Discovering object categories in image collections. in *Proceedings of the International Conference on Computer Vision*(2005)
16. L. Spinello, R. Triebel, D. Vasquez, K. Arras, R. Siegwart, Exploiting repetitive object patterns for model compression and completion. Eur. Conf. Comput. Vis. (2010)
17. R. Triebel, J. Shin, R. Siegwart, Segmentation and unsupervised part-based discovery of repetitive objects. in *Proceedings of Robotics: Science and Systems* (2010)
18. C. Westin, S. Peled, H. Gudbjartsson, R. Kikinis, F. Jolesz, Geometrical diffusion measures for mri from tensor basis analysis. in *Proceedings of ISMRM* '97(1997)

Probabilistic Collision Detection Between Noisy Point Clouds Using Robust Classification

Jia Pan, Sachin Chitta and Dinesh Manocha

Abstract We present a new collision detection algorithm to perform contact computations between noisy point cloud data. Our approach takes into account the uncertainty that arises due to discretization error and noise, and formulates collision checking as a two-class classification problem. We use techniques from machine learning to compute the collision probability for each point in the input data and accelerate the computation using stochastic traversal of bounding volume hierarchies. We highlight the performance of our algorithm on point clouds captured using PR2 sensors as well as synthetic data sets, and show that our approach can provide a fast and robust solution for handling uncertainty in contact computations.

1 Introduction

The problems of collision detection and proximity computation are widely studied in different areas, including robotics, physically-based modeling, haptics and virtual environments. In particular, reliable and fast collision detection algorithms are required for robot motion planning, grasping and dynamics simulation to enforce the non-penetration constraints with the environment.

Most of the prior work on collision detection assumes an exact geometric description of the objects in the scene, typically represented as a polygon mesh. However, these methods may not work well for robots operating in real-world environments, where only partial observations of the environment are possible based on robot sensors. For example, inaccurate motor control makes a robot

J. Pan (✉) · D. Manocha
The Department of Computer Science, UNC, Chapel Hill, USA
e-mail: panj@cs.unc.edu

D. Manocha
e-mail: dm@cs.unc.edu

S. Chitta
Willow Garage Inc., Menlo Park CA 94025, USA
e-mail: sachinc@willowgarage.com

H.I. Christensen and O. Khatib (eds.), *Robotics Research*,
Springer Tracts in Advanced Robotics 100, DOI 10.1007/978-3-319-29363-9_5

deviate from its exact configuration and the sensors tend to add noise to the environment measurements. Current robot sensors including cameras and LIDAR and new devices such as Kinect can easily generate detailed point cloud data of real-world environments. However, it is hard to directly use prior collision detection algorithms which perform a boolean query and compute a yes/no answer. Moreover, exact collision checking may not be suitable in terms of handling uncertainty in perception and control, which also causes uncertainty in collision results. For many robotics applications, such as grasping or motion planning, we need to reduce the risk of physical contacts between the robot and the environment that may result in damages. Hence, we need to develop methods that tend to minimize the probability of collisions. Our point cloud collision and proximity algorithm can also be used to improve many methods' feasibility and robustness in real world. For example, algorithms in tactile manipulation usually require an exact or approximated mesh model of manipulated objects (e.g., [21]) and our method can extend them to directly handle the point clouds provided by sensors (See Fig. 1).

Main Results: In this paper, we present a probabilistic collision detection algorithm that can handle environments with uncertainty. Our approach can handle noisy or inexact point data representations that are gathered using sensors. In order to handle point cloud data with noise, we reformulate the collision detection problem as a two-class classification problem, where points of different objects belong to different classes. The collision probability is directly related to the separability of the corresponding two-class problem, which can be elegantly and efficiently solved using support vector machines (SVMs). We accelerate the computation using bounding volume hierarchies and perform a stochastic traversal of the hierarchies that takes into account noise and uncertainty. These hierarchies are updated for dynamic scenes or when the robot head or the gripper moves. Our probabilistic collision algorithm also estimates the contact points and contact normals. We test

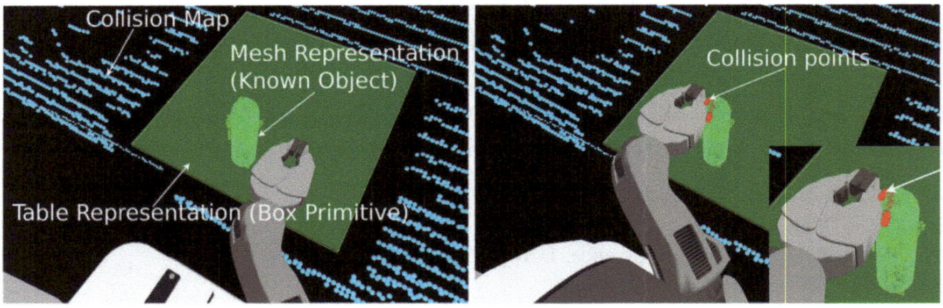

Fig. 1 A visual representation of the collision information generated by the sensors on the PR2 robot. (*Left*) The environment includes the points in a collision map (in *light blue*), mesh representations for known objects detected through visual sensing (*green cylindrical object* on table), and an exact geometric representation of the table surface (*green flat surface*). A detailed mesh model for the robot is also seen in the picture. (*Right*) A representation of the collision points (shown by *red spheres*) between the gripper and the object on the table

our algorithm on point clouds generated from PR2 sensors and synthetic data sets. Our method can provide robust results for probabilistic collision detection and its run-time performance is similar to that of hierarchy-based collision detection algorithms for triangle meshes (e.g., 500–1000 ms for 10 K points on a single CPU core).

The rest of the paper is organized as follows. We survey related work in Sect. 2. We introduce our notation and give an overview of the approach in Sect. 3. Section 4 shows how probabilistic collision detection computation is reduced to robust classification. We highlight the performance of our algorithm on different benchmarks in Sect. 5.

2 Previous Work

The field of probabilistic robotics provides a mathematical framework to handle the uncertainty that exists in the physical world [30]. It deals with representing uncertainty explicitly using the calculus of probability distribution and obtains robust control choices relative to the uncertainty in the robot system. Probabilistic robotics can handle perception uncertainty (or environment uncertainty) due to sensor and action errors. However, previous approaches tend to use simple methods to model environment uncertainty, such as feature-based methods or occupancy grid based methods [30]. These models can only provide a rough description of the environment while many robot actions (e.g., grasping) require more detailed information for robust computation.

2.1 Uncertainty of Point Cloud Data

Raw point cloud data obtained from sensor data can have a high degree of uncertainty, which results mainly from discretization error and noise. As a result, it is difficult to obtain robust estimation of high-order features like surface normals. This causes difficulty for many applications that require precise estimates of normal vectors at the boundary, such as grasping.

Many approaches consider uncertainty of point clouds implicitly. For example, [25, 28] encode surface uncertainty as a parameter tolerance for learning algorithms as they apply geometric operations (e.g., reconstruction) on the point clouds. However, without an explicit model of uncertainty, we can only consider a single uncertainty formulation for the overall surface, but may not be able to model varying uncertainty at different parts of the surface for local control.

There is recent work on explicitly modeling the uncertainty of point cloud data for different applications. Bae et al. [1] present a closed-form expression for the positional uncertainty of point clouds. Pauly et al. [20] propose two methods, confidence map and likelihood map, to analyze shape uncertainty in point clouds for resampling and reconstruction applications. Jenke et al. [11] describe a Bayesian model for point cloud uncertainty for surface reconstruction.

2.2 *Collision Detection*

Prior collision detection methods mainly focus on performing efficient and accurate contact computations between objects represented by triangulated primitives [17].

In terms of collision checking with point clouds, there are several simple methods. For example, we can first reconstruct triangle meshes from point clouds and then perform exact collision checking between the reconstructed surfaces. However, this approach suffers from inefficiency (>10 s for 10 K points) and robustness issues that arise in terms of using reconstruction algorithms (e.g., reconstruction quality, sensitiveness to parameter and noise, etc.). We can also simply expand every point as a sphere with suitable radius and approximate the object as a union of spheres [10] for collision checking. The main difficulty is in terms of automatically choosing different sphere radii for different points. Other direct collision checking methods for point cloud data are based on using bounding volume hierarchies [13, 27] and reconstructing implicit functions at the leaf nodes, which are prone to robustness issues. Minkowski sums of point clouds have also been used for collision queries [16]. Sucan et al. [29] describe a collision map data structure, which uses axis aligned cubes to model the point cloud and to perform collisions with a robot. Some applications, including virtual reality and haptics, need real-time collision checking, and use probabilistic criteria based on minimum distance computation between the point sets [15]. However, these methods do not take into account point cloud data's inherent shape uncertainty that arises from discretization or sampling [20].

There has been relatively little work in terms of handling uncertainty in collision detection. A special type of collision uncertainty is discussed in [7], which projects objects onto different image planes to perform collision culling using GPU-based computation. Guibas et al. [8] propose a method to compute the collision probability between 2D objects composed of line segments in a 2D environment with uncertainty. In order to estimate the collision uncertainty, this method models the endpoints of a line segment as probability distributions with a rectangular support region. Missiuro et al. [18] also try to model uncertainty in probabilistic roadmaps by using the collision probability of a configuration to bias the sampling process for roadmap computation.

3 Overview

In this section we introduce the notation used in the rest of the paper and give an overview of our approach.

The main pipeline of our system consists of three steps: (1) Obtain raw data from sensors and filter the point clouds to remove points on the robot and reduce the shadow effect [29]; (2) Compute the separating surface between two point clouds by estimating the noise from sensor parameters (Sects. 4.1–4.3); (3) Estimate the collision probability for each point and the overall collision probability between two point clouds (Sect. 4.4). Moreover, we use bounding volume hierarchies to accelerate the computation and recompute the hierarchies for dynamic environments (Sect. 4.5).

The inputs to our collision detection algorithm are the point clouds. In some cases, we need to perform the collision query between two different point clouds or between a point cloud and a polygonal object (e.g., when the mesh representation of a robot hand or gripper is available). We first present our approach for two different point clouds, and later show how it can be applied to a point cloud and a polygonal object.

Let the two point clouds be denoted as C_1 and C_2. We assume that each point cloud C is obtained from sensors and is a partial and noisy representation of the underlying exact surface S. There are two kinds of errors introduced in the generation of point clouds: *discretization errors* and *position errors* or *noise uncertainty*. Intuitively, the discretization error refers to how these point samples are distributed on the boundary of the surface and the position error measures the imprecision in the coordinates of each point. Formally, we assume C is generated from S according to the following process: first a series of n sample points $\mathbf{x}_i'$ is generated according to some sampling process and we use the symbol $p(\mathbf{x}_i'|S)$ to represent the distribution of coordinates for a random point $\mathbf{x}_i'$, i.e., it models the *discretization error*. Next, $\mathbf{x}_i$ is generated from $\mathbf{x}_i'$ according to some noise distribution $p(\mathbf{x}_i|\mathbf{x}_i'; \Sigma_i)$, i.e., it models the *position error*. Generally $p(\mathbf{x}_i'|S)$ is not given, but we can estimate it based on the observed point-cloud data with some assumptions about surface smoothness and sampling density. The symbol Σ_i is used to model point cloud's uncertainty due to noise, and is typically computed based on the sensor characteristics. For example, Σ_i may measure the level of noise that is a combination of sensing noise, motion uncertainty and deformation error. Then the overall uncertainty of a point $\mathbf{x}_i$ can be modeled as

$$\mathbf{x}_i|S \sim p(\mathbf{x}_i|S) = \int p(\mathbf{x}_i'|S)p(\mathbf{x}_i|\mathrm{x}_i'; \Sigma_i)d\mathbf{x}_i'. \tag{1}$$

In this formulation, we have an implicit assumption that the sensor is able to capture the features of the underlying surface. For example, more sample points $\mathbf{x}_i'$ are generated near the sharp features so that we can reconstruct the necessary features of the original model.

The output of the collision detection algorithm is a probability $\mathbb{P}_{C_1,C_2}$ that estimates whether two point clouds C_1 and C_2 are in-collision.

3.1 Separating Surface

Given a point cloud, we can possibly reconstruct a triangulated surface representation using Bayesian optimization. That is, the underlying surface should be the one with the maximum probability:

$$\hat{S} = \underset{S}{\operatorname{argmax}}\, p(S|\{\mathbf{x}_i\}_{i=1}^n) = \underset{S}{\operatorname{argmax}}\, p(S) \prod_i p(\mathbf{x}_i|S). \tag{2}$$

Next, we can perform collision checking based on reconstructed models. However, reconstruction process is only an estimation and the collision computation based reconstruction can be rather inaccurate. Our formulation is based on the theory of convex sets: two convex sets are non-intersecting if there exists an oriented *separating plane* P so that one set is completely in the positive (open) halfspace P^+ and the other completely in the negative (open) halfspace P^- [19]. For non-convex sets, we extend the concept of separating plane to the *separating surface*: two sets are non-intersecting (or *separable*) if and only if there exists a separating surface P between them. Previous work in collision detection [19, 22] is limited to the special case when P is composed of multiple planes.

We extend the idea of separating surfaces to handle point clouds. Given two point clouds $C_1 = \{\mathbf{x}_i^1\}_{i=1}^{n_1}$ and $C_2 = \{\mathbf{x}_i^2\}_{i=1}^{n_2}$ with n_1 and n_2 elements, respectively, a separating surface P is a surface that can separate the two sets completely with C_1 in P^+ and C_2 in P^-. In this case, P^+ and P^- represent a partition of the space $\mathscr{R}^3$ into two parts. Notice that here P should not be an arbitrary surface, i.e., it should not be a very complex function in terms of acting as a valid separating surface. Otherwise, even if P can completely separate the point clouds, it may not be able to separate the underlying surfaces. Such a problem is called *overfitting* in machine learning literature, i.e., the statistical model biases too much on the observed data and may not be able to predict the underlying model correctly. In order to avoid overfitting, we need to assume *regularity conditions* for P, which intuitively impose suitable smoothness constraints on the separating surface. For example, we represent P as a parameterized implicit surface $\{\mathbf{x} : f(\mathbf{x}; \theta) = 0\}$ with θ as its parameters. In this case, the regularity condition can limit the value of $f'(\mathbf{x}; \theta)$. Moreover, P^+ and P^- can be represented as $\{\mathbf{x} : f(\mathbf{x}; \theta) > 0\}$ and $\{\mathbf{x} : f(\mathbf{x}; \theta) < 0\}$, respectively. As a result, collision detection problem is reduced to finding the separating surface, i.e., deciding the parameter set θ, that can separate C_1 and C_2.

There is one major difference between point clouds and convex/non-convex sets.

In particular, for point cloud data, the existence of a separating surface is *not* a necessary or sufficient condition for non-intersection between the two sets. If two point clouds are noise-free and separable, their underlying surfaces may still be collision-free or in-collision, as shown in Fig. 2a, b. This is due to the discretization error from point-cloud sampling. The issue becomes more complicated when point clouds have position errors, as shown in Fig. 2c, d. This property of point cloud sets makes it difficult to perform exact collision checking, but is suitable for statistical

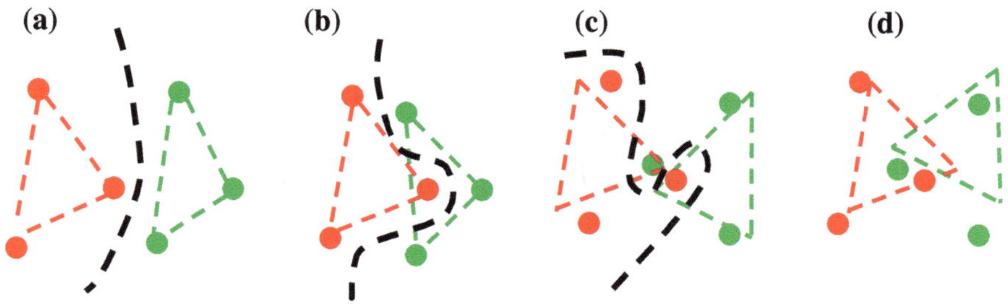

Fig. 2 Separating surface for point cloud sets. Point clouds in (**a**) and (**b**) are noise-free and are separable. However, due to discretization uncertainty, the underlying surfaces can be collision-free (**a**) or in-collision (**b**). Point clouds in (**c**) and (**d**) have some noise and may not be separable. And the underlying surfaces can be collision-free (**c**) or in-collision (**d**). Notice that we require suitable regularity or smoothness on the separating surface to avoid overfitting. For example, the separating surface provided in (**c**) has too large curvature and therefore is not valid. It in fact does not provide a good estimation for how to separate the underlying clouds. Collision result based on reconstructed meshes may not be reliable in all four cases due to discretization error (**a**, **b**) or position noise (**c**, **d**) or unsuitable parameters

learning approaches like SVM [3]. As a result, the probabilistic collision detection problem can be reduced to computing the optimal separating surface that minimizes the *separating error* for underlying surfaces: i.e., find θ that minimizes

$$\int_{\mathbf{x}\in S_1} 1_{\{\mathbf{x}\in P(\theta)^-\}} d\mathbf{x} + \int_{\mathbf{x}\in S_2} 1_{\{\mathbf{x}\in p(\theta)^+\}} d\mathbf{x}, \tag{3}$$

where S_1 and S_2 are the underlying surfaces for point clouds C_1 and C_2, respectively.

3.2 Probabilistic Model for Point Cloud Collision

We now present the probabilistic model for point cloud collision checking to compute the optimal separating surface. We rewrite $\mathbf{x}_i^l$ with $l \in \{1, 2\}$ as $(\mathbf{x}_i, c_i)$, where $\mathbf{x}_i = \mathbf{x}_i^l$ and $c_i = (-1)^{l+1} \in \{-1, 1\}$ denotes which object the point $\mathbf{x}_i$ belongs to. As a result, we have $n_1 + n_2$ elements in $\{(\mathbf{x}_i, c_i)\}$. As discussed in Sect. 3.1, collision checking between two point sets reduces to finding an optimal separating surface P. In machine learning terminology, this corresponds to finding an optimal classifier that can minimize the expected risk on the classification problem whose data is drawn from $\{\mathbf{x} : \mathbf{x} \in S_1 \bigcup S_2\}$ and its *training set* is $\{(\mathbf{x}_i, c_i)\}$. As a result, the collision detection problem is reduced to a machine learning problem. However, unlike typical machine learning algorithms which only deal with cases where $(\mathbf{x}_i, c_i)$ are specified exactly, we also need to take into account the noise in $\mathbf{x}_i$. Our solution is based on the maximum-likelihood (ML) scheme, i.e., the optimal surface should maximize the probability on the observed inputs $\{(\mathbf{x}_i, c_i)\}$.

Similar to Eq. (1), the joint probability for $(\mathbf{x}_i, c_i)$ can be expressed as

$$p(\mathbf{x}_i, c_i) = \int p(\mathbf{x}'_i, c_i; \theta) p(\mathbf{x}_i | \mathbf{x}'_i; \Sigma_i) d\mathbf{x}'_i. \tag{4}$$

Here θ is the parameter set used to represent the separating surface P. For example, P is $\{\mathbf{x} : \mathbf{w}^T \mathbf{x} + b = 0\}$ if P is a plane and $\theta = \{\mathbf{w}, b\}$. Or P is $\{\mathbf{x} : \mathbf{w}^T \Phi(\mathbf{x}) + b = 0\}$ if P is a hyper-plane in some high-dimensional inner product space $\mathscr{H}$ and Φ is the mapping $\Phi : \mathscr{R}^3 \mapsto \mathscr{H}$. The unknown surface parameter θ can be estimated from the point cloud data using ML:

$$\theta^* = \underset{\theta}{\operatorname{argmax}} \sum_{\theta} \operatorname{In} \int p(\mathbf{x}'_i, c_i; \theta) p(\mathbf{x}_i | \mathbf{x}'_i; \Sigma_i) d\mathbf{x}'_i \tag{5}$$

In practice, the integration over the unknown underlying surface sample $\mathbf{x}_i'$ makes it hard to compute the surface parameter. As a result, we consider an alternative form that is computationally more efficient. Specifically, we use an approximation to Eq. (5) based on a widely used heuristic for mixture estimation: we simply regard $\mathbf{x}_i'$ as a parameter of the model instead of a random variable. Then Eq. (5) reduces to:

$$\theta^* = \underset{\theta}{\operatorname{argmax}} \sum_{i} \ln \sup_{\mathbf{x}'_i} p(\mathbf{x}'_i, c_i; \theta) p(\mathbf{x}_i | \mathbf{x}'_i; \Sigma_i). \tag{6}$$

We present an algorithm to solve Eq. (6) in Sect. 4.

4 Probabilistic Collision Checking Between Point Clouds

In this section, we present our probabilistic algorithm for collision checking between point clouds using two-class classification. This reduces to computing the optimal separating surface that minimizes the function in Eq. (6).

4.1 Basic Formulation

For convenience, we first assume that the separating surface is a plane, i.e., $P = \{\mathbf{x} : \mathbf{w}^T \mathbf{x} + b = 0\}$. We also assume that the uncertainty due to noise can be described by a Gaussian distribution. We will relax these assumptions later. Based on these two assumptions, we have

$$p(\mathbf{x}'_i, c_i; \theta) \sim p(\mathbf{x}'_i) \exp(-\frac{(\mathbf{w}^T\mathbf{x}'_i + b - c_i)^2}{\sigma^2}) \quad \text{and} \\ p(\mathbf{x}_i|\mathbf{x}'_i; \Sigma_i) \sim \exp(-(\mathbf{x}_i|\mathbf{x}'_i)^T \Sigma_i^{-1}(\mathbf{x}_i - \mathbf{x}'_i)), \tag{7}$$

where σ and Σ_i are the covariance parameters of a Gaussian distribution.

As we will show in Sect. 5, the discretization uncertainty at $\mathbf{x}'_i$ can also be estimated as a Gaussian distribution with the observation $\mathbf{x}_i$ as mean. That is $p(\mathbf{x}'_i) \sim \exp(-(\mathbf{x}'_i - \mathbf{x}_i)^T \Psi_i^{-1}(\mathbf{x}'_i - \mathbf{x}_i))$, where Ψ_i is the covariance parameter for discretization uncertainty. Here we assume that the observed data $\mathbf{x}_i$ is fixed and the true value $\mathbf{x}'_i$ is subject to random errors. This is equivalent to the so-called *Berkson's model* in statistics literature [2]. Then Eq. (6) becomes

$$\theta^* = \underset{\theta}{\operatorname{argmax}} \sum_i \inf_{\mathbf{x}'_i} \left[\frac{(\mathbf{w}^T\mathbf{x}'_i + b - c_i)^2}{\sigma^2} + (\mathbf{x}_i - \mathbf{x}'_i)^T \widetilde{\Sigma}_i^{-1}(\mathbf{x}_i - \mathbf{x}'_i) \right], \tag{8}$$

where $\theta = \{\mathbf{w}, b\}$ and $\widetilde{\Sigma}_i^{-1} = \Sigma_i^{-1} + \Psi_i^{-1}$.

Moreover, notice that if $(\mathbf{x}_i - \mathbf{x}'_i)^T \widetilde{\Sigma}_i^{-1}(\mathbf{x}_i - \mathbf{x}'_i)$ is large, then $p(\mathbf{x}'_i, c_i; \theta)$ term will have a small value and can be ignored in the integration for $p(\mathbf{x}i, ci)$. As a result, we can constrain $\mathbf{x}_i$ to lie within the ellipsoid $\mathscr{E}_i = \{\mathbf{x}'_i : (\mathbf{x}_i - \mathbf{x}'_i)^T \widetilde{\Sigma}_i^{-1}(\mathbf{x}_i - \mathbf{x}'_i) \leq r_i^2\}$ and this will not influence the final result considerably. Also considering the regularity of separating surfaces, Eq. (8) can be approximated by an optimization formulation that is similar to support vector machine (SVM):

$$\begin{array}{lll} \underset{\mathbf{w},b,\xi_i}{\text{minimize}} & \frac{1}{2}\|\mathbf{w}\|^2 + \lambda \sum_{i=1}^{n} \xi_i & \\ \text{subject to} & c_i(\mathbf{w}^T\mathbf{x}'_i + b) \geq 1 - \xi_i, & \forall \mathbf{x}'_i \in \mathscr{E}_i \forall 1 \leq i \leq n; \\ & \xi_i \geq 0, & \forall 1 \leq i \leq n. \end{array} \tag{9}$$

The above formulation minimizes the upper bound on the classification error, which is equivalent to separating error in Eq. (3). Errors occur when $\xi_i \geq 1$, as $\mathbf{x}'_i$ lies on the wrong side of P. The quantity λ is the penalty for any data point $\mathbf{x}'_i$ that either lies within the margin on the correct side of P ($0 < \xi_i \leq 1$) or on the wrong side of P ($\xi_i > 1$). $\|\mathbf{w}\|$ is the regularization term which controls the smoothness of the separating surface.

It is easy to verify that $c_i(\mathbf{w}^T\mathbf{x}'_i + b)$ reaches its minimum at point $\mathbf{x}_i - r_i(\mathbf{w}^T\widetilde{\Sigma}_i\mathbf{w})^{1/2}\widetilde{\Sigma}_i\mathbf{w}$ and the minimum value is $c_i(\mathbf{w}^T\mathbf{x}_i + b) - r_i(\mathbf{w}^T\widetilde{\Sigma}_i\mathbf{w})^{1/2}$. As a result, Eq. (9) can be further written as:

$$\begin{array}{lll} \underset{\mathbf{w},b,\xi_i}{\text{minimize}} & \frac{1}{2}\|\mathbf{w}\|^2 + \lambda \sum_{i=1}^{n} \xi_i & \\ \text{subject to} & c_i(\mathbf{w}^T\mathbf{x}_i + b) \geq 1 - \xi_i + r_i\left\|\widetilde{\Sigma}_i^{1/2}\mathbf{w}\right\|, & \forall 1 \leq i \leq n; \\ & \xi_i \geq 0, & \forall 1 \leq i \leq n. \end{array} \tag{10}$$

Such optimization problems have been studied in the literature [26] and can be solved using second order cone programming (SOCP) methods. Once $\mathbf{w}$ and b are computed, we can compute $\xi_i = \max(0, 1 - c_i(\mathbf{w}^T\mathbf{x}_i + b) + r_i \left\| \widetilde{\Sigma}_i^{1/2}\mathbf{w} \right\|)$.

4.2 Non-Gaussian Uncertainty

The uncertainty of real-world sensors may not be accurately modeled using a Gaussian distribution. Our approach can also handle non-Gaussian uncertainty.

Shivaswamy et al. [26] point out that the ellipsoid radius r_i is related to the confidence of the classification result when the training data contains noise. Briefly, if we desire the underlying surface point $\mathbf{x}'_i$ with Gaussian distribution to lie on the correct side of the separating surface with a probability greater than κ_i

$$\mathbb{P}_{\mathbf{x}'_i \sim \mathcal{N}(\mathbf{x}_i, \widetilde{\Sigma}_i)}\left((c_i(\mathbf{w}^T\mathbf{x}'_i + b) \geq 1 - \xi_i\right) \geq \kappa_i, \tag{11}$$

then $r_i = \mathrm{cdf}^{-1}(\kappa_i)$, where $\mathrm{cdf}(u) = \frac{1}{\sqrt{2\pi}} \int_{-\infty}^{u} \exp\left(\frac{-s^2}{2}\right) ds$. Using multivariate Chebyshev inequality, this relationship between κ_i and r_i can be further extended to the case when $\mathbf{x}'_i$ follows non-Gaussian distribution. That is, if $\mathbf{x}'_i \sim (\mathbf{x}_i, \widetilde{\Sigma}_i)$ represents a family of distributions with a common mean and covariance given by $\mathbf{x}_\mathrm{i}$ and $\widetilde{\Sigma}_i$, and we want $\mathbf{x}_\mathrm{i}$ to lie on the correct side of the separating surface with a probability greater than κ_i

$$\sup_{\mathbf{x}'_i \sim (\mathbf{x}_i, \widetilde{\Sigma}_i)} \mathrm{P}_{\mathbf{x}'_i}\left((c_i(\mathbf{w}^T\mathbf{x}'_i + b) \geq 1 - \xi_i\right) \geq \kappa_i, \tag{12}$$

then $r_i = \sqrt{\frac{\kappa_i}{1 - \kappa_i}}$. This formulation implies that we can perform collision detection using Eq. (10) even when the uncertainty is non-Gaussian.

4.3 Non-linear Separating Surface

Linear separating surface is mainly limited to the case when all the underlying surfaces are convex. If any one of them is non-convex, a separating plane may not exist even when the surfaces are collision-free. Therefore, we need to extend our algorithm to non-linear P. Similar to typical SVM algorithms [31], we can remove the linear separating surface assumption by applying a *kernel trick* on the dual form of Eq. (10). Briefly, kernel trick is a method that transforms the Euclidean space $\mathscr{R}^n$

into another inner space $\mathscr{H}$ using mapping Φ and then replaces the inner product $\langle \mathbf{y}, \mathbf{z} \rangle \mathscr{R}^n$ by the new inner product $K(\mathbf{y}, \mathbf{z}) = (\Phi(\mathbf{y}), \Phi(\mathbf{z}))\mathscr{H}$ in space $\mathscr{H}$. Here $K(\cdot, \cdot)$ is called the *kernel function*. Usually a hyper-plane in $\mathscr{H}$ will correspond to a non-linear surface in $\mathscr{R}^n$, which is a popular way to construct non-linear classifiers in machine learning [9]. Some of the widely used kernel functions include linear ($K(\mathbf{y}, \mathbf{z}) = \mathbf{y}^T \mathbf{z}$) and Gaussian ($K(\mathbf{y}, \mathbf{z}) = \exp(-\gamma \|\mathbf{y} - \mathbf{z}\|^2)$).

Based on the kernel trick, the non-linear separating surface can be formulated as $P = \{\mathbf{x} : \mathbf{w}^T \Phi(\mathbf{x}) + b = 0\}$. To compute P, we first transform Eq. (10) into its dual form. Next, based on the Taylor-expansion technique [3], we replace $\mathbf{y}^T \mathbf{z} \partial \mathbf{z}$ by kernel function $K(\mathbf{y}, \mathbf{z})$ and replace $\mathbf{y}$ by the kernel gradient $\frac{\partial K(\mathbf{y}, \mathbf{z})}{\partial \mathbf{z}}$ and finally obtain the optimization formulation in non-linear case as

$$\begin{aligned}
\underset{\alpha_i, \mathbf{v}_i}{\text{maximize}} \quad & \sum_{i=1}^{n} \alpha_i - \frac{1}{2}\Big(\sum_{i=1}^{n}\sum_{j=1}^{n} \alpha_i \alpha_j c_i c_j K(\mathbf{x}_i, \mathbf{x}_j) + \sum_{i=1}^{n}\sum_{j=1}^{n} \alpha_i c_i (\widetilde{\Sigma}_j^{1/2} \frac{\partial K(\mathbf{x}_i, \mathbf{x}_j)}{\partial \mathbf{x}_j})^T \mathbf{v}_j \\
& + \sum_{i=1}^{n}\sum_{j=1}^{n} \alpha_j c_j (\widetilde{\Sigma}_j^{1/2} \frac{\partial K(\mathbf{x}_i, \mathbf{x}_j)}{\partial \mathbf{x}_j})^T \mathbf{v}_j + \sum_{i=1}^{n}\sum_{j=1}^{n} \mathbf{v}_i^T (\widetilde{\Sigma}_j^{1/2} \frac{\partial^2 K(\mathbf{x}_i, \mathbf{x}_j)}{\partial \mathbf{x}_i \partial \mathbf{x}_j} \widetilde{\Sigma}_j^{T/2}) \mathbf{v}_j\Big) \\
\text{subject to} \quad & \|\mathbf{v}_i\| \le r_i \alpha_i, \; 0 \le \alpha_i \le C, \forall 1 \le i \le n; \text{ and } \sum_{i=1}^{n} \alpha_i c_i = 0;
\end{aligned} \tag{13}$$

where C is a regularity term similar to λ in Eq. (10). Once α_i and $\mathbf{v}_i$ are computed, we can compute the formulation for the separating surface P

$$f(\mathbf{x}) = b + \sum_{j=1}^{n} \alpha_j c_j K(\mathbf{x}_j, \mathbf{x}) + \sum_{j=1}^{n} \mathbf{v}_j^T \widetilde{\Sigma}_j^{1/2} \frac{\partial K(\mathbf{x}_j, \mathbf{x})}{\partial \mathbf{x}_j} \tag{14}$$

and $\xi_i = \max(0, \xi_i')$, where

$$\xi_i' = 1 - c_i f(\mathbf{x}_i) + r_i \left\| \widetilde{\Sigma}_i^{1/2} f'(\mathbf{x}_i) \right\|. \tag{15}$$

Notice that the surface parameter b does not appear in the dual form, but it can be computed based on Karush–Kuhn–Tucker conditions [5]. We first choose i so that $0 < \alpha_i < C$, $\|\mathbf{v}_i\| < r_i \alpha_i$ and then set $\xi_i' = 0$ in Eq. (15) to obtain b. Moreover, notice that all the results for non-linear separating surface are consistent with those for linear separating surface, which use a linear kernel $K(\mathbf{y}, \mathbf{z}) = \mathbf{y}^T \mathbf{z}$.

4.4 Probabilistic Collision Decision

Based on the computed separating surface, we present a simple scheme to perform probabilistic collision detection between the point clouds. First, we compute the collision probability for each point, i.e., the probability that $\mathbf{x}_i'$ lies on the wrong side of separating surface:

$$\mathbb{P}_{\mathbf{x}_i' \sim \mathcal{N}(\mathbf{x}_i, \widetilde{\Sigma}_i)}(c_i f(\mathbf{x}_i') \leq 0) = \mathrm{cdf}(-c_i f(\mathbf{x}_i) / \left\| \widetilde{\Sigma}_i^{1/2} f'(\mathbf{x}_i) \right\|). \tag{16}$$

We denote this per-point probability as $\mathbb{P}(\mathbf{x}_i)$. Next, we need to use an appropriate metric to measure the collision probability between two point clouds. For two exact models, collision occurs if *any* subsets of them are in-collision. Therefore, for point clouds C_1 and C_2, it seems to be reasonable to define the collision probability between them as $1 - \prod_{\mathbf{x} \in \{C_1 \bigcup C_2\}} [1 - \mathbb{P}(\mathbf{x})]$. However, this metric may have some issues: when the number of points increases, its value will go to zero instead of converging to the real collision probability. The reason is that this metric does not consider the dependency between collision states of nearby points. Our approach for computing collision probability only involves far-away points with large per-point collision probability. First, we compute the maximum per-point collision probability $\max_{\mathbf{x}} \mathbb{P}(\mathbf{x})$. Next, we find all the points whose per-point collision probabilities are near the maximum value, e.g., more than $0.8 \max_{\mathbf{x}} \mathbb{P}(\mathbf{x})$. For points that are close to each other, we only use one of them in the whole body collision probability computation. The first rule filters out points whose collision probabilities are not large enough so as to improve the stability of collision results while the second rule filters out points that are closely correlated. Finally, we compute the collision probability between point clouds based on the left $m \ll n$ points $\{\tilde{\mathbf{x}}_i\} : \mathbb{P}_{C_1,C_2} = 1 - \prod_{i=1}^m [1 - \mathbb{P}(\tilde{\mathbf{x}}_i)]$. We can also use a simpler version of this metric which only considers the point with the maximum collision probability: $\mathbb{P}_{C_1,C_2} = \max_{\mathbf{x} \in C_1 \bigcup C_2} \mathbb{P}(\mathbf{x})$. For collision between exact models, the two metrics are equivalent, as $\mathbb{P}(\tilde{\mathbf{x}}_i) = \max_{\mathbf{x}} \mathbb{P}(\mathbf{x}) = 1$, for all i. The simpler metric can not distinguish the collision states when point clouds have one or more far-away points with large per-point collision probability, but it is more convenient to distinguish between collision-free and in-collision cases.

4.5 Acceleration Using Bounding Volume Hierarchies

We have reduced the problem of collision detection between two point clouds to a two-class classification problem and can solve it with SVM. However, performing collision detection by directly using Eq. (13) introduces some challenges. First, the timing complexity of SVM can be $\mathcal{O}(n^3)$, where $n = n_1 + n_2$ is the number of points in the two point clouds. As a result, the underlying algorithm can be slow for dense point clouds. Second, the two point clouds corresponding to different objects may have different numbers of points, which can result in unbalanced training data in terms of using machine learning algorithms. Moreover, if the two point clouds under consideration correspond to objects with different sizes (e.g., a large room and a small robot), it will cause the optimization algorithm to have a lower separating error for the large object and higher error for the small object.

We use bounding volume hierarchies (BVH) to overcome these problems. These hierarchies provide a quick identification of objects or parts of an object that can be easily culled away and therefore perform exact collision queries on relatively few primitives.

5 Implementation and Results

In this section, we describe some details of our implementation and highlight the performance on several benchmarks.

5.1 Implementation

First, we discuss how to estimate the distribution of the underlying surface sample $p(\mathbf{x}'_i)$. The mean of $p(\mathbf{x}'_i)$ is $\mathbf{x}_i$ due to our unbiased assumption. We estimate the covariance $\boldsymbol{\Psi}_i$ based on the formulation described in [20]:

$$\Psi_i = \frac{\Sigma_{j=1}^{n}(\mathbf{x}_j - \mathbf{x}_i)(\mathbf{x}_j - \mathbf{x}_i)^T \exp(-\|\mathbf{x}_i - \mathbf{x}_j\|^2/\tau_i^2)}{\Sigma_{j=1}^{n} \exp(-\|\mathbf{x}_i - \mathbf{x}_j\|^2/\tau_i^2)}, \tag{17}$$

where n is the total number of points and τ_i is a parameter used to remove the influence of points too far away from $\mathbf{x}_i$. We set $\tau_i = \tau \cdot \eta_i . \tau$ as a global scale parameter and the variable $\eta_i = \frac{r}{\sqrt{k}}$ denotes the local sample spacing estimated from k a k-neighborhood, where r is the radius of the enclosing sphere of the k-nearest neighbors of $\mathbf{x}_i$.

Our algorithm is based on machine learning techniques and includes some parameters that need to be tuned. Fortunately, we find that our method is not sensitive to the parameter choice if we preprocess the data by scaling it to $[0, 1]^3$ volume in 3D. Scaling is considered important in terms of robustness of SVM, especially for the non-linear case. Moreover, scaling also helps us in computing the parameters that are suitable for the point clouds with different sizes or configurations. In practice, scaling also changes the uncertainty of each point, so we need to update the noise level from $\widetilde{\Sigma}_i$ to $\mathbf{S}\widetilde{\Sigma}_i\mathbf{S}^T$, where $\mathbf{S} = \mathrm{diag}(s_1, s_2, s_3)$ is the scaling matrix.

We have used our algorithm on data captured using robot sensors. Note that our method is designed for noisy environments where the ground-truth for collision detection is unknown. In this case, exact collision algorithms are not applicable as we don't have an exact representation of the environment. Therefore, it is difficult to directly compare the quality or accuracy of our algorithm with prior methods. However, our method can guarantee: (1) For easy collision queries, i.e., when the distance between two collision-free objects is large or the two objects are in deep penetration, our method will give collision probability near 0 or 1. In this case, only very large noise can reverse the outcome of the query. However, our probabilistic algorithm would give the same result as the exact approach that first performs mesh reconstruction from the point clouds. (2) For challenging queries, i.e., when two objects are almost in-contact or have a small penetration, our method computes a collision probability near 0.5, because these configurations are more susceptible to noise. Exact collision algorithms will still provide a yes-no result, but the accuracy of the exact algorithm is governed by the underlying sampling and mesh reconstruction algorithm. If a yes-no collision answer is required, our algorithm uses two thresholds $A \geq 0.5 \geq B$: if collision probability $>A$, we report *collision-free*; if collision probability $<B$, we report *in-collision*; if collision probability is between A and B, we report *in-contact*. For example, when collision-avoidance is critical for the underlying applications, we can use large conservative value for A and small conservative value for B to achieve higher guarantees.

5.2 Results

We highlight the performance of our algorithm on real-world point clouds as well as synthetic data sets. We also compare its accuracy with prior collision detection techniques. The running time of our probabilistic algorithm is similar to that of exact collision detection algorithms and varies based on number of primitives and their relative configuration.

We evaluate the performance of our algorithm on a synthetic data set corresponding to a moving piano in a room with tables. We first generate a point cloud by sampling the polygons and adding some noise. Next, we use the PQP package to perform exact collision detection and separation distance query between the exact, triangulated model and compared the results with probabilistic collision detection on the resulting point cloud (see Fig. 3). We see a high correlation between our results and the actual separation distance, and it varies based on the level of noise. This shows that our approach is quite robust and even works well in degenerate configurations, e.g., when the two objects are barely touching or very close to each other.

Such configurations are more susceptible to noise and the exact collision detection algorithms are very sensitive to these configurations.

We have applied our probabilistic collision detection to the point cloud data generated for manipulation using the PR2 robot. Point cloud data on the PR2 robot is generated from a scanning laser range finder (*Hokuyo Top-URG(UTM-30LX)*) and a stereo camera (*WGE-100*), which is combined with an active texture projector to obtain good 3D data from untextured objects. The robot is placed in front of a table with multiple household objects (e.g., bowls, cans) on the table at a distance of

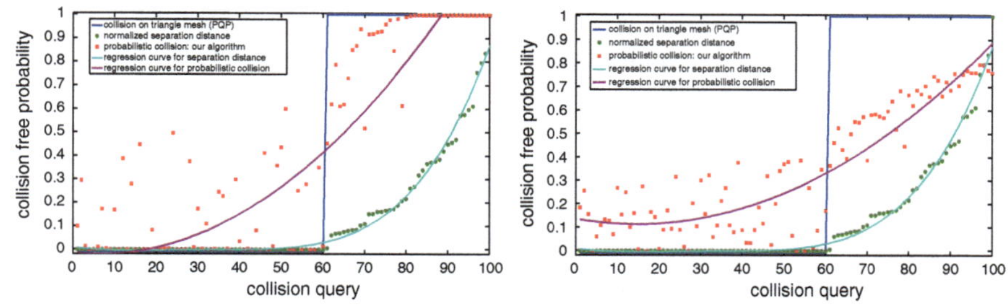

Fig. 3 Comparison between the results for 100 random queries between prior collision detection algorithms for exact triangle meshes and our algorithm on the point clouds (generated by sampling and adding noise). We show the results of exact collision detection and separation distance as well. If the noise in the point cloud is small (the *left* figure), our method returns 0 or 1 collision probability for most queries. When the queries correspond to a small separation distance or penetration depth (i.e., difficult cases), our algorithm computes collision-free probability close to 0.5. Furthermore, the collision-free probability is higher when the separation distance is large for non-overlapping objects. If the noise is large (the *right* figure), fewer queries return 0 or 1 collision probability. We see a good correlation between the regression curves computed by our algorithm and the exact queries on these synthetic datasets

about 1.5 m from the robot's sensors. The point clouds are a discretized (about ±1.5 cm in range) representation of the real environment and are generated periodically by each sensor. The data is noisy and exhibits speckles especially in the vicinity of boundaries of objects and boundaries of the field of view of the sensor. The sensors are calibrated with respect to each other and the arms using a known calibration pattern. The known position of the arms, measured using encoders, is used to filter out the points corresponding to the arms from the point clouds obtained by the sensors. Typical point clouds generated by the stereo sensors on the PR2 robot have more than 40,000 points and are generated at 20 Hz. Point clouds generated by the laser range scanners typically have about 10,000 points. The data from the point clouds is aggregated into a *collision map* representation. The collision map is a 3-dimensional occupancy grid maintained at a fixed resolution. The resulting collision maps are at 1 cm resolution and have about 2,000 occupied cells. A complete triangulated mesh representation of the robot, including the arms and the gripper, is also available as input for the collision checker.

There are very few algorithms or systems available for collision checking between noisy point clouds. As a result, we compare our algorithm with the implementation in ROS (based on ODE) and exact collision detection on reconstructed meshes.

The collision checking procedures used in ROS are currently based on the collision checking implementation in the ODE software package. The input to the collision checker is a combination of mesh models for the robot and objects in the environment and the collision map. The points in the collision map are represented as axis-aligned box primitives whose length is equal to the resolution at which the collision map is maintained. The current representation of the collision space considers every point in the collision map to be a potential obstacle. Thus, noise in the sensor data can frequently lead to false positives, i.e., the detection of potential collisions in parts of the environment where there are no obstacles. There is no robust criterion to compute the box size, e.g., a function of noise, so we can't compare all the features of our method with ODE collision checking. We also use a reconstruction algorithm to compute a triangle mesh from the point clouds and perform triangle-based collision as well as separation distance computation using PQP. In many ways, this formulation only provides an approximation of the ground truth and is used to evaluate the robustness of our algorithm.

As shown in Fig. 4a, our result matches well with the exact collision detection algorithm, especially with the separation distance computation. Furthermore, we notice that the collision probability of our approach changes slowly when the noise increases. It is more robust as compared to the yes-no result computed by ODE on the point clouds, which is likely to frequently switch between collision-free and in-contact configurations, when the noise level changes. We also apply our algorithm on a dataset generated using Kinect RGB-D cameras. This dataset corresponds to an indoor environment at the Intel Lab in Seattle captured using Kinect sensors. The results of our probabilistic collision detection on this dataset are shown in Fig. 4b.

Moreover, from Figs. 3 and 4, we observe that configurations with the same distances to the obstacles can have large spread in the computed collision probabilities. The reason is that distance is only a partial measurement of collision status while our collision probability is a more complete description about collision status and provides more detailed information about the relative configurations.

For one query, our method needs about 500–1000 ms for about 10,000 points on one Intel Core i7 3.2 GHz CPU, based on BVH acceleration. It is about 5–10 times slower than optimized collision packages on models with 10 K triangles (e.g., PQP can compute collisions in such situations in about 50–100 ms). However, the reconstruction algorithms take more than 10 s to compute the triangulated mesh from the point cloud. Moreover, our current implementation can be optimized in

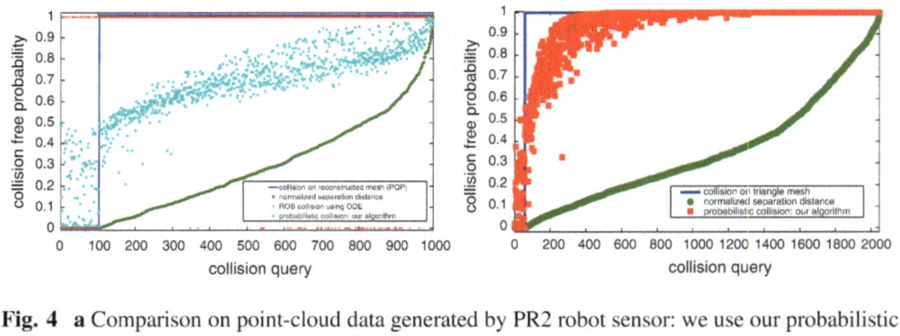

Fig. 4 **a** Comparison on point-cloud data generated by PR2 robot sensor: we use our probabilistic collision detection on the noisy point cloud versus results computed by ODE package used in ROS versus exact collision and distance queries on the reconstructed mesh model. Our results on the point cloud are more robust as compared to the ODE package. **b** Result of our algorithm on Kinect data

several ways, such as replacing the non-linear kernel by approximated linear kernel [23] and using more efficient SVM methods designed for large scale data [6]. We expect an optimized probabilistic collision method to have similar speed to the PQP algorithm. Furthermore, our approach can provide more detailed information and can be easily combined with planning/reasoning algorithms. For example, we can combine it with trajectory optimization algorithms (e.g., CHOMP [24], STOMP [12]) to find a smooth path that has a minimum probability of colliding with the obstacles.

6 Conclusions and Future Work

We have presented a novel and robust method for contact computation between noisy point cloud data using machine learning methods. We reformulate collision detection as a two-classification problem and compute the collision probability at each point using support vector machines. The algorithm can be accelerated by using bounding volume hierarchies and performing a stochastic traversal. We have tested the results on synthetic and real-world data sets and the preliminary results are promising.

There are many avenues for future work. We need to test the performance on different robotic systems and evaluate its performance on tasks such as planning and grasping. It would be useful to extend this approach to continuous collision checking, which takes into account the motion of the robot between discrete intervals along the path. Similar probabilistic methods can also be developed for other queries, including separation and penetration depth computation. Finally, we are interested in improving the algorithm to handle dynamic environments where points may change position or can be added or removed from the environment due to movement, occlusion or incremental data, based on incremental SVM [4] and BVH refitting techniques [14].

Acknowledgments This work was supported in part by ARO Contract W911NF-10-1-0506, NSF grants 0917040, 0904990, and 1000579, and Willow Garage. The dataset generated using Kinect RGB-D cameras was provided to us by Dieter Fox and Peter Henry at the University of Washington.

References

1. K.-H. Bae, D. Belton, D.D. Lichti, A closed-form expression of the positional uncertainty for 3D point clouds. Trans. Pattern Anal. Mach. Intell. **31**, 577–590 (2009)
2. J. Berkson, Are there two regressions? J. Am. Stat. Assoc. **45**(250), 164–180 (1950)
3. J. Bi, T. Zhang, Support vector classification with input data uncertainty, in *Advances in Neural Information Processing Systems* (2005), pp. 161–168

4. G. Cauwenberghs, T. Poggio, Incremental and decremental support vector machine learning, in *Advances in Neural Information Processing Systems* (2001)
5. C.-C. Chang, C.-J. Lin, *LIBSVM: A Library for Support Vector Machines* (2001)
6. R.-E. Fan, K.-W. Chang, C.-J. Hsieh, X.-R. Wang, C.-J. Lin, Liblinear: a library for large linear classification. J. Mach. Learn. Res. **9**, 1871–1874 (2008)
7. N. Govindaraju, M. Lin, D. Manocha, Fast and reliable collision culling using graphics hardware. Trans. Vis. Comput. Graph. **12**(2), 143–154 (2006)
8. L. Guibas, D. Hsu, H. Kurniawati, E. Rehman, Bounded uncertainty roadmaps for path planning. Algorithmic Found. Robot. VIII **57**, 199–215 (2009)
9. T. Hofmann, B. Schölkopf, A.J. Smola, Kernel methods in machine learning. Ann. Stat. **36**(3), 1171–1220 (2008)
10. P.M. Hubbard, Approximating polyhedra with spheres for time-critical collision detection. Trans. Graph. **15**, 179–210 (1996)
11. P. Jenke, M. Wand, M. Bokeloh, A. Schilling, W. Straßer, Bayesian point cloud reconstruction, in *Eurographics* (2006), pp. 379–388
12. M. Kalakrishnan, S. Chitta, E. Theodorou, P. Pastor, S. Schaal, Stomp: stochastic trajectory optimization for motion planning, in *International Conference on Robotics and Automation* (2011)
13. J. Klein, G. Zachmann, Point cloud collision detection, in *Eurographics* (2004), pp. 567–576
14. C. Lauterbach, Q. Mo, D. Manocha, gProximity: hierarchical gpu-based operations for collision and distance queries. Comput. Graph. Forum **29**(2), 419–428 (2010)
15. J.-K. Lee, Y.J. Kim, Haptic rendering of point set surfaces, in *EuroHaptics* (2007), pp. 513–518
16. J.-M. Lien, Point-based Minkowski sum boundary, in *Pacific Graphics* (2007), pp. 261–270
17. M. Lin, D. Manocha, Collision and proximity queries, in *Handbook of Discrete and Computational Geometry* (CRC Press, Inc., 2004), pp. 787–808
18. P.E. Missiuro, N. Roy, Adapting probabilistic roadmaps to handle uncertain maps, in *International Conference on Robotics and Automation* (2006), pp. 1261–1267
19. D.M. Mount, Geometric intersection, in *Handbook of Discrete and Computational Geometry* (CRC Press, Inc., 2004), pp. 857–876
20. M. Pauly, N.J. Mitra, L. Guibas, Uncertainty and variability in point cloud surface data, in *Symposium on Point-Based Graphics* (2004), pp. 77–84
21. A. Petrovskaya, O. Khatib, Global localization of objects via touch. Trans. Robot. **27**, 569–585 (2011)
22. M. Ponamgi, D. Manocha, M.C. Lin, Incremental algorithms for collision detection between solid models, in *Symposium on Solid Modeling and Applications* (1995), pp. 293–304
23. A. Rahimi, B. Recht, Random features for large-scale kernel machines, in *Advances in Neural Information Processing Systems* (2007)
24. N. Ratliff, M. Zucker, J.A.D. Bagnell, S. Srinivasa, Chomp: gradient optimization techniques for efficient motion planning, in *International Conference on Robotics and Automation* (2009)
25. B. Schölkopf, J. Giesen, S. Spalinger, Kernel methods for implicit surface modeling, in *Advances in Neural Information Processing Systems* (2005), pp. 1193–1200
26. P.K. Shivaswamy, C. Bhattacharyya, A.J. Smola, Second order cone programming approaches for handling missing and uncertain data. J. Mach. Learn. Res. **7**, 1283–1314 (2006)
27. D. Steinemann, M. Otaduy, M. Gross, Efficient bounds for point-based animations, in *Symposium on Point-Based Graphics* (2007), pp. 57–64
28. F. Steinke, B. Schölkopf, V. Blanz, Support vector machines for 3d shape processing, in *Eurographics* (2005), pp. 285–294
29. I.A. Sucan, M. Kalakrishnan, S. Chitta, Combining planning techniques for manipulation using realtime perception, in *International Conference on Robotics and Automation* (2010), pp. 2895–2901
30. S. Thrun, W. Burgard, D. Fox, *Probabilistic Robotics* (The MIT Press, 2005)
31. V.N. Vapnik, *The Nature of Statistical Learning Theory* (Springer, New York, 1995)

Active Classification: Theory and Application to Underwater Inspection

Geoffrey A. Hollinger, Urbashi Mitra and Gaurav S. Sukhatme

Abstract We discuss the problem in which an autonomous vehicle must classify an object based on multiple views. We focus on the active classification setting, where the vehicle controls which views to select to best perform the classification. The problem is formulated as an extension to Bayesian active learning, and we show connections to recent theoretical guarantees in this area. We formally analyze the benefit of acting adaptively as new information becomes available. The analysis leads to a probabilistic algorithm for determining the best views to observe based on information theoretic costs. We validate our approach in two ways, both related to underwater inspection: 3D polyhedra recognition in synthetic depth maps and ship hull inspection with imaging sonar. These tasks encompass both the planning and recognition aspects of the active classification problem. The results demonstrate that actively planning for informative views can reduce the number of necessary views by up to 80 % when compared to passive methods.

1 Introduction

Consider the following scenario, which occurs when observing an environment with an underwater vehicle: given a playback of imaging sonar data from the vehicle, the task is to determine which frames contain objects of interest (e.g., mines [23], explosives, ship wreckage, enemy submarines, marine life [20], etc.). We will

G.A. Hollinger (✉) · G.S. Sukhatme
Computer Science Department, Viterbi School of Engineering,
University of Southern California, Los Angeles, CA 90089, USA
e-mail: gahollin@usc.edu

G.S. Sukhatme
e-mail: gaurav@usc.edu

U. Mitra
Electrical Engineering Department, Viterbi School of Engineering,
University of Southern California, Los Angeles, CA 90089, USA
e-mail: ubli@usc.edu

H.I. Christensen and O. Khatib (eds.), *Robotics Research*,
Springer Tracts in Advanced Robotics 100, DOI 10.1007/978-3-319-29363-9_6

refer to these problems as *underwater inspection*, since an object is being inspected to determine its nature. We are interested in utilizing sensor data, such as depth map information, to determine the nature of a potential object of interest. Such problems are typically formulated as *passive* classification, where some data are given, and the goal is to determine the nature of this data.

While passive classification problems are challenging in themselves, what is often overlooked is that robotic applications allow for *active* decision making. In other words, an autonomous vehicle performing a classification task has control over how it views the environment. The vehicle could change its position, modify parameters on its sensor, or even manipulate the environment to improve its view. For instance, it may be difficult to determine the nature of an object when viewed from the top (due to lack of training data, lack of salient features, occlusions, etc.), but the same object may be easy to identify when viewed from the side. As an example, Fig. 1 shows an explosive device placed on a ship's hull viewed from two different angles with imaging sonar. The explosive is easier to identify when viewed from the side (left image) versus from above (right image) due to the reflective qualities of its material.

In addition to choosing the most informative views of the object, an autonomous vehicle is able to act adaptively by modifying its plan as new information from viewing the object becomes available. Consider an object of interest, such as an explosive, that has an identifiable feature on a particular side. If the vehicle receives a view that increases the likelihood of that object being in the frame, it would be advantageous to search for that identifiable feature to either exclude or confirm the identification of that object. A significant benefit from acting adaptively has been shown in the stochastic optimization and planning domains [7, 11].

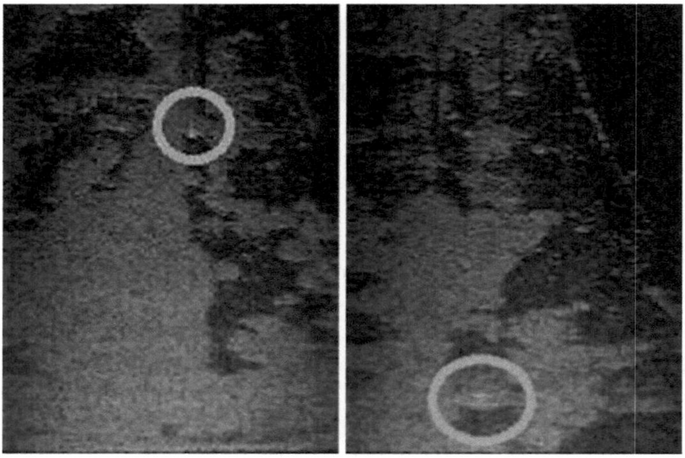

Fig. 1 An explosive device (*circled*) placed on a ship hull viewed using an imaging sonar. The explosive is easier to identify when viewed from the side (*left image*) than when viewed from above (*right image*). This difference motivates active planning to identify the object

In this paper, we apply the above insights to active inspection in the underwater domain. This paper makes three main contributions. We

1. **formalize** the active classification problem, combining classical work in sequential hypothesis testing with recent work in active learning,
2. **analyze** the benefit of adaptivity, leading to an information theoretic heuristic for planning informative paths for active classification, and
3. **apply and test** the approach to underwater classification in a simulated domain and using real-world data.

2 Related Work

The problem of active multi-view recognition has been studied extensively in the computer vision community, dating back to early work in active vision [1] and next-best view planning [6]. While early work primarily optimized views using a geometric approach, later work incorporated probabilistic models into active vision systems [2, 8, 19].[1] Such approaches have also been applied to depth maps in the context of medical imagery [25]. While different forms of information gain play a critical role in active vision, a key distinction in our work is the notion of adaptivity. In active classification problems, selecting the next best observation, or even an initial ordering of informative observations, may not result in overall performance optimization. It is in this regard that we provide new analysis of the benefit of adaptivity and make connections to performance guarantees in submodular optimization and active learning. Our analysis is complementary to prior computer vision work and could potentially be extended to many of these alternative frameworks.

In this paper, we connect two classical problems: active classification and sequential hypothesis testing. Sequential hypothesis testing arises when an observer must select a sequence of noisy experiments to determine the nature of an unknown [22]. A key distinction between sequential hypothesis testing and active classification is that the type of experiment does not change in sequential testing. One of the first applications of sequential hypothesis testing to sensor placement applications was due to Cameron and Durrant-Whyte [5]. They discuss a Bayesian selection framework for identifying 2D images with multiple sensor placements. This work provides a foundation for the formulation discussed in the current paper.

The active classification problem can be seen as an instance of informative path planning [18]. Informative path planning optimizes the path of a robot to gain the maximal amount of information relative to some performance metric. It has been shown in several domains, including sensor placement [14] and target search [13],

[1]There are a number of additional active vision works relevant to the present paper. We direct the interested reader to Roy et al. [17] for a survey.

that many relevant metrics of informativeness satisfy the theoretical property of *submodularity*. Submodularity is a rigorous characterization of the intuitive notion of diminishing returns that arises in many active planning application.

Recent advances in active learning have extended the property of submodularity to cases where the plan can be changed as new information is incorporated. The property of *adaptive submodularity* was introduced by Golovin and Krause [11], which provides performance guarantees in many domains that require adaptive decision making. Their recent work examines these theoretical properties in the context of a sequential hypothesis testing problem with noisy observations [12]. The idea of acting adaptively has also been examined in stochastic optimization and shown to provide increases in performance for stochastic covering, knapsack [7], and signal detection [16]. To our knowledge these ideas have not been formally applied to robotics applications.

In the underwater inspection and surveying domains, there has been limited work in utilizing multiple views to classify underwater mines. In some work, an assumption is made that all views provide the same amount of information [23], and in other work the focus is on designing high-level mission planning capabilities to ensure coverage of the sea floor [24]. The closest prior work to our own discusses active object recognition with imaging sonar using a Partially Observable Markov Decision Process [15]. The authors focus on the optimal algorithm, which grows infeasible as the number of possible viewing locations increases. This prior research does not provide a scalable approximate algorithm, and the authors do not analyze the benefit of adaptivity or possible performance guarantees.

3 Problem Formulation

We will now formulate the active classification problem within the sequential hypothesis testing framework [22]. The goal is to determine the class of an unknown object given a set of N possibilities $\mathscr{H} = \{h_1, \ldots, h_N\}$. Let H be a random variable equal to the true class of the object. In the simplest case, a binary classification task is considered (e.g., $H = h_0$ denotes an object of interest and $H = h_1$ denotes the lack of such an object). We can observe the object from a set of possible locations $\mathscr{L} = \{L_1, \ldots, L_M\}$, where the locations themselves are not informative.[2] There is a cost of moving from location L_i to location L_j, which we denote as d_{ij} . In robotics applications, this cost is determined by the kinematics of the vehicle and the dynamics of both the vehicle and environment.

A set of features $\mathscr{F} = \{F_1, \ldots, F_K\}$ is also given that distinguishes between objects. Each feature F_i is a random variable, which may take on some values

[2]We formulate the problem for the case of discrete locations. If continuous locations are available, an interpolation function can be used to estimate the informativeness of a location based on the discrete training data (see Sect. 6).

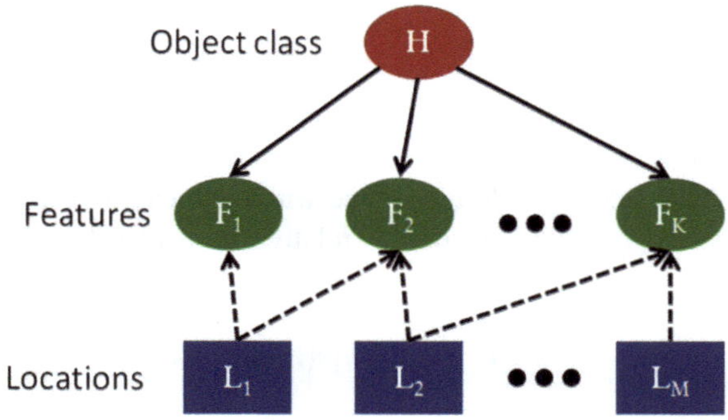

Fig. 2 Graphical model of an active classification problem. The goal is to determine the value of the hypothesis H by observing a subset of features $F_1, \ldots, F_K$. The features cannot be viewed directly, but must instead be viewed by moving to some locations $L_1, \ldots, L_M$. The solid lines denote stochastic dependence, and the dashed lines denote which features can be viewed by visiting each location. Dependencies between the features could also exist, which would break the conditional independence assumption

(e.g., binary, discrete, or continuous). Given one or more template images for each class n, we can calculate a function $G(L) : \mathscr{L} \to \mathscr{F}$ mapping viewing location L to the features for which realizations will be observed from that viewing location. In general, this mapping may be stochastic and dependent on the class. The mapping from location to features is a key characteristic of robotics applications that differentiates our problem from the more common problem where the features can be observed directly [12]. Figure 2 shows a graphical model of the resulting problem.

We assume knowledge of a prior distribution for each class $P(H)$, as well as a conditional probability for each feature given the class $P(F_k \mid H)$. The conditional distribution represents the probability of each feature taking on each of its possible values given the class. These probabilities can be estimated via training data. The features that have been viewed evolve as the robot moves from location to location. At a given time t, the robot is at location $L(t)$, and we observe realizations of some new features $\mathscr{F}_t \subset \mathscr{F}$. Let us define $\mathscr{F}_{1:t} := \bigcup_{i=1}^{t} \mathscr{F}_i$ as the features observed up until time t. If we assume that the features are conditionally independent given the class, we can calculate a distribution $b(t) = \{b_1, \ldots .b_N\}$ using standard recursive Bayesian inference [21]:

$$b(t) := P(H \mid \mathscr{F}_{1:t}) \tag{1}$$

$$= \eta\, b(t-1) \prod_{F \in \mathscr{F}_t} P(F \mid H), \tag{2}$$

where η is a normalizing constant.

The goal is to find a policy π that takes a belief distribution $b(t)$, current location $L(t)$, and observation history $\mathscr{F}_{1:t}$ and determines the next location from which to view the object. Note that the dependence on the observation history and current distribution allows the policy to be adaptive as new information becomes available.

3.1 Noiseless Case

Ideally, we would like to run the policy until we know the object's class. If the observations do not contain any noise, this goal is reachable. For each hypothesis h, a policy π will have a cost $c(\pi, h)$ associated with the locations the policy visits. We define the expected cost of this policy relative to a distribution on hypothesis $P(H)$ as:

$$c(\pi) := \mathbb{E}_H[c(\pi, h)]. \tag{3}$$

This equation represents the expected cost for the policy π . For the noiseless case, we assume that each hypothesis h has an associated vector $V_h = [f_1, \ldots, f_K]$ of feature values that *always* occur for that hypothesis. As a result, $P(F_1, \ldots, F_K \mid H)$ only takes on the values of one or zero. An incomplete feature vector V is said to be consistent with a hypothesis h if for all $f \in V$, we have $f \in V_h$.

Without observation noise, we may fully determine the hypothesis by observing some features (in some cases all features). Let $\mathscr{V}(V)$ represent the number of classes that are consistent with partial feature vector V (also referred to in prior work as the version space [12]). Let $V(\pi, h)$ be the feature vector that results from executing policy π with hypothesis h. The optimal policy is now the one that optimizes the equation below:

$$\pi^* = \underset{\pi}{\operatorname{argmin}}\; c(\pi) \text{ s.t. } \mathscr{V}(V(\pi, h)) = 1 \text{ for all } h \in \mathscr{H}. \tag{4}$$

Even in the noiseless case, there may be insufficient features to determine the exact class of the unknown object. In these cases, the goal would be to observe the fewest number of features that reduce the number of consistent classes as much as if all features were observed.

3.2 Noisy Observations

When the observations are noisy, it will likely be impossible to determine the class of an unknown object with certainty. However, as in the decision theory literature, we minimize the expected loss (also known as the Bayes' risk [22]) of the final classification decision. We will now formulate the problem of minimizing Bayes' risk for the case of noisy observations. With noisy observations, $P(F \mid H)$ takes on values other than one or zero. As a result, there is no longer a deterministic vector V_h associated with each hypothesis, and typically we cannot uniquely determine the hypothesis even by observing all features.

In the noisy observation case, we can generate a policy that minimizes a loss function $l(d, h)$ associated with making a decision d for that object (i.e., deciding on the object's class). For instance, if the object is an explosive, a false negative could incur a very high cost, but a false positive would be a lower cost. If we select the class with maximum a posteriori probability after running a policy π, we can calculate the expected loss for running that policy to completion:

$$l(\pi) := \mathbb{E}_H[l(d,h) \mid \pi]. \tag{5}$$

Let τ be an acceptable threshold on expected loss. A natural goal is to incur the lowest cost and achieve the same expected loss. The resulting optimization problem is given below:

$$\pi^* = \underset{\pi}{\text{argmin}}\ c(\pi) \text{ s.t. } l(\pi) \leq \tau. \tag{6}$$

4 Proposed Solution

The goal is to optimize the expected loss for a policy π. The expected loss is a function of the final belief *b(T)*, which represents $P(H \mid \mathscr{F}_{1:T})$. Calculating this loss on an infinite horizon would require examining an exponential number of paths in the horizon length. To make the computation feasible, we can use the truncated expected loss:

$$\pi^* = \underset{\pi \in \Pi(1:T)}{\text{argmin}}\ \mathbb{E}_H[l(d,h) \mid \pi(1:T)]. \tag{7}$$

A related measure of the quality of *b(T)* is the *information gain* of the class given the features observed: $IG(H; \mathscr{F}_{1:T}) = \mathbb{H}(H) - \mathbb{H}(H \mid \mathscr{F}_{1:T})$, where $\mathbb{H}$ is the entropy. We will motivate the use of information gain further in Sect. 5. A heuristic for solving the active classification problem using information gain can be formulated as below:

$$\pi^* = \underset{\pi \in \Pi(1:\mathrm{T})}{\text{argmax}}\ \mathbb{E}_H[IG(H; \mathscr{F}_{1:T}) \mid \pi(1:T)], \tag{8}$$

where $\Pi(1:T)$ is the set of all possible policies truncated at time *T*. If this optimization is performed on the receding horizon, it allows for adaptive decision making with a finite lookahead. The path costs can be implicitly incorporated by looking ahead to a "cost horizon." This approach has been shown to perform well in similar information gathering domains [13].

For some loss functions, the information gain objective is equivalent to minimizing the Bayes' risk. One such function for the binary hypothesis case is the standard 0/1 loss, where cost of one is incurred for an incorrect classification, and no cost is incurred for a correct classification.

5 Theoretical Analysis

We next relate the active classification problem to recent advances in active learning theory that allow us to analyze the performance of both non-adaptive and adaptive policies. Prior work in active vision does not provide tools for analyzing the performance of approximate solutions. With the goal of generating approximation guarantees for scalable algorithms, we provide a preliminary analysis of the theoretical properties of active classification objective functions.

Active classification can be seen as an instance of informative path planning [18]. Given some potential locations to make observations, the informative path planning problem is to maximize a function $F(A)$, where $A = \{L_1, L_2, \ldots, L_T\}$ is a set of locations visited by the vehicle up to an end time T. In most cases, the sets of possible locations to visit are constrained by obstacles, vehicle kinematics, or other factors. For the active classification problem, $F(A) = -\mathbb{E}_H[l(d,h) \mid A]$, the negative expected loss after observing along path A.

5.1 Performance Guarantees

A *non-adaptive policy* is one that generates an ordering of locations to view and does not change that ordering as features are observed. The non-adaptive policy will typically be easier to compute and implement, since it can potentially be computed offline and run without modification. Performance guarantees in the non-adaptive informative path planning domain are mainly dependent on the objective function (i.e., the informativeness of the views) being non-decreasing and submodular on the ground set of possible views. A set function is non-decreasing if the objective never decreases by observing more locations in the environment. A set function is submodular if it satisfies the notion of diminishing returns (see Singh et al. [18] for a formal definition).

Information gain has been shown to be both non-decreasing and submodular if the observations are conditionally independent given the class [14], as is assumed in this paper (see Sect. 3). Thus, if the loss function is equivalent to information gain (e.g., 0/1 loss with binary hypotheses), then the active classification problem optimizes a non-decreasing, submodular function. Let A^{IG} be the set of locations visited by the information gain heuristic with a one-step lookahead. For non-adaptive policies without path constraints (e.g., when traversal costs between locations are negligible compared to observation cost), we have the following performance guarantee: $F(A^{IG}) \geq (1 - 1/e)F(A^{opt})$ [14].

When path constraints are considered, the recursive greedy algorithm, a modification of greedy planning that examines all possible middle locations while constructing the path, can be utilized to generate a path A^{rg} [18]. Recursive greedy provides a performance guarantee of $F(A^{rg}) \geq F(A^{opt})/\log(|A^{opt}|)$, where $|A^{opt}|$ is the number of location visited on the optimal path. However, the recursive greedy algorithm requires pseudo-polynomial computation, which makes it infeasible for some application domains. To our knowledge, the development of a fully polynomial algorithm with performance guarantees in informative path planning domains with path constraints is still an open problem. Hence, we utilize a one-step heuristic in our experiments in Sect. 6.

The performance guarantees described above do not directly apply to adaptive policies. An *adaptive* policy is one that determines the next location to select based on the observations at the previously viewed locations. Rather than a strict ordering of locations, the resulting policy is a tree of locations that branches on the observation history from the past locations. As noted earlier, the concept of adaptive submodularity [11] allows for some performance guarantees to extend to adaptive policies as well. When the observations are noiseless, the information gain heuristic satisfies the property of adaptive submodularity. This result leads to a performance guarantee on the cost of the one-step information gain adaptive policies in sequential hypothesis testing domains without path constraints: $c(\pi_{IG}) \leq c(\pi^{opt})$ $(\ln(1/p_{min}) + 1)$, where $p_{min} := \min_{h \in \mathcal{H}} P(h)$. When noisy observation are considered, a reformulation of the problem is required to provide performance guarantees (i.e., information gain is not adaptive submodular). However, Golovin et al. [12] show that the related Equivalence Class Determination Problem (ECDP) optimizes an adaptive submodular objective function and yields a similar logarithmic performance guarantee. The direct application of ECDP to active classification is left for future work.

5.2 Benefit of Adaptivity

We now examine the benefit of adaptive selection of locations in the active classification problem. As described above, the non-adaptive policy will typically be easier to compute and implement, but the adaptive policy could potentially perform better. A natural question is whether we can quantify the amount of benefit to be gained from an adaptive policy for a given problem. To begin our analysis of adaptivity, we consider the problem of minimizing the expected cost of observation subject to a hard constraint on loss.[3]

Problem 1 Given hypotheses $\mathcal{H} = \{h_1, h_2, \ldots, h_N\}$, features $\mathcal{F} = \{F_1, F_2, \ldots, F_K\}$, locations $\mathcal{L} = \{L_1, \ldots, L_M\}$, costs $c(L_i, L_j) = d_{ij}$ for observing location

[3]Note that the related problem of minimizing expected loss subject to a hard constraint on budget is also relevant. While similar examples show that there is a benefit to acting adaptively in this case, we defer detailed analysis to future work.

i when at location j, and a loss function defined as $l(d, h)$ for selecting hypothesis d when the true hypothesis is h. We wish to select a policy π such that:

$$\pi^* = \underset{\pi}{\operatorname{argmin}}\ c(\pi) \text{ s.t. } l(\pi) \leq \tau, \tag{9}$$

where $l(\pi) := \mathbb{E}_H[l(d,h) \mid \pi], c(\pi) := \mathbb{E}_H[c(\pi, h)]$, and τ is a scalar threshold.

We now show that the optimal non-adaptive policy can require exponentially higher cost than an adaptive policy for an instance of this problem:

Theorem 1 *Let π_{adapt} be the optimal adaptive policy, and $\pi_{non-adapt}$ be the optimal non-adaptive policy. There is an instance of Problem 1 where $c(\pi_{adapt}) = \log(N)$ and $c(\pi_{non-adapt}) = N - 1$, where is N is the number of hypotheses.*

Proof We adopt a proof by construction. Let $\tau = 0$, i.e., the required expected loss is zero. Let the features be observed directly through the corresponding locations (i.e., $G(L_i) = F_i$ and $M = K$). Let there be N hypotheses and $M = N - 1$ features. Assign a cost $c(F) = 1$ for all features. The loss $l(d, h) = 1$ for all $d \neq h$ and $l(d, h) = 0$ for $d = h$.

Let $P(h) > 0$ for all $h \in \mathcal{H}$. Let $P(F_1 \mid h_i) = 1$ for all $i \in \{1, \ldots, N/2\}$ and $P(F_1 \mid h_i) = 0$ for all $i \in \{N/2+1, N\}$. That is, feature F_1 is capable of deterministically differentiating between the first half and second half of the hypotheses. $P(F_2 \mid h_i) = 1$ for all $i \in \{1, N/4\}$, $P(F_3 \mid h_i) = 0$ for all $i \in \{N/4+1, N/2\}$, and $P(F_2 \mid h_i) = 1/2$ for all $i \in \{N/2+1, N\}$. That is, feature F_2 is capable of deterministically differentiating between the first fourth and second fourth of the hypothesis space but gives no information about the rest of the hypotheses. Similarly, define $P(F_3 \mid h_i) = 1$ for all $i \in \{N/2+1, 3N/4\}$, $P(F_3 \mid h_i) = 0$ for all $i \in \{3N/4+1, N\}$, and $P(F_2 \mid h_i) = 1/2$ for all $i \in \{1, N/2\}$. The remaining features are defined that differentiate progressively smaller sets of hypotheses until each feature differentiates between two hypotheses.

The adaptive policy will select F_1 first. If F_1 is realized positive, it will select F_2. If F_1 is realized negative, it will select F_3. It will continue to do a binary search until $\log(N)$ features are selected. The true hypothesis will now be known, resulting in zero expected loss. In contrast, the non-adaptive policy must select all $N - 1$ features to ensure realizing the true hypothesis and reducing the expected loss to zero. □

The adaptivity analysis in Theorem 1 requires multiple hypotheses, and the potential benefit of adaptivity increases as the number of hypotheses increases. For the two hypothesis case, however, the benefit of adaptivity may be very small. In the binary examples we have examined, all cases showed little or no benefit from adaptivity. Furthermore, if there is a strict ordering on the informativeness of the viewing locations independent of the current distribution on the hypotheses, we conjecture that the benefit of acting adaptively will be zero [16].

6 Implementation and Experiments

In this section, we examine the active classification problem experimentally through the use of both synthetic images and data from imaging sonar during ship hull inspection. The results confirm the benefit of active view selection in these application domains as well as the benefit of adaptivity when more than two hypotheses are considered. For all experiments, we assume a simple 0/1 loss model, where a cost of one is incurred for a false classification, and a cost of zero is incurred for a correct classification.

6.1 Synthetic Images

The goal of our first experiments is to differentiate between possible polyhedra using depth maps from different views. The relevance of polyhedra recognition to underwater inspection is direct, as explosive devices are often cubic or pyramidal in shape [9]. This is a particularly challenging active recognition problem due to similarities and symmetries between polyhedra. These experiments are designed to (1) demonstrate the benefit of selecting the views with the highest potential for information about the unknown object, and (2) examine the benefit of acting adaptively when multiple possible objects are examined.

To identify the polyhedra, we utilize salient features extracted from the synthetic depth map. Training images were created from 24 different viewpoints around the objects, and the OpenCV [4] SURF feature extractor [3] was used to extract features for the different object and viewpoints viewpoints. Noisy test images were then created with Gaussian white noise ($\sigma = 0.23\ m$). Figure 3 shows SURF features extracted from synthetic depth maps of the polyhedra. The number of SURF features is greater when viewing the polyhedra vertices when compared to viewing the faces.

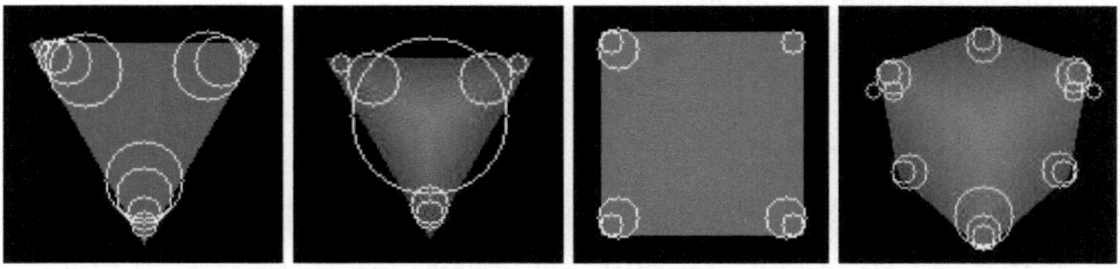

Fig. 3 SURF features extracted from depth maps of tetrahedron and cubes viewed from different angles. Viewing the vertices provides more distinguishing information than viewing the faces

6.1.1 Two Objects

The intuition is that it will be easier to identify the object in some viewpoints than in others, due to the presence of additional salient features. We analyze this claim by comparing informative view selection to random view selection on synthetic depth map data from a cube and tetrahedron. The information gain of each view was calculated based on the number of expected salient features corresponding to the true object minus the expected number of false correspondences. This calculation requires comparing all views to the corresponding views of each other object ($O(N^2)$ computation in the number of hypotheses). After the cross-correlations were computed, planning was completed in milliseconds. To apply adaptive view selection, we calculate the information gain from the current distribution over the features, which changes as new views are observed.

In these experiments, path constraints are not considered, though the view ordering could easily be used to generate a feasible path on the finite horizon. Figure 4 shows results comparing the information gain heuristic with random view orderings. Number of correct correspondences are reported based on the assumption that the object with the largest number of correspondences will be selected. Utilizing the information gain heuristic to determine the most informative views leads to as much as a 35 % increase in the number of correct feature correspondences with limited views. Adaptive view selection does not provide much benefit over the non-adaptive technique, as expected from the small adaptivity gap in the binary hypothesis case (see Sect. 5). Note that, for comparison, only 24 views are considered, and all methods will provide the same performance after seeing all these views.

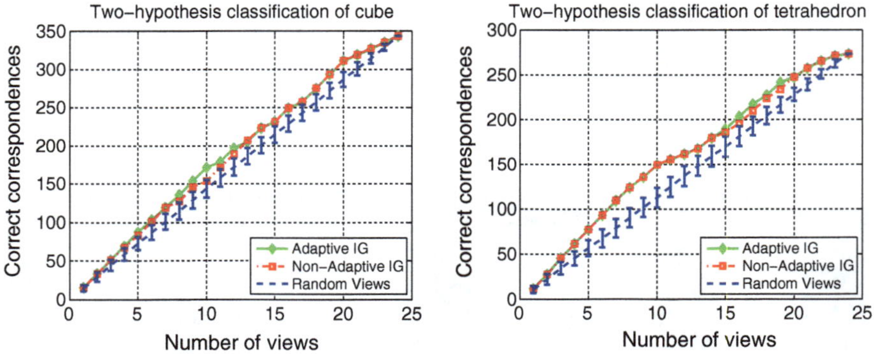

Fig. 4 Multi-view classification experiments with synthetic images of a cube and tetrahedron viewed from 24 different angles (best viewed in color). Utilizing the expected information gain of the next view improves the number of SURF feature correspondences when limited views are used. Random view results are averaged over 100 orderings; error bars are one standard deviation

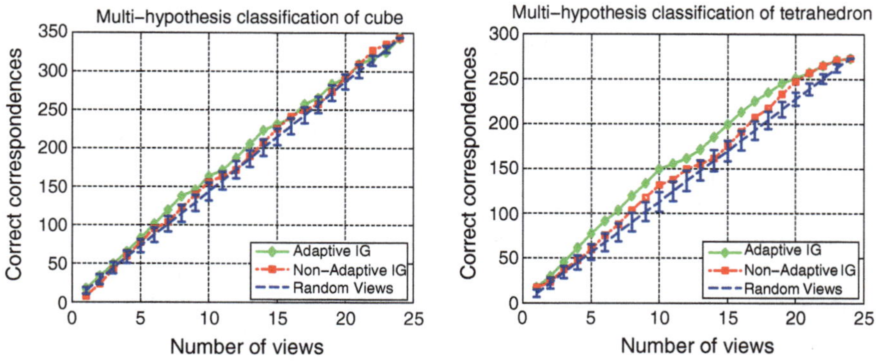

Fig. 5 Classification experiments with synthetic images of the five platonic solids (best viewed in color). The results for a cube and tetrahedron test object are shown. Adaptively selecting the most informative views based on past information tends to improve classification accuracy, and acting adaptively increases this benefit. Random view results are averaged over 100 random orderings; error bars are one standard deviation

6.1.2 Multiple Objects

The benefit of active classification is now examined for cases where more than two object classes are considered. In addition to the cube and tetrahedron, we include training images of the icosahedron, octahedron, and dodecahedron as possible object classes. The theoretical analysis in Sect. 5 suggests that acting adaptively should improve performance for the multi-hypothesis problem. Figure 5 shows results for classifying the cube and tetrahedron when additional hypotheses are considered for the other three platonic solids. The adaptive policy outperforms both random view selection and the non-adaptive policy the majority of the time. The difference is particularly significant for the tetrahedron. Note that the dominance of the adaptive policy is not true at all data points. These results suggest that adding additional hypotheses in some cases reduces the performance of active view selection.

6.2 Imaging Sonar Data

To examine the benefit of active classification on real-world data, we utilize data from imaging sonar depth maps taken from a ship hull inspection with an underwater vehicle. The vehicle has already executed a path, and we utilize the proposed framework to order the viewpoints based on informativeness. Such information could then be utilized to plan additional inspection paths of the object.

The goal of the inspection is to determine whether an explosive has been placed on the ship hull. The explosive appears as a small patch of bright pixels on the imaging sonar depth map. Since the sonar data is not dense enough to provide salient features, we take a simpler approach of using the brightness of the pixels as the feature base. A brightness threshold was learned by minimizing the number of misclassified pixels in labeled data. The performance metric is the total number of pixels correctly classified as part of the explosive device. We utilize this metric because images with a large number of corresponding pixels may provide additional information during post-processing or to a human operator.

A separate test set was held out of the labeled set to determine if the most informative views could be predicted using the learned threshold and expected view quality. There were 100 frames in the training and 75 frames in the test set. The training and test frames were from different trajectories, but with the same background. The frame rate was approximately 2 fps. The information gain in these experiments was calculated based on the expected number of pixels corresponding to the explosive in a given view, which was found using an average of the hand-labeled pixels in the training set images weighted by their distance (using data from a DVL sensor). A squared exponential weighting was used.

Figure 6 shows the results of running the information gain approach versus random views. We also compare to the initial (very poor) view ordering from the data as well as two simple ordering methods: sorting the views based on minimum distance to the object and sorting based on the maximum angle of view (see Fig. 1 for the intuition behind this method). The results show that actively choosing the views with the highest expected information improves classification performance. For example, choosing informative views reduces the number of views for 15 correct pixel identifications by nearly 80 % versus random selection (from 38 views to 8 views).

For visual reference, Fig. 7 shows images of decreasing expected pixel classifications. Intuitively, the images where the explosive stands out from the background should provide the most information. Despite some incorrect predictions, it is clearly beneficial to examine those viewpoints predicted to be informative. It should be noted that the informativeness of the images depends on the quality of the low-level sonar processing. With perfect low-level data processing, all images may have high informativeness, which would reduce the benefit of active classification.

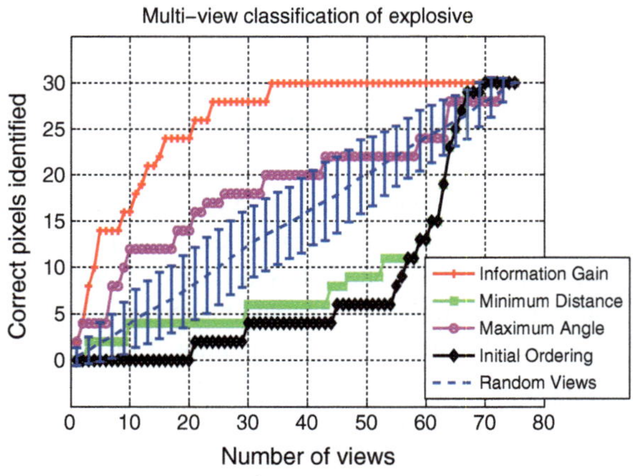

Fig. 6 Multi-view classification experiments with imaging sonar identifying an explosive on a ship hull. With limited views, utilizing information gain leads to a larger number of pixels correctly identified as part of the object of interest. Random view results are averaged over 100 random orderings; error bars are one standard deviation

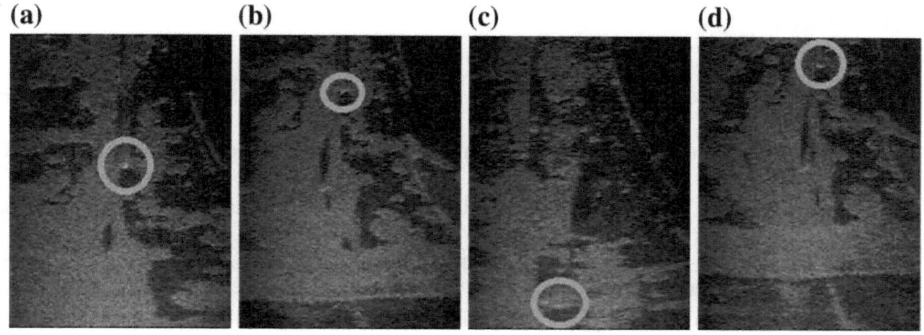

Fig. 7 Imaging sonar depth maps of an explosive device (*circled*) placed on a ship's hull. The depth maps are ordered based on the expected number of pixels in the image corresponding to a possible explosive. Note that the explosive is easy to identify in image (**a**), more difficult to identify in image (**b**), and very difficult to identify in image (**c**). Image **d** is expected to be a low information view, when in fact the explosive is relatively easy to identify. **a** Exp. gain: 3.7, **b** Exp. gain: 2.1, **c** Exp. gain: 1.5, **d** Exp. gain: 0.8

7 Conclusions and Future Work

This paper has shown that actively choosing informed views improves performance for inspection tasks in the example underwater domain. The experimental results demonstrate that depth map information can be utilized to recognize objects of interest, and that (compared to passive methods) up to 80 % fewer views need to be examined if the views are chosen based on their expected information content. In addition, acting adaptively by re-evaluating the most informed views as new information becomes available leads to improvement when more than two classes are considered. These results are consistent with theoretical analysis of the benefit of adaptivity.

Future work includes further theoretical analysis of performance guarantees, particularly in the case of path constraints. The analysis and tools developed in this paper can also be applied to related underwater inspection problems, such as reconstructing ocean floors or ship hulls for inspection [10]. Such tasks have typically been formulated as coverage problems, but the use of alternative objective functions based on uncertainty reduction could both improve performance and allow these problems to be analyzed in the context of active classification. Finally, the analysis in this paper has applications beyond underwater inspection. Tasks such as ecological monitoring, reconnaissance, and surveillance are just a few domains that would benefit from active planning for the most informed views. Through better control of the information we receive, we can improve the understanding of the world that we gain from robotic perception.

Acknowledgements The authors gratefully acknowledge Franz Hover and Brendan Englot at MIT for imaging sonar data and technical support while processing the data. This research has been funded in part by the following grants: ONR N00014-09-1-0700, ONR N00014-07-1-00738, NSF 0831728, NSF CCR-0120778, and NSF CNS-1035866.

References

1. Y. Aloimonos, I. Weiss, A. Bandopadhyay, Active vision. Int. J. Comput. Vision **1**(4), 333–356 (1987)
2. T. Arbel, F.P. Ferrie, Entropy-based gaze planning. Image Vis. Comput. **19**(11), 779–786 (2001)
3. H. Bay, A. Ess, T. Tuytelaars, L. Gool, SURF: speeded up robust features. Comput. Vis. Image Underst. **110**(3), 346–359 (2008)
4. G. Bradski, A. Kaehler, *Learning OpenCV: Computer Vision with the OpenCV Library* (O'Reilly Media, 2008)
5. A. Cameron, H. Durrant-Whyte, A Bayesian approach to optimal sensor placement. Int. J. Robot. Res. **9**(5), 70–88 (1990)
6. C. Connolly, The determination of next best views, in *Proceedings of the IEEE Conference on Robotics and Automation* (1985)
7. B. Dean, M. Goemans, J. Vondrak, Approximating the stochastic knapsack: the benefit of adaptivity. Math. Oper. Res. **33**(4), 945–964 (2008)
8. J. Denzler, C. Brown, Information theoretic sensor data selection for active object recognition and state estimation. IEEE Trans. Pattern Anal. Mach. Intell. **24**(2), 145–157 (2002)
9. G. Dobeck, J. Cobb, Fusion of multiple quadratic penalty function support vector machines (QPFSVM) for automated sea mine detection and classification, in *Proceedings of the SPIE* (2002)
10. B. Englot, F. Hover, Planning complex inspection tasks using redundant roadmaps, in *Proceedings of the International Symposium on Robotics Research* (2011)
11. D. Golovin, A. Krause, Adaptive submodularity: a new approach to active learning with stochastic optimization, in *Proceedings of the Conference on Learning Theory* (2010)
12. D. Golovin, D. Ray, A. Krause, Near-optimal Bayesian active learning with noisy observations, in *Proceedings of the Neural Information Processing Systems* (2010)
13. G. Hollinger, S. Singh, J. Djugash, A. Kehagias, Efficient multi-robot search for a moving target. Int. J. Robot. Res. **28**(2), 201–219 (2009)
14. A. Krause, C. Guestrin, Near-optimal nonmyopic value of information in graphical models, in *Proceedings of the Uncertainty in Artificial Intelligence* (2005)
15. V. Myers, D. Williams, A POMDP for multi-view target classification with an autonomous underwater vehicle, in *Proceedings of the IEEE OCEANS Conference*, Sept 2010
16. M. Naghshvar, T. Javidi, Active M-ary sequential hypothesis testing, in *Proceedings of the IEEE International Symposium on Information Theory* (2010)
17. S.D. Roy, S. Chaudhury, S. Banarjee, Active recognition through next view planning: a survey. J. Pattern Recognit. **37**(3), 429–446 (2004)
18. A. Singh, A. Krause, C. Guestrin, W. Kaiser, Efficient informative sensing using multiple robots. J. Artif. Intell. Res. **34**, 707–755 (2009)
19. M. Sipe, D. Casasent, Feature space trajectory methods for active computer vision. IEEE Trans. Pattern Anal. Mach. Learn. **24**(12),1634–1643 (2002)
20. D. Steinberg, S. Williams, O. Pizarro, M. Jakuba, Towards autonomous habitat classification using Gaussian mixture models, in *Proceedings of the IEEE/RSJ International Conference on Intelligent Robots and Systems* (2010)
21. S. Thrun, W. Burgard, D. Fox, *Probabilistic Robotics* (MIT Press, Cambridge, 2005)
22. A. Wald, Sequential tests of statistical hypotheses. Ann. Math. Stat. **16**(2), 117–186 (1945)
23. D. Williams, Bayesian data fusion of multiview synthetic aperture sonar imagery for seabed classification. IEEE Trans. Image Process. **18**(6), 1239–1254 (2009)
24. D. Williams, On optimal AUV track-spacing for underwater mine detection, in *Proceedings of the IEEE International Conference on Robotics and Automation* (2010)
25. X. Zhou, D. Comaniciu, A. Krishnan, Conditional feature sensitivity: a unifying view on active recognition and feature selection, in *Proceedings of the International Conference on Computer Vision* (2003)

The Importance of Structure

Carl Henrik Ek and Danica Kragic

Abstract Many tasks in robotics and computer vision are concerned with inferring a continuous or discrete state variable from observations and measurements from the environment. Due to the high-dimensional nature of the input data the inference is often cast as a two stage process: first a low-dimensional feature representation is extracted on which secondly a learning algorithm is applied. Due to the significant progress that have been achieved within the field of machine learning over the last decade focus have placed at the second stage of the inference process, improving the process by exploiting more advanced learning techniques applied to the same (or more of the same) data. We believe that for many scenarios significant strides in performance could be achieved by focusing on representation rather than aiming to alleviate inconclusive and/or redundant information by exploiting more advanced inference methods. This stems from the notion that; given the "correct" representation the inference problem becomes easier to solve. In this paper we argue that one important mode of information for many application scenarios is not the actual variation in the data but the rather the higher order statistics as the *structure* of variations. We will exemplify this through a set of applications and show different ways of representing the structure of data.

1 Introduction

A central question to solve when designing an artificial system is how to make it aware and capable of interaction with the environment. The level of usefulness of a robot is considered through its capability of reacting to and adjusting its behavior to changes in the environment. Today's robots, equipped with different sensors such as

C.H. Ek (✉)
University of Bristol, Bristol, UK
e-mail: carlhenrik.ek@bristol.ac.uk

D. Kragic
Royal Institute of Technology, Stockholm, Sweden
e-mail: dani@csc.kth.se

H.I. Christensen and O. Khatib (eds.), *Robotics Research*,
Springer Tracts in Advanced Robotics 100, DOI 10.1007/978-3-319-29363-9_7

cameras, microphones and depth sensors acquire information from the environment at very high precision and rate. Through this rapid development it is now possible to design artificial systems who's sensory systems are more capable than those of the human. However, despite getting more and more detailed observations of the environment, the progress in what we are able to infer through reasoning from this data have not seen the same rapid development. Our central argument in this paper is, *Given the "right" information about a domain inferring the correct answer becomes an easier problem.* The development of sensory systems have rather than focusing on providing the *"right"* information been aimed at simply acquiring *more* information. The justification for this has been the development of more and more advanced machine learning algorithms capable of dealing with larger amounts of data of more complicated distributions. However, the fact still remains that the progress in terms inference have not followed that of the sensory systems.

One of the strengths of human inference is its capability of being selective with the information it uses to reason [1]. During our development we construct strong (conditional) priors which helps us filter the enormous amount of information that our sensory systems acquires to only use a small subset of the data which is relevant for the task, as indicated by the concept of intentional blindness shown in [2]. Rather the opposite approach seems to be dominant when building artificial systems where we try to extract and model more and more of the variations in the sensory data and exploit more advanced learning algorithms for inference from a very complicated input domain. A describing example is object categorisation in computer vision where the dominant approach is to use local image descriptor such as SIFT [3] to model the sensory data. Clearly the information extracted by such features contains significant amounts of variance which is not relevant for the task which means that in order to be able to generalize within categories the inference algorithm needs to learn to ignore data and focus on the discriminating information. In many situations the discriminating information represents only a small part of the variance in the extracted representation which often means a significant challenge in terms of modeling and inference.

In this paper we argue that rather than focusing on building models capable of representing a larger amounts of the variance in the sensory, we should aim to carefully consider what information that is actually relevant. We argue for representations that focus on the structure of variations rather than accurate descriptions of the local variations in the data. Our motivation stems from the notion that the biggest challenge when it comes to inference is not discrimination per say but rather its complementary notion that of generalization. I.e. the key problem is not to extract variance that separates certain classes but rather avoid extracting variance that corresponds to within class variations. As an example, having observed a specific instance of a mug we can reasonably reliably detect that mug again, the big challenge is to create a system which is capable of generalizing over different mugs separating them from other objects.

We argue that the important questions are concerned with generalization on a level where the global structure is the dominant discriminating factor and not the local variations see Fig. 1. To that end we will describe a set of different scenarios

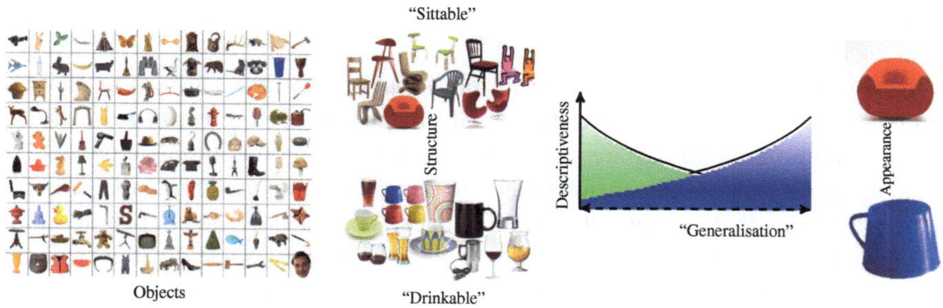

Fig. 1 The above figure tries to highlight the notion of the importance of structure that we try to convey in this paper. The example above shows a large data-base of objects to the far *left*. Of these we want find a representation in order to classify objects at a certain resolution. If the representation naturally generalizes, i.e. it does not reflect within class variance but only between this task is easy to solve. In this paper we argue that for a coarse scale task such as separating "sittable" from "drinkable" objects the discriminating variance is represented by the global structure. While for a high resolution task such as separating the "*red* felt comfy chair" or the "*blue* plastic mug" the discriminating information is contained in the appearance cues. We believe that in robotics we are generally interested in the first type of these two task why therefore find representations of global structures is important

where structural representations and models are of key importance. Through these examples we will show different approaches for exploiting global structure. However, we would like to point out that the purpose of this paper is not to provide a solution to a specific problem but rather exemplify our notion through a range of applications where structure is important to stimulate further discussion on the subject.

2 Structure and Generalization

There are three central concepts in this paper; those of *generalization*, *discrimination* and that of *structure*. To explain what we mean by these we use the example of object modeling as this provides a intuitive example of the concepts that we address in this paper. Object modeling is a necessary prerequisite for equipping a robot with the ability of detection, identification and manipulation. Dependent on the task, we wish to acquire a representation that generalize over specific classes and is able to discriminate between others. Formally this means that we wish to model the between class variance but not the within. Thus, the two concepts generalization and discrimination are complementary, but from a traditional representation point of view the biggest challenge is not to retain (the discriminative part) but rather to remove (the generalizing part) information. An example of this is representing object from visual data for the task of categorization. The main challenge is not to find a representation that separates for example mugs from glasses, as they look different the information is contained in the observations, but rather to remove the information that separates different mugs and different glasses from each other.

Any statistical method relies on the presumption that we can acquire enough samples of a space that can describe it well. Images are high-dimensional meaning that it is not possible to acquire such a data-set easily. The traditional approach has been to extract a low-dimensional feature representation assuming that we can acquire samples that describe the feature space. The most obvious approach is to extract this information from a local patch in the image as clearly this will per definition contain less variations. The central question is then: What level contains the desirable generalization and discrimination characteristics for a specific task? Clearly, on the most local level, being the colour of a pixel, we can model the information robustly and the assumption of sampling the feature space well is going to be forefilled by observing a single image. However, we also know that statistics of such local features will not contain discriminating information for other than the most simple task while it will generalize over a large range of different images. Here is an important notion: the more local a feature, the less discriminative it becomes. Thus, there is a trade-off here that needs to be considered, local enough to be robust and well sampled and global enough to be descriptive, see Fig. 2.

The traditional approach have been to try and use more and more descriptive local features by acquiring large (and growing!) training data sets and then exploit a supervised machine learning technique capable of learning a secondary representation, often through the use of kernel machines or metric learning, that achieves the desired balance between generalization and discrimination.

We argue that the there is a different paradigm where we could use less informative local descriptors while still being able to discriminate. That is to aim to create strong models of the *structure* between the local features and not stop at first

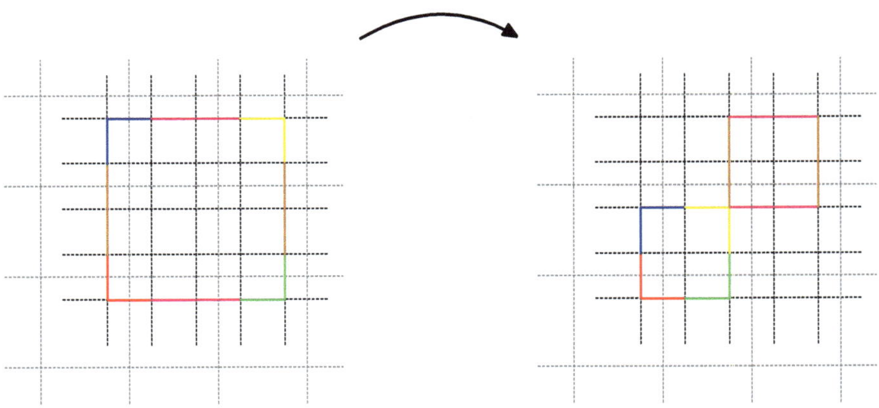

Fig. 2 The above figure shows two different objects with two different scales of local representation, *dotted* (fine) and *dashed* (coarse). First order statistics from the fine resolution will not be able to discriminate the two objects while at the coarser scale they will be different. However, using a coarser scale implies that each cell has a higher dimensionality requiring more samples in order to represent the space well

order statistics such as the so popular *Bag-or-words* techniques. However, how to encode structure is a non-trivial problem that we believe needs to be addressed with much more focus. We do not think that there is one single approach for representing structure but rather a large range of different tools and approaches. In the reminder of this paper we will show different applications and different intuitions and tools that are useful and provide insights into how to deal with different tasks by including a structural element. Our goal with this paper is rather to raise questions than provide specific solutions.

3 Temporal Structure

Many tasks in robotics deal with dynamical scenes where the relevant information is contained in the order of events. A goal of robotics is learning by demonstration [4] where the task is for a robot to extract the relevant notion of a task by observing a demonstrator. Various subproblems have been studied related to task planning and sequencing, detection of motion primitives, developing models for structured collections of actions [5]. The underlying question has been how to acquire a representation that in a sufficient manner generalizes the objective(s) of the task. Take for example the task of clearing a table. Here the appearance of both the objects and the table are irrelevant. Rather the important information that generalizes the task lies in the structure of the events not the actual events themselves. I.e. the task remains the same if the cutlery are cleared before the plates or vice-versa. In this section we describe different applications where we, through some model of temporal structure, manage to simplify an otherwise complicated inference task.

3.1 Interaction

Recently, [6] suggested a method for action classification by representing the temporal structure of the interactions that takes place in the scene. Using visual measurements from a camera the approach first segments the objects in the scene for each frame in a video sequence. The temporal structure is encoded by a graph representing each frame, every object being a node and connected component sharing an edge, see Fig. 3. This process removes all information associated with the appearance and identity of the objects leaving only the interaction. The final processing step is to remove the duration of the interactions and only retain the sequence of topologically different graphs. The intuition behind the representation is that for discriminating between actions the temporal structure of the interactions of objects independent of their identity contains sufficient information. This is significantly different to the more traditional approach for modeling actions such as [7–9] which extracts a representation that retains a significant amount of the

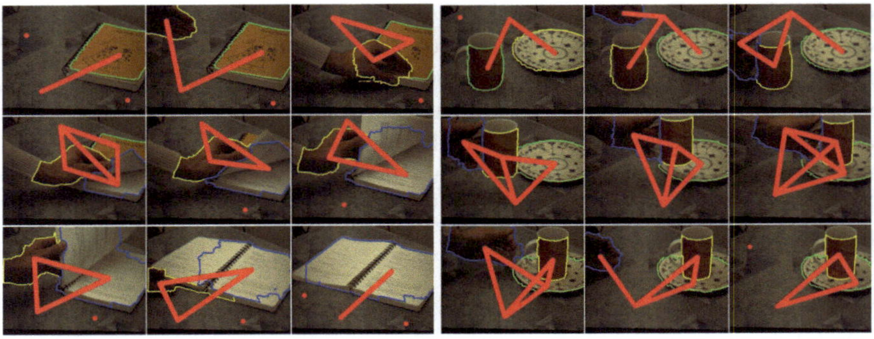

Fig. 3 The *left* example shows an instance of the **Opening Book** action while the *right* shows the **Moving Object**. In each of the images the result of the segmentation and its corresponding graph have been overlaid. Only the spatial relations between the segments are extracted and no identification of the objects is performed

variance related to appearance. This means that we have to learn the invariance related to appearance from data. This requires significantly larger amounts of training data and puts additional challenges on the learning machinery that needs to explain away this non-relevant variance and extract the important variance from the feature. In order to represent each frame the authors in [6] defines a specific semantic extracted from the node connectivity in the graphs and the alterations under this semantic over time is represented as a matrix. A simple distance measure is then defined to compare two different matrices which given a training data-set allows for action classification.

One of the major drawbacks of the approach suggested in [6] is that it is very sensitive to noise as it assumes that each node in the graph represents a single object. In order to circumvent this problem, we have developed a general framework for encoding the structure of variation in a semantic chain using a robust machinery derived from work in text representation [10]. We are motivated by the approach presented in [11] where a feature space representation of a string is presented. By deriving a vector space representation of a string independent of its length strings can be compared by standardized tools from statistical learning. The parametrisation is sensitive to both the order and the existence of letters in the string and does therefore encode both the structure and the appearance of the string. Being infeasible to compute for most typically sized data-sets the feature space is represented implicitly through the use of a kernel function [12]. More formally the feature space we use is spanned by all possible permutations of all lengths of the letters in the semantic alphabet, with an inner product defined as a function of the matching part of the overlap between two strings, see Fig. 4. Clearly the spaces is infinite dimensional but as any string of a shorter length compared to the basis are orthogonal the maximum dimensionality is bounded. Similarly to the original string kernel [11] an efficient recursive computation of the inner product can be formulated representing the feature space implicitly using by a kernel.

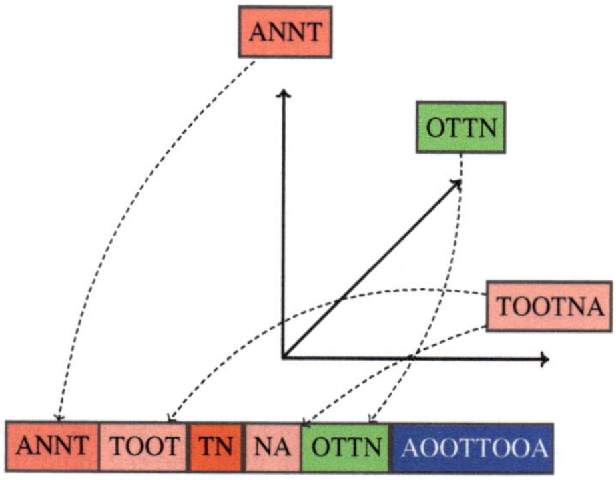

Fig. 4 For a the specific semantic alphabet, here defining the four different interaction relationships between objects: $\{A, N, T, O\}$, we above show a subspace of the feature space representing the sequence. The sequence **ANNT** (*red*) and **OTTN** (*green*) exists in order in the string and will therefore project parallel to the corresponding basis while the **TOOTNA** does not which will induce a non-zero angle between the string and the basis. This means that the representation will be sensitive to gaps in the string making it robust to noise

The above example completely removes all variance associated with appearance from the observations and only retains information about structure. For the task of discriminating between the different actions defined in [6] this contains sufficient information. However, it is easy to think of scenarios where this information is not sufficient for performing the task. However, the kernel based framework can easily be adapted to encode structure where the appearance is also retained as this is simply about defining a semantic that also encodes the appearance. As an example of such we will describe an approach for representing object categories that retain both the appearance and the structure of the object. An idea for the future is the integration of this approach with the probabilistic models for action encoding presented in [13] (Fig. 5).

3.2 *Object Detection*

A robot should be able to interact with it surroundings by applying actions to objects. Thus, a very important task is to identify and extract objects from sensory data. The visual domain contains a rich description of the environment and by segmenting objects from the background detailed models of individual objects can be built. Image segmentation is concerned with clustering "similar" pixels into segments and has attracted considerable interest in computer vision. There are many different approaches and assumptions used to define similarity between pixels. Because of computational limitations, but also due to the challenge of formulating general appearance models, the focus has been on local statistics such as colour distributions and gradients [14, 15]. This has meant that for all but the simplest

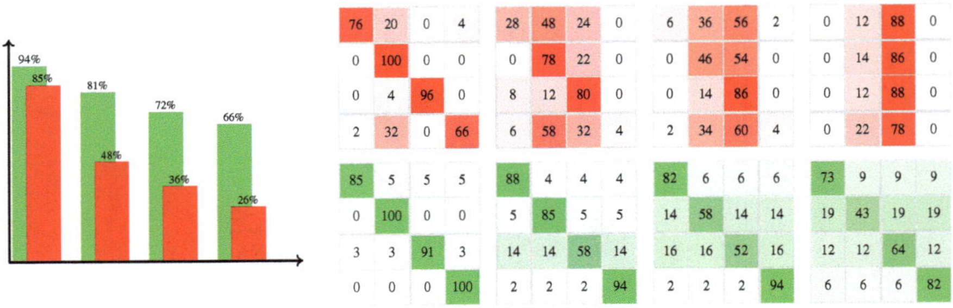

Fig. 5 *Left* The *bar plot* above shows the classification rate associated with increasing noise to the *right*. The *green bars* identifies our kernel approach while the *red* indicates the performance of the original method. *Right* Confusion matrices for increasing noise. The classes are ordered as **Moving Object**, **Making Sandwich**, **Opening Book** and **Filling Liquid**. The *red* matrices show the results for the original approach while the results of our method is shown in *green*. With increasing amount of noise the original measure is unable to disambiguate between the different actions classifying every action as belonging to opening book. For the same data the kernel approach is able to differentiate between the classes and the performance is reduced much more gracefully

objects it is quite unlikely that the clusters retained by an image segmentation approach will corresponds to actual objects in the scene.

The work in image segmentation shows the non-trivial nature of formulating consistent appearance cues based on local statistics that corresponds to objects in the image. This has meant that most successful approaches are interactive, requiring a human to refine and rectify the result produced in an iterative manner [15]. In an autonomous system we cannot rely on interaction to leverage and include human object priors for segmentation but rather need to create a self-contained system.

In [16] we presented an active system for object segmentation which exploits both traditional appearance based assumptions in collaboration with temporal cues in an active iterative manner. Image segmentation techniques are good at grouping pixels into consistent regions. This often mean that for all but the simplest objects this will result in an *over* segmentation where each object is divided into several different segments. Acknowledging the fact that it is a non-trivial task to create appearance models that encapsulates the long range pixel interactions that generalizes over objects we turn our attention to a different domain. In many applications we can assume that the objects of interest in the scene are rigid. Further, each local element or point on such an object moves according to simple rules of rigid motion. This means these rules *generalizes* over all points belonging to the same object. To that end we use the initial segmentation from the appearance cues as an hypothesis of the objects in the scene. In correspondence with this the robot introduces motion by interacting with the scene. Modeling the motion we can easily verify if the appearance segmentation is consistent with the rigid motion assumption. In [16] we describe which combines local appearance cues with a method for modeling rigid motion to use them in a complementary fashion. We show results for a common table-top scenario where and appearance based method used on its own would fail, Fig. 6.

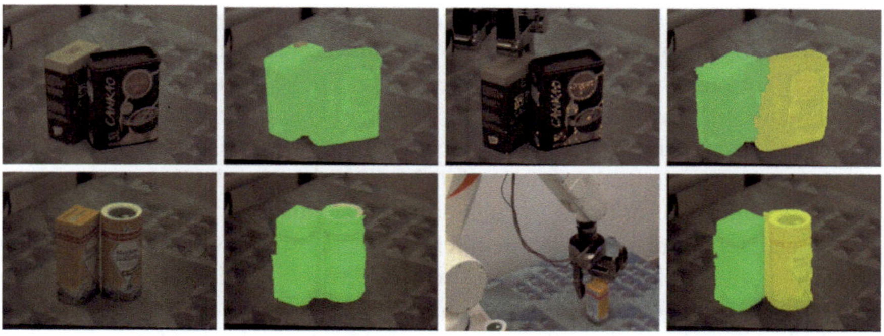

Fig. 6 The *left* most column shows two scenarios where two objects have been placed on a table *top*. Using a traditional appearance based image segmentation approach it is not possible to separate the objects. By introducing motion into the scene by letting the robot interact with the environment the motion can be modeled and the objects separated in the *right* most image

This approach shows how by exploiting a simple assumption we can actively introduce a variance corresponding to the level of generalization we are interested in such a manner that it can easily be extracted from the environment.

4 Spatial Structure

In previous section, we described applications and tasks exemplifying the importance of temporal structure. In this section we discuss structure on a different level namely the structure on a spatial level.

Similarly to the temporal case we argue that the interesting generalization for many tasks are represented by structural information. On example is our use of language, where we would use an structural adjective such as *striped* to discriminate on a coarse level while for identifying specific objects we would add local appearance descriptions such as *red and white*. The currently dominating approach is to use a local representation of each instance and hope that the inference procedure is capable of extracting the information that generalizes between the classes by observing enough examples. As we have previously stated this is a very challenging task from a learning perspective, as quite likely only a small portion, if any, of the variance in the local descriptor will contain generalizing information.

In this section we describe two different task where the generalizing information is contained in the spatial structure of the local appearance and not the local appearance itself.

4.1 Object Representation

Being able to discriminate between objects both on category and instance level is of key importance for a wide range of task in robotics. This requires an object representation that is capable of generalizing over the desired task dependent domain. In computer vision object categorisation has attracted a significant interest. Especially in recent years with the collection of public datasets and high profile competitions such as the Pascal VOC challenge [17]. A large range of different techniques have been applied to the problem where the dominating approach is to aim to extract discriminating information from local image descriptors by relying on the capabilities of different machine learning approaches.

Compared to computer vision researchers roboticists enjoy the luxury of being able to apply several different types of sensory streams in addition to cameras for extracting information of the environment. Recently with the introduction of affordable depth sensors has allowed us to consider dense depth information not as a specialized domain but rather something that can be assumed as readily available. In [18, 19] a robust *3D* feature is presented which represents each local patch of an object as belonging to a specific geometric class. In Fig. 7 the feature is shown extracted from a set of typical household items. Clearly, only describing the geometrical local structure on the object is not likely to provide discriminative information between a large range of different object why the global structure needs to be encoded. To that end in [20] the author presents an approach to encode the global structure by encoding the distribution of local patches along rays between each patch.

The results presented are impressive but modeling the distribution of geometrical classes between local patches is not going to retain the full structure of the object and in order to be able to scale in terms of the level of generalization we believe that a stronger representation is needed. In specific we do not think that rays are a good way of encoding the structure of a surface. The objective is to find a representative global statistics that encodes the structure of the object. What we mean in formal terms is that: an object is a two dimensional surface embedded in a three

Fig. 7 Object features representing the local geometrical class encoded by *colour* shown for three different objects, from the *left* box, citrus fruit and mug

dimensional space which encapsulate a non-empty volume. This implies that given a point on the object one can travel to any other point belonging to the object by traversing this enclosing surface. It is the shape of this surface is what we wish to represent. In this notion of a surface lies our objection towards the use of rays. The surface is a two dimensional object meaning that relating two points to each other requires two degrees of freedom. The position along a ray does not respect the shape of the surface but is rather a construction to create a simple measure of sampling the three dimensional volume along a single parameter. By defining a path respecting the surface of the object, such as the use of an approximate geodesic [21], this defines a distance between each point that reflects the shape of the surface of the object. This distance induces an ordering of each local patch and by representing this ordering rather than the non-surface respecting ordering induces by a ray we believe a more descriptive representation can be found.

Given that we can sample statistics of the object along paths that reflect the true global structure of the object the question remains what type of statistics should be encoded. The obvious approach would be to encode only first order statistics such as in [20] as it can be done in a robust manner and is less sensitive to difference is sampling resolution. However, we believe that the important information is in the ordering of the local patches not simply the distribution. To that end we wish to take a similar approach as in [10] and exploit robust and principled kernel approaches representation and inference. In specific, where the semantic in [6] does not reflects the local appearance we wish to exchange the semantical alphabet to use the local representation presented in [18]. Rather than modeling the interaction between segments in time we aim to model the interaction spatial, where the time domain is replaces with a distance measure along the object. We believe that this approach has the potential of improving object categorisation and classification in a similar manner as it improved action classification as shown in [10]. Our intuition why this will lead to improvement is two fold; only modeling the local structure we are likely to need a very detailed descriptor which is likely to be susceptible to noise. By using a less descriptive local feature as [18] we believe this can be avoided. Secondly, the generalization and discrimination will be encoded by using the robust string kernel approach developed in [10] allowing us to exploit principled and robust inference algorithms for classification.

5 Data Conditional Dependence and Factorization

The previous examples we have discussed have addressed representation of data for a specific problem where we argue that the global structure of the variations in the observations is the key component to model and represent not the actual variations themselves. In this section we will describe a more general case where we do not have a specific task in mind but rather want to acquire a complete model of the data and model its underlying distribution.

In many scenarios of robotics we are given observations of the environment in a factorised form. This can either be that the observations naturally factorises describing separate modalities or through the use of different sensors and or feature representations. Assuming that the observations of the environment $\mathbf{Y}$ factorises into k separate terms $[\mathbf{Y}_1,\ldots,\mathbf{Y}_k]$ this means that from a probabilistic view point the complete model of the environment is represented by the joint distribution, $P(\mathbf{Y}) = P(\mathbf{Y}_1,\ldots,\mathbf{Y}_k)$. However, for many scenarios in robotics the dimensionality of this distribution makes it intractable to learn. In order to proceed one can exploit conditional independence in the observations imposing a structure on the joint distribution such as,

$$P(\mathbf{Y}) = \prod_{i=1}^{k} P(\mathbf{Y}_k|\pi_k), \tag{1}$$

where π_k corresponds to the subspace of $\mathbf{Y}$ that induces a dependency on $\mathbf{Y}_k$ thereby imposing a structure on the observation.

Extracting dependency structures in data is a very hard problem with the number of possible structures growing super-exponentially with the number of variables or nodes. Recently significant strides have been made towards being able to treat structure learning in a principled manner through the development of structural priors such as the Chinese Restaurant Process [22–24] and Indian Buffet Process [25, 26]. However, the use of such priors introduces significant limitations on the individual factors in the model meaning that they are not applicable in the general scenario. This means that for many problems researchers have to resort to using heuristic or greedy approaches. Of specific success have been the application of such methods when the data is discrete. However, for most robotic applications we deal with continuous data which means that such approaches have in general been beyond us. As a result, for the general case we often have to assume the structure and or the factorization of the data to be known a priori [27].

In recent [28–30] work we have created a model which encodes the tradeoff between loss of precision as introduced by discretisation process and the benefit of learning the structure by exploiting the heuristic approaches developed for such data. The proposed method learns a continuous latent variable model of each observation space represented by a set of discrete key states. It does so by exploiting recent advances in probabilistic dimensionality reduction [31] and by introducing a specific prior who balances the trade-off between discretisation and representation in a principled manner. In Fig. 8 a schematic figure of the graphical model proposed in [28] and the learned intermediate representation used for clustering is shown. Application of proposed method has allowed us to learn the conditional structure from large collections of both discrete and continuous variables within the same model. In Fig. 9 the resulting learned structure for modeling a range of different sensor data for a grasping task is shown. This is an example of by enforcing a specific structure on a lower level allows us to learn the more global structure of the data which is often much less trivial to have a notion of. Even

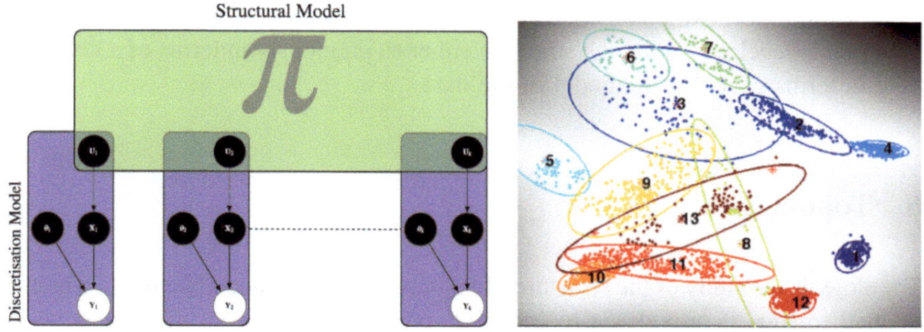

Fig. 8 The *left* image shows a schematical graphical model of the structure learning approach. For each continuous observation space $\mathbf{Y}_i$ we learn a low dimensional representation $\mathbf{X}_i$ with a functional relationship to the observed data parametrised by θ_i. Further, the low-dimensional space is represented using a set of discrete locations $\mathbf{U}_i$. Given that we have a completely discrete representation in terms of the $\mathbf{U}_i$ we can apply traditional heuristic methods for learning the structure π. The *right* image shows and example of the low-dimensional continuous representation and the discretisation colour coded. The separation between the clusters is controlled by a prior modeling the trade-off between discretisation and representation

Name	D	N	Description
task	-	3	Task Identifier
obcl	-	6	Object Class
size	3	8	Object Dimensions
cvex	1	4	Convexity Value [0, 1]
shcv	3	7	Shape Class Vector (Zernike Similarity)
fcon	20	20	Final Hand Configuration
dir	4	20	Approach Direction (Quaternion)
pos	3	14	Grasp Position
egpc	2	6	Eigengrasp Pre-Configuration
upos	3	11	Unified Spherical Grasp Position
fvol	1	6	Free Volume
gbvl	1	4	Volume of Grasped Boxes
pshcv	3	7	Part Shape Class Vector (Zernike Similarity)
pecce	1	3	Part Eccentricity [0, 1]
g1bx	1	2	Grasped-1-Box Value [0, 1]
qeps	1	5	Grasp Stability Measure (eps)
qvol	1	3	Grasp Stability Measure (vol)

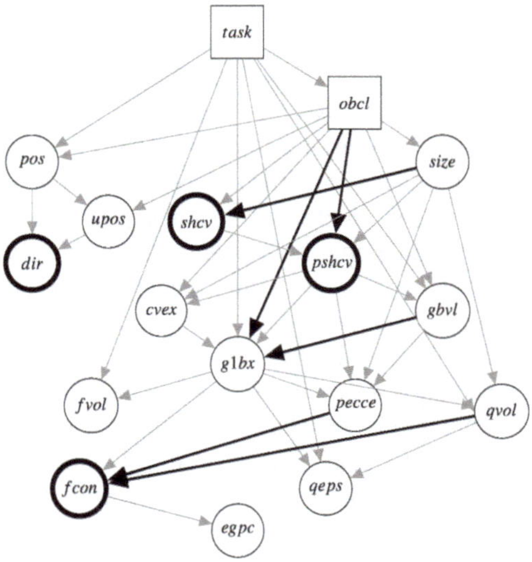

Fig. 9 Example of a learned factorised representation of 17 different observation spaces for a grasping scenario. To the *left* the different features are shown and to the *right* the resulting graphical model with the learned structure. The structure is very complicated and it is highly unlikely that we would be able to specify it a priori

though it might not be directly obvious this approach is not particularly different from the previous described methods as: on a lower level we enforce a structure, either in the case here as a discretisation or in the object category example by on the local feature level extracting specific structures such as edges or face normals, then on a global level we model the structure either as previously in terms of a task or as here in terms of a density model of the data.

6 Topology

Topology is the study of the structure of geometrical spaces and objects. As a branch of mathematics it provides a toolbox for extracting qualitative measurements of geometrical objects. We believe that tools from topology can provide a machinery to encode the global type of structure that we have argued throughout this paper being essential for acquiring a generalizable representation of the environment. However, topology as branch of pure mathematics was not aimed at analyzing uncertain scenarios where we measure the environment through sparse and potentially noisy samples as is often the case in robotics. In [32] the authors argue that by careful consideration of the problem setting, topological tools are applicable to the type of problems where statistical learning have usually been the dominating paradigm. The authors also argue that topological reasoning has the potential to eliviate some of the shortcomings fundamental to statistical learning. In specific, we like to highlight the following observations of statistical learning made in the paper; *Coordinates are rarely natural*, *Metrics are necessarily not justified* and *The need for large scale qualitative information*. The two first observations relate to the fact that as the dominant portion of statistical learning approaches work on vector spaces where the inner product is assumed to be naturally interpretable. However, observations are often "shoehorned" into vector spaces which are not natural in the sense that the inner product does not relate to the intrinsic structure of the data. In order to reason about the space we require some form of similarity measure between points providing a distance or an ordering of the space. If the data is represented in vectorial space the natural similarity measure is the use of a norm. However, if the vectorial representation per say is not a natural representation of the data neither will the distance be. Especially relationships at large scale are likely to be less informative compared to local. This is indicated by the success of approaches which relaxes the assumption about the parametrisation to only assume it to be locally metric such as simple nearest neighbor methods [33–35] and the success of kernel induced feature spaces based on radial basis functions which emphasizes the local structure in the data. This is also the foundation for the last intuition that we wish to highlight from [32] that of the need for a qualitative measure of the data.

We have throughout this paper argued the importance of understanding the global structure of data. Given that it is only at best on a local scale we can associate significance to the similarity measure, we need tools that can in a principled manner provide qualitative measure on the global structure of a set of data induced by a local measure. A set of data and its structure can be studied by creating a graph where a node represents each samples with paths connecting nodes according to some similarity measure. Assuming that, we can at least on a local scale derive a somewhat natural notion of similarity, this graph represents the structure of the whole dataset that is induced by this local measure. The field of algebraic topology defines a formalism for providing qualitative measures on such graphs. However, one central question remains: on what scale the local similarity measure is relevant? In order to reduce the effects of noise in the samples we wish to use as large range of interaction as possible, however if too large we run the risk of connecting non-related components. This problem is well known in machine learning for constructing local affinity matrices [21, 36, 37]. In order to circumvent this problem the idea of Persistent Homology has been introduced which studies how the qualitative measure changes by varying the range of the local interactions. Persistent homology provides tools which can potentially make algebraic topology applicable as a formalism for studying uncertain data.

We believe that a symbiosis between statistical learnings tools with its principles for modeling in scenarios with uncertainty and missing data together with the tools for qualitative measurements of structure provided by topology has the potential of achieving a synergic effect for merging local observations and global structure in a unified framework.

7 What Next?

Robots acting and interacting in realistic environments rely on perception, planning and control for motion generation. Although state of the art algorithms are capable of finding solutions that results in successful goal generation in some applications, they are still not able to flexibly make use of the gathered experience and use it for solving a similar/related problem on a future occasion. Extracting the semantics of the task is one of the major bottlenecks that still remain to be solved and we argued in this paper that this is in general dependent on using the *right* representation for the problem at hand. A good representation of data is one that except for being robust is capable of generalizing at the desired level.

In regard to motion generation, the classical approach operates in a complete configuration or state space represented at the level of generalized coordinates considering all joint angles and their 3D pose. This requires a computationally expensive state space optimization and randomized exploration in very large search spaces. In a EU funded project TOMSY (www.tomsy.eu) we study representations of actions and morphologies using topology-based abstractions in a layered manner and to implement dexterous manipulation on articulated and flexible objects using mappings between the topology-based abstract space, task space and joint space of metamorphic manipulators.

In this paper, have argued that one important mode of information for many application scenarios is not the actual variation in the data but the rather the higher order statistics as the structure of variations. We have exemplified this through a set of applications and show different ways of representing the structure of data, considering applications such as scene understanding, object recognition and data representation for grasping.

References

1. R. Rensink, J. ORegan, J. Clark, On the failure to detect changes in scenes across brief interruptions. Vis. Cogn. **7**(1), 127–145 (2000)
2. D.J. Simons, C.F. Chabris, Gorillas in our midst: sustained inattentional blindness for dynamic events. Perception **28**, 1059–1074 (1999)
3. D.G. Lowe, Distinctive image features from scale-invariant keypoints. Int. J. Comput. Vis. **60** (2), 91–110 (2004)
4. B.D. Argalla, S. Chernova, M. Veloso, B. Browning, A survey of robot learning from demonstration. Robot. Auton. Syst. **57**(5), 469–548 (2009)
5. V. Kruger, D. Kragic, A. Ude, C. Geib, The meaning of action: a review on action recognition and mapping. Adv. Robot. **21**(13), 1473–1501 (2007)
6. E. Aksoy, A. Abramov, F. Wörgötter, B. Dellen, Categorizing object-action relations from semantic scene graphs, in *IEEE International Conference on Robotics and Automation*, 2010, pp. 398–405
7. I. Laptev, P. Perez, Retrieving actions in movies, in *IEEE International Conference on Computer Vision*, 2007, pp. 1–8
8. I. Laptev, M. Marszalek, C. Schmid, B. Rozenfeld, Learning realistic human actions from movies, in *IEEE Conference on Computer Vision and Pattern Recognition*, 2008, pp. 1–8
9. H. Kjellstro¨m, J. Romero, D. Kragic, Visual object-action recognition: Inferring object affordances from human demonstration. Comput. Vis. Image Underst. **115**, 81–90 (2011)
10. G. Luo, N. Bergström, C.H. Ek, D. Kragic, Representing actions with kernels, in *International Conference of Intelligent Robots and Systems*, 2011
11. H. Lodhi, C. Saunders, J. Shawe-Taylor, N. Cristianini, C. Watkins, Text classification using string kernels. J. Mach. Learn. Res. **2**, 419–444 (2002)
12. N. Cristianini, J. Shawe-Taylor, *An introduction to Support Vector Machines and Other Kernel Based Learning Methods* (Cambridge University Press, 2006)
13. V. Kruger, D.L. Herzog, Sanmohan A. Ude, D. Kragic, Learning actions from observations. Robot. Autom. Mag. **17**(2), 30–43 (2010)
14. D. Comaniciu, P. Meer, Mean shift: a robust approach toward feature space analysis. IEEE Trans. Pattern Anal. Mach. Intell. **24**(5), 603–619 (2002)
15. Y. Boykov, M.-P. Jolly, Interactive graph cuts for optimal boundary & region segmentation of objects in N-D Images, in *IEEE International Conference on Computer Vision*, 2005, pp. 105–112
16. N. Bergström, C.H. Ek, M. Björkman, D. Kragic, Scene understanding through interactive perception, in *International Conference on Vision Systems*, 2011
17. M. Everingham, L. Van Gool, C. K. Williams, J. Winn, A. Zisserman, The PASCAL Visual Object Classes Challenge 2010 (VOC2010) (2010)
18. R. Rusu, N. Blodow, M. Beetz, Fast Point Feature Histograms (FPFH) for 3D registration, in *International Conference on Robotics and Automation*, 2009, pp. 3212–3217
19. R. Rusu, A. Holzbach, N. Blodow, M. Beetz, Fast geometric point labeling using conditional random fields, in *IEEE/RSJ International Conference on Intelligent Robots and Systems*, 2009, pp. 7–12
20. R.B. Rusu, Semantic 3D Object Maps for Everyday Manipulation in Human Living Environments, Ph.D. thesis, Technische Universität München (2009)
21. J.B. Tenenbaum, V. de Silva, J.C. Langford, A global geometric framework for nonlinear dimensionality reduction. Science **290**(5500), 2319–2323 (2000)

22. Y. Teh, M. Jordan, M. Beal, D. Blei, Hierarchical Dirichlet processes. J. Am. Stat. Assoc. **101** (476), 1566–1581 (2006)
23. J. Pitman, *Combinatorial Stochastic Processes* (Springer, St. Flour Summer School, Berlin, 2006)
24. H. Wallach, S. Jensen, L. Dicker, K. Heller, An alternative prior process for nonparametric bayesian clustering, in *International Conference on Artificial Intelligence and Statistics*
25. R. Adams, H. Wallach, Z. Ghahramani, Learning the Structure of Deep Sparse Graphical Models, in *International Conference on Artificial Intelligence and Statistics*, 2010
26. T.L. Griffiths, Z. Ghahrmani, Infinite latent feature models and the Indian buffet process, in: *Advances in Neural Information Processing*, 2006, pp. 475–482
27. D. Song, K. Huebner, V. Kyrki, D. Kragic, Learning task constraints for robot grasping using graphical models, in *IEEE/RSJ International Conference on Intelligent Robots and Systems*, 2010, pp. 1579–1585
28. C.H. Ek, D. Song, D. Kragic, Learning conditional structures in graphical models from a large set of observation streams through efficient discretisation, in *International Conference on Robotics and Automation, Workshop on Manipulation under Uncertainty*, 2011
29. D. Song, C.H. Ek, K. Huebner, D. Kragic, Embodiment-specific representation of robot grasping using graphical models and latent-space discretization, in *IEEE/RSJ International Conference on Intelligent Robots and Systems*, 2011, pp. 1–8
30. D. Song, C.H. Ek, K. Huebner, D. Kragic, Multivariate discretization for bayesian network structure learning in robot grasping, in *International Conference on Robotics and Automation*, 2011
31. M. Titsias, N. Lawrence, Bayesian gaussian process latent variable model, in *International Conference on Artificial Intelligence and Statistics*, 2010
32. G. Carlsson, Topology and data. Am. Math. Soc. **46**(2), 255–308 (2009)
33. G. Shakhnarovich, T. Darrell, P. Indyk, Nearest-Neighbor Methods in Learning and Vision (MIT Press, 2005)
34. G. Shakhnarovich, P. Viola, T. Darrell, Fast pose estimation with parameter-sensitive hashing, in *IEEE International Conference on Computer Vision*, 2003, pp. 750–757
35. O. Boiman, E. Shechtman, M. Irani, In defense of nearest-neighbor based image classification, in *Computer Vision and Pattern Recognition*, 2008, pp. 1–8
36. K.Q. Weinberger, F. Sha, L. K. Saul, Learning a kernel matrix for nonlinear dimensionality reduction, in *International Conference on Machine Learning*, 2004
37. S.T. Roweis, L.K. Saul, Nonlinear dimensionality reduction by locally linear embedding. Science **290**

Modular Design of Image Based Visual Servo Control for Dynamic Mechanical Systems

Robert Mahony

Abstract This paper presents a modular framework for design of image based visual servo control for fully actuated dynamic mechanical systems. The approach taken uses the formalism of port Hamiltonian systems to track energy exchanged between the mechanical system and virtual potentials or Hamiltonians associated with each image feature. Asymptotic stability of the system is guaranteed by injecting damping to the otherwise conservative system. A simple approach based on full state measurement is presented and then extended to deal with unmeasured relative depth of image features.

1 Introduction

Visual serving algorithms have been extensively developed in the robotics field over the last twenty years. *Image-Based Visual Servo* (IBVS) control regulates the dynamics of features in the image plane directly [1], resolving the cartesian motion planning task implicitly [2, 3]. IBVS control methods offer advantages in robustness to camera and target calibration errors, reduced algorithmic complexity and are easily extended multiple camera scenarios [4]. Visual servo control was first developed for rigid industrial serial-link robotic manipulators [4] where the dynamics of the system are easily compensated using computed torque (or high gain) control design. As a consequence, classical IBVS control is framed in the kinematic setting [1]. The last decade has seen a number of developments of visual servo control algorithms that consider the full dynamics of a mechanical system motivated by non-stiff robotic manipulators, such as those used in human safe environments, and other applications such as visual servo control of aerial vehicles. Kelly et al. considered the full dynamics of a robot and used a simple image based error feedback along with damping to prove asymptotic stability of the full system, firstly for planar robots [5], then later for full 6-DOF manipulators with known

R. Mahony (✉)
Australian National University, ACT, Canberra 0200, Australia
e-mail: Robert.Mahony@anu.edu.au

H.I. Christensen and O. Khatib (eds.), *Robotics Research*,
Springer Tracts in Advanced Robotics 100, DOI 10.1007/978-3-319-29363-9_8

image depth [6]. Zergeroglu et al. [7] used robust backstepping methods and adaptive estimation to deal with unknown calibration and system dynamics. Bishop et al. [8] considered a similar problem using non-linear PD control techniques. Hamel et al. [9] used an image centroid feature to ensure passivity properties of system and then applied robust backstepping to control a dynamic under-actuated model of an aerial robotic vehicle. Maruyama et al. [10] took an approach similar to that of Kelly [5, 6] and worked with a general Euler-Lagrange model of a robotic manipulator. A key issue in all visual servo control schemes is overcoming the loss of relative depth information associated with using an imaging system to observe a target point. Maruyama and Fujita further developed their work to include an observer that estimates the camera pose [11, 12], effectively estimating the unkown depths of the image points. An advantage of this approach is that a positive-definite control Lypunov function for the closed-loop system is available and provides an estimate of L_2 gain of the visual servo control loop [12]. Since this work incorporates an explicit estimate of the system pose it is natural to use a pose based error for the control criteria, leading to position based visual servo control [4]. Kawai et al. [13], however, showed that an image based regulation error can be used if desired; although the control algorithm still contains the complexity of a full pose estimator. Several authors have also considered using navigating functions [14–16] for visual servo control of dynamic systems.

In this paper, I present a novel modular framework for design of image based visual servo control for dynamic systems. The proposed approach uses the structure of port Hamiltonian systems represented graphically using the bondgraph formalism. Each image feature is associated with a separate branch of the bondgraph, ensuring a modularity and structural simplicity to the design framework. The natural pairing of generalized forces (efforts) and generalized velocities (flows) associated with the mechanical system are transformed into image flow and image effort in a power preserving modulated transformer. Image Hamiltonians are introduced in the image space that represent stored energy associated with the image variables. The control objective is assigned by shaping of the total potential energy of the complete bond-graph using the flexibility available in choosing the image Hamiltonian potentials. Asymptotic stability of the system is obtained by injecting damping into the otherwise conservative system. Since each image branch of the graph is independent of the others the approach is inherently modular and image points can be added or removed arbitrarily as long as care is taken that the total system energy is preserved and that the system potential is always shaped with a minimum at the desired pose. A further advantage of the approach is that it is straightforward to interface the proposed design framework with other control algorithms that have a port-Hamiltonian representation. For example, bilateral force-feedback teleoperation of the system can be achieved by simply connecting a external source port representing the master system to the bondgraph.

The initial results of the paper are presented under the assumption that all system variables are measured. In practice, the relative depth of each image feature is never directly measured due to the physical nature of imaging systems. The second part of the paper presents a modification of the design framework that incorporates an on-line estimate of the relative depth.

2 Classical Image Based Visual Servo Control

Classical Image Based Visual Servo (IBVS) control was developed in the kinematic setting for serial manipulator devices [4]. Let $\{A\}$ denote an inertial (base) frame of reference. Let $\{B\}$ denote the body-fixed (end-effector) frame of reference, and let ${}^A\xi_B$ (resp. ${}^AR_B$) denote the position (resp. orientation) of $\{B\}$ with respect to $\{A\}$. Note that ${}^A\xi_B \in \mathbb{R}^3$ while ${}^AR_B$ is a rotation matrix. I will consider a mechanical system with N degrees of freedom, such as, but not limited to, an N-link serial robotic manipulator. The generalized coordinates (joint variables) of the system are denoted $q \in \mathbb{R}^N$. Let ${}^BV = {}^B_AV_B$ and ${}^B\Omega = {}^B_A\Omega_B$ denote the linear and angular velocity of $\{B\}$ with respect $\{A\}$ expressed in $\{B\}$. Let ${}^BU = ({}^BV, {}^B\Omega)$ denote the combined spatial velocity of $\{B\}$ with respect to $\{A\}$ expressed in $\{B\}$. Then the Jacobian $J(q) \in \mathbb{R}^{6\times N}$ (velocity Jacobian of a robotic manipulator) gives the relationship between spatial and generalized velocities

$$ {}^BU = J(q)\dot{q}. \tag{1} $$

Any physical point p can be given coordinates either in the world frame ${}^Ap \in \{A\}$ or in the end-effector frame ${}^Bp \in \{B\}$. The mapping between these coordinate representations of p is given by the transformation mapping ${}^AH_B : {}^Bp \mapsto {}^Ap$,

$$ {}^Ap = {}^AH_B({}^Bp) = {}^AR_B\,{}^Bp + {}^A\xi_B. $$

with inverse ${}^Bp = {}^AH_B^{-1}({}^Ap) = {}^AR_B^\top({}^Ap - {}^A\xi_B)$. The time variation of ${}^Bp$ is given by

$$ {}^B\dot{p} = {}^Bp \times {}^B\Omega - {}^BV + {}^AR_B^{\top A}\dot{p}. \tag{2} $$

In the sequel, I will only consider the case where the point p is stationary in the world frame, that is ${}^A\dot{p} = 0$. In this case, it is convenient to write a matrix form for (2)

$$ {}^B\dot{p} = A({}^Bp){}^BU := (-I_3 \; {}^Bp_\times)\begin{pmatrix} {}^BV \\ {}^B\Omega \end{pmatrix} $$

where ${}^Bp_\times$ is the 3×3 matrix such that ${}^Bp_\times v = {}^Bp \times v$ for any vector $v \in \mathbb{R}^3$. The matrix $A({}^Bp) \in \mathbb{R}^{3\times 6}$ is a Jacobian that maps Euclidean velocity of the body-fixed frame into velocity of the point ${}^Bp$ associated with ego-motion of $\{B\}$.

Classical eye-in-hand IBVS control uses a set of features s_i observed by a camera attached to the end-effector frame of reference. Many different visual features have been considered in the past, however, for this paper I only consider point features, that is the image coordinates $s_i = (u_i, v_i)$ of the image of an observed point p_i for $i = 1,\ldots,n$. The principles presented in this paper, however, should generalize to any image based feature.

Let $(x_i, y_i, z_i)^\top = {}^Bp \in \mathbb{R}^3$ index the body-fixed frame coordinates of a set of point features in $\mathbb{R}^3$. The image coordinates of these features observed using a calibrated pinhole camera are

$$s_i := \begin{pmatrix} u_i \\ v_i \end{pmatrix} = \sigma \begin{pmatrix} \frac{x_i}{z_i} \\ \frac{y_i}{z_i} \end{pmatrix} \tag{3}$$

where $\sigma > 0$ is a scalar that represents the focal length of the camera. That is s_i is a non-linear function $f(x_i, y_i, z_i) = \sigma\left(\frac{x_i}{z_i}, \frac{y_i}{z_i}\right)$.

The image interaction matrix or image Jacobian, L_i, is the linear relationship between the instantaneous variation of the image feature and the generalized coordinates, (Remark 1 justifies why I write $L_i{:} = L_i(s_i, z_i, q)$)

$$\begin{aligned} L_i(s_i, z_i, q) &= Df({}^Bp_i)A({}^Bp_i)J(q) \\ &= \frac{1}{z_i}\begin{pmatrix} \sigma & 0 & -u_i \\ 0 & \sigma & -v_i \end{pmatrix}(-I_3({}^Bp_i)_\times)J(q) \in \mathbb{R}^{2\times N} \end{aligned} \tag{4}$$

recalling that $s_i = (u_i, v_i)$. The image interaction matrix is the linear mapping between generalized velocities of the manipulator and the velocity of the observed image point

$$\dot{s}_i = L_i(s_i, z_i, q)\dot{q}. \tag{5}$$

Let $s = (s_1^\top, \ldots, s_n^\top) \in \mathbb{R}^{2n}$ be the concatenated vector of stacked image features and define

$$L(s, z_1, \ldots, z_n, q) := \begin{pmatrix} L_1(s_1, z_1, q) \\ \vdots \\ L_n(s_n, z_n, q) \end{pmatrix} \in \mathbb{R}^{2n\times N} \tag{6}$$

Then the combined image kinematics can be written

$$\dot{s} = L(s, z_1, \ldots, z_n, q)\dot{q} \tag{7}$$

Remark 1 Recalling Eq. 4 it appears to be the case that $L_i := L_i(s_i, z_i, {}^Bp_i, q)$ depends on the full target position ${}^Bp_i$ as well as s_i, z_i and q. In fact the term $({}^Bp_i)_\times$

in the rightmost block of $A(^Bp_i)$ can be combined with the depth dependence of $Df(^Bp_i)$ to yield

$$\frac{1}{z_i}(^Bp_i)_\times = \begin{pmatrix} s_i \\ 1 \end{pmatrix}_\times.$$

Although this trick means that the rightmost 3×3 block of $Df(^Bp_i)A(^Bp_i)$ depends only on the measured image feature s_i, the z_i dependence of the leftmost block remains. Thus, to compute L_i one requires the variables s_i, z_i and q.

Let $s_i^* \in \mathbb{R}^2$ be a set of desired image coordinates. The task considered in IBVS control is to servo control the robot until the camera is in a position such that $s_i = s_i^*$ for $i = 1,\ldots,n$. The desired image points $\{s_i^*\}$ should be chosen to be feasible, that is that there should exist a physical pose of the camera in which the observed points match the desired $\{s_i^*\}$. Typically image based visual servo control is used to return to a pose that has been visited before and the desired image coordinates are computed directly from a reference image obtained at the desired pose, ensuring that the goal features are feasible. The image error is defined to be

$$\tilde{s} = s - s^* = \left((s_1 - s_1^*)^\top, \ldots, (s_n - s_n^*)^\top\right) \in \mathbb{R}^{2n}. \tag{8}$$

The control is chosen to stabilize the image error

$$\dot{q} = -kL^\dagger(s, z_1, \ldots, z_n, q)\tilde{s}$$

where $k > 0$ is a position scalar gain and $L^\dagger = (L^\top L)^{-1}L^\top$ is the matrix pseudo-inverse. The image features s and manipulator joint coordinates q are measured and are available and can be used in computing L. The z_i coordinates of ^Bp_i, however, are usually not available due to the nature of imaging devices. The simplest work around to this issue used in practice is to use approximations $(\hat{z}_1, \ldots, \hat{z}_n)$ to the depth and compute an estimate $\hat{L} := L(s, \hat{z}_1, \ldots, \hat{z}_n, q)$ of the true interaction matrix. There are plethora of schemes proposed in the literature to compute estimates of $\{\hat{z}_i\}$ or directly compute estimates of $\hat{L}$ or even its pseudo-inverse [4, 17–19].

3 Modular Image Based Visual Servo Control for a Dynamic Systems

This section introduces a modular port Hamiltonian frame work for design image based visual servo control algorithms. A key advantage of this framework is that each image feature is treated as a separate energy port attached to the system dynamics. Thus, the separate image features can be treated as detachable modules in the control design leading to a number of advantages that will be discussed in future sections. The theory is developed starting from a general Euler-Lagrange model of a mechanical system expressed with respect to generalized coordinates. Although this model is directly applicable to the standard dynamic models of robotic manipulators, it is not necessarily this class of systems that is of most interest as potential applications for the results. For example, the port Hamiltonian approach has recently been applied aerial robotic vehicles [20] and there are a large number of additional systems that can easily and naturally be modeled using this framework.

Let $q \in \mathbb{R}^N$ denote generalised coordinates for a fully actuated mechanical system with generalised inertia matrix $M(q)$ and potential function $U(q)$. Lagrange's equations yield dynamics

$$M(q)\ddot{q} = -C(q,\dot{q})\dot{q} - \frac{\partial U}{\partial q}(q) + \tau \tag{9}$$

where $\tau \in \mathbb{R}^N$ are the generalised forces. The Hamiltonian associated with the mechanical system is

$$H_0(q,p) := \frac{1}{2}p^\top M^{-1}(q)p + U(q), \quad p = M(q)\dot{q} \tag{10}$$

and the Euler-Lagrange dynamics (9) is equivalent to the standard port Hamiltonian equations with flow and effort variables

$$\begin{aligned} \dot{q} &= \frac{\partial H}{\partial p}(q,p), \\ \dot{p} &= -\frac{\partial H}{\partial q}(q,p) + \tau, \end{aligned} \tag{11}$$

where $(\tau,\ \dot{q})$ is the effort-flow pairing of the energy multi-port of the Hamiltonian. That is

$$\frac{d}{dt}H_0(q,p) = \langle \tau | \dot{q} \rangle = \tau^\top \dot{q} \tag{12}$$

where the $\langle \cdot | \cdot \rangle$ is the power in the port. In the graphical representation of bond-graphs[1] the port Hamiltonian system is represented as a single storage bond attached to a 1-junction as shown in Fig. 1. The causality stroke of the bond (horizontal line at the tip of the arrow) indicates that the effort τ is the input to the Hamiltonian system while the flow $\dot{q}$ is the output, as is natural in the differential Eq. (9).

[1]In Fig. 1 the arrow on the bond should be a half arrow with the hook on the side of the flow variable. However, I was unable to typeset this effectively, and I have chosen to make all bonds full arrows in this paper to make the notation consistent.

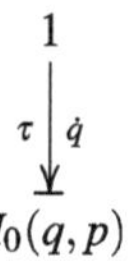

Fig. 1 Single bond storage element indicating the Hamiltionian dynamics associated with a mechanical system with Hamiltonian $H_0(q, p)$ and Dirac structure as specified in (11)

The approach taken is to treat each image feature as a separate bond interconnection in a bondgraph with the Hamiltonian dynamics of the system present as initial element. Thus, each image feature s_i will result in a bond attached to the 1-junction in Fig. 1 with an associated effort denoted τ_i as shown in Fig. 2. Following the rules of 1-junctions in bondgraphs the flow $\dot{q}$ in each of the bonds is equal while the forces add $\tau + \sum_{i=1}^{n} \tau_i = 0$.

In order to interface the *i*th image bond to the image feature a non-linear modulated transformer is used to transform the flow $\dot{q}$ to the image flow $\dot{s}_i$. The transformer relationship is defined by the relationship between image flow and the generalized velocity given by (5). That is the image interaction matrix $L_i(s_i, z_i, q)$ is the defining matrix for the transformer relationship

$$\dot{s}_i = L_i(s_i, z_i, q)\dot{q}, \quad \tau_i = L_i(s_i, z_i, q)^\top e_i \tag{13}$$

where e_i is the effort variable that will be associated with the flow $\dot{s}_i$. The relationship between image effort e_i and τ_i is implied by the principle of power conservation in the transformer. A key contribution of this paper is the identification of the *image effort* e_i as an independent concept from the efforts τ_i that are expressed as forces the generalized coordinates and the energy pairing of the dual variables e_i and $\dot{s}_i$ in image space. The image effort is a similar construct to the generalized *operational* force vectors introduced in Khatib's work [21], however, here since the

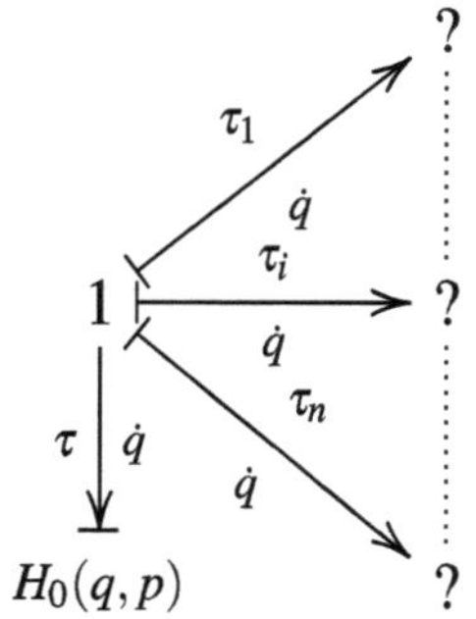

Fig. 2 Attachment of a multiple image feature bonds. Each image feature contributes an effort τ_i to the 1-junction that couples with the flow $\dot{q}$ of the mechanical system. These efforts sum to produce the IBVS control action. The *question marks* indicate that this part of the bondgraph remains to be defined

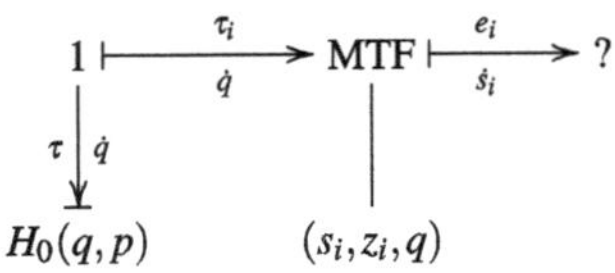

Fig. 3 Modulated transformer to relate image flow $\dot{s}_i$ with the flow $\dot{q}$ at the 1-junction

efforts live in the image space, they can only ever be associated with virtual forces. The resulting bondgraph has the form shown in Fig. 3 where only the ith image bond is shown and where the line without an arrowhead indicates exchange of information without flow of power. The transformer is modulated by the signals (s_i, z_i, q) via the dependence of $L_i(s_i, z_i, q)$ on these variables.

In Fig. 3 the causality of the bonds indicates that the effort variable τ_i depends on e_i (13), and that the effort variable e_i depends on the element placed at the question mark. The specification of the flow variable $\dot{s}_i$ indicates that the image coordinates s_i should act as the energy variables associated with any storage associated with the image effort e_i. The approach taken in this paper is to use an *image Hamiltonian*, that is a spring like storage element defined in the image space, to define the relationship between the image flow $\dot{s}_i$ and the image effort e_i. A natural first choice for an image Hamiltonian is the squared norm of image error. Assume that the target image coordinates s_i^* are known and that they are constant. Define an image storage element (or Hamiltonian) by

$$H_i(s_i) := \frac{1}{2} k_i \| s_i - s_i^* \|^2 \tag{14}$$

where k_i is a positive scalar. Treating this element in the classical manner then the natural interconnection (Dirac structure) for the bondgraph is to assign

$$e_i := \frac{\partial H_i}{\partial s_i} = k_i (s_i - s_i^*) \tag{15}$$

It follows that

$$\frac{d}{dt} H_i = \left(\frac{\partial H_i}{\partial s_i} \right)^\top \dot{s}_i = \langle e_i | \dot{s}_i \rangle = k_i (s_i - s_i^*)^\top \dot{s}_i$$

as expected for a storage element. The resulting bond graph is shown in Fig. 4.

The proposed image Hamiltonian $H_i(s_i)$ (14) is chosen such that the minimum energy in of the image Hamiltonians occurs when $s_i = s^*$, normally corresponding

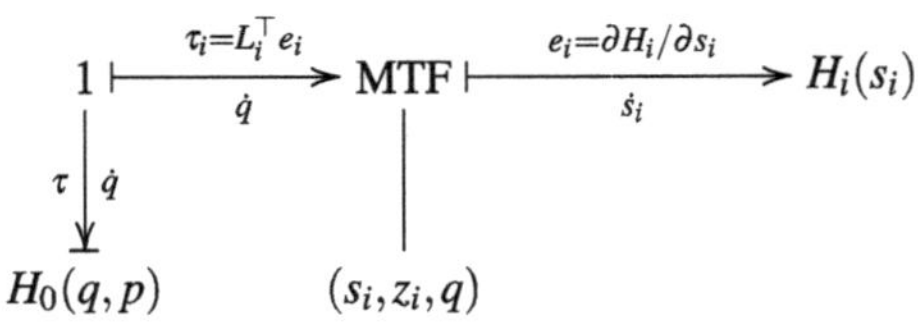

Fig. 4 Complete structure of the modular image branch of the proposed bondgraph design framework

to correct positioning of the camera. In more generality, the image Hamiltonians should be chosen to shape the energy of total system, including the potential energy in the mechanical system and any other storage elements that may be integrated into the system, in order that its minimum corresponds to the desired regulation point of the system. Shaping the energy of the total system may require additional insight into the system configuration or control task, and indeed may utilize more than just the image variables. In particular, in the present case where the mechanical system itself has potential energy then this must be balanced either by shaping the image Hamiltonians or by adding additional storage that cancels the potential $U(q)$ of the mechanical system. In the present development I will compensate the potential $U(q)$ directly by adding a separate Hamiltonian storage element $-U(q)$ attached the 1-junction that depends on the known generalized coordinates q, the upper bond to the 1-junction added in Fig. 5.

The final bondgraph for the closed-loop system (Fig. 5) is obtained by adding an additional dissipation term R, such that $\delta = R\dot{q}$ with $R > 0$ is a positive definite matrix. Without adding this term then the underlying system will be conservative and the closed-loop response of the system would be that of an oscillator. By choosing the damping suitably, the response of the system can be tuned to converge asymptotically to the minimum of the potential energy.

For the bondgraph given in Fig. 5 the generalized forces τ to the mechanical system are specified by the relationship implied by the 1-junction

$$\begin{aligned}\tau &= -\sum_{i=1}^{n} \tau_i - \delta + \frac{\partial U}{\partial q}(q) \\ &= -\sum_{i=1}^{n} k_i L_i(s_i, z_i, q)^\top (s_i - s_i^*) - R\dot{q} + \frac{\partial U}{\partial q}(q)\end{aligned} \tag{16}$$

The total energy of the closed-loop system is given by sum of all the storage elements

$$H(q,p,s_i) = H_0(q,p) + \sum_{i=1}^{n} H_i(s_i) - U(q) = \frac{1}{2} p^\top M^{-1}(q) p + \frac{1}{2} \sum_{i=1}^{n} k_i \|s_i - s_i^*\|^2 \tag{17}$$

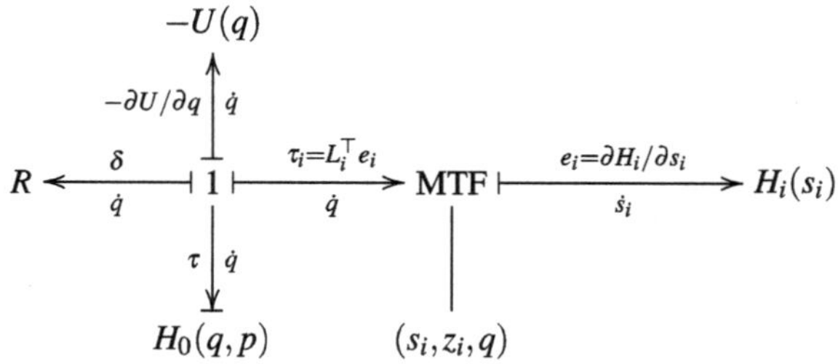

Fig. 5 Proposed modular framework for dynamic image based visual servo control

The theory of port-Hamiltonian system now guarantees that

$$\frac{d}{dt}H(q,p,s_i) = -\langle\delta|\dot{q}\rangle = -\dot{q}^\top R\dot{q}. \tag{18}$$

Lemma 1 *Consider the generalized mechanical system (9) with input τ given by (16). Assume that the total energy $H(q, p, s_i)$ (17) has a unique minimum at (q^*, p^*, s_i^*) and is radially unbounded in (q, $\dot{q}$). Assume furthermore that $L(s, z_1, \ldots, z_n, q)$ (6) is full rank for all q. Then, for any initial condition ($q(0)$, $\dot{q}$ (0)) the closed-loop trajectory ($q(t)$, $\dot{q}(t)$) $\to$ (q^*, 0).*

Proof Since $H(q, p, s_i)$ is radially unbounded in (q, $\dot{q}$) then (12) implies that trajectories of (q, $\dot{q}$) are bounded and exist for all time. From (18) along with Lyapunov's theorem and LaSalle's principle then (q, $\dot{q}$) converges to the largest forward invariant set containedin $\dot{q}^\top R\, \dot{q} = 0$ proving that $\dot{q}(t) \to 0$.

From (9) it follows that on LaSalle's invariant set $\tau = \partial U/\partial q$. Substituting from (16) one has

$$\sum_{i=1}^{n} k_i L_i(s_i, z_i, q)^\top (s_i - s_i^*) = L(s, z_1, \ldots, z_n, q)^\top \mathrm{diag}\,(k_1 I_2, \ldots, K_2 I_2)(s - s^*) = 0$$

where diag($k_1 I_2, \ldots, k_n I_2$) is the diagonal $2n \times 2n$ matrix with block diagonal entries $k_i I_2$ and ($s - s^*$) is given by Eq. 8. Since $L(s, z_1, \ldots, z_n, q)$ is full rank it follows that $\tilde{s} = 0$ and the uniqueness of the minimum of the Hamiltonian energy implies that $q(t) \to q^*$. *QED.*

The proposed control scheme provides a simple and intuitive design methodology for image based visual servo control of a mechanical system. A key advantage of the proposed control is that, unlike the classical IBVS approach, it is not a linearizing control design. In classical IBVS the pseudo inverse of the image Jacobian $L(s, z_1, \ldots, z_n, q)$ implicit in the control input may lead to significant stability issues in the closed-loop system if the estimate of L is poor. In contrast, the proposed scheme benefits from the natural robustness and passivity of the port Hamiltonian framework. Good asymptotic performance of the closed-loop response, however, will depend critically on tuning the damping and spring coefficients of the system.

The proposed approach is highly modular with each image feature dealt with as a separate branch of the bondgraph. This structure has considerable advantages in the practical implementation of image based visual servo control. In particular, the proposed framework allows one to develop heuristic schemes that can switch between features, drop old features, or add new features as long as the total energy of the system is conserved. For example, if at time t_1 it is wished to switch from image feature s_i to a new image feature $\bar{s}_i$ then by choosing $\bar{k}_i$ such that

$$\bar{k}_i = k_i \frac{\|s_i - s_i^*\|^2}{\|\bar{s}_i - \bar{s}_i^*\|^2}$$

the value of the total Hamiltonian $H(t)$ is continuous at time t_1. As a consequence the basic Lyapunov stability of the system is preserved and $\dot{q} \to 0$.

A modification of this idea can be used if an image feature is leaving the field of view. As an image feature i exits the field of view at time t_1, the energy in the ith image Hamiltonian

$$H_i(s_i(t_1)) = \frac{k_i}{2} \|s_i(t_1) - s_i^*\|^2$$

can be partitioned into $(n - 1)$ portions H_i^j for $j = 1,\ldots,n$ with $j \neq i$. Each $k_j^{\text{after}} := k_j^{\text{before}} + \Delta_j$ can then be augmented by

$$\Delta_j = 2\frac{H_i^j}{\|s_i - s_i^*\|^2}, \quad j = 1,\ldots,n, \quad j \neq i$$

that increases the energy in the image Hamiltonians $H_j(s_j)$ to compensate for the energy lost when $H_i(s_i)$ is removed from the bondgraph. A new image feature s_{n+1} can be added in a similar manner by taking energy from the existing image Hamiltonians, with the caveat that it may be necessary to choose the scaling k_{n+1} of the new feature sufficiently small to ensure there is sufficient energy available to cover its initial potential and that the energy taken from each image Hamiltonian must be less than its total available.

Another classical problem in image based visual servo control, that of ensuring that no image feature leaves the field of view during the evolution of the closed-loop system is also easily addressed. In this case the image Hamiltonian can be augmented by a barrier function

$$H_i(s_i) = \frac{k_i}{2} \|s_i - s_i^*\|^2 + \boldsymbol{\Phi}(s_i)$$

where $\boldsymbol{\Phi}$ is a positive definite function that is radially unbounded on the boundary of the image. It is still necessary to ensure that the total Hamiltonian of the system is shaped such that the minimum energy corresponds to the desired pose of the system.

A final advantage of the proposed approach is that it can be easily interfaced with other port-Hamiltonian control modalities. For example, exogenous user input such as a haptic interface can be added by including an additional bond to an exogenous source associated with the mode of input of the user. In this way it is straightforward to integrate bilateral force-feedback teleoperation into the image based visual servo control framework.

The proposed approach will still suffer from certain of the failings of classical IBVS control. In particular, the assumption that the total Hamiltonian $H(q, p, s_i)$ is radially unbounded in $(q, \dot{q})$ made in Lemma 1 is not satisfied in many classical configurations of IBVS, and without this assumption it is possible that the trajectory in q may become unbounded. For example, the now classical 'Chaumette conundrum' [22] has four image points located in a square configuration with the camera initial condition such that the observed points are exactly 180° rotated around the centre axis. In this case the proposed control will act to reduce the distance of all four points to the target vector causing them to contract into the centre of the image, corresponding to the camera moving infinitely far away along image axis, an unbounded motion in q. In this case the Hamiltonian is indeed decreasing along the closed-loop trajectories of the system, however, the potential is not radially unbounded in q.

4 On-Line Estimation of Image Depth

This section presents a extension of the development in Sect. 3 that incorporates an estimator for the unknown relative depth signals required in the implementation of (16).

The generalized forces τ_i (Fig. 5) associated with each individual image feature are given by the relationship

$$\tau_i = -k_i L_i(s_i, z_i, q)^\top (s_i - s_i^*).$$

In order to implement the proposed control exactly it is necessary to have measurements of the variables s_i, s_i^*, z_i and q. Of these signals it is only the relative depth z_i that is classically considered unknown. The image coordinates s_i and s_i^* are the primary measurements of the IBVS servo control problem, while the classical IBVS control problem is formulated in the context of serial robotic manipulator for which the generalized coordinates q are available to the control algorithm. In some of the more modern applications of IBVS control, such as those involving aerial robotic vehicles, the generalized coordinates q may be more difficult to measure, however, in the present paper I will focus on the classical problem formulation where the only unmeasured variable is the relative range z_i of an image feature.

The most common approach to dealing with unknown depth z_i in IBVS control is to use an estimate $\hat{z}_i$ in the algorithm implementation. The simplest approach is to choose the estimate $\hat{z}_i$ to be constant based on a best guess available given the expected environmental configuration [4]. Using the expected depth at the regulation point is known to provide good asymptotic behaviour of the closed-loop response. An alternative approach is to build an estimator that uses information obtained during the trajectory motion to estimate the relative depth z_i online [17]. Other approaches include partial pose estimation [18] and optimization based methods [19]. The approach taken in this paper is most closely coupled to an observer, or adaptive control [17] based approach. The novelty in the proposed approach is the integration of this idea into the port-Hamiltonian framework.

Consider the bondgraph shown in Fig. 6. The leftmost 1-junction, the Hamiltonian H_0 for the mechanical system, the $-U(q)$ potential, and the damping R are identical to the bondgraph in Fig. 5 and motivated in the same way. The difference in this section lies on the righthand side of the graph where the contributions from the translational and rotational motion of camera frame to the generalized mechanical system are separated into two modulated transformers. Recalling (1) it is clear that the relationship between rigid-body and generalized coordinate velocity can be decomposed as

$$\begin{pmatrix} {}^BV \\ {}^B\Omega \end{pmatrix} = \begin{pmatrix} J_V(q) \\ J_{\Omega(q)} \end{pmatrix} \dot{q},$$

and the matrices $J_V(q)\mathbb{R}^{3\times N}$ and $J_\Omega(q) \in \mathbb{R}^{3\times N}$ define the two modulated transformers in Fig. 6. The principle of conservation of power implies that

$$\tau_V = J_V^\top(q)F, \quad \tau_\Omega = J_\Omega^\top(q)\Gamma.$$

providing a decomposition of the generalized force τ into a component due to translational motion and one due to rotational motion of the camera frame. This decomposition is undertaken before the separate image branches of the bondgraph are created. The additional 1-junctions in the translational and rotational branch of the bondgraphs are added to provide a junction for the n image feature branches of the bondgraph, however, now each image feature generates a pair of image bonds, one associated with translational motion and one associated with rotational motion. Only the ith image bonds are shown in Fig. 6 and the dotted line is used to indicate that there are n modular pairs of image bonds attaching to the two 1-junctions. The

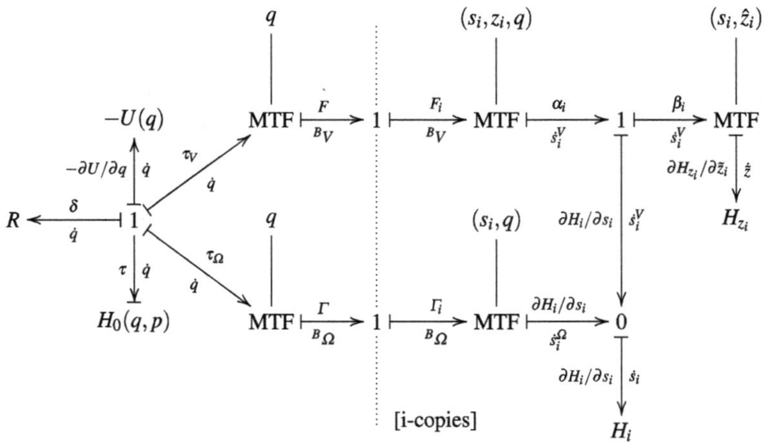

Fig. 6 The proposed architecture for modular image based visual servo control with depth estimation

transformation into image velocity is achieved using a pair of modulated transformers based on a decomposition of Eq. 4. Firstly, write (4) as

$$L_i(s_i, z_i, q) = \big(K_i^V(s_i, z_i, q) K_i^\Omega(s_i, q)\big) J(q)$$

where

$$K_i^V(s_i, z_i, q) = -\frac{1}{z_i}\begin{pmatrix} \sigma & 0 & -u_i \\ 0 & \sigma & -v_i \end{pmatrix}, \tag{19}$$

$$K_i^\Omega(s_i, q) = \begin{pmatrix} \sigma & 0 & -u_i \\ 0 & \sigma & -v_i \end{pmatrix}\begin{pmatrix} u_i \\ v_i \\ 1 \end{pmatrix}_\times . \tag{20}$$

Here (20) is obtained by factoring the $1/z_i$ dependence into the ${}^B p_\times$ to obtain the term $(u_i, v_i, 1)_\times$ as was discussed in Remark 1. Define two new image flow variables, $\dot{s}_i^V$ and $\dot{s}_i^\Omega$ for translational and rotational image flow respectively, by

$$\dot{s}_i^V = K_i^V(s_i, z_i, q)^B V, \quad \dot{s}_i^\Omega = K_i^\Omega(s_i, q)^B \Omega. \tag{21}$$

The two relationships (21) define the second set of modulated transformers (to the right of the dotted line) in the bondgraph Fig. 6. The reason for taking this approach is to separate the part of L_i that depends explicitly on the relative depth z_i into a single modulated transformer (the translational motion) branch that has a *scalar* dependence on the unknown variable. The modulated transformer in the rotational motion branch of the bondgraph depends only on known variables and can be implemented explicitly.

The translational and rotational image flows add to generate the full image flow

$$\dot{s}_i = \dot{s}_i^V + \dot{s}_i^\Omega. \tag{22}$$

This summation is implemented in the 0-junction in the bottom right of the bond-graph Fig. 6. Here the image effort $\partial H_i/\partial s_i$ is equal in each of the branches of the 0-junction while the image flows add in accordance to the directions of the bond arrows. This completes the separation of a modular image feature branch of the bondgraph into two branches that rejoin to implement the image Hamiltonian H_i in the bottom right of the bondgraph. The justification for this decomposition of the bondgraph is to enable the inclusion of an estimator for the unknown depth into each translational motion image feature sub-branch of the bondgraph. In modifying the bondgraph, the image flow $\dot{s}_i^V$ must be preserved in-order to implement the image Hamiltonian. The image effort associated with the translational motion, however, can be modified. Access to the relevant signal is achieved by introducing a 1-junction into the upper branch of the bondgraph just before the translational and rotational image velocities are recombined in the 0-junction. The 1-junction introduced preserves flow in each branch ensuring that $\dot{s}_i^V$ is preserved. The action of the 1-junction is to introduce a relationship between the three efforts associated with bonds on the 1-junction

$$\alpha_i = \beta_i + \frac{\partial H_i}{\partial s_i}. \tag{23}$$

Let $\hat{z}_i$ denote a new variable that is an estimate of the relative depth z_i. Define

$$\alpha_i := \frac{z_i}{\hat{z}_i} \frac{\partial H_i}{\partial s_i}.$$

This assignment fixes the effort β_i according to (23)

$$\beta_i = \frac{(z_i - \hat{z}_i)}{\hat{z}_i} \frac{\partial H_i}{\partial s_i} \tag{24}$$

and fully defines the three bonds attached to the 1-junction in the top right of Fig. 6. The reason for the choice of α_i can be seen by computing the relationship between F_i and α_i given by the modulated transformer

$$F_i = \frac{1}{z_i} \begin{pmatrix} \sigma & 0 \\ 0 & \sigma \\ -u_i & -v_i \end{pmatrix} \alpha_i = \frac{1}{\hat{z}_i} \begin{pmatrix} \sigma & 0 \\ 0 & \sigma \\ -u_i & -v_i \end{pmatrix} \left(\frac{\hat{z}_i}{z_i} \alpha_i \right) = \frac{1}{\hat{z}_i} \begin{pmatrix} \sigma & 0 \\ 0 & \sigma \\ -u_i & -v_i \end{pmatrix} \frac{\partial H_i}{\partial s_i}$$

Thus, the effort F_i can now be computed based on a 'virtual' modulated transformer (obtained by replacing the unknown depth z_i by its estimate $\hat{z}_i$) acting on the image effort $\partial H_i/\partial s_i$. All variables here are available and the control can be implemented.

It remains to terminate the bond $(\dot{s}_i^V, \beta_i)$ in the top right of the bondgraph. To save notational complexity I will assume the image Hamiltonian is given by $H_i(s_i) = k/2 \|s_i - s_i^*\|^2$ as was discussed in Sect. 3. This implies that $\partial H_i/\partial s_i = k(s_i - s_i^*)$.

Define a general form of dynamics for the observer $\hat{z}_i$

$$\dot{\hat{z}}_i = w_i \tag{25}$$

where w_i is an arbitrary driving term. The state $\hat{z}_i$ becomes an internal dynamic state in the implementation of the control algorithm. Let

$$\tilde{z}_i = \hat{z}_i - z_i. \tag{26}$$

denote the observer error.

Define a Hamiltonian for the relative depth error to be

$$H_{zi}(\tilde{z}_i) := \frac{\gamma}{2}\|\tilde{z}_i\|^2.$$

The natural flow variable associated with H_{zi} is $\dot{\tilde{z}}_i$ while the corresponding effort is

$$\frac{\partial H_{zi}}{\partial \tilde{z}}(\tilde{z}) = \gamma \tilde{z}_i. \tag{27}$$

The idea is to hypothesize a modulated transformer to relate the effort β_i with the effort $\partial H_{z_i}/\partial \tilde{z}$ and then implement this transformer by choosing the observer dynamics w_i to ensure that the flows $\dot{s}_i^V$ and $\dot{\tilde{z}}_i$ conform to the transformer relationship. Let $P(\cdot)$ denote the transformer relationship then

$$\beta_i = P(\cdot)\frac{\partial H_{z_i}}{\partial \tilde{z}}$$

and hence substituting from (24) and (27) and using $\partial H_i/\partial s_i = k_i(s_i - s_i^*)$ (14) one obtains

$$P(\hat{z}_i, s_i, s_i^*) := \frac{k}{\gamma \hat{z}_i}(s_i - s_i^*) \in \mathbb{R}^2 \tag{28}$$

Applying the principle of conservation of power then

$$\dot{\tilde{z}}_i = \frac{k}{\gamma \hat{z}_i}(s_i - s_i^*)^\top \dot{s}_i^V$$

Recalling (25) and differentiating (26) one has

$$\dot{\hat{z}} = w_i = \dot{z}_i + \dot{\tilde{z}}_i = {}^B V_z + \frac{k}{\gamma \hat{z}_i}(s_i - s_i^*)^\top \dot{s}_i^V \tag{29}$$

since the velocity $\dot{z}_i = {}^B V_z$ is just the z-axis velocity of the body fixed frame velocity, recalling that the target points are stationary. This process assigns observer dynamics to $\hat{z}_i$.

Space limitations prevent a formal statement of stability of the closed-loop system in this configuration. It is clear, however, that an analogous argument to that in Lemma 1 will guarantee firstly that $\dot{q} \to 0$ and secondly (using LaSalle's principle) that $q \to q^*$. There is no guarantee that the observer error $\tilde{z}_i$ actually converges to zero as the modulated transformer $P(\hat{z}_i, s_i, s_i^*)$ (28) decreases to zero as $s \to s_i^*$. Thus, the observer Hamiltonian H_{z_i} becomes decoupled from the full Hamiltonian system in the limit as $q \to q_i^*$ even if $\hat{z}_i \neq z_i$. Hence, even though the energy minimum of the total Hamiltonian occurs for $\tilde{z}_i = 0$, the modulated transformer can act to partition off energy in the observer Hamiltonian H_{z_i} that remains locked in storage as the system comes to equilibrium.

5 Conclusions

This paper used the port-Hamiltonian formalism to provide a design methodology for dynamic image based visual servo control. The approach is conceptually simple once the underlying bondgraph formalism is accepted. The graphical representation provides a powerful visualization of design framework and I believe it will provide considerable insight leading to the development of practical control algorithms for a range of relevant problems.

Acknowledgements This research was supported by the Australian Research Council through Future Fellowship FT0991771 "Foundations of Vision Based Control of Robotic Vehicles".

References

1. B. Espiau, F. Chaumette, P. Rives, IEEE Trans. Robot. Autom. **8**(3), 313 (1992)
2. C. Samson, M. Le Borgne, B. Espiau, *Robot Control: The Task Function Approach* (The Oxford Engineering Science Series (Oxford University Press, Oxford, U.K., 1991)
3. R. Pissard-Gibollet, P. Rives, in *Proceedings of the IEEE International Conference on Robotics and Automation, ICRA'95* (Nagasaki, Japan, 1995), pp. 166–171
4. S. Hutchinson, G. Hager, P. Corke, IEEE Trans. Robot. Autom. **12**(5), 651 (1996)
5. R. Kelly, IEEE Trans. Robot. Autom. **12**(5), 759 (1996)
6. R. Kelly, R. Carelli, O. Nasisi, B. Kuchen, F. Reyes, IEEE/ASME Trans. Mechatr. **5**(1), 39 (2000). doi:10.1109/3516.828588
7. E. Zergeroglu, D. Dawson, M. de Queiroz, A. Behal, IEEE/ASME Trans. Mechatron. **6**, 322 (2001)
8. B. Bishop, M. Spong, Control Eng. Pract. **7**, 423 (1999)
9. T. Hamel, R. Mahony, IEEE Trans. Robot. Autom. **18**(2), 187 (2002)
10. A. Maruyama, M. Fujita, Adv. Robot. **12**(14), 67 (1998)
11. A. Maruyama, H. Kawai, M. Fujita, in *Proceedings of 40th IEEE Conference on Decision and Control*, vol. 5 (2001), pp. 4415–4420. doi:10.1109/.2001.980897
12. M. Fujita, H. Kawai, M.W. Spong, IEEE Control Syst. Technol. **15**(1), 40 (2007). doi:10.1109/TCST.2006.883236
13. H. Kawai, T. Murao, M. Fujita, in *Proceedings of IEEE Computer Aided Control System Design IEEE International Conference on Control Applications IEEE International Symposium Intelligent Control* (2006), pp. 740–745. doi:10.1109/CACSD-CCA-ISIC.2006.4776738
14. N. Cowan, J. Weingarten, D. Koditschek, IEEE Trans. Robot. Autom. **18**(4), 521 (2002)
15. R. Kelly, J. Moreno, R. Campa, in *43rd IEEE Conference on Decision and Control* (Atlantis, Paradise Island, Bahama, 2004), pp. 4028–4033
16. T. Murao, H. Kawai, M. Fujita, in *IEEE International Conference on Control Applications* (Yokohama, Japan, 2010)

17. N. Papanikolopoulos, P.K. Khosla, T. Kanade, in *Proceedings of the American Control Conference* (1991), pp. 962–967
18. E. Malis, F. Chaumette, S. Boudet, IEEE Trans. Robot. Autom. **15**(2), 238 (1999)
19. J.A. Piepmeier, *A dynamic quasi-newton method for model independent visual servoing* (Phd, Georgia Institute of Technology, Atlanta, USA, 1999)
20. S. Stramigioli, R. Mahony, P. Corke, in *Proceedings of the IEEE International Conference on Robotics and Automation (ICRA)* (2010), pp. 5302–5308
21. O. Khatib, JSME Int. J. Ser. C: Dyn. Control Robot. Des. Manuf. **36**(3), 277 (1993)
22. F. Chaumette, in *Conflux of vision and control LNCIS*, vol. 237 (1998), pp. 66–78

Force Sensing by Microrobot on a Chip

Tomohiro Kawahara and Fumihito Arai

Abstract In this paper, we discuss a force sensing by microrobot called magnetically driven microtool (MMT) in a microfluidic chip. On-chip force sensor is fabricated by assembling layers to neglect the friction issue and it is actuated by permanent magnets, which supply mN order force to stimulate microorganisms. The displacement is magnified by designing beams on the force sensor and the sensor achieved 100 μN resolutions. We succeeded in on-chip stimulation and evaluation of *Pleurosira laevis* by developed MMT with force sensing structure.

1 Introduction

In bioscience field, the measurement of applied stimulation to a single cell is highly required to figure out functions and properties of cells. In particular, this approach is very useful for understanding a mechanism of microorganisms in terms of neurology [1] and bio-fuel technology [2, 3]. Recently, specific characteristics of algae living in fresh water have been investigated to increase a growth rate toward the realization of efficient bio-fuel. However, these growth mechanisms are still not fully understood especially cell response by mechanical stimulation, even though mechanical approach for single cell is well known as an effective way of increasing growth. In addition, after algae stimulation, precise observation of the cells with chemical reactions is highly required. Therefore, in order to evaluate the relationship between stimulation and a response of single cell, on-chip cell stimulation and evaluation device is required to be maintained in a stable experimental condition.

T. Kawahara (✉)
Kyushu Institute of Technology, 2-4 Hibikino, Wakamatsu-ku, Kitakyushu
Fukuoka 808-0196, Japan
e-mail: kawahara@lsse.kyutech.ac.jp

F. Arai
Nagoya University, Furo-cho, Chikusa-ku, Nagoya, Aichi 464-8603, Japan
e-mail: arai@mech.nagoya-u.ac.jp

H.I. Christensen and O. Khatib (eds.), *Robotics Research*,
Springer Tracts in Advanced Robotics 100, DOI 10.1007/978-3-319-29363-9_9

On the other hand, there have been many works that have used micro force sensors based on MEMS/NEMS technologies to measure the characteristics of microscale objects [4–13]. Sun et al. [5] proposed an SOI-based 2DOF force sensor that consists of spring structures to detect changes in capacitance. This sensor can sense a pushing force with a resolution of 0.01 μN. It was applied to the examination of the mechanical property of zona pellucida of oocyte [6]. Wacogne et al. [9] developed the SU-8 nega-photoresist-based force sensor having a cantilever structure. By applying a μN force to the human oocyte, they found that there are difference in the stiffness of the oocyte and zona pellucida. Nakajima et al. [10] developed nanoprobes fabricated by focused ion beam (FIB) etching to measure the stiffness of *C. elegans* by the atomic force microscope (AFM) cantilever-based force sensing. By using a nanomanipulator, the nanoprobe was used to squeeze *C. elegans*. Recently, Cappelleri et al. [11] proposed a μN force sensor for microrobotics. This sensor has a unique spring structure fabricated using polydimethylsiloxane (PDMS), which is a silicon-based organic elastic polymer with a Young's modulus of 360 kPa to 3 MPa. This sensor can measure the force of two directions by using the vision sensor. However, these sensors must be operated using micro-nano-manipulators to measure objects in a narrow space. This implies that it would be difficult to completely seal the measurement system to prevent contamination of solution and objects.

On the basis of this background, we discuss the on-chip force sensing by using a magnetically driven microtool (MMT) to evaluate the stimulant property of microorganisms, as illustrated in Fig. 1. By using a non-contact actuated microtool having a force sensing structure (i.e., a beam structure), we can apply mN-order force to a single cell, and then estimate the applied force from the deformation of the beam. Since a perfectly closed biochip is used, we avoid contamination and are able to maintain a stable experimental condition during measurements. This is

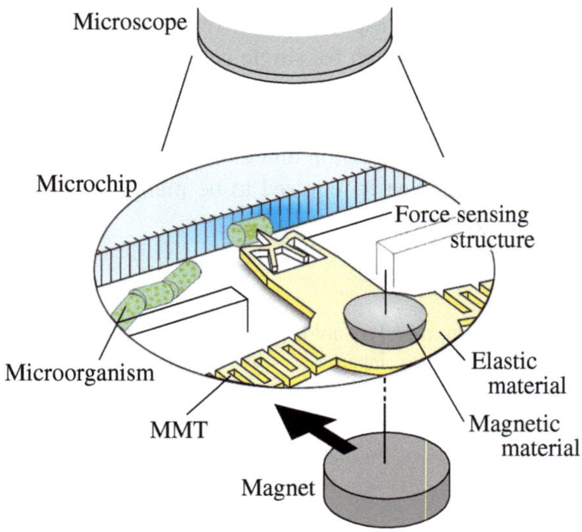

Fig. 1 Basic concept of force sensing by MMT; the microtool with a force sensing structure is composed of a magnetic material and an elastic material

important because it is necessary to continuously observe the response of the microorganisms to chemical reactions after stimulation. The remainder of this paper is organized as follows. In Sect. 2, we describe the basic concept and the prototype design of the on-chip force sensing. In Sect. 3, we describe the basic experiments performed using the developed microchip with microtools and evaluate the effectiveness of the proposed approach. Further, we show how the proposed approach is applied to the stimulation of *Pleurosira laevis* which is one of centric diatom. Finally, we present the concluding remarks and our future plans in Sect. 4.

2 Magnetically Driven Microtool with Force Sensing Structure

2.1 Concept

In previous works by our group, magnetically driven microtools (MMTs) have been proposed for automation of cell manipulation [14–16]. The MMT placed in a biochip is fabricated from a magnetic material, and it can be actuated by a permanent magnet from outside the microfluidic chip. Thus, four advantages of the MMT are as follows.

(1) **Powerful**: Optical tweezers and AFM are difficult to apply enough force to the microorganisms which has a stiff structure. The MMT controlled by permanent magnet can apply mN order stimulation to single cell in a microfluidic chip.
(2) **High dexterity**: Since the MMT has 2 degree of freedom, the microorganisms can be manipulated by MMT where we want it to move.
(3) **On-chip**: Since the perfectly closed microchip is used, we can avoid any contamination and keep stable condition during measurements. It contributes to the continuous observation of the response by the chemical of microorganisms after stimulation.
(4) **Cost**: The combination of the polymer based MMT structure and the camera based measurement achieves very low cost microsystem.

To elucidate the relation between response of cell and mechanical stimulation, quantitative evaluation of the applied force is highly required. Therefore, we add a force sensing structure to the tip of MMT, as shown in Fig. 1. Then, the deformation of the beam is measured by a microscope to achieve on-chip force sensing.

2.2 Basic Sensing Principle

In the first step of on-chip force sensing, we design a frame-shaped beam for the MMT. This structure is not a cantilever structure, and it is important to keep the posture of the MMT straight when single cell is pushed. In addition, to increase the sensitivity of the force sensing by camera, a magnification mechanism of the beam deformation is placed in the frame, as shown in Fig. 2a.

To design the parameters of the force sensing structure, we set the following mechanical model for the frame, shown in Fig. 2a. Here, F, δ, l, h, d, r, $I_1 = b_1 h_1^3/12$, $\mathrm{I}_2 = b_2 h_2^3/12$, and E are the force to be measured, displacement of the beam, width of the frame, height of the frame, width of the magnification mechanism, height of the magnification mechanism, second moment of area of the beam BC, second moment of area of the beam CD, and Young's modulus, respectively. The displacement of the magnification mechanism is

$$X = \alpha\delta, \tag{1}$$

where α is the magnification ratio.

Now, we suppose that half part of the frame structure and the point G is fixed on a wall, as shown in Fig. 2b, where P, MA, and MG are the horizontal stress for point A, moment around point A, and moment around point G, respectively. When the point A is pushed from the horizontal direction, its displacement is as follows:

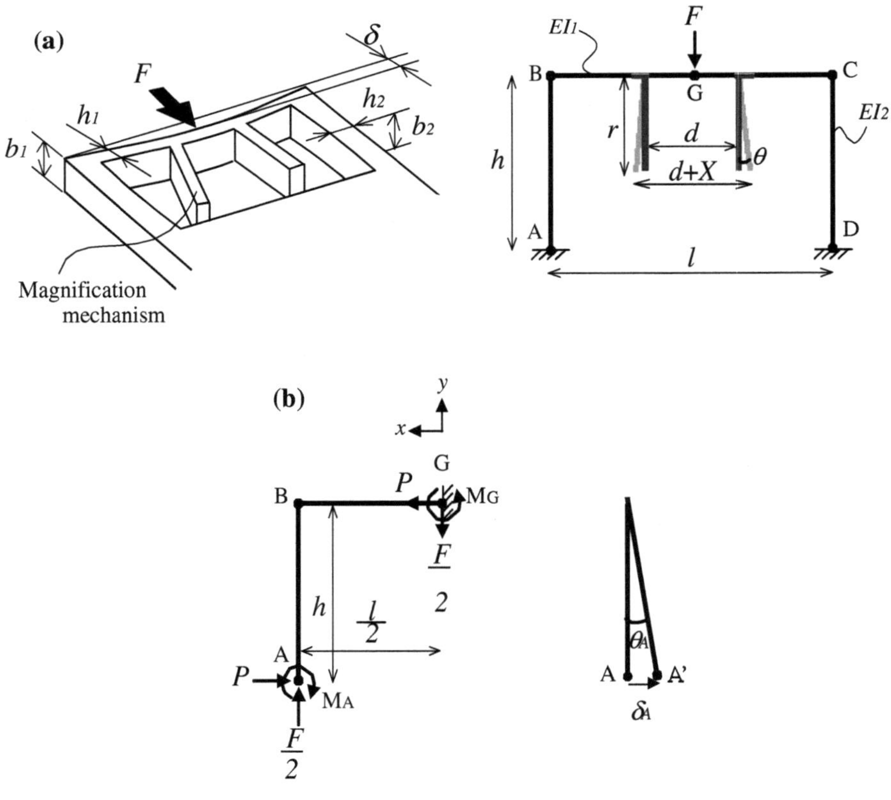

Fig. 2 Model of force sensing by MMT. **a** Design parameters of the force sensing structure. **b** Details of half part of the force sensing structure

$$\delta_A = \frac{Ph\frac{l}{2}}{EI_1}h - \frac{\frac{F}{2}\left(\frac{l}{2}\right)^2}{2EI_1}h - \frac{M_A\frac{l}{2}}{EI_1}h + \frac{Ph^3}{3EI_2} - \frac{M_A h^2}{2EI_2} = 0. \quad (2)$$

Then, the angle of point A is as follows:

$$\theta_A = \frac{Ph\frac{l}{2}}{EI_1} - \frac{\frac{F}{2}\left(\frac{l}{2}\right)^2}{2EI_1} - \frac{M_A\frac{l}{2}}{EI_1} + \frac{Ph^2}{2EI_2} - \frac{M_A h}{EI_2} = 0. \quad (3)$$

From (2) and (3), we obtain

$$M_A = \frac{I_2 F l^2}{8(I_1 h + 2I_2 l)}, \quad P = \frac{3I_2 F l^2}{8h(I_1 h + 2I_2 l)}.$$

Therefore, the displacement of the center of the beam BC is

$$\begin{aligned} \delta &= -\frac{Ph\left(\frac{l}{2}\right)^2}{2EI_1} + \frac{\frac{F}{2}\left(\frac{l}{2}\right)^3}{3EI_1} + \frac{M_A\left(\frac{l}{2}\right)^2}{2EI_1} \\ &= \frac{Fl^3}{96EI_1}\left(\frac{2I_1 h + I_2 l}{I_1 h + 2I_2 l}\right). \end{aligned} \quad (4)$$

On the other hand, from the balance of moment $(MA - Ph - MG + (1/4)Fl = 0)$,

$$M_G = \frac{Fl}{4}\left(\frac{I_1 h + I_2 l}{I_1 h + 2I_2 l}\right).$$

Here, we consider the moment around point G from the frame model, as shown in Fig. 2b,

$$EI_1\theta = M_G x - \frac{F}{4}x^2.$$

Then, the beam angle of the magnification mechanism at $x = d/2$ is

$$\theta = \frac{1}{EI_1}\left[M_G\frac{d}{2} - \frac{F}{4}\left(\frac{d}{2}\right)^2\right].$$

Finally, we can obtain the displacement of the magnification mechanism as follows:

$$\begin{aligned} X &= 2r\theta \\ &= \frac{Frd^2}{8EI_1}\left[\frac{2l(I_1 h + I_2 l)}{d(I_1 h + 2I_2 l)} - 1\right]. \end{aligned} \quad (5)$$

Therefore, from (5), we can estimate the applied force

$$F = kX, \tag{6}$$

where,

$$k = \frac{8EI_1}{rd^2}\left[\frac{2l(I_1h + I_2l)}{d(I_1h + 2I_2l)} - 1\right]^{-1}. \tag{7}$$

2.3 *Fabrication*

To achieve on chip force sensing, there are two problems that needs to be solved by fabrication.

(1) **Friction**: There is large friction between the MMT and the substrate of the microchip. It is critical issue for sensing because the effect of the friction is included in the measured force data.
(2) **Assembling**: Since the force sensing structure becomes very thin, it is difficult to assemble the MMT to the microchip without any damage.

To solve these problems, we introduce a layered fabrication technique which employs assembly of several layers, as shown in Fig. 3. The space between the MMT and the substrate is maintained by a microspacer; this prevents friction

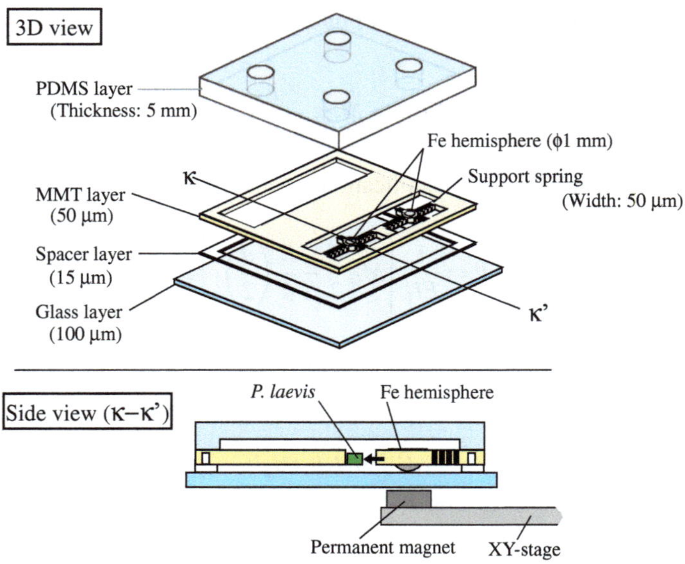

Fig. 3 Layered fabrication for on-chip force sensing

Table 1 Parameters for designed force sensing structure

Parameter	Description	Value
$h1$	Width of frame BC	0.02 mm
$h2$	Width of frame AB and CD	0.04 mm
$b1$	Height of frame BC	0.05 mm
$b2$	Height of frame AB and CD	0.05 mm
h	Length of frame AB and CD	0.60 mm
l	Length of frame BC	0.92 mm
r	Length of magnification mechanism	0.50 mm
d	Distance between magnification mechanism	0.50 mm
E	Young's modulus of frame	3.2 GPa

between the force sensing structure of the MMT and the substrate. By using a layered structure, the micropattern of the MMT is protected, and it is easy to assemble the microparts.

In terms of biocompatibility, we select an SU-8 negative photoresists for a material of the MMT to prevent the damage to microorganisms. As is well known, SU-8 is flexible (approx. 2–10 GPa), it can be easily used to fabricate a micropattern using photolithography technique The Young's modulus of SU-8 depends on its thickness and the amount of exposure [17, 18]. The designed parameters of force sensing structure are summarized, as shown in Table 1.

Figure 4 shows the results of the FEM analysis based on the designed parameters. To ensure enough displacement and rigidity of the MMT, we changed the width of the support spring in various parameters. Finally, we decided the spring with the width of 50 μm. From the results of the simulation, we confirmed that the support spring can be manipulated by the magnet actuation successfully, and consequently the force sensing structure works well.

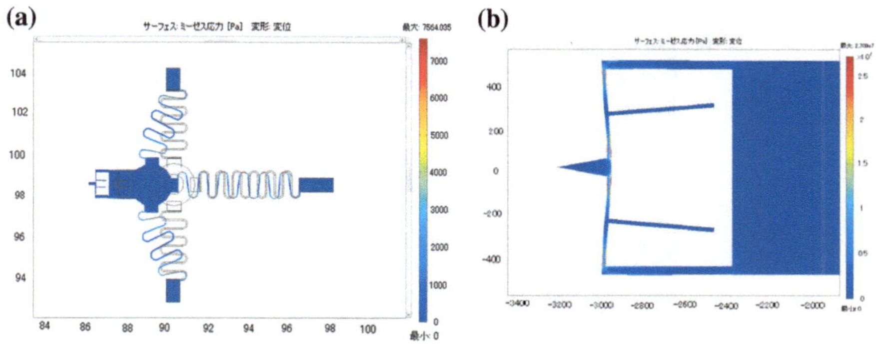

Fig. 4 Results of FEM simulation. **a** Support spring. **b** Force sensing mechanism

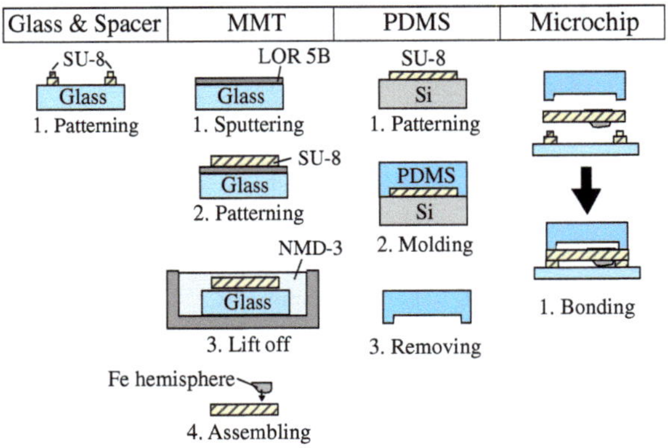

Fig. 5 Process flow of the microchip with MMTs

Figure 5 shows the fabrication process for the proposed layer type MMT with the frame structure. The microchip is composed of the glass substrate with the thickness of 100 μm, the microspacer with the thickness of 15 μm, the SU-8 layer with the thickness of 50 μm includes the MMTs and the microchannel, the iron

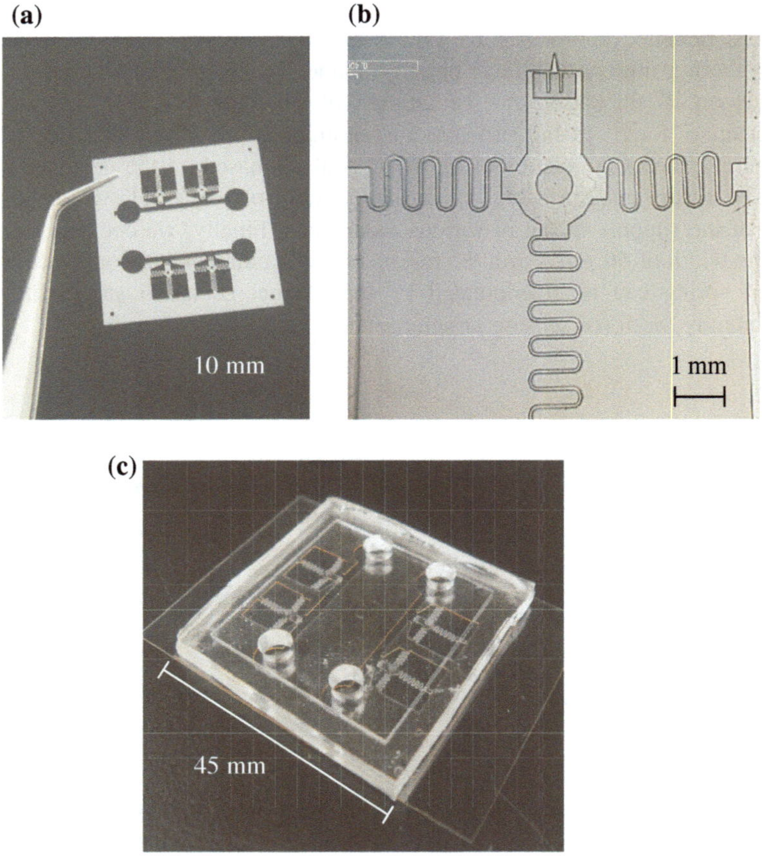

Fig. 6 Fabricated MMT within one layer and microchip. **a** MMT layer. **b** Overview of the MMT. **c** Assembled microchip with 4 measurement sites

hemispheres with the diameter of 1 mm to actuate the MMTs, and the PDMS cover with the thickness of 5 mm. Figure 6a shows the fabricated SU-8 layer, (b) shows the enlarged view of the fabricated MMT, and (c) shows the completely assembled microchip with 4 measurement chambers. To push a local point of small cell, the triangleshaped probe was added to the tip of MMT.

3 Experiments

3.1 Evaluation of the Fabricated MMT

Next, we evaluate the force measurement by the developed MMT. To measure the accurate force data, the MMT is pushed to the commercial force sensor with the accuracy of 10 μN by using the linear stage with the accuracy of 2 μm, as shown in Fig. 7. Then, we measure the deformation by the microscope with the CCD camera (resolution: 1 μm/pixel). From Fig. 8a, we confirmed that there is a high linearity between the displacement of frame δ and the displacement of magnification mechanism *X*, and the magnification ratio is 2.72. Therefore, we can use *X* to estimate the applied force. Figure 8b shows the measured force data by the commercial sensor and the estimated force data calculated from (6) and (7). From this result, it is confirmed that we can measure the accurate force data by using developed force sensing structure. Figure 8c shows the individual difference of MMTs. From this result, the variation among the MMTs is less than 10 %. This means that by calibrating the one of the MMT from the fabricated layer, we can use other MMTs without calibration.

Through these experiments, we obtain the minimum resolution of the force sensing as approximately 100 μN. The resolution of this type of force sensor is decided by camera resolution and sensitivity of frame deformation. As a first step, we think that this performance is reasonable to determine the force range to push cells.

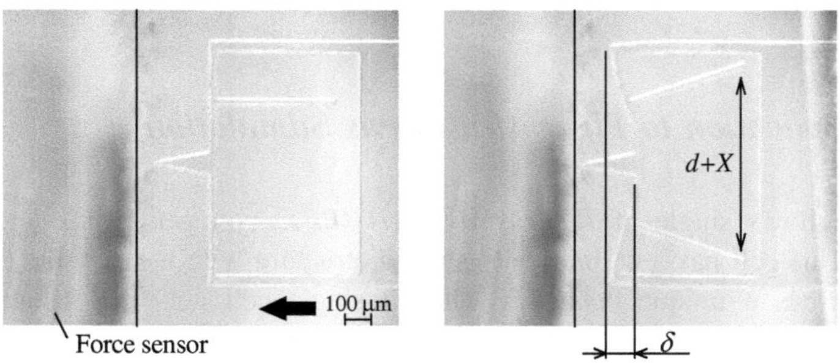

Fig. 7 Overview of experiment

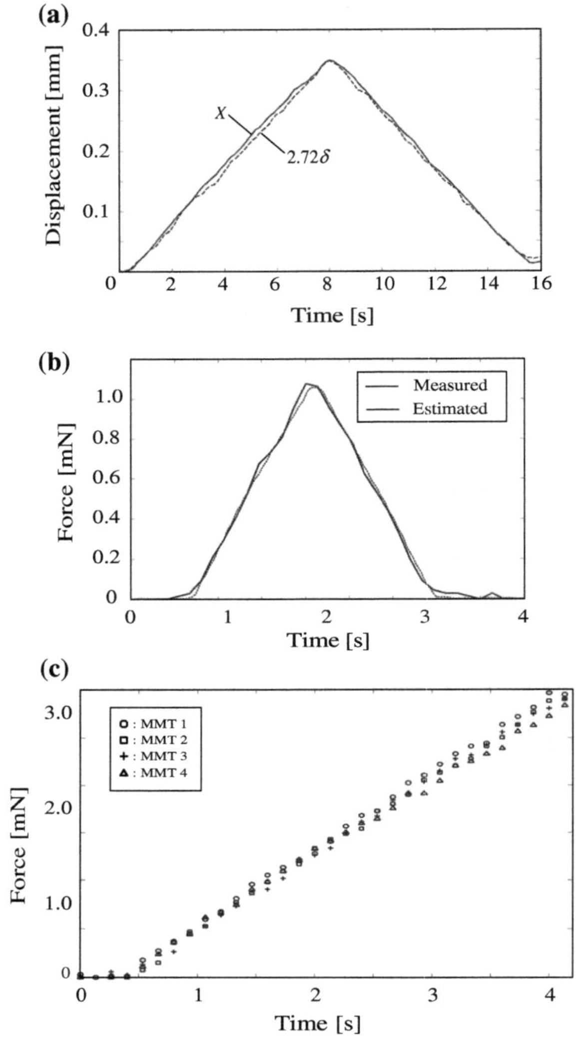

Fig. 8 Results of evaluation of force sensing structure. **a** Comparison between the displacement δ and the displacement X. **b** Force calibration. **c** Individual difference

3.2 Application to Pleurosira Laevis Stimulation

Figure 9 shows single of *Pleurosira laevis* (*P. laevis*) which is one of centric diatom. This cell has a cylindrically-shaped structure with a glass like outer wall. This cell has a unique behavior, when a single cell gets the stimulation, the chloroplasts gather to around the nucleus, as shown in Fig. 9c. Furthermore, this agglomeration is transmitted to other connected cells and also unconnected cells [19]. This mechanisms still not fully understood, because conventional approaches could not measure the applied force in a perfectly closed microchip [20]. Therefore, we applied the developed MMT with force sensing structure to stimulate and

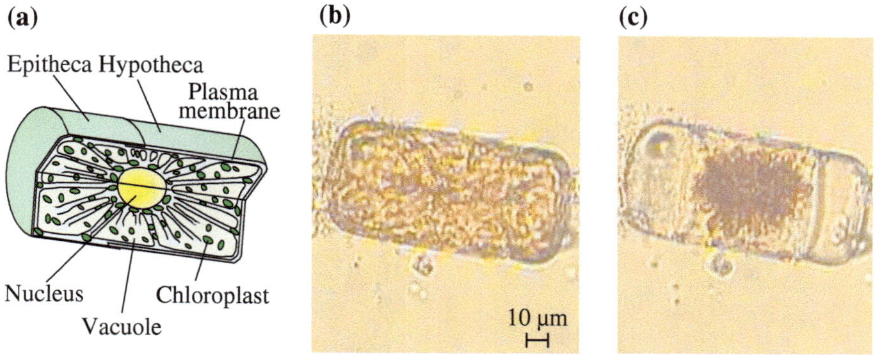

Fig. 9 Overview of Single cell of *Pleurosira laevis*. **a** Basic structure. **b** Normal condition. **c** After stimulation

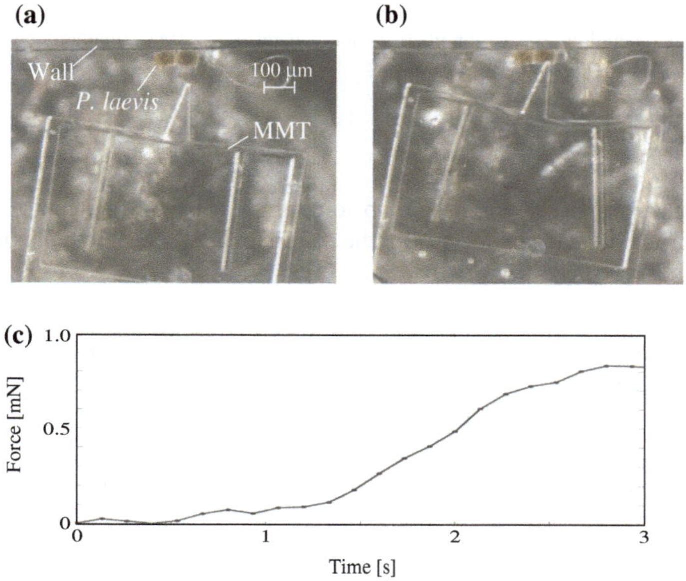

Fig. 10 Stimulation experiment for the *Pleurosira laevis*. **a** Before stimulation. **b** During stimulation. **c** Estimated force

evaluate *P. laevis*. As a result, we succeeded in the pushing of the *P. laevis* by the MMT, as shown in Fig. 9a, b. Then, the applied force is estimated as shown in Fig. 10c. The agglomeration propagation phenomenon of *P. laevis* is also observed by MMT stimulation, as shown in Fig. 11a–d.

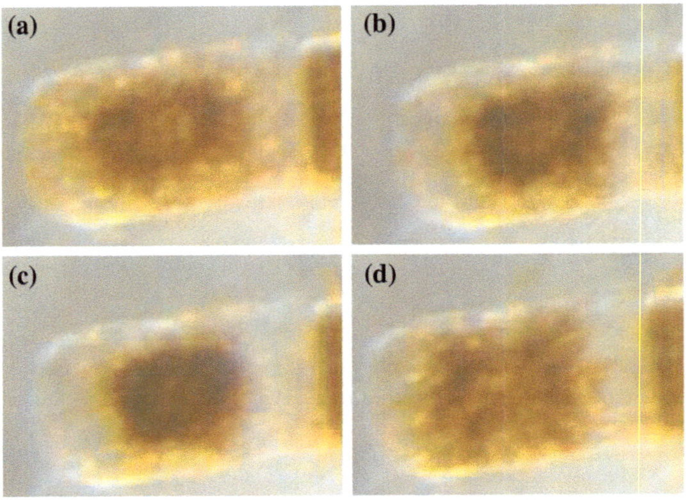

Fig. 11 Agglomeration propagation phenomenon generated by MMT stimulation, **a** 0 min later, **b** 10 min later, **c** 15 min later, **d** 25 min later

4 Conclusions

In this paper, we discussed an on-chip force sensing using a magnetically driven microtool (MMT) that can be used for the measurement of the stimulant property of microorganisms:

1. The frame-based force sensing structure with a displacement magnification mechanism is designed and fabricated on the basis of analytical approach.
2. The microchip consists of a microspacer, an MMT layer, an Fe hemisphere, and a PDMS cover. These parts are assembled by using a layer fabrication technique.
3. Through basic experiments, we have confirmed the performance of the force sensing with an accuracy of 100 μN.

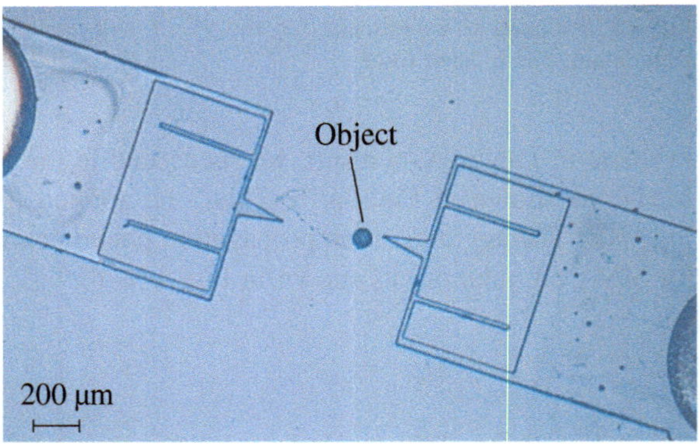

Fig. 12 Dual-arm MMT for microorganism manipulation

4. Through experiments for *P. laevis* stimulation, we confirmed that the proposed force sensing mechanism works well, and that the MMT has sufficient power to push and stimulate *P. laevis*.

We believe that the developed approach is useful for sensing of living cells in a biochip. The combination of a simple and disposable SU-8 based structure and vision-based measurement affords a very low cost force sensing system. In future work, the relationship between the applied force and the agglomeration phenomenon of *P. laevis* is investigated by the dual-arm MMT, as shown in Fig. 12.

Acknowledgments This work has been supported by the Ministry of Education, Culture, Sports, Science and Technology, Grant-in-Aid for Scientific Research (22860030) and the Nagoya University Global COE Program, "COE for Education and Research of Micro-Nano Mechatronics". Finally, we would like to acknowledge Dr. Hiroyuki Kawano, Dr. Ikuko Shihira-Ishikawa, and Dr. Atsushi Miyawaki, RIKEN Brain Science Institute for their great support on *P. laevis* experiments.

References

1. J.M.J. Racca, A. Philibert et al., A comparison between diatom-based pH inference models using artificial neural networks (ANN), weighted averaging (WA) and weighted averaging partial least squares (WA-PLS) regressions. J. Paleolimnol. **26**, 411–422 (2001)
2. V.T. Yadugiri, Milking diatoms—a new route to sustainable energy. Curr. Sci. **97**(6), 748–750 (2009)
3. T.V. Ramachandra, D.M. Mahapatra, B. Karthick, Milking diatoms for sustainable energy: biochemical engineering versus gasoline-secreting diatom solar panels. Ind. Eng. Chem. Res. **48**, 8769–8788 (2009)
4. M.E. Fauver, D.L. Dunaway, D.H. Lilienfeld et al., Microfabricated cantilevers for measurement of subcellular and molecular forces. IEEE Trans. Biomed. Eng. **45**(7), 891–898 (1998)
5. Y. Sun, B.J. Nelson, D.P. Potasek et al., A bulkmicrofabricated multi-axis capacitive cellular force sensor using transverse comb drives. J. Micromech. Microeng. **12**, 832–840 (2002)
6. Y. Sun, K.T. Wan, K.P. Roberts et al., Mechanical property characterization of mouse zona pellucida. IEEE Trans. Nanobiosci. **2**(4), 279–286 (2003)
7. K.H. Jeong, C. Keller, L. Lee, Direct force measurements of biomolecular interactions by nanomechanical force gauge. Appl. Phys. Lett. **86**, 193901-1–193901-3 (2005)
8. S. Koch, G. Thayer, A. Corwin et al., Micromachined piconewton force sensor for biophysics investigations. Appl. Phys. Lett. **89,** 173901-1–173901-3 (2006)
9. B. Wacogne, C. Pieralli, C. Roux et al., Measuring the mechanical behaviour of human oocytes with a very simple SU-8 micro-tool. Biomed. Microdevices **10**, 411–419 (2008)
10. M. Nakajima, M.R. Ahmad, S. Kojima et al., Local stiffness measurements of *C. elegans* by buckling nanoprobes inside an environmental SEM, in *Proceedings of the IEEE/RSJ International Conference on Intelligent Robots and Systems* (2009), pp. 4695–4700
11. D.J. Cappelleri, G. Piazza, V. Kumar, Two-dimensional, vision-based μN force sensor for microrobotics, in *Proceddings of the IEEE Inernational Conference on Robotics and Automation* (2009), pp. 1016–1021
12. K. Ikuta, F. Sato, K. Kadoguchi et al., Optical driven master-slave controllable nanomanipulator with real-time force sensing, in *Proceedings of the IEEE International Conference on Micro Electro Mechanical Systems* (2008), pp. 539–5421

13. M. Papi, L. Sylla, T. Parasassi et al., Evidence of elastic to plastic transition in the zona pellucida of oocytes using atomic force spectroscopy. Appl. Phys. Lett. **94**, 153902-1–153902-3 (2009)
14. Y. Yamanishi, S. Sakuma, K. Onda et al., Powerful actuation of magnetized microtools by focused magnetic field for particle sorting in a chip. Biomed. Microdevices **10**, 411–419 (2008)
15. M. Hagiwara, T. Kawahara, Y. Yamanishi et al., Driving method of microtool by horizontally arranged permanent magnets for single cell manipulation. Appl. Phys. Lett. **97**(0137011-97), 013701–013703 (2010)
16. M. Hagiwara, T. Kawahara, Y. Yamanishi et al., On-chip magnetically actuated robot with ultrasonic vibration for single cell manipulations. Lab. Chip **11**, 2049–2054 (2011)
17. H.S. Khoo, K.K. Liu, F.G. Tseng, Mechanical strength and interfacial failure analysis of cantilevered SU-8 microposts. J. Micromech. Microeng. **13**, 822–831 (2003)
18. D. Bachmann, B. Schoberle, S. Kuhne et al., Fabrication and characterization of folded SU-8 suspensions for MEMS applications. Sens. Actuators A **130–131**, 379–386 (2006)
19. N. Makita, I. Shihira-Ishikawa, Chloroplast assemblage by mechanical stimulation and its intercellular transmission in diatom cells. Protoplasma **197**, 86–95 (1997)
20. Y. Hanada, K. Sugioka, H. Kawano et al., Nano-aquarium for dynamic observation of living cells fabricated by femtosecond laser direct writing of photostructurable glass. Biomed. Microdevices **10**, 403–410 (2008)

Force Control and Reaching Movements on the iCub Humanoid Robot

Giorgio Metta, Lorenzo Natale, Francesco Nori and Giulio Sandini

Abstract This paper is about a layered controller for a complex humanoid robot: namely, the iCub. We exploited a combination of precomputed models and machine learning owing to the principle of balancing the design effort with the complexity of data collection for learning. A first layer uses the iCub sensors to implement impedance control, on top of which we plan trajectories to reach for visually identified targets while avoiding the most obvious joint limits or self collision of the robot arm and body. Modeling errors or misestimation of parameters are compensated by machine learning in order to obtain accurate pointing and reaching movements. Motion segmentation is the main visual cue employed by the robot.

1 Introduction

In this paper we consider a solution to the problem of reaching for a visually identified target in the context of the control of a humanoid robot platform, considering both potential forceful interactions with objects or people and gross mistakes due to miscalibration of the controller parameters. Our reference platform is the iCub [1], a humanoid robot shaped as a three and half years old child. The iCub, by design, only uses "passive" sensors as for example cameras, gyroscopes, pres-

G. Metta (✉) · L. Natale · F. Nori
iCub Facility, Istituto Italiano di Tecnologia, Via Morego 30, Genoa, Italy
e-mail: giorgio.metta@iit.it

L. Natale
e-mail: lorenzo.natale@iit.it

F. Nori
e-mail: francesco.nori@iit.it

G. Sandini
Robotics, Brain and Cognitive Sciences, Istituto Italiano di Tecnologia,
Via Morego 30, Genoa, Italy
e-mail: giulio.sandini@iit.it

H.I. Christensen and O. Khatib (eds.), *Robotics Research*,
Springer Tracts in Advanced Robotics 100, DOI 10.1007/978-3-319-29363-9_10

sure, force and contact sensors, microphones and so forth. We excluded the use of lasers, sonars and other esoteric sensing modalities.

In this conditions and in an unstructured environment where human can freely move and work (our laboratory space in the daily use of the iCub), it is unlikely that the robot obtains an accurate model of the environment for precise impact-free planning of movements. One common solution [2] is to control the robot mechanical impedance and, simultaneously, minimize impacts by using for example vision and trajectory planning. The possibility of impedance control lowers the requirements of vision and guarantees a certain degree of safety in case of contacts with the environment—though, strictly speaking, the robot can still be potentially dangerous and cause damage if it moves fast.

The control architecture described in this paper is not very different in principle from a standard computed torque approach [3]. A first layer compensates for the dynamics and linearizes the system. Because of the communication bus of the iCub controllers, of bandwidth requirements, and implementation constraints, it operates in joint space. A second layer subsequently plans trajectories starting from a description of the target position in extrinsic space and merging joint limits, a secondary task specification, inverse kinematics and singularity avoidance. We show in the remainder of the paper how this is implemented by mixing hand-coded models of the robot dynamics and kinematics together with machine learning.

Reaching and pointing is fundamental in learning about the environment enabling interaction with objects and their manipulation to achieve complex tasks. In this sense these are the basic building blocks of a complex cognitive architecture for the iCub.

2 Experimental Platform: The iCub

The iCub is one of the results of the RobotCub project, an EU-funded endeavor to create a common platform for researchers interested in embodied artificial cognitive systems [4].

The initial specifications of the robot aimed at replicating the size of a three and a half years old child. In particular, it was required that the robot be capable of crawling on all fours and possess fine manipulation abilities. For a motivation of why these features are important, the interested reader is referred to Metta et al. [5].

Dimensions, kinematic layout and ranges of movement were drafted by considering biomechanical models and anthropometric tables [6]. Rigid body simulations were used to determine the crucial kinematic features in order to perform the set of desired tasks and motions, i.e. reaching, crawling, etc. [7]. These simulations also provided joint torques requirements. Data were then used as a baseline performance indicator for the selection of the actuators. The final kinematic structure of the robot is shown in Fig. 1c. The iCub has 53 degrees of freedom (DoF). Its kinematics has several special features which are rarely found in other humanoid robots: e.g. the waist has three DoF which considerably increase the robot's

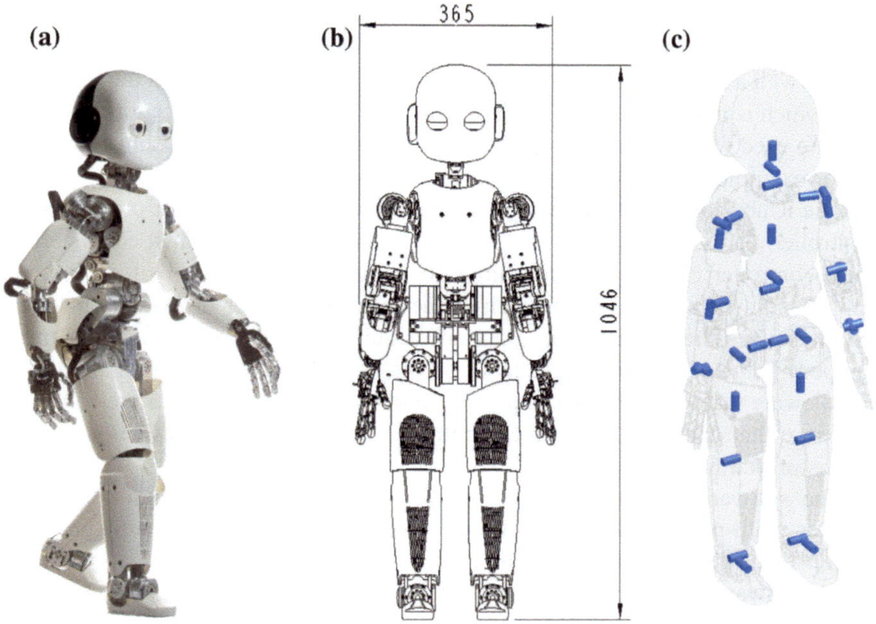

Fig. 1 The iCub platform: *panel a* a picture of the latest realization of the iCub; *panel b* approximate dimensions height × width; and *panel c* the kinematic structure of the major joints

mobility; the three DoF shoulder joint is constructed to have its axes of rotation always intersecting at one point.

To match the torque requirements we employed rotary electric motors coupled with speed reducers. We found this to be the most suitable choice in terms of robustness and reliability. Motor groups with various characteristics were developed (e.g. 40, 20 and 11 Nm) for different placements into the iCub. We used the Kollmorgen-DanaherMotion RBE type brushless frameless motor (BLM) and a CSD frameless Harmonic Drive as speed reducer. The use of frameless components allowed further optimization of space and reduced weight. Smaller motors for moving the fingers, eyes and neck are from Fulhaber in various sizes and reduction gear ratios.

Cable drives were used almost everywhere on the iCub. Most joints have relocated motors as for example in the hand, shoulder (besides one joint), elbow, waist and legs (apart from two joints). Cable drives are efficient and almost mandatory in order to optimize the motor locations and the overall "shape" of the robot. All joints in the hand are cable driven. The hand of the iCub has 20 joints which are moved by only 9 motors: this implies that some of the joints are under-actuated and their movement is obtained by means of the cable couplings. Similarly to the human body most of the hand actuation is in the forearm subsection. The head is another particular component of the iCub enabling independent vergence movements supported by a three DoF neck for a total of six DoF.

By design we decided to only use "passive sensors" and in particular cameras, microphones, gyroscopes and accelerometers, force/torque (FTS) and tactile sensors as well as the traditional motor encoders. Of special relevance is the sensorized skin which is not easily found in other platforms as well as the force/torque sensors that are used for force/impedance control (see later). No active sensing is provided as for example lasers, structured light projectors, and so forth.

The iCub mounts custom-designed electronics which consists of programmable controller cards, amplifiers, DACs and digital I/O cards. This ecosystem of microcontroller cards relies on multiple CAN bus lines (up to 10) for communication and synchronization and then connects with a cluster of external machines via a Gbit/s Ethernet network. Data are acquired and synchronized (and timestamped) before being made available on the network. We designed the software middleware that supports data acquisition and control of the robot as well as all the firmware that operates on the microcontrollers which eventually drive each single transistor that moves the motors.

The software middleware is called YARP [8]. YARP is a thin library that enables multi-platform and multi-IDE development and collaboration by providing a layer that shields the user from the quirks of the underlying operating system and robot hardware controllers. The complete design of the iCub (drawings, schematics, specifications) and its software (both middleware and controllers) is distributed according to the GPL or the LGPL licenses.

3 Dynamics

The first layer of the proposed architecture is based on computation of the body dynamics and implements joint position and velocity control on top of joint-level impedance. In the simplest possible version, the controller cards implement a 1 ms feedback loop relying on the error e defined as (Fig. 3):

$$e = \tau - \tau_d, \tag{1}$$

where τ is the vector of joint torques and τ_d its desired value. We do not know τ directly on the iCub but we have access to estimates through the force/torque sensors (FTSs). They are mounted as indicated in Fig. 2 in the upper part of the limbs and can therefore be used to detect wrenches at any location in the iCub limbs and not only at the end-effector as it is more typical for industrial manipulators.

We show that τ can be estimated from the FTS measurements of each limb (equations repeat identical for each limb). Let's indicate with w_s the wrench measured by the FTS and assume that it is due to an actual external wrench at a known location (e.g. at the end-effector) which we call w_e. We can estimate w_e by propagating the measurement on the kinematic chain of the limb (changing coordinates):

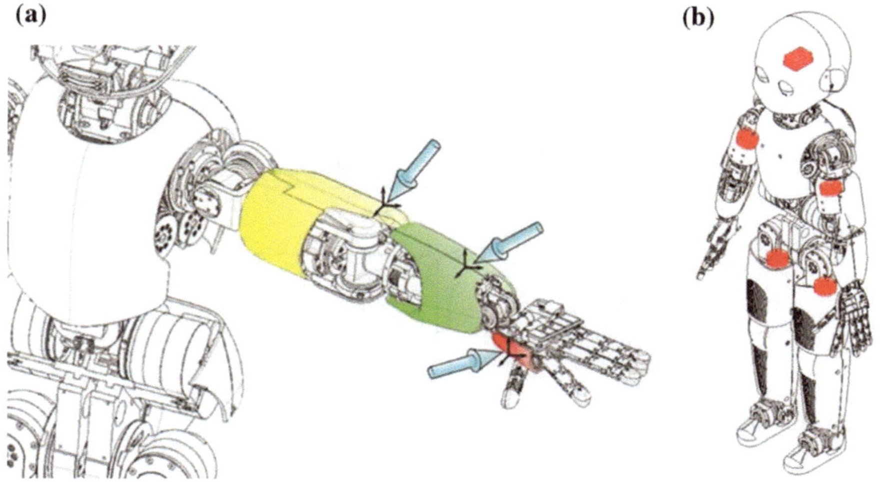

Fig. 2 In **a** a typical interaction of the iCub arm with the environment exemplified here with a number of wrenches at different locations and in **b** the location of the four FTSs of the iCub in the *upper part* of the limbs (proximal with respect to the reference frame of the robot kinematic chains) and of the inertial sensors mounted in the head

$$\hat{w}_e = \begin{bmatrix} I & 0 \\ -[\bar{r}_{se}]_\times & I \end{bmatrix} \cdot (w_s - w_i), \tag{2}$$

with $[\bar{r}_{se}]_\times$ the skew-symmetric matrix representing the cross product with the vector $\bar{r}_{se}$, $\hat{w}_e$ the estimate of w_e, and w_i the internal wrench (due to internal forces and moments). Note that $[\bar{r}_{se}]_\times$ is a function of q, the vector of joint angles. w_i can be estimated from the dynamics of the limb (either with the Lagrange or Newton-Euler formulation). To estimate τ_e we only need to project $\hat{w}_e$ to the joint torques using the transposed Jacobian, i.e.:

$$\hat{\tau}_e = J^T(q) \cdot \hat{w}_e. \tag{3}$$

We can then use this estimate in a control loop by defining the torque error e as:

$$e = \hat{\tau}_e - \tau_d, \tag{4}$$

where $\hat{\tau}_e$ is an estimate of τ regulated by a PID controller of the form:

$$u = k_p \cdot e + k_d \cdot \dot{e} + k_i \cdot \int e, \tag{5}$$

where k_p, k_d and k_i are the usual PID gains and u the amplifier output (the PWM duty cycle which determines the equivalent applied voltage at the motor). Similarly we can build an impedance controller in joint space by making τ_d of the form:

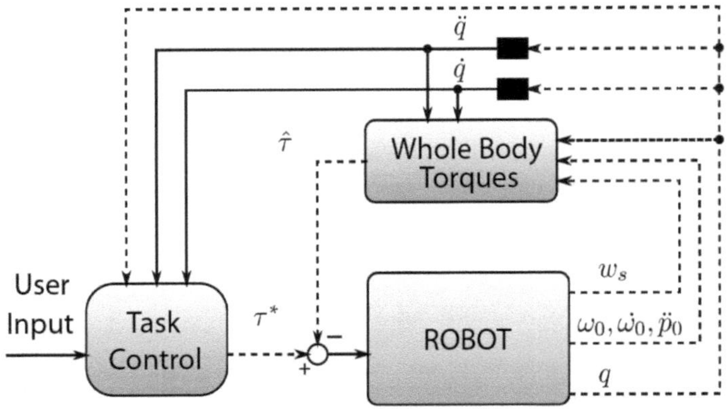

Fig. 3 The torque controller of the iCub. See text for details

$$\tau_d = K \cdot (q - q_d) + D \cdot (\dot{q} - \dot{q}_d), \tag{6}$$

which can be implemented at the controller card level if K and D are diagonal matrices. Furthermore, we can command velocity by making:

$$q_d(t) = q_d(t - \delta_t) + \dot{q}_d(t)\delta_t, \tag{7}$$

with δt the control cycle interval (1 ms in our case). This latter modality is useful when generating whole trajectories incrementally. The actual computation of the dynamics and kinematics is based on a graph representation which we detail in the following.

We start by considering an open (single or multiple branches) kinematic chain with n DoF composed of n + 1 links. Adopting the Denavit-Hartenberg notation [3], we define a set of reference frames (0), (1),..., (n), attached at each link. The ith link of the chain is described by a vertex v_i (sometimes called node), usually represented by the symbol ⓘ A hinge joint between the link i and the link j (i.e. a rotational joint) is represented by an oriented edge $e_{i,j}$ connecting v_i with v_j : ⓘ→ ⓙ . In a n DoF open chain, each vertex (except for the initial and terminal, v_0 and v_n respectively) has two edges. Therefore, the graph representation of the n-link chain is an oriented sequence of nodes v_i, connected by edges $e_{i-1,i}$. The orientation of the edges can be either chosen arbitrarily (it will be clear later on that the orientation simply induces a convention) or it can follow from the exploration of the kinematic tree according to the *regular numbering scheme* [9], which induces a parent-child relationship such that each node has a unique input edge and multiple output edges. We further follow the classical Denavit-Hartenberg notation, we assume that each joint has an associated reference frame with the z-axis aligned with the rotation axis; this frame will be denoted $\langle e_{i,j} \rangle$. In kinematics, an edge $e_{i,j}$ from v_i to v_j represents the fact that $\langle e_{i,j} \rangle$ is fixed in the ith link. In dynamics, $e_{i,j}$ represents the fact that the dynamic equations will compute (and make use of) $w_{i,j,}$ i.e. the wrench that the *i*th link exerts on the *j*th link, and not the equal and opposite reaction $-w_{i,j,}$ i.e. the wrench that the *j*th link exerts on the *i*th link. In order to simplify the computations of the inverse dynamics on the graph, kinematic and dynamic measurements have been explicitly represented. Specifically, the graph representation has been enhanced with a new set of graphical symbols: a triangle to represent kinematic quantities (i.e. velocities and acceleration of links— $\omega, \dot{\omega}, \dot{p}, \ddot{p}$), and a rhombus for wrenches (i.e. force sensors measurements on a link—f, μ). Moreover these symbols have been further divided into known quantities to represent sensors measurements, and unknown to indicate the quantities to be computed, as in the following:

- ▽: unknown kinematic information
- ▼: known (e.g., measured) kinematic information
- ◇: unknown dynamic information
- ◆: known (e.g., measured) dynamic information

In general, kinematic variables can be measured by means of gyroscopes, accelerometers, or simply inertial sensors. When attached on link *i*th, these sensors provide angular and linear velocities and accelerations $(\omega, \dot{\omega}, \dot{p} \text{ and } \ddot{p})$ at the specific location where the sensor is located. We can represent these measurement in the graph with a *black triangle* (▼) and an additional edge from the proper link where the sensor is attached to the triangle. As usual, the edge has an associated reference frame, in this case corresponding to the reference frame of the sensor. An unknown kinematic variable is represented by a *white triangle* (▽) with an associated edge going from the link (where the unknown kinematic variable is attached) to the triangle. Similarly, we introduce two new types of nodes with a rhomboidal shape: *black rhombus* (◆) to represent known (i.e. measured) wrenches, *white rhombus* (◇) to represent unknown wrenches which need to be computed. The reference frame associated to the edge will be the location of the applied or unknown wrench. The complete graph for the iCub is shown in Fig. 4.

From the graph structure, we can define the update rule that brings information across edges and by traversing the graph we therefore compute either dynamical or kinematic unknowns (◇ and ▽ respectively). For kinematic quantities this is:

$$\begin{aligned} \omega_{i+1} &= \omega_i + \dot{\theta}_{i+1} z_i, \\ \dot{\omega}_{i+1} &= \dot{\omega}_i + \ddot{\theta}_{i+1} z_i + \dot{\theta}_{i+1} \omega_i \times z_i, \\ \ddot{p}_{i+1} &= \ddot{p}_i + \dot{\omega}_i \times r_{i,i+1} + \omega_{i+1} \times (\omega_{i+1} \times r_{i,i+1}), \end{aligned} \tag{8}$$

where z_i is the z-axis of $\langle i \rangle$, i.e. we propagate information from the base to the end-effector visiting all nodes and moving from one node to the next following the edges. The internal dynamics of the manipulator can be studied as well: if the dynamical parameters of the system are known (mass m_i, inertia I_i, center of mass C_i), then we can propagate knowledge of wrenches applied to e.g. the end-effector

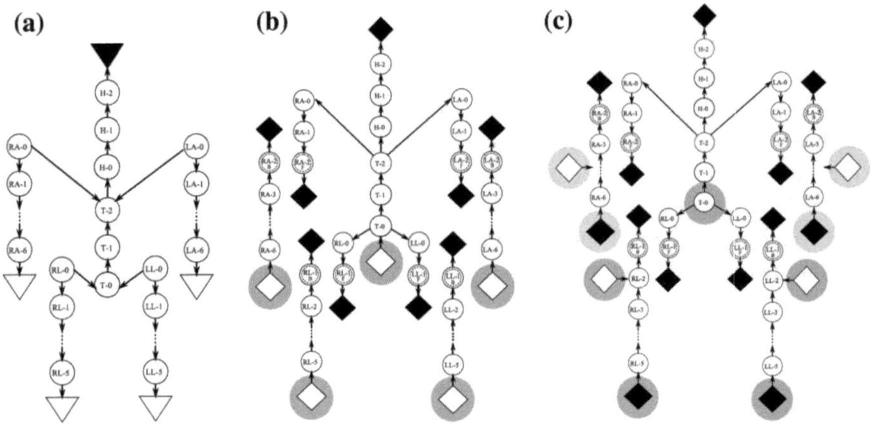

Fig. 4 Representation of iCub's kinematic and dynamic graph. In **a** iCub's kinematics. The inertial sensor measure (▼) is the unique source of kinematic information for the whole branched system. **b** iCub's dynamics when the robot is standing on the mainstay and moving freely in space. Given the four FTSs, the main graph is cut by the four links hosting the sensors, and a total of five sub-graphs are finally generated. The unknowns are the external wrenches at the end-effectors: if the robot does not collide with the environment, they are zero, whereas if a collision happens, then an external wrench arises. The displacement between the expected and the estimated wrenches allows detecting contacts with the environment under the hypothesis that interactions can only occur at the end-effectors. The external wrench on top of the head is assumed to be null. Notice that the mainstay is represented by a unknown wrench ◇. **c** iCub's dynamics when the robot is crawling (four points of contact with the ground). As in the previous case, five subgraphs are generated after the insertion of the four FTSs measurements, but unlike the free-standing case, here the mainstay wrench is removed, being the iCub on the floor. Specific locations for the contacts with the environment are given as part of the task: the unknown external wrenches (◇) are placed at wrists and knees, while wrenches at the feet and palms are assumed known and null (▼). Interestingly, while moving on the floor the contact with the upper part could be varying (e.g. wrists, palms, elbows), so the unknown wrenches could be placed in different locations than the ones shown in the graph

(f_{n+1} and μ_{n+1}) to the base frame of the manipulator so as to retrieve forces and moments f_i, μ_i:

$$\begin{aligned} f_i &= f_{i+1} + m_i \ddot{p}C_i, \\ \mu_i &= \mu_{i+1} - f_i \times r_{i-1}, C_i + f_{i+1} \times r_{r_i, C_i} + I_i \dot{\omega}_i + \omega_i \times (I_i \omega_i), \end{aligned} \tag{9}$$

where

$$\ddot{p}C_i = \ddot{p}_i + \dot{\omega}_i \times r_i, C_i + \omega_i \times (\omega_i \times r_i, C_i), \tag{10}$$

noting that these are the classical recursive Newton-Euler equations. Knowledge of wrenches enables the computation of w_i as needed in Eq. 2 or the corresponding joint torques from $\tau_i = \mu_i^T z_{i-1}$.

3.1 Validation and Further Improvements

In order to validate computation of the dynamics, we compared measurements from the FTSs with their model-based prediction. The wrenches w_s from the four six-axes FTSs embedded in the limbs are compared with the analogous quantities $\hat{w}_s$ predicted by the dynamical model, during unconstrained movements (i.e. null external wrenches). Kinematic and dynamic parameters are retrieved from the CAD model of the robot. Sensor measurements w_s can be predicted assuming known wrenches at the limbs extremities (hands or feet) and then propagating forces up to the sensors. In this case, null wrenches are assumed, because of the absence of contact with the environment. Table 1 summarizes the statistics of the errors $(w_s - \hat{w}_s)$ for each limb during a given, periodic sequence of movements, with the robot supported by a rigid metallic mainstay, and with the limbs moving freely without self collision or contact with the environment. Table 1 shows the mean and the standard deviation of the errors between measured and predicted sensor wrench during movement. Figure 5 shows a comparison between w_s and $\hat{w}_s$ for the left arm (without loss of generality, all limbs show similar results).

Subsequently we investigated methods to improve the estimates of the robot dynamics. In another set of experiments we thus compared various non-parametric learning methods with the rigid body model just presented. We refer the interested reader to Gijsberts et al. [10]. We report here only the main findings. The task of learning here is the estimation of the wrenches due to the internal dynamics (w_i) given the FTS readings (w_s) and the robot configuration $(q, \dot{q}, \ddot{q})$; we do not take into account inertial information.

We compared various methods from the literature as for example the widely used Local Weighted Projection Regression (LWPR), the Local Gaussian Process (LGP) and Gaussian Process Regression (GPR) as presented by Nguyen-Tuong et al. [11] with an incremental version of Kernel Ridge Regression (also known as

Table 1 Error in predicting FT sensor measurement (see text for details)

	ε_{f0}	ε_{f1}	ε_{f2}	$\varepsilon_{\mu 0}$	$\varepsilon_{\mu 1}$	$\varepsilon_{\mu 2}$
$\bar{\varepsilon}$	−0.3157	−0.5209	0.7723	−0.0252	0.0582	0.0197
σ_ε	0.5845	0.7156	0.7550	0.0882	0.0688	0.0364
Right arm: $\varepsilon \equiv \hat{w}_{s,RA} - w_{s,RA}$						
$\bar{\varepsilon}$	−0.0908	−0.4811	0.8699	0.0436	0.0382	0.0030
σ_ε	0.5742	0.6677	0.7920	0.1048	0.0702	0.0332
Left arm: $\varepsilon \equiv \hat{w}_{s,LA} - w_{s,LA}$						
$\bar{\varepsilon}$	−1.6678	3.4476	−1.5505	0.4050	−0.7340	0.0171
σ_ε	3.3146	2.7039	1.7996	0.3423	0.7141	0.0771
Right leg: $\varepsilon \equiv \hat{w}_{s,RL} - w_{s,RL}$						
$\bar{\varepsilon}$	0.2941	−5.1476	−1.9459	−0.3084	−0.8399	0.0270
σ_ε	1.8031	1.8327	2.3490	0.3365	0.8348	0.0498
Left leg: $\varepsilon \equiv \hat{w}_{s,LL} - w_{s,LL}$						

SI units: f: [N], μ: [Nm]

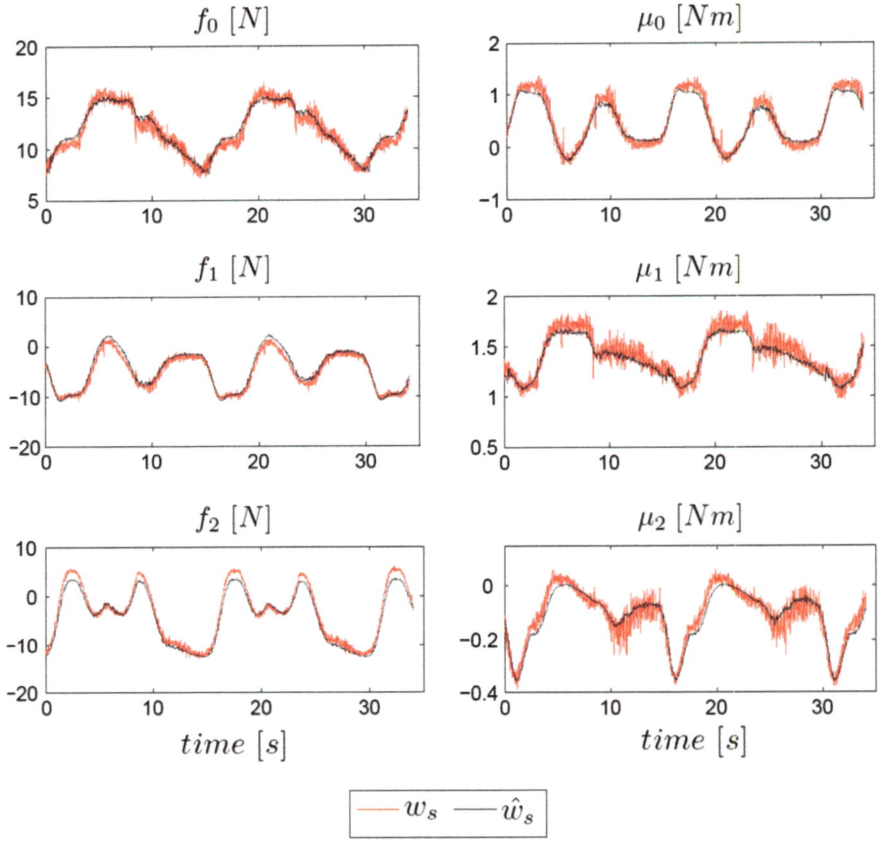

Fig. 5 Comparison between the wrench measured by the FT sensor and that predicted by the model, during a generic contact-free movement of the *left* arm. The three plots on the *left* are forces expressed in [N]; the three rightmost plots are the moments in [Nm]

Sparse Spectrum Gaussian Process) with the aim of maintaining eventually an incremental open-ended learner updating the estimation of the robot dynamics on-line. Our incremental method relies on an approximation of the kernel (see [12]) based on a random sampling of its Fourier spectrum. The more random features, the better the approximation. We considered approximations with 500, 1000, and 2000 features. In the following we call KRR the plain kernel ridge regression method and RFRR^D the random feature version for D features. Various datasets (e.g. Barret, Sarcos) were used from the literature (for comparison [11]) before applying the method to the iCub.

The results in Fig. 6 show that KRR often outperforms GPR by a significant margin, even though both methods have identical formulations for the predictive mean and KRR hyperparameters were optimized using GPR. These deviations indicate that different hyperparameter configurations were used in both experiments. This is a common problem with GPR in comparative studies: the marginal

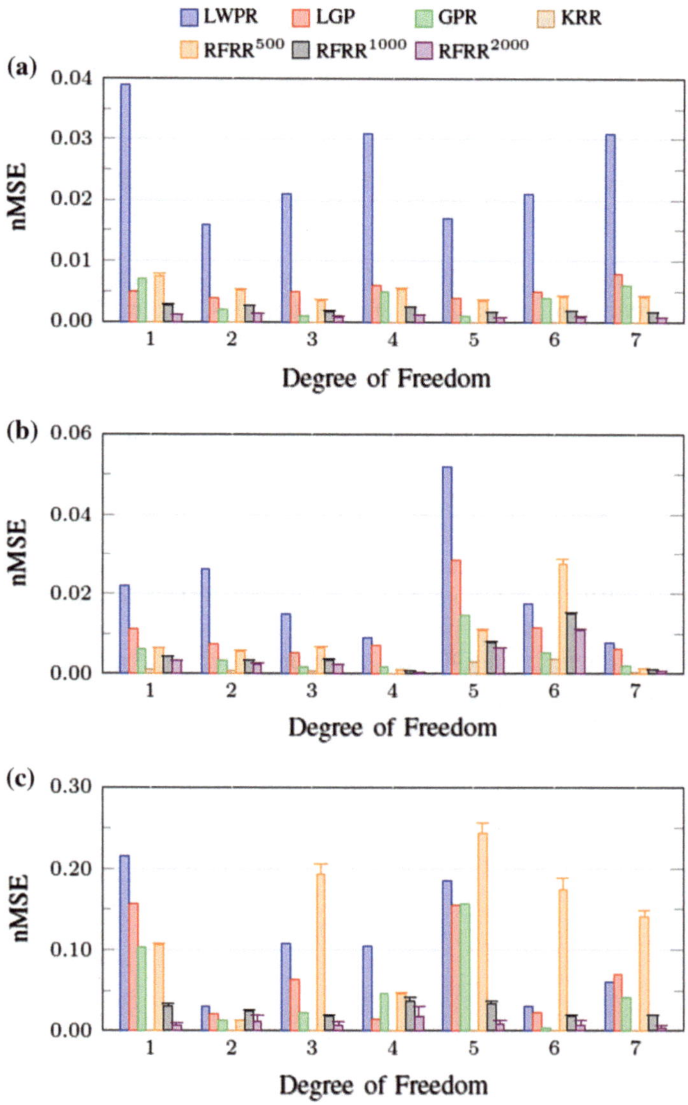

Fig. 6 Prediction error per degree of freedom for the **a** Simulated Sarcos, **b** Sarcos, and **c** Barrett datasets. The results for LWPR, GPR, and LGP are taken from Nguyen-Tuong et al. [11]. The mean error over 25 runs is reported for RFRR with $D \in 500, 1000, 2000$, whereas error bars mark a distance of one standard deviation. Note that in some cases the prediction errors for KRR are very close to zero and therefore barely noticeable

likelihood is non-convex and its optimization often results in a local optimum that depends on the initial configuration. Hence, we have to be cautious when interpreting the comparative results on these datasets with respect to generalization performance. The comparison between KRR and RFRR, trained using identical hyperparameters, remains valid and gives an indication of the approximation quality of RFRR. As expected, the performance of RFRR steadily improves as the number of random features increases. Furthermore, RFRR^{1000} is often sufficient to obtain satisfactory predictions on all datasets. RFRR^{500}, on the other hand, performs poorly on the Barrett dataset, despite using distinct hyperparameter configurations for each degree of freedom. In this case, RFRR^{1000} with a shared hyperparameter configuration is more accurate and requires overall less time for prediction.

Figure 7 shows how the average nMSE develops as test samples are predicted in sequential order using either KRR or RFRR. RFRR requires between 5000 and 10000 samples to achieve performance comparable to KRR. The performance of KRR, on the other hand, decreases over time. In particular on the iCub dataset it suffers a number of large errors, causing the average nMSE to show sudden jumps. This is a direct consequence of the unavoidable fact that training and test samples are not guaranteed to be drawn from the same distribution. Incremental RFRR, on the other hand, is largely unaffected by these changes and demonstrates stable predictive performance. This is not surprising, as RFRR is incremental and thus (1) it is able to adapt to changing conditions, and (2) it eventually has trained on significantly more samples than KRR. Furthermore, Fig. 7 shows that 200 random features are sufficient to achieve satisfactory performance on either dataset. In this case, model updates of RFRR require only 400 μs, as compared to 2 and 7 ms when using 500 or 1000 random features, respectively. These timing figures make incremental RFRR suitable for high frequency loops as needed in robot control tasks.

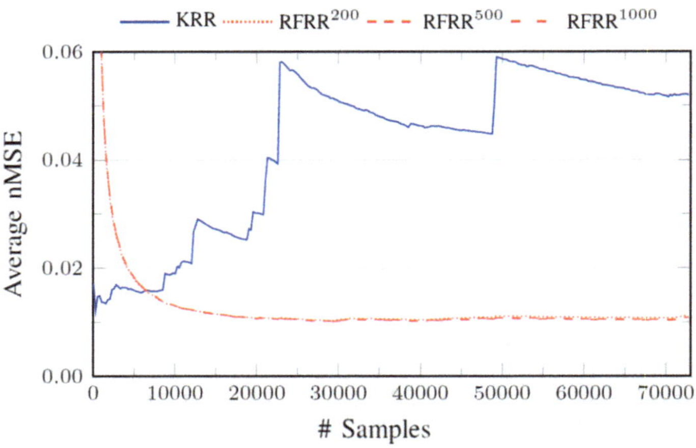

Fig. 7 Average prediction error with respect to the number of test samples of KRR and incremental RFRR with $D \in 200, 500, 1000$ on the iCub dataset. The error is measured as the nMSE averaged over the force and torque output components. The standard deviation over 25 runs of RFRR is negligible in all cases, for clarity we report only the mean without *error bars*

In conclusion, this shows that for a relatively complex robot like the iCub, good estimation of the internal dynamics is possible and that a combination of non-parametric and parametric methods can provide simultaneously good generalization performance, fast and incremental learning. Not surprisingly, lower errors are obtained with learning. In the next section we see how to build on this controller to reach for visually identified targets.

4 Kinematics

We consider the general problem of computing the value of joint angles q_d in order to reach a given position in space $x_d \in \mathbb{R}^3$ and orientation $\alpha_d \in \mathbb{R}^4$ of the end-effector (where α_d is a representation of rotation in axis/angle notation). Note that q_d can be directly connected to the input of the impedance controller described in Sect. 3. It is desired that the computed solution satisfies a set of additional constraints expressed as generic inequalities—we see later the reason for constraining the solution of the optimization problem. This can be stated as follows:

$$q_d = \operatorname{argmin}_{q \in \mathbb{R}^n}\left(\|\alpha_d - K_\alpha(q)\|^2 + \beta(q_{rest} - q)^T W(q_{rest} - q)\right),$$
$$s.t. \begin{cases} \|x_d - K_x(q)\|^2 < \varepsilon \\ qL < q < qU \end{cases}, \tag{11}$$

where K_x and K_α are the forward kinematic functions for the position and orientation of the end-effector for a given configuration q; q_{rest} is a preferred joint configuration, W is a diagonal weighting matrix, β a positive scalar weighting the influence of the terms in the optimization and ε a parameter for tuning the precision of the movement. Typically $\beta < 1$ and $\varepsilon \in [10^{-5}, 10^{-4}]$. The solution to Eq. 11 has to satisfy the set of additional constraints of joint limits $q_L < q < q_U$ with q_L, q_U the lower and upper bounds respectively. In the case of the iCub, we solved this problem for ten DoF—seven of the arm and three of the waist and we determined the value of q_{rest} so that the waist is as upright as possible. The left and right arm can be both controlled by switching from one or the other kinematic chain (e.g. as a function of the distance to the target).

We used an interior point optimization technique to solve the problem in 11. In particular we used IpOpt [13], a public domain software package designed for large-scale nonlinear optimization. This approach has the following advantages:

1. Quick convergence. IpOpt is reliable and fast enough to be employed in control loops at reasonable rates (tens of milliseconds), as e.g. compared to more traditional iterative methods such as the Cyclic Coordinate Descent (CCD) adopted in [14];

2. Scalability. The intrinsic capability of the optimizer to treat nonlinear problems in any arbitrary number of variables is exploited to make the controller structure easily scalable with the size of the joint space. For example, it is possible to change at run time from the control of the 7-DoF iCub arm to the complete 10-DoF structure inclusive of the waist or to any combination of the joints depending on the task;
3. Automatic handling of singularities and joint limits. This technique automatically deals with singularities in the arm Jacobian and joint limits, and can find solutions in virtually any working conditions;
4. Tasks hierarchy. The task is split in two subtasks: the control of the orientation and the control of the position of the end-effector. Different priorities can be assigned to the subtasks. In our case the control of position has higher priority with respect to orientation (the former is handled as a nonlinear constraint and thus is evaluated before the cost);
5. Description of complex constraints. It is easy to add new constraints as linear and/or nonlinear inequalities either in task or joint space. In the case of the iCub, for instance, we added a set of constraints that avoid reaching the limits of the tendons that actuate the three joints of the shoulder.

Once q_d is determined as described above, there is still the problem of generating a trajectory from the current robot configuration q to $q_{d.}$ Simultaneously, we would like to impose suitable smoothness constraints to the trajectory. This has been obtained by using the Multi-Referential Dynamical Systems approach [14], whereby two dynamical controllers, one in joint space and another in task space, evolve concurrently (Fig. 8). The coherence constraint, that is $\dot{x} = J\,\dot{q}$, with J the

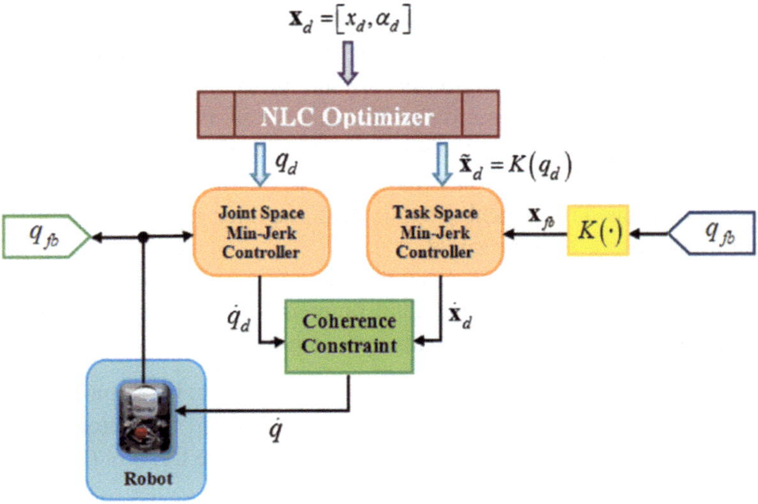

Fig. 8 The multi-referential scheme for trajectory generation. K is the forward kinematics map; q_{fb} is the vector of encoder signals

Jacobian of the kinematics map, guarantees that at each instant of time the trajectory is meaningful. This is enforced by using the Lagrangian multipliers method and can be tuned to modulate the relative influence of each controller (i.e. to avoid joint angles limits). The advantage of such a redundant representation includes the management of the singularities while maintaining a quasi-straight trajectory profile of the end-effector in the task space—reproducing a human-like behavior [15].

Differently from the work of Hersch and Billard, we designed a feedback trajectory generator instead of the VITE (Vector-Integration-To-Endpoint) method used in open loop. A complete discussion of the rationale of the modifications to the trajectory generation is outside the scope of this paper; the interested reader is referred to Pattacini et al. [16]. Reasons to prefer a feedback formulation include the possibility of smoothly connecting multiple pieces of trajectories and correcting on line for accumulation of errors due to the enforcement of the constraints of the multi-referential method.

4.1 Validation and Further Improvements

As earlier for the dynamics, we compared our method with other methods from the literature. The comparison with the method of Hersch et al. [14] was almost immediate since the work was developed on the iCub. This provides the multi-referential approach together with the VITE trajectory generation at no cost. Additionally, we included in the assessment another controller representing a more conventional strategy that uses the Damped Least-Squares (DLS) rule [17] coupled with a secondary task that comprises the joints angles limits by means of the gradient projection method [18]. This solution employs the third-party package Orocos [19], a tool for robot control that implements the DLS approach and whose public availability and compliance with real-time constraints justified its adoption as one of the reference controllers.

In the first experiment we put to test the three selected schemes in a point-to-point motion task wherein the iCub arm was actuated in the "7-DoF mode" and where the end-effector was controlled both in position and orientation. Results show that paths produced by our controller and by the DLS-based system are well restricted in narrow tubes of confidence intervals and are quite repeatable; conversely the VITE is affected by a much higher variability. Figure 9 highlights results for a set of 10 trials of a typical reaching task where the right hand is moved from a rest position to a location in front of the iCub with the palm directed downward.

Table 2 summarizes the measured in-target errors for the three cases: all the controllers behave satisfactory, but the DLS achieves lower errors because operates continuously on the current distance from the target $x_{d,}$ being virtually capable of canceling it at infinite time. On the contrary, strategies based on the interaction with an external solver bind the controller module to close the loop on an approximation $\tilde{x}_d$ of the real target that is determined by the optimization tolerances as in 11.

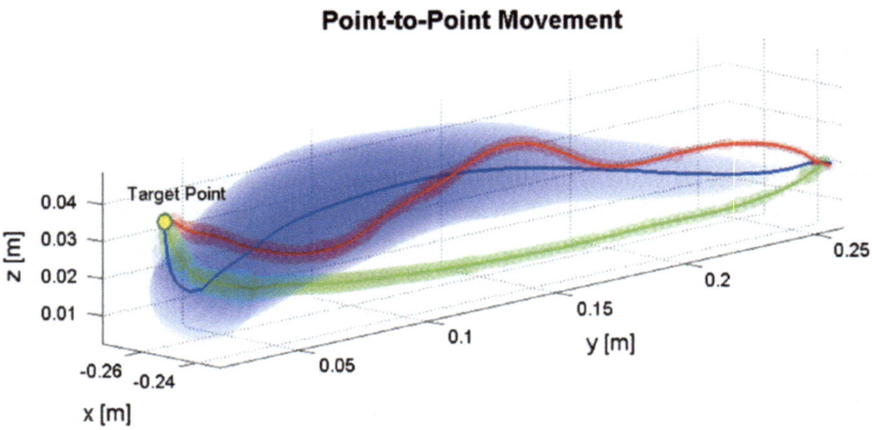

Fig. 9 Point-to-point Cartesian trajectories executed by the three controllers: the VITE-based method produces on average the *blue line*, the minimum-jerk controller result is in *green*, the DLS system using Orocos in *red*. Bands containing all the measured paths within a confidential interval of 95 % are drawn in corresponding colors. Controllers settings are: $T = 2.0$ s for the minimum-jerk system, $\alpha = 0.008$, $\beta = 0.002$, $K_P = 3$ for the VITE (see [14] for the meaning of the parameters), and $\mu = 10^{-5}$ for the damping factor of the DLS algorithm

Table 2 Mean errors along with the confidence levels at 95 % computed when the target is attained

Controller	Position error (mm)	Orientation error (rad)	Mean radius of the trajectory band (mm)
VITE	$1.3 \pm 1.4 \times 10^{-3}$	0.041 ± 0.05	10 ± 10.8
Min-jerk	$3.0 \pm 1.3 \times 10^{-3}$	0.048 ± 0.008	2.5 ± 1.5
DLS	$1.3 \pm 1.4 \times 10^{-3}$	0.016 ± 0.028	2.0 ± 1.36

An average measure of the variability of executed path is also given for the three controllers

Additional experiments tend to favor our method. For example, measuring the jerk of the resulting trajectory shows a gain of our method by 43 % from the VITE and of about 69 % from DLS. This turns out to be crucial for more complicated trajectories when speed factors make the minimum jerk controller even more advantageous.

Further improvements can be made on the quality of the inverse kinematic results by means of machine learning. As for the dynamics, we initially estimated the function K from the CAD models of the iCub. This is a good initial guess in need of refinement. The goal here is therefore to design a procedure that allows enforcing eye-hand coordination such that, whenever the robot reliably localizes a target in both cameras, it can also reach it. Here we further simplified the problem (from the visual point of view) and decided to learn only the position of the end-effector (x, y, z) since the orientation of the hand in the image is difficult to detect reliably. For this problem, the input space is defined by the position of the hand (or the target) in the two cameras (u_l, v_l) and (u_r, v_r) with respect to the current head configuration.

To sum up, having defined the input and the output space, the map M that is to be learned is:

$$(x, y, z)_H = M(u_l, v_l, u_r, v_r, T, V_s, V_g), \tag{12}$$

where $(u_l, v_l, u_r, v_r) \in \mathbb{R}^4$ represent the visual input of the position of the hand in the iCub cameras, whereas $(T, V_s, V_g) \in \mathbb{R}^3$ accounts for the proprioceptive part of the input designating the tilt, the pan and the vergence of the eyes; finally, $(x, y, z)_H \in \mathbb{R}^3$ is the Cartesian position of the hand expressed in the head-centered frame.

This map can be learned by a regression method if enough training samples are available and these can be in turn collected if we can measure (u_l, v_l, u_r, v_r) by means of vision (see Sect. 5). Some preliminary results by using a sigmoidal neural network from Matlab (Neural Network Toolbox) trained with backpropagation can be seen in Fig. 10. The training phase is carried out off-line. The neural network consists of 7 nodes in the linear input layer, 50 nodes for the hidden layer implemented with the ordinary hyperbolic tangent function and 3 nodes in the linear output layer: an overall number of 15000 samples has been employed for training and validation, whereas 5000 samples have been used for testing. The neural

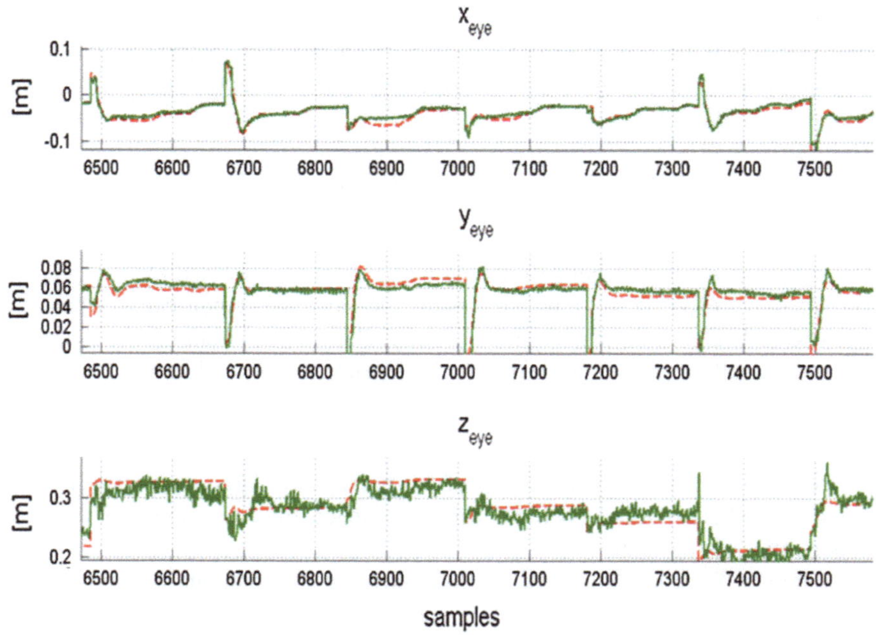

Fig. 10 The desired target (*dashed red*) and the corresponding outputs of the neural network (*green*) for the three Cartesian coordinates in the head centered frame

network provides a very good estimation of M as demonstrated by the testing phase. Notably, as expected, the z component estimation is the most affected by noise since it accounts principally for the distance of the hand from the head, a value that is not directly measured by the cameras but only indirectly from binocular disparity. The inspection of the mean and standard deviation supports this claim, i.e. mean error 0.00031 m and standard deviation of 0.0055 m for the x and y components and about twice as big for z.

In summary, it is relevant to outline here that an upcoming activity has been planned with the purpose to replace the off-line training phase with a fully online version that resorts to random features as in Gijsberts et al. [10] and will eventually make the robot learn the eye-hand coordination completely autonomously.

5 Vision

The remaining piece of information in this journey through the structure of the iCub controller is certainly vision. We strive to provide reliable estimates of object in space since this enables the control of action as presented earlier. One appealing visual cue is motion and we have been recently able to devise a method which provides motion segmentation independent from the movement of the cameras.

Our method is based on the analysis of failures of the standard Lucas-Kanade algorithm [20]. As a general rule, in order to verify that the instant velocity v of a point p has been correctly estimated, the patch W around that point in the image I_t is compared to the patch of the same size at $p + v$ in the new image I_{t+1} (where the original point is supposed to have moved). Given a suitable threshold $\Theta_{M,}$ the discrepancy measure

$$M(p) = \sum_{q \in W} (I_t(p+q) - I_{t+1}(p+v+q))^2, \tag{13}$$

is then used to evaluate whether tracking was correctly performed ($M(p) < \Theta_M$) or not ($M(p) \geq \Theta_M$) It is thus interesting to analyze empirically when the Lucas-Kanade algorithm tends to fail and why. Conclusions from this investigation will lead directly to a method to perform independent motion detection. The main empirical circumstances in which errors in the evaluation process of the optical flow arise are three:

- Speed. The instantaneous velocity of the point is too large with respect to the window where motion is being considered. Hence, the computation of temporal derivatives is difficult;
- Rotations. The motion around the point has a strong rotational component and thus, even locally, the assumption regarding the similarity of velocities fails;
- Occlusions. The point is occluded by another entity and obviously it is impossible to track it in the subsequent frame.

Tracking failures caused by high punctual speed depend exclusively on the scale of the neighborhood where optical flow is computed. This issue is usually solved by the so called pyramidal approach which applies the Lucas-Kanade method at multiple image scales. This allows evaluating iteratively larger velocities first and then smaller ones. Instead we determined empirically that when rotations cause failures in the tracking process, this is often a consequence of a movement independent from that of the observer. The third situation in which Lucas-Kanade fails, is caused by occlusions. In this context the main role in determining whether optical flow has been successfully computed is played by the speed at which such occlusion takes place.

We therefore look for points where tracking is likely to fail as soon as one of the conditions discussed is met, i.e. flow inconsistencies due to rotations or occlusions. In detail, we run Lucas-Kanade over a uniform grid on the image, perform the comparison indicated in Eq. 13 and then filter for false positives (isolated failures). The results is a set of independent moving blobs.

We tested the method both in controlled situations (a small robotic device moving linearly in front of the iCub) and, more generally, in tracking people and other moving objects in the laboratory. Figure 11 shows results of tracking with both stationary and moving cameras (therefore without and with ego-motion respectively). In the configuration considered, a linear speed of 10 cm/s corresponds to one pixel per frame in a 30 frames-per-second (fps) acquisition.

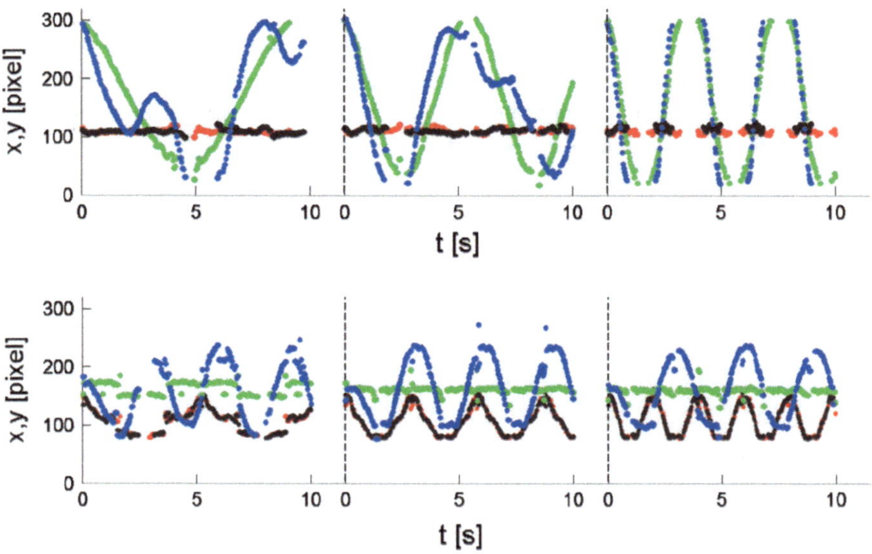

Fig. 11 Trajectories of the *x, y* coordinates of the center of mass of the areas detected as moving independently. The cart is moving parallel (*up*) or orthogonal (*down*) with respect to the image plane. The plots are reported for the following cart speeds: from *left* to *right* 20, 40, 100 cm/s. Colors legend: (1) *green* for *x* and *red* for *y* in the case of a static head; (2) *blue* for *x* and *black* for *y* in the case of a head rotating at 20 °/s

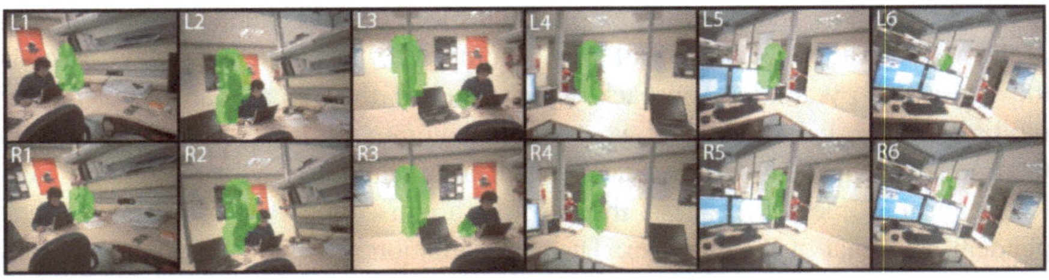

Fig. 12 A sequence of images recorded during the real time stereo tracking of a person walking in front of the iCub: six images are shown in temporal order from L1 to L6 (*left* camera) and R1 to R6 (*right* camera). The walking person is highlighted with a *green* blob using the result of proposed algorithm

Experiments were conducted up to 100 cm/s and with the iCub head adding movement up to 40 deg/s.

The sequence of images in Fig. 12 is an example of a more naturalistic tracking. In spite of the complexity of the background, it is evident from the images that our method produces robust detection of the moving target with a behavior that varies smoothly in time and is consistent with respect to the two different views acquired from the left and right cameras of the robot. In particular, the movement of the target is effectively tracked both when the person is far from the robot (frames 1 and 6) as well as when he gets closer to it (frames 2–5). Furthermore a substantial modification to light conditions exists with a maximum of brightness reached approximately at frame 4. The algorithm is robust to occlusions: this is visible at the frames in which pillars and posters cover the person. Notably, at frame 3 another person sitting at the table produces a secondary blob with his hand. This distractor is of limited size and it does not interfere with the task since the tracker is instructed to follow the largest blob in the sequence.

These are the data that at the moment the iCub uses for attention, for tracking and which are eventually passed to the reaching controller described earlier. We favored robustness to accuracy here in order to be able to run learning methods and exploration of the environment for considerable periods of time (e.g. as for collecting the 20000 samples mentioned in Sect. 4.1). Our experiments show that this goal has been fully achieved.

6 Conclusions

This paper deals with the problem of building a reliable architecture to control reaching in a humanoid robot where many degrees of freedom need to be coordinated. We have shown original solutions to vision (using motion), to kinematics (using robust optimization and a multi-referential trajectory formulation) and dynamics (by enabling impedance control from a set of FTSs). Although certain aspects of these methods are somewhat traditional, their specific application and combination is novel. We took particular care in testing all methods rigorously and comparing them with other methods in the literature.

Furthermore, the entire implementation of this software is available, following the iCub policies, as open source (GPL) from the iCub repository. These libraries and modules, besides running on the iCub, are available to the research community at large. The algorithms are almost always embedded in static libraries ready to be picked up by others.

The iCub repository can be found at http://www.icub.org and browsed on Source-Forge (http://www.sourceforge.net). Several videos of the iCub showing the methods described in this paper are available on the Internet and in particular at this site: http://www.youtube.com/watch?feature=playerprofilepage&v=LMGSok5tN4A.

Acknowledgments The research described in this paper is the result of work from several EU projects: CHRIS grant agreement number FP7-ICT-215805, Poeticon grant agreement number PF7-ICT-215843, Xperience grant agreement number PF7-ICT-270273 and EFAA grant agreement number FP7-ICT-270490. Many students and postdocs were involved in this research and are gratefully acknowledged: in particular Ugo Pattacini, Carlo Ciliberto, Vadim Tikhanoff, Matteo Fumagalli, Serena Ivaldi, Arjan Gijsberts, and Marco Randazzo.

References

1. G. Metta, L. Natale, F. Nori et al., The iCub humanoid robot: an open-systems platform for research in cognitive development. Neural Networks, special issue on Social Cognition: From Babies to Robots. **23**, 1125–1134 (2010)
2. L. Villani, J. De Shutteer, *Force control—Chapter 7*, eds. by B. *Siciliano,* O. Khatib. Springer Handbook of Robotics (Springer, 2008)
3. L. Sciavicco, B. Siciliano, *Modelling and Control of Robot Manipulators. Advance textbooks in Control and Signal Processing*, 2nd edn. (Springer, 2005)
4. G. Metta The iCub website (2010). http://www.iCub.org
5. G. Metta, D. Vernon, G. Sandini, The RobotCub approach to the development of cognition: implications of emergent systems for a common research agenda in epigenetic robotics. in *5th Epigenetic Robotics Workshop*, Nara, Japan, July 2005
6. A.R. Tilley, *The Measure of Man and Woman: Human Factors in Design* (Wiley Interscience, 2002)
7. N.G. Tsagarakis, G. Metta, G. Sandini et al., iCub: the design and realization of an open humanoid platform for cognitive and neuroscience research. Adv. Robot. **21**(10), 1151–1175 (2007)
8. P. Fitzpatrick, G. Metta, L. Natale, Towards long-lived robot genes. Robot. Auton. Syst. **56**(1), 29–45 (2008)
9. R. Featherstone, D.E. Orin, Dynamics—Chapter 2, in B. *Siciliano,* O. Khatib, Springer Handbook of Robotics (Springer, 2008)
10. A. Gijsberts, G. Metta, Incremental learning of robot dynamics using random features. in *IEEE International Conference on Robotics and Automation.* Shanghai, China, 9–13 May 2011
11. D. Nguyen-Tuong, J. Peters, M. Seeger, Computed torque control with nonparametric regression models. in *American Control Conference (ACC 2008)*, pp. 212–217, June 2008
12. A. Rahimi, B. Recht, Random features for large-scale kernel machines. Adv. Neural Inform. Process. Syst. **20**, 1177–1184 (2008)

13. A. Wätcher, L.T. Biegler, On the implementation of a primal-dual interior point filter line search algorithm for large-scale nonlinear programming. Math. Programm. **106**(1), 25–57 (2006)
14. M. Hersch, A.G. Billard, Reaching with multi-referential dynamical systems. Auton. Robots **25**(1–2), 71–83 (2008)
15. W. Abend, E. Bizzi, P. Morasso, Human arm trajectory formation. Brain **105**, 331–348 (1982)
16. U. Pattacini, F. Nori, L. Natale, G. Metta, G. Sandini, An experimental evaluation of a novel minimum-Jerk Cartesian controller for humanoid robots. in *IEEE/RSJ International Conference on Intelligent Robots and Systems*, pp. 1668–1674, Taipei, Taiwan, 18–22 Oct 2010
17. A.S. Deo, I.D. Walker, Robot subtask performance with singularity robustness using optimal damped least-squares. in *IEEE International Conference on Robotics and Automation* (1992)
18. H.Y. Lee, B.J. Yi, Y. Choi (2007) A realistic joint limit algorithm for kinematically redundant manipulators. in *IEEE International Conference on Control, Automation and Systems*
19. The Orocos website. http://www.orocos.org/kdl
20. B.D. Lucas, T. Kanade, An iterative image registration technique with an application to stereo vision. in *Proceedings of IJCAI*, pp. 674–679 (1981)

Analytical Least-Squares Solution for 3D Lidar-Camera Calibration

Faraz M. Mirzaei, Dimitrios G. Kottas and Stergios I. Roumeliotis

Abstract This paper addresses the problem of estimating the intrinsic parameters of the 3D Velodyne lidar while at the same time computing its extrinsic calibration with respect to a rigidly connected camera. Existing approaches to solve this nonlinear estimation problem are based on iterative minimization of nonlinear cost functions. In such cases, the accuracy of the resulting solution hinges on the availability of a precise initial estimate, which is often not available. In order to address this issue, we divide the problem into two least-squares sub-problems, and analytically solve each one to determine a precise initial estimate for the unknown parameters. We further increase the accuracy of these initial estimates by iteratively minimizing a batch nonlinear least-squares cost function. In addition, we provide the minimal observability conditions, under which, it is possible to accurately estimate the unknown parameters. Experimental results consisting of photorealistic 3D reconstruction of indoor and outdoor scenes are used to assess the validity of our approach.

1 Introduction and Related Work

As demonstrated in the DARPA Urban Challenge, commercially available high-speed 3D lidars, such as the Velodyne, have made autonomous navigation and mapping within dynamic environment possible. In most applications, however,

This work was supported by the University of Minnesota (DTC), and the National Science Foundation (IIS-0643680, IIS-0811946, IIS-0835637).

F.M. Mirzaei (✉) · D.G. Kottas · S.I. Roumeliotis
University of Minnesota, Minneapolis, MN 55455, USA
e-mail: faraz@cs.umn.edu

D.G. Kottas
e-mail: dkottas@cs.umn.edu

S.I. Roumeliotis
e-mail: stergios@cs.umn.edu

H.I. Christensen and O. Khatib (eds.), *Robotics Research*,
Springer Tracts in Advanced Robotics 100, DOI 10.1007/978-3-319-29363-9_11

another sensor is employed in conjunction with the 3D lidar to assist in localization and place recognition. In particular, spherical cameras are often used to provide visual cues and to construct photorealistic maps of the environment. In these scenarios, accurate extrinsic calibration of the six degrees of freedom (d.o.f.) transformation between the two sensors is a prerequisite for optimally combining their measurements.

Several methods exist for calibrating a 2D laser scanner with respect to a camera. The work of Zhang and Pless [17] relies on the observation of a planar checkerboard by both sensors. In particular, corners are detected in images and planar surfaces are extracted from the laser measurements. The image features are used to determine the normal vector and distance of the planes where the laser-scan endpoints lie. Using this geometric constraint, the estimation of the transformation between the two sensors is formulated as a non-linear least-squares problem and solved iteratively. A simplified linear least-squares solution is also provided to initialize the iterative nonlinear algorithm. More recently, Naroditsky et al. have presented a minimal approach for calibrating a 2D laser scanner with respect to a camera, using only six measurements of a planar calibration board [6]. The computed transformation is then used in conjunction with RANSAC to initialize an iterative least-squares refinement.

The existing 2D laser scanner-camera calibration methods are extended to 3D lidars in [7, 16]. In both works, a geometric constraint similar to that of [17] is employed to form a nonlinear least-squares cost function which is iteratively minimized to estimate the lidar-camera transformation. An initial estimate for the iterative minimization is determined based on a simplified linear least-squares method [16]. Specifically, the estimation of relative rotation and translation are decoupled, and then each of them is computed from a geometric constraint between the planar segments detected in the measurements of both the 3D lidar and the camera. An alternative 3D lidar-camera calibration approach is described in [11], where several point correspondences are manually selected in images and their associated lidar scans. Then, the PnP algorithm of [9] is employed to find the transformation between the camera and the 3D lidar based on these point correspondences. In a different approach, presented in [13], the structural edges extracted from 3D lidar scans are matched with the vanishing points of the corresponding 2D images to compute a coarse 3D lidar-camera transformation, followed by an iterative least-squares refinement.

The main limitation of the above methods is that they assume the 3D lidar to be intrinsically calibrated. If the lidar's intrinsic calibration is not available or sufficiently accurate, then the calibration accuracy as well as the performance of subsequent lidar-camera data fusion significantly degrades. In [7], this issue is partially addressed for the Velodyne 3D lidar by first calibrating only some of its intrinsic parameters. However, this intrinsic calibration procedure is also iterative, and no method is provided for initializing it. While several of the intrinsic parameters of a lidar may be initialized using the technical drawings of the device (if available), other parameters, such as the offset in the range measurements induced by the delay in the electronic circuits, cannot be determined in this way.

To address these limitations, in this work we propose a novel algorithm for joint estimation of both the intrinsic parameters of the Velodyne 3D lidar and the lidar-camera transformation. Specifically, we use measurements of a calibration plane at various configurations to establish geometric constraints between the lidar's intrinsic parameters and the lidar-camera 6 d.o.f. relative transformation. We process these measurement constraints to estimate the calibration parameters as follows: First, we analytically compute an initial estimate for the intrinsic and extrinsic calibration parameters in two steps. Subsequently, we employ a batch iterative (nonlinear) least-squares method to refine the accuracy of the estimated parameters.

In particular, to analytically compute an initial estimate, we relax the estimation problem by seeking to determine the transformation between the camera and each one of the conic laser scanners within the Velodyne, along with its intrinsic parameters. As a first step, we formulate a nonlinear least-squares problem to estimate the 3 d.o.f. rotation between each conic laser scanner and the camera, as well as a subset of the laser scanner's intrinsic parameters. The optimality conditions of this nonlinear least-squares form a system of polynomial equations, which we solve analytically using an algebraic-geometry approach to find all its critical points. Amongst these, the one that minimizes the least-squares cost function corresponds to the global minimum and provides us with the initial estimates for the relative rotation and the first set of intrinsic lidar parameters. In the next step, we use a linear least-squares algorithm to compute the initial estimate for the relative translation between the camera and the conic laser scanners, and the remaining intrinsic parameters.

Once all initial estimates are available, we finally perform a batch iterative joint-optimization of the lidar-camera transformation and the lidar's intrinsic parameters. As part of our contributions, we also study the observability properties of the problem and present the minimal necessary conditions for concurrently estimating the lidar's intrinsic parameters and the lidar-camera transformation. These observability conditions provide a guideline for designing high-accuracy calibration procedures. Our experimental results demonstrate that our proposed method significantly improves the accuracy of the intrinsic calibration parameters of the Velodyne lidar, as well as, the lidar-camera transformation.

2 Problem Formulation

The Velodyne HDL-64E lidar consists of 64 conic laser scanners mounted on a rotating head so that they span a 360° panoramic (azimuth) view (see Fig. 1). Each laser scanner has a horizontal offset from the axis of rotation, and a vertical offset from adjacent laser scanners. Additionally, each laser scanner points to a different elevation angle, such that, collectively, all the laser scanners cover a 27° vertical field of view. Therefore, once the lidar's head completes a full rotation, each laser scanner has swept a cone in space specified by its elevation angle. Let {L} be the

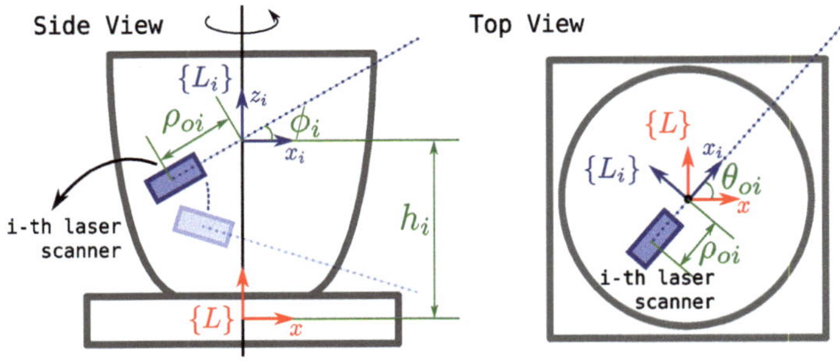

Fig. 1 Internal structure of the Velodyne 3D lidar. The total number of laser beams is 64 and they span a vertical field of view of 27 The intrinsic parameters of the Velodyne describe the measurements of each laser scanner in its coordinate frame, $\{L_i\}$, and the transformation between the Velodyne's fixed coordinate frame, $\{L\}$, and $\{L_i\}$. Note that besides the physical offset of the laser scanners from the axis of rotation, the value of ρ_{oi} depends on the delay in the electronic circuits of the lidar

lidar's fixed frame of reference whose z-axis is the axis of rotation of the sensor's head (see Fig. 1). Also, let $\{L_i\}$, i = 1,…,64, be the coordinate frame corresponding to the i-th laser scanner, such that its origin is at the center of the associated cone on the z-axis of {L} with vertical offset hi from the origin of {L}, its z-axis aligned with that of {L}, and its x-axis defining an angle θ_{oi} with the x-axis of {L}. We determine the direction of the *k*-th shot of the *i*-th laser beam from its corresponding elevation angle, ϕ_i, and azimuth measurement, θ_{ik}, and denote it with[1]:

$$^{Li}\bar{p}_k \triangleq \begin{bmatrix} \cos\phi_i \cos\theta_{ik} \\ \cos\phi_i \sin\theta_{ik} \\ \sin\phi_i \end{bmatrix}. \tag{1}$$

The distance measured by the *k*-th shot of the *i*-th laser scanner is represented by ρ_{ik}. The real distance to the object that reflects the *k*-th shot of the *i*-th laser beam is $\alpha_i(\rho_{ik} + \rho_{oi})$, where α_i is the scale factor, and ρ_{oi} is the range offset due to the delay in the electronic circuits of the lidar and the horizontal offset of each laser scanner from its cone's center. In this way, the position of the *k*-th point measured by the *i*-th laser scanner is described by

$$^{Li}\bar{p}_k = \alpha_i(\rho_{ik} + \rho_{oi})\,^{Li}\bar{p}_k. \tag{2}$$

The transformation between $\{L_i\}$ and $\{L\}$ (i.e., h_i and θ_{oi}), the scale α_i, offset ρ_{oi}, and elevation angle ϕ_i, for $i = 1, \ldots, 64$, comprise the intrinsic parameters of the

[1]Throughout this paper, $^A\mathbf{p}$ denotes the expression of a vector $\mathbf{p}$ with respect to frame $\{A\}$ and $\bar{\mathbf{p}}$ is the corresponding unit vector. $^A_B\mathbf{C}$ is the rotation matrix rotating vectors from frame $\{B\}$ to frame $\{A\}$, and $^A\mathbf{t}_B$ is the position of the origin of $\{B\}$ in $\{A\}$. $\mathbf{I}_n$ is the $n \times n$ identity matrix, and $\mathbf{0}_{m\times n}$ is $m \times n$ matrix of zeros.

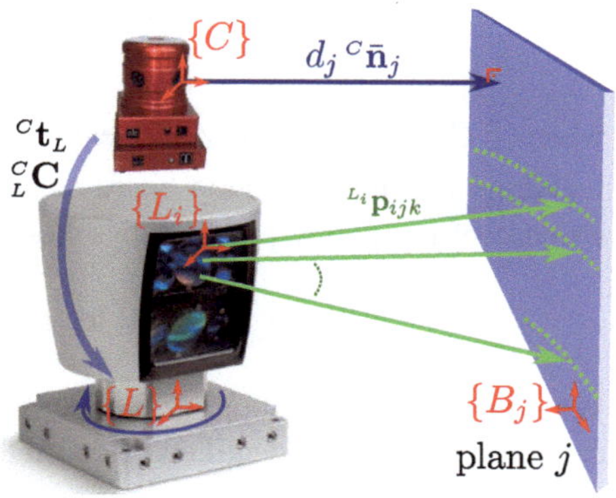

Fig. 2 Geometric constraint between the j-th plane, the Ladybug $\{C\}$, and the i-th laser scanner, $\{L_i\}$. Each laser beam is described by a vector ${}^{Li}\mathbf{p}_{ijk}$. The plane is described by its normal vector ${}^c\bar{\mathbf{n}}_j$ and its distance d_j both expressed with respect to the Ladybug

lidar that must be precisely known for any application, including photorealistic reconstruction of the surroundings. Since the intrinsic parameters supplied by the manufacturer may not be accurate,[2] in this work we estimate them along with the transformation with respect to a camera.

The Ladybug2 spherical vision system consists of six rigidly-connected calibrated cameras equipped with wide-angle lenses (see Fig. 2). The extrinsic transformations between the different cameras are provided by the manufacturer with high accuracy. Therefore, the measurements from any of the cameras can be easily transformed to the Ladybug's fixed frame of reference, denoted as $\{C\}$.

The Ladybug is rigidly connected to the lidar, and our objective is to determine the 6 d.o.f. relative transformation between the two, as well as the intrinsic parameters of the lidar. For this purpose, we employ a planar calibration board with fiducial markers, at M different configurations to establish geometric constraints between the measurements of the lidar and the Ladybug, their relative transformation, and the lidar's intrinsic parameters.

Specifically, at the j-th plane configuration, $j = 1,\ldots,M$, the fiducial markers whose positions are known with respect to the calibration board's frame of reference $\{B_j\}$, are first detected in the Ladybug's image. The 6 d.o.f. transformation between $\{C\}$ and $\{B_j\}$ is then computed using a PnP algorithm [9], from which the normal vector and the distance of the target plane in the Ladybug's frame are extracted as:

$$ {}^C\bar{\mathbf{n}}_j \triangleq {}^C_{Bj}C[0 \quad 0 \quad -1]^T \tag{3} $$

[2]Note that when the technical drawings of the lidar are available, an initial estimate for h_i, $_{oi}$, and ϕ_i can be readily obtained. Computing an initial estimate for ρ_{oi} and α_i, however, is significantly more challenging even for the manufacturer, since their values do not solely depend on the physical dimensions of the device.

$$d_j \triangleq {}^{C}\bar{\mathbf{n}}_{jt}^{TC}\mathbf{t}_{Bj} \tag{4}$$

where ${}^{A}_{B_j}\mathbf{C}$ and ${}^{c}\mathbf{t}_{B_j}$ represent the relative rotation and translation between the Ladybug and the calibration board at the j-th configuration. Consequently, in the absence of noise, any point ${}^{C}\mathbf{p}$ that lies on the j-th plane satisfies:

$${}^{C}\bar{\mathbf{n}}_{jt}^{TC}\mathbf{p} - d_j = 0. \tag{5}$$

We now turn our attention to the lidar point measurements reflected from the j-th calibration plane and identified based on the depth discontinuity. Let us denote such points as ${}^{L_i}\mathbf{p}_{ijk}$, $k = 1, \ldots, N_{ij}$, measured by the lidar's i-th laser scanner [see (2)]. Transforming these points to the Ladybug's frame, and substituting them in (5) yields:

$${}^{C}\bar{\mathbf{n}}_j^T\left({}^{C}_{Li}\mathbf{C}^{Li}\mathbf{P}_{ijk} + {}^{C}\mathbf{t}_{Li}\right) - d_j = 0 \tag{6}$$

$$\underset{\Rightarrow}{(2)}\ \alpha_i\left(\rho_{ijk} + \rho_{oi}\right){}^{C}\bar{\mathbf{n}}_{j}^{T}{}^{C}_{L_i}C^{Li}\bar{\mathbf{p}}_{ijk} + {}^{C}\bar{\mathbf{n}}_j^{T}{}^{C}\mathbf{t}_{Li} - d_j = 0 \tag{7}$$

where ${}^{C}_{L_i}\mathbf{C}$ and ${}^{C}\mathbf{t}_{L_i}$ are the relative rotation and translation between the Ladybug and the i-th laser scanner.

In addition to the camera and laser scanner measurements, the following constraints can also be used to increase the accuracy of the calibration process. Specifically, since the z-axis of $\{L_i\}$ is aligned with the z-axis of $\{L\}$, while their x-axes form an angle θ_{oi}, the following constraint holds for all ${}^{C}_{L_i}\mathbf{C}$:

$${}^{C}_{Li}C = {}^{C}_{L}CC_z(\theta_{oi}) \tag{8}$$

where $\mathbf{C}_z(\theta_{oi})$ represents a rotation around the z-axis by an angle θ_{oi}. Additionally, the origin of each laser-scanner frame lies on the z-axis of $\{L\}$ with vertical offset of h_i from the origin of $\{L\}$, resulting in the following constraint:

$${}^{C}_{L}\mathbf{C}^T\left({}^{C}\mathbf{t}_{L_i} - {}^{C}\mathbf{t}_L\right) = \begin{bmatrix}0 & 0 & h_i\end{bmatrix}^T \tag{9}$$

In the presence of noise, the geometric constraint in (7) is not exactly satisfied. Therefore, to estimate the unknown parameters, we form a constrained nonlinear least-squares cost function from the residuals of this geometric constraint over all point and plane observations [see (22)]. In order to minimize this least-squares cost, one has to employ iterative minimizers such as the Levenberg-Marquardt [8], that require a precise initial estimate to ensure convergence. To provide accurate initialization, in the next three sections we present a novel analytical method to estimate the lidar-Ladybug transformation and all intrinsic parameters of the lidar (except the elevation angle φ_i which is precisely known from the manufacturer). In order to reduce the complexity of the initialization process, we temporarily drop the constraints in (8) and (9) and seek to determine the transformation between the camera and each of the laser scanners (along with each scanner's intrinsic parameters) independently. Once an accurate initial estimate is computed, we lastly perform an iterative non-linear least-squares refinement that explicitly considers (8) and (9), and increases the calibration accuracy (see Sect. 2.4).

2.1 Analytical Estimation of Offset and Relative Rotations

Note that the term ${}^{C}\bar{\mathbf{n}}_j^{T\,C}\mathbf{t}_{L_i} - d_j$ in (7) is constant for all points k of the i-th laser scanner that hit the calibration plane at its j-th configuration. Therefore, subtracting two constraints of the form (7) for the points ${}^{Li}\mathbf{p}_{ijk}$ and ${}^{Li}\mathbf{p}_{ijl,}$ and dividing the result by the nonzero scale, α_i, yields:

$$ {}^{C}\bar{\mathbf{n}}_{j}^{T}{}^{C}_{L_i}\mathbf{C}\left(\mathbf{u}^i_{jkl} + \rho_{oi}\mathbf{v}^i_{jkl}\right) = 0 \tag{10} $$

where $\mathbf{u}^i_{jkl} \triangleq \rho^{Li}_{ijk}\bar{\mathbf{p}}_{ijk} - \rho^{Li}_{ijl}\bar{\mathbf{p}}_{ijl}$ and $\mathbf{v}^i_{jkl} \triangleq {}^{Li}\bar{\mathbf{p}}_{ijk} - {}^{Li}\,\bar{\mathbf{p}}_{ijl}$. Note that the only unknowns in this constraint are the relative rotation of the i-th laser scanner with respect to the Ladybug, ${}^{C}_{L_i}\mathbf{C}$, and its offset, ρ_{oi}. Let us express the former, ${}^{C}_{L_i}\mathbf{C}$, using the Cayley-Gibbs-Rodriguez (CGR) parameterization [12], i.e.,

$$ {}^{C}_{Li}\mathbf{C}(\mathbf{s}) = \frac{\bar{\mathbf{C}}(\mathbf{s}_i)}{1+\mathbf{s}_i^T\mathbf{s}_i}, \quad \bar{\mathbf{C}}(\mathbf{s}_i) \triangleq \left((1-\mathbf{s}_i^T\mathbf{s}_i)\mathbf{I}_3 + 2\lfloor \mathbf{s}_i \times \rfloor + 2\mathbf{s}_i\mathbf{s}_i^T\right) \tag{11} $$

where $\mathbf{s}_i^T = [s_{i1}\ s_{i2}\ s_{i3}]$ is the vector of CGR parameters that represent the relative orientation of the i-th laser scanner with respect to the Ladybug, and $\lfloor \mathbf{s} \times \rfloor$ is the corresponding skew-symmetric matrix [12]. Substituting (11) in (10), and multiplying both sides with the nonzero term $1+\mathbf{s}_i^T\mathbf{s}_i$ yields:

$$ {}^{C}\bar{\mathbf{n}}_j^T\bar{\mathbf{C}}(\mathbf{s}_i)\left(\mathbf{u}^i_{jkl} + \rho_{oi}\mathbf{v}^i_{jkl}\right) = 0 \tag{12} $$

This algebraic constraint holds exactly in the absence of noise. In that case, the method presented in Sect. 2.5 can be employed to recover the unknowns given the minimum number of measurements. In the presence of noise, however, (12) becomes:

$$ {}^{C}\bar{\mathbf{n}}_j^T\bar{\mathbf{C}}(\mathbf{s}_i)\left(\mathbf{u}^i_{jkl} + \rho_{oi}\mathbf{v}^i_{jkl}\right) = \eta^i_{jkl} \tag{13} $$

where η^i_{jkl} is a nonzero residual. In this case, we estimate $\mathbf{s}_i$ and ρ_{oi} by solving the following nonlinear least-squares problem:

$$\hat{\mathbf{s}}_i, \hat{\rho}_{oi} = \min_{\mathbf{s}_i, \rho_{oi}} J_i, \quad J_i \triangleq \frac{1}{2} \sum_{j=1}^{M} \sum_{k=1}^{\frac{N_{ij}}{2}} \sum_{l=\frac{N_{ij}}{2}+1}^{N_{ij}} \left({}^{C}\bar{\mathbf{n}}_j^T \bar{\mathbf{C}}(\mathbf{s}_i) \left(\mathbf{u}_{jkl}^i + \rho_{oi} \mathbf{v}_{jkl}^i \right) \right)^2 \tag{14}$$

where, without loss of generality, we have assumed N_{ij} is even. Note that the N_{ij} points from the i-th laser scanner, and the j-th configuration of the calibration plane are divided into two mutually exclusive groups so as to ensure that each point appears in the least-squares cost only once and hence avoid noise correlations.

When a sufficient number of plane configurations are observed, we employ a recently proposed algebraic method to directly solve this nonlinear least-squares problem without requiring initialization [14]. Specifically, we first form the following polynomial system describing the optimality conditions of (14):

$$f_{i\ell} = \frac{\partial J_i}{\partial s_{i\ell}} = 0, \quad \ell = 0, \ldots, 3 \quad \text{and} \quad s_{i0} \triangleq \rho_{oi}. \tag{15}$$

Note that the cost function in (14) is a polynomial of degree six in the elements of $\mathbf{s}_i$ and ρ_{oi}. Therefore, (15) consists of four polynomials of degree five in four variables. This polynomial system has 243 solutions that comprise the critical points of the least-squares cost function J_i, and can be computed using the eigenvalue decomposition of the so-called *multiplication matrix* (see Sect. 2.2). The globally optimal solution of the least-squares problem is the critical point that minimizes (14), and it is selected through direct evaluation of the cost function J_i. We point out that the computational complexity of solving (15) and finding the global minimum does not increase with the addition of measurements, since the degree and number of polynomials expressing the optimality conditions are fixed regardless of the number of calibration-plane configurations and laser-scanner points reflected from them. Moreover, computing the contribution of all points to the coefficients of the polynomials $f_{i\ell}, \ell = 0, \ldots, 3$, increases only linearly with the number of measurements.

2.2 Polynomial System Solver

In order to solve the polynomials describing the optimality conditions of (15), we compute the *multiplication matrix*, a generalization of the companion matrix to systems of multivariate polynomial equations, whose eigenvalues are the roots of the associated polynomial system [1]. In the following, we briefly describe an efficient method for computing the multiplication matrix. For a detailed discussion on solving systems of polynomial equations, we refer the interested reader to [4].

Let us denote a *monomial* in $\mathrm{x} = [x_1 \cdots x_n]^T$ by $x^{\gamma} \triangleq x_1^{\gamma_1} x_2^{\gamma_2} \cdots x_n^{\gamma_n}$, $\gamma_i \in \mathbb{Z}_{\geq 0}$, with degree $\sum_{i=1}^{n} \gamma_i$. A polynomial of degree d in $\mathbf{x}$ is denoted by $f = \mathbf{c}^T \mathbf{x}_d$ where $\mathbf{x}_d$ is the $\binom{n+d}{n}$-dimensional vector of monomials of degree up to and including d, and

$\mathbf{c}$ is the vector of coefficients of equal size. We assume that the given system of equations has n polynomials, denoted by $f_i = \mathbf{c}_i^T \mathbf{x}_{di} = 0, i = 1, \ldots, n$, each of them with degree d_i. The total degree of the polynomial system is $d \triangleq \max_i d_i$. By padding the coefficient vectors of f_i's with zeros, and stacking them together in $\mathbf{C}$, we can present the polynomial system in the matrix form of $\mathbf{C}\mathbf{x}_d = \mathbf{0}$.

A system of polynomial equations defines an *ideal* I as the set of all the polynomials that can be generated as $\sum_i f_i h_i$ where h_i is any polynomial in $\mathbf{x}$. Clearly the elements of the ideal become zero at the solutions of the original (generator) polynomial system. The Gröbner basis $G \triangleq \langle g_1, \ldots g_t \rangle$ of an ideal is a finite subset of the ideal such that (i) the remainder of the division of any polynomial to it is unique, (ii) any polynomial whose division by the Gröbner basis results in zero remainder, is a member of the associated ideal. The first property can be expressed as: $\varphi(\mathbf{x}) = \mathrm{r}(\mathbf{x}) + \sum_{i=1}^{t} g_i(\mathbf{x}) h_i(\mathbf{x})$ where φ is any polynomial in $\mathbf{x}$, h_i's are the quotient polynomials, and r is the unique remainder. We hereafter use the name "remainder" as the remainder of the division of a polynomial by the Gröbner basis. The Gröbner basis for an ideal generated from polynomials with integer or rational numbers can be computed using implementations of the so-called Buchberger's algorithm [4] on symbolic software packages such as Macaulay2 or Maple. Computation of the Gröbner basis for polynomials with floating-point coefficients is much more difficult due to quick accumulation of round-off errors in the Buchberger's algorithm.

The remainders of the polynomials that are not in an ideal are instrumental in finding the solutions (i.e., variety) of that ideal. It can be shown that all such remainders can be expressed as a linear combination of a specific (unique) group of monomials that comprise the so-called *normal set* [4]. The normal set can be easily obtained from the Gröbner basis of an ideal, and under mild conditions,[3] its cardinality equals the number of solutions (real and complex) of the ideal, and it will contain the monomial 1 [4, p. 43]. The important point here is that the normal set is generically fixed across different instantiations of the polynomials. Therefore, we can compute the normal set of an instance of the problem (e.g., integer or rational coefficients) and use it when the coefficients are floating point.

Let us assume that the cardinality of the normal set is s, and represent its monomials in a vector form $\mathbf{x}_B$. Then multiplication of $\mathbf{x}_B$ with a generic polynomial $\varphi(\mathbf{x})$ yields:

$$\varphi(\mathbf{x}) \cdot \mathbf{x}_B = \mathbf{M}_\varphi \mathbf{x}_B + \begin{bmatrix} h_{11} & \cdots & h_{1t} \\ \vdots & & \vdots \\ h_{s1} & \cdots & h_{st} \end{bmatrix} \begin{bmatrix} g_1 \\ \vdots \\ g_t \end{bmatrix} \tag{16}$$

[3]These conditions are: (i) the ideal must be radical, (ii) its variety must be non-empty and zero dimensional [3]. These conditions are generally satisfied for the current problem.

where h_{ij}'s are polynomials in $\mathbf{x}$, and g_i's are the elements of the Gröbner basis. In this expression, $\mathbf{M}_\varphi$ is called the *multiplication matrix* associated with φ. This relationship holds since the remainder of any polynomial (including $x^\gamma \varphi(\mathbf{x}), x^\gamma \in \mathbf{x}_B$) can be written as a linear combination of $\mathbf{x}_B$. Now, if we evaluate (16) at $\mathbf{x} = \mathbf{p}$, a solution of the ideal, all g_i's become zero, and we get $\varphi(\mathbf{p}) \cdot \mathbf{x}_B = \mathbf{M}_\varphi \mathbf{x}_B$, where $\mathbf{p}_B$ is $\mathbf{x}_B$ evaluated at $\mathbf{p}$. Clearly, $\mathbf{p}_B$ is an eigenvector of M_φ, and $\varphi(\mathbf{p})$ is the associated eigenvalue. Therefore, if we select $\varphi(\mathbf{x})$ equal to one of the variables (e.g., x_i), we can read off the x_i-coordinate of the solutions as the eigenvalues of M_φ. Further-more, depending on the ordering of the monomials when computing the Gröbner basis, $\mathbf{x}_B$ may include all first-order monomials x_1, ..., x_n. In that case, one can simultaneously read off all the coordinates of the solutions, for an arbitrary choice of φ, as long as it is nonzero and distinct at each solution of the ideal.

When the Gröbner basis is available (such as in polynomial systems with integer coefficients), one can use it directly to compute remainders of $\varphi(\mathbf{x}) \cdot \mathbf{x}_B$, and construct $\mathbf{M}_\varphi$.. This is not possible, however, when working with polynomials with floating-point coefficients. Therefore we employ the method proposed in [2] to compute $\mathbf{M}_\varphi$. We first note that some of the monomials of $\phi(\mathbf{x})$ $\mathbf{x}_B$ remain in $\mathbf{x}_B$, while some others do not. We form the vector $\mathbf{x}_R$ from the latter monomials, and write:

$$\varphi(\mathbf{x}) \cdot \mathbf{x}_B = \mathbf{M}'_\varphi \begin{bmatrix} \mathbf{x}_R \\ \mathbf{x}_B \end{bmatrix} \tag{17}$$

where $\mathbf{M}'_\varphi$ is called the unreduced multiplication matrix. Our objective is to express the remainders of $\mathbf{x}_R$ as a linear combination of $\mathbf{x}_B$ without using the Gröbner basis. For this purpose, we expand each original polynomial f_i by multiplying it with all the monomials up to degree $\ell - d_i$ (ℓ to be determined later). Clearly all these new expanded polynomials belong to the ideal generated by the original polynomials, and they have monomials up to degree ℓ. Thus, we can write them collectively in matrix form as $\mathbf{C}_e \mathbf{x}_\ell = 0$. We reorder $\mathbf{x}_\ell$ and $\mathbf{C}_e$ as:

$$\mathrm{C}_e \mathbf{x}_\ell = \begin{bmatrix} \mathbf{C}_E & \mathbf{C}_R & \mathbf{C}_B \end{bmatrix} \begin{bmatrix} \mathbf{x}_E \\ \mathbf{x}_R \\ \mathbf{x}_B \end{bmatrix} = 0 \tag{18}$$

where $\mathbf{x}_E$ are the monomials that belong neither to $\mathbf{x}_R$ nor to $\mathbf{x}_B$. Multiplying (18) with $\mathbf{N}^T$, the left null space of $\mathbf{C}_E$, and decomposing $\mathbf{N}^T \mathbf{C}_R = \mathbf{QR} = [\mathbf{Q}_1 \mathbf{Q}_2][\mathbf{R}_1^T \mathbf{0}]^T$ using QR factorization, yields:

$$[\mathbf{N}^T \mathbf{C}_R \mathbf{N}^T \mathbf{C}_B] \begin{bmatrix} \mathbf{x}_R \\ \mathbf{x}_B \end{bmatrix} = \mathbf{Q} \begin{bmatrix} \mathbf{R}_1 & \mathbf{Q}_1^T \mathbf{N}^T \mathbf{C}_B \\ 0 & \mathbf{Q}_2^T \mathbf{N}^T \mathbf{C}_B \end{bmatrix} \begin{bmatrix} \mathbf{x}_R \\ \mathbf{x}_B \end{bmatrix} = 0. \tag{19}$$

If ℓ is selected sufficiently large, $\mathbf{R}_1$ will be full rank [10], which allows us to solve (19) and find $\mathbf{x}_R$ as a function of $\mathbf{x}_B$, i.e., $\mathbf{x}_R = -\mathbf{R}_1^{-1}\mathbf{Q}_1^T N^T \mathbf{C}_B \mathbf{x}_B$. Substituting this relationship in (17) yields the multiplication matrix:

$$\mathbf{M}_\varphi = \mathbf{M}'_\varphi \begin{bmatrix} \mathbf{I}_s \\ -\mathbf{R}_1^{-1}\mathbf{Q}_1^T \mathbf{N}^T \mathbf{C}_B \end{bmatrix}. \tag{20}$$

For solving Eq. (15), we had to expand the polynomials up to degree $\ell = 15$ and arrived at a multiplication matrix $\mathbf{M}_\varphi$ of dimensions 243×243. Finally, we mention that it is possible to compute the multiplication matrix without explicit computation of the normal set. Further details on this subject and also on possible numerical instabilities and their remedies are given in [2, 10, 15].

2.3 Analytical Estimation of Scale and Relative Translation

Once the relative rotation, ${}^C_{L_i}\mathbf{C}$, and the offset, ρ_{oi}, of each laser scanner, $i = 1, \ldots, 64$ are computed, we use linear least-squares to determine the relative translation and scale from (7). Specifically, we stack together all the measurement constraints on the i-th laser scanner's scale and relative translation (from different points and calibration-plane configurations), and write them in a matrix form as:

$$\begin{bmatrix} {}^C\bar{\mathbf{n}}_1^T & (\rho_{i11} + \rho_{oi})\,{}^C\bar{\mathbf{n}}_j^T\,{}^C_{L_i}\mathbf{C}\,{}^{L_i}\bar{\mathbf{p}}_{i11} \\ {}^C\bar{\mathbf{n}}_1^T & (\rho_{i12} + \rho_{oi})\,{}^C\bar{\mathbf{n}}_j^T\,{}^C_{L_i}\mathbf{C}\,{}^{L_i}\bar{\mathbf{p}}_{i12} \\ \vdots & \vdots \\ {}^C\bar{\mathbf{n}}_M^T & (\rho_{iMN_{iM}} + \rho_{oi})\,{}^C\bar{\mathbf{n}}_j^T\,{}^C_{L_i}\mathbf{C}\,{}^{L_i}\bar{\mathbf{p}}_{iMN_{iM}} \end{bmatrix} \begin{bmatrix} {}^C\mathbf{t}_{L_i} \\ \alpha_i \end{bmatrix} = \begin{bmatrix} d_1 \\ d_2 \\ \vdots \\ d_M \end{bmatrix} \tag{21}$$

Under the condition that the coefficient matrix on the left-hand side of this equality is full rank (see Sect. 2.5), we can easily obtain the i-th laser scanner's scale factor, α_i, and relative translation, ${}^C\mathbf{t}_{L_i}$, by solving (21).

2.4 Iterative Refinement

Once the initial estimates for the transformation between the Ladybug and the laser scanners, and the intrinsic parameters of the lidar are known (Sects. 2.1–2.3), we employ an iterative refinement method to enforce the constraints in (8) and (9). Specifically, we choose the coordinate frame of one of the laser scanners (e.g., the 1-st laser scanner) as the lidar's fixed coordinate frame, i.e., $\{L\} = \{L_1\}$. Then for $\{L_i\}$, $i = 2, \ldots, 64$, we employ the estimated relative transformation with respect to the Ladybug (i.e., ${}^C_{L_i}\mathbf{C}$ and ${}^C\mathbf{t}_{L_i}$) to obtain the relative transformations between $\{L_i\}$ and $\{L\}$. From these relative transformations, we only use the z component of the translation to initialize each laser scanner's vertical offset, h_i [see (9)], and the *yaw* component of the rotation to initialize each laser scanner's θ_{oi} [see (8)].

We then formulate the following constrained minimization problem to enforce (8) and (9):

$$\begin{aligned} &\min \sum_{i,j,k} \left[\alpha_i\left(\rho_{ijk} + \rho_{oi}\right){}^{C}\bar{\mathbf{n}}_1^{T}{}^{C}_{L_i}\mathbf{C}^{L_i}\bar{\mathbf{p}}_{ijk} + {}^{C}\bar{\mathbf{n}}_j^{T}{}^{C}\mathbf{t}_{L_i} - d_j\right]^2 \\ &\text{s.t.}\ {}^{C}_{L_i}\mathbf{C} = {}^{C}_{L}\mathbf{C}\mathbf{C}_z(\theta_{oi}),\quad {}^{C}_{L}\mathbf{C}^T\left({}^{C}\mathbf{t}_{L_i} - {}^{C}\mathbf{t}_{L_L}\right) = [0\,0\,h_i]^T \end{aligned} \tag{22}$$

where the optimization variables are α_i, ρ_{oi}, θ_{oi}, h_i, $i = 2, \ldots, 64$, α_1, ρ_{o1}, ${}^{C}\mathbf{t}_L$, and ${}^{C}_{L}\mathbf{C}$.[4] Note that the constraints in (22) should be taken into account using the method of Lagrange multipliers. For the implementation details of the Levenberg-Marquardt algorithm we refer the interested reader to [8].

2.5 Observability Conditions

In this section, we examine the conditions under which the unknown lidar-Ladybug transformation and the intrinsic parameters of the lidar are observable, and thus can be estimated using the algorithms in Sects. 2.1–2.4.

2.5.1 Observation of One Plane

Suppose we are provided with lidar measurements that lie only on one plane whose normal vector is denoted as ${}^{C}\bar{\mathbf{n}}_1$. In this case, it is easy to show that the measurement constraint in (6) does not change if ${}^{C}_{L_i}\mathbf{C}$ is perturbed by a rotation around ${}^{C}\bar{\mathbf{n}}_1$, represented by the rotation matrix $\mathbf{C}'$:

$${}^{C}\bar{\mathbf{n}}_1^T\mathbf{C}'\mathbf{C}^{C}_{L_i}\mathbf{p}_{i1k} + {}^{C}\bar{\mathbf{n}}_1^{T}{}^{C}\mathbf{t}_{L_i} - d_1 = 0 \Rightarrow {}^{C}\bar{\mathbf{n}}_1^{T}{}^{C}_{L_i}\mathbf{C}^{L_i}\mathbf{p}_{i1k} + {}^{C}\bar{\mathbf{n}}_1^{T}{}^{C}\mathbf{t}_{L_i} - d_1 = 0 \tag{23}$$

The second equation is obtained from the first, since ${}^{C}\bar{\mathbf{n}}_1$ is an eigenvector of $\mathbf{C}'$, thus ${}^{C}\bar{\mathbf{n}}_1^T C' = {}^{C}\bar{\mathbf{n}}_1^T$. Therefore, when observing only one plane, any rotation around the plane's normal vector is unobservable. Similarly, if we perturb ${}^{C}\mathbf{t}_{Li}$ by a translation parallel to the plane, represented by $\mathbf{t}'$, the measurement constraint does not change:

[4] In general, the optimization should be performed over φ_i as well. However, in our experiments, we observed that the provided value of φ_i by the manufacturer is sufficiently accurate, and thus excluded it from the calibration.

$$^C\bar{\mathbf{n}}_1^{T}\,{}^C_{L_i}\mathbf{C}\,{}^{L_i}\mathbf{p}_{i1k} + {}^C\bar{\mathbf{n}}_1^{\mathrm{T}}\left({}^C\mathbf{t}_{L_i} + \mathbf{t}'\right) - d_1 = 0 \Rightarrow {}^C\bar{\mathbf{n}}_1^{T}\,{}^C_{L_i}\mathbf{C}\,{}^{L_i}\mathbf{p}_{i1k} + {}^C\bar{\mathbf{n}}_1^{\mathrm{T}}\,{}^C\mathbf{t}_{L_i} - d_1 = 0. \tag{24}$$

This relationship holds since ${}^C\bar{\mathbf{n}}_1^{\mathrm{T}}\mathbf{t}' = 0$. Therefore, when observing only one plane, any translation parallel to the plane's normal is unobservable.

2.5.2 Observation of Two Planes

Consider now that we are provided with measurements from two planes, described by ${}^C\bar{\mathbf{n}}_1$, d_1, ${}^C\bar{\mathbf{n}}_2$, d_2. If we perturb the laser scanner's relative translation with $\mathbf{t}'' \propto {}^C\bar{\mathbf{n}}_1 \times {}^C\bar{\mathbf{n}}_2$ [see (24)], none of the measurement constraints will change, since ${}^C\bar{\mathbf{n}}_1^T\mathbf{t}'' = {}^C\bar{\mathbf{n}}_2^T\mathbf{t}'' = 0$. Therefore, we conclude that the relative translation cannot be determined if only two planes are observed.

2.5.3 Observation of Three Planes

In this section, we prove that when three planes with linearly independent normal vectors are observed, we can determine all the unknowns. For this purpose, we first determine the relative orientation ${}^C_{L_i}\mathbf{C}$ and the offset ρ_{oi} and then find the scale α_i and relative translation ${}^C\mathbf{t}_{L_i}$. Let us assume that the i-th laser scanner has measured four points on each plane, denoted as $\left(\rho_{ijk}, {}^{L_i}\bar{\mathbf{p}}_{ijk}\right)$, $j = 1, 2, 3$, $k = 1, \ldots, 4$. Each of these points provides one constraint of the form (7). We first eliminate the unknown relative translation and scale, by subtracting the constraints for point $k = 1$ from $k = 2$, point $k = 2$ from $k = 3$, and point $k = 3$ from $k = 4$, and obtain:

$$^C\bar{\mathbf{n}}_j^{T}\,{}^C_{L_i}\mathbf{C}\left(\mathbf{u}^i_{j12} + \rho_{oi}\mathbf{v}^i_{j12}\right) = 0 \tag{25}$$

$$^C\bar{\mathbf{n}}_j^{T}\,{}^C_{L_i}\mathbf{C}\left(\mathbf{u}^i_{j23} + \rho_{oi}\mathbf{v}^i_{j23}\right) = 0 \tag{26}$$

$$^C\bar{\mathbf{n}}_j^{T}\,{}^C_{L_i}\mathbf{C}\left(\mathbf{u}^i_{j34} + \rho_{oi}\mathbf{v}^i_{j34}\right) = 0 \tag{27}$$

where $\mathbf{u}^i_{jkl} \triangleq \rho^{L_i}_{ijk}\bar{\mathbf{p}}_{ijk} - \rho^{L_i}_{ijl}\bar{\mathbf{p}}_{ijl}$, $\mathbf{v}^i_{jkl} \triangleq {}^{L_i}\bar{\mathbf{p}}_{ijk} - {}^{L_i}\bar{\mathbf{p}}_{ijl}$, and $j = 1, 2, 3$. Note that ${}^{L_i}\bar{\mathbf{p}}_{ijk}$ and ${}^{L_i}\bar{\mathbf{p}}_{ijl}$ lie on the intersection of the unit sphere and the cone specified by the beams of the i-th laser scanner. Since the intersection of a co-centric unit sphere and a cone is always a circle, we conclude that all $\mathbf{v}^i_{jkl}$ for a given i given belong to a plane and have only two degrees of freedom. Thus, we can write $\mathbf{v}^i_{j34}$ as a linear combination of $\mathbf{v}^i_{j12}$ and $\mathbf{v}^i_{j23}$, i.e.,

$$\mathbf{v}^i_{j34} = a\mathbf{v}^i_{j12} + b\mathbf{v}^i_{j23} \tag{28}$$

for some known scalars a and b. Substituting this relationship in (27), and using (25)–(26) to eliminate the terms containing ρ_{oi} yields:

$$^C\bar{\mathbf{n}}^T_j {}^C_{L_i}\mathbf{C}\left(\mathbf{u}^i_{j34} - a\mathbf{u}^i_{12} - b\mathbf{u}^i_{j23}\right) = 0 \tag{29}$$

for j = 1, 2, 3. The only unknown in this equation is the relative orientation ${}^C_{L_i}\mathbf{C}$ of the i-th laser scanner. These equations are identical to those for orientation estimation using line-to-plane correspondences, which is known to have at most eight solutions that can be analytically computed when $^C\bar{\mathbf{n}}_j$, j = 1, 2, 3, are linearly independent [3]. Once ${}^C_{L_i}\mathbf{C}$ is known, we can use any of (25)–(27) to compute the offset ρ_{oi}. Finally, the scale and the relative translation can be obtained from (21).

3 Experiments

In order to validate the proposed calibration method, we conducted a series of experiments. Specifically, we rigidly connected a Velodyne 3D lidar and a Ladybug2 spherical vision system, and recorded measurements of a 36” × 40” calibration plane with 16 fiducial markers at 40 different configurations. By processing the Ladybug’s images using a PnP algorithm followed by a least-square refinement [9], we computed the normal vector and the distance of the calibration plane at each configuration. We then identified the approximate location of the calibration plane in the lidar scans based on a coarse prior estimate for the relative rotation of the Velodyne and the Ladybug. Within these approximate locations, we detected the lidar data points reflected from the calibration plane, based on their depth discontinuity.

Once the Velodyne’s measurements for each configuration of the calibration plane were available, we used the methods described in Sects. 2.1–2.4 to accurately estimate the lidar’s intrinsic parameters and the lidar-camera transformation. Note, however, that in order to increase the robustness of our algorithm to outliers, we did not directly use the raw laser points measured by the lidar. Instead, for each laser scanner, we fit small line segments to the intersection of the laser scanner’s beam and the calibration plane, and used the endpoints of these line segments as the lidar’s measurements.[5]

[5]Note that in general the intersection of the cone induced by the laser scanner’s beam with a plane results in a conic section, and not a straight line. However, in practical situation this conic section can be approximated with a sequence of straight line segments.

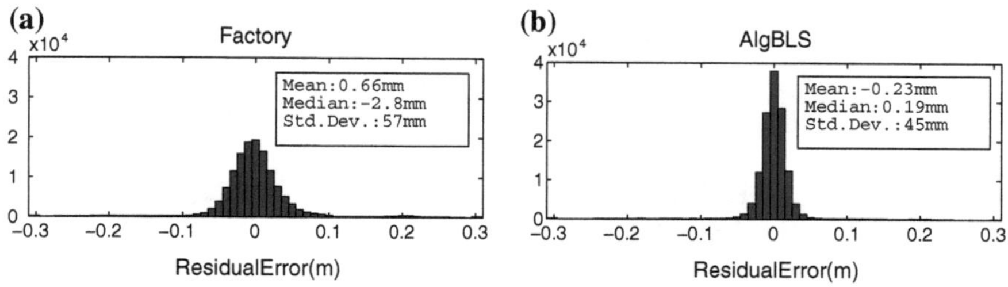

Fig. 3 Histograms of the residual error for the lidar points belonging to 40 planar surfaces using the intrinsic lidar parameters from: **a** the manufacturer; **b** the proposed method

We compare the accuracy of the estimated intrinsic lidar parameters with those provided by the manufacturer. For this purpose, we transform the raw lidar measurements to the lidar's Euclidean frame [see (2)] using both the estimated and the manufacturer-provided intrinsic parameters. We then fit planes to the lidar points belonging to the calibration planes and use the residual fitting error to evaluate the accuracy of the estimated intrinsic parameters. In Fig. 3, the histogram of these residual errors for the manufacturer-provided and the estimated intrinsic parameters are shown, clearly demonstrating the superior accuracy of our method.

To further evaluate the performance of our calibration algorithm, we created photorealistic reconstructions of several indoor and outdoor scenes from the University of Minnesota campus (see Fig. 4). For each scene, the raw measurements of the lidar are first transformed to Euclidean coordinates using the estimated intrinsic parameters of the lidar and then they are expressed in the Ladybug's frame of reference. In the next step, the lidar points are overlaid on the spherical image provided by the camera to associate them with an image pixel. Note, however, that after this step many of the image pixels will not be associated with any lidar points due to the low resolution of the lidar scans compared to the Ladybug's images. We assigned such "orphan" pixels a 3D point obtained through linear interpolation of the surrounding lidar points. The final result is a set of image pixels with 3D coordinates (i.e., 3D pixels). Finally, we converted the 3D pixels to 3D surfaces using Delaunay triangulation [5]. In Fig. 4, a selection of the reconstructed surfaces are shown for indoor and outdoor scenes. Note that white gaps in the reconstructed surfaces result from missing lidar measurements due to occlusion or specular reflection of the laser beams from glass and shiny surfaces.

4 Conclusions and Future Work

In this paper, we presented a novel method for intrinsic calibration of a Velodyne 3D lidar and extrinsic calibration with respect to a camera. Specifically, we developed an analytical method for computing a precise initial estimate for both the lidar's intrinsic parameters and the lidar-camera transformation. Subsequently, we

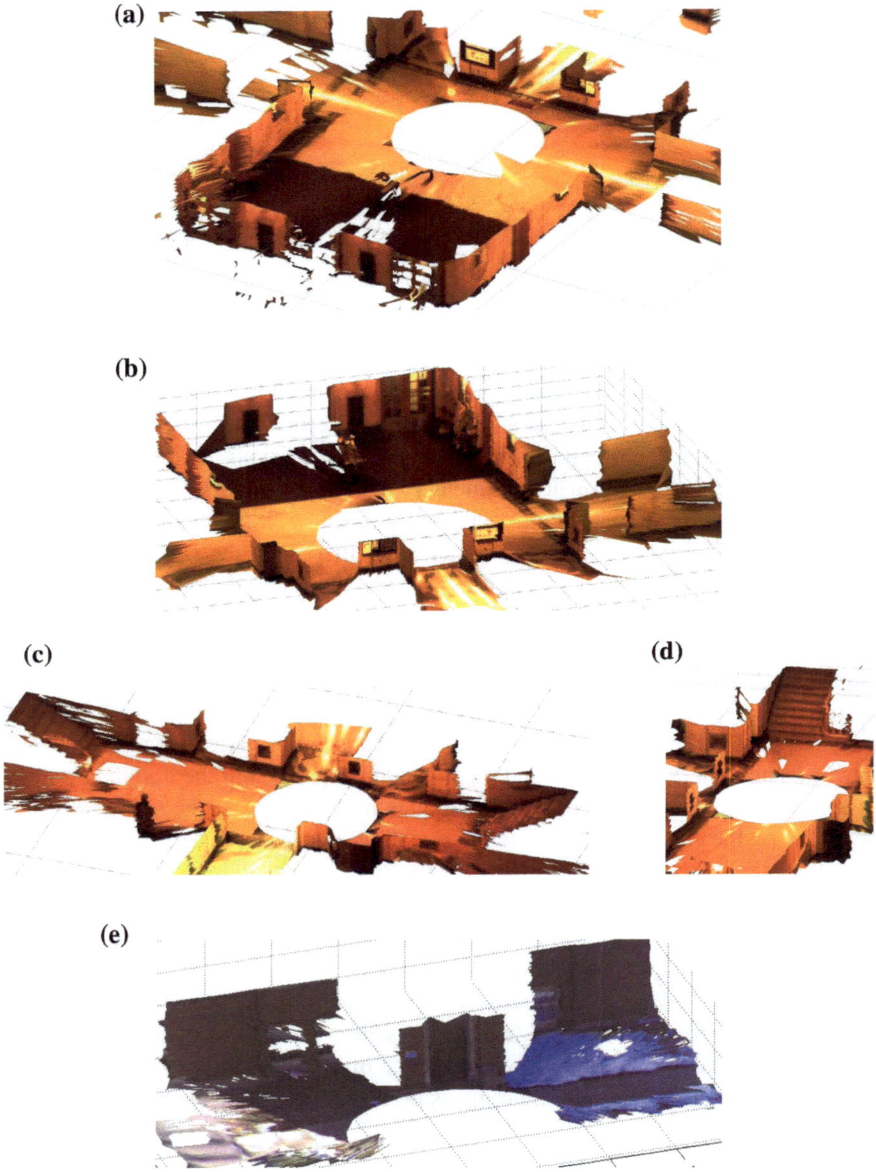

Fig. 4 A selection of the results from photorealistic reconstruction of indoor and outdoor scenes (best viewed in *color*). The *white gaps* are the regions where at least one of the sensors did not return meaningful measurements (e.g., due to occlusion, specular reflections, or limited resolution and field of view). Note that the depth of the scene can be inferred from the dotted grids. **a**, **b** Center of a building with several corridors, viewed from different directions; **c**, **d** A indoor scene containing two stair cases, viewed from two different directions; **e** An outdoor scene with snow on the ground

used these estimates to initialize an iterative nonlinear least-squares refinement of all the calibration parameters. Additionally, we presented an observability analysis to determine the minimal conditions under which it is possible to estimate the calibration parameters. Experimental results from both indoor and outdoor scenes are used to demonstrate the achieved accuracy of the calibration process by photorealistic reconstruction of the observed areas. Optimally combining multiple images and lidar scans over consecutive time steps for mapping large areas while at the same time increasing the 3D points' resolution and revealing occluded areas, is part of our ongoing research work.

References

1. W. Auzinger, H.J. Stetter, An elimination algorithm for the computation of all zeros of a system of multivariate polynomial equations, in *Proceedings of International Conference on Numerical Mathematics*. Singapore (1988), pp. 11–30
2. M. Byröd, K. Josephson, K. Åström, A column-pivoting based strategy for monomial ordering in numerical Gröbner basis calculations, in *Proceedings of the European Conference on Computer Vision*, Marseille, France (2008), pp. 130–143.
3. H.H. Chen, Pose determination from line-to-plane correspondences: existence condition and closed-form solutions. IEEE Trans. Pattern Anal. Mach. Intell. **13**(6), 530–541 (1991)
4. D. Cox, J. Little, D. O'Shea, *Using Algebraic Geometry* (Springer, 2004)
5. M. de Berg, O. Cheong, M. van Kreveld, M. Overmars, *Computational Geometry: Algorithms and Applications* (Springer, 2008)
6. O. Naroditsky, A. Patterson IV, K. Daniilidis, Automatic alignment of a camera with a line scan lidar system, in *Proceedings of the IEEE International Conference on Robotics and Automation*. Shanghai, China (2011)
7. G. Pandey, J. McBride, S. Savarese, R. Eustice, Extrinsic calibration of a 3d laser scanner and an omnidirectional camera, in *Proceedings of the IFAC Symposium on Intelligent Autonomous Vehicles*. Lecce, Italy (2010)
8. W.H. Press, S.A. Teukolsky, W.T. Vetterling, B.P. Flannery, *Numerical Recipes in C* (Cambridge University Press, Cambridge, 1992)
9. L. Quan, Z.D. Lan, Linear n-point camera pose determination. IEEE Trans. Pattern Anal. Mach. Intell. **21**, 774–780 (1999)
10. G. Reid, L. Zhi, Solving polynomial systems via symbolic-numeric reduction to geometric involutive form. J. Symb. Comput. **44**(3), 280–291 (2009)
11. D. Scaramuzza, A. Harati, R. Siegwart, Extrinsic self calibration of a camera and a 3d laser range finder from natural scenes, in *Proceedings of the IEEE/RSJ International Conference on Intelligent Robots and Systems*. San Diego, USA (2007), pp. 4164–4169
12. M.D. Shuster, A survey of attitude representations. J. Astronaut. Sci. **41**(4), 439–517 (1993)
13. I. Stamos, L. Liu, C. Chen, G. Wolberg, G. Yu, S. Zokai, Integrating automated range registration with multiview geometry for the photorealistic modeling of large-scale scenes. Int. J. Comput. Vision **78**(2), 237–260 (2008)
14. N. Trawny, S.I., Roumeliotis, On the global optimum of planar, range-based robot-to-robot relative pose estimation, in *Proceedings of the IEEE International Conference on Robotics and Automation*. Anchorage, AK (2010)
15. N. Trawny, X.S. Zhou, S.I. Roumeliotis, 3D relative pose estimation from six distances, in *Proceedings of the Robotics*: *Science* and *Systems*. Seattle, WA (2009)

16. R. Unnikrishnan, M. Hebert, Fast extrinsic calibration of a laser rangefinder to a camera. Technical report, Carnegie Mellon University, Robotics Institute (2005)
17. Q. Zhang, R. Pless, Extrinsic calibration of a camera and laser range finder (improves camera calibration), in *Proceedings of the IEEE/RSJ International Conference on Intelligent Robots and Systems*. Sendai, Japan (2004), pp. 2301–2306

Tactile Object Recognition and Localization Using Spatially-Varying Appearance

Zachary Pezzementi and Gregory D. Hager

Abstract In this work, we present a new method for doing object recognition using tactile force sensors that makes use of recent work on "tactile appearance" to describe objects by the *spatially-varying appearance* characteristics of their surface texture. The method poses recognition as a localization problem with a discrete component of the state representing object identity, allowing the application of sequential state estimation techniques from the mobile robotics literature. Ideas from geometric hashing approaches are incorporated to enable efficient updating of probabilities over object identity and pose. The method's strong performance is demonstrated experimentally both in simulation and using physical sensors.

1 Introduction

Haptic object recognition has long been a goal of robotics research, and the important role of tactile information has been recognized for decades [3]. Nonetheless, most existing haptic recognition techniques use tactile information only for localizing contact points and estimating local surface normals or curvature to constrain the object geometry [1, 4–6, 9, 11]. Several researchers have used sequential state estimation techniques to localize the pose of a known object using touch sensing [7, 10, 13]. In recent work [16], we added object identity to the state being estimated in such an approach; this led to a geometry-based object recognition method that used occupancy grid maps as the underlying object representation. This method works very well in 2D, but there are computational challenges in scaling it to 3D, so we wished to incorporate other sources of information.

Only recently (e.g., in [15, 17]) has the potential to use tactile force sensors to characterize surface textural properties been realized, giving a notion of "tactile

Z. Pezzementi (✉) · G.D. Hager
Johns Hopkins University, Baltimore, MD, USA
e-mail: zap@cs.jhu.edu

G.D. Hager
e-mail: hager@cs.jhu.edu

H.I. Christensen and O. Khatib (eds.), *Robotics Research*,
Springer Tracts in Advanced Robotics 100, DOI 10.1007/978-3-319-29363-9_12

appearance" inspired by appearance-based recognition techniques from the computer vision literature. The work in this paper builds upon the idea of recognition as localization and incorporates appearance information into a new method that characterizes the *spatially-varying appearance* (SVA) characteristics of object surfaces for recognition.

Our approach was inspired in part by geometric hashing techniques, a review of which is provided in [20]. In standard geometric hashing, a basis for 3D space is formed from a set of three 3D points. Our measurements are much more informative than just contact point locations though, so we can take advantage of this extra information to greatly improve efficiency. Though two contact points are not sufficient to define a basis in a 3D space, they can be used to constrain a transformation up to one degree of freedom of uncertainty; our contact points also have associated surface normal estimates, which can in many cases be used to constrain this last degree of freedom. Additionally, we have a tactile image associated with each point, giving it an appearance signature that can be used to distinguish individual points. We therefore extend the geometric hashing algorithm by incorporating this additional information as probabilistic constraints, maintaining the advantage of fast lookup times while reducing space requirements from $O\left(\binom{N}{3}\right)$ to $O\left(\binom{N}{2}A^2\right)$, where A is a factor of the ambiguity of appearance of a surface patch, explained in Sect. 3.2.

1.1 Sensing Tactile Appearance

This work used a capacitive sensor system made by Pressure Profile Systems, consisting of a 6×6 array of individual pressure sensors, each of which is square with 2 mm sides, shown in Fig. 1. The physical sensors and a simulation thereof were used in the experiments presented here. Further details on the sensors and the simulator are available in [14].

In [15], the simulation environment from [14] was extended to full robotic exploration using touch sensing, which required the use of a set of surface contact controllers to collect consistent sensor readings. The goal of these controllers is to produce as consistent a tactile image as possible each time the same region of an object surface is sensed, regardless of the angle of approach. The controllers developed for this task control the location and orientation of the sensor up to rotation about the sensor normal (since this rotation can not be controlled for without knowing the object pose), leaving invariance to this final degree of freedom to the appearance representation. In brief, these controllers alternate between pressing the sensor against the object surface under PD control to achieve a target average pressure and reorienting the sensor normal to align with a local estimate of the object surface normal; these steps are iterated until convergence.

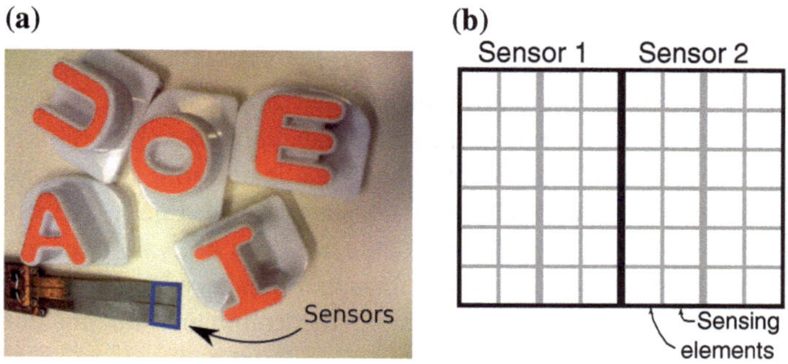

Fig. 1 **a** shows set of raised letters used in the geometry experiments alongside our tactile sensor system, with the sensing area highlighted in *blue*. **b** shows the layout of sensor elements within that highlighted area. Only the central 6×6 element region was used for recognition. **a** Letters and sensors. **b** Sensor layout

Descriptors are extracted from the sensor readings to describe the object surface's local appearance properties. This work uses two descriptors from [15] that account for the aforementioned required invariance to rotation. The first, called moment-normalized (MN), uses spatial moments to determine a major axis for the image; then the image is rotated to bring this major axis to a canonical orientation, and the resulting image is used as a descriptor. The second descriptor, moment-normalized translation-invariant (MNTI), begins with the same steps as MN. Then the result is padded, the 2D Fourier transform is taken, and the magnitudes of the low-frequency coefficients are used as the descriptor. Due to the invariances introduced by discarding phase information, MNTI is more robust to small translations of the sensor with respect to the object surface. More details on the exploration process and the descriptors' formulation and rationale can be found in [15].

2 Spatially Varying Appearance (SVA) and Bayes Filters

A Bayes filtering approach is used to maintain estimates of the unknown object's state, x_t, which consists of the identity and pose of the unknown object, at each time step. During each time step, a command, u_t, is sent to the robot, and a sensor measurement, z_t, is received as a result. Following the notation of [19], the "belief" of the state, $bel(x)$, is updated as

$$\overline{bel}(x_t) = \int \Pr(x_t|u_t, x_{t-1}, \mathbf{M}) bel(x_{t-1}) dx_{t-1} \tag{1}$$

$$bel(x_t) = \eta \frac{\Pr(x_t|z_t, \mathbf{M})}{\Pr(x_t, \mathbf{M})} \overline{bel}(x_t) \tag{2}$$

$$= \eta \ \Pr(x_t|z_t, \mathbf{M})\overline{bel}(x_t) \quad (3)$$

where η is a (different) normalization constant in Eqs. 2 and 3. Equation 2 is obtained by applying Bayes rule to the standard belief update [19]. In this particular application, the prior on object identity and pose is uniform, so the $\Pr(x_t, \mathbf{M})$ term can be folded into the normalization constant to get Eq. 3. We consider the explored object to be fixed in space, so the state does not vary over time. Each u_t therefore contributes only kinematic information; since we consider the resolved end effector position to be available in z_t, the commands do not contribute to recognition.

Single sensor readings provide only weak constraints on the possible pose of the object. The pose can be fully constrained by triplets of sensor readings, but this would add extreme space requirements. $\Pr(x_t|z_t, \mathbf{M})$ is therefore estimated from the constraints imposed by pairs of sensor readings received so far. The mapping from measurements to states is evaluated as

$$\Pr(x_t|z_t, \mathbf{M}) = \Pr(x_t|\{\text{pair}(z_i, z_t), \text{pair}(z_t, z_i);\ i = 1, \ldots t-1\}, \mathbf{M}) \quad (4)$$

$$= \prod_{i=1,\ldots,t-1} \Pr(x_t|\text{pair}(z_t, z_i), \mathbf{M}) \prod_{i=1,\ldots,t-1} \Pr(x_t|\text{pair}(z_i, z_t), \mathbf{M}) \quad (5)$$

where Eq. 5 follows from Eq. 4 by applying Bayes' rule and assuming the constraints of all pairs are conditionally independent given the state. Estimation of the individual probabilities in Eq. 5 is driven by our maps, $\mathbf{M}$, which contain information about surface patch pairs acquired during training. The training phase consists of collecting a large number of sensor readings that cover the surface of each object to be recognized. The map contains characterizations of pairs of surface patches from each object along with the identity of that object and their location on its surface. During testing, the identity and pose of the unknown object are then constrained by matching observed pairs of surface regions to map pairs, denoted $\mathbf{m}_{[\cdot]}$:

$$\begin{aligned} &\Pr(x_t|\text{pair}(z_{a1}, z_{b1}), \mathbf{M}) \\ &= \sum_{\mathbf{m}_{a2}, \mathbf{m}_{b2} \in \mathbf{M}} \underbrace{\Pr(x_t|\text{match}(\text{pair}(z_{a1}, z_{b1}), \text{pair}(\mathbf{m}_{a2}, \mathbf{m}_{b2}))).}_{\text{match constraint}} \\ &\qquad \underbrace{\Pr(\text{match}(\text{pair}(z_{a1}, z_{b1}), \text{pair}(\mathbf{m}_{a2}, \mathbf{m}_{b2})))}_{\text{match likelihood}} \end{aligned} \quad (6)$$

First the match likelihood term will be discussed in Sect. 3, then the distributions imposed on the state space by the match constraint term are covered in Sect. 4.

3 Mapping Spatially-Varying Appearance

The purpose of the SVA map is to evaluate Pr(match(pair(z_{a1}, z_{b1}), pair(z_{a2}, z_{b2}))), the probability that two pairs of sensor readings correspond to the same pair of regions on an object surface. This matching, since it makes use of appearance information, depends on the characterization of local appearance to be consistent over a local region; locations nearby a given point on the object surface are assumed to have a similar appearance characterization. This effectively means that the object surface must be sampled densely enough during training for the appearance characterization used to be consistent across regions whose size corresponds to the sampling resolution in the neighborhood.

3.1 Using Surface Patch Pair Statistics

Each sensor reading consists of a tactile image and a 3D translation and orientation of the sensor in the robot frame. Readings are collected using the controllers described in Sect. 1.1, so rotation about the sensor normal is not controlled for. The geometry of a pair of surface patches a and b is therefore described by the positions of the two patches, p_a and p_b, and their estimated surface normals, n_a and n_b, shown in Fig. 2.

Tactile appearance is described through association with appearance classes in the form of clusters $\mathbf{c}_i$ in the space of appearance descriptors (described in Sect. 1.1). Each cluster corresponds to an appearance class of physical surfaces that gives rise to measurements with certain characteristics picked out by the descriptor being used. We can evaluate the likelihood of a measurement belonging to appearance class i as $\Pr(z_t|\mathbf{c}_t)$. $\mathbf{c}(\mathbf{v}_j)$ will be taken to represent the set of those likelihoods for the appearance feature $\mathbf{v}_j$ extracted from measurement z_j.

Unknown objects may be encountered in any pose, so the map of surface patch pairs is indexed by quantities independent of pose. Let $\mathbf{v}_a$ and $\mathbf{v}_b$ be the features describing each patch's appearance. Regardless of the pose of the object, these values and the distance between the points, δ_{ab}, should not change. Pairs are then indexed by $\mathbf{c}(\mathbf{v}_a)$, $\mathbf{c}(\mathbf{v}_b)$, and δ_{ab}, ordered to distinguish pair(z_a, z_b) from pair(z_b, z_a).

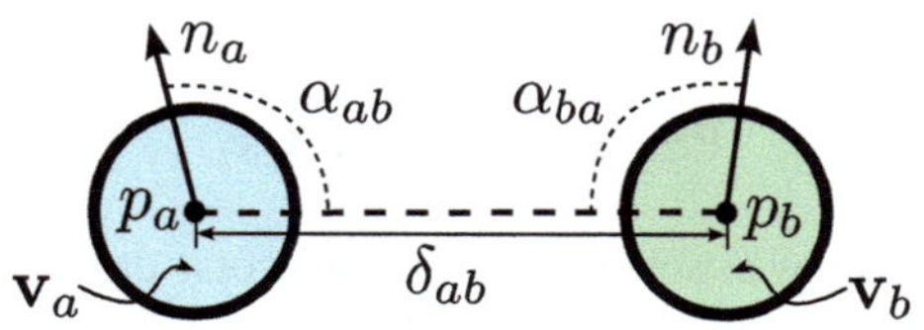

Fig. 2 Illustration of the relevant geometry for dealing with a pair of surface patches, labeled a and b: Each has a centroid $p_{[\cdot]}$, appearance feature $\mathbf{v}[\cdot]$ (visualized by *color*), and surface normal estimate $n_{[\cdot]}$. δ_{ab} denotes the distance between the patch centroids. The angle between each patch's surface normal and the vector between the patches (*dotted line*) is marked as $\alpha_{[\cdot]}$

3.2 Appearance Class Likelihoods

Now we wish to model $\Pr(z_t|\mathbf{c}_t)$, the surface patch measurement likelihoods associated with each appearance class, where appearance classes are defined by clusters in the space of appearance features. A hard clustering method, such as *k*-means, is excessively restrictive, as the appearance class of many inputs may be legitimately ambiguous, so we opt for a soft clustering approach that only associates a feature with the most likely clusters.

Let the affinity between features $\mathbf{v}_a$ and $\mathbf{v}_b$, $aff(\mathbf{v}_a, \mathbf{v}_b)$ be given by their inner product, $\langle \mathbf{v}_a, \mathbf{v}_b \rangle$. We use Partitioning Around Medoids [12] to form n_C clusters from the set of all features acquired in training using $aff(\cdot, \cdot)$, each represented by a medoid med_i, such that each feature $\mathbf{v}_j$ is associated with the nearest medoid by its membership m_j. Affinities of members of a cluster to the medoid were assumed to be distributed roughly as a Gaussian with mean 1 and standard deviation, ψ_i, computed for each cluster $\mathbf{c}_i$ as

$$\psi_i = \frac{\sum_j (1 - aff(\mathbf{v}_j, med_i))^2 \mathrm{Ind}(m_j, i)}{\sum_j \mathrm{Ind}(m_j, i)} \tag{7}$$

where Ind(i, j) is an indicator function equal to 1 if $i = j$ and 0 otherwise. Then the appearance class likelihoods of each feature are given initially by

$$\Pr(\mathbf{v}_j|\mathbf{c}_i) = \frac{1}{\sqrt{\pi\psi_i}} exp\left(-\frac{(1 - aff(\mathbf{v}, med_i))^2}{2\psi_i} \right) \tag{8}$$

$$\Pr(\mathbf{c}_i|\mathbf{v}) = \eta \frac{\Pr(\mathbf{v}_j|\mathbf{c}_i)}{\sum_j \Pr(\mathbf{v}_j|\mathbf{c}_j)} \tag{9}$$

where η is a normalization constant. Unlikely matches are then pruned away by setting

$$\mathrm{best}_{i,j} = \max_i \Pr(\mathbf{v}_j|\mathbf{c}_i) \tag{10}$$

$$\mathrm{Prune}(\mathbf{v}_j|\mathbf{c}_i) = \begin{cases} \Pr(\mathbf{c}_i|\mathbf{v}_j) & \Pr(\mathbf{c}_i|\mathbf{v}_j) > T_a \mathrm{best}_{i,j} \\ 0 & \text{otherwise} \end{cases} \tag{11}$$

$$\Pr(z_j|\mathbf{c}_i) = \eta \, \mathrm{Prune}(\mathbf{v}_j|\mathbf{c}_i) \tag{12}$$

In our experiments, T_a was set to 0.75. In the worst case, e.g. if all appearance classes have the same likelihood, this procedure can produce a match to every class, making the appearance ambiguity mentioned in Sect. 1, A, equal n_C in the worst case. In practice, however, many fewer matches are common.

3.3 Matching Distances

Distances were matched using kernelized histograms. The range of possible values was discretized into a set of n_{DB} uniform regions with distance bin centers

$$\mathrm{bin}D_i = \mathrm{minDist} + (\mathrm{maxDist} - \mathrm{minDist})\frac{i+0.5}{n_{DB}} \tag{13}$$

and δ_{ab} was associated with the nearest bins through linear interpolation to give degrees of association with each bin, $\Pr(\mathrm{bin}D_i|\delta_{ab})$.

3.4 Putting it all Together

Let CA and CB be random variables corresponding to the appearance classes of surface patches a and b respectively for both the prospectively matching pairs and D be another random variable corresponding to the distance between points in the pairs. Marginalizing over appearance and distance classes gives

$$\begin{aligned} &\Pr(\mathrm{match}(\mathrm{pair}(z_{a1}, z_{b1}), \mathrm{pair}(z_{a2}, z_{b2}))) \\ &\quad = \sum_{i,j,k} \Pr(\mathrm{pair}(z_{a1}, z_{b1}), \mathrm{pair}(\mathbf{m}_{a2}, \mathbf{m}_{b2}), CA = \mathbf{c}_i, CB = \mathbf{c}_j, D = \mathrm{bin}D_k \end{aligned} \tag{14}$$

Since the observed measurements are independent of the map measurements given the appearance and distance classes, each of which is independent of the others, this can be manipulated into the form

$$\begin{aligned} &\Pr(\mathrm{match}(\mathrm{pair}(z_{a1}, z_{b1}), \mathrm{pair}(\mathbf{m}_{a2}, \mathbf{m}_{b2}))) \\ &\quad = \sum_{i=1}^{n_C} (\Pr(z_{a1}|CA = \mathbf{c}_i) \Pr(\mathbf{m}_{a2}|CA = \mathbf{c}_i) \Pr(CA = \mathbf{c}_i). \\ &\qquad \sum_{j=1}^{n_C} (\Pr(z_{b1}|CB = \mathbf{c}_j) \Pr(\mathbf{m}_{b2}|CB = \mathbf{c}_j) \Pr(CB = \mathbf{c}_j). \\ &\qquad \sum_{k=1}^{n_{DB}} (\Pr(\delta_{a1b1}|D = \mathrm{bin}D_k) \Pr(\delta_{a2b2}|D = \mathrm{bin}D_k) \Pr(D = \mathrm{bin}D_k))) \end{aligned} \tag{15}$$

This provides a way to evaluate the probability of each pair of observed points corresponding to pairs of regions in the object maps. The appearance likelihoods can be computed by evaluating Eq. 12, and the distance likelihoods can be obtained from the joint distribution of Sect. 3.3 as $\Pr(\delta_{ab}, \mathrm{bin}D_k) = \Pr(\delta_{ab}|\mathrm{bin}D_k)$ $\Pr(\mathrm{bin}D_k)$. In our experiments, the class priors, $\Pr(\mathbf{c}_i)$ for CA and CB (these distributions are taken to be equal), were assumed to be uniform. The mapping can be efficiently implemented using a hash multimap, indexed by bin numberings, to support fast lookups without using excessive storage when the space is sparsely covered (particularly, e.g., when there are many appearance classes).

4 Recognition and Localization from SVA Maps

Given a matched pair of surface patches, pair(z_{a1}, z_{b1}) and pair(z_{a2}, z_{b2}) we now wish to estimate the set of rigid transformations that would align them.

4.1 Initial Alignment of Surface Patch Pairs

We begin with a version of the method of [2] to align 3D point clouds, simplified to the case of two points. Continuing with the notation of Sect. 3, this procedure gives a rotation, R_1, that is effective for aligning the points of contact, p_{a1} with p_{a2} and p_{b1} with p_{b2}, but it leaves rotations about the axis between the points in the pair, $\text{axis}_{a,b} = (p_b - p_a)/\|p_b - p_a\|^2$, unconstrained. The surface normals associated with each patch are next used to further constrain the aligning transformations.

The normals are limited in their ability to be used in this respect, though, in two ways: Each normal only constrains rotation about $\text{axis}_{a,b}$ if it is not colinear with $\text{axis}_{a,b}$, and the observed sensor surface normals themselves may be unconstrained if the object surface normal in the area is not well-defined, e.g. in the case of an edge or corner. Because of these factors, we have considerably more confidence in the point locations than in the surface normals, so we are comfortable parameterizing the aligning transformation as a rigid transformation based on point locations followed by a rotation about $\text{axis}_{a1,b1}$ by an angle β with uncertainty.

We will first discuss our handling of constraints on the surface normals in Sect. 4.2, then this will be incorporated with the colinearity issue into our full estimate of the axis-angle rotation portion of the transformation with its associated uncertainty in Sect. 4.3.

4.2 Estimating Constraints on Sensor Normals

The surface normal at a surface patch can only be used to constrain rotation if it is itself constrained, so our goal here is to estimate the level of constraint the object surface imposed upon the sensor normal when a reading z_i was taken by looking at the associated tactile image, $\mathbf{I}_i$. Our approach is to infer a rough set of contact points of the sensor with the object surface from sensor elements with non-zero responses. In order for the sensor normal to be well-constrained, the surface should make contact with the sensor in at least three well-separated, non-colinear locations. Algorithm 1 quantifies a way of measuring the degree of fulfillment of this requirement, returning *normConf*(n_i).

A set of 3D contact points are estimated from $\mathbf{I}_i$. Their centroid is subtracted, giving a set of relative positions, which are assembled into a matrix, A, whose singular value decomposition is computed. For edge or point contacts, there should be less than two significant singular values, and in this case the function returns a confidence of zero. Otherwise it returns a value that approaches one as the two largest singular values approach each other, i.e. as the contact type approaches fully planar.

Algorithm 1 Estimate Normal Constraint

```
pts ← ∅
avgPt ← point3D(0,0,0)
for all sensor elements i do
    if val(i) > contactThresh then
        p ← point3D(getX(i), getY(i), estimateDepth(val(i)))
        add p to pts
        avgPt ← avgPt + p
    end if
end for
if sizeOf(pts) < 3 then
    return 0
end if
avgPt ← avgPt/sizeOf(pts)
r ← 1
A ← matrix(sizeOf(pts), 3)
for all points p in pts do
    rowOf(A, r) ← p − avgPt
    r ← r + 1
end for
S = svd(A) {Returns sorted vector of singular values}
return S[2]/S[1] {Ratio of two largest singular values}
```

4.3 Formulating Axis-Angle Uncertainty

Finally, we incorporate the constraint imposed by the normals on about-axis rotation with the uncertainty in the normals themselves to estimate a distribution over possible axis-angle rotations to complete the alignment of the two patch pairs.

First the surface normals associated with the first patch pair, n_{a1} and n_{b1}, and the axis between them, $\text{axis}_{a1,b1}$ are rotated according to the transformation obtained in Sect. 4.1 to be in the same coordinate system as n_{a2} and n_{b2}. Then a projection is computed to project each pair's normals onto the plane normal to $R_1\text{axis}_{a1,b1}$, giving projected normals $\{pn_{[\cdot]}\}$. Next, a rotation is computed to align these projected normals once again based on the method of [2]:

$$H = pn_{a1}pn_{a2}^T + pn_{b1}pn_{b2}^T \tag{16}$$

$$USV^T = \text{svd}(H) \tag{17}$$

$$R_\beta = VU^T \tag{18}$$

The projection has the effect of scaling each vector by the degree to which it is perpendicular to $\text{axis}_{a1,b1}$ in the rotated space, thereby also scaling its contribution to the least squares error being minimized in the fit. If the determinant of R_β is negative, this generally means S is rank defficient and the sign of one column of U is unconstrained, so R_β is reset to $V\,diag(1,\,-1)U^T$, giving a valid rotation.

Finally, the angle of rotation is extracted from R_β to get $\hat{\beta}$ our estimate of β. The overall confidence in this value is estimated as

$$q_a = \|pn_{a1}\|normConf(n_{a1})\|pn_{a2}\|normConf(n_{a2}) \tag{19}$$

$$q_b = \|pn_{b1}\|normConf(n_{b1})\|pn_{b2}\|normConf(n_{b2}) \tag{20}$$

$$alignConf = \frac{q_a + q_b}{2} \tag{21}$$

The distribution of possible true values of β is then estimated as a Gaussian with mean $\hat{\beta}$ and standard deviation given by $\sigma_{\text{init}}/alignConf$. In our experiments, σ_{init} was conservatively set to 0.1 radians.

In practice, the distribution over x_t is maintained as a sparse histogram. Except at initialization, when the distribution can be implicitly assumed equal to the prior (e.g., uniform), the likelihood in most bins will be zero. The data structure can be efficiently implemented using a hash map indexed by the histogram bin numbers.

5 Experiments

The approach was tested on 3D models in our simulation environment and on a set of raised letter shapes using our physical sensor system. The simulation experiments are described in Sect. 5.1, then those on physical sensors follow in Sect. 5.2.

5.1 3D Simulation Experiments

To test the algorithm, a set of 10 objects from the Princeton Shape benchmark [18] was used, shown in Fig. 3. The sample objects were selected to span a variety of shapes and local surface characteristics, and all were scaled to the same size.

A set of 1000 sensor readings of each object was collected for training, and a separate 100 readings of each object were collected for testing. All sensor readings were collected by the following process: A location p on the object surface was chosen uniformly at random. The position of the tactile sensor was set to $q = p + dn$, a small distance d away in the direction of the local surface normal, n. The sensor was oriented so that its surface normal was in the direction $-n$ plus a small random perturbation. Then the sensor moved in the direction of the object until the controllers described in Sect. 1.1 converged.

As described in Sect. 3, appearance descriptors were extracted from each sensor reading and these were grouped into 25 clusters using k-medoids, then an SVA map was built of all the objects. The map used 40 bins to discretize the space of interpatch distances that covered a range from 30 to 160 mm. The objects themselves were scaled 80 mm in their largest dimension.

Recognition performance was measured as a function of the number of sensor readings seen so far by averaging results over a number of trials. In each trial, readings were selected uniformly at random from the set of test readings for the selected unknown object. An object pose was generated and the pose of each sensor reading was transformed according to this unknown pose before it was presented to the recognition algorithm. This pose consisted of an arbitrary 3D rotation and a translation in the range [−200, 200] mm in each direction. One test repetition consisted of one trial of recognition on each object from the set. Performance was averaged over three test repetitions to get the final results shown in Fig. 4. Accuracy of the SVA approach is shown alongside that of the appearance-only approach of [15] for comparison.

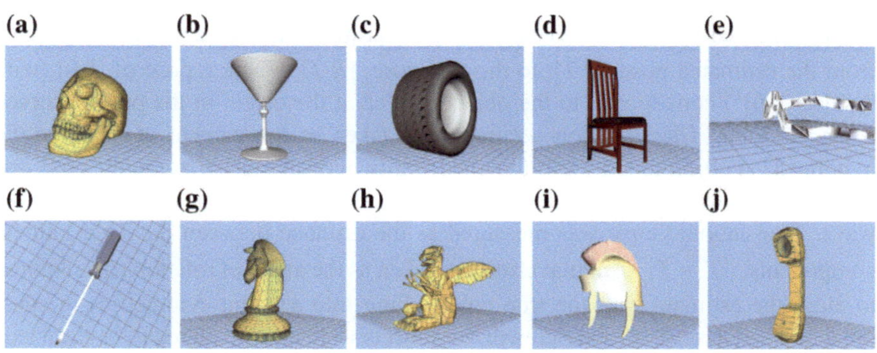

Fig. 3 The set of models from the Princeton Shape Benchmark [18] used for testing. **a** Skull. **b** Glass. **c** Tire. **d** Chair. **e** Pliers. **f** Screwdriver. **g** Knight. **h** Dragon. **i** Helmet. **j** Phone

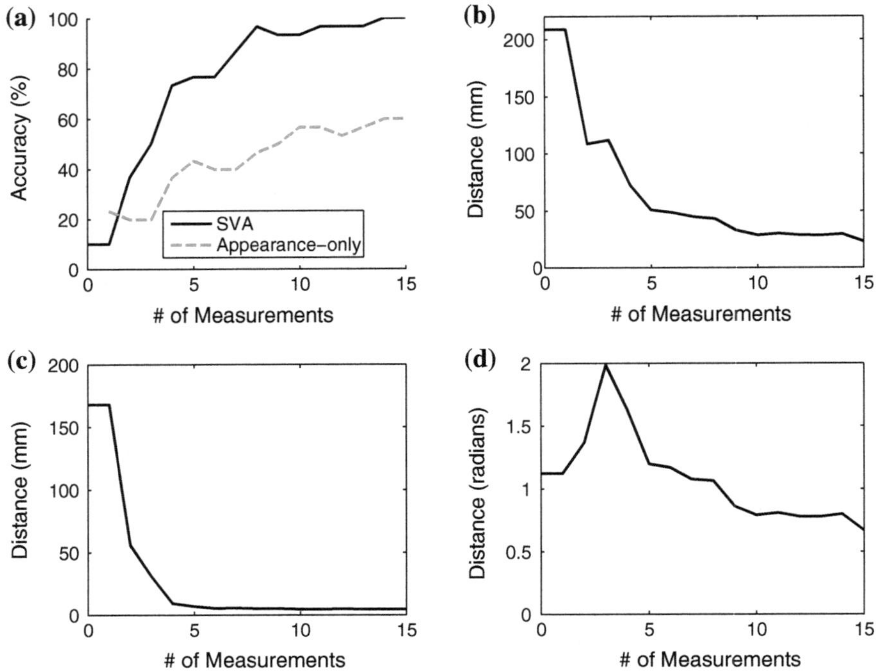

Fig. 4 SVA recognition and localization results on Princeton set. **a** Classification accuracy. **b** Inertial error. **c** Translational error. **d** Angular error

The virtual histogram used to maintain pose estimates used 50 bins for each translation dimension and represented rotation by a 3-dimensional vector in Rodrigues form (with magnitude encoding rotation angle) divided into 9 bins per dimension. At each time step, the hypothesized object identity was taken as the object with the most probability weight, summed over all possible poses. The hypothesized object pose was taken as the centroid of the histogram bin with the highest weight.

Performance was measured in terms of classification accuracy and of distance from the estimated pose, $[\hat{R}\,\hat{T}]$, to the true pose, $[R\ T]$, where a pose of $[\mathbf{I}\ \mathbf{0}]$ (with $\mathbf{0} = [0, 0, 0]^T$) corresponds to the object located at the origin in the pose observed during training. Classification accuracy was taken as the percent of the time the hypothesized object identity was the true identity over all trials. Error in the pose was recorded only when the estimated identity was correct, and it was measured in three ways: Translational error was measured as the distance between the Translational components, $\left\|\hat{T} - T\right\|$. Angular error was taken as the angle of rotation, ϕ_e, required to align the estimate with the true pose, taking into account symmetries in some objects. The Glass, Tire, and Screwdriver objects were all considered to have an axis of rotational symmetry. Let this axis be denoted ζ; then we have

$$\phi_e = \begin{cases} \min_\gamma \left\| unskew(logm(R_{R\zeta}(\gamma)R\hat{R}^T)) \right\| & \text{if symmetric about } \zeta \\ \left\| unskew(logm(R\hat{R}^T)) \right\| & \text{otherwise} \end{cases} \tag{22}$$

where *logm* denotes the matrix logarithm, *unskew* extracts the vector v from $sk(v)$, its corresponding skew-symmetric matrix, and $R_\zeta(\gamma)$ is a rotation about ζ by γ. Inertial distance was measured according to the metric of [8, Eq. 4], where each object's mass and moment of inertia was approximated by a solid sphere of radius 40 mm. This metric combines translational and angular error into a measurement of the energy required to align the two transformations. Localization accuracy was not measured for the appearance-only approach, since it does not estimate object pose. The MNTI descriptor was used to characterize appearance due to its invariance properties' robustness to small translations. Figure 4 graphs all of the error metrics above: Fig. 4a shows recognition accuracy. Figure 4b–d show inertial, translational, and angular error respectively. Classification accuracy climbs to 100 % with thirteen sensor readings. Translational error drops to slightly above 4 mm, the lowest expected attainable error using histogram bins of width 8 mm. Angular error remains high, however, most likely due to near-symmetries in the objects; e.g., the knight's base is axially symmetric and the phone nearly has two-fold rotational symmetry. As a result of the angular error, inertial error decreases more slowly.

5.2 2D Physical Sensor Experiments

The SVA mapping approach was also tested using physical sensors on capital vowels from a child's set of raised letters (from a Leap Frog "Fridge phonics" magnetic alphabet set), shown in Fig. 1 along with our sensors.

The letters were approximately 2.5 cm per side, so less than a quarter of the letter was visible in any single reading. In order to cover the entire object, readings were collected at 16 planar positions arranged in a 4×4 grid with a spacing of 6.8 mm, oriented at 12 evenly-spaced angles at each location for a total of 192 readings. A mechanical system was constructed to position the letters coplanar with the sensors and press them down with a consistent force.

Two readings were taken at each of the 192 poses. The first set of readings at each pose was used for training, while the second set was used for testing.

As before, readings from the unknown object were transformed according to a randomly selected object pose for each trial. This pose consisted of a translation in x and y in the range $[-10, 10]$ mm in each direction and an arbitrary rotation in the plane. This pose space was discretized using 21 bins for each dimension of translation and 9 bins for each dimension of the Rodrigues vector representing rotation. The map used 100 bins for distance, covering a range of 3–20 mm. Since there was a discrete set of contact locations, invariance to translation was less of a concern, so the Moment-Normalized descriptor was used.

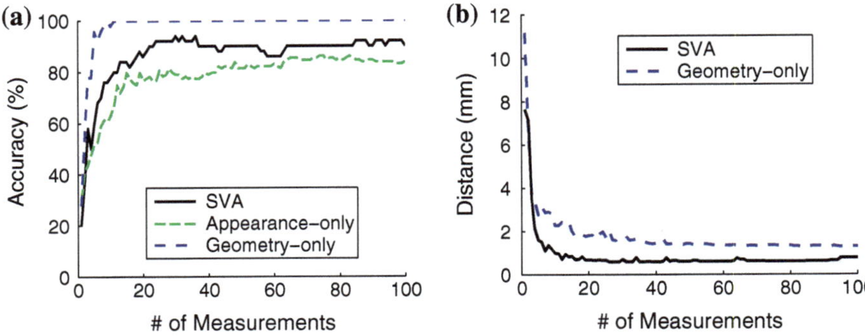

Fig. 5 SVA recognition and localization results on raised letters. **a** Classification accuracy. **b** Inertial error

Classification accuracy and inertial distance are shown in Fig. 5. Again, performance of the SVA and appearance-only approach [15] are shown as well as that of the geometry-only approach of [16]. Classification accuracy quickly climbs above 90 % within about 30 sensor readings. Inertial distance was measured taking into account symmetries in the letters, so that angular error was measured with respect to the closest of the true pose and its 180° rotation for "I"; "O" was considered fully symmetric, so that angular error was always zero. This metric drops below 1 mm within about 20 sensor readings. This is once again close to the expected optimum given the virtual histogram resolution. While SVA does not achieve the recognition rates of the geometric method in this 2D case, it gives better localization.

6 Discussion

We have presented a method that makes use of both the appearance content of tactile force sensor readings and the geometric information associated with each. The method was demonstrated on both simulated and real tactile data sets, exhibiting strong performance both in recognition accuracy and pose estimation. Performance was not perfect, however, so we provide some analysis of why that may be, ideas for improvement, and guidance on how to apply the method in different situations.

6.1 Analysis of Failure Modes

It is interesting to compare the experimental results of Sect. 5.2 to those of the purely geometric approach of [16] on the same data. The purely geometric approach eventually achieved 100 % classification accuracy when using a histogram with 10,000 bins for the pose space, but with a localization accuracy of just over 2 mm; the SVA approach did not quite reach 100 % accuracy on the letter set, but its localization accuracy was below 1 mm. Higher pose accuracy is achievable because a higher-resolution histogram can be maintained using a forward mapping from sensor readings to pose and a sparse representation of the probability space. This same difference also makes the SVA approach much more amenable to extension to full 3D. A natural question would be why the SVA method does not achieve perfect classification accuracy. It is instructive to address this question in some detail, as it gives guidance on considerations for applying the method in other situations.

A classification failure must result from the true pose not being among those considered able to explain a patch pair. If the bin of the true pose is assigned zero probability, then the optimal solution will not be found unless all probabilities go to zero and the distribution is reseeded. There are a few ways this might occur:

1. A sensor reading may be of part of the object surface not observed in training. There would then be no valid match in the map for pairs containing that point.
2. A patch pair may not be matched to the nearest corresponding regions in the map because the estimated appearance and distance do not match well enough.
3. The distribution of aligning transformations computed for a correct (or close) patch pair match may not place significant probability on the true object pose. This might be due to a misestimation of the surface patch's locations or (more likely) their surface normals.
4. A combination of the factors above, each acting in part, could push the probability of the true pose below machine precision.

The first situation would not occur with the raised letter data, since sensor readings were taken at set locations, and those locations were the same in training as in testing. The third failure mode is also not likely on the letter data for the same reason, and since the surface normals were all known and equal. The fourth failure mode also did not seem to come into play in our experiments with our chosen parameters. The most likely source of classification failures therefore seems to be the second item, in the appearance classification of surface patch pairs.

6.2 *Extensions*

Like the geometric method, for which each particle or histogram bin calculation is independent, this approach has great potential for parallelization. With this method, the computations for each surface patch pair can be carried out in parallel. Additionally, the computations for each prospective match for a surface patch pair are also independent, leading to another level of parallelization.

Although this method was developed with tactile force sensors as the intended source of information, it should be noted that it is equally applicable to other sensing modalities. For instance, a stereo vision system could also be used to acquire surface patch information comprising location, surface normal estimates, and appearance. A particularly interesting extension would be to examine what appearance properties can be characterized by both vision and touch; then cross-modality models could be built from one modality and then used in another or with both modalities.

Acknowledgements This work was supported by NSF grants MRI-0722943 and IIS-0748338, and a Link Foundation Fellowship for Simulation and Training.

References

1. P. Allen, P. Michelman, Acquisition and interpretation of 3-d sensor data from touch. IEEE Trans. Robot. Autom. **6**(4), 397–404 (1990)
2. K.S. Arun, T.S. Huang, S.D. Blostein, Least-squares fitting of two 3-d point sets. IEEE Trans. Pattern Anal. Mach. Intell. **9**, 698–700 (1987). http://portal.acm.org/citation.cfm?id=28809.28821
3. R. Bajcsy, What can we learn from one finger experiments?, in *International Symposium on Robotics Research* (Bretton Woods, NH, 1984), pp. 509–527
4. J. Bay, Tactile shape sensing via single and multifingered hands, in *International Conference on Robotics and Automation*, vol. 1 (Scottsdale, AZ, 1989), pp. 290–295
5. S. Caselli, C. Magnanini, F. Zanichelli, E. Caraffi, Efficient exploration and recognition of convex objects based on haptic perception, in IEEE *International Conference on Robotics and Automation*, vol. 4 (Minneapolis, MN, 1996), pp. 3508–3513
6. S. Casselli, C. Magnanini, F. Zanichelli, On the robustness of haptic object recognition based on polyhedral shape representations, in *IEEE/RSJ International Conference on Intelligent Robots and Systems*, vol. 2 (1995), p. 2200
7. S. Chhatpar, M. Branicky, Localization for robotic assemblies using probing and particle filtering, in *2005 IEEE/ASME International Conference on Advanced Intelligent Mechatronics. Proceedings* (2005), pp. 1379–1384
8. G. Chirikjian, S. Zhou, Metrics on motion and deformation of solid models. J. Mech. Des. **120**, 252 (1998)
9. R. Fearing, Tactile sensing mechanisms. Int. J. Robot. Res. **9**(3), 3–23 (1990)
10. K. Gadeyne, H. Bruyninckx, Markov techniques for object localization with force-controlled robots. Int. Conf. Adv. Robot. 91–96 (2001) (citeseer)
11. W. Grimson, T. Lozano-Perez, Model-based recognition and localization from tactile data, in *IEEE International Conference on Robotics and Automation*, vol. 1, (Atlanta, GA, 1984), pp. 248–255
12. L. Kaufman, P. Rousseeuw, *Finding Groups in Data: An Introduction to Cluster Analysis*, vol. 5 (Wiley, 1990)
13. A. Petrovskaya, O. Khatib, S. Thrun, A. Ng, Touch based perception for object manipulation, in *Robotics Science and Systems, Robot Manipulation Workshop* (2007)
14. Z. Pezzementi, E. Jantho, L. Estrade, G.D. Hager, Characterization and simulation of tactile sensors, in *Haptics Symposium* (Waltham, MA, USA, 2010), pp. 199–205
15. Z. Pezzementi, E. Plaku, C. Reyda, G.D. Hager, Tactile object recognition from appearance information. IEEE Trans. Rob. **27**(3), 473–487 (2011)
16. Z. Pezzementi, C. Reyda, G.D. Hager, Object mapping, recognition, and localization from tactile geometry, in *IEEE International Conference on Robotics and Automation*, Shanghai, China, 2011, pp. 5942–5948

17. A. Schneider, J. Sturm, C. Stachniss, M. Reisert, H. Burkhardt, W. Burgard, Object identification with tactile sensors using bag-of-features, in *IROS 2009. IEEE/RSJ International Conference on Intelligent Robots and Systems, 2009* (2009), pp. 243–248
18. P. Shilane, P. Min, M. Kazhdan, T. Funkhouser, The Princeton shape benchmark, in *Shape Modeling International*, Genova, Italy, 2004, pp. 167–178
19. S. Thrun, W. Burgard, D. Fox, *Probabilistic Robotics* (Intelligent Robotics and Autonomous Agents) (The MIT Press, 2005)
20. H.J. Wolfson, I. Rigoutsos, Geometric hashing: an overview. IEEE Comput. Sci. Eng. **4**, 10–21 (1997)

The Antiparticle Filter—An Adaptive Nonlinear Estimator

John Folkesson

Abstract We introduce the *antiparticle filter*, *AF*, a new type of recursive Bayesian estimator that is *unlike* either the extended Kalman Filter, *EKF*, unscented Kalman Filter, *UKF* or the particle filter *PF*. We show that for a classic problem of robot localization the AF can substantially outperform these other filters in some situations. The AF estimates the posterior distribution as an auxiliary variable Gaussian which gives an analytic formula using no random samples. It adaptively changes the complexity of the posterior distribution as the uncertainty changes. It is equivalent to the EKF when the uncertainty is low while being able to represent non-Gaussian distributions as the uncertainty increases. The computation time can be much faster than a particle filter for the same accuracy. We have simulated comparisons of two types of AF to the EKF, the iterative EKF, the UKF, an iterative UKF, and the PF demonstrating that AF can reduce the error to a consistent accurate value.

1 Introduction

Non-linear estimation is crucial to robotics as well as many other fields. Fundamentally one needs to describe uncertainty in some 'state' based on indirect and noisy observations of it. This state is known to evolve in time and that evolution can introduce uncertainty. When both the observation and the evolution are described by linear system equations wrt the state estimate and the noise is white Gaussian then the Kalman filter is the optimal estimator, [1]. The problem becomes harder when the system is nonlinear. One approach is to linearize the system equations wrt the state around the current estimated state as in the extended Kalman Filter *EKF* [2, 3].

J. Folkesson (✉)
Centre for Autonomous Systems, Royal Institute of Technology (KTH), Stockholm, Sweden
e-mail: johnf@csc.kth.se

H.I. Christensen and O. Khatib (eds.), *Robotics Research*,
Springer Tracts in Advanced Robotics 100, DOI 10.1007/978-3-319-29363-9_13

The EKF has led to the development of many variations. These mostly address the issue of inconsistency. For the EKF inconsistency arises as a consequence of the linearization. The EKF is a Gaussian estimator which implies that the posterior distribution is parametrized by a state mean and covariance. When the covariance is too small we say that the estimate is overconfident [4]. This can lead to divergence of the estimate or to bad inferences based on it. In order improve filter stability methods of better estimating or simple adjusting the covariance have been tried, such as, the robust extended Kalman filter [5, 6], or the method used for robot localization in [7].

A better estimate of the covariance is obtained by computing the estimate based on mapping a selected set of points through the non-linearity, allowing the estimate to reflect more than a single point. This approach led to the unscented Kalman filter, *UKF* [8–10], linear regression Kalman filters *LRKF* [11–13], the shifted Rayleigh filter [14], and the filter in [15].

Divergence or unstable behavior often occurs while incorporating observations that require large changes to the estimated mean. The iterative Kalman filter, *IEKF* [16, 17] can help this problem by applying the EKF formula and linearizations repeatedly until convergence to a stable solution.

All of the above estimators use a Gaussian posterior. As the uncertainty in the estimates becomes more significant wrt the non-linearity, the Gaussian distribution can no longer adequately represent the posterior. This situations can be addressed by the particle filter *PF*, [18–21]. The PF can estimate any distribution and makes no requirements on the form of the noise. It does so by sampling the distribution. It essentially carries out many simulations, *particles*, of possible evolutions of the system, re-weighting the particle set based on the observations. The PF is very popular and powerful. It has one drawback: the estimate depends on having many particles near the true state. It is important to keep the particles spread evenly and densely. Unfortunately so called particle deprivation (or depletion) is inevitable (eventually) using the PF. Particle deprivation is the condition of having too few different particles. This can be made less significant by increasing the number of particles but this leads to higher computational costs.

Particle deprivation is even more of a problem in higher dimensions. This can sometimes be solved by Rao-Blackwellized particles filters, such as, FastSLAM [22, 23]. Alternatively, a novel Gaussian particle filter is proposed in [24] which does not require re-sampling but does assume a Gaussian posterior.

Other solutions to representing non-Gaussian distributions are the sum of Gaussians as used in [25] on the simultaneous localization and mapping, problem. Or the Gaussian sum quadrature filter as presented in [13]. Exact batch methods and some of the graphical algorithms [26–30] can solve the exact system by successive approximation. In [31] a method that combines graphical belief propagation, particle filters and uses auxiliary variables is presented.

2 Overview of This Article

We will developed a new class of estimators which we call *antiparticle filters AF* [32]. We shall describe two members of that class which we call the quadratic and the trigonometric AF, *QAF*, and *TAF*. We compare these filters to the EKF, IEKF, UKF, IUKF[1] and PF.

We start by discussing the AF posterior distribution. We show how it can change complexity adaptively as the uncertainty changes by introducing and removing auxiliary variables. We then discuss how to estimate this distribution with prediction and update steps.

This is then followed by our simulations which show how for a robot localization problem with large uncertainty the AF is more accurate, stable and consistent than the other types of filters. The PF can be made more accurate and consistent by increasing the number of particles. We show the PF for a number of particles that gives similar computation times as our filter and one that is significantly slower but still not as consistent or accurate.

3 The Auxiliary Variable Gaussian Distribution

As state estimates evolve in time the uncertainty increases during the prediction step which moves the estimate forward in time using the dynamic equations. When observations are made the uncertainty is reduced in an update step. If no observations are made for a long time the uncertainty can become too large, making the nonlinearities significant. This uncertainty is often highly correlated. So that even if all the covariance matrix elements are growing this growth is really only in one (or few) problem directions. We propose to factor out part of this problem direction and model its uncertainty in a n^{λ} dimensional auxiliary variable vector called λ. So that the posterior distribution takes the form:

$$p(\mathrm{x}) = \int_{-\infty}^{\infty} G(\mathrm{x} - \mathbf{m}^{\lambda}, P) G(\lambda, C) (d\lambda)^{n\lambda} \tag{1}$$

where $G(\lambda, C)$ is the zero mean Gaussian distribution with covariance C. We see that the mean of the state estimate $\mathbf{m}^{\lambda}$ is conditionally parametrized by λ. This $\mathbf{m}^{\lambda}$ is a differentiable function of λ. By assuming different forms for this we can generate different shapes for the distribution and different filter types. We will later use this fact:

$$E[\lambda\lambda^T] - E[\lambda]E[\lambda^T] = C. \tag{2}$$

[1]The iterative UKF is an iterative version of the UKF similar to the IEKF.

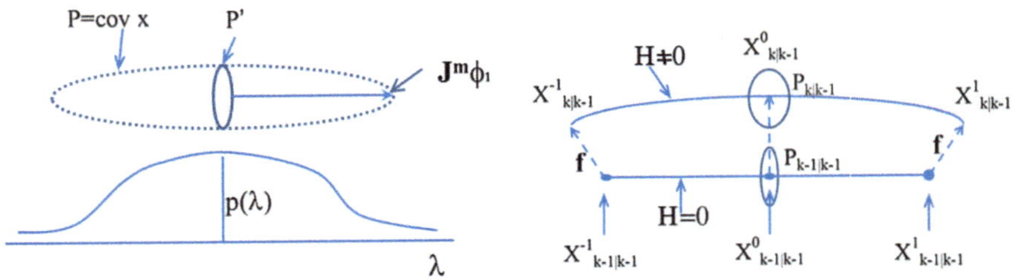

Fig. 1 The *left* illustrates that the **x** covariance can be exactly moved between P and (P', $\mathbf{J}^m$, C). The *right* shows the subsequent predict phase

We can derive some results by using the Taylor series expansion of $\mathbf{m}^\lambda$:

$$\mathbf{m}^\lambda = \mu + \mathbf{J}^m \lambda + \frac{1}{2}\lambda^T \mathbf{H}^m \lambda + O(\lambda^3) \tag{3}$$

where $\mu = \mathbf{m}^0$, $\mathbf{H}^m$ is an **x** space vector of symmetric λ space Hessian matrices and the Jacobian matrix $\mathbf{J}^m$ is a matrix from λ to **x** space.

$$E[\mathbf{x}] = \int_{-\infty}^{\infty} \mathbf{m}^\lambda G(\lambda, C)(d\lambda)^{n\lambda} \approx \mu + \frac{1}{2}\sum_{i,j} \mathbf{H}^m_{ij} C_{ij}$$

$$E[\mathbf{x}\mathbf{x}^T] - E[\mathbf{x}]E[\mathbf{x}^T] \approx P + \mathbf{J}^m C (\mathbf{J}^m)^T + \frac{1}{2}\sum_{i,j,k,l} \mathbf{H}^m_{ij} C_{ik} C_{jl} (\mathbf{H}^m_{kl})^T$$

These two equations are used when creating and destroying auxiliary variables with minimal changes to the mean and covariance in **x**. This is done adaptively in response to the changes in uncertainty. When the uncertainty is low the distribution can be simplified by reducing the number of auxiliary dimensions and vice versa as shown schematically in Fig. 1.

Destroying Auxiliary Dimensions

If we chose a basis in the λ space so that $C = I$ then we can eliminate dimension q from λ by making these adjustments to the remaining parameters:

$$\mu \leftarrow \mu + \frac{1}{2}\mathbf{H}^m_{qq}, \quad P_{ij} \leftarrow P_{ij} + \Delta P_{ij} = P_{ij} + J^m_{iq} J^m_{jq} + \sum_k H^m_{iqk} H^m_{jqk} - \frac{1}{2} H^m_{iqq} H^m_{jqq}$$

In general, when the trace of ΔP above is below a threshold for some q we will eliminate that dimension by making the adjustments above and setting $\lambda_q = 0$.[2] This ΔP decreases during updates and increases during prediction so we normally do this check after updating with new observations.

[2] We use the threshold 0.01 in our simulations.

Creating Auxiliary Dimensions

We can create new auxiliary dimensions by moving some of the uncertainty in unit direction $\mathbf{u}$, from P into a new λ dimension, q. For this we must chose the new parameters to meet these constraints with $C_{qq} = 1$ and $\mathbf{H}_{qk} = 0$:

$$\mathbf{J}_q^m = \sigma\sqrt{1-\delta}\mathbf{u}, \quad P = P - \mathbf{u}(1-\delta)\sigma^2\mathbf{u}^T,$$
$$P^{-1} = P^{-1} + \frac{P^{-1}\mathbf{u}(1-\delta)\sigma^2(P^{-1}\mathbf{u})^T}{\delta}$$

where $\sigma^{-2} = \mathbf{u}^T P^{-1}\mathbf{u}$ and δ is a small number to keep P positive definite. In practice we use the largest eigenvalue of P for σ^2 and the eigen vector for $\mathbf{u}$. If the eigenvalue is above some threshold[3] then we create a new dimension q.

Parameterizing the Distribution—The Antiparticles

We will assume that the vector functions $\mathbf{m}^\lambda$ can be parametrized in two ways. One is by some canonical parametrization, π which can be used to directly compute $\mathbf{m}^\lambda$ for any lambda. The second is the values of $\mathbf{m}^\lambda$ at a set of $\lambda = \varphi_i$ from which the canonical parameters can be derived. This set of values with be notated by $\mathbf{x}^i = m^{\varphi i}$. The φ_i are chosen distributed around the mean $\lambda = 0$ value using the C in a manner reminiscent of how the regression (sigma) points are chosen in the LRKF (UKF).

We will call the set $\{\mathbf{x}^i\}$ the antiparticles as they will function similarly to particles in the PF but are not random samples. We assume that we can go from the canonical parametrization to the antiparticles and back as required.

$$\{\mathbf{x}^i\} \leftrightarrow \{\pi\} \tag{4}$$

The antiparticles will be sufficient for doing prediction and are the result of the update. We will find it necessary to use the canonical parameters for updates. They are also used when creating and destroying auxiliary dimensions.

4 Estimation of the Non-linear Process

The starting point of all estimators is the process model:

$$\mathbf{x}_k = \mathbf{f}(\mathbf{x}_{k-1}, \mathbf{u}_k) + \omega_k, \quad \omega_k \sim G(0, Q_k) \tag{5}$$

$$\mathbf{z}_k = \mathbf{h}(\mathbf{x}_k) + v_k, \quad v_k \sim G(0, R_k) \tag{6}$$

Here $\mathbf{x}_k$ is the state, $\mathbf{z}_k$ are the measurements, $\mathbf{u}_k$ are the control signals, and Q/R are the white noise covariances. We will use subscripts $k-1|k-1$ and $k|k-1$ for the parameters before and after the kth prediction respectively.

[3]We use 1.0 for this threshold.

Prediction

Prediction is done by marginalizing over $\mathbf{x}_{k-1}$:

$$\int p(\mathbf{x}_k|\mathbf{x}_{k-1}, \mathbf{u}_k)p_{k-1|k-1}(\mathbf{x}_{k-1})d\mathbf{x}_{k-1} = p_{k|k-1}(\mathbf{x}_k) \quad (7)$$

For the above process and our assumed form for the distribution this becomes

$$\int G(\mathbf{x}_k - \mathbf{m}^{\lambda}_{k|k-1}, p_{k|k-1})G(\lambda, C_{k|k-1})(d\lambda)^{n\lambda} \approx \propto$$
$$\iint G(\mathbf{x}_k - f(\mathbf{x}_{k-1}, u_k), Q_k)G(\mathbf{x}_{k-1} - \mathbf{m}^{\lambda}_{k-1|k-1}, P_{k-1|k-1})G(\lambda, C_{k-1|k-1})d\mathbf{x}_{k-1}(d\lambda)^{n\lambda}$$

We linearize around an estimated state $\mathbf{x}_k = \mathbf{m}^{\lambda}_{k-1|k-1}$ just as in the EKF.

$$\mathbf{f}(\mathbf{x}_{k-1}, \mathbf{u}_k) \approx \bar{\mathbf{x}}^{\lambda} + \mathbf{J}^f_{\lambda}(\mathbf{x}_{k-1} - \mathbf{m}^{\lambda}_{k-1|k-1}), \quad \mathbf{J}^f_{\lambda} \equiv \frac{\partial \mathrm{f}(\mathbf{x}_{k-1}, \mathbf{u}_k)}{\partial \mathbf{x}_{k-1}}\Big|_{\mathbf{x}_{k-1}=\mathbf{m}^{\lambda}_{k-1|k-1}}$$
$$\bar{\mathbf{x}}^{\lambda} \equiv \mathrm{f}(m^{\lambda}_{k-1|k-1}, \mathbf{u}_k), \qquad P^{\lambda} \equiv Q_k + \mathbf{J}^f_{\lambda}P_{k-1|k-1}(\mathbf{J}^f_{\lambda})^T.$$

Using these substitutions we can do the marginalization above and find:

$$\int G(\mathbf{x}_k - \mathbf{m}^{\lambda}_{k-1|k-1}, P_{k|k-1})G(\lambda, C_{k|k-1})(d\lambda)^{n\lambda} \approx \propto \int \sqrt{|P^{\lambda}|}G(\mathbf{x}_k - \bar{\mathbf{x}}^{\lambda}, P^{\lambda})G(\lambda, C_{k-1|k-1})(d\lambda)^{n\lambda}$$

In light of this equation we make the following prediction rules:

$$P_{k|k-1} = P^{\lambda}|_{\lambda=0}, \quad C_{k|k-1} = C_{k-1|k-1}, \quad (8)$$

$$\mathbf{m}^{\lambda}_{k|k-1} = \bar{\mathbf{x}}^{\lambda} \Rightarrow \{\mathbf{x}^{\mathbf{i}}_{\mathbf{k}|\mathbf{k}-\mathbf{1}}\} = \{\mathbf{f}(\mathbf{x}^{i}_{k-1|k-1})\} \quad (9)$$

The approximations are that P^{λ} does not vary with λ and that by mapping the antiparticles through the nonlinear **f**, the $\mathbf{m}^{\lambda}_{k|k-1}$ will follow $\bar{\mathbf{x}}^{\lambda}$ for all λ.

We end up with a simple predict formula that uses the EKF prediction for P, leaves C unchanged and maps the antiparticles through the dynamics in the same way that the mean is treated by the EKF, illustrated in Fig. 1. If no update is to be done we can feed the antiparticles directly into the next predict step without converting to the canonical form.

Update

During update the observation measurements are used to form the posterior distribution. The most important part of the update is locating the maximum likelihood estimate, *MLE*, for both $\mathbf{x}_k$ and λ. The canonical parametrization is used for this. Once this MLE is found a new set of antiparticles is found around this MLE. These then are ready for the next prediction step.

Setting $\Delta\mathbf{z}(\mathbf{x}_k) = \mathbf{z}_k - \mathbf{h}(\mathbf{x}_k)$ and Bayes rule gives us the posterior:

$$G(\mathbf{x}_k - m^{\lambda}_{k|k}, P_{k|k})G(\lambda, C_{k|k}) \propto G(\mathbf{x}_k - m^{\lambda}_{k|k-1}, P_{k|k-1})G(\Delta z(\mathbf{x}_k), R)G(\lambda, C_{k|k-1})$$

Taking the exponents on the right hand side above we see

$$(\mathbf{x}_{MLE}, \lambda_{MLE}) = \arg \min_{\mathbf{x}_k,\lambda} g(\mathbf{x}_k, \lambda) \tag{10}$$

where g is given by:

$$\begin{aligned} g(\mathbf{x}_k, \lambda) = [(\mathbf{x}_k - \mathbf{m}^{\lambda}_{k|k-1})^T P^{-1}_{k|k-1}(\mathbf{x}_k \\ - \mathbf{m}^{\lambda}_{k|k-1}) + \Delta z(\mathbf{x}_k)^T R^{-1} \Delta z(\mathbf{x}_k) + \lambda^T C^{-1}_{k|k-1}\lambda]/2. \end{aligned}$$

We will repeatedly make use of linearization of the function h:

$$\Delta\mathbf{z}(\mathbf{x}_k) = \mathbf{z}_k - \mathbf{h}(\mathbf{x}_k) \approx \mathbf{z}_k - \mathbf{h}(\mathbf{x}) + J^h(\mathbf{x}_k - \mathbf{x}) \tag{11}$$

We do a three phase update which is shown schematically in Fig. 2. Phase 1 varies only λ while holding $\mathbf{x}_k - \mathbf{m}^{\lambda}_{k|k-1}$. This moves along the $\mathbf{m}^{\lambda}$ curves to minimize Eq. (10). Phase 2 varies both $\mathbf{x}_k$ and λ. Phase 3 holds $\lambda = \varphi_i$ while varying $\mathbf{x}_k$ to find the updated antiparticle set.

Update Phase 1—Getting near the MLE

We begin by converting the antiparticle set to the canonical parametrization. We seek a point λ on the $\mathbf{m}^{\lambda}_{k|k-1}$ curves to best explain the measurements.

$$\lambda_0 = \arg \min_{\lambda} g(\mathbf{m}^{\lambda}_{k|k-1}, \lambda) \tag{12}$$

We use the Gauss-Newton method starting from $\lambda_0 = 0$,

$$\nabla g = (C^{-1}_{k|k-1}\lambda_0 - (J^h \mathbf{J}^m_{k|k-1})^T R_k^{-1} \Delta\mathbf{z})|_{\mathbf{x}=\mathbf{m}^{\lambda_0}_{k|k-1}} \tag{13}$$

$$\Omega \equiv C^{-1}_{k|k-1} + (J^h \mathbf{J}^m_{k|k-1})^T R_k^{-1} J^h \mathbf{J}^m_{k|k-1}|_{\mathbf{x}=\mathbf{m}^{\lambda_0}_{k|k-1}} \tag{14}$$

$$\Delta\lambda_0 = -\Omega^{-1}\nabla g \tag{15}$$

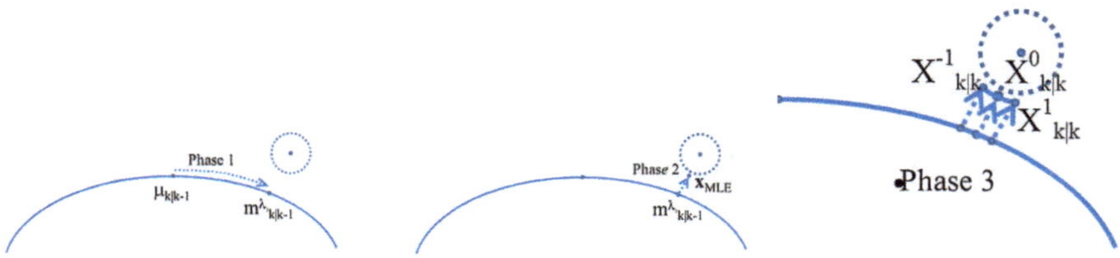

Fig. 2 Here we show the three update phases

where $\mathbf{J}^m_{k|k-1}$ denotes the Jacobian of $\mathbf{m}^{\lambda}_{k|k-1}$ wrt. λ We will use this $\Delta\lambda_0$ to move iteratively towards a solution to Eq. (12).

Update Phase 2—Finding the MLE

In the second phase we move to the solution to Eq. (10) by varying both $\mathbf{x}_k$ and λ. We use the Gauss-Newton method starting at $\mathbf{x}_k = \mathbf{m}^{\lambda 0}$ and $\lambda = \lambda_0$.

$$\omega_\lambda \equiv (\mathbf{J}^m_{k|k-1})^T P^{-1}_{k|k-1}(\mathbf{x}_k - \mathbf{m}^{\lambda}_{k|k-1}) - C^{-1}_{k|k-1}\lambda \tag{16}$$

$$\mathbf{w}_x \equiv (J^h)^T R_k^{-1}\Delta\mathbf{z}(\mathbf{x}_k) - P^{-1}_{k|k-1}(\mathbf{x}_k - \mathbf{m}^{\lambda}_{k|k-1}) \tag{17}$$

$$W_{\lambda\lambda} \equiv (\mathbf{J}^m_{k|k-1})^T P^{-1}_{k|k-1}\mathbf{J}^m_{k|k-1} + C^{-1}_{k|k-1} \tag{18}$$

$$W_{x\lambda} \equiv P^{-1}_{k|k-1}\mathbf{J}^m_{k|k-1} \tag{19}$$

$$W_{xx} \equiv (J^h)^T R_k^{-1} J^h + P^{-1}_{k|k-1} \tag{20}$$

$$\begin{pmatrix} \Delta\lambda \\ \Delta\mathbf{x}_k \end{pmatrix} = W^{-1}\mathrm{w} = \begin{pmatrix} W_{\lambda\lambda} & W^T_{x\lambda} \\ W_{x\lambda} & W_{xx} \end{pmatrix}^{-1} \begin{pmatrix} \omega\lambda \\ \mathbf{w}_x \end{pmatrix} \tag{21}$$

Update C and P

We set $P_{k|k} = (W_{xx})^{-1}|_{MLE}$ just as the IEKF. This gives the right shape for the distribution near the MLE. For the C update we note that $C_{k|k}$ is the covariance of λ for the posterior distribution, Eq. (2). By estimating this using the expansion around the MLE we find:

$$C_{k|k} \approx \int_{-\infty}^{\infty} (\lambda - \lambda_{MLE})^2 G((\lambda - \lambda_{MLE}, \mathbf{x} - \mathbf{x}_{MLE}), W^{-1})(d\mathbf{x})^n (d\lambda)^{n\lambda} \tag{22}$$

$$C^{-1}_{k|k} = [W_{\lambda\lambda} - W^T_{x\lambda}(W_{xx})^{-1}W_{x\lambda}]|_{x_k}, \lambda = MLE \tag{23}$$

$$= C^{-1}_{k|k-1} + (J^h\mathbf{J}^m)^T (J^h P_{k|k-1}(J^h)^T + R)^{-1} J^h\mathbf{J}^m \tag{24}$$

Update Phase 3—Creating a New Antiparticle Set

For phase 3 we need to create a new antiparticle set that reflects the measurements. The λ_{MLE} and C are used to produce new values for φ_i. These then produce a new antiparticle set by minimizing g subject to $\lambda = \varphi_i$.

$$\mathbf{x}^i_{k|k} = \arg\min_{\mathbf{x}_k} g(\mathbf{x}_k, \varphi_i) \tag{25}$$

$$\Delta\mathbf{x}^i_{k|k} = (W_{xx})^{-1} w_x|_{\mathbf{x}_k = \mathbf{x}^i_{k|k}, \lambda = \varphi_i} \tag{26}$$

We again use the Gauss-Newton method this time starting from $\mathbf{x}^i = \mathbf{m}^{\varphi_i}_{k|k-1}$.

Gauss-Newton Iterations

In each of the three update phases we compute a Δ increment. This will in general have the wrong direction and the wrong magnitude although it will be approximately correct. We allow for the direction being wrong by applying the method iteratively stopping when the change in g is not significant. We allow for the magnitude being wrong by doing a line search in the direction of Δ and finding the minimum g value along the line. This is done by computing g at N points along the line between 0 and 2Δ. We then interpolate to the minimum using a quadratic fit to the minimum and the points to either side.[4]

5 QAF and TAF

The quadratic antiparticle filter, QAF, uses this parametrization:

$$\mathbf{m}^{\lambda} = \mu + \Lambda\lambda + \frac{1}{2}\lambda^{T}\Gamma\lambda \tag{27}$$

where $\{\pi\} = \{\mu + \Lambda, \Gamma\}$.[5] We chose C to be diagonal then:

$$\varphi_0 = 0, \quad \varphi_i = (0, \ldots, \sqrt{C_{ii}}, \ldots, 0)^T, \quad \varphi_{-i} = -\varphi_{i,}$$
$$\varphi_{i,j} = (\varphi_i + \varphi_j)/\sqrt{2}, \quad i, j \in \{1, 2, \ldots n_\lambda\}, j < i.$$

This leads to a generalization of the linear correlations in the Gaussian distribution to parabolic ones. Parabolas can not model very larger circular nonlinearities. This lead us to try circular parametrizations. The trigonometric antiparticle filter, TAF, uses this parametrization:

$$m_\theta^\lambda = \mu_\theta + \sum_j \alpha_{\theta j}\lambda_j \tag{28}$$

$$m_i^\lambda = \mu_i + \rho_i \sin\sum_j \alpha_{ij}\lambda_j + \kappa_i(\cos\sum_j \beta_{ij}\lambda_j - 1), \quad i \neq \theta \tag{29}$$

$\{\pi\} = \{\mu, \rho, \alpha, \kappa, \beta\}$.[6] For the TAF we use an additional $2n^\lambda$ values $\varphi_{\pm 2i} = \varphi_{\pm i}/2$, for $i = 1, 2, \ldots, n^\lambda$.

[4]We used N = 21.

[5]$\mathbf{\Lambda}$ is a matrix from the $\lambda \to \mathbf{x}$ spaces and x vector $\mathbf{\Gamma}$ is a symmetric matrix wrt λ.

[6]α and β are matrices from the $\lambda \to \mathbf{x}$ spaces.

6 Experiments

We did simulations of a robot moving in 2D.

$$\mathbf{f}(\mathbf{x}, \mathbf{u}) = \begin{pmatrix} x \\ y \\ \theta \end{pmatrix} + \begin{pmatrix} \cos\theta & 0 \\ \sin\theta & 0 \\ 0 & 1 \end{pmatrix} \begin{pmatrix} \Delta s \\ \Delta\theta \end{pmatrix} \tag{30}$$

With $Q_k = J_u Q J^T + 10^{-10} I$ where J_u is the Jacobian of **f** wrt **u** and Q is a 2×2 diagonal matrix. The observations were range and bearing to point features with known data associations. Our UKF and PF[7] implementations were as in [33] and the IUKF was an iterative version.

For the first experiment we did 800 simulations for each of 3 values of Q, small, medium and large.[8] The robot started at the center of a circle of features and moved with odometery reading steady increments $(0.2, 0, 0)^T$ and the true path having noise added to that according to Q. The experiment was designed to have the robot first see one feature giving partial pose information and then later a second feature giving the exact pose. The consistency and accuracy was analyzed at the point before the first update, at the update, before the second feature update, at the second feature update, and then at selected times after the second feature was observed.

An example of one run is shown in Fig. 3. One can see how the AF is able to approximately model the crescent shaped distribution. Figure 4 shows the root mean square xy error as it evolved after the first update. We see that the QAF and the TAF converged more rapidly than the other methods. Notice that the PF tends to gets stuck at a fixed error depending on the density of particles while the other methods improve the pose more rapidily. Figure 4 also shows the number of runs with errors outside a region of ± 1, ± 1, ± 0.1 in (x, y, θ) error.[9] This indicates the divergent estimates as the measurement of two features should bring the pose into that region. In order to evaluate consistency we computed a test statistic based on the mahalanobis distance

$$d^2 = (\mathbf{x} - \mathbf{x}_{true})^T (Cov_x)^{-1} (\mathbf{x} - \mathbf{x}_{true}) \tag{31}$$

The cumulative χ_3^2 distribution of d^2 was recorded at each time. This would be uniformly distributed if the error were Gaussian and the estimates were consistent. The errors should become approximately Gaussian after sufficient updates are performed. By sorting these values and plotting them we can compare the empirical distribution to the ideal straight line in the top left of Fig. 5. It is not possible to show all these plots but we can summarize the consistency by plotting the

---

[7]As in Thrun pp. 220–228 (UKF), 250–252 (PF) and 110 (low variance sampling).

[8]Q was 10^{-5}, 10^{-4}, or 10^{-3} along the diagonal.

[9]The simulated distance units are arbitrary but can be thought of as meters while the angles are in radians.

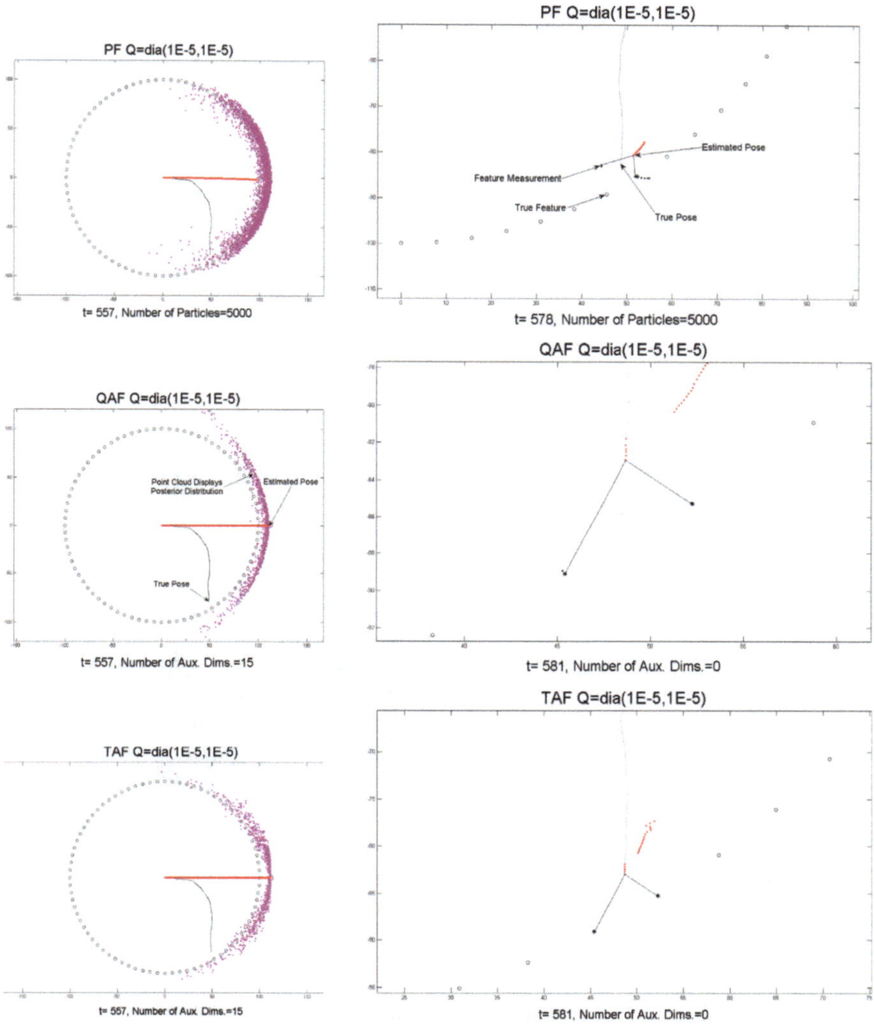

Fig. 3 Here we show the posterior distribution for the PF top (here we show 5,000 particles, while we later decided to use 2,000 and 20,000 for the experiments) QAF center and the TAF bottom. The *left* is just before observing the first feature and the right is 5 steps after observing the second features. The QAF and TAF posterior are displayed by sampling from it. The estimate is the *thick* (*red*) *line* while the *thin* (*black*) track is the true path. We see that for this outlier simulation on the *left* the PF is more consistent in its shape and the QAF is the least consistent. The better ends of the crescent for the TAF help it to update more accurately than the QAF as seen by the slight error in the first updates of the second feature shown as dots in the *right column figures*. In the *bottom right figure* these *dots* all lie on *top* of the true feature location (displayed as the ring of *small circles*) indicating that all the estimates explained the measurements. The QAF had the first of update with the second feature not quite line up the measurement of the feature with its true location. Both TAF and QAF could correct the pose accurately after seeing the second feature two times. The PF is hopelessly off as it had no particle near the correct pose. The same run for the 4 Gaussian estimators EKF, IEKF, UKF, and IUKF had the robot pose outside of the *circle* of features heading in. Clearly a Gaussian ellipse can not represent this situation

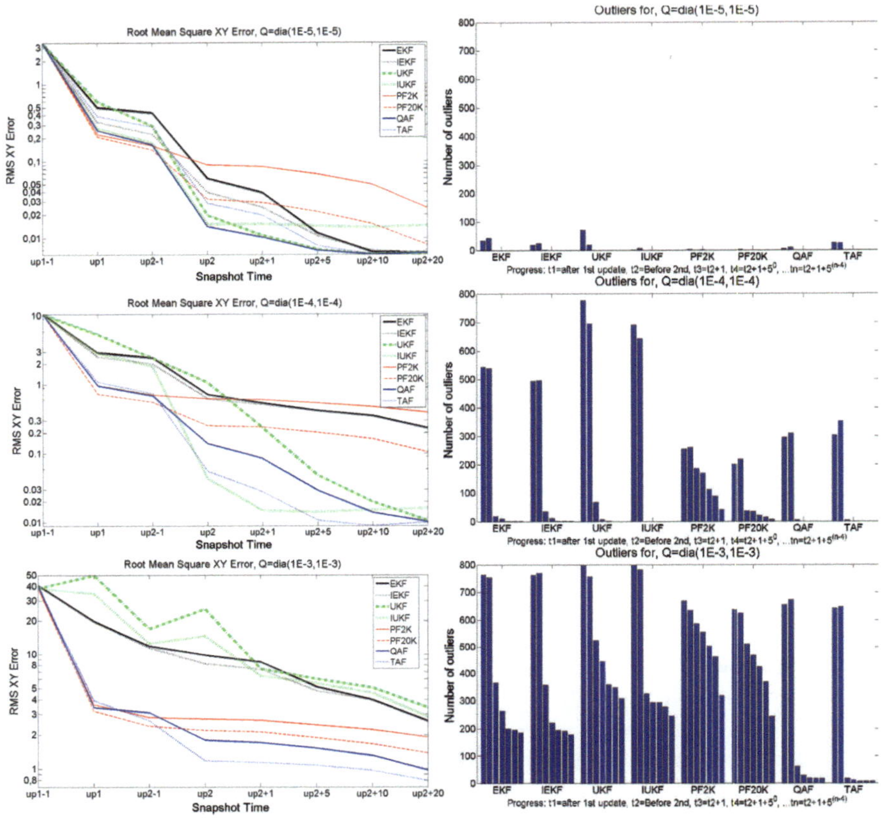

Fig. 4 *Left* is the root mean square error in x-y distance for small, medium and big Q. On the *right* are the number of simulations with pose error outside a box ± 1, ± 1, ± 0.1 for times: the first update, before second feature, the second feature update, then 1, 5, 10 and 20 after the second feature update. The AF are both seen to be more accurate than any of the other estimators

Kolmogorov-Smirnov test statistic for all plots over time relative to the updates (Fig. 5). This statistic is the maximum vertical distance between the empirical and ideal curves. The effect of particle depletion is seen by the very poor PF consistency shown here. The TAF has lower values in general and reaches a 'consistent' Gaussian posterior sooner than the others.

Table 1 summarizes the average computation times for experiment 1. This is not very informative as it depends on the matlab implementation details and the difficulty of the simulation but one can get some idea of the time trade-offs. We can see

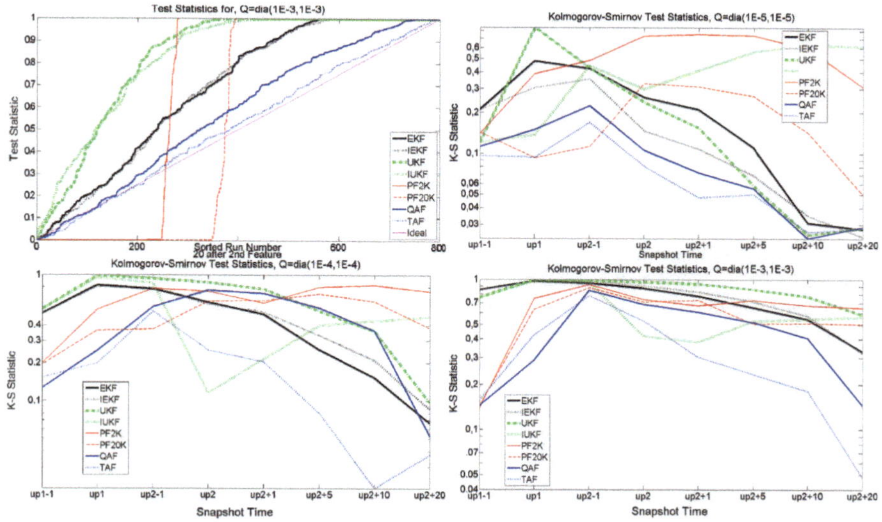

Fig. 5 The *top left* shows and example of the computation of the Kolmogrov-Smirnov, KS, statistic for the case of the big Q 20 updates after observing the second feature. The *curves* should lie on the *straight line* from 0 to 1 if the estimator was consistent. The KS-statistic is the maximum vertical distance between this *line* and the *curve*. In this case the AF estimators are much closer to consistent than the others. The evolution of the KS-statistic is shown in the other three plots for the three Q values. The TAF seems to generally be the least inconsistent estimator. The PF with 2,000 and 20,000 particles both become inconsistent

Table 1 The mean computation times for the various filters for experiment 1 in s

Q(2,2)	EKF	IEKF	UKF	IUKF	PF2K	PF20K	QAF	TAF
1E-5	0.65	0.78	0.40	1.22	1.04	8.09	1.92	1.69
1E-4	0.65	0.80	0.40	1.22	1.03	7.91	2.17	2.11
1E-3	0.57	0.81	0.35	1.05	0.94	7.24	4.55	2.38

that the TAF seemed to be 2 times faster than the QAF when the nonlinearities were large (up to 13 auxiliary dimensions for the largest Q). This is most likely due to more easily moving to the MLE for the TAF parametrization. We can see a time factor of about 2–5 for the AF with moderate nonlinearities over the Gaussian estimators.

We then looked at how the errors behave after a series of updates each with insufficient information to fully localize the robot. We did simulations with a square 4 feature map and a true path that observed each feature in turn. We can see in Fig. 6 that the AF were less sensitive to the build up of errors around the square.

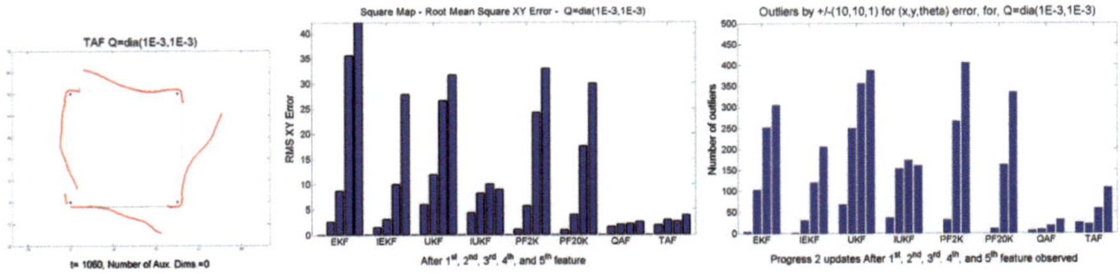

Fig. 6 The *left* shows and example from the 500 simulations in the second experiment. The TAF estimated path is shown as the *thicker* (*red*) *curve* while the true path is the *thin black line*. The four features are observed in turn starting and ending with the *left bottom corner*. We see that the estimate is able to mostly correct the angular error each time. *Center* we show the root mean square xy error 2 steps after each feature is used to correct the pose. The 2 steps is to let the estimate settle after the large change of first update. We see how the errors are amplified by all the estimators but the AF have much lower errors. The *right* shows the number of estimates outside a box ± 10, ± 10, ± 1 for (x, y, θ) errors at these same times. The PF with 2,000 and 20,000 particles are especially poor at this test as the particle depletion gets worse each time. The IUKF was considerable better than any of the other Gaussian estimators. The QAF was the best, slightly better than the TAF. That is probably because the distance with no updates was half that of the experiment 1. This lead to smaller angular error and less difference in shape from parabolic

7 Conclusion

We have investigated a completely new type of nonlinear filter the AF. The AF methods can be significantly more accurate than the other popular recursive non-linear estimators. The TAF was in some ways slightly better than the QAF on this problem for which it was tailored. The QAF is a more general filter. We also showed the new estimators were significantly more consistent than the EKF, IEKF, UKF, IUKF, and PF on a robot localization problem in 2D. In general the PF will be able to represent more shapes than the AF and so it is a more general solution. In particular it can estimate multimodal distributions. For problems that can be modeled approximately by an AF, the AF has strengths making it a viable choice of estimator.

Acknowledgements This work was supported by the SSF through its Centre for Autonomous Systems (CAS).

References

1. R. Kalman, A new approach filtering and prediction problems. Trans. ASME J. Basic Eng. **82** (1), 35–45 (1960)
2. L. Ljung, Asymptotic behavior of the extended kalman filter as a parameter estimator for linear systems. IEEE Trans. Autom. Control **24**(1), 36–51 (1979)
3. S. Huang, G. Dissanayake, Convergence and consistency analysis for extended kalman filter based slam. IEEE Trans. Robot. Autom. **23**(5), 1036–1050 (2007)

4. K. Reif, S. Gunther, E. Yaz, R. Unbehauen, Stochastic stability of the discrete-time extended kalman filter. IEEE Trans. Autom. Control **44**(4), 714–728 (1999)
5. K. Xiong, H. Zhang, L. Liu, Adaptive robust extended kalman filter for nonlinear stochastic systems. Control Theory Appl. IET **2**(3), 239–250 (2008)
6. W. Zhang, B.S. Chen, C. Tseng, Robust H_∞ filtering for nonlinear stochastic systems. IEEE Trans. Signal Process. **53**(2), 589–598 (2005)
7. L. Jetto, S. Longhi, G. Venturini, Development and experimental validation of an adaptive extended kalman filter for the localization of mobile robots. IEEE Trans. Robot. Autom. **15**(2), 219–229 (1999)
8. S.J. Julier, J.K. Uhlmann, Unscented filtering and nonlinear estimation, in *Proceedings of the IEEE*, vol. 92 (2004), pp. 401–422
9. E. Wan, R. Van Der Merwe, The unscented kalman filter for nonlinear estimation, in *Adaptive Systems for Signal Processing, Communications, and Control Symposium 2000. AS-SPCC. The IEEE 2000*, vol. 1, no. 1 (2000), pp. 153–158
10. K. Xiong, H. Zhanga, C. Chan, Performance evaluation of UKF-based nonlinear filtering. Automatica **42**(2), 261–270 (2006)
11. K. Ito, K. Xiong, Gaussian filters for nonlinear filtering problems. IEEE Trans. Autom. Control **45**(5), 910–927 (2000)
12. T. Lefebvre, H. Bruyninck, J.D. Schutter, Kalman filters for nonlinear systems: a comparison of performance. Int. J. Control **77**(7), 639–653 (2004)
13. I. Arasaratnam, S. Haykin, R.J. Elliott, Discrete-time nonlinear filtering algorithms using gauss hermite quadrature, in *Proceedings of the IEEE*, vol. 95 (2007), pp. 953–977
14. J. Clark, R. Vinter, M. Yaqoob, Shifted rayleigh filter: a new algorithm for bearings-only tracking. IEEE Trans. Aerosp. Electr. Syst. **43**(4), 1373–1384 (2007)
15. M. Norgaard, N.K. Poulsen, O. Ravn, New developments in state estimation for nonlinear systems. Automatica **36**, 1627–1638 (2000)
16. B. Bell, F. Cathey, The iterated kalman filter update as a gauss-newton method. IEEE Trans. Autom. Control **38**(2), 294–297 (1993)
17. Y. Bar-Shalom, T. Fortmann, *Tracking and Data Association* (Academic Press, New York, NY, 1987)
18. N. Gordon, D. Salmond, A. Smith, Novel approach to nonlinear/nongaussian bayesian state estimation, in *Proceedings of the IEEE for Radar and Signal Processing*, vol. 140 (1993), pp. 107–113
19. M.S. Arulampalam, S. Maskell, N. Gordon, T. Clapp, A tutorial on particle filters for online nonlinear/non-gaussian bayesian tracking. IEEE Trans. Signal Process. **50**(2), 174–188 (2002)
20. J. Carpenter, P. Clifford, P. Fearnhead, Improved particle filters for nonlinear problems. IEEE Proc. Radar Sonar Navig. **146**(1), 2–7 (1999)
21. F. Dellaert, D. Fox, W. Burgard, S. Thrun, Monte carlo localization for mobile robots, in *Proceedings of the IEEE International Conference on Robotics and Automation* (1999), pp. 1322–1328
22. M. Montemerlo et al., FastSLAM: A factored solution to the simultaneous localization and mapping problem, in *Proceedings of the National Conference on Artificial Intelligence (AAAI-02)* (Edmonton, Canada, 2002)
23. G. Grisetti, C. Stachniss, W. Burgard, Improving grid-based slam with rao-blackwellized particle filters by adaptive proposals and selective resampling, in *Proceedings of the IEEE Int. Conference on Robotics and Automation* (2005), pp. 2432–2437
24. J. Kotecha, P. Djuric, Gaussian particle filtering. IEEE Trans. Signal Process. **51**(10), 2592–2601 (2003)
25. H. Durrant-Whyte, S. Majumder, S. Thrun, M. de Battistal, S. Scheding, A bayesian algorithm for simultaneous localisation and map building, in *Springer Tracts in Advanced Robotics*, vol. 6 (2003), pp. 49–60
26. Y. Bresler, A. Macovski, Exact maximum likelihood parameter estimation of superimposed exponential signals in noise. IEEE Trans. Acoust. Speech Signal Process. **34**(5), 1081–1089 (1986)

27. K.R. Muske, J.B. Rawlings, *Nonlinear Moving Horizon State Estimation* (Kluwer, 1995)
28. C. Rao, J. Rawlings, D. Mayne, Constrained state estimation for nonlinear discrete-time systems: stability and moving horizon approximations. IEEE Trans. Autom. Control **48**(2), 246–258 (2003)
29. J. Folkesson, H.I. Christensen, Graphical SLAM: a self-correcting map, in *Proceedings of the IEEE International Conference on Robotics and Automation (ICRA04)*, vol. 1 (2004)
30. F. Dellaert, Square root sam: Simultaneous location and mapping via square root information smoothing, in *Robotics: Science and Systems* (2005)
31. M. Briers, A. Doucet, S. Singh, Sequential auxiliary particle belief propagation, in *Proceedings of the IEEE International Conference on Information Fusion*, vol. 1 (2005), pp. 1–8
32. J. Folkesson, Robustness of the quadratic antiparticle filter for robot localization, in *Proceedings of the European Conference on Mobile Robots*, vol. 1 (2011)
33. S. Thrun, W. Burgard, D. Fox, *Probalistic Robotics* (MIT Press, 2005)

Visual Odometry and Mapping for Autonomous Flight Using an RGB-D Camera

Albert S. Huang, Abraham Bachrach, Peter Henry, Michael Krainin, Daniel Maturana, Dieter Fox and Nicholas Roy

Abstract RGB-D cameras provide both a color image and per-pixel depth estimates. The richness of their data and the recent development of low-cost sensors have combined to present an attractive opportunity for mobile robotics research. In this paper, we describe a system for visual odometry and mapping using an RGB-D camera, and its application to autonomous flight. By leveraging results from recent state-of-the-art algorithms and hardware, our system enables 3D flight in cluttered environments using only onboard sensor data. All computation and sensing required for local position control are performed onboard the vehicle, reducing the dependence on unreliable wireless links. We evaluate the effectiveness of our system for stabilizing and controlling a quadrotor micro air vehicle, demonstrate its use for constructing detailed 3D maps of an indoor environment, and discuss its limitations.

A.S. Huang (✉) · A. Bachrach · N. Roy
Computer Science and Artificial Intelligence Laboratory, Massachusetts Institute of Technology, Cambridge, MA 02139, USA
e-mail: albert@csail.mit.edu

A. Bachrach
e-mail: abachrac@csail.mit.edu

N. Roy
e-mail: nickroy@csail.mit.edu

P. Henry · M. Krainin · D. Fox
Department of Computer Science and Engineering, University of Washington, Seattle, WA, USA
e-mail: peter@cs.washington.edu

M. Krainin
e-mail: mkrainin@cs.washington.edu

D. Fox
e-mail: fox@cs.washington.edu

D. Maturana
Department of Computer Science, Pontificia Universidad Católica de Chile, Santiago, Chile
e-mail: dimatura@puc.cl

H.I. Christensen and O. Khatib (eds.), *Robotics Research*,
Springer Tracts in Advanced Robotics 100, DOI 10.1007/978-3-319-29363-9_14

1 Introduction

Stable and precise control of an autonomous micro air vehicle (MAV) demands fast and accurate estimates of the vehicle's pose and velocity. In cluttered environments such as urban canyons, under a forest canopy, and indoor areas, knowledge of the 3D environment surrounding the vehicle is additionally required to plan collision-free trajectories. Navigation systems based on wirelessly transmitted information, such as Global Positioning System (GPS) technologies, are not typically useful in these scenarios due to limited range, precision, and reception. Thus, the MAV must estimate its state and plan collision-free trajectories using only its onboard sensors.

RGB-D cameras capture RGB color images augmented with depth data at each pixel. A variety of techniques can be used for producing the depth estimates, such as time-of-flight imaging, structured light stereo, dense passive stereo, laser range scanning, etc. While many of these technologies have been available to researchers for years, the recent application of structured light RGB-D cameras to home entertainment and gaming [32] has resulted in the wide availability of low-cost RGB-D sensors well-suited for robotics applications. In particular, the Microsoft Kinect sensor, developed by PrimeSense, provides a 640×480 RGB-D image at 30 Hz. When stripped down to its essential components, the Kinect weighs 115 g—light enough to be carried by a small MAV.

Previously, we have developed algorithms for MAV flight in cluttered environments using LIDAR [3] and stereo cameras [1]. LIDAR sensors currently available in form factors appropriate for use on a MAV are very high precision, but only provide range measurements along a plane around the sensor. Since they can only detect objects that intersect the sensing plane, they are most useful in environments characterized by vertical structures, and less so in more complex scenes.

Structured light RGB-D cameras are based upon stereo techniques, and thus share many properties with stereo cameras. The primary differences lie in the range and spatial density of depth data. Since RGB-D cameras illuminate a scene with an structured light pattern, they can estimate depth in areas with poor visual texture, but are range-limited by their projectors.

This paper presents our approach to providing an autonomous micro air vehicle with fast and reliable state estimates and a 3D map of its environment by using an on-board RGB-D camera and inertial measurement unit (IMU). Together, these allow the MAV to safely operate in cluttered, GPS-denied indoor environments. The control of a micro air vehicle requires accurate estimation of not only the position of the vehicle but also the velocity—estimates that our algorithms are able to provide. Estimating a vehicle's 3D motion from sensor data typically consists of estimating its relative motion at each time step by aligning successive sensor measurements such as laser scans or RGB-D frames, a process most often known as "visual odometry" when comparing camera or RGB-D images. The primary contribution of this paper is to provide a systematic experimental analysis of how the best practices in visual odometry using an RGB-D camera enable the control of a micro air vehicle.

Given knowledge of the relative motion of the vehicle from sensor frame to sensor frame, the 3D trajectory of the vehicle in the environment can be estimated by integrating the relative motion estimates over time. Given knowledge of the vehicle position in the environment, the locations of obstacles in each sensor frame can also be used to construct a global map. However, while often useful for local position control and stability, visual odometry methods suffer from long-term drift and are not suitable for building large-scale maps. To solve this problem, we also demonstrate how our previous work on RGB-D Mapping [14] can be incorporated to detect loop closures, correct for accumulated drift and maintain a representation of consistent pose estimates over the history of previous frames. We describe our overall system, justify the design decisions made, provide a ground-truth evaluation, and discuss its capabilities and limitations.

2 Related Work

Visual odometry refers to the process of estimating a vehicle's 3D motion from visual imagery alone, and dates back to Moravec's work on the Stanford cart [25]. The basic algorithm used by Moravec and others since then is to identify features of interest in each camera frame, estimate depth to each feature (typically using stereo), match features across time frames, and then estimate the rigid body transformation that best aligns the features over time. Since then, a great deal of progress has been made in all aspects of visual odometry. Common feature detectors in modern real-time algorithms include Harris corners [12] and FAST features [33], which are relatively quick to compute and resilient against small viewpoint changes. Methods for robustly matching features across frames include RANSAC-based methods [18, 22, 28] and graph-based consistency algorithms [17]. In the motion estimation process, techniques have ranged from directly minimizing Euclidean distance between matched features [16], to minimizing pixel reprojection error instead of 3D distance [28]. When computation constraints permit, bundle adjustment has been shown to help reduce integrated drift [22].

Visual odometry estimates local motion and generally has unbounded global drift. To bound estimation error, it can be integrated with simultaneous localization and mapping (SLAM) algorithms, which employ loop closing techniques to detect when a vehicle revisits a previous location. Most recent visual SLAM methods rely on fast image matching techniques [26, 35] for loop closure. As loops are detected, a common approach is to construct a pose graph representing the spatial relationships between positions of the robot during its trajectory and environmental features, creating constraints that link previous poses. Optimization of this pose graph results in a globally aligned set of frames [10, 19, 30]. For increased visual consistency, Sparse Bundle Adjustment (SBA) [37] can be used to simultaneously optimize the poses and the locations of observed features.

In the vision and graphics communities, a large body of work exists on alignment and registration of images for 3D modeling and dense scene reconstruction (e.g., Polleyfeys et al. [31]). However, our focus is on primarily on scene modeling for robot perception and planning, and secondarily for human situational awareness (e.g., for a human supervisor commanding the MAV).

The primary focus in the visual odometry communities has been on ground vehicles, however, there has been significant amount of research on using visual state estimation for the control of MAVs. For larger outdoor helicopters, several researchers have demonstrated various levels of autonomy using vision based state estimates [6, 20]. While many of the challenges for such vehicles are similar to smaller indoor MAVs, the payload and flight environments are quite different. For smaller MAVs operating in indoor environments, a number of researchers have used monocular camera sensors to control MAVs [2, 5, 8, 36]. However, these algorithms require specific assumptions about the environment (such as known patterns) to obtain the unknown scale factor inherent in using a monocular camera. Previous work in our group used a stereo camera to stabilize a MAV in unknown indoor environments [1], but the computation had to be performed offboard, and no higher level mapping or SLAM was performed. Finally, there has been considerable work in using laser range finders for MAV navigation and control [3, 11, 13, 34] with the limitations discussed earlier in this paper.

3 Approach

The problem we address is that of a quadrotor helicopter navigating in an unknown environment. The quadrotor must use the onboard RGB-D sensor to estimate its own position (local estimation), build a dense 3D model of the environment (global simultaneous localization and mapping) and use this model to plan trajectories through the environment.

Our algorithms are implemented on the vehicle shown in Fig. 1. The vehicle is a Pelican quadrotor manufactured by Ascending Technologies GmbH. The vehicle has a maximal dimension of 70 cm, and a payload of up to 1000 g. We have mounted a stripped down Microsoft Kinect sensor and connected it to the onboard flight computer. The flight computer, developed by the Pixhawk project at ETH Zurich [24], is a 1.86 GHz Core2Duo processor with 4 GB of RAM. The computer is powerful enough to allow all of the real-time estimation and control algorithms to run onboard the vehicle.

Following our previous work, we developed a system that decouples the real-time local state estimation from the global simultaneous localization and mapping (SLAM). The local state estimates are computed from visual odometry (Sect. 3.1), and to correct for drift in these local estimates the estimator periodically

Fig. 1 Our quadrotor micro air vehicle (MAV). The RGB-D camera is mounted at the base of the vehicle, tilted slightly down

incorporates position corrections provided by the SLAM algorithm (Sect. 3.2). This architecture allows the SLAM algorithm to use much more processing time than would be possible if the state estimates from the SLAM algorithm were directly used to control the vehicle.

3.1 Visual Odometry

The visual odometry algorithm that we have developed is based around a standard stereo visual odometry pipeline, with components adapted from existing algorithms. While most visual odometry algorithms follow a common architecture, a large number of variations and specific approaches exist, each with its own attributes. The contribution of this paper is to specify the steps of our visual odometry algorithm and compare the alternatives for each step. In this section we specify these steps, and in Sect. 4 we provide the experimental comparison of each step in the visual odometry pipeline. Our overall algorithm is most closely related to the approaches taken by Mei et al. [23] and Howard [17].

1. **Image Preprocessing**: An RGB-D image is first acquired from the RGB-D camera (Fig. 2). The RGB component of the image is converted to grayscale and smoothed with a Gaussian kernel of $\sigma = 0.85$, and a Gaussian pyramid is constructed to enable more robust feature detection at different scales. Each level of the pyramid corresponds to one octave in scale space. Features at the higher scales generally correspond to larger image structures in the scene, which generally makes them more repeatable and robust to motion blur.
2. **Feature Extraction**: Features are extracted at each level of the Gaussian pyramid using the FAST feature detector [33]. The threshold for the FAST detector is adaptively chosen using a simple proportional controller to ensure a

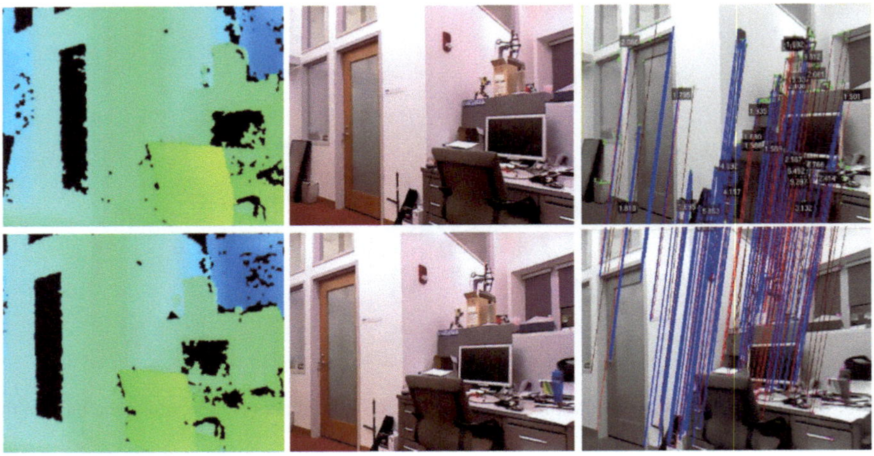

Fig. 2 The input RGB-D data to the visual odometry algorithm alongside the detected feature matches. The *top row* images are from time t, the *bottom row* images are from time $t + 1$. The *left column* is the depth image, and the *middle column* is the corresponding RGB image. The *right hand column* shows the pixels that are matched between frames, where inlier feature matches are drawn in *blue* and outliers are drawn in *red*

sufficient number of features are detected in each frame. The depth corresponding to each feature is also extracted from the depth image. Features that do not have an associated depth are discarded. To maintain a more uniform distribution of features, each pyramid level is discretized into 80×80 pixel buckets, and the 25 features in each bucket with the strongest FAST corner score are retained.

3. **Initial Rotation Estimation**: For small motions such as those encountered in successive image frames, the majority of a feature's apparent motion in the image plane is caused by 3D rotation. Estimating this rotation allows us to constrain the search window when matching features between frames. We use the technique proposed by Mei et al. [23] to compute an initial rotation by directly minimizing the sum of squared pixel errors between downsampled versions of the current and previous frames.

   One could also use an IMU or a dynamics model of the vehicle to compute this initial motion estimate, however the increased generality of the image based estimate is preferable, while providing sufficient performance. An alternative approach would be to use a coarse-to-fine motion estimation that iteratively estimates motion from each level of the Gaussian pyramid, as proposed by Johnson et al. [18].
4. **Feature Matching**: Each feature is assigned an 80-byte descriptor consisting of the brightness values of the 9×9 pixel patch around the feature, normalized to zero mean and omitting the bottom right pixel. The omission of one pixel results in a descriptor length more suitable for vectorized instructions. Features are then matched across frames by comparing their feature descriptor values using a

mutual-consistency check [28]. The match score between two features is the sum-of-absolute differences (SAD) of their feature descriptors [17], which can be quickly computed using SIMD instructions such as Intel SSE2. A feature match is declared when two features have the lowest scoring SAD with each other, and they lie within the search window defined by the initial rotation estimation. Once an initial match is found, the feature location in the newest frame is refined to obtain a sub-pixel match. Refinement is computed by minimizing the sum-of-square errors of the descriptors, using ESM to solve the iterative nonlinear least squares problem [4]. We also use SIMD instructions to speed up this process.

5. **Inlier Detection**: Although the constraints imposed by the initial rotation estimation substantially reduce the rate of incorrect feature matches between frames, an additional step is necessary to further prune away bad matches. We follow Howard's approach of computing a graph of consistent feature matches, and then using a greedy algorithm to approximate the maximal clique in the graph [15, 17]. The graph is constructed according to the fact that rigid body motions are distance-preserving operations—the Euclidean distance between two features at one time should match their distance at another time. Thus, each pair of matched features across frames is a vertex in the graph, and an edge is formed between two such pairs of matched feature if the 3D distance between the features does not change substantially from the prior frame to the subsequent frame. For a static scene, the set of inliers make up the maximal clique of consistent matches. The max-clique search is approximated by starting with an empty set of matched feature pairs and iteratively adding matched feature pairs with greatest degree that is consistent with all matched feature pairs in the clique (Fig. 2). Overall, this algorithm has a runtime quadratic in the number of matched feature pairs, but runs very quickly due to the speed of the consistency checking. In Sect. 4, we compare this approach to RANSAC-based methods [22, 28].
6. **Motion Estimation**: The final motion estimate is computed from the matched features in three steps. First, Horn's absolute orientation method provides an initial estimate by minimizing the Euclidean distances between the inlier feature matches [16]. Second, the motion estimate is refined by minimizing feature reprojection error using a nonlinear least-squares solver [4]. This refinement step implicitly accounts for the increase in depth uncertainty with range due to the fact that the depth estimates are computed by stereo matching in image space. Finally, feature matches exceeding a fixed reprojection error threshold are discarded from the inlier set and the motion estimate is refined once again.
   To reduce short-scale drift, we additionally use a keyframe technique. Motion is estimated by comparing the newest frame against a reference frame. If the camera motion relative to the reference frame is successfully computed with a sufficient number of inlier features, then the reference frame is not changed. Otherwise, the newest frame replaces the reference frame after the estimation is finished. If motion estimation against the reference frame fails, then the motion

estimation is tried again with the second most recent frame. This simple heuristic serves to eliminate drift in situations where the camera viewpoint does not vary significantly, a technique especially useful when hovering.

3.2 Mapping

Visual odometry provides locally accurate pose estimates; however global consistency is needed for metric map generation and navigation over long time-scales. We therefore integrate our visual odometry system with our previous work in RGBD-Mapping [14]. This section focuses on the key decisions required for real-time operation; we refer readers to our previous publication for details on the original algorithm that emphasizes mapping accuracy [14].

Unlike the local pose estimates needed for maintaining stable flight, map updates and global pose updates are not required at a high frequency and can therefore be processed on an offboard computer. The MAV transmits RGB-D data to an offboard laptop, which detects loop closures, computes global pose corrections, and constructs a 3D log-likelihood occupancy grid map. For coarse navigation, we found that a grid map with 10 cm resolution provided a useful balance between map size and precision. Depth data is downsampled to 128×96 prior to a voxel map update to increase the update speed, resulting in spacing between depth pixels of approximately 5 cm at a range of 6 m. Incorporating a single frame into the voxel map currently takes approximately 1.5 ms.

As before, we adopt a keyframe approach to loop closure—new RGB-D frames are matched against a small set of keyframes to detect loop closures, using a fast image matching procedure [14]. New keyframes are added when the accumulated motion since the previous keyframe exceeds either 10° in rotation or 25 cm in translation. When a new keyframe is constructed, a RANSAC procedure over FAST keypoints [33] compares the new keyframe to keyframes occurring more than 4 s prior. As loop closure requires matching non-sequential frames, we obtain putative keypoint matches using Calonder randomized tree descriptors [7]. A new keypoint is considered as a possible match to an earlier frame if the L2 distance to the most similar descriptor in the earlier frame has a ratio less than 0.6 with the next most similar descriptor. RANSAC inlier matches establish a relative pose between the frames, which is accepted if there are at least 10 inliers. Matches with a reprojection error below a fixed threshold are considered inliers. The final refined relative pose between keyframes is obtained by solving a two-frame sparse bundle adjustment (SBA) system, which minimizes overall reprojection error.

To keep the loop closure detection near constant time as the map grows, we limit the keyframes against which the new keyframe is checked. First, we only use keyframes whose pose differs from the new frame (according to the existing estimates) by at most 90° in rotation and 5 m in translation. We also use Nistér's vocabulary tree approach [29], which uses a quantized "bag of visual words" model to rapidly determine the 15 most likely loop closure candidates. Keyframes that pass these tests are matched against new frames, and matching is terminated after the first successful loop closure. On each successful loop closure, a new constraint is added to a pose graph, which is then optimized using TORO [9]. Pose graph optimization is typically fast, converging in roughly 30 ms. Corrected pose estimates are then transmitted back to the vehicle, along with any updated voxel maps.

Greater global map consistency can be achieved using a sparse bundle adjustment technique that optimizes over all matched features across all frames [21]. However, this is a much slower approach and not yet suitable for real-time operation.

3.3 State Estimation and Control

To control the quadrotor, we integrated the new visual odometry and RGB-D Mapping algorithms into our system previously developed around 2D laser scan-matching and SLAM [3]. The motion estimates computed by the visual odometry are fused with measurements from the onboard IMU in an Extended Kalman Filter. The filter computes estimates of both the position and velocity, which are used by the PID position controller to stabilize the position of the vehicle.

We keep the SLAM process separate from the real-time control loop, instead having it provide corrections for the real-time position estimates. Since these position corrections are delayed significantly from when the measurement upon which they were based was taken, we must account for this delay when we incorporate the correction by retroactively modifying the appropriate position estimate in the state history. All future state estimates are then recomputed from this corrected position, resulting in globally consistent real-time state estimates.

By incorporating the SLAM corrections after the fact, we allow the real-time state estimates to be processed with low enough delay to control the MAV, while still incorporating the information from SLAM to ensure drift free position estimation.

4 Experiments

This section presents results that compare our design decisions with other approaches, especially with respect to the ways these decisions affect autonomous flight. First, we compare our approach to visual odometry and mapping with alternatives. In some cases, computational speed is preferred over accuracy. Second, we present results using the RGB-D camera to stabilize and control a MAV. We characterize the performance of the system as a whole, including its limitations.

4.1 Visual Odometry

There are a variety of visual odometry methods, and the existing literature is often unclear about the advantages and limitations of each. We present results comparing a number of these approaches and analyze their performance. As is true in many domains, the tradeoffs can often be characterized as increased accuracy at the expense of additional computational requirements. In some cases, the additional cost is greatly offset by the improved accuracy.

We conducted a number of experiments using a motion capture system that provides 120 Hz ground truth measurements of the MAV's position and attitude. The motion capture environment can be characterized as a single room approximately 11 m $\times$ 7 m $\times$ 4 m in size, lit by overhead fluorescent lights and with a wide variation of visual clutter—one wall is blank and featureless, and the others have a varying number of objects and visual features (see Fig. 3). While this is not a large volume, it is representative of many confined, indoor spaces, and provides the opportunity to directly compare against ground truth.

We recorded a dataset of the MAV flying various patterns through the motion capture environment. Substantial movement in X, Y, Z, and yaw were all recorded, with small deviations in roll and pitch. We numerically differentiated the motion capture measurements to obtain the vehicle's ground truth 3D velocities, and compared them to velocities and trajectories as estimated by the visual odometry and mapping algorithms.

Table 1 shows the performance of our integrated approach, and its behavior when adjusting different aspects of the algorithm. Each experiment varied a single aspect from our approach. We present the mean velocity error magnitude, the overall computation time per RGB-D frame, and the *gross failure rate*. We define a gross failure to be any instance where the visual odometry algorithm was either unable to produce a motion estimate (e.g., due to insufficient feature matches) or where the error in the estimated 3D velocities exceeded a fixed threshold of 1 m/s. Timing results were computed on a 2.67 GHz laptop computer.

The dataset was designed to challenge vision-based approaches to the point of failure, and includes motion blur and feature-poor images, as would commonly be encountered indoors and under moderate lighting conditions. Our algorithm had a mean velocity error of 0.387 m/s and a 3.39 % gross failure rate, and is unlikely to

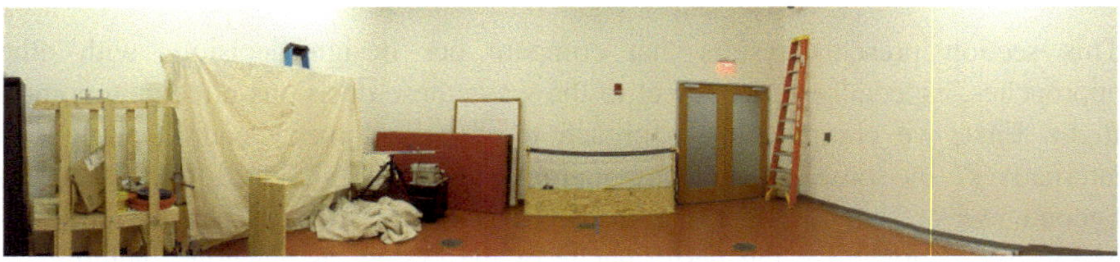

Fig. 3 Panorama photograph of the motion capture room used to conduct our ground-truth experiments. Visual feature density varies substantially throughout this room

Table 1 Comparison of various approaches on a challenging dataset

	Velocity error (m/s)	% gross failures	Total time (ms)
Our approach	0.387 ± 0.004	3.39	14.7
Inlier detection			
RANSAC	0.412 ± 0.005	6.05	15.3
Preemptive RANSAC	0.414 ± 0.005	5.91	14.9
Greedy max-clique—*our approach*	0.387 ± 0.004	3.39	14.7
Initial rotation estimate			
None	0.388 ± 0.004	4.22	13.6
Gaussian pyramid levels			
1	0.387 ± 0.004	5.17	17.0
2	0.385 ± 0.004	3.52	15.1
3—*our approach*	0.387 ± 0.004	3.39	14.7
4	0.387 ± 0.004	3.50	14.5
Reprojection error minimization			
Bidir. Gauss-Newton	0.387 ± 0.004	3.24	14.7
Bidir. ESM—*our approach*	0.387 ± 0.004	3.39	14.7
Unidir. Gauss-Newton	0.391 ± 0.004	3.45	14.6
Unidir. ESM	0.391 ± 0.004	3.47	14.6
Absolute orientation only	0.467 ± 0.005	10.97	14.4
Feature window size			
3	0.391 ± 0.004	5.96	12.8
5	0.388 ± 0.004	4.24	13.7
7	0.388 ± 0.004	3.72	14.2
9—*our approach*	0.387 ± 0.004	3.39	14.7
11	0.388 ± 0.004	3.42	15.7
Subpixel feature refinement			
No refinement	0.404 ± 0.004	5.13	13.1
Adaptive FAST threshold			
Fixed threshold (10)	0.385 ± 0.004	3.12	15.3
Feature grid/bucketing			
No grid	0.398 ± 0.004	4.02	24.6

Error is computed using a high resolution motion capture system for ground truth

have been capable of autonomously flying the MAV through the entire recorded trajectory. In contrast, in environments with richer visual features, we have observed mean velocity errors of 0.08 m/s, with no gross failures, significantly lower than the values reported in Table 1.

Inlier detection RANSAC based methods [28] are more commonly used than the greedy max-clique approach. We tested against two RANSAC schemes, traditional RANSAC and Preemptive RANSAC [27]. The latter attempts to speed up

RANSAC by avoiding excessive scoring of wrong motion hypotheses. In our experiments, when allocated a comparable amount of computation time (by using 500 hypotheses), greedy max-clique outperformed both.

Initial rotation estimation A good initial rotation estimate can help constrain the feature matching process and reduce the number of incorrect feature matches. Disabling the rotation estimate results in slightly faster runtime, but more frequent estimation failures.

Gaussian pyramid levels Detecting and matching features on different levels of a Gaussian pyramid provides provides resilience against motion blur and helps track larger features.

Reprojection error We compared undirectional motion refinement, which minimizes the reprojection error of newly detected features onto the reference frame, with bidirectional refinement, which additionally minimizes the reprojection error of reference features projected onto the new frame. We additionally compared a standard Gauss-Newton optimization technique with ESM. Bidirectional refinement does provide slightly more accuracy without substantially greater cost, and we found no significant difference between Gauss-Newton and ESM.

Feature window size As expected, larger feature windows result in more successful motion estimation at the cost of additional computation time. Interestingly, a very small window size of 3×3 yielded reasonable performance, a behavior we attribute to the constraints provided by the initial rotation estimate.

Subpixel refinement, adaptive thresholding, and feature bucketing We found the accuracy improvements afforded by subpixel feature refinement outweighed its additional computational cost. While the lighting in the motion capture experiments did not substantially change, adaptive thresholding still yielded a lower failure rate. We would expect the accuracy difference to be greater when flying through more varied lighting conditions. Finally, without feature bucketing, the feature detector often detected clusters of closely spaced features, which in turn confused the matching process and resulted in both slower speeds and decreased accuracy.

Timing On the 2.6 GHz laptop computer used for comparisons, our algorithm requires roughly 15 ms per frame. The timing per stage is as follows. Preprocessing: 2.1 ms, feature extraction: 3.1 ms, initial rotation estimation: 1.0 ms, feature matching: 6.0 ms, inlier detection: 2.2 ms, and motion estimation required less than 0.1 ms. Runtimes for the computer onboard the MAV are roughly 25 ms per frame due to the slower clock speed (1.86 GHz), but are still well within real-time.

4.2 Mapping and Autonomous Flight

In addition to evaluating the visual odometry algorithms against motion capture results, we also conducted a number of autonomous flight experiments in the motion capture system and in larger environments. In these experiments, the vehicle

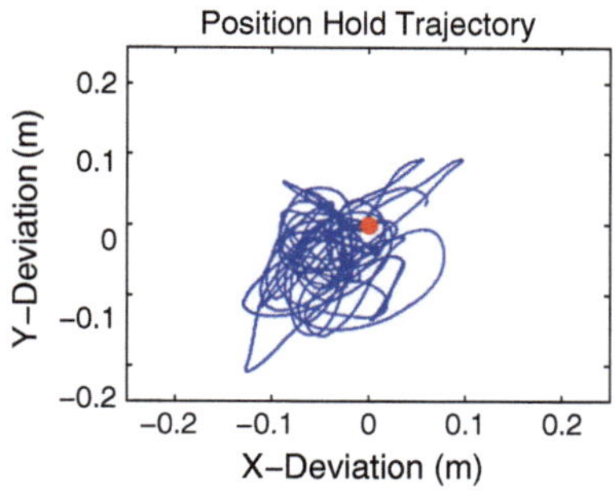

Metric	
Duration	90 s
Mean speed	0.10 m/s
Mean pos. deviation	6.2 cm
Max pos. deviation	19 cm

Fig. 4 A plot showing the ground truth trajectory of the vehicle during position hold. The *red dot* near the center is the origin around which the vehicle was hovering. The vehicle was controlled using visual odometry, and its position measured with a motion capture system

flew autonomously with state estimates provided by the algorithms presented in this paper. The vehicle was commanded through the environment by a human operator selecting destination waypoints using a graphical interface.

Figure 4 shows an example trajectory where the MAV was commanded to hover at a target point, along with statistics about how well it achieved this goal. The ground truth trajectory and performance measures were recorded with the motion capture system.

In addition to the flights performed in the small motion capture environment, we have flown in a number of locations around the MIT campus, and at the Intel Research office in Seattle. Two such experiments are shown in Fig. 5.

As the MAV covers greater distances, the RGB-D mapping algorithm limits the global drift on its position estimates by detecting loop closures and correcting the trajectory estimates. The trajectory history can then be combined with the RGB- D sensor data to automatically generate maps that are useful both for a human operator's situational awareness, and for autonomous path planning and decision

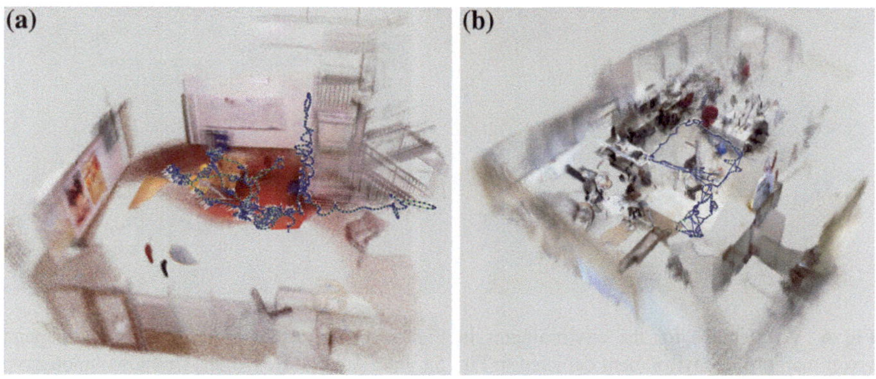

Fig. 5 Trajectories flown by the MAV in two navigation experiments

making. While the ground truth position estimates are not available, the quality of the state estimates computed by our system is evident in the rendered point cloud. A video demonstrating autonomous flight and incremental mapping is available at http://groups.csail.mit.edu/rrg/isrr2011-mav.

4.3 Navigation

Figure 6a shows an occupancy voxel map populated using the dense depth data provided by the RGB-D sensor. These occupancy maps can be used for autonomous path planning and navigation in highly cluttered environments, enabling flight through tight passageways and in close proximity to obstacles. Figure 6b shows a rendering of the MAV's internal state estimate as it flew through the environment depicted in Fig. 5b, and a path planned using the occupancy map and a simple dynamic programming search strategy. While these renderings are not necessary for obstacle avoidance, they would serve to provide a human operator with greater situational awareness of the MAV's surrounding environment.

5 Discussion and Future Work

The system described in this paper enables autonomous MAV flight in many unknown indoor environments. However, there remain a great number more challenging situations that would severely tax our system's abilities. Motion estimation algorithms based on matching visual features, such as ours and virtually all

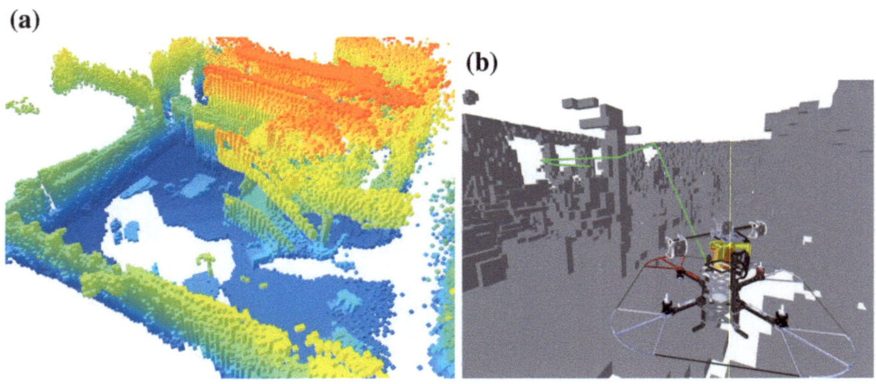

Fig. 6 Voxel maps for the environments in Fig. 5. **a** Dense maximum-likelihood occupancy voxel map of the environment depicted in Fig. 5a, false-colored by height. Unknown/unobserved cells are also tracked, but not depicted here. **b** A voxel map of the environment in Fig. 5b allows the vehicle to plan a collision-free 3D trajectory (*green*)

other visual odometry techniques, do not perform as well in regions with few visual features. In large open areas, the visible structure is often far beyond the maximum range of the Kinect. As a result, the system actually performs better in cluttered environments and in close quarters than it does in wide open areas. Handling these challenges will likely require the integration of other sensors such as conventional stereo cameras or laser range-finders. As these sensors have different failure modes, they serve to complement each other's capabilities. Additional sensing modalities can reduce, but not eliminate, state estimation failures. Further robustness can be gained by designing planning and control systems able to respond appropriately when the state estimates are extremely uncertain, or to plan in ways that minimize future uncertainty [13].

Our state estimation algorithms assume a static environment, and assume that the vehicle moves relatively slowly. As the vehicle flies faster, the algorithms will need to handle larger amounts of motion blur, and other artifacts resulting from the rolling shutter in the Kinect cameras. Larger inter-frame motions resulting from greater speeds may in turn require more efficient search strategies to retain the real-time estimation capabilities required to control the vehicle. Relaxing the static environment assumptions will likely require better ways of detecting the set of features useful for motion estimation. When moving objects comprise a substantial portion of the visible image, the maximal clique of consistent feature matches may not correspond to the static environment.

Further work is also required to improve the accuracy and efficiency of the presented algorithms. Currently, the visual odometry, sensor fusion, and control algorithms are able to run onboard the vehicle; however, even with the modifications discussed in Sect. 3.2, the loop closing and SLAM algorithms are not quite fast enough to be run using the onboard processor. In other cases, we have actively traded estimation accuracy for computational speed. Figure 7 shows the mapping accuracy possible with further processing time, using more computationally intensive techniques presented in our previous work [14].

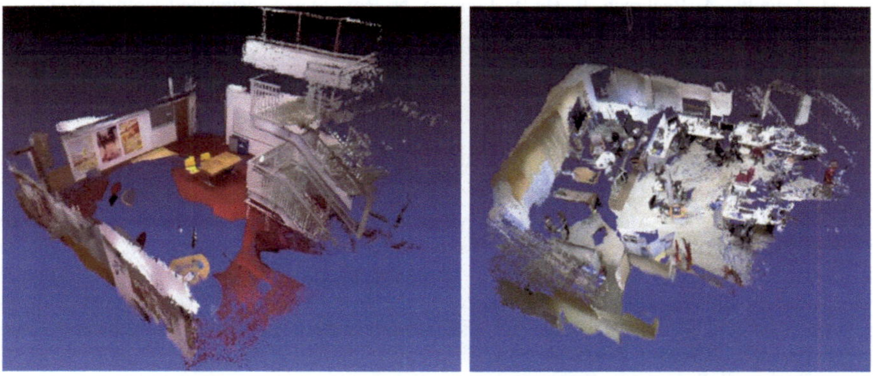

Fig. 7 Textured surfaces generated offline using sparse bundle adjustment, with data collected from autonomous flights

While the maps presented in this paper are fairly small, the methods presented scale to much larger environments. We have previously demonstrated building-scale mapping with a hand-collected data set [14], although autonomous map construction of very large spaces will require exploration algorithms that keep the vehicle well localized (e.g., in visually rich areas).

6 Conclusion

This paper presents an experimental analysis of our approach to enabling autonomous flight using an RGB-D sensor. Our system combines visual odometry techniques from the existing literature with our previous work on autonomous flight and mapping, and is able to conduct all sensing and computation required for local position control onboard the vehicle. Using the RGB-D sensor, our system is able to plan complex 3D paths in cluttered environments while retaining a high degree of situational awareness. We have compared a variety of different approaches to visual odometry and integrated the techniques that provide a useful balance of speed and accuracy.

Acknowledgements This research was supported by the Office of Naval Research under MURI N00014-07-1-0749, Science of Autonomy program N00014-09-1-0641 and the Army Research Office under the MAST CTA. D.M. acknowledges travel support from P. Universidad Cato′lica's School of Engineering. P.H. and D.F. are supported by ONR MURI grant number N00014-09-1-1052, and by the NSF under contract number IIS-0812671, as well as collaborative participation in the Robotics Consortium sponsored by the U.S Army Research Laboratory under Agreement W911NF-10-2-0016.

References

1. M. Achtelik, A. Bachrach, R. He, S. Prentice, N. Roy, Stereo vision and laser odometry for autonomous helicopters in GPS-denied indoor environments, in *Proceedings of the SPIE Unmanned Systems Technology XI*, vol. 7332, Orlando, F, 2009
2. S. Ahrens, D. Levine, G. Andrews, J.P. How, Vision-based guidance and control of a hovering vehicle in unknown, GPS-denied environments, in *IEEE International Conference on Robotics and Automation*, 2009, pp. 2643–2648
3. A. Bachrach, R. He, N. Roy, Autonomous flight in unknown indoor environments. Int. J. Micro Air Veh. **1**(4), 217–228 (2009)
4. S. Benhimane, E. Malis. Improving vision-based control using efficient second-order minimization techniques, in *IEEE International Conference on Robotics and Automation*, 2004
5. M. Blösch, S. Weiss, D. Scaramuzza, R. Siegwart, Vision based MAV navigation in unknown and unstructured environments, in *IEEE International Conference on Robotics and Automation*, 2010, pp. 21–28
6. G. Buskey, J. Roberts, P. Corke, G. Wyeth. Helicopter automation using a low-cost sensing system. Comput. Control Eng. J. **15**(2), 8–9 (2004)

7. M. Calonder, V. Lepetit, and P. Fua. Keypoint signatures for fast learning and recognition, in European Conference on Computer Vision (2008), pp. 58–71
8. K. Celik, Soon J. Chung, and A. Somani. Mono-vision corner SLAM for indoor navigation, in IEEE International Conference on Electro/Information Technology (2008), pp. 343–348
9. G. Grisetti, S. Grzonka, C. Stachniss, P. Pfaff, and W. Burgard. Estimation of accurate maximum likelihood maps in 3D, in IEEE International Conference on Intelligent Robots and Systems (2007)
10. G. Grisetti, C. Stachniss, S. Grzonka, and W. Burgard. A tree parameterization for efficiently computing maximum likelihood maps using gradient descent, in Proceedings of Robotics: Science and Systems (2007)
11. S. Grzonka, G. Grisetti, W. Burgard, Towards a navigation system for autonomous indoor flying, in IEEE International Conference on Robotics and Automation (Kobe, Japan, 2009)
12. C. Harris, M. Stephens, A combined corner and edge detector, in *Alvey Vision Conference* (1988), pp. 147–151
13. R. He, S. Prentice, N. Roy, Planning in information space for a quadrotor helicopter in a GPS-denied environment, in *IEEE International Conference on Robotics and Automation* (Los Angeles, CA, 2008), pp. 1814–1820
14. P. Henry, M. Krainin, E. Herbst, X. Ren, D. Fox, RGB-D mapping: using depth cameras for dense 3d modeling of indoor environments, in *International Symposium on Experimental Robotics* (2010)
15. H. Hirschmuller, P.R. Innocent, J.M. Garibaldi, Fast, unconstrained camera motion estimation from stereo without tracking and robust statistics, in *Proceedings of International Conference on Control, Automation, Robotics and Vision*, vol. 2, 2002, pp. 1099–1104
16. B.K.P. Horn, Closed-form solution of absolute orientation using unit quaternions. J. Opt. Soc. Am. **4**(4), 629–642 (1987)
17. A. Howard, Real-time stereo visual odometry for autonomous ground vehicles, in *IEEE International Conference on Intelligent Robots and Systems* (2008)
18. A.E. Johnson, S.B. Goldberg, Y. Cheng, L.H. Matthies, Robust and efficient stereo feature tracking for visual odometry, in *IEEE International Conference on Robotics and Automation* (Pasadena, CA, 2008)
19. M. Kaess, A. Ranganathan, F. Dellaert, iSAM: incremental smoothing and mapping. IEEE Trans. Robot. (TRO) **24**(6), 1365–1378 (2008)
20. J. Kelly, G.S. Sukhatme, An experimental study of aerial stereo visual odometry, in *Proceedings of Symposium on Intelligent Autonomous Vehicles* (Toulouse, France, 2007)
21. K. Konolige, Sparse bundle adjustment, in *Proceedings of the British Machine Vision Conference (BMVC)* (2010)
22. K. Konolige, M. Agrawal, J. Sola, Large-scale visual odometry for rough terrain, in *International Symposium on Robotics Research* (Hiroshima, Japan, 2007)
23. C. Mei, G. Sibley, M. Cummins, P. Newman, I. Reid, A constant time efficient stereo SLAM system, in *British Machine Vision Conference* (2009)
24. L. Meier, P. Tanskanen, F. Fraundorfer, M. Pollefeys, Pixhawk: a system for autonomous flight using onboard computer vision, in *IEEE International Conference on Robotics and Automation* (2011)
25. H. Moravec, *Obstacle avoidance and navigation in the real world by a seeing robot rover*. PhD thesis, Stanford University (1980)
26. P. Newman, G. Sibley, M. Smith, M. Cummins, A. Harrison, C. Mei, I. Posner, R. Shade, D. Schröter, L. Murphy, W. Churchill, D. Cole, I. Reid, Navigating, recognising and describing urban spaces with vision and laser. Int. J. Robot. Res. **28**(11–12) (2009)
27. D. Nistér, Preemptive RANSAC for live structure and motion estimation. Mach. Vis. Appl. **16**, 321–329 (2005)
28. D. Nistér, O. Naroditsky, J. Bergen, Visual odometry, in *Computer Vision and Pattern Recognition* (Washington, D.C., 2004), pp. 652–659
29. D. Nistér, H. Stewenius, Scalable recognition with a vocabulary tree, in *Computer Vision and Pattern Recognition* (2006)

30. E. Olson, J. Leonard, S. Teller, Fast iterative optimization of pose graphs with poor initial estimates, in *IEEE International Conference on Robotics and Automation* (2006), pp. 2262–2269
31. M. Pollefeys, D. Nister, J.-M. Frahm, A. Akbarzadeh, P. Mordohai, B. Clipp, C. Engels, D. Gallup, S.-J. Kim, P. Merrell, C. Salmi, S. Sinha, B. Talton, L. Wang, Q. Yang, H. Stewenius, R. Yang, G. Welch, H. Towles, Detailed real-time urban 3D reconstruction from video. Int. J. Comput. Vis. **72**(2), 143–167 (2008)
32. PrimeSense. http://www.primesense.com
33. E. Rosten, T. Drummond, Machine learning for high-speed corner detection, in *European Conference on Computer Vision* (2006)
34. S. Shen, N. Michael, V. Kumar, Autonomous multi-floor indoor navigation with a computationally constrained MAV, in *IEEE International Conference on Robotics and Automation* (Shanghai, China, 2011)
35. N. Snavely, S. Seitz, R. Szeliski, Photo tourism: exploring photo collections in 3D, in *ACM Transactions on Graphics (Proceedings of SIGGRAPH)* (2006)
36. B. Steder, G. Grisetti, C. Stachniss, W. Burgard, Visual SLAM for flying vehicles. IEEE Trans. Robot. **24**(5), 1088–1093 (2008)
37. B. Triggs, P. McLauchlan, R. Hartley, A. Fitzgibbon, Bundle adjustment—a modern synthesis. Vis. Algor.: Theory Pract. 153–177 (2000)

Efficient Planning in Non-Gaussian Belief Spaces and Its Application to Robot Grasping

Robert Platt, Leslie Kaelbling, Tomas Lozano-Perez and Russ Tedrake

Abstract The limited nature of robot sensors make many important robotics problems partially observable. These problems may require the system to perform complex information-gathering operations. One approach to solving these problems is to create plans in *belief-space*, the space of probability distributions over the under-lying state of the system. The belief-space plan encodes a strategy for performing a task while gaining information as necessary. Most approaches to belief-space planning rely upon representing belief state in a particular way (typically as a Gaussian). Unfortunately, this can lead to large errors between the assumed density representation of belief state and the true belief state. This paper proposes a new sample-based approach to belief-space planning that has fixed computational complexity while allowing arbitrary implementations of Bayes filtering to be used to track belief state. The approach is illustrated in the context of a simple example and compared to a prior approach. Then, we propose an application of the technique to an instance of the grasp synthesis problem where a robot must simultaneously localize and grasp an object given initially uncertain object parameters by planning information-gathering behavior. Experimental results are presented that demonstrate the approach to be capable of actively localizing and grasping boxes that are presented to the robot in uncertain and hard-to-localize configurations.

R. Platt (✉) · L. Kaelbling · T. Lozano-Perez · R. Tedrake
Computer Science and Artificial Intelligence Laboratory,
Massachusetts Institute of Technology, 32 Vassar St., Cambridge, MA, USA
e-mail: rplatt@csail.mit.edu

L. Kaelbling
e-mail: lpk@csail.mit.edu

T. Lozano-Perez
e-mail: tlp@csail.mit.edu

R. Tedrake
e-mail: russt@csail.mit.edu

H.I. Christensen and O. Khatib (eds.), *Robotics Research*,
Springer Tracts in Advanced Robotics 100, DOI 10.1007/978-3-319-29363-9_15

1 Introduction

A fundamental objective of robotics is to develop systems that can perform tasks robustly even in unstructured environments. One way to achieve this is to create a planner capable of simultaneously localizing the state of the system and of reaching a particular goal state. It is common to model control problems such as these as partially observable Markov decision processes (POMDPs). However, in general, finding optimal solutions to POMDPs has been shown to be PSPACE complete [1]. Even many approximate approaches are computationally complex: the time complexity of standard point-based algorithms, such as HSVI and SARSOP, is exponential in the planning horizon [2–4]. These algorithms calculate policies in *belief-space*, the space of probability distributions over the underlying state space. Very few of these algorithms can handle continuous state and action spaces [5, 6].

In an effort to avoid the computational complexity of creating policies, a new set of approaches have recently been proposed which create plans based on expected information content. In one class of approaches, large numbers of candidate trajectories in the underlying state space are evaluated in terms of the information that is likely to be gained during execution [7–9]. Trajectories are selected that optimize information content or minimize the likelihood of collisions. These approaches work well in scenarios where the likelihood of generating information-gathering trajectories by sampling the underlying space is high. A different class of approaches create plans in a parametrization of belief-space [10–12]. These approaches are potentially better positioned to generate complex information-gathering plans, but since they plan directly in the belief-space, the dimensionality of the planning problem is potentially very large. With the exception of [12], the planning approaches listed above assume that Bayes filtering will be performed using a Gaussian density function [7–11]. However, the popularity of the particle filter relative to the extended Kalman filter or unscented Kalman filter suggests that in many robot problems, belief state is not well-represented as a Gaussian. Furthermore, simply extending an approach such as in [10, 11] to non-Gaussian distributions quickly results in an intractable planning problem because of the high dimensionality of typical non-Gaussian parametrizations.

This paper proposes an approach to planning in high-dimensional belief-spaces that tracks belief state using an accurate, high-dimensional filter, but creates plans using a fixed-dimensional sampled representation of belief. We leave the implementation of the high-dimensional filter as a design choice, but expect that it will be a histogram filter or a particle filter. In order to create a new plan, the high-dimensional belief state is projected onto a hypothesis in the underlying state space and a set of sampled competing states. Plans are created that generate observations that differentiate the hypothesis from the other samples while also reaching a goal state. During execution, we monitor KL divergence between the actual (high- dimensional) belief-space trajectory and a belief-space trajectory associated with the plan. If divergence exceeds a threshold, we halt execution and create a new plan starting from the current belief (this re-planning approach is similar to that taken in [10, 11]). In a technical report that expands upon this paper, we have shown that if each new plan found has a below-threshold cost, then the algorithm eventually localizes the true state of the system and reaches a goal region with probability one [13]. We illustrate the approach in the context of a one-dimensional manipulation problem and compare it to the approach proposed in [10]. Then, we show that the approach can be used to solve a version of the grasp synthesis problem where the robot must simultaneously localize and grasp an object. The algorithm generates robot arm trajectories that gain information by "scanning" the boxes using a laser scanner and pushing one of the boxes as necessary in order to gain information. The algorithm terminates in a pre-grasp configuration that is likely to lead to a successful grasp. The approach is tested over a range of randomly selected box configurations.

2 Problem Statement

This paper is concerned with the problem of reaching a desired goal state when the initial state is uncertain and may only be estimated based on partial or noisy observations. Consider a discrete-time system with continuous non-linear deterministic[1] process dynamics,

$$x_{t+1} = f(x_t, u_t), \tag{1}$$

where state, $x \in \mathbb{R}^n$, and action, $u \in \mathbb{R}^l$, are column vectors. At each time step, the system makes an observation, $z \in \mathbb{R}^m$, that is a non-linear stochastic function of state:

$$z_t = h(x_t) + v_t, \tag{2}$$

where $v_t \sim N(0, Q)$ is zero-mean Gaussian noise with variance Q.

Bayesian filtering can be used to estimate state based on actions taken and observation perceived. The state estimate is represented by a probability distribution function, $\pi(x; b)$, that is a function of the parameter vector, $b \in \mathcal{B}$. We will refer to b, (and sometimes the probability distribution, $\pi(x; b)$) as the *belief state*. Suppose that at time t, the system starts in belief state, b_t, takes action, u_t, and perceives observation, z_{t+1}. Then, belief state can be updated to incorporate the new information using the Bayesian filter update equation. For deterministic process dynamics, it is:

$$\pi(f(x, u_t); b_{t+1}) = \frac{\pi(x; b_t) P(z_{t+1} | x, u_t)}{P(z_{t+1})}, \tag{3}$$

[1]Although we have formally limited ourselves to the case of zero process noise, we find in Sect. 4 that empirically, our algorithm performs well in environments with bounded process noise.

where we implicitly assume that $P(z_{t+1}) \neq 0$. Although, in general, it is impossible to implement Eq. 3 exactly using a finite-dimensional parametrization of belief-space, a variety of approximations exist in practice [14].

The objective of belief-space planning is to achieve task objectives with a given minimum probability. Specifically, we want to reach a belief state, b, such that

$$\Theta(b, r, x_g) = \int_{x \in B_n(r)} \pi(x + x_g; b) > \omega, \tag{4}$$

where $B_n(r) = \{x \in \mathbb{R}^n, x^T x \leq r^2\}$ denotes the r-ball in $\mathbb{R}^n$, for some $r > 0$, $x_g \in \mathbb{R}^n$ denotes the goal state, and ω denotes the minimum probability of success. It is important to notice the similarities between this problem and the more general partially observable Markov decision process (POMDP) framework. Both problems are concerned with controlling partially observable systems. However, whereas in the POMDP formulation, the objective is to minimize the expected cost, in our problem, the objective is to reach a desired region of state space with a guaranteed minimum probability of success.

3 Algorithm

This paper extends the approach proposed in [10] to non-Gaussian belief spaces. Our algorithm iteratively creates and executes a series of belief-space plans. A replanning step is triggered when, during plan execution, the true belief state diverges too far from the nominal trajectory.

3.1 *Creating Plans*

The key to our approach is a mechanism for creating horizon-T belief-space plans that guarantees that new information is incorporated into the belief distribution on each planning cycle. The basic idea is as follows. Given a prior belief state, b_1, define a "hypothesis" state to be at the maximum of the distribution,

$$x^1 = \arg\max_{x \in \mathbb{R}^n} \pi(x; b_1).$$

Then, sample $k - 1$ states from the prior distribution,

$$x^i \sim \pi(x; b_1),\ i \in [2, k], \tag{5}$$

such that the pdf at each sample is greater than a specified threshold, $\pi(x^i; b_1) \geq \varphi > 0$, and there are at least two unique states (including x^1). We search for a sequence of actions, $\mathbf{u}_{1:T-1} = (u_1, \ldots, u_{T-1})$, that result in as wide a margin as possible between the observations that would be expected if the system were in the hypothesis state and the observations that would be expected in any other sampled state. As a result, a good plan enables the system to "confirm" that the hypothesis state is in fact the true state or to "disprove" the hypothesis state. If the hypothesis state is disproved, then the algorithm selects a new hypothesis on the next re-planning cycle, ultimately causing the system to converge to the true state.

To be more specific, consider that if the system starts in state x, and takes a sequence of actions $\mathbf{u}_{1:t-1}$, then the most likely sequence of observations is:

$$\mathbf{h}_t(x, \mathbf{u}_{1:t-1}) = \left(h(x)^T, h(f(x, u_1))^T, h(F_3(x, \mathbf{u}_{1:2}))^T, \ldots, h(F_t(x, \mathbf{u}_{1:t-1}))^T\right)^T,$$

where $F_t(x, \mathbf{u}_{1:t-1})$ denotes the state at time t when the system begins in state x and takes actions, $\mathbf{u}_{1:t-1}$. We are interested in finding a sequence of actions over a planning horizon T, $\mathbf{u}_{1:T-1}$, that maximizes the squared observation distance

$$\left\|\mathbf{h}_T\left(x^i, \mathbf{u}_{1:T-1}\right) - \mathbf{h}_T\left(x^1, \mathbf{u}_{1:T-1}\right)\right\|_{\mathbb{Q}}^2,$$

summed over all $i \in [2, k]$, where $\|a\|_A = \sqrt{a^T A^{-1} a}$ denotes the Mahalanobis distance and $\mathbb{Q} = diag(Q, \ldots, Q)$ denotes a block diagonal matrix of the appropriate size composed of observation covariance matrices. The wider the observation distance, the more accurately Bayes filtering will be able to determine whether or not the true state is near the hypothesis in comparison to the other sampled states.

Notice that the expression for observation distance is only defined with respect to the sampled points. Ideally, we would like a large observation distance between a region of states about the hypothesis state and regions about the other samples. Such a plan would "confirm" or "disprove" regions about the sampled points - not just the zero-measure points themselves. We incorporate this objective to the first order by minimizing the Frobenius norm of the gradient of the measurements,

$$\mathbf{H}_t(x, \mathbf{u}_{1:t-1}) = \frac{\partial \mathbf{h}_t(x, \mathbf{u}_{1:t-1})}{\partial x}.$$

These dual objectives, maximizing measurement distance and minimizing the Frobenius norm of the measurement gradient, can simultaneously be optimized by minimizing the following cost function:

$$\bar{J}\left(x^1, \ldots, x^k, u_{1:T-1}\right) = \frac{1}{k}\sum_{i=2}^{k} e^{-\Phi(x^i, \mathbf{u}_{1:T-1})}, \tag{6}$$

where

$$\Phi\left(x^i, u_{1:T-1}\right) = \left\|\mathbf{h}_T\left(x^i, \mathbf{u}_{1:T-1}\right) - \mathbf{h}_T\left(x^1, \mathbf{u}_{1:T-1}\right)\right\|_{\Gamma(x^i, \mathbf{u}_{1:T-1})}^2.$$

The weighting matrix (i.e. the covariance matrix) in the metric above is defined

$$\begin{aligned}\Gamma(x, \mathbf{u}_{1:T-1}) = 2\mathbb{Q} + \mathbf{H}_T(x, \mathbf{u}_{1:T-1}) V \mathbf{H}_T(x, \mathbf{u}_{1:T-1})^T \\ + \mathbf{H}_T(x^1, \mathbf{u}_{1:T-1}) V \mathbf{H}_T(x^1, \mathbf{u}_{1:T-1})^T,\end{aligned} \quad (7)$$

where $V \in \mathbb{R}^{n \times n}$ is a diagonal weighting matrix.

In order to find plans that minimize Eq. 6, it is convenient to restate the problem in terms of finding paths through a parameter space. Notice that for any positive semi-definite matrix, A, and vector, x, we have $x^T Ax \geq x^T \bar{A} x$, where $\bar{A}$ is equal to A with all the off-diagonal terms set to zero. Therefore, we have the following lower-bound,

$$\Phi(x^i, \mathbf{u}_{1:t-1}) \geq \sum_{t=1}^{T} \phi(F_t(x^i, \mathbf{u}_{1:t-1}), F_t(x^1, \mathbf{u}_{1:t-1})),$$

where

$$\begin{aligned}\phi(x, y) &= \frac{1}{2} \|h(x) - h(y)\|^2_{\gamma(x,y)}, \\ \gamma(x, y) &= 2Q + H(x)H(x)^T + H(y)H(y)^T,\end{aligned}$$

and $H(x) = \partial h(x)/\partial x$. As a result, we can upper-bound the cost, $\bar{J}$ (Eq. 6), by

$$\begin{aligned}\bar{J}(x^1, \ldots x^k, \mathbf{u}_{1:T-1}) &\leq \frac{1}{k} \sum_{i=1}^{k} e^{-\sum_{t=1}^{T} \phi(F_t(x^i, \mathbf{u}_{1:t-1}), F_t(x^1, \mathbf{u}_{1:t-1}))} \\ &\leq \frac{1}{k} \sum_{i=1}^{k} \prod_{t=1}^{T} e^{-\phi(F_t(x^i, \mathbf{u}_{1:t-1}), F_t(x^1, \mathbf{u}_{1:t-1}))}\end{aligned}. \quad (8)$$

As a result, the planning problem can be written in terms of finding a path through a parameter space, $(x_t^{1:k}, w_t^{1:k}) \in \mathbb{R}^{2k}$, where x_t^i denotes the state associated with the ith sample at time t and the weight w_t^i, denotes the "partial cost" associated with sample i. This form of the optimization problem is stated as follows.

Problem 1

$$\text{Minimize} \quad \frac{1}{k} \sum_{i=1}^{k} (w_T^i)^2 + \alpha \sum_{t=1}^{T-1} u_t^2 \quad (9)$$

$$\text{subject to} \quad x_{t+1}^i = f(x_t^i, u_t), i \in [1, k] \quad (10)$$

$$w_{t+1}^i = w_t^i e^{-\phi\left(x_t^i, x_t^1\right)},\ i \in [1, k] \tag{11}$$

$$x_1^i = x^i, w_1^i = 1,\ i \in [1, k] \tag{12}$$

$$x_T^1 = x_g \tag{13}$$

Problem 1 should be viewed as a planning problem in $\left(x^{1:k}, w^{1:k}\right) \in \mathbb{R}^{2k}$ where Eqs.12 and 13 set the initial and final value constraints, Eqs. 10 and 11 define the "belief space dynamics", and Eq. 9 defines the cost. Notice that we have incorporated a quadratic cost into the objective in order to cause the system to favor short paths. Problem 1 can be solved using a number of planning techniques such as rapidly exploring random trees [15], differential dynamic programming [16], or sequential quadratic programming [17]. We use sequential quadratic programming to solve the direct transcription of Problem 1. The direct transcription solution will be denoted

$$\mathbf{u}_{1:T-1} = \text{DIRTRAN}\left(\mathrm{x}^{1:\mathrm{K}}, \mathrm{x}_{\mathrm{G}}, \mathrm{T}\right), \tag{14}$$

for the sample set, $x^{1:k}$, goal state constraint, x_g, and time horizon, T. Sometimes, we will call DIRTRAN without the final value goal constraint (Eq. 13). This will be written, $\mathbf{u}_{1:T-1} = \text{DIRTRAN}(\mathrm{x}^{1:\mathrm{K}}, \mathrm{T})$. It is important to recognize that the computational complexity of planning depends only on the number of samples used (the size of k in step 3 of Algorithm 1) and not strictly on the dimensionality of the underlying space. This suggests that the algorithm can be efficient in high-dimensional belief spaces. In fact, we report results in [13] from simulations that indicate that the algorithm can work well when very few samples (as few as three or four) are used.

3.2 Re-planning

After creating a plan, our algorithm executes it while tracking belief state using an accurate, high-dimensional filter chosen by the system designer. We denote this Bayesian filter update as

$$b_{t+1} = G(b_t, u_t, z_{t+1}).$$

If the true belief state diverges too far from a nominal trajectory derived from the plan, then execution stops and a new plan is created. The overall algorithm is outlined in Algorithm 1. Steps 2 and 3 sample k states from the distribution with the hypothesis state, $x^1 = \arg\max_{x \in \mathbb{R}^n} \pi(x; b)$, located at the maximum of the prior distribution. The prior likelihood of each sample is required to be greater than a minimum threshold, $1 > \varphi \geq 0$. In step 4, CREATEPLAN creates a horizon-T plan,

$\mathbf{u}_{1:T-1}$, that solves Problem 1 and generates a nominal belief-space trajectory, $\bar{b}_{1:T}$. Steps 6 through 12 execute the plan. Step 8 updates the belief state given the new action and observation using the Bayes filter implementation chosen by the designer. Step 9 breaks plan execution when the actual belief state departs too far from the nominal trajectory, as measured by the KL divergence, $D_1[\pi(x; b_{t+1}), \pi(x; \bar{b}_{t+1})] > \theta$. The second condition, $\bar{J}(x, \ldots, x^k, \mathbf{u}_{1:t-1}) < 1 - \rho$, guarantees that the *while* loop does not terminate before a (partial) trajectory with cost $\bar{J} < 1$ executes. The outer *while* loop terminates when there is a greater than ω probability that the true state is located within r of the goal state, $\Theta(b, r, x_g) > \omega$ (Eq. 4). In the technical report that expands upon this paper [13], we show that if, for each iteration of the *while* loop, the two conditions in step 9 are met on some time step, $t < T$, then it is possible to guarantee that Algorithm 1 will eventually localize the true state of the system and the *while* loop will terminate.

Input : initial belief state, b, goal state, x_g, planning horizon, T, and belief-state update, G.

1 **while** $\Theta(b, r, x_g) \leq \omega$ **do**
2    $x^1 = \arg\max_{x \in \mathbb{R}^n} \pi(x; b)$;
3    $\forall i \in [2, k], x^i \sim \pi(x; b) : \pi(x^i; b) \geq \varphi$;
4    $\bar{b}_{1:T}, \mathbf{u}_{1:T-1} =$ `CreatePlan`$(b, x^1, \ldots, x^k, x_g, T)$;
5    $b_1 = b$;
6   **for** $t \leftarrow 1$ **to** $T-1$ **do**
7   execute action u_t, perceive observation z_{t+1};
8      $b_{t+1} = G(b_t, u_t, z_{t+1})$;
9   **if** $D_1\left[\pi(x; b_{t+1}), \pi(x; \bar{b}_{t+1})\right] > \theta$ **and** $\bar{J}(x^1, \ldots, x^k, \mathbf{u}_{1:t-1}) < 1 - \rho$ **then**
10   break
11   **end**
12   **end**
13    $b = b_{t+1}$;
14 **end**

Algorithm 1: Belief-space re-planning algorithm

Algorithm 2 shows the CreatePlan procedure called in step 4 of Algorithm 1. Step 1 calls DirTran parametrized by the final value constraint, x_g. If DirTran fails to satisfy the goal state constraint, then the entire algorithm terminates in failure. Step 2 creates a nominal belief-space trajectory by integrating the user-specified Bayes filter update over the planned actions, assuming that observations are generated by the hypothesis state. If this nominal trajectory does not terminate in a belief state that is sufficiently confident that the true state is located within r of the hypothesis, then a new plan is created in steps 4 and 5 that is identical to the first except that it does not enforce the goal state constraints. This allows the algorithm to gain information incrementally in situations where a single plan does not lead to a sufficiently "peaked" belief state. When the system gains sufficient information, this branch is no longer executed and instead plans are created that meet the goal state constraint.

Input : initial belief state, b, sample set, $x^1,\ldots,x^k$, goal state, x_g, and time horizon, T.
Output: nominal trajectory, $\bar{b}_{1:T}$ and $\mathbf{u}_{1:T-1}$

```
u_{1:T-1} = DirTran(x^1,...,x^k, x_g, T);
b̄_1 = b; ∀t ∈ [1 : T − 1], b̄_{t+1} = G(b̄_t, u_t, h(x_t^1));
if Θ(b, r, x_g) ≤ ω then
    u_{1:T-1} = DirTran(x^1,...,x^k, T);
    b̄_1 = b; ∀t ∈ [1 : T − 1], b̄_{t+1} = G(b̄_t, u_t, h(x_t^1));
end
```

Algorithm 2: CREATEPLAN procedure

3.3 Illustration

Figures 1 and 2 illustrate the process of creating and executing a plan in a robot manipulation scenario. Figure 1 shows a horizontal-pointing laser mounted to the end-effector of a two-link robot arm. The end-effector is constrained to move only vertically along the dotted line. The laser points horizontally and measures the range from the end-effector to whatever object it "sees". There are two boxes and a gap between them. Box size, shape, and relative position are assumed to be perfectly known along with the distance of the end-effector to the boxes. The only uncertain variable in this problem is the vertical position of the end-effector measured with respect to the gap position. This defines the one-dimensional state of the system and is illustrated by the vertical number line in Fig. 1. The objective is to localize the vertical end-effector with respect to the center of the gap (state) and move the end-effector to the center of the gap. The control input to the system is the vertical velocity of the end-effector.

Figure 2a illustrates an information-gathering trajectory found by DIRTRAN that is expected to enable the Bayes filter to determine whether the hypothesis state is

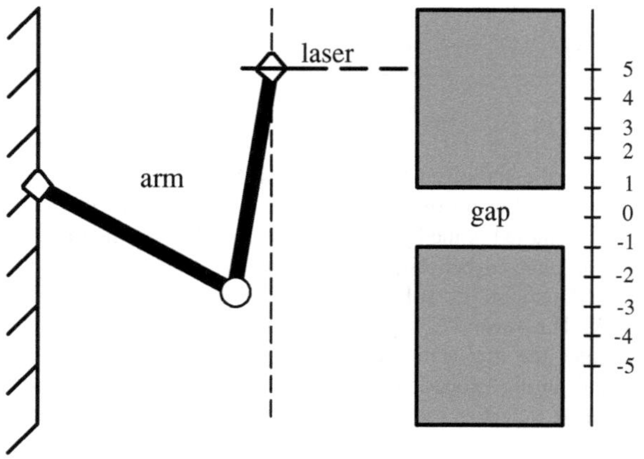

Fig. 1 Localization scenario. The robot must simultaneously localize the gap and move the end-effector in front of the gap

indeed the true state while simultaneously moving the hypothesis to the goal state (end-effector at the center of the gap). The sample set used to calculate the trajectory was $x^1, \ldots, x^k = 5, 2, 3, 4, 6, 7, 8$, with the hypothesis sample located at $x^1 = 5$. The action cost used while solving Problem 1 was $\alpha = 0.0085$. DirTran was initialized with a random trajectory. The additional small action cost smooths the trajectory by pulling it toward shortest paths without changing information gathering or goal directed behavior much. The trajectory can be understood intuitively. Given the problem setup, there are two possible observations: range measurements that "see" one of the two boxes and range measurements that "see" through the gap. The plan illustrated in Fig. 2a moves the end effector such that different sequences of measurements would be observed depending upon whether the system were actually in the hypothesis state or in another sampled state.

Figure 2b, c show the nominal belief-space trajectory and the actual trajectory, respectively, in terms of a sequence of probability distributions superimposed on each other over time. Each distribution describes the likelihood that the system started out in a particular state given the actions taken and the observations perceived. The nominal belief-space trajectory in Fig. 2b is found by simulating the

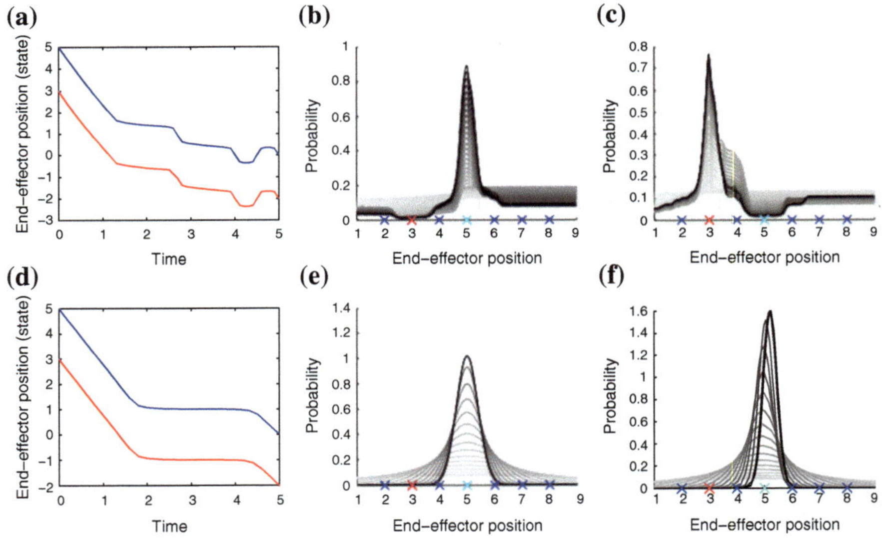

Fig. 2 Illustration of CreatePlan. **a** An information-gathering trajectory (state as a function of time) found using direct transcription. *Blue* denotes the trajectory that would be obtained if the system started in the hypothesis state. *Red* denotes the trajectory obtained starting in the true state. **b** The planned belief-space trajectory illustrated by probability distributions superimposed over time. Distributions early in the trajectory are *light gray* while distributions late in the trajectory are dark. The seven "X" symbols on the horizontal axis denote the positions of the samples (*red* denotes the true state while *cyan* denotes the hypothesis). **c** The actual belief-space trajectory found during execution. **d–f** Comparison with the EKF-based method proposed in [10]. **d** The planned trajectory. **e** The corresponding nominal belief-space trajectory. **f** Actual belief-space trajectory

belief-space dynamics forward assuming that future observations will be generated by the hypothesis state. Ultimately, the planned trajectory reaches a belief state distribution that is peaked about the hypothesis state, x_1 (the red “X”). In contrast, Fig. 2c illustrates the actual belief-space trajectory found during execution. This trajectory reaches a belief state distribution peaked about the true state (the cyan “X”). Whereas the hypothesis state becomes the maximum of the nominal distribution in Fig. 2b, notice that it becomes a minimum of the actual distribution in Fig. 2c. This illustrates the main idea of the algorithm. Figure 2b can be viewed as a trajectory that “trusts” that the hypothesis is correct and takes actions that confirm this hypothesis. Figure 2c illustrates that even when the hypothesis is wrong, the distribution necessarily gains information because it “disproves” the hypothesis state (notice the likelihood of the region about the hypothesis is very low).

Figure 2d–f compares the performance of our approach with the extended Kalman filter-based (EKF-based) approach proposed in [10]. The problem setup is the same in every way except that cost function optimized in this scenario is:

$$J(u_{1:T-1}) = \frac{1}{10}\left(\sigma_T^2\right)^T \sigma_T^2 + 0.0085 u_{1:T-1}^T u_{1:T-1},$$

where σ_T^2 denotes covariance. There are several differences in performance Notice that the two approaches generate different trajectories (compare Fig. 2a, d). Essentially, the EKF-based approach maximizes the EKF reduction in variance by moving the maximum likelihood state toward the edge of the gap where the gradient of the measurement function is large. In contrast, our approach moves around the state space in order to differentiate the hypothesis from the other samples in regions with a small gradient. Moreover, notice that since the EKF-based approach is constrained to track actual belief state using an EKF Bayes filter, the tracking performance shown in Fig. 2f is very bad. The EKF innovation term actually makes corrections in the wrong direction. However, in spite of the large error, the EKF covariance grows small indicating high confidence in the estimate.

4 Simultaneous Localization and Grasping

In real-world grasping problems, it is just as important to localize an object to be grasped as it is to plan the reach and grasp motions. We propose an instance of the grasp synthesis problem that we call *simultaneous localization and grasping* (SLAG) where the localization and grasp planning objectives are combined in a single planning problem. In most robot implementations, the robot is able to affect the type or quality of information that it perceives. For example, many robots have the ability to move objects of interest in the environment or move a camera or LIDAR through the environment. As a result, SLAG becomes an instance of the planning under uncertainty problem. The general SLAG problem is important because good solutions imply an ability to grasp objects robustly even when their position or shape is uncertain.

4.1 Problem Setup

Our robot, *Paddles*, has two arms with one paddle at the end of each arm (see Fig. 3a). Paddles may grasp a box by squeezing the box between the two paddles and lifting. We assume that the robot is equipped with a pre-programmed "lift" function that can be activated once the robot has placed its two paddles in opposition around the target box. Paddles may localize objects in the world using a laser scanner mounted to the wrist of its left arm. The laser scanner produces range data in a plane parallel to the tabletop over a 60° field of view.

We use Algorithm 1 to localize the planar pose of the two boxes parametrized by a six-dimensional underlying metric space. The boxes are assumed to have been placed at a known height. We reduce the dimensionality of the planning problem by introducing an initial perception step that localizes the depth and orientation of the right box using RANSAC [18]. From a practical perspective, this is a reasonable simplification because RANSAC is well-suited to finding the depth and orientation of a box that is assumed to be found in a known region of the laser scan. The remaining (four) dimensions that are not localized using RANSAC describe the horizontal dimension of the right box location and the three-dimensional pose of the left box. These dimensions are localized using a Bayes filter that updates a histogram distribution over the four-dimensional state space based on laser measurements and arm motions measured relative to the robot. The histogram filter is comprised of 20000 bins: 20 bins (1.2 cm each) describing right box horizontal position times 10 bins (2.4 cm each) describing left box horizontal position times 10 bins (2.4 cm each) describing left box vertical position times 10 bins (0.036 radians each) describing left box orientation. While it is relatively easy for the histogram filter to localize the remaining four dimensions when the two boxes are separated by a gap (Fig. 3b), notice that this is more difficult when the boxes are

(a) (b) (c)

Fig. 3 Illustration of the grasping problem (**a**). The robot must localize the pose and dimensions of the boxes using the laser scanner mounted on the *left* wrist. This is relatively easy when the boxes are separated as in (**b**) but hard when the boxes are pressed together as in (**c**)

pressed together (Fig. 3c). In this configuration, the laser scans lie on the surfaces of the two boxes such that it is difficult to determine where one box ends and the next begins. Note that it is difficult to locate the edge between abutting boxes reliably using vision or other sensor modalities in general this is a hard problem.

Our implementation of Algorithm 1 used a set of 15-samples including the hypothesis sample. The algorithm controlled the left paddle by specifying Cartesian end-effector velocities in the horizontal plane. These Cartesian velocity commands were projected into the joint space using standard Jacobian Pseudoinverse techniques [19]. The algorithm was parametrized by process dynamics that modeled arms motions resulting from velocity commands and box motions produced by pushes from the arm. Box motions were modeled by assuming zero slip while pushing the box and assuming the center of friction was located at the center of the area of the box "footprint". The observation dynamics described the set of range measurements expected in a given paddle-box configuration. For planning purposes, the observation dynamics were simplified by modeling only a single forward-pointing scan rather than the full 60° scan range. However, notice that since this is a conservative estimate of future perception, low cost plans under the simplified observation dynamics are also low cost under the true dynamics. Nevertheless, the observation model used for *tracking* (step 8 of Algorithm 1) accurately described measurements from all (100) scans over the 60° range. The termination threshold in Algorithm 1 was set to 50 % rather than a higher threshold because we found our observation noise model to overstate the true observation noise.

Our hardware implementation of the algorithm included some small variations relative to Algorithm 1. Rather than monitoring divergence explicitly in step 9, we instead monitored the ratio between the likelihood of the hypothesis state and the next most probable bin in the histogram filter. When this ratio fell below 0.8, plan execution was terminated and the *while* loop continued. Since the hypothesis state must always have a maximal likelihood over the planned trajectory, a ratio of less than one implies a positive divergence. Second, rather than finding a non-goal directed plan in steps 3–5 of Algorithm 2, we always found goal-directed plans.

Figure 4 illustrates an example of an information-gathering trajectory. The algorithm begins with a hypothesis state that indicates that the two boxes are 10 cm apart (the solid blue boxes in Fig. 4a). As a result, the algorithm creates a plan that scans the laser in front of the two boxes under the assumption that this will enable the robot to perceive the (supposed) large gap. In fact, the two boxes about each other as indicated by the black dotted lines in Fig. 4a. After beginning the scan, the histogram filter in Algorithm 1 recognizes this and terminates execution of the initial plan. At this point, the algorithm creates the pushing trajectory illustrated in Fig. 4b. During execution of the push, the left box moves in an unpredicted way due to uncertainty in box friction parameters (this is effectively process noise). This eventually triggers termination of the second trajectory. The third plan is created based on a new estimate of box locations and executes a scanning motion in front of the boxes is expected to enable the algorithm to localize the boxes with high confidence.

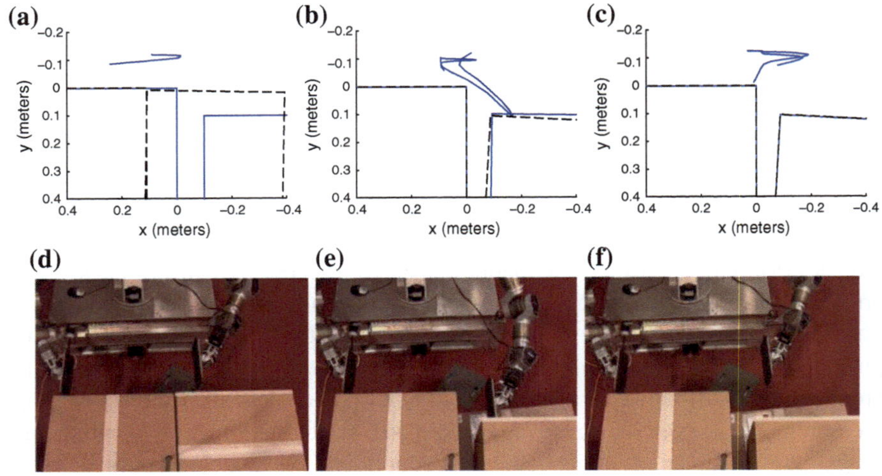

Fig. 4 Example of a box localization task. In **a**, **d**, the robot believes the gap between the boxes is large and plans to localize the boxes by scanning this gap. In **b**, **e**, the robot has recognized that the boxes abut each other and creates a plan to increase gap width by pushing the *right box*. In **c**, **f**, the robot localizes the boxes by scanning the newly created gap

4.2 Localization Performance

At a high level, the objective of SLAG is to robustly localize and grasp objects even when the pose or shape of those objects is uncertain. We performed a series of experiments to evaluate how well this approach performs when used to localize boxes that are placed in initially uncertain locations. On each grasp trial, the boxes were placed in a uniformly random configuration (visualized in Fig. 5a, c). There were two experimental contingencies: "easy" and "hard". In the easy contingency, both boxes were placed randomly such that they were potentially separated by a gap. The right box was randomly placed in a 13 $\times$ 16 cm region over a range of 15°. The left box was placed uniformly randomly in a 20 $\times$ 20 cm region over 20° measured with respect to the right box (Fig. 5a). In the hard contingency, the two boxes were pressed against each other and the pair was placed randomly in a 13 $\times$ 16 cm region over a range of 15° (Fig. 5b).

Figure 5c, d show right box localization error as a function of the number of updates to the histogram filter since the trial start. 12 trials were performed in each contingency. Each blue line denotes the progress of a single trial. The termination of each trial is indicated by the red "X" marks. Each error trajectory is referenced to the ground truth error by measuring the distance between the final position of the paddle tip and its goal position in the left corner of the right box using a ruler. There are two results of which to take note. First, all trials terminate with less than 2 cm of error. Some of this error is a result of the coarse discretization of possible right box positions in the histogram filter (note also the discreteness of the error plots). Since the right box position bin size in the histogram filter is 1.2 cm, we would expect a

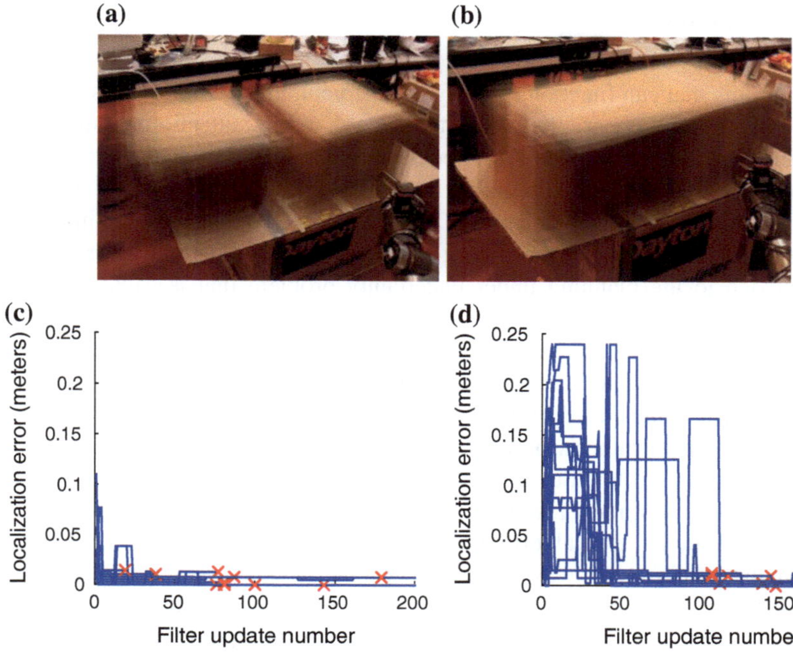

Fig. 5 "Easy" and "hard" experimental contingencies. **a** Shows images of the 12 randomly selected "easy" configurations (both box configurations chosen randomly) superimposed on each other. **b** Shows images of the 12 randomly selected "hard" configurations (boxes abutting each other). **c**, **d** Are plots of error between the maximum a posteriori localization estimate and the true box pose. Each *line* denotes a single trial. The *red* "X" marks denote localization error at algorithm termination

maximum error of at least 1.2 cm. The remaining error is assumed to be caused by errors in the range sensor or the observation model. Second, notice that localization occurs much more quickly (generally in less than 100 filter updates) and accurately in the easy contingency, when the boxes are initially separated by a gap that the filter may used to localize. In contrast, accurate localization takes longer (generally between 100 and 200 filter updates) during the hard contingency experiments. Also error prior to accurate localization is much larger reflecting the significant possibility of error when the boxes are initially placed in the abutting configuration. The key result to notice is that even though localization may be difficult and errors large during a "hard" contingency, all trials ended with a small localization error. This suggests that our algorithm termination condition in step 1 of Algorithm 1 was sufficiently conservative. Also notice that the algorithm was capable of robustly generating information gathering trajectories in all of the randomly generated configurations during the "hard" contingencies. Without the box pushing trajectories found by the algorithm, it is likely that some of the hard contingency trials would have ended with larger localization errors.

5 Conclusions

Creating robots that can function robustly in unstructured environments is a central objective of robotics. We argue that it is not enough to limit attention to developing better perception algorithms. Robust localization of relevant state in real-world scenarios is not always possible unless the robot is capable of creating and executing information-gathering behaviors. One avenue toward achieving this is the development of algorithms for planning under uncertainty. Our paper proposes a new approach to the planning under uncertainty problem that is capable of reasoning about trajectories through a non-Gaussian belief-space. This is essential because in many robot problems it is not possible to track belief state accurately by projecting onto an assumed density function (Gaussian or otherwise).

The approach is illustrated in the context of a grasping task. We propose a new setting of the grasp synthesis problem that we call simultaneous localization and grasping (SLAG). We test our algorithm using a particular instance of a SLAG problem where a robot must localize two boxes that are placed in front of it in unknown configurations. The algorithm generates information gathering trajectories that move the arm in such a way that a laser scanner mounted on the end-effector is able to localize the two boxes. The algorithm potentially pushes the boxes as necessary in order to gain information. Several interesting questions remain. First, our algorithm focuses primarily on creating information gathering plans. However, this ignores the need for "caution" while gathering the information. One way to incorporate this into the process is to include *chance constraints* into Problem 1 [20]. One approximation that suggests itself is to place constraints on the sample set used for planning. However, since we use a relatively small sample set, this may not be sufficiently conservative. Alternatives should be a subject for future work.

Acknowledgments This work was supported in part by in part by the NSF under Grant No. 0712012, in part by ONR MURI under grant N00014-09-1-1051 and in part by AFOSR grant AOARD- 104135.

References

1. C. Papadimitriou, J. Tsitsiklis, Math. Oper. Res. **12**(3), 441 (1987)
2. T. Smith, R. Simmons, in *Proceedings Uncertainty in Artificial Intelligence* (2005)
3. H. Kurniawati, D. Hsu, W.S. Lee, in *Proceedings of Robotics: Science and Systems (RSS)*
4. S. Ross, J. Pineau, S. Paquet, B. Chaib-draa, J. Mach. Learn. Res. **32**, 663 (2008)
5. H. Bai, W. Hsu, D. Lee, A. Ngo, in *Workshop on the Algorithmic Foundations of Robotics (WAFR)* (2010)
6. J. Porta, N. Vlassis, M. Spaan, P. Poupart, J. Mach. Learn. Res. **7**, 2329 (2006)
7. J. Van der Berg, P. Abbeel, K. Goldberg, in *Proceedings of Robotics: Science and Systems (RSS)* (2010)
8. S. Prentice, N. Roy, in *12th International Symposium of Robotics Research* (2008)
9. N. Du Toit, J. Burdick, in *IEEE International Conference on Robotics and Automation (ICRA)* (2010)

10. R. Platt, R. Tedrake, L. Kaelbling, T. Lozano-Perez, in *Proceedings of Robotics: Science and Systems (RSS)* (2010)
11. T. Erez, W. Smart, in *Proceedings of the International Conference on Uncertainty in Artificial Intelligence* (2010)
12. K. Hauser, in *Workshop on the Algorithmic Foundations of Robotics (WAFR)* (2010)
13. R. Platt, L. Kaelbling, T. Lozano-Perez, R. Tedrake, A hypothesis-based algorithm for planning and control in non-gaussian belief spaces. Technical Report CSAIL-TR-2011-039, Massachusetts Institute of Technology (2011)
14. A. Doucet, N. Freitas, N. Gordon (eds.), *Sequential Monte Carlo Methods in Practice* (Springer, 2001)
15. S. LaValle, J. Kuffner, Int. J. Robot. Res. **20**(5), 378 (2001)
16. D. Jacobson, D. Mayne, *Differential Dynamic Programming* (Elsevier, 1970)
17. J. Betts, *Practical Methods for Optimal Control Using Nonlinear Programming* (Siam, 2001)
18. M. Fischler, R. Bolles, Commun. ACM **24**, 381 (1981)
19. L. Sciavicco, B. Siciliano, *Modelling and Control of Robot Manipulators* (Springer, 2000)
20. L. Blackmore, M. Ono, in *Proceedings of the AIAA Guidance, Navigation, and Control Conference* (2009)

Pose Graph Compression for Laser-Based SLAM

Cyrill Stachniss and Henrik Kretzschmar

Abstract The pose graph is a central data structure in graph-based SLAM approaches. It encodes the poses of the robot during data acquisition as well as spatial constraints between them. The size of the pose graph has a direct influence on the runtime and the memory requirements of a SLAM system since it is typically used to make data associations and within the optimization procedure. In this paper, we address the problem of efficient, information-theoretic compression of such pose graphs. The central question is which sensor measurements can be removed from the graph without loosing too much information. Our approach estimates the expected information gain of laser measurements with respect to the resulting occupancy grid map. It allows us to restrict the size of the pose graph depending on the information that the robot acquires about the environment. Alternatively, we can enforce a maximum number of laser scans the robot is allowed to store, which results in an any-space SLAM system. Real world experiments suggest that our approach efficiently reduces the growth of the pose graph while minimizing the loss of information in the resulting grid map.

1 Introduction

Maps of the environment are needed for a wide range of robotic applications including transportation and delivery tasks, search and rescue, or efficient automated vacuum cleaning robots. The capability of building an appropriate model of the environment allows for designing robots that can efficiently operate in complex environments only based on their on-board sensors and without relying on external

C. Stachniss (✉) · H. Kretzschmar
Institute for Computer Science, University of Freiburg, Freiburg im Breisgau, Germany
e-mail: cyrill.stachniss@igg.uni-bonn.de

H. Kretzschmar
e-mail: kretzsch@informatik.uni-freiburg.de

C. Stachniss
Institute of Geodesy and Geoinformation, University of Bonn, Bonn, Germany

H.I. Christensen and O. Khatib (eds.), *Robotics Research*,
Springer Tracts in Advanced Robotics 100, DOI 10.1007/978-3-319-29363-9_16

reference systems. In the past, several effective approaches to robot mapping have been developed. A popular approach to address the simultaneous localization and mapping (SLAM) problem models the poses of the robot as nodes in a graph. Spatial constraints between poses resulting from observations and odometry are encoded as edges. Often, graph-based approaches marginalize out features or local grid maps and reduce the problem to trajectory estimation without prior map knowledge, followed by mapping with known poses.

Most of the SLAM approaches assume that map learning is carried out as a preprocessing step and that the robot then uses the acquired model for tasks such as localization and path planning. A robot that has to extend the map of its environment during long-term operation cannot apply most of the existing graph-based mapping approaches since their complexity grows with the length of the robot's trajectory. The reason for this is that standard graph-based approaches constantly add new nodes to the graph. As a result, memory and computational requirements grow over time, preventing long-term mapping applications. A constantly growing graph slows down graph optimization and makes it more and more costly to find constraints between the current pose and former poses, i.e., to identify loop closures. There exist also incremental methods for online corrections that perform partial optimizations. These methods are mostly orthogonal to our contribution.

The contribution of this paper is an information-theoretic approach to lossy pose graph compression to allow graph-based SLAM systems to operate over extended periods of time. Figure 1 depicts a motivating example. The top image shows the pose graph and the resulting map obtained by a standard graph-based approach to SLAM. The bottom image displays the corresponding pose graph along with the map resulting from our information-theoretic compression approach. As can be seen, significantly less nodes are required to provide a comparable mapping result. We present an approach to select laser scans for removal such that the expected loss of information with respect to the map is minimized. Our unbiased selection applies the information-theoretic concept of mutual information to determine the laser scans that should be removed. In order to keep the pose graph compact, the corresponding pose node needs to be eliminated from the pose graph. This is achieved by applying an approximate marginalization scheme that maintains the natural sparsity pattern that is observed in the context of SLAM. Our approach is highly relevant to long-term mapping, particularly when the robot frequently re-traverses already visited areas. It allows us to build an any-space SLAM system that aims at minimizing the expected loss of information.

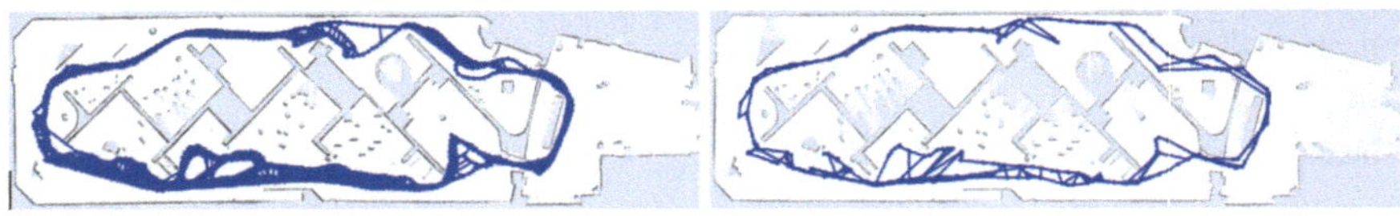

Fig. 1 The goal of our work is to compress the SLAM pose graph (*left*) to a sparse pose graph (*right*), while minimizing the loss of information in the graph and the resulting map

2 Related Work

A large variety of graph-based SLAM approaches have been proposed [5, 7, 8, 11, 15, 17, 19]. Most of these approaches to SLAM do not provide means to effectively prune the graph. Instead, they add more and more nodes to the graph over time. Some approaches group nodes into rigid local sub-maps [7] or subdivide the map into connected frames that contain maps that capture the local environment [1]. Typically, these methods do not discard nodes that store information about the environment and therefore do not prevent the graph from growing.

One way to reduce the number of nodes in the graph is to sample the trajectory of the robot at an appropriate spatial decimation [6]. A similar method is to only add a new node to the graph if it is not spatially close to any existing node [11]. Konolige and Bowman [12] presented an approach to lifelong mapping that uses a single stereo camera and that is able to update the map when the environment changes. Their method discards views based on a least recently used algorithm. The above mentioned techniques do not rely on information-theoretic concepts to determine which measurements to discard. Similar to that, hierarchical techniques [5, 8, 17] have been employed to bound the computational requirements by optimizing only higher levels of the hierarchy.

In contrast to that, Davison [3] analyzes mutual information, particularly in the case of Gaussian probability distributions, to guide image processing. In the vision community, Snavely et al. [20] aim to find a skeletal subgraph with the minimum number of interior nodes that spans the full graph while achieving a bound on the full covariance. Their technique is used for reconstructing scenes based on large, redundant photo collections.

Kaess and Dellaert [10] consider the information gain of measurements in the state estimate within the iSAM framework. In contrast to that, our approach estimates the mutual information of laser scans and the occupancy grid map, thus considering the effect on the resulting grid map explicitly. Ila et al. [9] propose to only incorporate non-redundant poses and informative constraints based on the relative distance between poses in information space and the expected information gain of candidate loop closures. As opposed to our maximum-likelihood approach to SLAM based on pose graphs, their method applies an information filter and does not marginalize out poses that were already added. Recently, Eade et al. [4] presented a view-based monocular SLAM system that reduces the complexity of the graph by marginalization and subsequent suppression of edges incident to nodes of high degrees. Their heuristic discards the constraints that most agree with the current state estimate. This, however, introduces a bias into the system.

When discarding laser range scans, our approach will marginalize out the corresponding pose node from the pose graph. Exact marginalization, however, would result in a dense pose graph and thus we apply an approximate marginalization scheme [14] that is based on local Chow-Liu trees. Other robotics researchers also use Chow-Liu trees for approximating probability distributions, e.g., [2].

3 Brief Introduction to Graph-Based SLAM

Graph-Based approaches to SLAM model the poses of the robot as nodes in a graph. The edges of the graph model spatial constraints between the nodes. These constraints arise from odometry measurements and from feature observations or scan matching. The so-called SLAM front-end interprets the sensor data to extract the constraints. The so-called SLAM back-end typically applies optimization techniques to estimate the configuration of the nodes that best matches the spatial constraints.

Our laser-based front-end uses correlative scan matching to estimate a constraint between the current node and the previous node. Our method also generates loop closure hypotheses by matching the current laser scan against a set of scans that is determined by the relative positional uncertainties and then rejects false hypotheses using the spectral clustering approach described by Olson [18]. Our method incrementally optimizes the pose graph while adding the poses and the constraints to it. Once the poses are estimated, the laser scans are used to render an occupancy grid map of the environment. The robot therefore stores the laser scans that correspond to the pose nodes in the pose graph.

The back-end aims at finding the spatial configuration $\mathbf{x}^*$ of the nodes that maximizes the log likelihood of the observations. Let $\mathbf{x} = (x_1^T, \ldots, x_n^T)^T$ be a vector where x_i describes the pose of node i, and let z_{ij} and $\mathbf{\Omega}_{ij}$ be the mean and the information matrix of an observation of node j seen from node i assuming Gaussian noise. Furthermore, let $\mathbf{e}_{ij}(\mathbf{x})$ be an error vector which expresses the difference between an observation and the current configuration of the nodes and let $\mathscr{C}$ be the set of pairs of nodes for which a constraint exists. Assuming the constraints to be independent, we have

$$\mathbf{x}^* = \arg\min_{\mathbf{x}} \sum_{\langle i,j\rangle \in \mathscr{C}} \mathbf{e}_{ij}(\mathbf{x})^T \Omega_{ij} \mathbf{e}_{ij}(\mathbf{x}). \tag{1}$$

Our approach applies the technique proposed in [8], which uses sparse Cholesky factorization to efficiently solve the system of linearized equations that is obtained from Eq. (1).

4 Discarding Laser Scans by Information-Theoretic Means

The main contribution of this paper is an approach to select the laser scans that are most informative with respect to the map estimate. Our technique aims at minimizing the expected loss of information in the resulting map without introducing a bias during the selection of the laser scans. Such a technique is important to allow for long-term robot mapping since a robot that keeps all scans will run out of resources at some point. In addition to that, our method can be used to directly implement an any-space SLAM system. Whenever the memory limit is reached, our algorithm discards the laser scans that are expected to be least informative about the map and marginalizes out the corresponding pose nodes.

4.1 Finding the Most Informative Subset of Laser Scans

We define the map M as a random variable describing the state of the world. It is highly correlated to the random variables $Z_{1:t}$ describing the laser scans $z_{1:t}$ recorded at the poses $\mathrm{x}_{1:t}$. We use Z^j to refer to an individual beam of laser scan Z_i. To estimate the state of the world m, we consider the posterior probability distribution of the map M given the laser measurements $z_{1:t}$. In this section, we are interested in finding the subset $Z^* \subseteq Z_{1:t}$ of at most n laser measurements that is expected to result in the smallest uncertainty about the map M.

Following the notation of MacKay [16], the average reduction in the uncertainty of the map M due to a set Z of laser measurements is given by the mutual information

$$I(M;Z) = H(M) - H(M|Z), \tag{2}$$

where H is the entropy. Hence, we want to find the subset $Z^* \subseteq Z_{1:t}$ of at most n laser measurements such that the mutual information of the map M and the subset Z^* is maximized, i.e.,

$$Z^* = \underset{Z \subseteq Z_{1:t}, |Z| \leq n}{\arg\max} \; H(M) - H(M|Z). \tag{3}$$

The conditional entropy $H(M|Z)$ of the map M given the set Z of measurements is the expected value, over the space of all possible measurements, of the conditional entropy of the map given the individual measurements z:

$$H(M|Z) = \int_z p(z) H(M|Z = z) dz \tag{4}$$

4.2 Efficiently Estimating Mutual Information

Unfortunately, computing the conditional entropy given in Eq. (4) is infeasible without approximations since integrating over the space of all possible combinations of up to n laser measurements is practically impossible. In addition to that, computing the entropy $H(M|Z = z)$ of a map given a set of measurements z typically requires model assumptions about the world.

To efficiently compute $H(M|Z)$, we make the following assumptions. We assume the laser measurements and the individual laser beams to be independent. Furthermore, we model the map M as a standard occupancy grid map, i.e., a grid of independent discrete binary random variables C that take the values Val (C) = {"free", "occupied"}. The entropy of an occupancy grid map M given a set of measurements z is then given by

$$H(M|Z = z) = \sum_{C \in M} H(C|Z = z) \tag{5}$$

$$= -\sum_{C \in M} \sum_{c \in \mathrm{Val}(C)} P(C = c|z) \log P(C = c|z). \tag{6}$$

To efficiently compute Z^*, we additionally ignore the distribution over $x_{1:t}$ and operate on the most likely estimate $x^*_{1:t}$, which is given in Eq. (1). Furthermore, similar to most works on robot localization, we assume the likelihood of sensing a specific object to decrease with range. The a priori probability of the jth beam of a range measurement z_i, denoted as Z_i^j, without any knowledge of the map M can be described by the exponential distribution

$$p\left(z_i^j\right) = \begin{cases} \eta \lambda e^{-\lambda z_i^j} & z_i^j \leq z_{\max}, \\ 0 & z_i^j \leq z_{\max}, \end{cases} \tag{7}$$

where $z_{\max}$ denotes the maximum range of the scanner, λ is a parameter of the measurement model, and η is a normalizing constant.

There are three possible outcomes of a measurement of a laser beam with respect to a particular grid cell that is located along the ray of the beam and given no prior map information. The laser beam either traverses the cell and thus observes the cell as free, the laser beam ends in the cell and thus observes the cell as occupied, or the laser beam does not observe the cell. The probability distribution of the outcome can be computed by integrating over the density $p(z^j)$. For the three cases, namely, the probabilities that (i) the beam Z^j does not reach a particular grid cell C that is located along the ray of the beam, (ii) the beam ends in that cell (measured as occupied), and (iii) the beam passes through that cell (measured as free) are given by

$$P\left(z_i^j \text{ does not observe } C\right) = \int_0^{d_1(x_i^*, C)} p\left(z_i^j\right) dz_i^j \tag{8}$$

$$P\left(z_i^j \text{ observes } C \text{ as occupied}\right) = \int_{d_1(x_i^*,C)}^{d_2(x_i^*,C)} p\left(z_i^j\right) dz_i^j \tag{9}$$

$$P\left(Z_i^j \text{ observes } C \text{ as free}\right) = \int_{d_2(x_i^*,C)}^{zmax} p\left(z_i^j\right) dz_i^j, \tag{10}$$

where $d_1\left(x_i^*, C\right)$ is the distance between the pose x_i^* (see Eq. (1)) from which the laser scan Z_i is taken and the closest border of the grid cell C in the direction of the jth beam (the border where the beam enters the cell). Similarly, $d_2\left(x_i^*, C\right)$ is the distance to the border of the grid cell C in the direction of the jth beam that is furthest away from x_i^* (the border where the beam leaves the cell).

By exploiting Eqs. (8)–(10), we can avoid integrating over all potential measurements as in Eq. (4). Instead, we can sum over all potential measurement outcomes. This results in the mutual information

$$I(C; Z) = H(C) - \sum_{z' \in \mathscr{A}_Z} P(z')H(C|z') \tag{11}$$

of the grid cell C and the set Z of laser measurements. Here, we consider the set $\mathscr{A}_Z$ of all possible measurement outcomes z' with respect to the grid cell C of all k laser scans that are recorded close enough to potentially measure the cell.

In general, the number of possible combinations of grid cell measurement outcomes is exponential in k as it is illustrated in the left image of Fig. 2. It is therefore practically infeasible to enumerate all the combinations in a tree. In our approach, we use a standard inverse measurement model, $p(c|z_i^j)$, for laser range scanners that updates each cell using one of the three values l_{free}, l_{occ}, and l_0. Since the effect of a set of observations on a particular cell does not depend on the *order* in which the measurements were obtained, this model allows us to efficiently combine nodes in the tree of all possible combinations. In fact, the result only depends on the *number* of free and occupied observations, i.e., the histogram of measurement outcomes, see Fig. 2 (right) for an illustration.

By ignoring the order, the number of histograms that we have to compute is quadratic in k:

$$\#outcomes(k) = \sum_{i=0}^{k} \left(\sum_{j=0}^{k-i} 1 \right) = \sum_{i=0}^{k} (k - i + 1) \tag{12}$$

$$= (k+1)^2 - \sum_{i=0}^{k} i = (k+1)^2 - \frac{(k+1)k}{2} \tag{13}$$

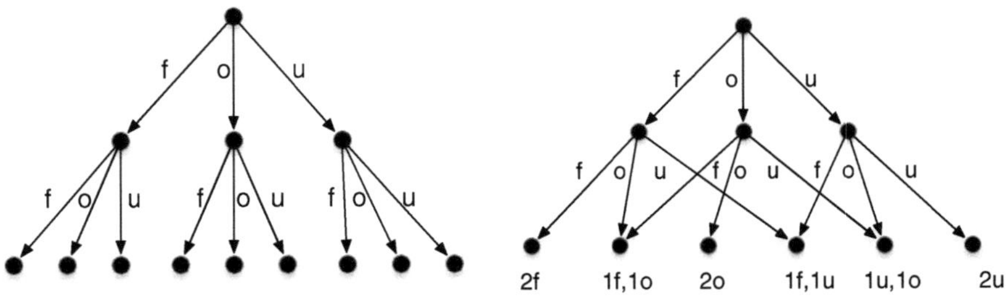

Fig. 2 *Left* all combination of measurement combinations that can occur for n beams, here with $n = 2$. In the figure, f = free, o = occupied, and u = unknown. The number of combinations is 3^n. *Right* Since the order of the measurements is irrelevant, the number of *possible outcomes* is quadratic in n

$$= \frac{k^2}{2} + \frac{3}{2}k + 1 \in \mathcal{O}(k^2) \tag{14}$$

If the individual probabilities of obtaining free/occupied/unknown measurements were equal for all laser scans, the probabilities of these outcomes could be computed by a multinomial distribution.

However, this is not the case here due to our model given in Eq. (7), which specifies that the likelihood of observing a cell depends on the distance from the view point to the cell under consideration. In our case, computing the probabilities of all outcomes requires cubic time in k. This can be achieved by using a hash table that is indexed by the number of free, occupied, and unknown measurements and that stores the accumulated probability mass for the corresponding outcome. By traversing the graph of possible outcomes (see right image of Fig. 2), from top to bottom, the computations can be performed as shown in Algorithm 1 (where $P(\langle \cdot,\cdot,\cdot \rangle)$ is implemented via a hash table) and the number of probabilities that need to be considered is

$$3\sum_{i=1}^{k} \#outcomes(k) = \frac{3}{2}k^3 + \frac{9}{2}k^2 + 3k \in \mathcal{O}(k^3). \tag{15}$$

Thus, to compute the probability for each possible outcome is cubic in k, i.e. the number of measurements that can observe the grid cell C.

Fortunately, the number k of scans that the algorithm has to consider is typically bounded: First, the maximum measurement range of laser scanners restricts the set of scans that have to be considered. Second, our technique discards laser scans online while building the graph and thus k typically stays small during mapping. We can efficiently further reduce the computational burden by only considering at most l laser scans when computing the histograms. One good way of choosing the l laser scans is selecting the ones with the highest likelihood of measuring C. This likelihood is given by $1 - P(Z_i^j \text{ does not observe } C)$, see Eq. (8). Thus, this approximation yields a linear complexity in k (for selecting the l scans out of k laser scans).

Finally, the mutual information $I(M; Z)$ of the map M and the set Z of laser scans is given by

$$I(M;Z) = \sum_{C \in M} I(C;Z). \tag{16}$$

All terms needed to compute Z^* in Eq. (3) are specified and can be computed or approximated efficiently.

4.3 Discarding Laser Scans Online

Our approach can be used in two ways. First, by introducing a bound on the total number of laser scans, our method results in an any-space SLAM system. Second, setting a threshold for the expected information gain of laser scans, our algorithm only keeps scans that are expected to provide at least a certain amount of information about the map.

Algorithm 1 Compute probabilities for all measurement outcomes

```
Require: Z : set of laser measurements, C : cell
Ensure: P(⟨·,·,·⟩) : the probabilities of all outcomes (free, occupied, not observed)
  P(⟨0,0,0⟩) = 1
  for r = 1 ... k do
    for all ⟨f,o,u⟩ with f+o+u == r do
      P(⟨f+1,o,u⟩) += P(⟨f,o,u⟩)P(Z_i^j observes C as free);
      P(⟨f,o+1,u⟩) += P(⟨f,o,u⟩)P(Z_i^j observes C as occupied);
      P(⟨f,o,u+1⟩) += P(⟨f,o,u⟩)P(Z_i^j does not observe C);
    end for
  end for
  return P(⟨·,·,·⟩)
```

Computing the subset Z^* of n laser measurements that most reduces the uncertainty about the map has been shown to be at least NP-hard [13]. Fortunately, the problem is submodular. Hence, greedily selecting measurements results in obtaining a set of measurements that is at most a constant factor (≈0.63) worse than the optimal set. Motivated by this insight, our approach estimates the subset Z^* by successively discarding laser scans. In each step, it discards the laser scan that is expected to be least informative.

5 Maintaining a Sparse Pose Graph

A pose graph can be seen as a Gaussian Markov random field (GMRF) that models the belief of the robot. In this view, each pose is a random variable that is represented as one node in the GMRF and each constraint between two poses in the pose graph is a binary potential between the nodes in the GMRF. Marginalizing out a pose node from the graph implies summarizing the information stored in the edges that connect that node in the edges between nodes that are kept. The main problem of exact marginalization is that it introduces new edges between all pairs of variables that are directly related to the eliminated variables, adding a so-called elimination clique to the graph, see for example [6]. This, unfortunately, destroys the natural sparsity pattern that is typical to SLAM problems.

Therefore, we apply an approximate marginalization scheme to maintain sparsity, which is important for long-term mapping tasks. The key idea is to replace the elimination clique, which is created when marginalizing a node, by a tree-shaped approximation of the clique. The optimal tree-shaped approximation with respect to the Kullback-Leibler divergence is given by the Chow-Liu tree. Here, the Chow-Liu tree is the maximum-weight spanning tree of the mutual information graph of the clique. This tree can be computed by assigning the mutual information of each two variables to the corresponding edges belonging to the elimination clique and then applying Kruskal's algorithm. The mutual information is computed according to Davison [3], which describes an efficient solution for the Gaussian case. We refer the reader to [14] for more details on the approximate marginalization scheme.

It should be noted that only the elimination clique is transformed to a tree structure, not the whole pose graph. Otherwise, all loop closing information would be lost.

6 Experimental Evaluation

To evaluate the presented approach, we carried out several experiments using a real ActivMedia Pioneer-2 robot equipped with a SICK laser range finder. In addition to that, we applied our method to a series of popular benchmark datasets. We compare our approach with the performance when no scans are discarded (referred to as "standard approach"). The experiments are designed to show that our approach to informed pose graph compression is well suited for long-term mobile robot mapping as well as for standard SLAM problems.

6.1 Mapping Results

Four different datasets have been considered for the evaluation. We used one self recorded dataset in which the robot traveled in our lab environment for an extended period of time (Fig. 3), a previously recorded dataset from a Freiburg computer science building (Fig. 4), as well as the Intel Research Lab (Fig. 5), and the FHW dataset (Fig. 1), both provided by Dirk Hähnel.

Figure 3 shows four maps built during our experiments with a fixed limit to 200 nodes. The first one depicts the pose graph obtained with the standard approach. The second one shows the state of our approach before the robot entered the left side of the corridor. Therefore, the limit of 200 nodes is used to model the right part only. The third image shows the pose graph modeling the entire environment. Note how our approach redistributed the nodes in the environment, still complying with the 200 node limit. Finally, the fourth image shows the map when setting a threshold on the mutual information.

Further mapping results showing the results of our approach in contrast to the standard approach are depicted in Figs. 4 and 5 as well as in the motivating example in Fig. 1. By visual inspection, the obtained grid maps look similar, although only a fraction of the original laser scans have been used to build the grid maps.

To compare the output of the SLAM algorithm more quantitatively, we also analyzed the estimate of the pose graph for the resulting nodes. We especially analyzed the mean and uncertainty estimates for the individual poses of the robot and compared them to the corresponding ones built without compressing the pose graph. Figure 6 depicts the 3σ covariance ellipses of the poses in the graphs. Our approach keeps less than 9 % of the edges of the original graph (349 of 3916) but only 2.8 % of the probability mass of the original pose graph is not covered by our approximation. The covariance estimates of our approach are typically more conservative (in this experiment by 41 %) since less information is used during mapping.

6.2 Memory and Runtime Requirements

In this section, we analyze the memory requirements of our approach in terms of the size of the resulting pose graph. In the first experiment, the robot moved around in our lab environment for an extended period of time (Fig. 3). The plots in Figs. 7 and 8 clearly suggest that the experiment leads to an explosion in terms of memory requirements when using the standard approach. This has a direct influence on the computational complexity: First, the optimization technique scales with the number of edges, which grows roughly quadratically since the robot moves in the same environment and is not exploring new terrain. Second, the loop closing component of the SLAM front-end, which uses a scan matcher to find constraints between the

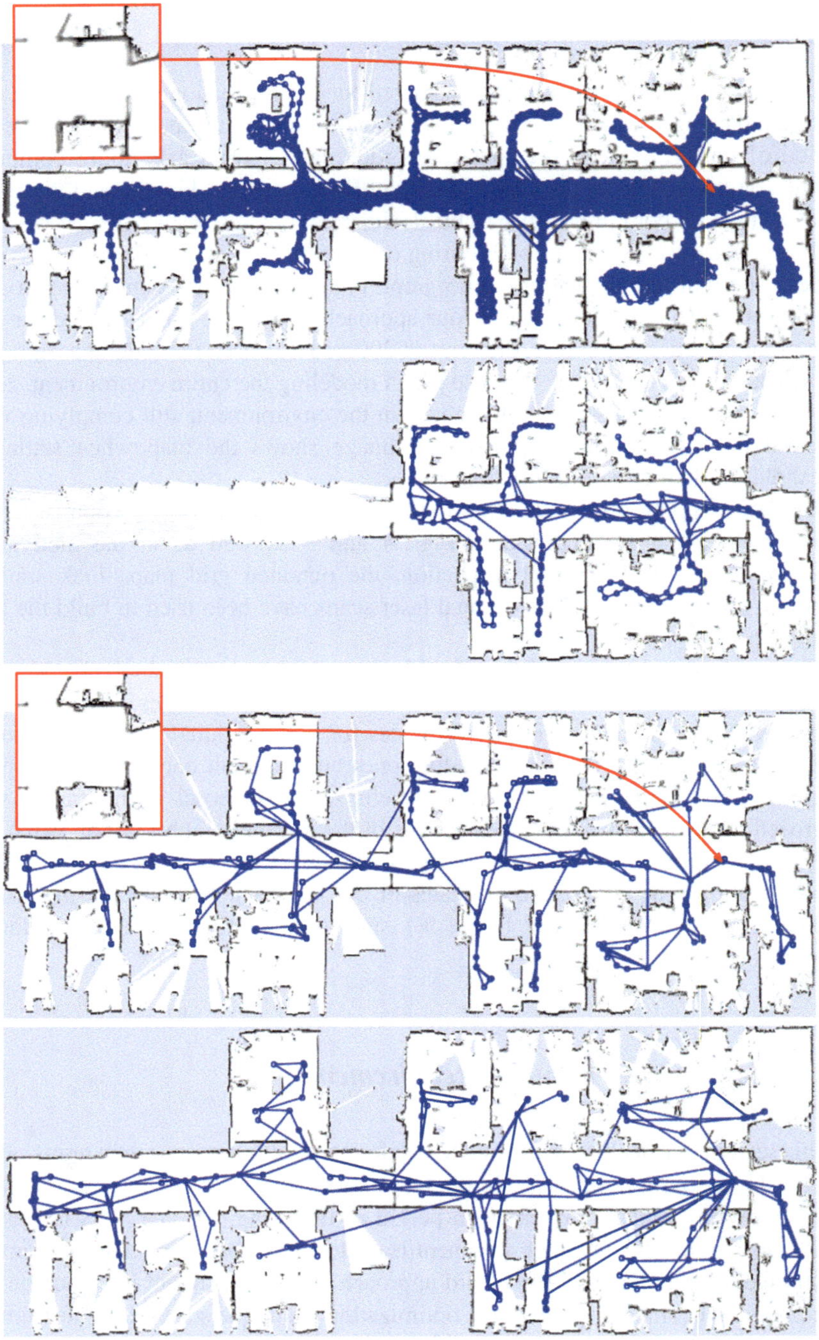

Fig. 3 The robot moved around in an office environment for an extended period of time, visiting the rooms and the corridor many times. *First* Standard approach. 2597 laser scans, 15695 edges *Second* Our approach at an intermediate time step, 200 laser scans, 264 edges. *Third* Our approach, 200 laser scans, 315 edges. *Fourth* Our approach when setting a threshold for the mutual information, 148 laser scans, 250 edges

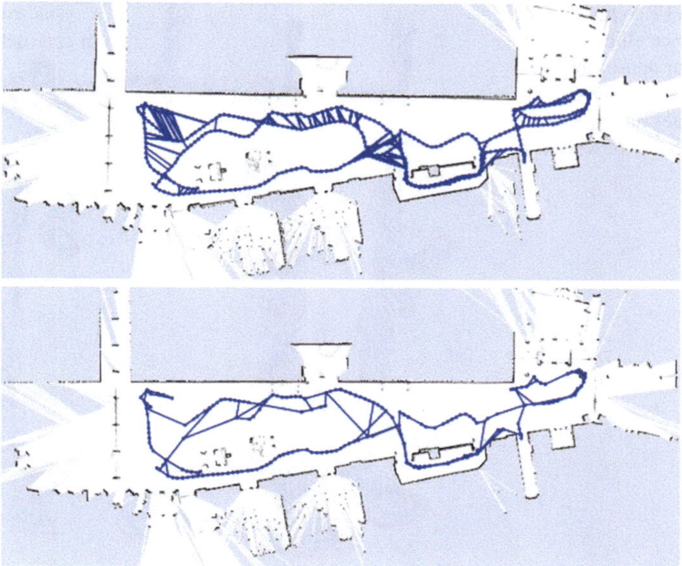

Fig. 4 Obtained map and pose graph for the FR101 dataset (*top* standard approach, *bottom* our approach). Since the robot does not frequently re-traverse known terrain, few scans were discarded (200 vs. 408 nodes and 246 vs. 723 edges)

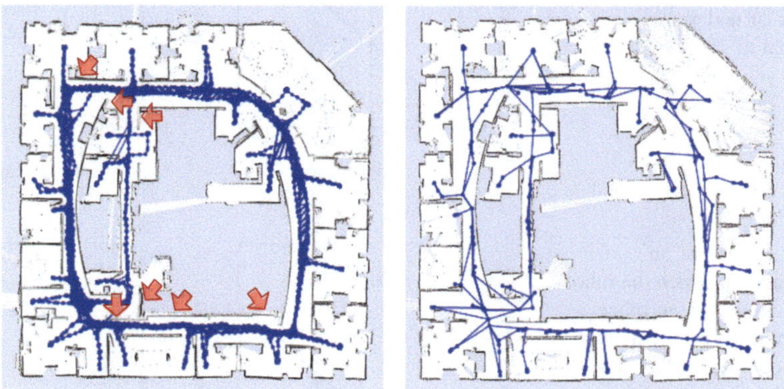

Fig. 5 Intel Research Lab. *Left* Standard approach, 1802 laser scans, 3916 edges. *Right* Our approach preserves the sparsity of the pose graph, 250 laser scans, 349 edges. *Arrows* indicate locally blurred areas or small alignment errors in the map obtained by the standard approach. In contrast to that, exact marginalization would result in 250 laser scans as well, but in 13052 edges

current scan and all former scans that were recorded in the vicinity of the robot, has to consider an increasing number of nodes in each step. In contrast to the standard approach, our approach compresses the pose graph such that the number of nodes in the graph remains constant (in Fig. 7 the threshold was set to 200 nodes). If we set a

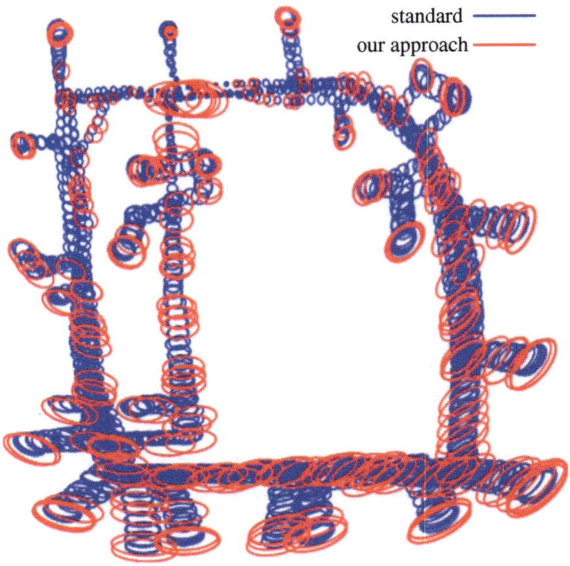

Fig. 6 Intel Research Lab.: 3σ covariance ellipses of the poses of our approach (*red*) and the standard one (*blue*). Our estimates are typically larger since less observations are used

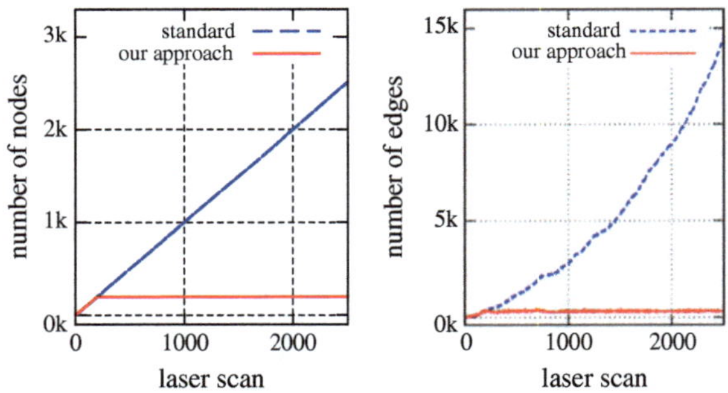

Fig. 7 Results of an experiment in which the robot moved around in an office environment for an extended period of time. The total number of nodes was restricted to 200

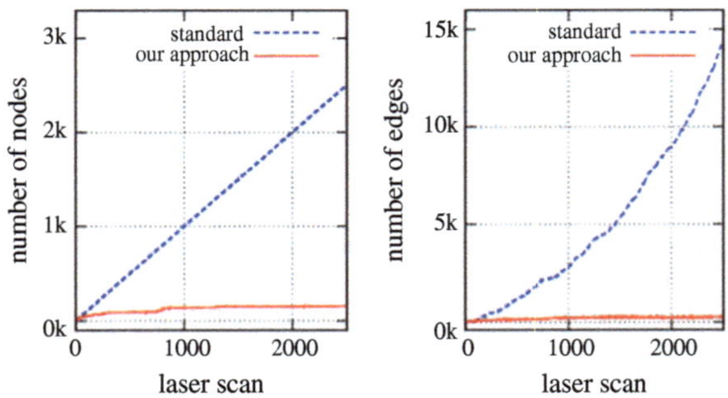

Fig. 8 Results of an experiment in which the robot moved around in an office environment for an extended period of time. In this experiment, our algorithm discarded all laser scans whose expected information gain was below a threshold

threshold for the mutual information instead of an upper bound for the number of nodes, the complexity does not grow as long as the robot does not explore new territory (see Fig. 8).

Our approach saves computational time as mentioned above but also introduces an overhead through the information-theoretic node selection. This overhead, however, is typically bounded since the number of nodes that have to be considered is bounded since our algorithm runs online and constantly discards nodes. Our current implementation of the information-theoretic laser scan selection is not optimized for speed—significant improvements could be obtained by caching results. Depending on the chosen parameters (particularly l, see Sect. 4.2) and on the environment that is mapped, the speed of our compression approach approximately ranges from running twice as fast as the standard approach to running four times slower than the standard approach. Our approach is beneficial when the robot frequently re-traverses already mapped areas. There is no gain if the robot mainly explores new territory.

6.3 *Effects on the Most Likely Occupancy Grid Map*

We furthermore analyzed the effects of our pruning technique on the resulting occupancy grid maps. We therefore compared the maps at a resolution of 10 cm and counted the number of cells that changed their most likely state (free, occupied, unknown) due to our pruning technique.

When mapping the Intel Research Lab, our pruning approach retained 349 of 1802 laser scans. As a consequence of this, 0.9 % of the cells changed. In the long-term experiment, our method kept 148 of 2597 laser scans and 1.6 % of the cells changed.

When mapping the FHW, our approach maintained 250 of 2049 scans and 1.2 % of the cells changed. Hence, the changes in the most likely maps are small.

6.4 *Scan Alignment and Map Quality*

We furthermore evaluated the effects of our pose graph compression technique that is applied during mapping on the quality of the resulting grid maps. In theory, the more observations are available, the better is the estimate. Ignoring measurements will lead to a belief with higher uncertainty. However, todays occupancy grid-based mapping systems typically involve some form of scan alignment or scan matching (e.g. to extract constraints). Such systems have the following disadvantage when it comes to long-term map learning. Whenever the robot obtains a measurement, the scan matcher aims at aligning the new scan with existing scans in order to solve the data association problem. The probability that the scan matcher thereby makes a small alignment error is nonzero. A scan which is incorporated at a slightly wrong position blurs the map. As a result, the probability that the scan matcher misaligns subsequent scans increases since scan matching is performed with misaligned scans. Hence, the probability of making alignment errors increases with the number of incorporated scans. In the long run, the map tends to become increasingly blurred and the mapping approach is likely to diverge.

Figures 5 and 3 depict the maps and graphs obtained from the Intel Research Lab dataset and the long-term experiments conducted in our office environment. The grid maps generated by the standard approach exhibit visibly more blur in several parts of the maps (see the arrows and the zoomed map view in the corresponding images). In general, the more often the robot re-traverses already visited terrain, the more blur is added to the maps. In contrast to the standard approach, our method discards scans and thus produces maps with sharp obstacle boundaries even in cases in which the robot frequently re-traverses already visited places. Although we do not claim that such a sharp map is a better estimate of the world, it better supports the scan matcher and reduces the risk of divergence in the mapping process.

7 Conclusion

In this paper, we presented a method for information-theoretic compression of pose graphs in graph-based SLAM, which is an important step towards long-term mapping. Our approach seeks to select the most informative set of laser scans and allows for restricting the size of the pose graph either based on a memory limit, resulting in an any-space mapping system, or based on a threshold on the minimum amount of information that a laser scan is expected to provide. Our approach estimates the expected information gain of laser measurements with respect to the resulting occupancy grid map. Real world experiments illustrate the effectiveness of our method for computing compressed pose graphs in the context of graph-based SLAM.

Acknowledgement This work has partly been supported by the DFG under SFB/TR-8, by the EC under FP7-231888-EUROPA and FP7-260026-TAPAS, and by Microsoft Research, Redmond. We would like to thank Giorgio Grisetti for his contribution to the marginalization as well as Maximilian Beinhofer for fruitful discussions. Furthermore, thanks to Dirk Hähnel for providing the FHW and the Intel Research Lab datasets.

References

1. M. Bosse, P.M. Newman, J.J. Leonard, S. Teller, An ATLAS framework for scalable mapping, in *Proceedings of the IEEE International Conference on Robotics and Automation (ICRA)* (2003), pp. 1899–1906
2. M. Cummins, P. Newman, FAB-MAP: probabilistic localization and mapping in the space of appearance. Int. J. Robot. Res. **27**(6), 647–665 (2008)
3. A.J. Davison, Active search for real-time vision, in *Proceedings of the International Conference on Computer Vision (ICCV)*, vol. 1 (2005)

4. E. Eade, P. Fong, M.E. Munich, Monocular graph SLAM with complexity reduction, in *Proceedings of the IEEE/RSJ International Conference on Intelligent Robots and Systems (IROS)*, Taipei, Taiwan (2010), pp. 3017–3024
5. C. Estrada, J. Neira, J.D. Tardós, Hierachical SLAM: real-time accurate mapping of large environments. IEEE Trans. Robot. **21**(4), 588–596 (2005)
6. R. Eustice, H. Singh, J.J. Leonard, Exactly sparse delayed-state filters for view-based SLAM. IEEE Trans. Robot. **22**(6), 1100–1114 (2006)
7. J. Folkesson, P. Jensfelt, H. Christensen, Vision SLAM in the measurement subspace, in *Proceedings of the IEEE International Conference on Robotics and Automation (ICRA)* (2005), pp. 325–330
8. G. Grisetti, R. Kümmerle, C. Stachniss, U. Frese, C. Hertzberg, Hierarchical optimization on manifolds for online 2D and 3D mapping, in *Proceedings of the IEEE International Conference on Robotics and Automation (ICRA)*, Anchorage, AK (2010)
9. V. Ila, J.M. Porta, J. Andrade-Cetto, Information-based compact pose slam. IEEE Trans. Robot. **26**(1), 78–93 (2010)
10. M. Kaess, F. Dellaert, Covariance recovery from a square root information matrix for data association. J. Robot. Auton. Syst. (RAS) **57**, 1198–1210 (2009)
11. K. Konolige, M. Agrawal, FrameSLAM: from bundle adjustment to realtime visual mapping. IEEE Trans. Robot. **24**(5), 1066–1077 (2008)
12. K. Konolige, J. Bowman, Towards lifelong visual maps, in *Proceedings of the IEEE/RSJ International Conference on Intelligent Robots and Systems (IROS)*, St. Louis, MO, USA (2009), pp. 1156–1163
13. A. Krause, C. Guestrin, Near-optimal nonmyopic value of information in graphical models, in *Proceedings of Uncertainty in Artificial Intelligence (UAI)* (2005)
14. H. Kretzschmar, C. Stachniss, G. Grisetti, Efficient information-theoretic graph pruning for graph-based SLAM with laser range finders, in *Proceedings of the IEEE/RSJ International Conference on Intelligent Robots and Systems (IROS)*, San Francisco, CA, USA (2011)
15. F. Lu, E. Milios, Globally consistent range scan alignment for environment mapping. Auton. Robot. **4**, 333–349 (1997)
16. D.J.C. MacKay, *Information Theory, Inference, and Learning Algorithms* (Cambridge Universit Press, 2003)
17. K. Ni, D. Steedly, F. Dellaert, Tectonic SAM: exact; out-of-core; submap-based slam, in *Proceedings of the IEEE International Conference on Robotics and Automation (ICRA)*, Rome, Italy (2007)
18. E. Olson, *Robust and Efficient Robotic Mapping*. Ph.D. thesis, MIT, Cambridge, MA, USA (2008)
19. P. Pfaff, R. Triebel, C. Stachniss, P. Lamon, W. Burgard, R. Siegwart, Towards mapping of cities, in *Proceedings of the IEEE International Conference on Robotics and Automation (ICRA)*, Rome, Italy (2007)
20. N. Snavely, S.M. Seitz, R. Szeliski, Skeletal graphs for efficient structure from motion, in *Proceedings of the IEEE Conference on Computer Vision and Pattern Recognition (CVPR)*, Anchorage, AK (2008), pp. 1–8

Part III
Planning

Demonstration-Guided Motion Planning

Gu Ye and Ron Alterovitz

Abstract We present demonstration-guided motion planning (DGMP), a new frame-work for planning motions for personal robots to perform household tasks. DGMP combines the strengths of sampling-based motion planning and robot learning from demonstrations to generate plans that (1) avoid novel obstacles in cluttered environments, and (2) learn and maintain critical aspects of the motion required to successfully accomplish a task. Sampling-based motion planning methods are highly effective at computing paths from start to goal configurations that avoid obstacles, but task constraints (e.g. a glass of water must be held upright to avoid a spill) must be explicitly enumerated and programmed. Instead, we use a set of expert demonstrations and automatically extract time-dependent task constraints by learning low variance aspects of the demonstrations, which are correlated with the task constraints. We then introduce multi-component rapidly-exploring roadmaps (MC-RRM), a sampling-based method that incrementally computes a motion plan that avoids obstacles and optimizes a learned cost metric. We demonstrate the effectiveness of DGMP using the Aldebaran Nao robot performing household tasks in a cluttered environment, including moving a spoon full of sugar from a bowl to a cup and cleaning the surface of a table.

1 Introduction

Over 10 million people need assistance with activities of daily living (ADLs) such as cooking, cleaning, and dressing in the United States. And this number is expected to grow dramatically worldwide as the number of elderly individuals continues to increase rapidly. Providing assistance to these individuals for ADLs is

G. Ye (✉) · R. Alterovitz
Department of Computer Science, The University of North Carolina, Chapel Hill, NC 27599, USA
e-mail: guye@cs.unc.edu

R. Alterovitz
e-mail: ron@cs.unc.edu

H.I. Christensen and O. Khatib (eds.), *Robotics Research*,
Springer Tracts in Advanced Robotics 100, DOI 10.1007/978-3-319-29363-9_17

currently very labor intensive and often requires moving an individual from their home to an impersonal institution, which comes at a significant cost to society. Recently developed personal robots have the potential to assist individuals with a variety of ADLs. However, the algorithms to control these robots for autonomous, safe assistance with household tasks is still a work in progress in the robotics research community.

A key challenge is the development of algorithms that enable personal robots to plan motions in unstructured environments to accomplish tasks currently performed by humans. These tasks often involve significant constraints on motion that humans are aware of from context and intuition. For example, when carrying a plate of food from the kitchen to the dining room, a person knows that tilting the plate sideways, while feasible, is undesirable. The robot must be aware of such task constraints and at the same time avoid unforeseen obstacles in unstructured environments.

We present a new approach that unifies ideas from two fields: robot motion planning and robot learning from demonstrations. The motion planning community has developed highly successful sampling-based methods that efficiently compute feasible plans that avoid obstacles. However, motion planning methods typically do not consider any task constraints unless explicitly programmed. As shown in Fig. 1, direct application of standard motion planning algorithms to a robot transferring a spoonful of sugar from a bowl to a cup will result in the robot spilling the sugar unless the robot is explicitly programmed to keep the spoon level. While such task constraints can be innocuous to program individually, requiring expert programmers to anticipate and program all such constraints for robots operating in human environments would be prohibitive. In contrast, the learning from demonstration

Fig. 1 We consider the household task of using a spoon to transfer sugar from a bowl to a cup on a table. Using standard motion planning algorithms without explicitly programming task constraints will fail since the spoon will not be kept level, causing the sugar to be spilled on the table (*left*). Methods based on learning from demonstrations are often unable to compute collision-free trajectories when novel obstacles are introduced in new locations not considered in the demonstrations (*middle*). Our method effectively combines motion planning with learning from demonstrations to avoid obstacles and keep the spoon level to successfully transfer the sugar (*right*)

community has developed highly effective approaches for extracting from multiple demonstrations the key motions required to accomplish a task. Such methods can explicitly or implicitly learn task constraints, such as the fact that spoons full of sugar should be kept level to avoid spillage. However, methods based on learning from demonstrations often falter when the robot must accomplish the task in a new, cluttered environment where unforeseen obstacles compel the robot to move outside the range of motions included in the input demonstrations. Our new approach, *Demonstration-Guided Motion Planning* (DGMP), combines the strengths of methods in motion planning and in learning from demonstration to both (1) avoid novel obstacles in cluttered environments, and (2) learn and maintain critical aspects of the motion required to successfully accomplish a task.

To teach the robot the motions necessary to assist in a household task, we use kinesthetic demonstrations; the robot's joints are placed in a passive mode and the demonstrator moves the robot's arms and torso through each step of the task. Kinesthetic demonstrations are ideally suited for teaching household assistance tasks. They do not require the human expert (e.g. an occupational therapist) to learn complicated computer/robot programming. Instead, they rely only on physical, realworld demonstrations that are an intuitive and user-friendly way for human experts to teach robots to perform new tasks.

The emphasis of DGMP is not on learning high-level task decompositions or low-level controls from the demonstrations, but rather to learn the constraints on the robot's motion that are required to accomplish a repeated task. From a set of demonstrations, we apply statistical analysis to learn low and high variance components of the motion. Low variance regions correspond to implicit constraints that should be satisfied when the motion plan is executed. We use these variances to define a cost metric over the configuration space of the robot.

After the learning phase, DGMP uses a new motion planning algorithm, a multi-component rapidly-exploring roadmap (MC-RRM). The planner guarantees obstacle avoidance and incrementally improves the plan, guaranteeing convergence to the optimal feasible solution for the cost metric as computation time is allowed to increase.

We demonstrate the effectiveness of DGMP using the Aldebaran Nao robot performing household tasks in a cluttered environment, including transferring sugar from a bowl to a cup using a spoon and wiping a table. Our results demonstrate that unlike using motion planning or learning from demonstration methods in isolation, our unified approach converges to an optimal solution and offers a significantly higher success rate in accomplishing tasks that involve both learning of task constraints as well as obstacle avoidance.

2 Related Work

Our framework bridges robot motion planning with robot learning from demonstrations. Learning from demonstration methods have been highly successful in enabling robots to learn task constraints and imitate task motions [4, 6]. Motion planning methods have been effective at computing feasible motions from a start configuration to a goal configuration while avoiding obstacles [10, 16].

Demonstrations can provide examples of the motion required to accomplish a task, and these demonstrations can be used to computationally learn a control policy that will enable a robot to autonomously execute the task motion subject to real-world noise and disturbances. Inverse reinforcement learning has been used to estimate the unknown objective function of a control policy from demonstrations in environments with complex dynamics. This approach, sometimes called apprenticeship learning, has been applied to learn control policies for car driving [2], helicopter acrobatics [12], and robotic knot tying [24]. Another approach models the variations across demonstrated motion trajectories using a Gaussian Mixture Model (GMM) and then uses Gaussian Mixture Regression (GMR) to estimate the ideal trajectory and a corresponding controller [7]. Our approach builds on the GMM/GMR work-flow from Calinon et al. for extracting local trajectories expressed in coordinate systems relative to objects in the environment [7]. The GMM/GMR approach has been applied to manipulation tasks such as moving chess pieces or feeding a doll and is robust to movement of obstacles included in the demonstrations [8]. However, these methods lack the ability to avoid novel obstacles that were not explicitly considered during the demonstrations, which is critical for motion planning in household environments.

Recent methods have used learning from demonstration methods to consider previously unseen obstacles, but existing methods are limited either to low dimension spaces, place limitations on the locations of obstacles, or do not allow for time-dependent task-space constraints. Prior work has investigated using global search methods such as A* or D* where path costs are learned from demonstrations. This approach has been successfully applied to navigating cars in a parking lot [1], maneuvering off-road vehicles in complex terrain [23], and generating natural motions for animated characters [17]. However, these methods do not consider time-dependent constraints and they discretize the state space, which does not scale well to higher degree of freedom systems like some personal robots. Another approach models demonstration trajectories as fluid currents in the task space and performs fluid dynamic simulation to derive a control policy [19]. Fluid simulation considers obstacles, but performance is unclear if obstacles pass through the demonstration trajectories.

In contrast to robot learning methods that have focused on extracting meaningful data from demonstrations, motion planning focuses on computing feasible plans that avoid obstacles. In particular, sampling based methods have been very successful for a wide variety of problems involving high-DOF robots [10, 16]. Prior work has investigated the use of RRTs combined with learned metrics to generate paths, but these methods are guaranteed to converge to a suboptimal solution [15]. One approach extends RRTs to sample only inside a provided number of standard deviations of a mean demonstrated trajectory, but may not find a feasible solution even if one exists [11]. RRTs have also been used in conjunction with task-based symbolic constraints [14]. Transition-based RRT (T-RRT) [13] is a sampling-based motion planner over cost maps that biases expansion of the tree to low cost regions of the configuration space. T-RRT can be used to generate natural motions by using a predefined human robot interaction cost metric [18]. GradienT-RRT extends T-RRT to use the gradients of the cost function to locally optimize the trajectory and facilitate finding solutions through narrow chasms [5]. GradienT-RRT can be used in conjunction with GMM to generate more natural motions and can also be used to constrain robot motion using explicit task space constraints. However, this prior work has not considered automatic learning of time-dependent task constraints from demonstrations and using that learned information to guide planning. Learning-based methods can also incorporate consideration of obstacles using potential fields [20], but this approach is sensitive to local minima.

In contrast to prior work, DGMP combines learning from demonstration with a sampling-based motion planner that incrementally refines the solution. Our DGMP framework learns task-space constraints and expresses them in a cost map, considers time-dependent criteria by aligning the robot's trajectory to demonstrations, avoids novel obstacles not included in demonstrations, works for high DOF robots, and generates an optimal solution as computation time is allowed to increase.

3 The Demonstration-Guided Motion Planning Framework

Our Demonstration-Guided Motion Planning (DGMP) framework consists of two major phases: learning and execution. The learning phase only needs to be performed once for a particular task and can then be applied to multiple task executions in different environments. In the DGMP learning phase, human experts perform several demonstrations of the task, which are encoded into a learned cost metric. In the DGMP execution phase, the robot performs the task in a new environment using the derived cost metric for guidance. An overview of the approach is shown in Fig. 2.

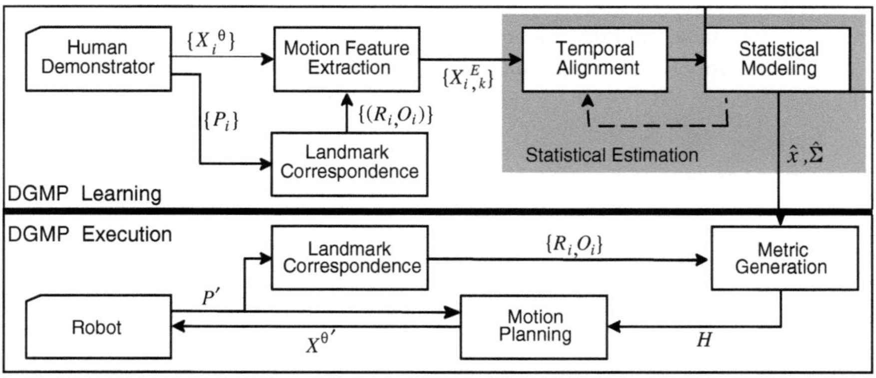

Fig. 2 The DGMP framework consists of a learning phase that is performed once per task and an execution phase that is performed each time the task is executed in a new environment

The DGMP learning phase requires that human experts control the robot to perform N demonstrations of the task. Although the framework allows demonstrations to be provided in any manner, our implementation assumes that we use kinesthetic demonstrations and that the robot includes position-controlled joints. For each demonstration, the robot's joints are placed in a passive mode and a human expert manually moves the robot's limbs to perform the task. We assume the robot has encoders at every joint, allowing the robot to "sense" its own motion and record joint angles as a function of time for each demonstration. The data obtained from demonstration i will be a sequence of joint angles X_i^θ as well as other sensor input Pi of the environment, such as images (from camera sensors) and point cloud data (from laser range finders or stereo image reconstructions).

The DGMP learning phase begins by aligning the demonstrations and transforming the demonstration data into a set of motion features expressed as a function of time. We consider motion features defined by the angles of the robot's joints as well as the position of points on the robot (e.g. the end effector and a grasped object) relative to landmarks in the environment (e.g. the sugar bowl or cup in Fig. 1). For a given time, the mean $\left\{\hat{x}^{(k)}\right\}$ and covariance matrix $\left\{\hat{\sum}^{(k)}\right\}$ of each motion feature k across multiple demonstrations reflect the task constraints intended by the human demonstrator. The lower the variance of a motion feature across demonstrations at a given time, the higher the consistency of the demonstrations with respect to that motion feature. Higher consistency implies the mean value of a motion feature should be followed more closely when performing the task. In contrast, high variance motion features likely do not need to be closely reproduced during execution for the robot to succeed in accomplishing the task.

In the DGMP execution phase, the robot senses its environment to (1) determine the landmarks' correspondences in the current environment, and (2) collect sufficient data to perform collision detection for motion planning. Combining the landmark correspondences with the means $\left\{\hat{x}^{(k)}\right\}$ and covariance matrices $\left\{\hat{\sum}^{(k)}\right\}$ of all the motion features, we define our cost metric H, which estimates the degree to which a a candidate plan matches the intent of the expert demonstrator. We then execute our new motion planning algorithm, MC-RRM, to search for a collision-free robot motion that minimizes the cost metric.

3.1 DGMP Learning Phase

The DGMP learning phase takes as input N demonstrations of the task. During each demonstration i, we record a time sequence of the robot's configuration $X_i^\theta = \left\{x_{i,t}^\theta\right\}_{t=1}^{T_i}$ where T_i is the length of the demonstration and $x_{i,t}^\theta$ is the vector joint angles at time t. We also record sensor input $\{P_i, i = 1, \ldots, N\}$ of the environment, such as images and point cloud data. From the sensed data, we require that M landmarks be identified, either manually or automatically using computer vision algorithms, that are present across all demonstrations and will likely be present in the execution environment. The landmarks serve as correspondence points in the environment across the demonstrations, and they may or may not be directly related to the task. In our implementation, we also explicitly define the robot's torso to be a landmark. For each demonstration, we denote the poses (orientations and positions) of the landmarks by $\{(R_{ji}, o_{ji}), j = 1, \ldots, M, i = 1, \ldots, N\}$. For example, in the task shown in Fig. 1a landmark was sensed on the sugar bowl.

3.1.1 Extracting Motion Features from Demonstrations

We expect that the key task constraints for a problem are satisfied across all the demonstrations. To automatically extract these task constraints, we consider a set of motion features that are designed to help identify aspects of the robot motions that are consistent across demonstrations but that may be hidden in the raw demonstration data. We consider configuration motion features and landmark-based motion features. In a *configuration motion feature*, we consider the robot's joint angles at a particular time. For a personal robot with many redundant degrees of freedom, this data helps in learning "natural" motions that are lost when only considering end effector motions. We denote the configuration motion feature trajectory as $X_i^{(0)} = X_i^{\theta}$. In a *landmark-based motion feature*, we consider the location of one or more points attached to moving parts of the robot (e.g. end effector, manipulator arm, and grasped objects) relative to sensed landmarks in the environment (e.g. a cup or bowl). For the vector of relevant points attached to moving parts of the robot, we define a "local" trajectory of that vector as the coordinates of the relevant points with respect to each landmark. The landmarks essentially serve as local coordinate systems for defining the trajectory of points on the robot. Landmark-based motion features can be used to find important consistencies across demonstrations, such as the position of the end effector relative to a relevant object for the task. We compute the end effector trajectory in the local coordinate system of each landmark $x_{i,t}^{(j)} = R_{ji}\left(x_{i,t}^{(1)} - o_{ji}\right), i = 1, \ldots, N, j = 1, \ldots, M$, where $x_{i,t}^{(1)}$ is the end effector position relative to the robot's torso (i.e. landmark #1). We represent a local trajectory as $X_i^{(j)} = \left\{x_{i,t}^{(j)}\right\}_{t=1}^{T_i}$. Combining the configuration and landmark-based motion features, we have $L = M + 1$ motion feature trajectories represented as $X_i^E = \left\{x_i^{(j)}\right\}_{j=0}^{L}$.

3.1.2 Computing the Cost Metric Using Statistical Modeling

Our objective is to identify low variance and high variance aspects of the motion feature trajectories across demonstrations in order to create a cost metric that will guide a motion planner to ensure that task constraints are satisfied as best as possible given the locations of obstacles in the execution environments.

The motion feature trajectories are time dependent signals and are obtained from different demonstrations with different time scales (e.g. due to varying demonstration speed). To correctly encode the task constraints across different demonstrations, the trajectories X_i^E must be temporally aligned. That is, for each time step t, the set of motion features for the ith demonstration $X_{i,t}^E = \left\{x_{i,t}^{(j)}\right\}_{j=0}^{M+1}$ is assigned with an aligned time step $g_i(t)$. Therefore, the aligned trajectory of X_i^E is represented as $X_{i,t}^A = \left\{X_{i,t'}^E | g_i(t') = t, t' = 1, \ldots, T_i, i = 1, \ldots, M\right\}$, $t = 1, \ldots, T$. We use linear interpolation resampling to ensure one observation per time slot. To compute g_i, we use dynamic time warping (DTW), which has been used in speech recognition [22] and robot learning [9, 12]. DTW uses dynamic programming to compute the optimal time alignment based on a distance function between points. We initially use a Euclidean distance function and then iteratively refine our solution using the output of the statistical modeling as discussed below.

Given the time aligned motion feature trajectories, our next objective is statistical modeling: to estimate the expected mean and covariance of the motion features across demonstrations. For a given time t, each motion feature has N observations, one from each demonstration. For notation simplification, let $x_{i,t}^{(k)}$ denote the motion feature k from demonstration i at time t after temporal alignment. We calculate the time dependent mean $\hat{x}$ and covariance matrix $\hat{\Sigma}$ for each time slice:

$$\hat{x}_t^{(k)} = \frac{1}{N}\sum_{i=1}^{N} x_{i,t}^{(k)}, \hat{\Sigma}_t^{(k)} = \frac{1}{N-1}\sum_{i=1}^{N}\left(x_{i,t}^{(k)} - \hat{x}_t^{(k)}\right)\left(x_{i,t}^{(k)} - \hat{x}_t^{(k)}\right)^T.$$

Figure 3 shows the statistical modeling result for two motion feature trajectories extracted for the table cleaning task described in Sect. 4.

We use the results of the statistical modeling to iteratively improve the temporal alignment. Rather than treating each DOF of the motion feature trajectories as having equal weight, we instead consider the similarities identified by the covariance computation. We generate weights for each DOF using the inverse of the covariance matrix, which results in similar motion features having a shorter distance in DTW. We re-execute DTW with the revised distance metric, re-evaluate the means and covariances, and loop until the result converges.

In the execution phase described below, we will search for an optimal trajectory in the robot's joint space that follows the modeled constraints, which we represent using a cost metric. The cost metric function depends on the landmark poses in the execution environment, where R_i and o_i define the transformation of the local

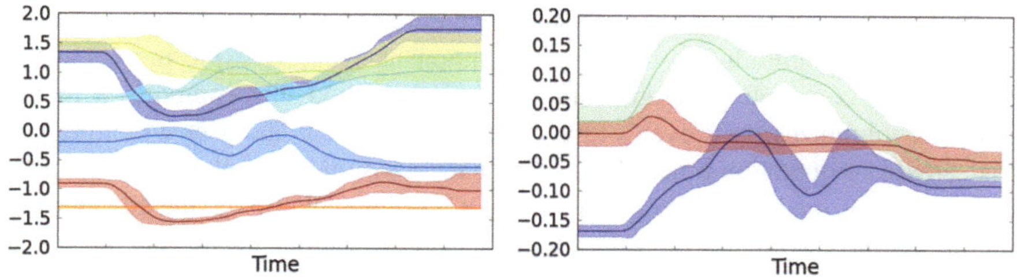

Fig. 3 The mean and covariance of the two motion feature trajectories, the 6D joint angle trajectories (*left*) and 3D end effector trajectory (*right*), extracted for the table cleaning task described in Sect. 4. The *red line* in the *right* plot, which corresponds to the *vertical* coordinate of the end effector trajectory, has low variance in the *middle* section; this indicates the end effector with the paper towel should be kept on the table surface

coordinate system of the ith landmark. With the input landmark poses provided during task execution, the cost of a given joint configuration θ at time t can be computed by the cost metric H as:

$$H(\theta, t) = (\theta - \hat{\theta}_t)^T W_t^{\theta} (\theta - \hat{\theta}_t) + \sum_{i=1}^{M} \left[\left(K(\theta) - \hat{x}_t^{\prime(i)} \right)^T W_t^{x(i)} \left(K(\theta) - \hat{x}_t^{\prime(i)} \right) \right],$$

where

$$\hat{\theta} = \hat{x}^{(M+1)}, \hat{\Sigma}^{\theta} = \hat{\Sigma}^{(M+1)}, W_t^{\theta} = \left(\hat{\Sigma}_t^{\theta}\right)^{-1}, \hat{x}_t^{\prime(i)} = R_i \hat{x}_t^{(i)} + o_i, W_t^{x(i)} = \left(R \hat{\Sigma}_t^{x(i)} R^T\right)^{-1}$$

and K is the forward kinematics function mapping the joint configuration to end effector position. The time-dependent cost metric function H is the output of the learning phase.

3.2 DGMP Execution Phase

We use the results of the learning phase to enable the robot to execute the task in new, cluttered environments. The robot must sense its environment to determine the locations of obstacles (e.g. the volume of the workspace that is not free) as well as sense landmarks and establish correspondences to the landmarks used in learning. To compute a motion plan that avoids obstacles while satisfying task constraints, DGMP takes advantage of two useful pieces of information. First, with knowledge of the landmark locations, the robot can compute the learned cost metric defined in the subsection above for any path. Second, it can compute a guiding path defined by the mean of the motion feature trajectories, which is equivalent to following the valley of the cost metric. This guiding path is not guaranteed to be collision free.

To harness these pieces of information and efficiently solve the motion planning problem, we introduce the multi-component rapidly-exploring roadmap (MC-RRM), a sampling based motion planning method that computes a motion plan that minimizes costs over a time-dependent cost map and uses a guiding path to decrease computation time. The guiding path is assumed to follow a valley of the cost map. This method is ideally suited for DGMP by explicitly taking advantage of the information available from the motion feature trajectory means and variances.

MC-RRM combines the benefits of PRMs and RRTs for motion planning problems over time-dependent cost maps in which a guiding path is provided. A traditional PRM with uniform sampling could solve this problem to optimality as the number of samples increases, but this approach is prohibitively slow for high dimensional configuration spaces in which most of the configuration space is not relevant for the task and due to challenges in temporally aligning a PRM plan to the demonstrations for computing the metric. On the other hand, RRT provides a fast incremental sampling-based algorithm for high dimensional spaces but is guaranteed to return a suboptimal solution [15]. MC-RRM builds on ideas from RRG and RRT* [15] and uses the guiding path to incrementally build a roadmap that enables finding an optimal solution as computation time is allowed to increase.

3.2.1 MC-RRM

MC-RRM takes as input a configuration space (C-space), a cost metric over the C-space, a set of obstacles in the C-space, a collision detector, and a guiding path $\tilde{x} = \{\tilde{x}_1, \ldots, \tilde{x}_T\}$ that traverses a local valley in the cost metric from the initial state to a goal state. The cost metric is defined by $H_e(x_1, x_2) > 0$ for an edge between a pair of arbitrary points x_1 and x_2 in C-space. MC-RRM returns a collision free trajectory such that the sum of the edge costs over the path is minimized.

As shown in Sect. 3.2.2, we can compute the guiding path directly based on the cost metric. With the newly observed obstacles in the execution environment, the ideal trajectory defined by this guiding path may not be feasible because parts of the trajectory collide with the obstacles. When the obstacles intercept the trajectory sparsely, the collision-free part of the ideal trajectory is still likely to be the part of the optimal path. Hence, the key challenge is to find optimal alternative pathways for the in-collision sections of the guiding path.

In MC-RRM, the points along the guiding path are regarded as guide points, whether in collision or not. At initialization of the MC-RRM, the collision-free guide points are added as nodes to a global graph (i.e. the roadmap) and adjacent points are connected by an edge if the edge is collision free. Due to obstacles, the global graph may contain multiple disconnected components. Figure 4 shows an example with two disconnected components. We denoted these components as C_i, which are represented as subgraphs of the global graph. For each C_i, a node set is maintained. The node set is initialized with the guide points from the corresponding disconnected component. In the sampling process, we sample new nodes in a manner similar to RRT for each of the C_i: a random sample is created; nearest

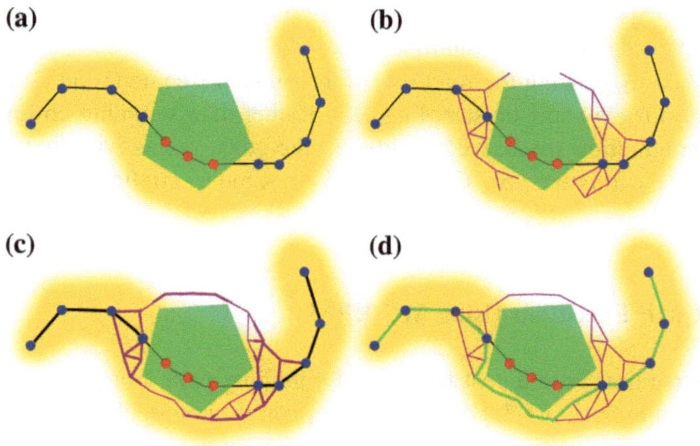

Fig. 4 MC-RRM in a 2D C-space where the guiding path intersects an obstacle (*green polygon*). **a** The guiding path is added to the graph, resulting in two disconnected components shown in *blue*. **b**, **c** Samples are drawn from a GMM (*yellow*), both components are extended, and nearest nodes are connected (*purple*), forming a graph. **d** Lowest cost path (*green*) computed by Dijkstra's algorithm

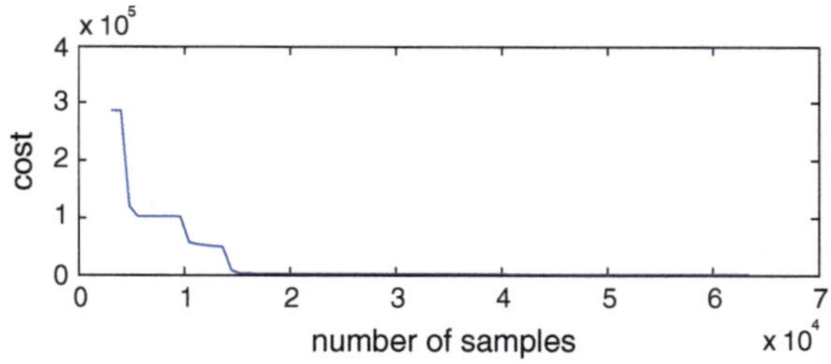

Fig. 5 Minimal path cost versus number of samples for one of the test cases. With traditional PRM, the minimal cost path is larger than 7×10^6 with 100,000 samples

neighbors in each C_i are selected; and each component is extended toward the random sample up to a distance $d_{\max}$ where a new node is added. The newly added nodes are connected by edges to the corresponding C_i as well as to the global graph. Every time a new node is added to the global graph, it is connected to all nearby nodes within some distance (as in RRG [15]), which could belong to any C_i. Until the allowed computation is reached, we periodically run Dijkstra's shortest path algorithm to compute the path with lowest cost (Fig. 5).

We note that each C_i only maintains a node set and does not need to store edges. Edges are only maintained in the global graph. As the sampling progresses, edges in the global graph may connect nodes in one component with nodes in another disconnected component. We maintain the components separately in order to maintain the expansion bias from each component that was originally separated by an obstacle. This ensures that gaps in the guiding path introduced by obstacles are incrementally explored from multiple directions.

In addition to growing the roadmap from collision-free nodes along the guiding path, we also use a heuristic sampling bias. Rather than sampling uniformly in the robot's configuration space, we consider all the guide points with their covariance matrices as a mixture of Gaussians. We sample configurations from this distribution.

3.2.2 DGMP Execution Using MC-RRM

MC-RRM requires both a cost metric and a guiding path. For the cost metric, we use H defined in Sect. 3.1.2. For the guiding path, we ignore obstacles and compute the robot configuration (i.e. the joint angles) that minimizes the cost metric at each t, $t = 1, \ldots, T$. To ensure satisfaction of the robot's kinematic constraints, we minimize H at each t using Lagrange optimization as in Calinon et al. [9]. For the sampling based motion planning algorithm, we compute the edge cost using our cost metric: $H_e(x_1, x_2) = \int_{x_1}^{x_2} H(x, k(x))dx$ where the integral is taken along the line segment from x_1 to x_2 in C-space and k is a time alignment function that estimates the corresponding time index of x in the original demonstration. In our current implementation, k returns the time index of the nearest point on the guiding path. This may introduce problems when the guiding path involves loops, which will be addressed in future work. To efficiently discretize the line integral for $H_e(x_1, x_2)$, we approximate the function as $H_e(x_1, x_2) = \sum_{t=t_1+1}^{t_2} H\left(\frac{(x_2-x_1)t}{t_2-t_1} + x_1, t\right)$, where $t_1 = k(x_1)$, $t_2 = k(x_2)$ the edge from x_1 to x_2, we assume the transition is performed in constant speed and interpolate configurations along the edge.

4 Results

We applied DGMP to the Aldebaran Nao robot [3] to perform two household tasks. The first task was to transfer sugar from a bowl to a cup in the presence of obstacles as shown in Figs. 6 and 7. The second task was to clean the surface (Fig. 8) of a table in the presence of obstacles as shown in Fig. 9.

The Aldebaran Nao includes 26 total degrees of freedom, including 5 DOF in each arm, 1 DOF for bending at the hip, 1 DOF for opening and closing each 3-finger gripper, and the remaining DOFs for the legs and neck. In our experiments, we used 6 DOF for each task: 5 DOF in the right arm and 1 DOF at the hip. We implemented DGMP using the Python programming language. All computation was performed on a 2.4 GHz Intel Xeon e5620 PC running 64-bit Linux.

In our first experiment, a sugar transfer task, the goal was to evaluate the ability of DGMP to (1) learn task-specific constraints that are critical to the success of the task, and (2) avoid novel obstacles in the environment. The robot was seated at a

Fig. 6 Execution of DGMP for the sugar transfer task. The robot successfully keeps the spoon level while avoiding the jar, a novel obstacle not included in the demonstrations

Fig. 7 Execution of DGMP for the sugar transfer task. The robot successfully keeps the spoon level while avoiding the canister, a novel obstacle not included in the demonstrations

table with a sugar bowl and a cup, and, in some cases, other items. The task was to scoop some sugar using a spoon and transfer the sugar to the cup without bumping the bowl, the cup, or any other items that may be on the table. Other than the demonstrations, we did not provide the method any explicit task-specific

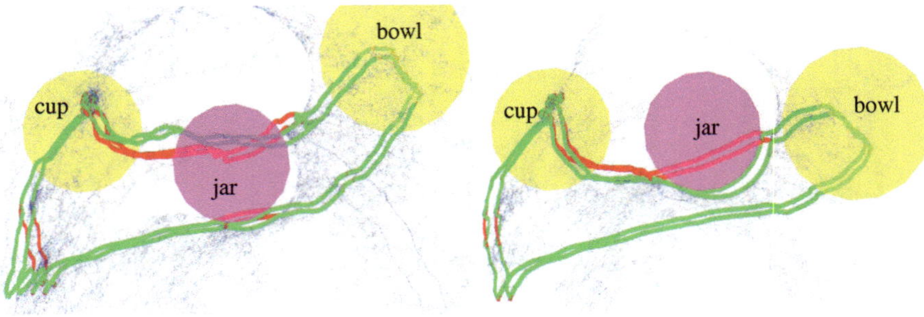

Fig. 8 A *top-down* view of the workspace for two test cases. The demonstration mean (*red*) and the MC-RRM plan (*green*) for the spoon motion, where two points near the spoon tip are tracked. MC-RRM samples projected into the workspace are shown as *blue dots*. Depending on the placement of the jar, the robot may detour above the jar (*left*) or to the side of the jar (*right*) in order to keep the spoon level while still satisfying the kinematic constraints the robot

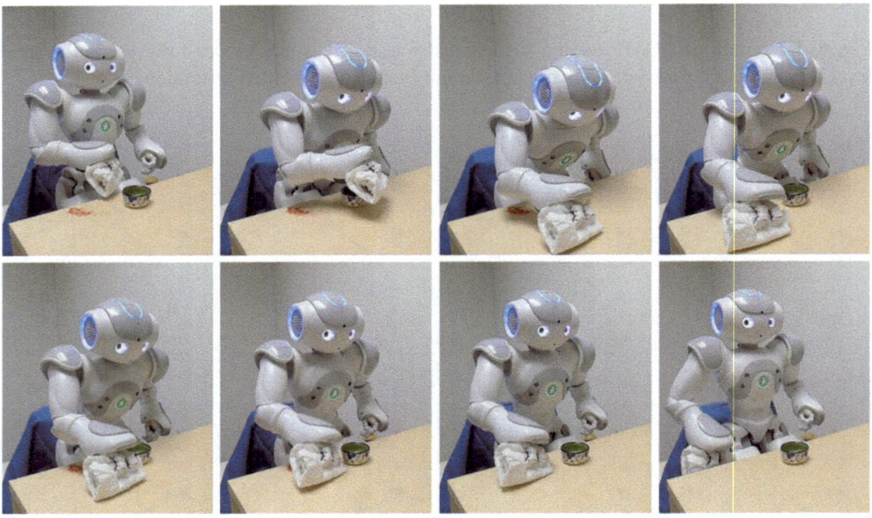

Fig. 9 Execution of DGMP for the table cleaning task. The robot successfully learned to keep the paper towel on the table while avoiding the bowl, which was not included in any of the demonstrations

information; e.g. we never explicitly constrained the spoon to be level when moving sugar from the bowl to the cup to prevent spillage. To successfully accomplish this task, the robot needed to (1) automatically learn that the spoon should be kept level as it moves from the bowl to the cup on order to prevent spilling the sugar, and (2) successfully avoid obstacles on the table when performing the task.

We conducted 7 kinesthetic demonstrations. In each demonstration, only the sugar bowl and cup were on the table. We varied the placement of the sugar bowl randomly within a 5 inch radius on the table across the demonstrations. We used 3 environment landmarks which were located on the chair, the cup, and the bowl. Each end-effector trajectory contained the x, y, z coordinates of two points attached to the robot (the end effector and the tip of the spoon), so each end-effector vector is 6 dimensional. Two points on the robot were chosen such that the appropriate tilt and level of the spoon could be learned (via the covariance matrices) for different stages of the task.

To evaluate performance of our method, we created 27 test cases. In each test case, the location of the sugar bowl and of the obstacle were randomly determined. The sugar bowl's location varied within a 4 inch range and the obstacle was placed in a 2.5 inch range. We provided the shape and locations of the bowl and obstacle (as would be extracted from a vision system) to the motion planner and then computed a plan for each test case. We then executed each plan on the Nao robot in the experimental setup. After completing the demonstrations, the learning phase required 42 s of computation time. In our unoptimized implementation, computing a motion plan then required 1.5 s when no obstacle was present and an average of 410 s when an obstacle was present. We show the convergence of the method in Fig. 5 and note that the cost metric of the path obtained by MC-RRM was orders of magnitude lower than for standard PRM at equivalent computation times.

A test case was considered *successful* if the robot (1) scoops sugar from the bowl and transfers it to the cup without spilling on the table or obstacle, and (2) does not contact the obstacle, bowl, or cup. We considered a test case to be *feasible* if it was possible for the robot to successfully accomplish the task given its kinematic constraints. Of the 27 test cases, 3 were not feasible due to the obstacles being too close to the robot's rest pose.

DGMP succeeded in 22 of the test cases, resulting in a success rate of 92 % of the 24 feasible test cases. In the two failure cases, the obstacle was very close to the robot, which results in a narrow passage in the robot's configuration space and the MC-RRM could not find a solution within 100,000 samples. For the same test cases, we also applied a pure learning-based approach in which we executed the mean time-aligned trajectory. Due to obstacle collisions, this approach resulted in only 3 successes, a success rate of 13 % of the feasible test cases. RRT resulted in zero successes because it always spills the sugar due to lack of knowledge of task constraints.

We also tested our method on a table cleaning task. As in the sugar transfer task, the Nao robot was seated at a table. The objective was to wipe the table clean using a grasped clump of paper towel. Six demonstrations were recorded of the robot wiping the table in an S-shaped curve without any obstacles. During task execution, a new obstacle was put on the table. To be successful, the robot needed to learn that the paper towel needed to be kept on the surface of the table and avoid obstacles. In all our task executions, DGMP was able to find a detouring path and successfully complete the wiping task, as shown in the example in Fig. 9.

Videos of a Nao robot performing these tasks using DGMP are available at: http://robotics.cs.unc.edu/DGMP.

5 Conclusion and Future Work

We presented demonstration-guided motion planning (DGMP), a new framework for planning motions for personal robots to perform household tasks. DGMP combines the strengths of sampling-based motion planning and robot learning from demonstrations to generate plans that (1) avoid novel obstacles in cluttered environments, and (2) learn and maintain critical aspects of the motion required to successfully accomplish a task. Sampling-based motion planning methods are highly effective at computing paths from start to goal configurations that avoid obstacles, but task constraints must be explicitly enumerated and programmed. Instead, we used a set of expert demonstrations and automatically extract task constraints by learning low variance aspects of the demonstrations. We then introduced multi-component rapidly-exploring roadmaps (MC-RRM), a sampling-based method that incrementally computes a motion plan that avoids obstacles and optimizes the learned cost metric. We demonstrated the effectiveness of DGMP using the Aldebaran Nao robot performing household tasks in a cluttered environment, including moving a spoon full of sugar from a bowl to a cup and cleaning the surface of a table. By effectively combining motion planning with learning from demonstration, our robot using DGMP accomplished its task successfully in over 90 % of its test cases for which a solution was feasible.

In the future work, we plan to extend our method to handle more general tasks with loops and pauses. This could be done by improving the time alignment function in the edge cost metric computation. We also plan to consider real-time sensing (including scene and object recognition) and moving obstacles, which may involve integrating perception-guided motion planning [21] with DGMP. We hope to build on DGMP to enable personal robots to assist humans with a larger class of tasks.

Acknowledgment This research was supported in part by the National Science Foundation (NSF) under awards #IIS-0905344 and #IIS-1117127 and by the National Institutes of Health (NIH) under grant #R21EB011628. The authors thank Jenny Womack from the Dept. of Allied Health Sciences for her input and John Thomas and Herman Towles from the Dept. of Computer Science for their help with the experiment setup.

References

1. P. Abbeel, D. Dolgov, A. Ng, S. Thrun, Apprenticeship learning for motion planning with application to parking lot navigation, in *Proceedings of IEEE/RSJ International Conference on Intelligent Robots and Systems (IROS)* (Sept. 2008), pp. 1083–1090
2. P. Abbeel, A.Y. Ng, Apprenticeship learning via inverse reinforcement learning, in *Proceedings of International Conference on Machine Learning (ICML)* (2004)
3. Aldebran Robotics, Aldebran Robotics NAO for education (2010), http://www.aldebaran-robotics.com/en/naoeducation
4. B.D. Argall, S. Chernova, M. Veloso, B. Browning, A survey of robot learning from demonstration. Robot. Auton. Syst. **57**, 469–483 (2009)

5. D. Berenson, T. Simeon, S.S. Srinivasa, Addressing cost-space chasms in manipulation planning, in *Proceedings of IEEE International Conference on Robotics and Automation (ICRA)* (May 2011), pp. 4561–4568
6. S. Billard, R. Calinon, S. Dillmann, Schaal, Robot programming by demonstration, in *Handbook of Robotics*, ed. by B. Siciliano, O. Khatib (Springer, 2008), ch. 59, pp. 1371–1394
7. S. Calinon, *Robot Programming by Demonstration*, 1st edn. (CRC Press Inc, Boca Raton, 2009)
8. S. Calinon, F.D'halluin, D.G. Caldwell, A.G. Billard, Handling of multiple constraints and motion alternatives in a robot programming by demonstration framework, in *9th IEEE—RAS International Conference on Humanoid Robots* (Dec. 2009), pp. 582–588
9. S. Calinon, F. Guenter, A. Billard, On learning, representing, and generalizing a task in a humanoid robot. IEEE Trans. Syst. Man Cybern. Part B **37**(2), 286–298 (2007)
10. H. Choset, K.M. Lynch, S. Hutchinson, G. Kantor, W. Burgard, L.E. Kavraki, S. Thrun, *Principles of Robot Motion: Theory, Algorithms, and Implementations* (MIT Press, 2005)
11. J. Claassens, An RRT-based path planner for use in trajectory imitation, in *Proceedings of IEEE International Conference on Robotics and Automation (ICRA)* (May 2010), pp. 3090–3095
12. A. Coates, P. Abbeel, A. Ng, Learning for control from multiple demonstrations, in *Proceedings of 25th International Conference on Machine Learning* (2008), pp. 144–151
13. L. Jaillet, J. Cortes, T. Simeon, Sampling-based path planning on configuration-space costmaps. IEEE Trans. Robot. **26**(4), 635–646 (2010)
14. R. Jakel, S.R. Schmidt-Rohr, M. Losch, R. Dillmann, Representation and constrained planning of manipulation strategies in the context of programming by demonstration, in *Proceedings of IEEE International Conference on Robotics and Automation (ICRA)* (May 2010), pp. 162–169
15. S. Karaman, E. Frazzoli, Incremental sampling-based algorithms for optimal motion planning, in *Proceedings of Robotics: Science and Systems* (2010)
16. S.M. LaValle, *Planning Algorithms* (Cambridge University Press, Cambridge, U.K., 2006)
17. S.J. Lee, Z. Popović, Learning behavior styles with inverse reinforcement learning. ACM Trans. Graph. (SIGGRAPH 2010) **29**(4), 122:1–122:7 (2010)
18. J. Mainprice, E. Sisbot, L. Jaillet, J. Cortés, R. Alami, T. Siméon, Planning human-aware motions using a sampling-based costmap planner, in *Proceedings of IEEE International Conference on Robotics and Automation (ICRA)* (May 2011), pp. 5012–5017
19. H. Mayer, I. Nagy, A. Knoll, E.U. Braun, R. Lange, R. Bauernschmitt, Adaptive control for human-robot skilltransfer: trajectory planning based on fluid dynamics, in *Proceedings of IEEE International Conference on Robotics and Automation (ICRA)* (April 2007), pp. 1800–1807
20. P. Pastor, H. Hoffmann, T. Asfour, S. Schaal, Learning and generalization of motor skills by learning from demonstration, in *Proceedings of IEEE International Conference on Robotics and Automation (ICRA)* (May 2009), pp. 763–768
21. R.B. Rusu, I.A. Sucan, B.P. Gerkey, S. Chitta, M. Beetz, L.E. Kavraki, Real-time perception-guided motion planning for a personal robot, in *Proceedings of IEEE/RSJ International Conference on Intelligent Robots and Systems (IROS)* (Oct. 2009), pp. 4245–4252
22. H. Sakoe, S. Chiba, Dynamic programming algorithm optimization for spoken word recognition. IEEE Trans. Acoust. Speech Sig. Process. **26**(1), 43–49 (1978)
23. D. Silver, J.A. Bagnell, A. Stentz, Learning from demonstration for autonomous navigation in complex unstructured terrain. Int. J. Robot. Res. **29**(12), 1565–1592 (2010)
24. J. van den Berg, S. Miller, D. Duckworth, H. Hu, A. Wan, X.-Y. Fu, K. Goldberg, P. Abbeel, Superhuman performance of surgical tasks by robots using iterative learning from human-guided demonstrations, in *Proceedings of IEEE International Conference on Robotics and Automation (ICRA)* (May 2010), pp. 2074–2081

Learning from Experience in Manipulation Planning: Setting the Right Goals

Anca D. Dragan, Geoffrey J. Gordon and Siddhartha S. Srinivasa

Abstract In this paper, we describe a method of improving trajectory optimization based on predicting good initial guesses from previous experiences. In order to generalize to new situations, we propose a paradigm shift: predicting *qualitative attributes* of the trajectory that place the initial guess in the basin of attraction of a low-cost solution. We start with a key such attribute, the choice of a goal within a goal set that describes the task, and show the generalization capabilities of our method in extensive experiments on a personal robotics platform.

1 Introduction

We are interested in pushing the boundaries of trajectory optimization for robots in complex human environments. Optimization techniques have been well-documented to struggle in these domains by getting stuck in high-cost local minima. We envision two possible means of alleviating the problem. The first is to improve the optimizer itself by widening the basin of attraction of low-cost minima. The second is to improve initialization and start with trajectories that are in the basin of attraction of low-cost minima. In our previous work [1], we made progress on the former by widening the basin of attraction of an efficient trajectory optimizer

A.D. Dragan (✉) · S.S. Srinivasa
The Robotics Institute, Carnegie Mellon University, Pittsburgh, USA
e-mail: adragan@cs.cmu.edu

S.S. Srinivasa
e-mail: siddh@cs.cmu.edu

G.J. Gordon
Machine Learning Department, Carnegie Mellon University, Pittsburgh, USA
e-mail: ggordon@cs.cmu.edu

H.I. Christensen and O. Khatib (eds.), *Robotics Research*,
Springer Tracts in Advanced Robotics 100, DOI 10.1007/978-3-319-29363-9_18

CHOMP [2] by taking advantage of the often available flexibility in goal selection. In this work, we focus on the latter: learning to generate initial trajectories that enable the optimizer to converge to low-cost minima.

So how does the robot acquire this trajectory-generating oracle? In designing the oracle, we take advantage of three key features: the optimization process itself, the repetition in the tasks, and the structure in the scenes. The optimization process relieves us from the need to produce low-cost initial trajectories. *The cost of the trajectory is irrelevant, as long as it lies in the basin of attraction of a low-cost trajectory.* Repetition or similarity in tasks allows the oracle to *learn from previous experience* how to produce trajectories. Finally, structure in the scenes suggests that we can use *qualitative attributes* to describe trajectories. For example, in a kitchen, we could say "go left of the microwave and grasp the object from the right." These attributes provide a far more compact representation of trajectories than a sequence of configurations. This work combines all three features and proposes a learning algorithm that, given a new situation, can generate trajectories in the basin of attraction of a low-cost trajectory by predicting the values of qualitative attributes that this trajectory should posses. As a consequence, instead of focusing on every single voxel of a scene at once, we first make some key decisions based on previous experience, and then refine the details during the optimization.

The idea of using previous experience to solve similar problems is not new. In Artificial Intelligence, it is known as Case-Based Reasoning [3, 4], where the idea is to use the solution to the most similar solved problem to solve a new problem. In the MDP domain, Konidaris and Barto [5] looked at transferring the entire value function of an MDP to a new situation. Stolle and Atkeson constructed policies for an MDP by interpolating between trajectories [6], and then used local features around states to transfer state-action pairs to a new problem [7, 8]. In motion planning, learning from experience has included reusing previous collision-free paths [9] or biasing the sampling process in randomized planners [10] based on previous environments.

Jetchev and Toussaint [11] explored trajectory prediction in deterministic and observable planning problems. They focused on predicting globally optimal trajectories: given a training dataset of situations and their globally optimal trajectories, predict the globally optimal trajectory for a new situation. Much like Case-Based Reasoning, their approach predicted an index into the training dataset of trajectories as the candidate trajectory [11, 12] or clustered the trajectories and predicted a cluster number [11, 13]. Since prediction is not perfect, a post-processing stage, where a local optimizer is initialized from the prediction is used to converge to the closest local minimum.

Our approach differers in two key ways. First, we take advantage of the necessity of the optimization stage, and focus on the easier problem of predicting trajectories that fall in the basin of attraction of low-cost minima. Second, by predicting low-dimensional attributes instead of whole past trajectories, we are able to generate trajectories beyond the database of previous experience, allowing us to generalize further away from the training set.

Although the dataset-indexing techniques are a promising start in the field of learning from experience for trajectory optimization, they are limited: they are reminiscent of earlier works in computer vision (e.g. [14]), where one way to classify an image is to find the closest image in the training set according to some features and predict its label (or find a set of closest images and verify their predictions in post-processing). In 2006, the vision community started thinking about learning the distance metric between images [15], and this is the state at which trajectory prediction is now. In 2009 however, the object recognition community started changing this classification paradigm and shifting towards a much more general way of recognizing objects based on a simple idea: predict attributes of the object instead of the object itself, and then use the attributes to predict the object [16, 17]. This not only improved recognition of known objects, but also *allowed learners to recognize objects they had never seen before*. A similar technique was used in [18] to recognize from brain scans words that a subject was thinking, by using physical attributes of the words as an intermediate representation. We propose to do the same for trajectory prediction: rather than predicting trajectories directly, we predict qualitative attributes of the trajectories first, such as where their goal point is or which side of an obstacle they choose, and then map these qualitative attributes into initial guesses for a local optimizer.

In this work, after providing the motivation for the idea of using attributes in trajectory prediction, we focus on one key such attribute of the trajectory: its end point. Most manipulation tasks are described by an entire region of goals rather than a single goal configuration, and our previous work [1] has shown that the choice of a goal significantly impacts the outcome of the optimizer. Therefore, by getting better at selecting good goals, we are able to initialize the optimizer in better basins of attraction. We take advantage of the fact that the robot has easy access to locally optimal trajectories though its local optimizer, and can obtain a rich dataset of multiple trajectories for a situation along with their performance. We compare several methods of learning to predict this attribute value, from predicting if a value is the best choice or not, to learning to rank these values and selecting the highest-ranking one as the prediction. We show that all these algorithms perform best when they take into account additional data about sub-optimal performance (the "error" in "trial-and-error"). Using this information, learners predict goals that achieve costs within 8–9 % of the minimum cost in a test suite of reaching tasks with different starting points, target object poses and clutter configurations. We also study the generalization capabilities of the method, by evaluating its robustness to differences between training and test distributions; and, we show examples where relying on a library of previously executed trajectories is not suitable. We see this work as the foundation for a learning framework where reinforcement from past experience can be harnessed to guide trajectory optimizers in making the right decisions.

2 Framework

2.1 Trajectory Optimization

Although our work can use any trajectory optimizer that produces consistent solutions, we will revise here the details of a particular optimizer that has proven to be very efficient in a wide range of manipulation tasks. In recent work [1], we introduced Goal Set CHOMP, an optimizer that can bend trajectories out of collision while remaining smooth, and that exploits the entire goal set of configurations allowed by the task in order to find better solutions. This algorithm was an improvement on the CHOMP optimizer [2], widening the basins of attraction of low-cost solutions by allowing trajectories to adapt their goals.

CHOMP optimizes a functional that trades off between a smoothness and an obstacle cost:

$$U[\xi] = \lambda f_{prior}[\xi] + f_{obs}[\xi] \quad \text{s.t. } h(\xi) = 0 \tag{1}$$

with the prior measuring a notion of smoothness such as sum squared velocities or accelerations along the trajectory ξ, the obstacle cost pushing all parts of the robot away from collision, and h capturing constraints on the trajectory.

We optimize the first-order Taylor Series expansion of U and h around ξ_t within a trust region shaped by a Riemannian metric A in the space of trajectories Ξ . This could be, for example, the Hessian of the prior cost, which will prefer smooth deformations as opposed to small deformations in the Euclidean norm. The resulting trajectory update rule looks like:

$$\xi_{t+1} = \underset{\xi \in \Xi}{\operatorname{argmin}}\, U(\xi_t) + g_t^T(\xi - \xi_t) + \frac{\eta_t}{2}\|\xi - \xi_t\|_A^2 \quad \text{s.t. } h(\xi_t) + h'(\xi_t)(\xi - \xi_t) = 0 \tag{2}$$

A convenient representation of trajectories for CHOMP is as a vector of way-points: $\xi = (\xi[1], \ldots, \xi[n])$. In this case, a typical constraint for CHOMP is a fixed goal: $\xi[n] = q_{\text{goal}}$. Goal Set CHOMP relaxes this assumption, and replaces the constraint by $h_n(\xi[n]) = 0$: the goal is restricted to a constraint surface instead of fixed.

2.2 Trajectory Attribute Prediction

The term "trajectory prediction" refers to the problem of mapping situations S (task descriptions) to a set of trajectories Ξ that solve them:

$$\tau : S \rightarrow \Xi \tag{3}$$

Previous work [11, 13] proposed solving this problem by learning to index into a dataset of examples. This approach is limited by the dataset of previously executed trajectories, much like, for example, the earlier work in object recognition was limited by labeled images it used. In our work, we combine the idea of using a lower dimensional representation of trajectories rather than the full dimensional representation with the ability to predict new trajectories that generalize to more different situations.

Our approach to solving the problem takes advantage of the capabilities of the optimizer. Since this optimizer is local, it will not produce the globally optimal trajectory independent of initialization, but it can produce various local minima with different costs. The training data set therefore contains not only the best trajectory found for the scene, but it can also include various other local optima. We also emphasize that trajectory prediction serves as an initialization stage for the optimizer, which leads to the following crucial observation: *In order to predict the optimal trajectory, we can predict any trajectory in its basin of attraction, and let the optimizer converge.*

So can we leverage this observation in such a way that we have the ability to predict new trajectories (rather than only the ones in the dataset) while avoiding the high dimensionality of the output space Ξ ? We propose that there often exist some lower-dimensional trajectory attributes such that predicting these attribute values, rather than a full-dimensional trajectory, places the optimizer in the desired basin of attraction. The insight is that in producing a trajectory, a planner is faced with a few key decisions that define the topology of the trajectory. Once the right decisions are made, producing a good trajectory comes down to local optimization from any initialization that satisfies those decisions. This implies that we can reduce the problem of predicting a good trajectory to that of predicting these core attributes, and then mapping these core attributes to a trajectory. We will discuss each of these two subproblems in turn.

2.3 Attributes

To explain the idea of attribute prediction, we start with the toy world from Fig. 1: a point robot needs to get from a start to a goal while minimizing the cost in (1). If we run CHOMP in this world, we get two solutions depending on the initial trajectory: a low and a high cost one. In order to converge to the low-cost trajectory, we can start with any trajectory to the right of the obstacle. Predicting the optimal trajectory reduces to predicting a single bit of information: right versus left of the obstacle.

In higher dimensional problems, there are many basins of attractions and instead of globally optimal trajectories we can talk about good local minima versus high-cost and sometimes infeasible local minima. In this setting, it is often the case that the

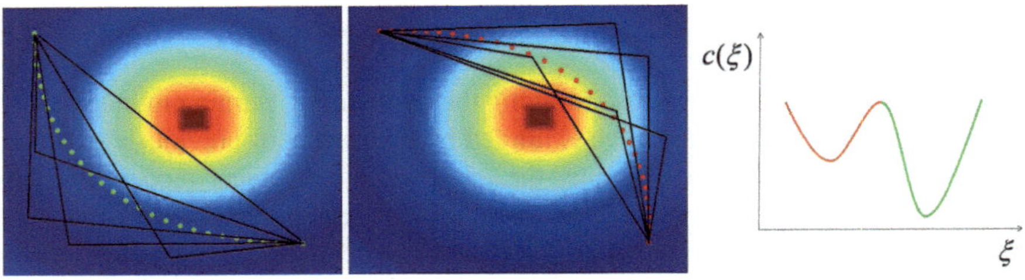

Fig. 1 A toy example that exemplifies the idea of attributes: there are two basins of attraction, and a simple attribute (the decision of going *right* vs. *left*) discriminates between them

lower-cost basins are still described by simple decisions (i.e. low-dimensional, even discrete, trajectory attributes). Figure 2 shows an example where going above an obstacle versus around it will determine whether the optimizer converges to a low cost trajectory versus a high cost one. In this case, a single bit of information will place the optimizer in a good basin of attraction. An optimizer like CHOMP can be initialized with a simple trajectory that satisfies this property, such as the one in Fig. 3, and, as exemplified in the same figure, will bend it out of collision to a low-cost trajectory.

Based on this observation, we propose changing the trajectory prediction paradigm to a trajectory attributes prediction problem where we first predict key attributes that a good trajectory should have:

$$\tau : S \rightarrow A(\Xi, S) \tag{4}$$

Here, $A(\Xi, S)$ denotes the trajectory attributes, which are conditioned on the situation, e.g. "in front of the shelf" or "elbow up around the cabinet". These attributes implicitly define a subset of trajectories $\Xi_A \subseteq \Xi$, and as a second step the optimizer is initialized from any trajectory $\xi \in \Xi_A$. The overall framework is

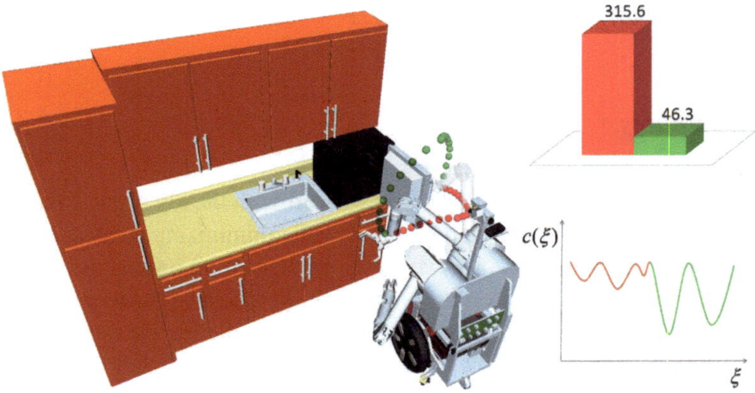

Fig. 2 High-dimensional problems are described by many basins of attraction, but there are often attributes of the trajectory that can discriminate between low cost basins and high cost basins. In this case, such an attribute is around versus above the fridge door

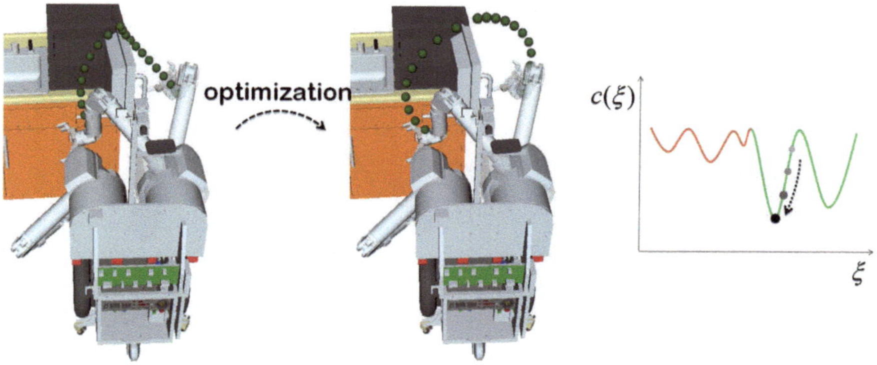

Fig. 3 Once the right choice is made (above the fridge door), we can easily create a trajectory that satisfies it. This trajectory can have high cost, but it will be in the basin of attraction of a low-cost solution, and running a local optimizer (e.g. CHOMP) from it produces a successful trajectory

$$S \rightarrow A(\Xi, S) \rightarrow \xi \in \Xi_A \rightarrow \xi^*$$

with ξ^* the locally optimal trajectory in the basin of attraction of ξ.

Constructing a trajectory from a set of attributes ($A(\Xi, S) \rightarrow \xi \in \Xi_A$) can be seen as solving a simple constrained optimization problem: starting from a straight line trajectory, we want to keep it short while satisfying certain constraints on a few of its way-points. Since this problem is convex, generating a trajectory from attributes is very fast. As an example of such a problem, "above X and then to the left of Y" translates into two constraints on two way-points of a piecewise linear trajectory. The example from Fig. 3 is an instantiation of that, with one constraint on one mid-point, above the fridge door, which generates two straight line segments in configuration space. Similarly, a goal attribute will be a constraint on the final end-point of the trajectory.

3 Learning to Select Good Goals

Most manipulation tasks are described by an entire region of goals rather than one particular configuration that the robot needs to be in such that it can achieve the task. Goal sets appear in reaching for an object, placing it on a surface, or handing it off to a person. In our previous work, we introduced Goal Set CHOMP, a trajectory optimizer that can take advantage of the goal set in order to obtain lower-cost solutions. However, this optimizer is still local, and the initial goal choice (the goal the initial trajectory ends at) still has a high impact on the final cost of the trajectories. Figure 4 plots this final cost for a variety of initial choices, in the problem of reaching for a target object in a small amount of clutter. Because of the large difference illustrated in the figure, the choice of a goal is a crucial component in the

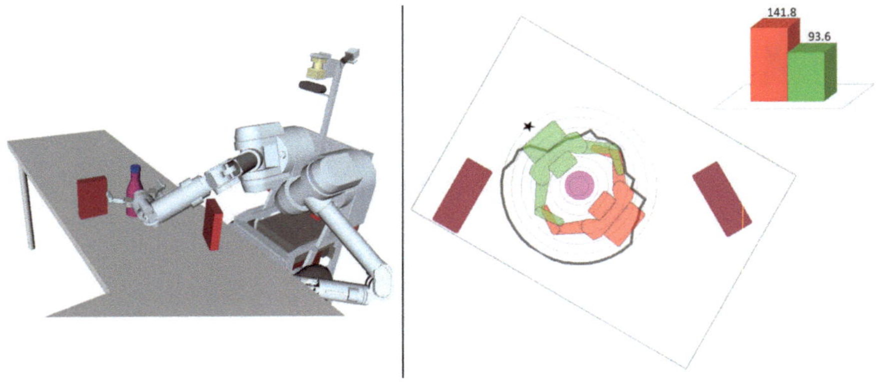

Fig. 4 *Left* the robot in one of the goal configurations for grasping the bottle. *Right* for the same scene, the *black* contour is a polar coordinate plot of the final cost of the trajectory Goal Set CHOMP converges to as a function of the goal it starts at; goals that make it hard to reach the object are associated with higher cost; the bar graph shows the difference in cost between the best goal (shown in *green* and marked with *) and the worst goal (shown in *red*)

optimizer's initialization process. Once a choice is made, a trajectory $\xi \in \Xi_A$ that satisfies this attribute value can be, for example, the straight line trajectory from the start to that goal. For any given situation, we can run an optimizer like Goal Set CHOMP with a straight line trajectory to each of the goals in a discretization of the goal set. In this section, we describe several different ways of taking advantage of this data in order to learn to predict what goal to initialize the optimizer with in order to minimize cost.

3.1 Some Words on Features

To enable learning, we designed features that capture potential factors in deciding how good a goal is. These are indicators of how much free space there is around the goal and how hard it is to reach it. A subset of these features are depicted in Fig. 5.

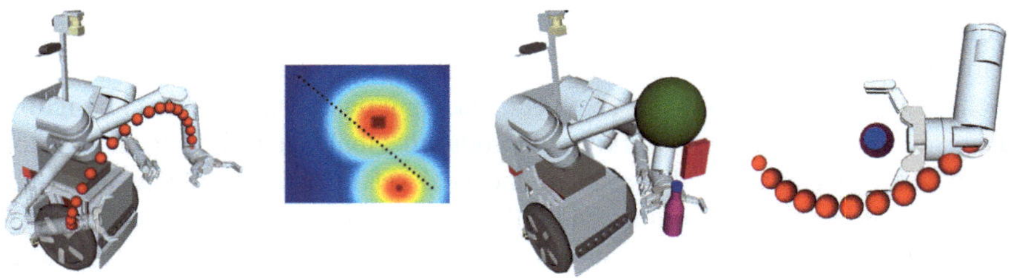

Fig. 5 Five of the features we are using. From *left* to *right* length of the straight line trajectory, cost of the goal and of the straight line trajectory, free space radius around the elbow and collision with the target object

We constructed these indicators with simplicity in mind, as a test of what can be done with very little input. We do however believe that much higher performance is achievable with a larger set of features, followed perhaps by a feature selection approach. We are also excited about the possibility of producing such features from a much rawer set using feature learning, although important questions, such as informing the algorithm about the kinematics of the robot, are still to be teased out.

3.1.1 A Minimal Set of Features

- The distance in configuration space from the starting point to the goal: $||\xi[n] - \xi[0]||$. Shorter trajectories tend to have lower costs, so minimizing this distance can be relevant to the prediction.
- The obstacle cost of the goal configuration: the sum of obstacle costs for all body points on the robot, $\sum_b c(x_b(\xi_n))$, with c an obstacle cost in the work-space and x_b the forward kinematics function at body point b.
- The obstacle cost of the straight-line trajectory from the start to the goal $\bar{\xi} : \sum_i \sum_b c\left(x_b\left(\bar{\xi}[i]\right)\right)$. If the straight line trajectory goes through the middle of obstacles, it can potentially be harder to reach a collision-free solution.
- The goal radius: a measure of the free space around the goal in terms of how many goals around it have collision-free inverse kinematics solutions. For example, the goal set of grasping a bottle can be expressed as a Workspace Goal Region [19] with a main direction of freedom in the yaw of the end effector (this allows grasping the bottle from any angle, as in Fig. 4). In this case, the feature would compute how many goals to the left and to the right of the current one have collision-free inverse kinematics solutions, and select the minimum of those numbers as the goal radius. The closer the clutter will be to the goal, the smaller this radius will be. It has the ability to capture the clutter at larger distances than the second feature can.
- The elbow room: the maximum radius of a collision-free sphere located at the elbow, indicating how much free space the elbow has around it for that particular goal configuration. Configurations that restrict the motion of the elbow are potentially harder to reach.
- The target collision amount: the percent of the last m configurations of the initial trajectory that are colliding with the target object $\frac{1}{m}\sum_{i=N-m+1}^{N} collides(\xi[i]^{sg})$. Here, $collides(q) = 1$ when q is in collision with the target object and 0 other-wise. This feature is another factor in how easy it is to reach the goal—if the initial trajectory passes through the target object, bending it out of collision could be too difficult.

3.1.2 Domain Adaptation

Among the features, the distance from the start as well as the initial trajectory cost can differ substantially between different scenes, and so may cause difficulty for generalization. A classical approach to deal with this problem is standardization, which we can not do directly because of the large difference between our training and test set statistics. The test set contains some scenes that are considerably harder, and some that are far easier than any in the training set: training data will never capture the entire diversity of the situations the robot will face. We still need to generalize to these situations, so we normalize the distance and cost features in a situation—this makes all situations have the same range of costs, allowing the learner to distinguish among them. We then add in the mean values of these two features, to give the learner access to how difficult the scene is, and only then standardize. More sophisticated domain adaptation strategies (e.g., [20]) are an area of future work.

3.2 Learners

We are comparing several learners, differing in the model they use (linear vs. nonlinear), how they use the data and whether they focus on predicting the best cost or on fitting the cost in general.

3.2.1 Classification

(a) *The Vanilla Version*: The easiest way to approach the problem of deciding which goal is optimal is to directly predict if a goal will be optimal or not. For every situation, we assign the goal corresponding to the minimum final cost the value 1, and 0 to all the other goals.
We can now train a standard classifier, such as a Support Vector Machine, to predict optimality of a goal. In a new scene, given a set of goal configurations, this classifier will select any number of goals to be optimal, and we will select a random one of these as the initial guess for the optimizer. If the classifier predicts that none of these goals are optimal, then we select randomly among all goals, i.e. the classifier has not given the optimizer any information.

(b) *The Data-Efficient Version*: Since we have access to costs and not just to the binary decision of "is optimal", another approach is to allow the classifier to predict any goal within a certain percent of the minimum cost. This can help by softening the data for the classifier, but there is of course a trade-off with predicting higher cost goals. We determined the value for this trade-off (the percent cutoff) on a validation set.

3.2.2 Inverse Optimal Control

(a) *The Vanilla Version*: A different way to look at the problem is to treat the best goals as expert demonstrations. In Inverse Optimal Control, we want to create a cost function that explains why the experts are optimal—in our case, we want a cost function c_{IOC} in feature space such that the best goal does have the best cost in every situation. Once we have this function, we can apply it to the goals in a new scene and choose the goal $g^* = \text{argmin}_g\ c_{IOC}\ (f_g)$ (here f_g denotes the features associated with goal g).
Taking the Maximum Margin Planning approach introduced in [21], we want to find a cost function $c_{IOC} = w^T f$ that makes the optimal goal have the lowest cost by some margin. To improve generalization, we will require a larger margin for goals that are farther away from the expert: in particular, we define $l(g, g')$ to be the *structured margin*, which is zero when $g = g'$ and large when g and g' are far apart. Then saying that some goal g is optimal means $w^T f_g \leq w^T f_{g'} \quad \forall g'$. Adding in our structured margin, penalizing constraint violations with a slack variable, and regularizing w, we have:

$$\min_w \sum_s \left(w^T f_{g^s_{exp}} - \min_i (w^T f_{g^s_i} - l(g^s_i, g^s_{exp})) \right) + \frac{\lambda}{2} \|w\|^2 \qquad (5)$$

where g^s_i denotes goal i in situation s, and $l(g, g') = \left\| f_g - f'_g \right\|^2$ is the structured margin which penalizes solutions from being far away in *feature space* from the expert in situation s. Overall, w pays a penalty for allowing non-expert goals to have low costs.
Taking the subgradient of (5) yields the following update rule:

$$w \leftarrow w - \alpha \left(\sum_s \left(f_{g^s_{exp}} - f_{g^{s*}} \right) + \lambda w \right) \qquad (6)$$

$$\text{where } g^{s*} = \arg\min_{g_i} (w^T f_{g^s_i} - l(g^s_i, g^s_{exp})) \qquad (7)$$

This algorithm is targeted at identifying the minimum cost goal (7), ignoring the costs associated with all other goals. It gains efficiency as it does not waste resources trying to explain what happens with other goals. Whether this focus on the expert pays off or not will be established in Sect. 4.

(b) *The Data-Efficient Version*: With IOC, there exists a way of introducing the true cost information (which we do have, unlike typical IOC problems which are only given expert examples), without losing the focus on the expert. By changing the margin l_s to be the true cost difference between the goal and the expert goal rather than the distance in features, $l(g_i^s, g_{exp}^s) = U(\xi_{g_{exp}}^{final}) - U(\xi_{g_i}^{final})$, the algorithm will ensure that the minimum with respect to its new cost is close in true cost to the expert, i.e. has low cost. In future work, we are interested in combining these two distance metrics and using a cutoff on cost difference as in the next Sect. 3.2.3.

3.2.3 Regression

(a) *The Vanilla Version*: A third way to predict the minimum cost goals is to predict the final cost associated to each of the goals:

$$f_{g_i}^s \rightarrow U(\xi_{g_i}^{final})$$

with $\xi_{g_i}^{final}$ the final trajectory obtained by initializing the optimizer with the straight line to goal g_i, and choose the best one:

$$g^* = \arg\min_{g_i} U(\xi_{g_i}^{final})$$

This is sometimes referred to as arg min-regression. We looked at three different regressors:

- Linear Regression: $w = F^{\dagger}C$, with F a matrix concatenating every feature vector on every situation, one per row, and C a vector concatenating all the final costs obtained by Goal Set CHOMP, one per row.
- Gaussian Process: A wide Gaussian radial basis kernel performed the best, since we need far knowledge transfers.
- Neural Network: We used the Back-Propagation Neural Network with one hidden layer. We determined the number of nodes in this layer, as well as the weight decay coefficient based on performance on a validation set.

(b) *The Data-Efficient Version*: Looking at the initial performance of Linear Regression on the training set (Fig. 6, left), it becomes apparent that there are a lot of data points with very high cost, and predicting that cost accurately is not only unnecessary, but leads to not being able to identify the good solutions from the mediocre ones. This suggests that even these regressors should not use all the data, but rather focus their efforts on discriminating among the lower-cost solutions by truncating the cost at some threshold. We selected this threshold based on a validation set as shown in Fig. 6 (right). The plot shows that a very low threshold degrades performance by confusing the learner to pay attention to the high-cost outliers, and a very high threshold also degrades performance by starving the learner of data. Figure 6 (center) portrays the new

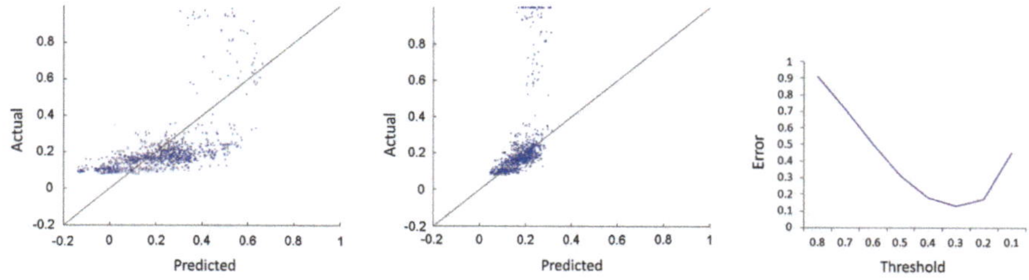

Fig. 6 From *left* to *right* the actual versus predicted cost without thresholding, the actual versus predicted cost with thresholding, and the dependence of the fit error of a validation set of medium and low cost examples on the threshold (on the *left* of the minimum, the regressors pays too much attention to high costs, on the *right* it uses too little data)

predictions based on the learned threshold, forming a much better fit for the solutions we are interested in, while keeping the high-cost predictions sufficiently high. We also tried to do the thresholding per scene instead of on the entire training data, but this did not cause a significant improvement, because the effect on how well the regressors can fit the data is minimal.

4 Experimental Results

4.1 Generalization from Limited Data

In a first experiment, we wanted to test how well we can generalize to new situations, going beyond the exemplars already executed. We used only two scenes for training, shown in Fig. 7, where the goal was the grasp the bottle while avoiding the table holding the object, as well as the box placed next to the target. We ran CHOMP to each goal in a discretization of the goal set, and recorded the final cost. Figure 7 shows the goals that produced the best cost for each of the scenes. We then trained a neural network to predict this cost given only the first three features from Sect. 3.1.1.

For testing, we moved the object to a very different location than in the training examples, also shown in Fig. 7. With a Nearest-Neighbor approach, the robot would identify one of the training scenes as closest, and initialize the optimizer from the best final trajectory for that scene. In this case, all the trajectories go to a goal that is sub-optimal or even colliding with the environment. The trajectory attributes approach, however, allows us to go beyond these previously executed trajectories. The learner predicts that goal shown on the right of Fig. 7 will produce the best cost. This goal has never been optimal in the training examples, yet because it stays away from clutter while maintaining a short distance from the starting configuration, the learner will recognize it as better than the other choices. Indeed, when initializing the optimizer from the straight line trajectory to that goal, the final

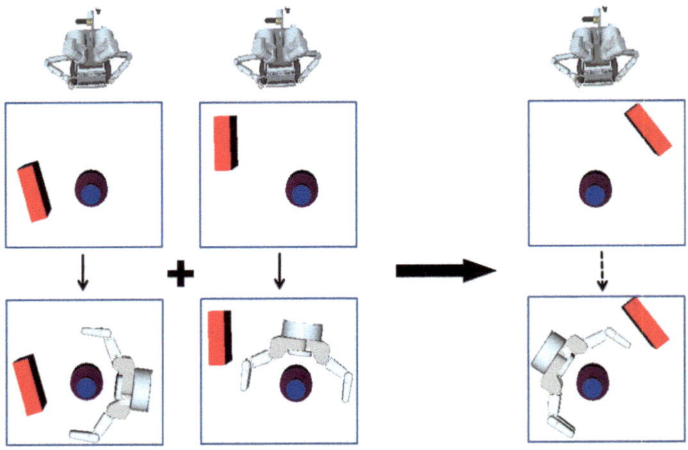

Fig. 7 Two training situations along with their corresponding best goal, and a test situation in which the correct goal is predicted. If the learner were constrained to the set of previously executed trajectories, it would not have been able to generalize to this new scene

cost is only 1 % higher than the best path we were able to find using multiple initializations of Goal Set CHOMP to the different goals.

4.2 Generalization Dependence on Train-Test Similarity

In this next experiment, we were interested in testing how far away from the training data we can transfer knowledge to. We created one testing situation, and trained two of the regressors (the Neural Network and the Gaussian Process) on situations that are more and more different from the testing one. In Fig. 8, we plot the performance in these cases as the percent of degradation of cost over the minimum that Goal Set CHOMP can reach—the final cost corresponding to initializing the optimizer with a straight line trajectory to the best goal. These performances, averaged across 15 different clutter configurations, are compared with our baseline: what happens if we randomly choose a collision-free goal, without any learning?

In the first setting, we train and test on the same dataset. Both the Neural Network and the GP perform drastically better than the no-learning baseline. We then change the situation slightly: first the clutter configurations change, then the target object position changes by approx. 20 cm, followed by the starting configuration of the robot. In the last but one test, we change all these situation descriptors drastically, and the performance decreases significantly, although the learning algorithms still outperform the baseline. Finally, we show that more variety in the training set can lead to better generalization. When we increase the number of examples in the training set—we still train on very different situations, but we

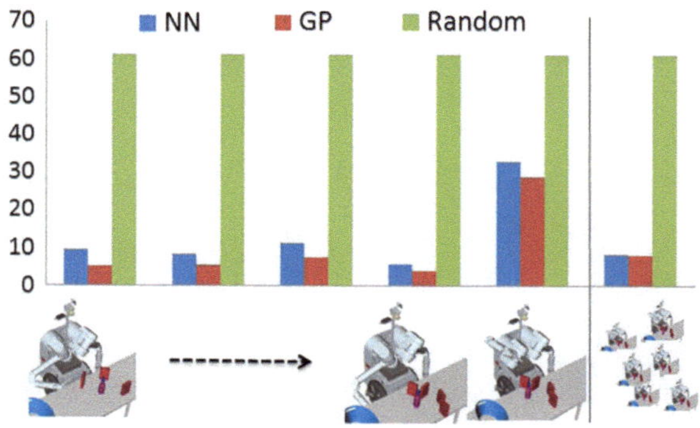

Fig. 8 The loss over the minimum cost on the same test set when training on scenes that are more and more different, until everything changes drastically in the scene and performance drops significantly. However, the loss decreases back to around 8 % when training on a wide range of significantly different scenes, showing that the algorithm can do far transfers if given enough variety in the training data

provide a wider range with more possible starting configurations and target poses—we notice that the performance again improves to about 8 % for both regressors. The random choice baseline does of course not take into account this data and performs the same, around 62 % degradation over the minimum cost.

4.3 Main Experiment

We are also interested in a realistic evaluation of the day-to-day performance of our system, as well as establishing which learning approach from Sect. 3.2 is most suitable for our problem. Should the learner focus on just the optimal goal or or should it also focus on the sub-optimal goals and their performance?

We created a large set of training examples, comprising of 90 situations varying in the starting configuration, target object pose, and clutter distribution. In each situation, we ran Goal Set CHOMP starting from the straight line trajectory to each of the collision-free goals in the discretized goal set (a total of 1154 examples) and recorded the final cost. We also created a test set of 108 situations (1377 examples) that differ in all three components from the training data.

Figure 9 (Left) shows the percentage of cost degradation over the minimum, averaged across all testing situations, for the five approaches from Sect. 3.2. The solid bars are the data-efficient versions of the algorithms: the regressors use thresholds established on a separate validation set, IOC uses the cost distance for the structured margin, and the classifier predicts goals close to the minimum as well. The vanilla versions of these methods, shown with transparent bars, always perform worse than their data-efficient counterparts.

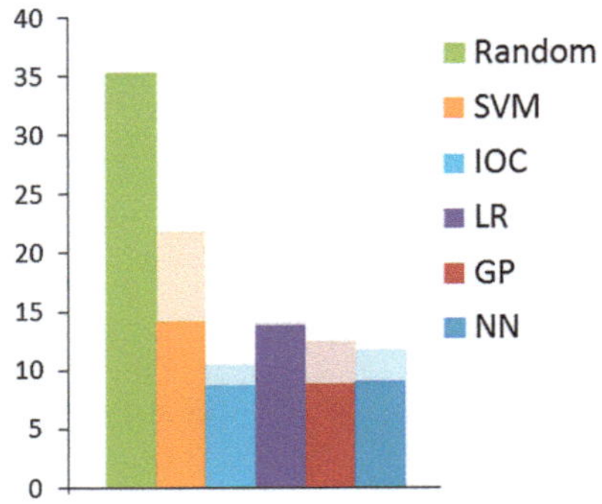

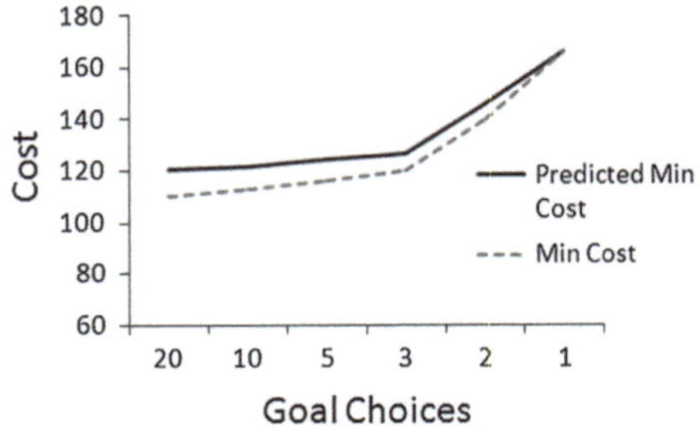

Fig. 9 *Left* Percentage loss over the best cost for all the methods. *Solid bars* are the data-efficient versions, and transparent bars are the vanilla algorithms, which perform worse. *Right* The predicted minimum cost versus the true minimum cost as function of the number of choices considered

The best performer is our version of data-efficient IOC—this algorithm focuses on predicting the expert rather than fitting cost, while taking into account the true cost and ensuring that non-expert predictions have low cost. Although both IOC and LR are linear, the advantage of IOC over LR is its expert prediction focus. The non-linear regressors have similar performance as IOC, and their advantage is a better fit of that data. The SVM is focusing on low-costs with a linear kernel, so its performance is, as expected, close to LR.

In these experiments, we had a fairly fine discretization of the goal set per scene. It makes sense to ask if we could get away with fewer choices. Figure 9 (Right) indicates that the answer is yes: with 5 goals, for example, we can predict the minimum cost better, and we this minimum is not a lot larger than the one considering, say, 20 goals.

5 Conclusion

In this paper, we proposed moving away from the learning from experience paradigm of predicting trajectories from a library. We proposed instead to predict the important attributes of trajectories that will place the initial guess for a trajectory optimizer in a good basin of attraction. We presented a first step towards this prediction paradigm by focusing on a very important trajectory attribute: the choice of a goal. We showed that the learner can generalize well by predicting this attribute, and presented results emphasizing the importance of our learning framework in practice. The next step is identifying the set of attributes that are required in order to differentiate between basins of attraction, based on both the optimizer and the current situation the robot is in. In parallel, we must improve the optimizer itself, to allow more flexibility in the prediction. We see this as an exciting challenge for machine learning and manipulation that will pave the road towards a more semantic and hierarchical way of planning, based on the robot's prior experience.

Acknowledgments This material is based upon work supported by DARPA-BAA-10-28, NSF-IIS-0916557, and NSF- EEC-0540865. Thanks to Chris Atkeson and the members of the Personal Robotics Lab for comments and fruitful discussions.

References

1. A.D. Dragan, N. Ratliff, S.S. Srinivasa, Manipulation planning with goal sets using constrained trajectory optimization, in *IEEE International Conference on Robotics and Automation* (2011)
2. N. Ratliff, M. Zucker, J.A. Bagnell, S. Srinivasa, CHOMP: gradient optimization techniques for efficient motion planning, in *IEEE ICRA* (2009), pp. 489–494
3. M.M. Veloso, Ph.D. Learning by analogical reasoning in general problem solving (CMU-CS-92-174) (1992)
4. R. Lopez De Mantaras, D. Mcsherry, D. Bridge, D. Leake, B. Smyth, S. Craw, B. Faltings, Retrieval, reuse, revision and retention in case-based reasoning. Knowl. Eng. Rev. **20**(03), 215 (2006)
5. G. Konidaris, A. Barto, Autonomous shaping: knowledge transfer in reinforcement learning, in *Proceedings of the 23rd International Conference on Machine Learning* (2006), pp. 489–496
6. M. Stolle, C.G. Atkeson, Policies based on trajectory libraries, in *IEEE ICRA* (May) (2006), pp. 3344–3349
7. M. Stolle, C.G. Atkeson, Knowledge transfer using local features, in *IEEE International Symposium on Approximate Dynamic Programming and Reinforcement Learning. ADPRL 2007* (2007)
8. M. Stolle, H. Tappeiner, J. Chestnutt, C.G. Atkeson, Transfer of policies based on trajectory libraries, in *IEEE/RSJ International Conference on Intelligent Robots and Systems* (2007), pp. 2981–2986
9. M. Branicky, R. Knepper, J. Kuffner, Path and trajectory diversity: theory and algorithms, in *IEEE International Conference on Robotics and Automation (ICRA)* (2008), pp. 1359–1364
10. S. Martin, S. Wright, J. Sheppard, Offline and online evolutionary bi-directional RRT algorithms for efficient re-planning in dynamic environments, in *IEEE CASE* (2007), pp. 1131–1136
11. N. Jetchev, M. Toussaint, Trajectory prediction, in *Proceedings of the 26th Annual International Conference on Machine Learning—ICML 09* (ACM Press, New York, USA, 2009), pp. 1–8
12. D. Dey, T. Liu, B. Sofman, J. Bagnell, Efficient Optimization of Control Libraries. Technical Report (CMU-RI-TR-11-20) (2011)
13. N. Jetchev, M. Toussaint, Trajectory prediction in cluttered voxel environments, in *IEEE International Conference on Robotics and Automation—ICRA2010*, Anchorage, AK, USA (2010)
14. D.G. Lowe, Object recognition from local scale-invariant features, in *Proceedings of the Seventh IEEE International Conference on Computer Vision*, vol. 2(8), pp. 1150–1157 (1999)
15. A. Frome, Y. Singer, J. Malik, Image retrieval and classification using local distance functions, in *Advances in Neural Information Processing Systems (NIPS)* (2006)
16. C. Lampert, H. Nickisch, S. Harmeling, Learning to detect unseen object classes by between-class attribute transfer, in *IEEE Conference on Computer Vision and Pattern Recognition* (2009), pp. 951–958

17. A. Farhadi, I. Endres, D. Hoiem, D. Forsyth, Describing objects by their attributes, in *IEEE Conference on Computer Vision and Pattern Recognition* (2009), pp. 1778–1785
18. M. Palatucci, G. Hinton, D. Pomerleau, T.M. Mitchell, Zero-shot learning with semantic output codes. Adv. Neural Inf. Process. Syst. **22**, 1–9 (2009)
19. D. Berenson, S.S. Srinivasa, D. Ferguson, A. Collet, J.J. Kuffner, Manipulation planning with workspace goal regions, in *Proceedings of the IEEE ICRA* (2009), pp. 618–624
20. J. Blitzer, H. Daume, ICML Tutorial on Domain Adaptation, http://adaptationtutorial.blitzer.com/
21. N. D. Ratliff, J.A. Bagnell, M.A. Zinkevich, Maximum margin planning, in *Proceedings of the 23rd International Conference on Machine Learning ICML 06* (2006), pp. 729–736

Planning Complex Inspection Tasks Using Redundant Roadmaps

Brendan Englot and Franz Hover

Abstract The aim of this work is fast, automated planning of robotic inspections involving complex 3D structures. A model comprised of discrete geometric primitives is provided as input, and a feasible robot inspection path is produced as output. Our algorithm is intended for tasks in which 2.5D algorithms, which divide an inspection into multiple 2D slices, and segmentation-based approaches, which divide a structure into simpler components, are unsuitable. This degree of 3D complexity has been introduced by the application of autonomous in-water ship hull inspection; protruding structures at the stern (propellers, shafts, and rudders) are positioned in close proximity to one another and to the hull, and clearance is an issue for a mobile robot. A global, sampling-based approach is adopted, in which all the structures are simultaneously considered in planning a path. First, the state space of the robot is discretized by constructing a roadmap of feasible states; construction ceases when each primitive is observed by a specified number of states. Once a roadmap is produced, the set cover problem and traveling salesman problem are approximated in sequence to build a feasible inspection tour. We analyze the performance of this procedure in solving one of the most complex inspection planning tasks to date, covering the stern of a large naval ship, using an a priori triangle mesh model obtained from real sonar data and comprised of 100,000 primitives. Our algorithm generates paths on a par with dual sampling, with reduced computational effort.

B. Englot (✉) · F. Hover
Massachusetts Institute of Technology, 77 Massachusetts Ave,
Cambridge, MA 02139, UK
e-mail: benglot@stevens.edu

F. Hover
e-mail: hover@mit.edu

H.I. Christensen and O. Khatib (eds.), *Robotics Research*,
Springer Tracts in Advanced Robotics 100, DOI 10.1007/978-3-319-29363-9_19

1 Introduction

A variety of autonomous surveillance, inspection, and distribution tasks can be solved using coverage path planning. Given an accurate model of the environment, a path is designed in which an agent sweeps its geometric footprint over 100 % of a required surface area. Manufacturing operations, security and maintenance inspections, painting, plowing, cleaning, environmental monitoring and mine-sweeping are a few of the many applications in which coverage path planning enables faster task completion compared with greedy or next-best-view strategies [9, 32].

In 2D workspaces with obstacles, cellular decomposition methods divide the free space into simple, easily-covered pieces [8, 10], allowing a full sweep of the open area. Alternatively, some applications call for the inspection of structure boundaries, and both deterministic (using Voronoi diagrams) [14, 35] and randomized (sampling-based) approaches [13, 18] have been used.

In 3D workspaces, the inspection task is typically one of boundary coverage. A structure is represented by a two-dimensional closed surface embedded in $\mathfrak{R}^3$, and the sensor must sweep over 100 % of the interior or exterior surface area. This problem is often solved by partitioning a 3D structure and planning individual inspection paths for separate components. In a 2.5D approach, the workspace is divided into 2D cross-sections, and planned paths over these cross-sections are assembled into a full 3D inspection [4, 18]. If a complex structure is comprised of distinct 3D components, one can plan individual inspection paths for each of them, assuming there is no risk of collision with neighboring components. This approach has been applied to painting the exterior surfaces of a car [5] and inspecting buildings in an urban environment [4]. In the former case, a segmentation algorithm automatically partitioned the car into topologically simple surfaces, and each was covered individually using a lawnmower-type trajectory [6]. In the latter case, each building was treated as an individual planning problem, and neighboring buildings were ignored (this required sufficient clearance between buildings). The key enabler for these modular approaches is that the plan for covering any one partition, component, or cross-section can be developed with no knowledge of the others.

Our coverage application is the autonomous in-water inspection of a ship hull, a 3D structure with challenging complexity at the stern due to shafts, propellers, and rudders in close proximity to one another and to the hull. The Bluefin-MIT Hovering Autonomous Underwater Vehicle [21], pictured in Fig. 3, is tasked with inspecting 100 % of the surface area at the stern using a forward-looking bathymetry sonar. Our vehicle is fully actuated and precision-maneuverable, but it cannot fit into the spaces between the component structures at the stern. If a 2.5D approach is adopted for coverage planning, it will need to be augmented with special, out-of-plane views in this problem to grant visibility of confined areas that are occluded in-plane. If a 3D modular approach is implemented, paths planned for component structures are at risk of collision with neighboring structures.

In consideration of these factors, we take a global optimization approach, in which all 3D protruding structures are considered simultaneously. The constraints are determined by the geometry of the 3D model provided as input. We use a triangle mesh, typically comprised of thousands of primitives, to accurately model a ship's running gear. Rather than explicitly optimizing robot configurations over the thousands of collision and visibility constraints posed by such geometry, sampling-based planning is used to find feasible means for the robot to peer into the low-clearance areas from a distance [25].

Sampling-based planning was first applied to a coverage problem by Gonzalez-Baños and Latombe [18, 19], who used random sampling to construct a solution to the 2D art gallery problem [33]. This method was recently utilized to achieve 2D view-planning for a laser-equipped wheeled robot [7]. The method was also extended to path planning in work by Danner and Kavraki, who approximated the traveling salesman problem (TSP) over the solution to the art gallery problem, planning inspections for complex 2D structures and for 3D cubes and pyramids [13].

We extend this work in several ways to enable sampling-based coverage path planning over complex, real-world 3D structures. We construct and analyze computationally a *redundant roadmap*, in which each geometric primitive is observed by multiple robot states. To enable fast planning over a large roadmap, tools from multi-robot [31] and multi-goal [30] planning are utilized to enable lazy collision-checking. The roadmap construction and collision-checking procedures are discussed in Sect. 2.

In Sect. 3 we discuss the methods by which the set cover problem (SCP) and TSP are approximated in sequence to build an inspection tour from a redundant roadmap. We compare two fast SCP approximation algorithms, the greedy algorithm [22, 28] and the linear programming rounding algorithm [20], with the dual sampling method of Gonzalez-Baños and Latombe.

In Sect. 4 we examine algorithm performance over ensembles of Monte Carlo trials in which randomly-sampled primitives must be inspected by a point robot in a 3D workspace. For simplicity, this workspace is devoid of obstacles. Finally, in Sect. 5 we apply the inspection-planning algorithm to a large-scale, real-world task, planning the inspection of a ship hull by the HAUV.

2 Sampling-Based Planning Procedure

In developing an inspection path, we employ two sampling-based routines. First, roadmap construction, which samples robot configurations and catalogs their sensor observations, creates a discrete state space from which the inspection path will be made. Second, a point-to-point planner, capable of finding collision-free paths for multi-degree-of-freedom robots in obstacle-filled workspaces, finds feasible paths joining the configurations on the roadmap. A stateflow diagram summarizing the coverage path planning procedure from start to finish is given in Fig. 1.

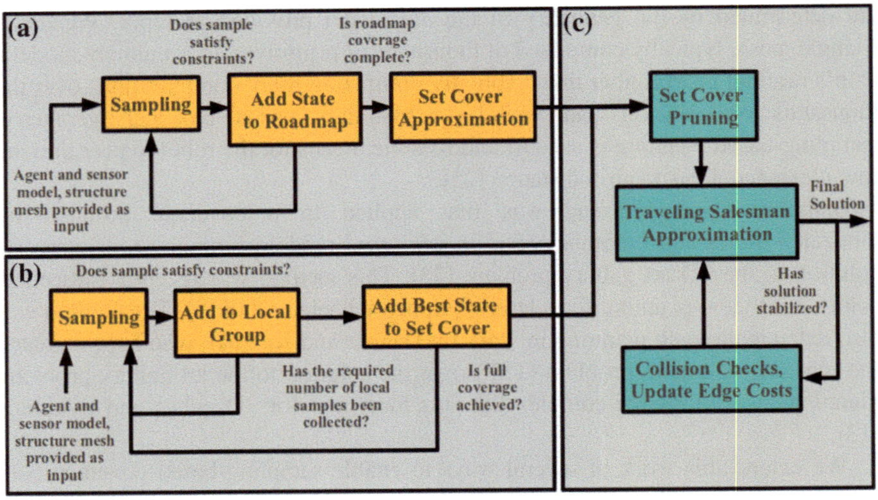

Fig. 1 A stateflow diagram illustrating the coverage path planning procedure from start to finish. **a** Steps of the procedure unique to the redundant roadmap method, including roadmap construction and solution of the set cover problem over the roadmap. **b** Steps of the procedure unique to the dual sampling method, including the sampling of local groups and iterative assembly of a set cover solution. **c** Steps of the procedure that we apply in both methods, including set cover pruning and the iterative solution of the traveling salesman problem

2.1 Roadmap Construction

Our algorithm adds configurations to a roadmap until each geometric primitive is observed a requisite number of times, which we term the *redundancy* of the roadmap. Construction begins with the selection of a geometric primitive which has not been observed the required number of times. Robot configurations are sampled uniformly at random in a local neighborhood of this primitive, avoiding exhaustive sampling in empty portions of the workspace. A configuration is added to the roadmap if it collects at least one required observation, and if the configuration is free of collisions and occlusions. In addition to collision-checking, this requires ray shooting; casting a line segment between the robot's sensor and each of the primitives inside the sensor footprint to ensure the line of sight is clear. After a configuration is added to the roadmap, another primitive is selected, and the procedure repeats until the redundancy requirement is satisfied. The full roadmap construction procedure is detailed in Algorithm 1.

Increased redundancy is intended to create a finely discretized state space from which a smaller covering subset of robot states is chosen. This procedure stands in contrast to that of prior work in sampling-based coverage [13, 18, 19], in which the final set of configurations used in the inspection is pieced together one-by-one. The *dual sampling* method, in each iteration, selects a geometric primitive that has not been observed, and samples in a local neighborhood of this primitive. Samples are drawn locally until a group of sufficient size is collected in which each sample observes at least one required primitive. Unlike the redundant roadmap method, the sample which contributes the largest quantity of new sensor information is immediately added to the set cover, the rest of the group is discarded, and local sampling continues elsewhere, until every primitive is observed at least once. The key tunable parameter of this procedure is the size of the locally-sampled group, which we term the *number of local samples*. We will compare the computational efficiency of our approach with that of the dual sampling method. A stateflow diagram of the dual sampling procedure to be implemented is given in Fig. 1.

Algorithm 1 $ConfigList = BuildRoadmap(Primitives, Obstacles, Redundancy)$

```
IncompletePrimitives ← Primitives
while IncompletePrimitives ≠ ∅ do
  SeedPrimitive ← ChooseRandomEntry(IncompletePrimitives)
  NewConfig ← FeasibleSample(SeedPrimitive, Obstacles)
  NewSightings ← Sensor(NewConfig, Primitives, Obstacles)
  NeededSightings ← NewSightings ∩ IncompletePrimitives
  if NeededSightings ≠ ∅ then
    ConfigList.add(NewCfg, NewSightings)
    for i ∈ NeededSightings do
      NeededSightings[i].incrementNumSightings()
      if NeededSightings[i].numSightings = Redundancy then
        IncompletePrimitives ← IncompletePrimitives \ NeededSightings[i]
      end if
    end for
  end if
end while
return ConfigList
```

2.2 *Lazy Point-to-Point Planning*

Efficient computation along roadmap edges is achieved with a lazy algorithm. As the roadmap is constructed, an adjacency matrix is maintained in which all entries represent the Euclidean norms among roadmap nodes. Computation of a Euclidean norm is far simpler than collision-checking and observation-checking along every edge of the roadmap. An initial inspection tour is computed over this naive adjacency matrix, and only the edges selected in the tour are collision-checked, not every edge of the roadmap. The bi-directional rapidly-exploring random tree (RRT) is utilized as the point-to-point planner [24]. Presumably, the computation of RRTs over the edges of the inspection tour increases the lengths of some edges. To address this, an iterative improvement procedure, similar to that of [30], is utilized. After the first set of feasible paths is obtained, the costs in the adjacency matrix are updated, and the inspection tour is recomputed using the new costs. This procedure is repeated, and goal-to-goal costs are iteratively updated, until there is no further improvement in the length of the returned path. Instead of requiring $O(n2)$ calls to the RRT, this approach requires $O(C * n)$ calls, where C is the number of iterations in which a new tour is computed. This procedure is detailed in Algorithm 2.

Algorithm 2 $RobotTour = LazyTourAlgorithm(Nodes, Obstacles)$

```
AdjMat ← EuclideanDistances(Nodes)
UnclearedEdges ← GetEdgePairs(Nodes)
ClearedEdges ← ∅
while NewTourCost ≠ PreviousTourCost do
   PreviousTourCost ← NewTourCost
   NewTourCost ← 0
   LazyTour ← ComputeTour(AdjMat)
   for Edge_ij ∈ LazyTour do
      if Edge_ij ∈ UnclearedEdges then
         FeasiblePath_ij ← RRT(Edge_ij, Obstacles)
         ClearedEdges ← ClearedEdges ∪ Edge_ij
         UnclearedEdges ← UnclearedEdges \ Edge_ij
         AdjMat(i, j) ← PathCost(FeasiblePath_ij)
      end if
      NewTourCost ← NewTourCost + AdjMat(i, j)
   end for
end while
RobotTour ← LazyTour
return RobotTour
```

3 Constructing an Inspection Tour

An efficient implementation of the RRT subroutine is only useful if computations over the the adjacency matrix are fast and efficient. However, the exact problem we aim to solve, finding the shortest path that collects an element from every set (where the sets are observations of primitives obtained at each roadmap node) is an instance of the generalized traveling salesman problem (GTSP), which is NP-hard and has no constant-factor approximation. Although branch-and-cut algorithms [16] and reduction to a non-metric asymmetric TSP [26, 29] have been characterized, these are not suitable for an iterative, real-time procedure (neither is solved by an approximation algorithm with a guaranteed termination time). As of this writing, a constant-factor approximation can only be obtained if each roadmap node is limited to exactly two primitive sightings, in which case the problem reduces to a Tour Cover [3]. We have found that stripping sensor information out of the roadmap to achieve an equivalent Tour Cover undoes any benefit of a constant-factor approximation.

Our approach is similar to that suggested by Current and Schilling [12], in which a GTSP (referred to in their work as the Covering Salesman Problem, a special geometric case of the GTSP [17]) is solved by posing, in sequence, a SCP subproblem and a Euclidean TSP subproblem. Both the SCP and Euclidean TSP can be approximated to within a constant factor of optimality using fast, polynomial-time algorithms. Recent work on penalizing both viewing and traveling costs [34] addresses the possibility of an arbitrarily bad result if a global optimization is broken into separate SCP and TSP subproblems. Although this would be possible for a sensor model with infinite range, the inspection problems for which our algorithms are intended involve robots with decidedly finite sensing radii. Specifically, our application of interest employs a bathymetry sonar with a 4 m sensing radius. The workspace is much larger than this, and thus we believe there is a strong correlation between the minimum-cardinality set cover and the minimum-cost GTSP.

3.1 Set Cover Subproblem

To solve the set cover subproblem, we rely on polynomial-time approximation algorithms that find solutions within guaranteed factors of optimality. We consider two such algorithms, a greedy algorithm and a linear programming (LP) rounding algorithm. The greedy algorithm [22, 28] simply adds to the set cover, on each iteration, the roadmap node with the largest number of observed primitives not yet in the cover. This algorithm solves the SCP within a factor of optimality that is bounded above by $\ln(m) + 1$, where m is the number of primitives required in the inspection. The rounding algorithm [20] solves the LP relaxation of the SCP, and then rounds the fractional solution according to a simple rule: if f is the largest number of roadmap nodes which share sightings of a primitive, then any roadmap node whose fractional decision variable is greater than or equal to $1/f$ is included in the cover. This method is guaranteed to return a solution within a factor f of optimality.

In the ship hull inspection example to be presented below, there are more than 105 primitives required in the inspection, giving a greedy algorithm approximation factor of about 12.5. At the same time, a typical value of f on a representative roadmap for this task is about twenty. Since these are both fast algorithms, and the approximation factors are of the same order, we will compare the two to assess their performance in practice.

Although both algorithms produce feasible solutions, these can often be pruned to yield feasible solutions of smaller size. Our pruning procedure, which runs in $O(n^2m)$ time, identifies configurations in the set cover which observe no geometric primitives uniquely, and in each iteration one of these configurations is randomly selected and pruned from the cover. The procedure repeats until every configuration in the cover is the unique observer of at least one geometric primitive.

3.2 Traveling Salesman Subproblem

To solve the TSP subproblem, we rely on another polynomial-time approximation. The algorithm of Christofides [11] computes the minimum spanning tree

(MST) over a graph, and then a minimum-cost perfect matching over the odd-degree nodes of the MST, achieving an approximation factor of 1.5 when the triangle inequality holds over the roadmap. Although our lazy computation procedure may occasionally violate the triangle inequality, RRT post-optimization smoothing ensures that there are no paths from a roadmap node i to a roadmap node k such than an alternate path from i to some node j to k is dramatically shorter. This assumption has proven successful in MST-only variants (with factor-2) for single and multi-agent coverage planning [13, 15], as well as pure multi-goal planning [30].

The Christofides approximation gives a good starting point for the TSP, but we also utilize a post-optimization improvement heuristic. Heuristics such as the Lin-Kernighan algorithm [27], which iteratively improves a TSP solution by swapping groups of edges, have succeeded in finding fast, high-quality solutions to very large TSP instances in practice [2]. We apply the chained Lin-Kernighan improvement procedure [1] for a short period of time after computation of each inspection tour.

4 Point Robot Test Case

First, we evaluate the performance of our inspection planning procedure on a point robot test case. This problem addresses algorithm performance as a function of the number of primitives, independent of collision and occlusion-checking. The unit cube is populated with a designated number of randomly sampled points, and the robot must plan a tour which observes them. Mimicking the HAUV inspection problem, the point robot has a four-dimensional state, comprised of three spatial coordinates, x, y, and z, and a yaw angle, θ. The sensor footprint is a cube, centered at the robot's location and designed to occupy one percent of the workspace volume, which is again representative of the parameters of a typical HAUV inspection planning problem. There are no obstacles in the point robot's workspace.

For several quantities of required primitives, ranging from 10^2 to 10^5, 100 instances of the planning procedure were run for each of three solution methods: redudant roadmaps with a greedy set cover, redundant roadmaps with LP rounding, and the dual sampling method. For the redundant roadmap cases, five different redundancies were tested, ranging from 1 to 50. For the dual sampling cases, five different numbers of local samples were tested, ranging from 10 to 1000. The chained Lin-Kernighan improvement heuristic was applied for 0.5 s after each computation of the Christofides TSP approximation. All trials were run on a Lenovo T400 laptop with a 2.53 GHz Intel Centrino 2 processor and 3 GB of RAM. To ensure the planning procedure was implemented using the very best data structures and algorithm implementations available, a variety of high-performance open-source software tools were utilized. A list of these software tools is included in the Appendix.

To sample in the local neighborhood of a primitive, a random configuration is constructed in a spherical coordinate system centered at the primitive. A range value is sampled uniformly at random between the minimum and maximum viewing range of the robot, and corresponding azimuth and elevation angles are randomly sampled as well. This places the robot at a position from which the primitive is in viewing range. Finally, the yaw angle is selected deterministically, such that a relative bearing of zero exists between the primitive and the robot. For a higher-dimensional vehicle state, a closed-form solution for angular orientation may not be available, and a Jacobian pseudoinverse method can be used to choose a robot orientation.

Figure 2 displays the results of this series of point-robot simulations. Increasing the redundancy of the coverage roadmap improved the quality of the greedy SCP solution and the LP rounding solution, but the relative quality of the LP rounding solution begins to worsen just short of a 1000-primitive inspection (for redundancies greater than one). In addition, the LP rounding algorithm, for large numbers of primitives, chooses much larger sets than the greedy algorithm. As a result, the pruning of sets became prohibitively expensive and LP set covers were not solved for large numbers of primitives.

Increasing the number of local samples in a dual sampling scheme improved the quality of the solution, which was comparable with the results for redundant roadmaps solved by the greedy set cover algorithm. As further basis for comparison, the length of the optimal "lawnmower" path for the point-robot's cube sensor to achieve 100 % coverage of the continuous workspace is plotted alongside the

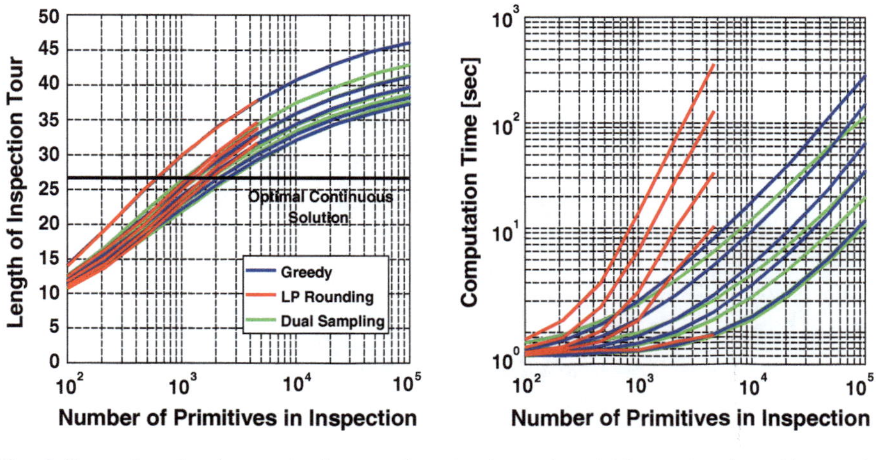

Fig. 2 Inspection planning results from a point robot in an obstacle-free, unit-cube workspace, in which the cube sensor is 1 % of the workspace volume. Inspection tour cost and computation time are plotted as a function of the number of required primitives; each data point represents the mean over 100 simulations. On *left*, LP rounding and greedy algorithm *lines* represent increasing roadmap redundancy [1, 5, 10, 25, 50] *downward* on the *vertical axis*. Dual sampling *lines* have increasing numbers of local samples, [10, 25, 100, 250, 1000] also moving downward. Roadmaps on *right plot* refer to computation times for the same trials, with redundancy and numbers of local samples increasing *upward* on the *vertical axis*. Due to prohibitively high computation time, larger quantities of primitives were not tested using the LP rounding algorithm, indicated by the end of the *red lines*

tour costs in Fig. 2. For a low number of primitives, the sensor does not have to cover the entire volume.

The two top-performing strategies, dual sampling and a redundant roadmap with the greedy SCP algorithm, encounter a similar asymptotic performance barrier as the local sampling and roadmap discretization is set finer and finer. Despite the neighborhood optimization of the dual method, and the global scope of the redundant roadmap, a bias persists due to the greedy basis for both methods.

5 AUV Inspection Test Case

Our procedure is next applied to a real-world problem, the inspection of the stern of a ship by the HAUV. Inspections are planned for the *SS Curtiss* (Fig. 3), a 200 m aviation logistics support ship, and the *USCGC Seneca*, an 80 m Coast Guard Cutter. The complex structures are large; the *Curtiss* has a single propeller 7 m in diameter and a shaft that is 1.5 m in diameter, while the *Seneca* has two shafts with propellers that are 2.5 m in diameter.

We first surveyed the ships with vertical and horizontal lawnmower patterns at safe distances of around 8 m. The preliminary surveys, although they did not

Fig. 3 An HAUV survey in progress at the *SS Curtiss*, a 200 m aviation logistics support ship. At *bottom right*, an annotated diagram of the HAUV, model 1B

achieve 100 % coverage of all structures, were intended to build a polygonal mesh model of each ship's stern suitable for planning a detailed inspection. For this, the Poisson reconstruction algorithm [23], which is typically applied to laser point clouds, was used to build watertight 3D meshes from acoustic range data, pictured in Figs. 4 and 5. Each mesh shown has been discretized such that no triangle edge is longer than 0.1 m, a resolution sufficient to identify a mine on the hull if all vertices are observed. Also in Figs. 4 and 5, the sensor footprint represents the sonar field of view when the sonar nods up and down through its full 180° range of rotation. Although the sonar can only produce a single range scan at a time, we assume that in this planned inspection, the vehicle, at each configuration, will nod the sonar over its full range of angular motion to obtain a larger field of view. Paths for the vehicle will be planned, as before, in x, y, z, and yaw angle θ.

Full-coverage roadmaps of varying redundancy were constructed, with 100 separate trials for each setting, using the same computer as the point-robot test case and an identical local sampling procedure. Both the LP rounding and the greedy SCP algorithm were again tested on these roadmaps, and the chained Lin-Kernighan improvement heuristic was applied in the same manner as in the point robot test case. Because a ship mesh comprises a large, non-convex obstacle, and the HAUV is not a point robot, the inspection planning procedure was used in

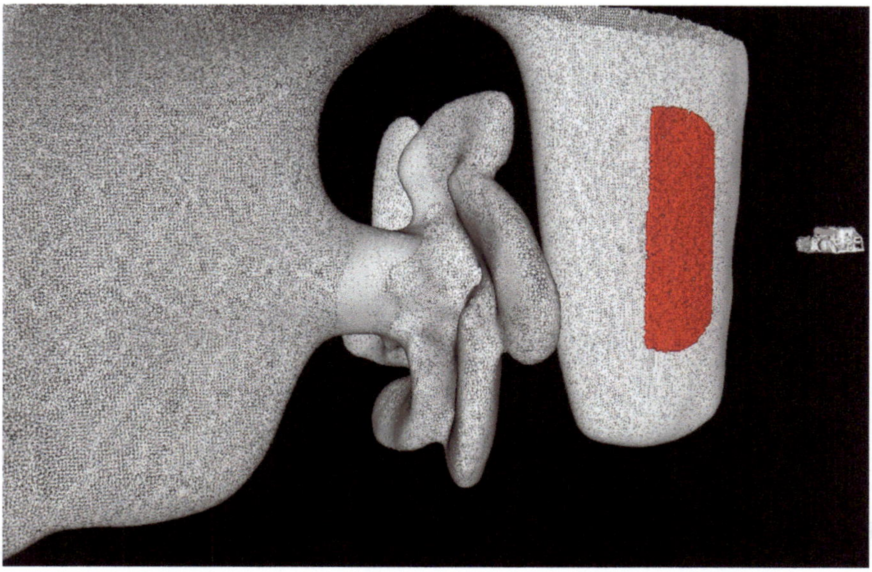

Fig. 4 A polygonal mesh obtained from our original, safe-distance survey of the *SS Curtiss* is depicted. The HAUV is illustrated in a configuration from which it observes a portion of the ship's rudder. The *red patch* shows mesh points imaged at a desired sensor range between 1 and 4 m, as the sonar sweeps through 180° pitch. The ship mesh contains 107, 712 points and 214, 419 triangular faces. The propeller is approximately 7 m in diameter

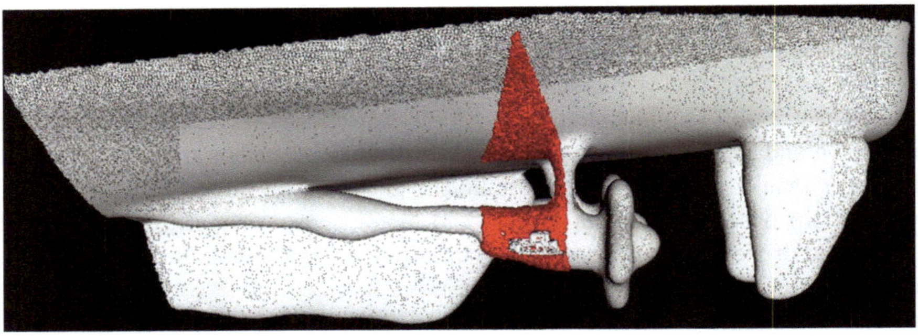

Fig. 5 A polygonal mesh obtained from our original, safe-distance survey of the *USCGC Seneca* is depicted. The HAUV is illustrated in a configuration from which it observes a portion of a shaft and propeller strut. The *red patch* shows mesh points imaged at a desired sensor range between 1 and 4 m, as the sonar sweeps through 180° pitch. The ship mesh contains 131, 657 points and 262, 173 triangular faces. Each propeller is approximately 2.5 m in diameter

its entirety, including collision-checking of sampled configurations and use of the bi-directional RRT to perform lazy inquiries of point-to-point paths.

As illustrated in Fig. 6, increased roadmap redundancy leads to an improvement in tour length, but the size of the improvement evidently diminishes as the redundancy increases. For solution of the set cover sub-problem, the LP rounding algorithm was inferior to the greedy algorithm for all redundancy settings greater than one, producing tours of greater length and also yielding set covers made up of much larger sets. As roadmap redundancy increased, the LP rounding algorithm required prohibitive amounts of time to eliminate unnecessary configurations from

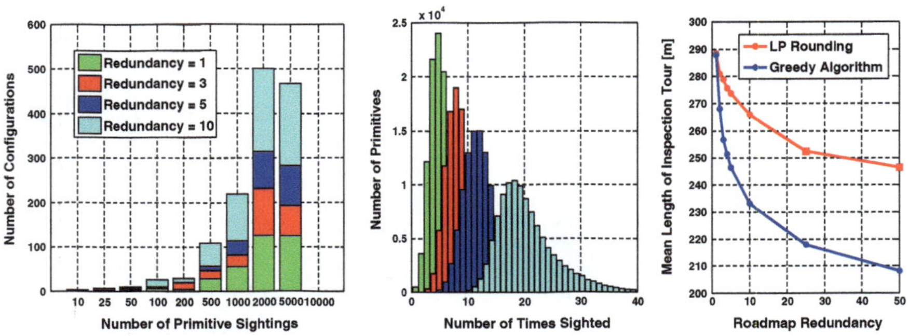

Fig. 6 Histograms display the coverage topology of a typical set of roadmaps for an instance of the *SS Curtiss* ship hull inspection task. The quantities of geometric primitives observed by roadmap configurations are illustrated at *left*, and the quantities of shared sightings of geometric primitives are illustrated at *center*. At *right*, a comparison of the LP Rounding Algorithm and Greedy Algorithm for inspection of the *SS Curtiss* is depicted. A variety of roadmap redundancies are examined, and the data for each redundancy represents the mean tour length over 100 trials (except for points marked with a *square*, which were limited to 10 trials only due to exhaustive computation time)

the set cover, making this pruning step the single most time-consuming step of the planning procedure. For problem instances using the greedy algorithm, runtime was instead dominated by the roadmap construction procedure.

To provide a clearer picture of the impact of increased redundancy, Fig. 6 also displays histograms showing roadmap coverage topology. It is clear that increased redundancy both increases the size of the roadmap and increases the mean and variance of the number of times a primitive is sighted.

In applying dual sampling to this planning problem, only three quantities of local samples were tested: five, ten, and twenty. Beyond a value of twenty, computation time grew prohibitively high. For each quantity of local samples, 100 separate trials of the dual sampling algorithm were run. We then identified redundant roadmaps with equivalent-or-better solution quality, and compared the computational performance of the two methods.

Tables 1 and 2 compare the performance of these algorithms, for the *SS Curtiss* and *USCGC Seneca*, respectively, in terms of mean tour length, mean computation time, and mean number of ray shooting calls. The dual sampling algorithms, in general, required more computation time than redundant roadmaps with greedy set covers, for equivalent-or-better mission costs. In particular, dual sampling required between 50 and 130° more ray shooting calls than redundant roadmaps. This data is evidence that the selectivity of the dual sampling method becomes a burden for 3D

Table 1 Selected tour costs over 100 HAUV inspection planning trials, *SS curtiss*

	Length of tour (m) mean (Min., Max.)	Comp. time (s)	R.S. Calls
Dual sampling with 5 local samples	263.8 (245.9, 281.0)	62.0	4.4×10^6
Roadmap of redundancy 3 (Greedy Alg.)	256.9 (241.9, 273.8)	54.1	2.5×10^6
Dual sampling with 10 local samples	250.7 (237.5, 266.9)	94.4	7.6×10^6
Roadmap of redundancy 5 (Greedy Alg.)	246.5 (229.7, 262.5)	72.3	3.5×10^6
Dual sampling with 20 local samples	240.1 (226.6, 256.1)	158.8	13.9×10^6
Roadmap of redundancy 10 (Greedy Alg.)	233.2 (213.6, 248.9)	118.1	6.0×10^6

Table 2 Selected tour costs over 100 HAUV inspection planning trials, *USCGC Seneca*

	Length of tour (m) mean (Min., Max.)	Comp. time (s)	R.S. calls
Dual sampling with 5 local samples	347.0 (326.5, 372.7)	243.2	7.4×10^6
Roadmap of redundancy 2 (Greedy Alg.)	346.4 (326.7, 366.6)	219.6	4.8×10^6
Dual sampling with 10 local samples	330.3 (298.5, 348.2)	346.1	11.7×10^6
Roadmap of redundancy 3 (Greedy Alg.)	330.6 (311.1, 351.1)	285.4	6.5×10^6
Dual sampling with 20 local samples	316.0 (296.8, 344.4)	555.5	19.7×10^6
Roadmap of redundancy 5 (Greedy Alg.)	314.9 (297.0, 333.0)	434.9	9.8×10^6

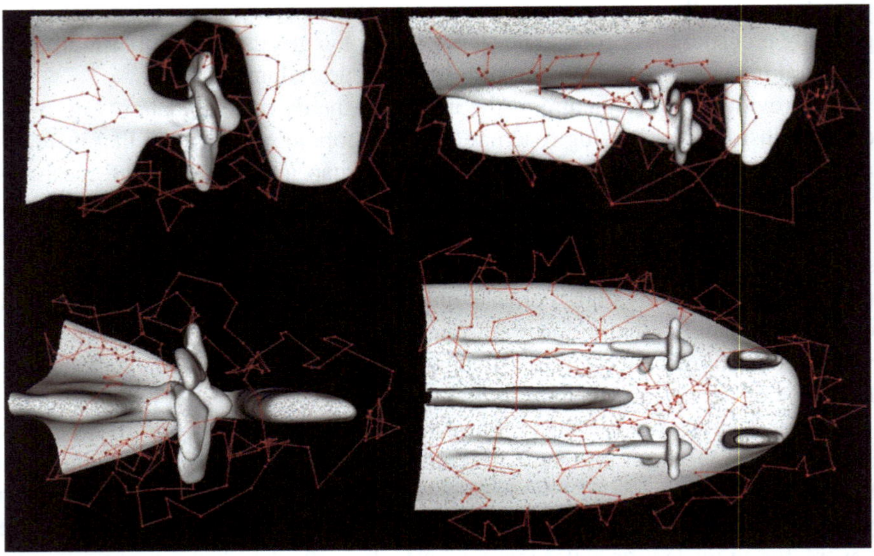

Fig. 7 Representative examples of planned HAUV inspection paths, with views from beneath the ships and from the side. The *left half* of the figure represents a roadmap of redundancy ten solved using the greedy SCP algorithm for inspection of the *SS Curtiss*. The tour depicted is 234 m in length and contains 167 distinct configurations. The *right half* of the figure represents the same computation for the *USCGC Seneca*. The tour depicted is 297 m in length and contains 259 distinct configurations

problems in which every sampled configuration must be checked against a large number of geometric primitives. The difference in overall computation time, less dramatic than the difference in ray shooting calls, was tempered by a significant cost common to both algorithms, the iterative solution of the TSP using lazy collision-checking. Representative HAUV inspection paths are depicted in Fig. 7.

6 Conclusion

In this work we presented an algorithm which plans fast, feasible inspection paths giving 100 % sensor coverage of required geometric primitives. Key developments are redundancy in a roadmap for coverage path planning, and the implementation of an integrated solution procedure for sampling-based coverage planning over complex 3D structures with several hundred thousand primitives (using highly developed, open-source routines wherever possible).

Redundancy improves the resolution of an initial covering roadmap, and we then sequentially apply practical set cover and traveling salesman algorithms, with lazy, point-to-point sampling-based planning. We have identified that this redundant roadmap method, in comparison to a dual sampling procedure, yields a consistent computational advantage in a large-scale, real-world coverage problem. Less time is spent checking and cataloging sensor observations, and a comparable if not superior planned path is produced as a result.

Future work entails the use of this method as a starting point for an iterative improvement procedure, which will guide the redundant roadmap solution toward a local optimum using continued sampling and the local replacement of configurations within the inspection tour.

Acknowledgments We would like to thank Dr. J. Vaganay and K. Shurn of Bluefin Robotics for essential field testing support in gathering the ship hull datasets. This work was supported by the Office of Naval Research under Grant N00014-06-10043, monitored by Dr. T.F. Swean.

Appendix

We give a table of open-source software resources used in our coverage path planning implementation (See Table 3).

Table 3 Resources used for coverage path planning software implementation

Software	Use	Link
OpenSceneGraph	KD-tree data structure for triangle mesh, ray shooting	http://www.openscenegraph.org
FLANN	KD-tree data structure for nearest-neighbor queries	http://www.cs.ubc.ca/~mariusm/index.php/FLANN/FLANN
OMPL	RRT implementation	http://ompl.kavrakilab.org/index.html
PQP	Collision checking	http://gamma.cs.unc.edu/SSV
Boost graph library	Minimum spanning tree	http://www.boost.org/doc/libs/1_46_1/libs/graph/doc/index.html
Blossom IV	Min-cost perfect matching	http://www2.isye.gatech.edu/~wcook/blossom4
Concorde	Lin-Kernighan TSP heuristic	http://www.tsp.gatech.edu/concorde.html
Meshlab	Processing and meshing of acoustic data	http://meshlab.sourceforge.net
Point cloud library	Viewer for rendering of paths and meshes	http://pointclouds.org

References

1. D. Applegate, W. Cook, A. Rohe, Chained Lin-Kernighan for large traveling salesman problems. INFORMS J. Comput. **15**(1), 82–92 (2003)
2. D. Applegate, R. Bixby, V. Chvatal, W. Cook, *The Traveling Salesman Problem: a Computational Study* (Princeton University Press, Princeton, 2006)
3. E.M. Arkin, M.M. Halldorsson, R. Hassin, Approximating the tree and tour covers of a graph. Inform. Process. Lett. **47**, 275–282 (1993)
5. P. Atkar, A.L. Greenfield, D.C. Conner, H. Choset, A. Rizzi, Uniform coverage of automotive surface patches. Int. J. Robot. Res. **24**(11), 883–898 (2005)
6. P. Atkar, A.L. Greenfield, D.C. Conner, H. Choset, A. Rizzi. Hierarchical segmentation of surfaces embedded in $\Re^3$ for auto-body painting, in *Proceedings of IEEE International. Conference on Robotics and Automation* (2005), pp. 572–577
7. P.S. Blaer, P.K. Allen, View planning and automated data acquisition for three-dimensional modeling of complex sites. J. Field Robot. **26**(11–12), 865–891 (2009)
4. P. Cheng, J. Keller, V. Kumar, Time-optimal UAV trajectory planning for 3D urban structure coverage, in *Proceedings of IEEE/RSJ International Conference on Intelligent Robots and Systems* (2008), pp. 2750–2757
8. H. Choset, P. Pignon, Coverage path planning: the boustrophedon decomposition, in *Proceedings of International Conference on Field and Service Robotics* (1997)
9. H. Choset, Coverage for robotics—a survey of recent results. Ann. Math. Artif. Intell. **31**, 113–126 (2001)
10. H. Choset, K.M. Lynch, S. Hutchinson, G. Kantor, W. Burgard, L.E. Kavraki, S. Thrun, *Principles of Robot Motion: Theory, Algorithms, and Applications* (MIT Press, Cambridge, 2005)
11. N. Christofides, Worst-case analysis of a new heuristic for the traveling salesman problem. Technical Report CS-93-13, Carnegie Mellon University (1976)
12. J.R. Current, D.A. Schilling, The Covering Salesman Problem. Transp. Sci. **23**(3), 208–213 (1989)
13. T. Danner, L. Kavraki, Randomized planning for short inspection paths, in *Proceedings of IEEE International Conference on Robotics and Automation*, vol 2 (2000), pp. 971–976
14. K. Easton, J. Burdick, A coverage algorithm for multi-robot boundary inspection, in *Proceedings of IEEE International Conference on Robotics and Automation* (2005), pp. 727–734
15. P. Fazli, A. Davoodi, P. Pasquier, A.K. Mackworth, Complete and robust cooperative robot area coverage with limited range, in *Proceedings of IEEE/RSJ International Conference on Intelligent Robots and Systems* (2010), pp. 5577–5582
16. M. Fischetti, J.J.S. Gonzalez, P. Toth, A branch-and-cut algorithm for the symmetric generalized traveling salesman problem. Oper. Res. **45**(3), 378–394 (1997)
17. M. Gendreau, G. LaPorte, F. Semet, The covering tour problem. Oper. Res. **45**(4), 568–576 (1997)
18. H. Gonzalez-Baños, J.-C. Latombe, Planning robot motions for range-image acquisition and automatic 3D model construction, in *Proceedings of AAAI Fall Symposium* (1998)
19. H. Gonzalez-Baños, J.-C. Latombe, A randomized art gallery algorithm for sensor placement, in *Proceedings of 17th Annual ACM Symposium on Computational Geometry* (2001), pp. 232–240
20. D.S. Hochbaum, Approximation algorithms for the set covering and vertex cover problems. 21.SIAM J. Comput. **11**(3) (1982)
21. F. Hover et al., A vehicle system for autonomous relative survey of in-water ships. J. Mar. Technol. Soc. **41**(2), 44–55 (2007)
22. D.S. Johnson, Approximation algorithms for combinatorial problems. J. Comput. Syst. Sci. **9**, 256–278 (1974)

23. M. Kazhdan, M. Bolitho, H. Hoppe, Poisson surface reconstruction, in *Proceedings of Fourth Eurographics Symposium on Geometry* (2006)
24. S. LaValle, J. Kuffner, Rapidly-exploring random trees: progress and prospects, in *Proceedings of Workshop on the Algorithmic Foundations of Robotics* (2000), pp. 293–308
25. S. LaValle, *Planning Algorithms* (Cambridge University Press, Cambridge, UK, 2006)
26. Y.N. Lien, E. Ma, Transformation of the generalized traveling salesman problem into the standard traveling salesman problem. Inf. Sci. **74**, 177–189 (1993)
27. S. Lin, B.W. Kernighan, An effective heuristic algorithm for the traveling salesman problem. Oper. Res. **21**(2), 498–516 (1973)
28. L. Lovasz, On the ratio of optimal integral and fractional covers. Discrete Math. **13**, 383–390 (1975)
29. C.E. Noon, J.C. Bean, An efficient transformation of the generalized traveling salesman problem. Technical Report 89-36, Department of Industrial and Operations Engineering (The University of Michigan, Ann Arbor, 1989)
30. M. Saha, T. Roughgarden, J.-C. Latombe, G. Sanchez-Ante, Planning tours of robotic arms among partitioned goals. Int. J. Robot. Res. **25**(3), 207–223 (2006)
31. G. Sanchez, J.-C. Latombe, On delaying collision checking in PRM planning: application to multi-robot coordination. Int. J. Robot. Res. **21**(1), 5–26 (2002)
32. W. Scott, G. Roth, J. Rivest, View planning for automated three-dimensional object reconstruction and inspection. ACM Comput. Surv. **35**(1), 64–96 (2003)
33. T. Shermer, Recent results in art galleries. Proc. IEEE **80**(9), 1384–1399 (1992)
34. P. Wang, R. Krishnamurti, K. Gupta, View planning problem with combined view and traveling cost, in *Proceedings of IEEE International Conference on Robotics and Automation* (2007), pp. 711–716
35. K. Williams, J. Burdick, Multi-robot boundary coverage with plan revision, in *Proceedings of IEEE International Conference on Robotics and Automation* (2006), pp. 1716–1723

Path Planning with Loop Closure Constraints Using an Atlas-Based RRT

Léonard Jaillet and Josep M. Porta

Abstract In many relevant path planning problems, loop closure constraints reduce the configuration space to a manifold embedded in the higher-dimensional joint ambient space. Whereas many progresses have been done to solve path planning problems in the presence of obstacles, only few work consider loop closure constraints. In this paper we present the AtlasRRT algorithm, a planner specially tailored for such constrained systems that builds on recently developed tools for higher-dimensional continuation. These tools provide procedures to define charts that locally parametrize manifolds and to coordinate them forming an atlas. AtlasRRT simultaneously builds an atlas and a Rapidly-Exploring Random Tree (RRT), using the atlas to sample relevant configurations for the RRT, and the RRT to devise directions of expansion for the atlas. The new planner is advantageous since samples obtained from the atlas allow a more efficient extension of the RRT than state of the art approaches, where samples are generated in the joint ambient space.

1 Introduction

In the recent years, there has been a growing interest around the problem of path planning with loop closure constraints [3, 6, 10, 20, 27, 31]. The reason behind this interest is that this problem appears in many relevant problems in Robotics such as coordinated manipulation [19], motion planning with parallel robots [34], robot grasping [25], constraint-based object positioning [24], or surgery robots [1]. This problem is also crucial in Biochemistry, when searching for conformational changes in molecular loops [37].

L. Jaillet · J.M. Porta (✉)
Institut de Robòtica i Informàtica Industrial, CSIC-UPC,
Llorens i Artigas 4-6, Barcelona, Spain
e-mail: porta@iri.upc.edu

L. Jaillet
e-mail: leonard.jaillet@inria.fr

H.I. Christensen and O. Khatib (eds.), *Robotics Research*,
Springer Tracts in Advanced Robotics 100, DOI 10.1007/978-3-319-29363-9_20

The mainstream of research in path planning in the two last decades [4, 16] has focused on variants of sampling-based path planners [14, 17] to efficiently solve the problem of robots with open loop kinematics operating in environments cluttered with obstacles. Obstacles induce a set of inequality constraints and planning motions that respect these constraints can be a very tough task, especially when it requires passing through narrow passages. Here, we address a more challenging situation where, beside the obstacles, the problem includes loop closure constraints represented by a set of equalities that must be fulfilled. Such constraints reduce the configuration space to a manifold embedded in the higher-dimensional ambient space defined by the joint variables involved in the problem.

The efficiency of sampling-based path planning approaches such as the Rapidly-Exploring Random Trees (RRTs) relies in the so called Voronoi exploration bias [17] which can only be obtained if the space to explore can be properly sampled. Thus, ideally, these approaches would require an isometric parametrization of the configuration space from which a uniform distribution of samples can be generated. Whereas for non-constrained systems such parametrization is straightforward, this is not the case when the loop closure constraints reduce the dimensionality of the configuration space.

Distance-based formulations [9, 31] can provide a global parametrization of the constrained configuration space for some particular families of mechanism with kinematic loops. Other approaches try to infer the parametrization from large sets of samples on the manifold [10], and task-space planners assume that a subset of variables related to the end-effector are enough to parametrize the configuration space [3, 27, 40].

In the absence of a global parametrization, Kinematics-PRM [8] samples a subset of joint variables and uses inverse kinematics to find values for the remaining ones. Unfortunately, this strategy is only valid for a limited class of mechanisms, and although some improvements have been proposed [5], the probability of generating invalid samples is significant. An alternative strategy to get valid configurations is to sample in the joint ambient space and to converge to the configuration space after each tree extension using numerical iterative techniques, either implementing random walks [38], or the more efficient Jacobian pseudoinverse method [2, 6, 30]. Despite being probabilistically complete [3], a uniform distribution of samples in the ambient space does not necessarily translate to a uniform distribution in the configuration space, which reduces the efficiency of these approaches. This problem is illustrated in Fig. 1 where the configuration space to be explored is a torus-like manifold of diameter four times smaller than the ambient space width. Figure 1a shows a RRT built from points sampled in the ambient space that has a poor coverage of the manifold. With the AtlasRRT presented in this paper, the process of diffusion is largely independent of the configuration space shape and of the ambient space bounds which improves the coverage of the manifold, as shown in Fig. 1b.

To improve the quality of the sampling, one can focus on a subset of the ambient space around the configuration space [43]. However, with this method points are still sampled in the ambient space, which can be of much higher dimensionality

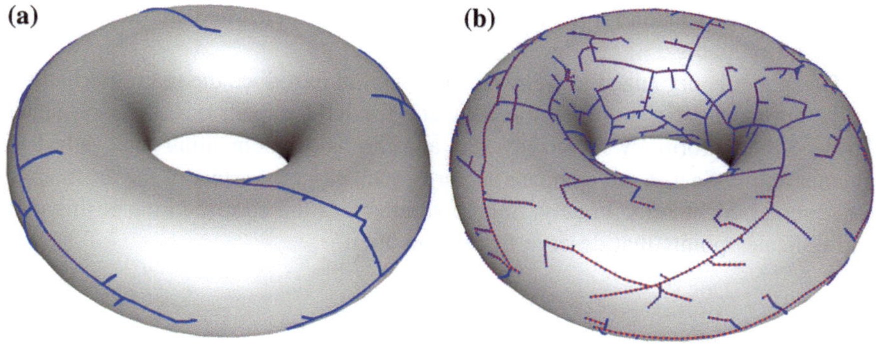

Fig. 1 Two RRTs built on a torus-like manifold after throwing 500 samples. **a** With an ambient space sampling, the exploration is focused on the outer parts of the torus and many samples do not produce any tree extension. **b** With an AtlasRRT, the diffusion process is largely independent of the ambient space which improves the coverage

than the configuration space. Um et al. [35] sketch a lazy RRT scheme where loosely coordinated RRTs are built on tangent spaces that locally approximates the manifold and that have the same dimensionality as the configuration space. However, the fact that the subtrees in different tangent spaces overlap affects the quality of the resulting RRT.

From Differential Geometry, it is well known that a manifold can be described by a collection of local parametrizations called charts, that can be coordinated within an atlas [22]. Higher-dimensional continuation techniques provide principled numerical tools to compute the atlas of an implicitly defined manifold starting from a given point, whereas minimizing the overlap between neighboring charts [11, 12]. One-dimensional continuation methods, have been strongly developed in the context of Dynamical Systems [15], whereas in Robotics, they have been mainly used for solving problems related to Kinematics [26, 29]. To the best of our knowledge, higher-dimensional continuation methods have been only used in Robotics to evaluate the dexterity of mechanisms [39]. In a previous work [20], we introduced a resolution complete path planner on manifolds based on higher-dimensional continuation tools. Despite its efficiency, this planner relies on a discretization of the manifold and the exploration could be blocked in the presence of narrow corridors, unless using a fine resolution with the consequent loose in performance. Moreover, the number of charts generated with this planner scales exponentially with the dimension of the configuration space. To overcome these limitations, we propose here a probabilistic complete planner based on RRTs with the consequent gain of efficiency, specially for high dimensional configuration spaces. The new method called AtlasRRT is based on a coordinated construction of an atlas and a bidirectional RRT. On the one hand, the atlas is used to adequately sample new configurations and thus, to retain the RRT Voronoi bias, despite exploring a non Euclidean configuration space. On the other hand, the RRT is used to determine directions of expansion for the atlas, so that the charts generated are those useful to find solution paths.

This paper is organized as follows. Next section introduce the main mechanisms of the approach, showing how the atlas and the RRT expansions are coordinated. Then, Sect. 3 formally describes the algorithms implementing the AtlasRRT planner and in Sect. 4 we compare its performance to other state of the art methods for several benchmarks. Finally, Sect. 5 summarizes the contributions of this work and indicates points that deserve further attention.

2 Building an RRT on a Manifold

In this section, we first describe the elements describing a chart and how to build a RRT within it. Next, we describe how to define an atlas properly coordinating the charts to obtain a RRT covering the whole configuration space manifold.

2.1 RRT on a Chart

Let us consider a n-dimensional joint ambient space and a k-dimensional configuration space implicitly defined by a set of constraints

$$\mathbf{F}(\mathbf{x}) = 0, \tag{1}$$

with $\mathbf{F} : \mathbb{R}^n \rightarrow \mathbb{R}^{n-k}$, and $n > k > 0$. Note that we adopt the standard convention in Kinematics [18] where the configuration space is defined as the set of points fulfilling the constraints (this is sometimes called constrained configuration space) that is embedded in the ambient space of the joint variables (called configuration space in some approaches). Moreover, we assume that the configuration space is manifold everywhere, without considering the presence of singularities.

A chart, C_i, locally parametrizes the k-dimensional manifold around a given point $\mathbf{x}_i$ with a bijective map, $\mathbf{x}_j = \psi_i\,(\mathbf{u}_j^i)$, between points $\mathbf{u}_j^i$ in $\mathbb{R}^k$ and points $\mathbf{x}_j$ on the manifold, with $\psi_i(\mathbf{0}) = \mathbf{x}_i$. Following [23], this mapping can be implemented using the k-dimensional space tangent at $\mathbf{x}_i$ (see Fig. 2). An orthonormal basis for this tangent space is given by the $m \times k$ matrix, Φ_i, satisfying

$$\begin{bmatrix} \mathbf{J}(\mathbf{x}_i) \\ \Phi_i^\top \end{bmatrix} = \Phi_i \begin{bmatrix} \mathbf{0} \\ \mathbf{I} \end{bmatrix}, \tag{2}$$

with $\mathbf{J}(\mathbf{x}_i)$ the Jacobian of $\mathbf{F}$ evaluated at $\mathbf{x}_i$, and $\mathbf{I}$, the identity matrix. Using this basis, the mapping ψ_i is computed by first computing the mapping ϕ_i from points in the tangent space to coordinates in the joint ambient space,

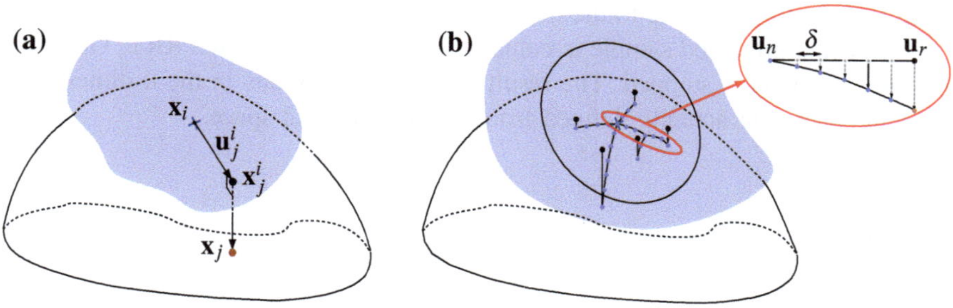

Fig. 2 **a** A chart C_i is implemented as a mapping $\mathbf{x}_j = \psi_i(\mathbf{u}^i_j)$ between the tangent space at $\mathbf{x}_i$ and the manifold. **b** Growing a RRT on the manifold from a single chart. *Black dots* are the samples thrown in the tangent space and *blue dots* are points on the manifold forming the RRT branches. The *inset* shows the process of generating a branch by linear interpolation in the tangent space with steps of size δ and successive projection

$$\mathbf{x}^i_j = \phi_i\left(\mathbf{u}^i_j\right) = \mathbf{x}_i + \Phi_i \mathbf{u}^i_j, \tag{3}$$

and then, orthogonally projecting this point on the manifold to obtain $\mathbf{x}_j$. This projection can be computed by solving the system

$$\begin{cases} \mathbf{F}(\mathbf{x}_j) = \mathbf{0}, \\ \Phi_i^\top\left(\mathbf{x}_j - \mathbf{x}^i_j\right) = \mathbf{0}, \end{cases} \tag{4}$$

using a Newton procedure where $\mathbf{x}_j$ is initialized to $\mathbf{x}^i_j$ and is iteratively updated by the $\Delta\mathbf{x}_j$ increments fulfilling

$$\begin{bmatrix} \mathbf{J}(\mathbf{x}_j) \\ \Phi_i^\top \end{bmatrix} \Delta\mathbf{x}_j = -\begin{bmatrix} \mathbf{F}(\mathbf{x}_j) \\ \Phi_i^\top\left(\mathbf{x}_j - \mathbf{x}^i_j\right) \end{bmatrix}, \tag{5}$$

until the error is negligible or for a maximum number of iterations.

The inverse mapping ψ_i^{-1} can be computed as the projection of a point on the tangent subspace

$$\mathbf{u}^i_j = \psi_i^{-1}(\mathbf{x}_j) = \Phi_i^\top\left(\mathbf{x}_j - \mathbf{x}_i\right). \tag{6}$$

Using the mapping provided by a chart, C_j, we can define a RRT on the part of the manifold covered by this chart, as shown in Fig. 2b. This can be achieved using the standard RRT exploration mechanism and projecting to the manifold whenever necessary. Thus, the tree is initialized at $\mathbf{x}_i$, a random point, $\mathbf{u}_r$, is drawn in a ball of radius R in $\mathbb{R}^k$ and their coordinates in ambient space are obtained as $\mathbf{x}_r = \phi_i\ (\mathbf{u}_r)$. Then, the point, $\mathbf{x}_n$, already in the RRT and closer to $\mathbf{x}_r$ is determined. The tree is extended from $\mathbf{x}_n$ by interpolating between $\mathbf{u}_n = \psi_i^{-1}\ (\mathbf{x}_n)$ and $\mathbf{u}_r$, using steps of a small size δ and projecting to the manifold after each step, as shown in the inset of Fig. 2b. If the projected sample is collision free, it is added to the RRT. Otherwise the tree extension is stopped. The result is a tree with points in the n-dimensional ambient space, but actually defined from a k-dimensional space.

2.2 RRT on an Atlas

Unless for particularly simple problems, a single chart is not enough to parametrize the whole configuration space. The validity area, $\mathcal{V}_i V_i$, for a chart, $\mathcal{C}_i$, is defined as the set of points $\mathbf{u}_j^i$ in the tangent space associated with $\mathcal{C}_i$ such that

$$\left\|\mathbf{u}_j^i\right\| \leq \beta R, \tag{7}$$

$$\left\|\mathbf{x}_j - \phi\left(\mathbf{u}_j^i\right)\right\| \leq \varepsilon, \tag{8}$$

$$\left\|\Phi_i^\top \Phi_j\right\| \leq 1 - \varepsilon, \tag{9}$$

where $0 < \beta < 1$, R is the radius of the sampling ball, $\mathbf{x}_j = \psi_i\,(\mathbf{u}_j^i)$ is the projection of $\mathbf{u}_j^i$ on the manifold, and Φ_j is the basis of the tangent space at $\mathbf{x}_{j.}$

The first condition in the definition of $\mathcal{V}_i$ ensures that new charts are eventually created if the boundaries of the search area are reached. The second condition bounds the error between the tangent space and the manifold. The third condition ensures a smooth curvature in the part of the manifold covered by $\mathcal{C}_i$ and a smooth transition between charts. These two last conditions bound the distorsion introduced by the chart map and, thus, ensure that a uniform distribution of samples in $\mathcal{V}_i$ translates to an approximately uniform distribution of configurations on the manifold.

The validity area of a chart is not precomputed, but discovered as the RRT grows. Whenever the RRT reaches a point outside the validity area of a given chart $\mathcal{C}_i$, a new chart is added to the atlas on the last point still in $\mathcal{V}_i$. To avoid the overlap between the validity areas of neighboring charts we introduce a mechanism to reduce the validity areas, similar to that in [11] (see Fig. 3). We associate a set of inequalities, $\mathcal{F}_i$, to each chart. This set is initially empty and when two charts $\mathcal{C}_i$ and $\mathcal{C}_j$ are neighbors (i.e., when the center of one of the charts are inside the validity area of the other chart) the following inequality is added to $\mathcal{F}_i$

$$2\mathbf{u}^\top \mathbf{u}_j^i \leq \left\|\mathbf{u}_j^i\right\|^2, \tag{10}$$

with $\mathbf{u}_j^i = \psi_i^{-1}(\mathbf{x}_j)$ This reduces the validity area to a half space defined in the tangent space associated to $\mathcal{C}_j$ given by the plane orthogonally bisecting vector $\mathbf{u}_i^j$.

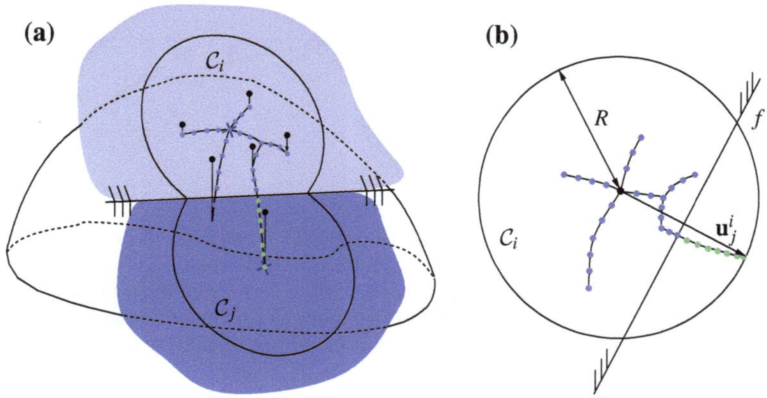

Fig. 3 **a** When a RRT leaves the validity area of chart $\mathcal{C}_i$, a new chart, $\mathcal{C}_j$, is created and their respective sampling areas are coordinated so that they do not overlap. **b** $\mathcal{C}_j$ adds an inequality, f, to $\mathcal{C}_i$ reducing its actual validity area. Note that after adding f, the nodes in green move from $\mathcal{C}_i$ to $\mathcal{C}_j$

Similarly, the validity area of $\mathcal{C}_i$ is cropped with the inequality defined from $\mathbf{u}_i^j = \psi_i^{-1}(\mathbf{x}_i)$, the projection of $\mathbf{x}_i$ on the tangent space associated with $\mathcal{C}_j$. When a given chart $\mathcal{C}_i$ is fully surrounded by neighboring charts, the intersection of the half spaces defined by the inequalities in $\mathcal{F}_i$ conform a polytope that conservatively bounds the actual validity area for the chart, taking into account the presence of neighboring charts.

The construction of an RRT on an Atlas proceeds as follows. First, a point $\mathbf{u}_r$ is sampled on the atlas. For this purpose, a chart $\mathcal{C}_r$ is randomly selected and a point is sampled on the ball of radius R defined on the tangent space associated with this chart. Random points not in the actual validity area (i.e., points not fulfilling the inequalities in $\mathcal{F}_r$) are rejected and the sampling process is repeated. Thus, the probability of generating a valid random point in a chart is proportional to the volume of its actual validity area and therefore, the sampling process selects points uniformly distributed within the area covered by the entire atlas.

The coordinates of sample $\mathbf{u}_r$ in ambient space, $\mathbf{x}_r$, are computed using ϕ_r. Then, $\mathbf{x}_n$, the node already in the RRT closer to $\mathbf{x}_r$, is determined, the random point $\mathbf{x}_r$ is projected to the tangent space of the chart $\mathcal{C}_c$ including $\mathbf{x}_n$, and the tree extension proceeds in this chart as detailed in Sect. 2.1. The extension only stops if the random point is reached or if the path is blocked by an obstacle. However, during the tree extension the growing branch can leave $\mathcal{C}_c$. If one of the inequalities in $\mathcal{F}_c$ is not fullfilled by the new point to be added to the RRT, the extension entered in the validity area of the neighboring chart that generated the violated inequality in $\mathcal{F}_c$. In this case, the tree extension continues in this neighboring chart by projecting $\mathbf{x}_r$ on it. If the RRT extension leaves the validity area of $\mathcal{C}_c$ but the new point fulfills all the inequalities in $\mathcal{F}_c$ (if any), a new chart is generated and the branch extension continues on it (see Fig. 4). Note that after creating a chart, some of the nodes assigned to neighboring charts might move to the area of validity of the new chart.

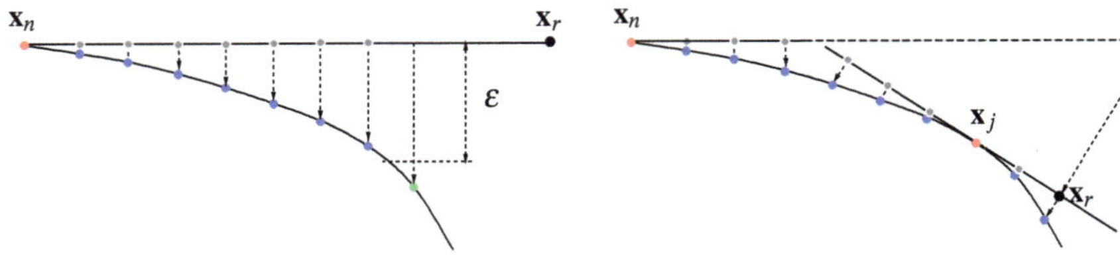

Fig. 4 RRT extension on a one-dimensional cut. The RRT is extended towards $\mathbf{x}_r$, the randomly selected point, from $\mathbf{x}_n$, the closest point already in the tree. In this case, $\mathbf{x}_n$ is the center of a chart. When the extension reaches a point (the *green dot*) where the error with respect to the chart at $\mathbf{x}_n$ is larger than ε, a new chart is created from the previous node already in the tree and $\mathbf{x}_r$ is projected to the new chart. Then, the RRT extension continues in the new chart until the projected $\mathbf{x}_r$ is reached

These nodes are directly identified since they are the ones that do not fulfill the inequality introduced to by the new chart.

Algorithm 1: The AtlasRRT algorithm.

AtlasRRT($\mathbf{x}_s, \mathbf{x}_g, \mathbf{F}$)
input : The query configurations, $\mathbf{x}_s$ and $\mathbf{x}_g$, a set of constraints, $\mathbf{F}$.
output: A path connecting $\mathbf{x}_s$ and $\mathbf{x}_g$.

$T_s \leftarrow$ INITRRT($\mathbf{x}_s$)
$T_g \leftarrow$ INITRRT($\mathbf{x}_g$)
$A \leftarrow$ INITATLAS($\mathbf{F}, \mathbf{x}_s, \mathbf{x}_g$)
DONE $\leftarrow$ FALSE
while not DONE **do**
    $\mathbf{x}_r \leftarrow$ SAMPLEONATLAS(A)
    $n_r \leftarrow$ NEARESTNODE($T_s, \mathbf{x}_r$)
    $\mathbf{x}_l \leftarrow$ EXTENDTREE($T_s, A, n_r, \mathbf{x}_r$)
    $n_l \leftarrow$ NEARESTNODE($T_g, \mathbf{x}_l$)
    $\mathbf{x}'_l \leftarrow$ EXTENDTREE($T_g, A, n_l, \mathbf{x}_l$)
    if $\|\mathbf{x}_l - \mathbf{x}'_l\| < \delta$ **then**
        DONE $\leftarrow$ TRUE
    else
        SWAP(T_s, T_g)
RETURN(PATH($T_s, \mathbf{x}_l, T_g, \mathbf{x}'_l$))

In the absence of other constraints, parameter β in Eq. (7) gives the ratio of sampled points that are outside the validity area and, thus, the ratio of points that will trigger the creation of new charts. Therefore, attending only to this criterion, the probability of creating new charts is

$$p = 1 - \beta^k. \quad (11)$$

3 AtlasRRT Algorithm

Algorithm 1 gives the pseudo-code for the AtlasRRT planner implementing the path planning approach described in the previous section. The algorithm takes $\mathbf{x}_s$ and $\mathbf{x}_g$ as start and goal configurations, respectively, and tries to connect them with a path on the manifold implicitly defined by a given set of constraints $\mathbf{F}$. The algorithm implements a bidirectional search method. To this end, two RRTs are initialized (lines 1 and 2), with the start and the goal configurations as respective root nodes. An atlas is also initialized with two charts centered at each one of these points (line 3). Next, the algorithm iterates trying to connect the two trees (lines 5–14). First, a configuration is sampled using the atlas and one of the RRT is extended as much as possible towards this random sample from the nearest node already in the tree, measured using the Euclidean distance. The other RRT is then extended towards the last node added to the first tree, from the nearest node already in this second tree. If this second extension reaches its objective, the trees are connected and the nodes in the RRTs are used to reconstruct a path between $\mathbf{x}_s$ and $\mathbf{x}_g$. Otherwise, the two RRTs are swapped and the extension process is repeated. Note that this top level search algorithm is the same as that in [3]. The differences appears in how to sample random points and how to add branches to the RRTs.

Algorithm 2: Sampling on an atlas.

SampleOnAtlas(A)
input : The atlas, A.
output: A sample on the atlas.

repeat
  $r \leftarrow$ RANDOMCHARTINDEX(A)
  $\mathbf{u}_r \leftarrow$ RANDOMONBALL(R)
until $\mathbf{u}_r \in \mathcal{F}_r$
RETURN($\phi_r(\mathbf{u}_r)$)

In the sampling process we take advantage of the atlas, as shown in Algorithm 2. A chart is selected at random with uniform distribution and then, a point is sampled within the ball of radius R bounding the sampling area of this chart. The process is repeated until a point is inside the actual validity area associated with its selected chart, i.e., until it fulfills all the inequalities associated with the chart, if any. Finally, the sample returned is formed by the ambient space coordinates for the selected point computed using the mapping ϕ_r for the selected chart, r.

The addition of a branch to a tree T is done following the steps detailed in Algorithm 3. This procedure operates in the chart $\mathcal{C}_c$ including the node to be extended, that is initially the chart associated to the nearest node n (line 1). The sample to reach in the tree extension is projected on $\mathcal{C}_c$ (lines 2 and 3) and the branch extension is iteratively performed while the branch is not blocked and $\mathbf{x}_r$ is not reached (lines 5–23). At each iteration, a node is added to the tree. To define the node to add, a small step of size δ is performed in the parameter space of $\mathcal{C}_c$ from the parameters of the current node $\mathbf{u}_n$ towards the parameters for the goal sample, $\mathbf{u}_r$. The resulting parameters $\mathbf{u}_j$ are projected to the manifold to obtain the configuration $\mathbf{x}_j$ (line 7). If the projection is not too far away from $\mathbf{x}_n$ and if it is collision free, the point is added to the tree. First, however, we verify if the new point is still inside $\mathcal{C}_c$. On the one hand, if the parameters for the point are outside the validity area of this chart (line 12), a new chart is added to the atlas (line 13) which becomes the chart where to operate. Procedure NEWCHART implements the chart creation, the detection of neighboring charts, and the reassignment of tree nodes, if necessary. On the other hand, if the parameters for the point are inside the validity area, but outside the area defined by the inequalities associated with $\mathcal{C}_c$, the point is in the validity area of a neighboring chart. Procedure MOVETOCHART (line 17) identifies this chart. Whenever the current chart changes (line 19), we compute the parameters for the current sample and for the goal one in the new $\mathcal{C}_c$. Note that this affects the computation of the next point, at line 6. Finally, the new point is added to the tree and associated with the current chart, $\mathcal{C}_c$ (line 23).

Algorithm 3: Adding a branch to the AtlasRRT.

ExtendTree (T, A, n, x_r)

input : A tree, T, an atlas, A, the index of the nearest node in the tree, n, and the random sample, x_r.

output : The last node added to the tree or x_n if no extension was performed.

1 $c \leftarrow$ CHARTINDEX(n)
2 $\mathrm{u}_r \leftarrow \psi_c^{-1}(\mathrm{x}_r)$
3 $\mathrm{x}_r \leftarrow \phi_c(\mathrm{u}_r)$
4 BLOCKED $\leftarrow$ FALSE
5 while not BLOCKED and $\|\mathrm{u}_n - \mathrm{u}_r\| > \delta$ do
6     $\mathrm{u}_j \leftarrow \mathrm{u}_n + (\mathrm{u}_r - \mathrm{u}_n)\,\delta / \|\mathrm{u}_r - \mathrm{u}_n\|$
7     $\mathrm{x}_j \leftarrow \psi_c(\mathrm{u}_j)$
8   if $\|\mathrm{x}_j - \mathrm{x}_n\| > 2\,\delta$ or COLLISION(x_j) then
9   BLOCKED $\leftarrow$ TRUE
10   else
11   NEW $\leftarrow$ FALSE
12   if $\mathrm{u}_j \notin \mathcal{V}_c$ then
13             $c \leftarrow$ NEWCHART(A, x_n)
14   NEW $\leftarrow$ TRUE
15   else
16   if $\mathrm{u}_j \notin \mathcal{F}_c$ then
17                 $c \leftarrow$ MOVETOCHART$(\mathrm{x}_n, \mathrm{u}_j)$
18   NEW $\leftarrow$ TRUE
19   if NEW then
20             $\mathrm{u}_j \leftarrow \psi_c^{-1}(\mathrm{x}_j)$
21             $\mathrm{u}_r \leftarrow \psi_c^{-1}(\mathrm{x}_r)$
22             $\mathrm{x}_r \leftarrow \phi_c(\mathrm{u}_r)$
23         $n \leftarrow$ ADDNODE(T, A, c, x_j)
24 RETURN(x_n)

Concerning the algorithm complexity and not considering the cost of collision detection, the most expensive steps in the algorithm are the search for nearest nodes in the tree (lines 7 and 9 of Algorithm 1), the identification of neighboring charts when adding a chart to the atlas (line 13 of Algorithm 3), and the computation of the mapping ψ_c in line 7 of Algorithm 3. The first two operations can be implemented using hierarchical structures reducing their cost to be logarithmic in the number of nodes of the corresponding tree and in the number of charts in the atlas, respectively. The cost of computing the mapping ψ_c scales with $O(n^3)$ since it is implemented as a Newton process with a bounded number of iterations where at each iteration a QR decomposition is used.

The algorithm basically uses four parameters: R, ε, δ, and p that appears in Eq. (11). R is a parameter defining the sampling area around a given point and can be safely set to a large value since the other criteria defining the validity area of a chart will eventually trigger the chart creation. ε regulates the amount of charts in the atlas. To address problems with a configuration space of moderate to large dimensionality it should not be set too small. A small δ should be used to avoid undetected collision between consecutive nodes in the RRT and sudden changes in the manifold curvature. Finally, p regulates the exploration since the larger this parameter the stronger the bias towards unexplored regions. In our experience adjusting the parameters is not an issue since the same values give good results for a large set of problems.

The probabilistic completeness of the RRT generated on a single chart over its sampling area is the same as that of a RRT defined in a k-dimensional Euclidean space. Thus, any point in the collision free region including the root of the tree will be eventually reached by the tree. In particular, points out of its validity area will be eventually reached and, consequently, new charts will be generated. The probabilistic completeness over the area covered by a new chart is also equivalent to that of a plain RRT. Assuming that the transition between charts is continuous and smooth, the local probabilistic completeness for each chart implies a global probabilistic completeness for the overall algorithm.

4 Experiments

We implemented the AtlasRRT planner described through Sects. 2 and 3 in C. The planner was integrated as a module of our position analysis toolbox [32], using SOLID [33] as collision detector, the GNU Scientific Library [7] for the linear algebra operations, and the kd-tree described in [42] for the nearest-neighbor queries. In principle, simple formulations are advantageous for continuation methods [36] and, thus, our position analysis toolbox is based on a formulation with redundant variables that yields a system of simple equations only containing linear, bilinear, and quadratic monomials [21]. This is a particularly challenging situation since the manifold here arises not only because of the loop closure constraints, but also due to the equations necessary to compensate for the redundancy in the formulation. Due to this redundancy, the planning system introduced in [35] can not be directly applied to this case. Thus, for the sake of comparison we use the HC-planner [20] and an adaptation of the planner introduced in [3] that is called here CB-RRT. The HC-planner is a resolution completed planner on manifolds that is based on a greedy best first search strategy on the graph implicitly defined by the centers of the charts in the atlas. The CB-RRT planner shares the bi-directional search strategy with AtlasRRT, but randomly samples in the joint ambient space and uses the Jacobian pseudo-inverse procedure to converge to points on the manifold when necessary. Note, however, that in our implementation some aspects of the planner in [3] are not considered (i.e., the Task Space Regions and the direct sampling using them) since they are neither included in the AtlasRRT. The comparison with the HC-Planner is used to evaluate the AtlasRRT search strategy whereas the comparison with the CB-RRT is used to assess the advantage of the atlas-based sampling strategy.

Figure 5 shows the four benchmarks used in this paper, ordered by increasing configuration space dimensionality. The first one is a problem where a small sphere

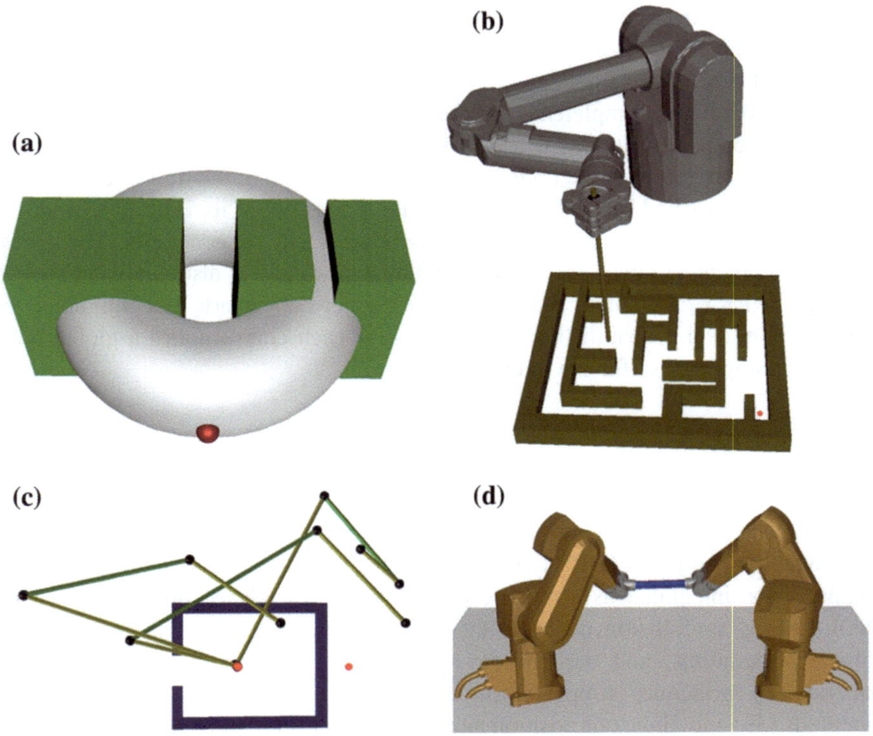

Fig. 5 The four benchmarks used to test the AtlasRRT planner. **a** A small sphere moving on a torus with obstacles and a narrow corridor. **b** The Barret arm solving a maze problem. **c** A planar parallel manipulator moving a peg out of a cul-de-sac. **d** Two Stäubli Rx60 industrial arms collaborating to move a cylindrical bar

has to pass through a narrow corridor, while staying on the surface of a torus. This example is used to emphasize the fact that AtlasRRT can be advantageous even with simple manifolds. The second example is the Barret arm solving a maze where the start configuration is depicted in the figure and the goal is marked with a red spot.

The stick moved by the arm has to stay in contact with the maze plane and perpendicular to it. Note, however, that the rotation around the stick is not blocked. This problem is specially challenging due to the obstacles. The third example is a planar mechanism similar to that used in [28] except that here, obstacles are considered since the manipulator has to move a peg attached to the point marked in red out of a cul-de-sac. The goal pose of the end effector is also marked with a red spot. This example is of mixed difficulty with respect to the arrangements of obstacles and to the dimensionality of the configuration space. Finally, the fourth example is a manipulation task with two Stäubli Rx60 industrial arms. In this case, collisions are not considered and the difficulty of the task arises only from the loop closure constraints and from the dimensionality of the configuration space.

Table 1 shows the performance comparison between the HC-planner, the CB-RRT and the AtlasRRT, averaged over 25 runs and for a maximal execution time of 600 s on a Intel Core i7 at 2.93 Ghz running Mac OS X with parameters set to $R = 0.75$, $\varepsilon = 0.5$, $\delta = 0.05$, and $p = 0.9$ for all the experiments. For each benchmark, the table gives the dimensionality of both the configuration space, k, and the ambient space, n. It also provides for each planner, the percentage of success (in the Succ. column), and for the successful runs, the execution times (in the Time column), as well as the number of charts and nodes required (given in the Charts and Nodes columns, respectively).

The results show that the AtlasRRT has always a better (or the same) success ratio than the HC-planner. For low dimensionality configuration spaces the HC-planner, when successful, is significantly fast. However, as the complexity of the obstacles grows, the probability of the HC-planner to fail also increases. The reason behind these failures is that when the HC-planner tries to enter a narrow corridor using too large charts, the atlas extension can get blocked as shown in Fig. 6. In these situations, there is not straight line between the center of the charts at the entrance of the corridor (marked with red dots) and the vertexes defining the frontier of expansion of the atlas. These failures can be avoided using a smaller R, but this will suppose an increase in memory use and a decrease in performance since more charts will be generated for the same problem. In the same situation, AtlasRRT will eventually grow a branch passing through the narrow passage, without adjusting the chart size. As the dimensionality of the configuration space increases, the HC-planner is also less efficient than AtlasRRT, even in problems without obstacles as for the manipulation problem with the two Rx60.

AtlasRRT is faster and at least as reliable as CB-RRT in all cases. Since both methods share the same search strategy, the better performance of the AtlasRRT can be explained by the higher quality of the samples obtained from the atlas. Note that, in general, AtlasRRT generates more nodes with a lower computational cost than CB-RRT since it does not suffer from unfruitful extensions, as when sampling

Table 1 Dimension of the configuration and ambient spaces, success rates, execution times, and number of nodes/charts for the three methods compared in this paper

HC-Planner						CB-RRT			AtlasRRT			
Benchmark	k	n	Succ.	Time	Charts	Succ.	Time	Nodes	Succ.	Time	Charts	Nodes
Torus	2	7	0.76	0.03	77	1.0	26.07	3177	1.0	0.10	57	1456
Barret	4	84	0.36	279.98	2204	0.04	584.66	7563	0.76	356.36	477	14413
Star	5	18	0.68	49.08	2779	1.0	5.43	13989	1.0	4.70	1028	22394
Two Rx60	6	108	1.0	90.77	158	–	–	–	1.0	14.24	20	366

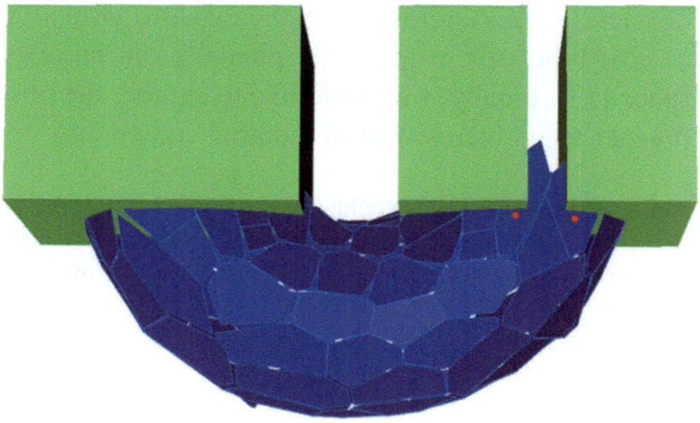

Fig. 6 For a given resolution, the HC-planner can get blocked when trying to enter in a narrow corridor since it only considers motions from the centers of chart (the *red* points in the figure) at the borders of the atlas

in the ambient space. Finally, the advantage of our method is particularly significant in the Barret example where CB-RRT only succeeded once out of 25 attempts and in the manipulation task problem with the two Rx60 where CB-RRT fails in all cases.

5 Conclusions

In this paper, we presented the AtlasRRT algorithm, an approach that uses an atlas to efficiently explore a configuration space manifold implicitly defined by loop closure constraints. The atlas is a collection of charts that locally parametrize the manifold. Samples are thrown uniformly on the charts and, since the error from the chart to the manifold is bounded, the distribution of samples on the manifold is close to be uniform. These samples are then used to efficiently grow a RRT connecting two given configurations and avoiding collisions with obstacles. This strategy contrasts with state of the art approaches that generate samples on the configuration space from uniformly distributed samples in the ambient space.

Since defining the full atlas for a given manifold is an expensive process, the AtlasRRT algorithm intertwines the construction of the atlas and the RRT: the partially constructed atlas is used to sample new configurations for the RRT, and the RRT is used to determine directions of expansion for the atlas. The approach retains the Voronoi exploration bias typical of RRT approaches in the sense that exploration is strongly pushed towards yet unexplored regions of the configuration space manifold.

The results included in this paper shows that our approach is more efficient than existing state of the art approaches. A more thoughtful evaluation must be carried, though, to fully characterize the performance of the new algorithm. Moreover, several modifications to the basic AtlasRRT algorithm can be devised. In particular, it might be useful to exploit the atlas structure to obtain a more meaningful distance between samples than the Euclidean one. Moreover, the presented approach can be combined with existing strategies to improve path planning in the presence of obstacles such as, for instance, the dynamic domain sampling [41] or the integration of cost functions to focus the planning on the relevant parts of the configuration space [13]. Finally, we would like to explore the possible extension of the proposed planner to problems with differential constraints.

Acknowledgments We would like to thank L. Ros for pointing us to the higher-dimension continuation tools and for fruitful discussions during the elaboration of this work. This work has been partially supported by the Spanish Ministry of Science and Innovation under projects DPI2010-18449 and DPI2014-57220-C2-2-P.

References

1. G. Ballantyne, F. Moll, The da Vinci telerobotic surgical system: virtual operative field and telepresence surgery. Surg. Clin. North Am. **83**(6), 1293–1304 (2003)
2. D. Berenson, S.S. Srinivasa, D. Ferguson, J.J. Kuffner, Manipulation planning on constraint manifolds, in *IEEE International Conference on Robotics and Automation*, pp. 1383–1390 (2009)
3. D. Berenson, S.S. Srinivasa, J.J. Kuffner, Task space regions: a framework for pose-constrained manipulation planning. Int. J. Robot. Res. **30**(12), 1435–1460 (2011)
4. H. Choset, K. Lynch, S. Hutchinson, G. Kantor, W. Burgard, L. Kavraki, S. Thrun, *Principles of Robot Motion: Theory, Algorithms, and Implementations* (MIT Press, 2005)
5. J. Cortés, T. Siméon, J.P. Laumond, A random loop generator for planning the motions of closed kinematic chains using PRM methods, in *IEEE International Conference on Robotics and Automation*, pp. 2141–2146 (2002)
6. S. Dalibard, A. Nakhaei, F. Lamiraux, J.P. Laumond, Whole-body task planning for a humanoid robot: a way to integrate collision avoidance, in *IEEE-RAS International Conference on Humanoid Robots*, pp. 355–360 (2009)
7. M. Galassi, et al., GNU Scientific Library Reference Manual. Network Theory Ltd. (2009)
8. L. Han, N.M. Amato, A kinematics-based probabilistic roadmap method for closed chain systems, in *Algorithmic and Computational Robotics—New Directions (WAFR2000)*, pp. 233–246 (2000)
9. L. Han, L. Rudolph, Inverse kinematics for a serial chain with joints under distance constraints, in *Robotics: Science and Systems II*, pp. 177–184 (2006)
10. I. Havoutis, S. Ramamoorthy, Motion synthesis through randomized exploration of submanifolds of configuration spaces, in *RoboCup 2009: Robot Soccer World Cup XIII*. Lecture Notes in Artificial Intelligence, vol. 5949, pp. 92–103 (2009)
11. M.E. Henderson, Multiple parameter continuation: computing implicitly defined k-manifolds. Int. J. Bifurc. Chaos **12**(3), 451–476 (2002)
12. M.E. Henderson, *Numerical Continuation Methods for Dynamical Systems: Path Following and Boundary Value Problems, Chap. Higher-Dimensional Continuation* (Springer, Berlin, 2007)
13. L. Jaillet, J. Cortés, T. Siméon, Sampling-based path planning on configuration-space costmaps. IEEE Trans. Rob. **26**(4), 635–646 (2010)
14. L.E. Kavraki, P. Svestka, J.C. Latombe, M.H. Overmars, Probabilistic roadmaps for path planning in high-dimensional configuration spaces. IEEE Trans. Robot. Autom. **12**, 566–580 (1996)
15. B. Krauskopf, H.M. Osinga, J. Galán-Vioque, Numerical Continuation Methods for Dynamical Systems: Path Following and Boundary Value Problems. Springer (2007)
16. S.M. LaValle, *Planning Algorithms* (Cambridge University Press, New York, 2006)

17. S.M. LaValle, J.J. Kuffner, Rapidly-exploring random trees: Progress and prospects, in *Algorithmic and Computational Robotics—New Directions (WAFR2000)*, pp. 293–308 (2000)
18. R.J. Milgram, J. Trinkle, The geometry of configuration spaces for closed chains in two and three dimensions. Homology, Homotopy Appl. **6**(1), 237–267 (2004)
19. C. Ott, O. Eiberger, W. Friedl, B. Bauml, U. Hillenbrand, C. Borst, A. Albu-Schafer, B. Brunner, H. Hirschmuller, G. Hirzinger, A humanoid two-arm system for dexterous manipulation, in *IEEE-RAS International Conference on Humanoid Robots*, pp. 276–283 (2006)
20. J.M. Porta, L. Jaillet, Path planning on manifolds using randomized higher-dimensional continuation, in *9th International Workshop on the Algorithmic Foundations of Robotics* pp. 337–353 (2010)
21. J.M. Porta, L. Ros, F. Thomas, A linear relaxation technique for the position analysis of multiloop linkages. IEEE Trans. Rob. **25**(2), 225–239 (2009)
22. A. Pressley, *Elementary Differential Geometry* (Springer, 2001)
23. W.C. Rheinboldt, MANPACK: a set of algorithms of computations on implicitly defined manifolds. Comput. Math Appl. **32**(12), 15–28 (1996)
24. A. Rodŕıguez, L. Basañez, E. Celaya, A relational positioning methodology for robot task specification and execution. IEEE Trans. Robot. **24**(3), 600–611 (2008)
25. C. Rosales, L. Ros, J.M. Porta, R. Suárez, Synthesizing grasp configurations with specified contact regions. Int. J. Robot. Res. **30**(4), 431–443 (2011)
26. B. Roth, F. Freudenstein, Synthesis of path-generating mechanisms by numerical methods. ASME J. Eng. Ind. **85**, 298–307 (1963)
27. A. Shkolmik, R. Tedrake, Path planning in 1000 + dimensions using a task-space Voronoi bias, in *IEEE International Conference on Robotics and Automation*, pp. 2892–2898 (2009)
28. N. Shvlab, G. Liu, M. Shoham, J.C. Trinkle, Motion planning for a class of planar closed-chain manipulators. Int. J. Robot. Res. **26**(5), 457–473 (2007)
29. A.J. Sommese, C.W. Wampler, *The Numerical Solution of Systems of Polynomials Arising in Engineering and Science* (World Scientific, 2005)
30. M. Stilman, Task constrained motion planning in robot joint space, in *IEEE/RSJ International Conference on Intelligent Robots and Systems*, pp. 3074–3081 (2007)
31. X. Tang, S. Thomas, P. Coleman, N.M. Amato, Reachable distance space: efficient sampling-based planning for spatially constrained systems. Int. J. Robot. Res. **29**(7), 916–934 (2010)
32. The CUIK project web page, http://www.iri.upc.edu/cuik
33. The SOLID web page, http://www.dtecta.com
34. L.W. Tsai, *Robot Analysis: The Mechanics of Serial and Parallel Manipulators* (Wiley, 1999)
35. T.T. Um, B. Kim, C. Suh, F.C. Park, Tangent space RRT with lazy projection: an efficient planning algorithm for constrained motions, in *Advances in Robot Kinematics*, pp. 251–260 (2010)
36. C.W. Wampler, A. Morgan, Solving the 6R inverse position problem using a generic-case solution methodology. Mech. Mach. Theory **26**(1), 91–106 (1991)
37. W.J. Wedemeyer, H. Scheraga, Exact analytical loop closure in proteins using polynomial equations. J. Comput. Chem. **20**(8), 819–844 (1999)
38. J.H. Yakey, S.M. LaValle, L.E. Kavraki, Randomized path planning for linkages with closed kinematic chains. IEEE Trans. Robot. Autom. **17**(6), 951–959 (2001)
39. F.C. Yang, E.J. Haug, Numerical analysis of the kinematic dexterity of mechanisms. J. Mech. Des. **116**, 119–126 (1994)
40. Z. Yao, K. Gupta, Path planning with general end-effector constraints: Using task space to guide configuration space search, in *IEEE/RSJ International Conference on Intelligent Robots and Systems*, pp. 1875–1880 (2005)
41. A. Yershova, L. Jaillet, T. Siméon, S.M. LaValle, Dynamic-domain RRTs: efficient exploration by controlling the sampling domain, in *IEEE International Conference on Robotics and Automation*, pp. 3856–3861 (2005)

42. A. Yershova, S.M. LaValle, Improving motion planning algorithms by efficient nearest neighbor searching. IEEE Trans. Rob. **23**(1), 151–157 (2007)
43. A. Yershova, S.M. LaValle, Motion planning for highly constrained spaces, in *Robot Motion and Control*. Lecture Notes on Control and Information Sciences, vol. 396, pp. 297–306 (2009)

Decentralized Control for Optimizing Communication with Infeasible Regions

Stephanie Gil, Samuel Prentice, Nicholas Roy and Daniela Rus

Abstract In this paper we present a decentralized gradient-based controller that optimizes communication between mobile aerial vehicles and stationary ground sensor vehicles in an environment with infeasible regions. The formulation of our problem as a MIQP is easily implementable, and we show that the addition of a scaling matrix can improve the range of attainable converged solutions by influencing trajectories to move around infeasible regions. We demonstrate the robustness of the controller in 3D simulation with agent failure, and in 10 trials of a multi-agent hardware experiment with quadrotors and ground sensors in an indoor environment. Lastly, we provide analytical guarantees that our controller strictly minimizes a nonconvex cost along agent trajectories, a desirable property for general multi-agent coordination tasks.

1 Introduction

Decentralized control of robotic systems has enabled complex group behaviors such as rendezvous, formation keeping and coverage to be applied to a wide range of engineering problems; however, the absence of centralized computation increases the demands on communication quality [1–4]. This paper focuses on the problem of optimizing communication quality between a multi-agent network of mobile robots and stationary sensors. In previous work [5] we developed a decentralized gradient-based controller that provably optimizes communication quality amongst

S. Gil (✉) · S. Prentice · N. Roy · D. Rus
Massachusetts Institute of Technology, Cambridge, MA, USA
e-mail: sgil@mit.edu

S. Prentice
e-mail: prentice@mit.edu

N. Roy
e-mail: nickroy@mit.edu

D. Rus
e-mail: rus@mit.edu

H.I. Christensen and O. Khatib (eds.), *Robotics Research*,
Springer Tracts in Advanced Robotics 100, DOI 10.1007/978-3-319-29363-9_21

Fig. 1 Multi-agent field test environment with Ascending Technology Pelican quadrotors (*solid outlines*) and stationary ground sensors (*dashed outlines*). Infeasible regions include the wall in the center of the room, an open staircase that is partially visible on the *right*, and a table (out of view)

the network, but this approach is limited to environments where the entire space is feasible. In practical scenarios, such as the indoor environment shown in Fig. 1, there often exist regions of space that are hazardous or untraversable. Such obstacles make designing the controller difficult for two main reasons: (1) the goal state, or optimal communication configuration, is unknown a priori and (2) the presence of infeasible regions introduce many constrained local minima that may be less satisfactory solutions. This work uses nonlinear optimization techniques to derive a decentralized controller that is easy to implement and addresses the problem of communication optimization in the presence of infeasible regions.

The introduction of infeasible regions raises many challenges. The cost for the communication optimization problem is nonconvex which is a necessary property of many interesting distributed tasks [6]. Therefore we aim for simple-to-implement controllers that have the desired properties of scalability, reliance only on local information, and that descend the cost along the trajectories of each agent. Gradient-based controllers are thus ideally suited. However, the presence of infeasible regions breaks up the free space into several sets over which we must optimize, introducing challenges both for convergence, and for the quality of the converged solution. As an example, an aerial vehicle may get "stuck" behind a wall that it could easily fly around if the direction of steepest descent of the cost happens to be perpendicular to the obstacle edge as illustrated in Fig. 2b, and so we need to consider a wider range of descent directions to avoid these scenarios. As a result of gradient-based optimization over a nonconvex environment, achievable convergence is either to a critical point of the cost in the best case, or to a point of improved cost on the edge of an infeasible region. Our aim is to derive a controller such that agents descend the cost along the generated trajectories, and where these

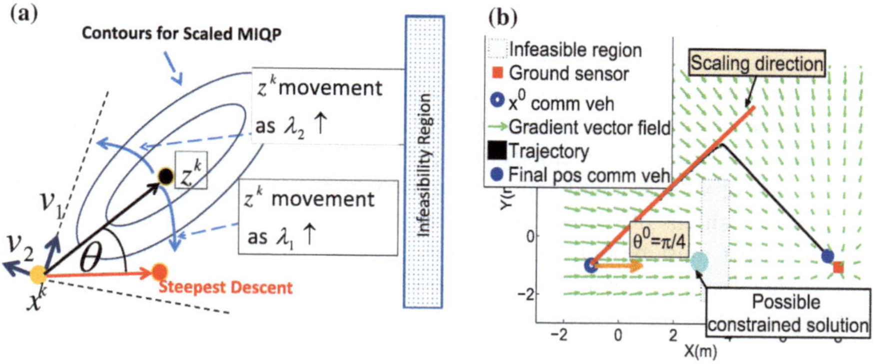

Fig. 2 Schematic showing the scaled direction and heading angle θ and the change of heading direction as the eigenvalues of S^k are changed in Fig. 3a. Simulation of basic scenario showing the utility of scaling to avoid getting stuck behind the wall as in Fig. 3b. Here, a constant scaling matrix $S^k = S$ is used. When the scaled direction reaches a perpendicular angle to the gradient, the trajectory moves along the steepest descent direction as discussed in Sect. 3.2.1

trajectories are biased towards directions that avoid infeasible regions and thus have a larger range of attainable improved cost solutions.

The use of gradient projection methods from nonlinear optimization [7] allows us to formulate our nonconvex problem as a simple quadratic program where the constraint set is a convex subset of the free space in the environment. We use the solution of a mixed integer program to effectively select the convex feasible region over which we optimize our cost and the result is a mixed integer quadratic program (MIQP) that can be solved efficiently for each agent. We show that the addition of a scaling matrix, that preserves the quadratic and thus efficiently solvable attributes of the problem, allows the designer to influence vehicle trajectories to move around infeasible regions and improve the range of attainable converged solutions. In particular, in Sect. 3.2.1 we derive analytical results relating the heading angle to the steepest descent direction for a chosen scaling matrix, and we show that we retain descent of the cost. Theorem 3 shows the existence of a sequence of scaling matrices such that our algorithm produces trajectories reaching unconstrained local minima of the communication cost, if such a trajectory exists for the given initial positions and environment. Although the derivation of a sequence of scaling matrices that guarantees convergence to unconstrained local minima of the cost remains an open question, in the Results Section we provide a heuristic selection of scaling matrices that demonstrates good performance in simulation and hardware experiments.

Section 4.1 presents our communication optimization algorithm and demonstrates the performance of the controller and its robustness in the case of agent failure. Lastly, in Sect. 4.2 we demonstrate our control method on real aerial vehicles that must navigate through an indoor environment (Fig. 1) to optimize communication amongst a network of three stationary ground vehicles.

1.1 Related Work

Artificial potential fields for obstacle avoidance as in [8–10], decomposition of the environment using different notions of a graph through which to search the space [11, 12], and shortest path methods as in [13], represent active areas of research for the problem of vehicle coordination in environments with obstacles. In the current work, final positions of the agents are local minima of the communication cost and since this cost must be optimized iteratively, these local minima are *unknown* a priori. Therefore we cannot assume knowledge of final goal states that we can navigate towards, and we cannot disallow minima from being inside of infeasible regions.

The characteristic that the optimization problem itself defines the agent trajectories makes this problem particularly challenging. The paper [14] also addresses a multi-agent optimization problem but for coverage of a 2D environment and uses a clever mapping inversion. Methods similar in spirit to our work are Mixed Integer Methods as in [15, 16], although these methods are different in that they also consider navigation to known goal states. For our work we must also descend the cost along agent trajectories as convergence to local minima of the cost and maintenance of connectivity for the network hinge on this requirement. Thus a strong motivation for this work is to ensure that descent of the cost is achieved at each iteration. The requirement of provably descending the cost along vehicle trajectories is common for many coordination tasks and thus illustrates the generality of the communication optimization problem to other multi-agent tasks [2–4].

2 Problem Formulation

In previous work [5], we derived a cost function that optimizes communication quality among aerial vehicles and ground sensors. This cost uses a Signal to Interference ratio (SIR) to weigh communication strength of a pair of vehicles against interference from neighboring vehicles and is a weighted sum of two terms where the first term maximizes the SIR of each individual link and the second term equalizes SIR over all links. The resulting behavior of the controller is designed by increasing or decreasing ρ, which is a scalar that assigns more or less weight to the second term in the cost.

The cost $H : \mathbb{R}^{(p \times N)} \to \mathbb{R}$ is defined over all vehicle positions $x_i^k \in \mathbb{R}^p$ for N vehicles at iteration k as:

$$H(x_1^k, \ldots, x_N^k) = \sum_{i}^{N} \sum_{j \neq i}^{N} -SIR_{ij} + \frac{\rho}{SIR_{ij} + \delta} \tag{1}$$

where i and j is shorthand for vehicles with positions x_i and x_j respectively, and δ is an arbitrarily small positive number to allow $SIR_{ij} = 0$. The Signal-to-Interference Ratio $(SIR) : \mathbb{R}^p \rightarrow \mathbb{R}$ is given by $SIR_{ij} = \frac{f_{ij}}{N_i + \sum_{k \in N_{i \setminus j}} f_{ik}}$ and the signal strength between two communicating agents i and j is given by $f_{ij} : \mathbb{R}^p \rightarrow \mathbb{R}$. The signal strength is given by $f_{ij} = \frac{P_0}{d_{ij}^{\beta}}$. All vehicles in the neighborhood of i not including j is denoted $N_{i \setminus j}, P_0 \in \mathbb{R}$ is a given maximum signal strength, β is a given dropoff parameter and often $\beta = 2$, and $d_{ij} = \|x_i - x_j\|$

2.1 Communication Optimization as a MIQP

We wish to move N agents along trajectories that descend the cost $H(\mathbf{x}^k)$ from (1), where $\mathbf{x}^k \in \mathbb{R}^{p\mathrm{x}N}$ is the vector of all vehicle positions at time k, while constraining this trajectory to remain outside of infeasible regions for all time. For each vehicle i with position $\mathbf{x}_i^k \in \mathbb{R}^p$ at iteration k, we wish to move an amount $s^k > 0$ along the direction of steepest descent, $-\nabla H(x^k)$, but we must enforce the constraint to stay within free space. The value $\nabla_i H(x^k) \in \mathbb{R}^p$ is the gradient of the cost H with respect to the position of vehicle x_i at time k and we note that although the cost is global, the derivative $\nabla_i H(x^k)$ depends only on *local* information and is distributed in this sense. We subsequently drop the subscript i to simplify notation so that x^k is the position of vehicle i and $\nabla H(x^k)$ refers to the gradient of H with respect to the position of agent i.

Gradient projection methods from nonlinear optimization allow us to formulate descent for our nonconvex cost while maintaining the constraint of staying a convex set. Because the free space of the environment is almost never convex we must divide the free space set into the intersection of many convex sets, which is possible in particular for environments with convex polygonal infeasible regions which is the case that we consider. We take advantage of the fact that each agent needs to optimize H only over its local environment and employ a mixed integer program to activate a local convex subset of the free space over which we can perform gradient projection. The result is a Mixed Integer Quadratic Program (MIQP):

$$\begin{aligned} &\min_{x,t} \left\| x - (x_i^k - s^k \nabla_i H(x^k)) \right\| \\ &s.t. A_l x \leq b_l + t_l M, \quad \forall l \in \{1, \ldots, L\} \\ &\sum_{j=1}^{E_l} t_{l_j} \leq E_l - 1, \quad \forall l \in \{1, \ldots, L\} \\ &t_{l_j} \in \{0,1\} \; \forall j, l \end{aligned} \tag{2}$$

where s^k is a scalar >0, L is the number of polygonal infeasible regions in the environment, E_l is the number of edges for infeasible region l, M is a sufficiently large scalar, and $t_l \in \mathbb{R}^{E_l}$ is a binary column vector returned by the MIQP for each infeasible region, and $A_l \in \mathbb{R}^{(E_l \times p)}, b_l \in \mathbb{R}^{E_l}$ describe the convex, polygonal, infeasible regions as defined next. We now provide the mathematical descriptions of infeasible regions and free space sets returned as solutions from the MIQP:

Definition 1 (*Infeasible Regions and Free Space Sets*) Infeasible regions are convex, polygonal sets that are the intersection of E_l halfspaces $\bigcap_{i=1}^{E_l} (A_i x \geq b_i)$. A vehicle may not move through an infeasible region but we assume communication strength is not affected. The binary column vectors from (2) encode feasible region constraints and thus a particular solution of binary variables $t^* \in \mathbb{R}^{LE_l}$ effectively “activates” one or more edges of each infeasible region such that these selected edges are the valid constraints enforced in solving the MIQP. The intersection of the halfspaces corresponding to the activated obstacle edges is always a closed and convex set denoted

$$X_{Ft^*} = X_F(t^*) = \{x | A_l x \leq b_l + t_l^* M\} \tag{3}$$

where $X_F(t^*)$ is the closed, convex free space subset corresponding to the binary variable solution t^* of Eq. (2).

The intuition for the formulation in Eq. (2) is that each vehicle moves as far along the direction of steepest descent of the cost H as possible while staying within feasible space. The problem with this formulation however, is that if the direction of steepest descent becomes perpendicular to the edge of an infeasible region then it is possible to get stuck behind this edge even in the case where the vehicle may be able to easily go around the obstacle by moving along a descent direction that is not that of steepest descent, see Fig. 2b. We address this problem in the formulation of the next section.

2.2 *Use of Scaling to Avoid Regions of Infeasibility*

The MIQP formulation from the last section can be solved efficiently using off-the shelf optimizers, and results in a very simple form of a controller but suffers from the limitation of always following the steepest descent direction, even in the case where this direction is obstructed by an infeasible region. Thus, we wish to improve the range of attainable solutions while conserving the simplicity of the MIQP from the previous section. To this aim we propose use of the *scaled* gradient projection method.

In nonlinear optimization theory the scaled gradient projection method is often used to improve rate of convergence [7]. Our objective, however, is to influence the vehicle trajectory towards directions that are not perpendicular to active constraint edges. In addition, the scaled gradient projection method amounts to the addition of a term that is quadratic in the optimization variable x and thus is also a quadratic program as in the previous case and can be easily solved. We define a new problem whose optimization results in a feasible waypoint $\bar{x}_S^k$ for agent i (where i subscripts are dropped):

$$\begin{aligned} \bar{x}_S^k = \underset{x}{\operatorname{argmin}} & \|x - z^k\|_{S^k} \\ \text{s.t.} & \, A_l x \leq b_l + t_l M \; \forall l \\ & \sum_{j=1}^{E_l} t_{l_j} \leq E_l - 1 \; \forall l \in 1, \ldots, L \\ & t_{l_j} \in \{0, 1\} \; \forall j, l \end{aligned} \tag{4}$$

where the matrix $S^k \in \mathbb{R}^{p \times p}$ is a positive definite matrix, and we use the notation $\|q\|_{S^k} = q' S^k q, \quad \forall q \in \mathbb{R}^p$ to represent the scaled norm. For this scaled formulation, the desired waypoint is

$$z^k = x^k - s^k (S^k)^{-1} \nabla H(x^k) \tag{5}$$

A more compact definition of (4) can be written using the representation of the free space set from (3) for each vector of binary variables t^* that solve (4):

$$\bar{x}_S^k = \arg \min_{x \in X_F(t^*)} \|z^k - x\|_{s^k} \tag{6}$$

The position update rule for x^{k+1} is given by:

$$x^{k+1} = x^k + \alpha^k (\bar{x}_S^k - x^k) = x^k + \alpha^k d^k \tag{7}$$

where the stepsizes α^k and s^k satisfy Assumption 1:

Assumption 1 There exist stepsizes $\alpha^k > 0$ and $\mathrm{s}^k > 0$ that are sufficiently small such that given a descent direction d^k, a step along this direction will not intersect an obstacle and will provide sufficient decrease of the cost in the sense of the Armijo, or limited minimization rule that are standard in Nonlinear Programming [7]. We assume that α^k and s^k satisfy these conditions throughout the paper.

A first order Taylor series expansion of the cost around the current point x^k shows that descent of the cost is possible for small enough stepsize along a valid descent direction d^k. In the case that the current iterate is at the edge of an obstacle, the step-size would necessarily be zero to avoid intersecting the obstacle and the method will stop. The requirement that S^k is positive definite is necessary to maintain descent of the cost $H(\mathbf{x}^{k+1}) < H(\mathbf{x}^k)$. In effect, our next waypoint x^{k+1} will minimize distance in the sense of the scaled norm to our desired waypoint z^k [7]. See Fig. 2.

We define the descent direction $d^k = \bar{x}_S^k - x^k$. The advantage is that now we can steer our trajectory to any heading relative to the direction of steepest descent $-\nabla H(x^k)$, as long as this direction satisfies $d^k = \left\{ (\bar{x}_S^k - x^k) | (\bar{x}_S^k - x^k)' \nabla H(x^k) < 0 \right\}$ where $x'y$ is the dot product of a vector x and a vector y, and the achieved relative heading angle θ depicted in Fig. 2 is defined as:

$$\theta = \arccos\left\{ \frac{(-\nabla H(x^k))' d^k}{\|\nabla H(x^k)\| \|d^k\|} \right\} \tag{8}$$

We use this flexibility to assign preference to paths that entirely clear regions of infeasibility that are in the direction of the negative gradient. In Sect. 3 we derive an analytical relationship between S^k and θ.

3 Analysis

3.1 Analysis of the Unscaled Controller

We show that the sequence of vehicle positions produced by the MIQP, in combination with the update rule from (7) produces strict descent directions such that $H(\mathbf{x}^{k+1}) < H(\mathbf{x}^k)$ for all k for stepsizes satisfying Assumption 1. Proving descent of the cost, such that $H(\mathbf{x}^{k+1}) - H(\mathbf{x}^k) < 0$, is made challenging by the general non-uniqueness of the solution for the binary variables t in Eq. (2). This in turn means that the convex subset over which we perform optimization may not be unique and may not contain the current iterate x^k which makes the classical descent proof for gradient projection methods not applicable.

From the result asserting that the cost is reduced at each iteration, and the fact that the local minima of H are finite as shown in previous work, [5], we expect convergence to a fixed point. This fixed point can either be at the edge of an infeasible region where the projection $\bar{x}^k = x^k$ (stationary) as defined in Lemma 1.4 or can be a critical point of the cost H. In the following section we show how scaling can be used to decrease the likelihood of getting "stuck" at the side of an infeasible region.

We use the concept of a vector d being *gradient-related*.

Definition 2 (*Gradient Related*) A bounded direction sequence $\{d^k\}_{k\in K}$ is gradient-related to any subsequence $\{x^k\}_{k\in K}$ that converges to a nonstationary point, if:

$$\lim_{k\to\infty} \sup_{k\in K} \nabla H(x^k)' d(x^k) < 0 \, \forall k \in K.$$

We use the following properties of projection from [7]:

Lemma 1 (Properties of the projection onto a convex set *X*) *Let X be nonempty, closed and convex, and let* $[z]^+$ *denote the projection of* $z \in \mathbb{R}^p$ *onto X:*

1. The projection of $z \in \mathbb{R}^p$ exists, is unique, and minimizes $\|z - x\|$ over $x \in X$.
2. *It must hold that* $(z - [z]^+)'(x - [z]^+) \leq 0, \forall x \in X$
3. *The projection function is continuous.*
4. *We have* $\tilde{x} = [\tilde{x} - s\nabla H(\tilde{x})]^+$ *for all* $s > 0$ *iff* $\tilde{x}$ *is stationary.*

We now seek to show that the d^k produced by the solution to (2) are gradient related for all k and thus for stepsizes satisfying Assumption 1 we have $H(\mathbf{x}^{k+1}) - H(\mathbf{x}^k) < 0$.

Theorem 1 *For the cost H that is differentiable everywhere, the sequence of directions* $\{d^k\}$ *produced by solving the MIQP formulation and using the provided update rule from* (7) *are directions of descent of the cost such that they satisfy the gradient related property for points* x^k *that are not stationary points of the cost.*

Proof The proof is identical to that of Theorem 2 with the scaling matrix set to the identity $S^k = I$. □

3.2 Analysis of the Scaled Gradient Projection Method for Avoidance of Infeasible Regions

We provide analytical results for three main problems related to the scaled version of the MIQP (4). First, we relate the scaling matrix S^k to the relative heading angle θ where this direction is relative to the direction of steepest descent. Second, we show that the use of a scaling matrix generates trajectories for each agent over which the cost is descended at each iteration. Lastly, we show that there exists a sequence $\{S^k\}$ of scaling matrices such that our formulation from (4) generates a trajectory converging to the more desirable *unconstrained* local minima of H if such a trajectory exists for the environment. Although the problem of deriving such a sequence of scaling matrices remains open, in the Results section we provide a heuristic method of generation of scaling matrices that circumvent regions of infeasibility but do not guarantee convergence to unconstrained local minima of H.

3.2.1 Controlling the Relative Heading Angle to Avoid Regions of Infeasibility

In this section we gain insight on how to design the scaling matrix to achieve the desired relative heading angle, θ. From our discussion in Sect. 2.2, we require that the scaling matrix S^k must be a positive definite matrix and with an orthonormal choice of eigenvectors $v_i \in \mathbb{R}^p$ we can write S^k in a decomposed form $S^k = V\Lambda V^T \cdot V = [v_1, \ldots, v_p]$ is a matrix of eigenvectors of S^k and $\Lambda \in \mathbb{R}^{p \times p}$ is a diagonal matrix of eigenvalues $[\lambda_1, \ldots, \lambda_p]$ of S^k. Furthermore we know that that all $\lambda_i > 0$ since S^k is positive definite. Therefore we can write any vector in $\mathbb{R}^p$, in particular the negative gradient vector $-\nabla H(x_k)$, as a linear combination of the v_i's. In particular, we can write $-\nabla H(x_k) = \sum_{i=1}^{p} \zeta_i v_i$ where ζ_i are scalars representing the component of $-\nabla H(x_k)$ in the direction of v_i, and we consider normalized eigenvectors such that $\|v_i\| = 1$. By the Pythagorean theorem, and the fact that the v_i are orthogonal, we have that

$$\|\nabla H(x_k)\|^2 = \sum_{i=1}^{p} (\zeta_i v_i)^T (\zeta_i v_i) = \sum_{i=1}^{p} (\zeta_i)^2 \tag{9}$$

We denote the unprojected heading direction $\tilde{d}^k$, and note that this is $\tilde{d}^k = (z^k - x^k)$ for the scaled gradient projection. From (5) we see that this is simply $s^k (S^k)^{-1} \nabla H(x_k)$. If we again use Pythagorean Theorem to write the expression for $\|\tilde{d}^k\|$, and the dot product $\nabla H(x^k)' \tilde{d}^k$ we get:

$$\|\tilde{d}^k\|^2 = \left\| s^k \sum_{i=1}^{p} \frac{1}{\lambda_i} \zeta_i v_i \right\|^2 = (s_k)^2 \sum_{i=1}^{p} \left(\frac{1}{\lambda_i}\right)^2 \zeta_i^2 \tag{10}$$

$$\nabla H(x^k)' \tilde{d}^k = s_k \sum_{i=1}^{p} \left(\frac{1}{\lambda_i}\right) \zeta_i^2 \tag{11}$$

Using the definition of the dot product and the definitions (10), (11), and (9), we get an expression relating the relative heading angle θ to the scaling matrix S^k via its eigenvectors and eigenvalues:

$$\cos(\theta) = \frac{(-\nabla H(x_k))' \tilde{d}^k}{\|\nabla H(x_k)\| \|\tilde{d}^k\|} = \frac{\sum_{i=1}^{p} \left(\frac{1}{\lambda_i}\right) \zeta_i^2}{\sqrt{\left(\sum_{i=1}^{p} \left(\frac{1}{\lambda_i}\right)^2 \zeta_i^2\right)\left(\sum_{i=1}^{p} (\zeta_i)^2\right)}} \tag{12}$$

This expression shows that the scaling matrix can be designed to achieve a specific relative heading angle by careful choice of its eigenvectors and eigenvalues. In particular we notice that if $S^k = I$ where I is the identity matrix and $\lambda_i = 1 \forall i$, then $\cos(\theta) = 1$, the heading angle is zero and we move in the direction of steepest descent as expected. Alternatively, putting a larger weight on the eigenvalues λ_i of S^k such that $\lambda_j \gg \lambda_i$, $\forall j \neq i$, will achieve the effect of causing the heading direction $\tilde{d}^k$ to align itself most with the component of $-\nabla H(x^k)$ along v_i See Fig. 2 for a schematic of a two dimensional case. However, as the ratio of λ_i gets larger, rate of convergence becomes slower so in general the heading angle θ should not be made larger than necessary.

Lastly, a result of (12) is that the direction $\tilde{d}^k$ can never be perpendicular to the negative gradient for a positive definite scaling matrix S^k. As the eigenvector v_i approaches the perpendicular direction to $-\nabla H(x^k)$, the component of the negative gradient vector along this direction $\zeta_i \to 0$ and thus, as seen from Eq. (10), $\tilde{d}^k$ cannot be made to move in a direction that is perpendicular to $-\nabla H(x^k)$.

3.2.2 Analysis for the Scaled Gradient Projection Method

In the last section we showed that the addition of a scaling matrix S^k allows us the flexibility to design the relative heading angle to avoid regions of infeasibility in the environment. We now show that the resulting directions $d^k = \bar{x}^k_S - x^k$ where $\bar{x}^k_S$ is solved for from Eq. (4), are descent directions such that the cost is reduced over agent trajectories. As in the unscaled case, finding descent directions d^k for our problem is made challenging due to the general non-uniqueness of the binary t variables in (4). From the definition of gradient relatedness, the desired property we wish to show is that $\nabla H(x^k)'d^k < 0$ for all k and all solutions d^k at iteration k. The intuition for our proof method is to use the solution d^k_1 which is defined for the convex subset containing the current iterate and which can be shown to always be gradient related, to bound all other solutions d^k_t that result from different solutions for the binary variables. From here we can show that $\nabla H(x^k)'d^k < 0$ for all d^k_t and k. Using the result of d^k being gradient related for all k, combined with Assumption 1, we get descent of the cost at each iteration such that $H(\mathbf{x}^{k+1}) < H(\mathbf{x}^k)$.

To avoid cumbersome notation, we subsequently drop the S subscript from $\bar{x}^k_S$ and the reader should assume all projections $\bar{x}^k$ in this section are scaled projections. We refer to a point x^k as not stationary if it is not equal to its projection such that $\bar{x}^k \neq x^k$.

Theorem 2 *For the cost $H : \mathbb{R}^{(p\times N)} \to \mathbb{R}$ that is differentiable everywhere, denoting the sequence of vehicle positions $\{x^k\}$ produced by solving the scaled MIQP formulation in (4) and using the provided update rule from (7), we have that all directions d^k are directions of descent of the cost such that they satisfy the gradient related property for all x^k not stationary.*

Proof We denote set containing the current iterate x^k as X_{F_1}, and the projection of z^k onto X_{F_1} as $\bar{x}^k_1$ and note that this is a solution of the scaled MIQP (6) over the set. From Lemma 1.2 generalized to scaled projections, it holds that $\left(z^k - \bar{x}^k_1\right)' S^k \left(x^k - \bar{x}^k_1\right) \le 0$ for $x^k \in X_{F_1}$ and $\bar{x}^k_1 \in X_{F_1}$. Expanding out this property and using the definition of z^k, continuity of the projection, and the fact that we are considering projection onto a single set X_F containing the current iterate x^k, we have that for all k (see [7]):

$$s^k \nabla H(x^k)'(\bar{x}_1^k - x^k) + \|\bar{x}_1^k - x^k\|_{S^k}^2 \leq 0,\ \forall k \tag{13}$$

□

The term $\|x^k - \bar{x}_1^k\|_{S^k}^2 > 0$ for all S^k positive definite and x^k, nonstationary such that $x^k \neq \bar{x}_1^k$. Thus from (13), and Lemma 1.4, we have:

$$\|x^k - \bar{x}_1^k\|_{S^k}^2 > 0 \Rightarrow s^k \nabla H(x^k)'(\bar{x}_1^k - x^k) < 0 \tag{14}$$

So that the direction d_t^k is always gradient related for x^k and $\bar{x}_1^k$ in the same convex subset, where x^k is non-stationary. We will use this inequality again later. We aim to prove that all directions $d_t^k = \bar{x}_1^k - x^k$ produced from the solutions of Eq. (4)

$$\nabla H(x^k)'(\bar{x}_t^k - x^k) < 0\, \forall k \tag{15}$$

Because $\bar{x}_1^k$ is a valid solution to the projection of z^k onto X_{F_1}, we know that any solution, $\bar{x}_1^k$, to the scaled MIQP in (4) must be within the elliptical set defined by $\bar{x}_1^k$:

$$\mathscr{E}_1 = \left\{c | (c - z^k)' S^k (c - z^k) \leq r_S\right\},\ r_S := (\bar{x}_1^k - z_k) S^k (\bar{x}_1^k - z^k) \tag{16}$$

Now we can write the gradient related condition that we wish to prove as:

$$-f^* = \min_{\bar{x}_t} (-\nabla H(x^k))'(\bar{x}_t - x^k) \tag{17}$$

$s.\ t.\ \bar{x}_t \in \mathscr{E}_1$

where our desired condition is that $-f^* > 0$ which ensures that the direction $(\bar{x}_t - x^k)$ is gradient related. The minimization problem written above is well-posed in that f is a continuous function minimized over a compact set $\mathscr{E}_1$ and thus there exists a minimum. Furthermore, this problem can be solved in closed form using Lagrange multipliers to yield the condition:

$$-f^* = -r_S + \|s^k \nabla H(x^k)\|_{S^{k-1}}^2 > 0 \tag{18}$$

If we take this a step further and substitute in the definition for r_S from (16), multiply through by (-1), expand, and simplify we get a new form for the inequality condition that we wish to prove:

$$2s^k \nabla H(x^k)'(\bar{x}_1^k - x^k) + (\bar{x}_1^k - x^k)' S^k (\bar{x}_1^k - x^k) < 0\, \forall k \tag{19}$$

We compare to the condition (14). From the reasoning shown in (14), we know that $2s^k\nabla H(x^k)'(\bar{x}_1^k - x^k) < 0$ for $\bar{x}_1^k \neq x^k$ which is true by the nonstationary assumption. Thus we have that this desired inequality always holds and all produced d^k are descent directions as desired and this completes the proof.

To gain more intuition notice that the condition in Eq. (19) is equivalent to requiring that $r_S = (\bar{x}_1^k - z_k)S^k(\bar{x}_1^k - z^k) < r_{S_{cr}} = \|s^k\nabla H(x^k)\|^2_{S^{k-1}}$. Intuitively what this means is that $\bar{x}_1^k$ is a valid projection of the desired waypoint z^k where the distance to z^k is smaller from $\bar{x}_1^k$ than from x^k in the scaled norm sense, such that $(\bar{x}_1^k - z^k)S^k(\bar{x}_1^k - z^k) < (x^k - z^k)S^k(x^k - z^k)$.

Because we attain descent of the cost at each iteration, and we are optimizing a continuous function over a compact set so that minima are well defined as shown in [5], we therefore expect convergence to a fixed point. This point can be at the edge of an infeasible region or at a critical point of the cost, although the use of scaling aims to circumvent those infeasible regions which do not contain local minima in their interiors.

3.3 *Existence of Optimal Sequence of Scaling Matrices*

Because we optimize a nonconvex cost, we target convergence to local minima. For the case where these local minima are reachable in feasible space, we consider a sequence of scaling matrices $\{S^k\}$ to be "optimal" if the controller resulting from using Algorithm 1 generates trajectories for all vehicles that converge to an unconstrained local minimum of H. The existence problem is to assert that if there exists such a trajectory for the given environment, then there also exists a sequence of scaling matrices such that the trajectory generated by Algorithm 1 is optimal. We do not find such a sequence, this remains an interesting open question. Instead, we prove the positive result for the existence problem.

Theorem 3 *If* $\exists\{g^k\} \rightarrow x^*_{unc}$, *where* $\{g^k\}$ *is a valid sequence of waypoints for each vehicle that converges to an unconstrained local minimum,* x^*_{unc}, *of* H *given* $\mathbf{x}^0$, *then* $\exists\{S^k\}$ *s.t.* $\{x^k\} \rightarrow x^*_{unc}$, *where* $\{x^k\}$ *is the trajectory sequence generated by using Algorithm 1 for each vehicle. A sequence* $\{g^k\}$ *is valid if* $H\{g^{k+1}\} - H\{g^k\} < 0$ *for all k,* $g^k \in X_F$, $\forall k$, $\forall k$ *where* X_F *is the entire feasible region of the environment, and the stepsize between any consecutive points* g^k, g^{k+1} *satisfies Assumption 1 and physical vehicle limits.*

Proof From Proposition 2 we must satisfy $-\nabla H(x^k)'d^k > 0$ for all k. From (11) we see that $\lambda_i > 0$ in order to satisfy this condition. We can write $g^{k+1} - g^k = \sum_{i=1}^{p} a_i v_i$ for some appropriate a_i since $g^{k+1} - g^k \in \mathbb{R}^p$ and the eigenvectors of S^k span $\mathbb{R}^p$. Since by the descent requirement on $\{g^k\}$ we have $-\nabla H(x^k)'(g^{k+1} - g^k) > 0$, $\forall k$ we can choose orthonormal basis vectors v_i of S^k such that $a_i > 0$ and $\zeta_i > 0$ for all i, where ζ_i are from (9) and thus the choice of $\lambda_i = \frac{s^k}{a_i}\zeta_i$ satisfies $\lambda_i > 0$, $\forall i$ and from the definition of d_k from (10) we see that we can always achieve $d_k = g^{k+1} - g^k, \forall k$ for this choice of λ Since S_k is fully determined through its eigenvectors and eigenvalues as $S_k = V\Lambda V'$ and we have shown that there exists a sequence $\{S_k\}$ for which $\{d_k\} = \{g^{k+1} - g^k\}$ and thus the resulting sequence of agent positions $\{x_k\}$ reaches the unconstrained local minimum of H if $\{g_k\}$ reaches the unconstrained local minimum from given initial positions. □

4 Results

4.1 Algorithm and Simulation Example

In this section we summarize our control method in Algorithm 1 and suggest a heuristic method for choosing an appropriate scaling matrix S^k for each vehicle. We demonstrate our algorithm and the suggested method for finding S^k via a Matlab simulation for four communication vehicles and eight ground sensor vehicles in three-dimensional space (Fig. 3).

Algorithm 1 Decentralized Control for Optimized Comms (for agent i)

```
x^k = x^0, k = 0.
while k == 0 OR |x^{k+1} - x^k| ≥ tol do
    k ← k+1
    {Compute scaling S^k using environment topology, see Algorithm 2.}
    {Compute gradient using neighbors of agent i:} ∇_i H(x^k)
    {Compute desired waypoint:} z^k ← x^k − s^k (S^k)^{-1} ∇_i H(x^k)
    {Compute: } x̄_S^k ←soln to (4)
    {Compute feasible stepsize α^k satisfying Assumption 1.}
    {Compute new point x^{k+1} for agent i using stepsize α^k:} x^{k+1} = x^k − α^k (x̄_S^k − x^k).
end while
```

4.1.1 Heuristic Selection of Scaling Matrix S^k

We suggest one possible method for choosing a scaling matrix S^k for each vehicle that is easily implemented and relies solely on map topology that is *local* to each agent. We show via simulation, the performance of the resulting optimization and its adaptive capabilities in the case of agent failures. For each agent, we draw a line along the direction of steepest descent which is plotted as a blue line in Fig. 3a, call this line gL. Let $\mathcal{O}$ be the first infeasible region intersected by gL. We wish to compute S^k such that we move around $\mathcal{O}$, so we compute the projection of the intersection point onto each of the L edges of $\mathcal{O}$ and choose the point such that the

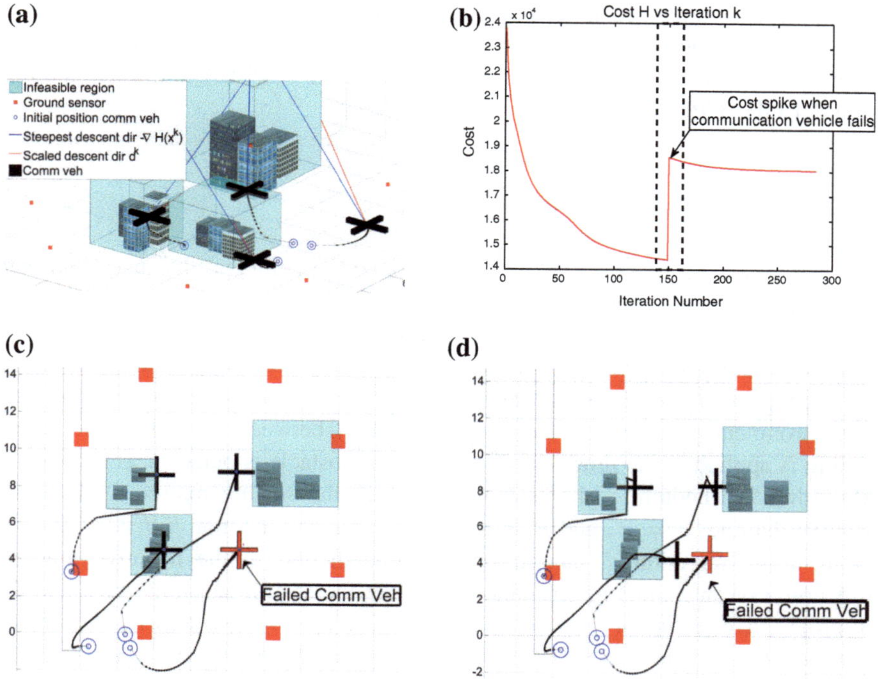

Fig. 3 **a** Scenario showing infeasible regions and 4 communication vehicles and 8 ground sensors. **b–d** Adaptive behavior when one communication vehicle fails (*red* quadrotor): remaining vehicles change trajectories to compensate. **b** Cost is always decreased along agent trajectories

chord from the current position x^k to the edge point e^* has the largest dot product $-\nabla H(x^k)'(e^* - x^k)$. This represents a direction that is as close to the direction of steepest descent as possible but that circumvents the infeasible region obstructing this direction. This chord is plotted in red for each agent whose steepest descent direction intersects an infeasible region in Fig. 3a. We use this chord to compute the first eigenvector of S^k so that $v_1 = (e^* - x^k)/\left\|(e^* - x^k)\right\|$, then v_2 and v_3 are simply any other unit vectors that are othornormal to each other and to v_1. Finally, we can set the eigenvalues $\lambda_2, \lambda_3 >> \lambda_1$ to attain a direction d^k closest to the v_1 direction, see discussion in Sect. 3.2.1 and Fig. 3a. We warn however that choosing the eigenvalue ratios too large will inversely effect convergence rate and thus this should not, in practice, be made larger than necessary. The matrix S^k is then computed via its eigenvectors and eigenvalues as $S^k = V\Lambda V'$ where V and Λ are defined in Sect. 3.2.1. If there is no infeasible region obstructing the direction gL for that vehicle, or there exists no such edge point e^* so that the dot product $-\nabla H(x^k)'(e^* - x^k) > 0$ (this is the case where no circumventing direction produces descent in the cost H), we simply set $S^k = I$, where I is the identity matrix. This algorithm is summarized in 2. For the simulation in Fig. 3 we set $\lambda_1 = 1, \lambda_2 = 50, \lambda_3 = 50$ and achieved satisfactory convergence in an average of 150 iterations where each iteration took on the order of 0.7 s using the CPLEX for Matlab toolbox on a 2.4 GHz CPU laptop.

4.1.2 Discussion on When to Use Scaling

The use of scaling is most effective when applied at sufficient distance from path obstructing infeasible regions. Since any descent direction d^k must be less than perpendicular to the negative gradient direction, as a vehicle gets closer to the edge of an obstructing infeasible region, the range of descent directions that can clear the obstructing region becomes smaller. Therefore we expect scaling to perform better in environments where there are larger distances between obstacles, and where scaling is applied at the time that an obstructing obstacle is detected as outlined in Algorithm 2. Theorem 2 shows that as long as the scaling matrix S^k is strictly positive definite, $\bar{x}^k \neq x^k$, and x^k is not a critical point such that $\nabla H(x^k) \neq 0$, then the resulting direction d^k can never be perpendicular to the negative gradient direction. For an intuitive explanation, consider the two dimensional case and the un-projected direction $\tilde{d}^k$ from (10). As one of the eigenvectors of S^k, say v_1, becomes perpendicular to $-\nabla H(x^k)$, the component of the negative gradient in the direction of v_1 approaches zero, $\varsigma_1 \to 0$ and $-\nabla H(x^k) \to \varsigma_2 v_2$. Therefore the direction $\tilde{d}^k = s^k(\frac{1}{\lambda_1}\varsigma_1 v_1 + \frac{1}{\lambda_2}\varsigma_2 v_2) \to s^k \frac{1}{\lambda_2}\varsigma_2 v_2$ which is exactly the negative gradient direction scaled by $s^k \frac{1}{\lambda_2}$. This means that even if scaling is applied incorrectly (almost perpendicular to the negative gradient), the resulting direction can never be perpendicular and in fact will align with the negative gradient direction, although, if λ_2 is a very large number it is seen that progress along this direction becomes very slow and convergence rate suffers as discussed in Sect. 3.2.1. Also if the current position is at a stationary point where the projection $\bar{x}^k$ is equal to x^k which may occur at the side of an obstacle, or at a critical point of the cost where $-\nabla H(x^k) = 0$, the resulting direction is zero even if nonzero scaling is applied. This can be seen easily from the update equation $x^{k+1} = x^k + \alpha^k d^k$ where $d^k = (\bar{x}^k - x^k)$ which is zero if x^k is stationary, or in free space $d^k = -s^k(S^k)^{-1}\nabla H(x^k) = 0$ at a critical point where $\nabla H(x^k) = 0$. Therefore the observations that (1) d^k can never be perpendicular to the direction of steepest descent (and actually approaches the steepest descent direction if scaling is applied perpendicular to the negative gradient), and (2) that the direction d^k is zero such that the method stops at stationary points or critical points even for positive scaling $S^k \neq I$, and finally that (3) scaling is more effective when applied at larger distances from path obstructing infeasible regions, motivate our recommendation of applying scaling for any path obstructing infeasible region within the vehicle sensing radius.

Algorithm 2 Heuristic Selection of Scaling Matrix S^k in 3D Using Local Information (for agent i)

{Define D: Sensing radius within which infeas. regions can be detected for vehicle i.}
{Define $\mathcal{O}$: closest infeas. region in steepest descent direction. }
{Define $rot(\pi/2)$: Rotation matrix by $\pi/2$.}
{Compute a line in direction of negative grad: }$gL = x^k - D\nabla H(x^k)$.
{ $ip \leftarrow$ intersection point of gL with closest face of infeasible region ($\mathcal{O}$)}
if $ip \neq \{0\}$ **then**
  {$EP \leftarrow$ all points on edges of $\mathcal{O}$ with smallest distance from ip}
  $e^* = \max_{e \in EP}(e - x^k)'gL$
  {Compute first orthonormal eigvec of S^k:} $v_1 \leftarrow (e^* - x^k)/(\|e^* - x^k\|)$
  $v_2 \leftarrow rot(\pi/2) * v_1$
  $v_3 \leftarrow (v_1 \times v_2)/(\|v_1 \times v_2\|)$
  $V \leftarrow [v_1\ v_2\ v_3]$
  {Set $\lambda_1 << \lambda_2, \lambda_3$ as discussed in Section 3.2.1}
  $\Lambda \leftarrow diag(\lambda_1, \lambda_2, \lambda_3)$
  $S^k \leftarrow V \Lambda V^T$
else
  {No obstacles in steepest descent direction, set scaling to identity: $S^k \leftarrow I$}
end if

4.2 Hardware Experiments

The algorithm was validated in a decentralized hardware experiment with two mobile quadrotor helicopters and three stationary ground sensors. This evaluation was performed in a known GPS-denied indoor environment with obstacles (the second floor atrium in the Stata Center at MIT). The hardware platform consisted of Ascending Technologies Pelican quadrotors,[1] each outfitted with a Hokuyo[2] UTM-30LX laser range-finder and 1.6 GHz Intel Atom processor (for details see [17]). Each vehicle performs onboard state estimation and control enabling completely autonomous flight. For practical purposes, each quadrotor communicates via WiFi with a corresponding ground station laptop, where human input and planning processes are run. The communication channel between the mobile and ground sensors is simulated. The environment and vehicles are shown in Fig. 1.

Ten trials were run, each starting at the initial configuration shown in Fig. 4 (labeled (x_{V1}^0, x_{V2}^0) for vehicles 1 and 2, respectively). The obstacle positions are overlayed on the gridmap in pink, and a solid outline denotes the configuration space boundaries, or *infeasible* regions. These regions do not impede communication; rather, they represent unsafe or untraversable regions. In this environment these obstacles were an open staircase, a thin wall, and a table. The quadrotors share

[1]Ascending Technologies GmbH. http://www.asctec.de.
[2]Hokuyo UTM-30LX Laser. http://www.hokuyo-aut.jp.

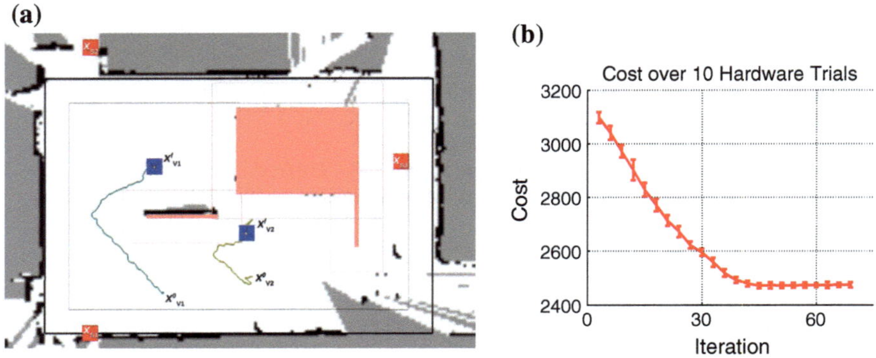

Fig. 4 **a** Overhead view of field test scenario ($\sim 11\,\mathrm{m} \times 7.5\,\mathrm{m}$). Obstacles (*pink*) and configuration space boundaries (*solid lines*) overlay a gridmap of the environment. Stationary ground sensors (x_{S1}, x_{S2}, x_{S3}) are shown as *red squares*. The 2 quadrotor trajectories are shown in *teal* and *yellow*, with initial positions (x^0_{V1}, x^0_{V2}) and final positions (x^f_{V1}, x^f_{V2}) highlighted by *blue squares*. **b** Average trial cost and standard deviation averaged over 10 trials

real-time pose information and at each control iteration 10 Hz compute their next waypoint according to Algorithm 1. The control commands were artificially throttled at 1 Hz by the waypoint executor. Figure 4 shows the trajectory of each vehicle during one trial, and the resulting local minima configuration to which they converge. Note that vehicle 1 moved around the wall. Vehicle 2 initially moved towards the wall, then converged to a point along the obstacle boundary distributed between sensors 1 and 3 and vehicle 1. The average duration over all trials was 65 s until convergence.

Video footage: http://groups.csail.mit.edu/drl/wiki/images/7/74/ISRR_v3d.mp4 or https://youtu.be/h-5Oiz7SM7o.

5 Discussion

We have presented a method for communication optimization in a heterogeneous network of aerial and ground vehicles in an environment with infeasible regions using the communication cost function from previous work [5]. We pursue extension to the general nonsmooth case, and study of the effect of obstacles on communication strength in future work. We have demonstrated both analytically and through simulation and hardware experiments, the utility of using a sequence of scaling matrices to improve the range of converged solutions by moving along trajectories that avoid infeasible regions.

Acknowledgements The authors acknowledge the MAST Project under ARL Grant W911NF-08-2-0004, the SMART Future Urban Mobility Project, and the NSFGRFP.

References

1. A. Jadbabaie, J. Lin, A.S. Morse, Coordination of groups of mobile autonomous agents using nearest neighbor rules. IEEE Trans. Autom. Control **48**, 988–1001 (2003)
2. J. Cortes, S. Martinez, T. Karatas, F. Bullo. Coverage control for mobile sensing networks. IEEE Trans. Robot. Autom. **20**, 243–255 (2004)
3. S. Martinez, J. Cortes, F. Bullo, Motion coordination with distributed information. IEEE Control Syst. Mag. (2007)
4. M Schwager, B Julian, D Rus, Optimal coverage for multiple hovering robots with downward-facing cameras, in *Proceedings of ICRA* (2009)
5. S. Gil, M. Schwager, B. Julian, D. Rus. Optimizing communication in air-ground robot networks using decentralized control, in *Proceedings of ICRA* (2010)
6. M. Schwager, J. J. Slotine, and D. Rus, Unifying geometric, probabilistic, and potential field approaches to multi-robot coverage control, in *Proceedings of ISRR* (2009)
7. Dimitri Bertsekas, *Nonlinear Programming* (Athena Scientific, Belmont, MA, 2008)
8. O. Khatib, Real-time obstacle avoidance for manipulators and mobile robots. Int. J. Robot. Res. **5** (1986)
9. C.W. Warren, Multiple robot path coordination using artificial potential fields, in *IEEE Internal Conference on Robotics and Automation* (1990)
10. S.S. Ge, Y.J. Cui, Dynamic motion planning for mobile robots using potential field method. Auton. Robots **13**, 207–222 (2002)
11. N. Ayanian, V. Kumar, Decentralized feedback controllers for multiagent teams in environments with obstacles. IEEE Trans. Robotics **26** (2010)
12. J. Cores, A. Ganguli, F. Bullo, Distributed deployment of asynchronous guards in art galleries, in *Proceedings of ACC* (2006)
13. N. Motee, A. Jadbabaie, G. Pappas, Path planning for multiple robots: an alternative duality approach, in *American Control Conference (ACC)* (2010)
14. C. Caicedo, M. Zefran, A coverage algorithm for a class of non-convex regions, in *Proceedings of CDC* (2008)
15. E. Stump, A. Jadbabaie, V. Kumar. Connectivity management in mobile robot teams, in *IEEE International Conference on Robotics and Automation*, pp. 1525–1530 (2008)
16. T. Schouwenaars, B. DeMoor, E. Feron, J. How, Mixed integer programming for multi-vehicle path planning, in *European Control Conference* (2001)
17. A. Bachrach, A. de Winter, R. He, G. Hemann, S. Prentice, N. Roy, *RANGE—Robust autonomous navigation in gps-denied environments* (In Proc, ICRA, 2010)

Pre-image Backchaining in Belief Space for Mobile Manipulation

Leslie Pack Kaelbling and Tomás Lozano-Pérez

Abstract There have been several recent approaches to planning and control in uncertain domains, based on online planning in a determinized approximation of the belief-space dynamics, and replanning when the actual belief state diverges from the predicted one. In this work, we extend this approach to planning for mobile manipulation tasks with very long horizons, using a hierarchical combination of logical and geometric representations. We present a novel approach to belief-space preimage backchaining with logical representations, an efficient method for on-line execution monitoring and replanning, and preliminary results on mobile manipulation tasks.

1 Introduction

As robots become more physically robust and capable of sophisticated sensing, navigation, and manipulation, we want them to carry out increasingly complex tasks. A robot that helps in a household must plan over the scale of hours or days, considering abstract features such as the desires of the occupants of the house, as well as detailed geometric models that support locating and manipulating objects. The complexity of such tasks derives from very long time horizons, large numbers of objects to be considered and manipulated, and fundamental uncertainty about properties and locations of those objects.

We have developed an approach to integrated task and motion planning that integrates geometric and symbolic representations in an aggressively hierarchical planning architecture, called HPN [10]. The hierarchical decomposition allows efficient solution of problems with very long horizons and the symbolic representations support abstraction in complex domains with large numbers of objects and are integrated effectively with the detailed geometric models that support motion

L.P. Kaelbling (✉) · T. Lozano-Pérez
CSAIL, MIT, Cambridge, MA 02139, USA
e-mail: lpk@csail.mit.edu

T. Lozano-Pérez
e-mail: tlp@csail.mit.edu

H.I. Christensen and O. Khatib (eds.), *Robotics Research*,
Springer Tracts in Advanced Robotics 100, DOI 10.1007/978-3-319-29363-9_22

planning. In this paper, we extend the HPN approach to handle two types of uncertainty: *future-state* uncertainty about what the outcome of an action will be, and *current state* uncertainty about what the current state actually is. Future-state uncertainty is handled by planning in approximate deterministic models, performing careful execution monitoring, and replanning when necessary. Current-state uncertainty is handled by planning in *belief space*: the space of sets or probability distributions over possible underlying world states.

There have been several recent approaches to integrated task and motion planning [3, 13, 14] but they do not address uncertainty. The use of belief space (and information spaces [11]) is central to decision-theoretic approaches to uncertain domains [9].

2 Hierarchical Pre-image Backchaining

Most planning problems require time on the order of the minimum of: $|S|$ and $|A^h$, where S is the size of the state space, $|A|$ is the size of the action space and h is the horizon (the length of the solution). Our approach to making planning tractable is to construct a temporal hierarchy of short-horizon problems, thus reducing the complexity of the individual planning problems we have to solve. The hierarchical approach will not always produce optimal plans; it is, however, complete in domains for which the goal is reachable from every state.

We formalize the effects of the robot's actions in a hierarchy of increasingly abstract operator descriptions; this hierarchy is constructed by postponing the consideration of preconditions of an action until more concrete levels of abstraction. Figure 1 shows part of a hierarchical plan for washing an object and putting it into storage. The operations are first considered abstractly, making a two-step, high-level plan. Then, the plan to wash the object is elaborated into two steps, of placing the object into the washer and then washing it. At the next level of abstraction down, we plan first to pick the object, and then to place it in its destination.

Even short-horizon plans are difficult to find when the branching factor is high. Searching forward from an initial, completely detailed, state description typically has a very high (or infinite) branching factor, and it can be difficult to prune actions heuristically and maintain the effectiveness of the planner. We address this problem by using *pre-image backchaining* [12], also known as *goal regression* [21], to search backward from the goal description. We presume that the goal has a simple abstract description, for example, requiring *can A* to be in *cupboard B*. That description denotes an infinite set of possible world states (varying according to the exact position of *can A*, the robot's pose, the locations and states of other objects in the world, the weather, etc.). Chaining backward from the goal, using symbolic operator descriptions together with the goal-driven construction of geometric preconditions, allows planning to proceed with a relatively low branching factor. Plan 10 in Fig. 1 shows a case in which, in order to place an object in storage, the

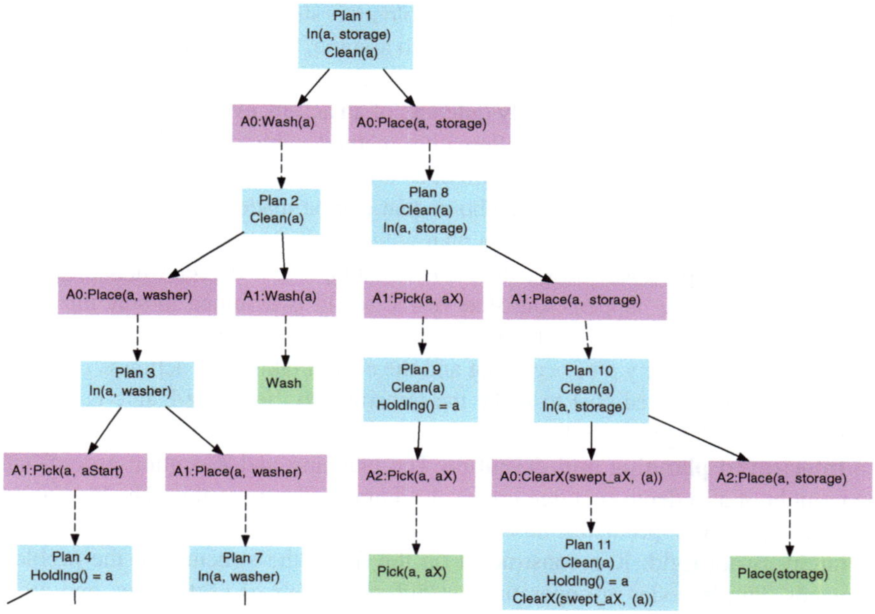

Fig. 1 Example of part of a hierarchical plan and execution tree

planner determines that it must first clear out the region of space through which the object must be moved, and then place the object. A geometric description of that region of space was computed on demand to specify the pre-image of placing the object into storage.

In the next two sections, we describe how to extend the HPN approach to handle uncertainty. First, we consider the case in which the robot knows its current state, but in which the world dynamics are stochastic. Then, we extend the method to the case in which there is uncertainty about the current state of the world.

3 Future-State Uncertainty

The decision-theoretic optimal approach to planning in domains with probabilistic dynamics is to make a *conditional plan*, in the form of a tree, supplying an action to take in response to any possible outcome of a preceding action [22]. For efficiency and robustness, our approach to stochastic dynamics is to construct a deterministic approximation of the dynamics, use the approximate dynamics to build a plan, execute the plan while perceptually monitoring the world for deviations from the expected outcomes of the actions, and replan when deviations occur. This method has worked well in control applications [4, 6, 15, 20] as well as symbolic planning domains [23].

Determinization There are several potential strategies for constructing a determinized model. A popular approach is to assume, for the purposes of planning, that the most likely outcome is the one that will actually occur. This method can never take advantage of a less-likely outcome of an action, even if it is the only way to achieve a goal. We pursue an alternative method, which considers all possible outcomes, but rather than modeling them as a randomized choice that is made by nature, instead modeling them as a choice that can be made by the agent [1]. This method integrates the desire to have a plan with a high success probability with the desire to have a plan with low action cost by adopting a model where, when an undesirable outcome happens, the state of the world is assumed to stay the same, allowing the robot to repeat that action until it has the desired result. If the desired outcome has probability p of occurring and the cost of taking the action is c, then in this model the expected cost to make the transition to the desired state is c/p. We will search for the plan that has the least cost under this model.

Interleaved planning and execution The planning and execution process can be thought of as a depth-first tree traversal. Figure 1 illustrates this: The blue nodes represent planning problems at different levels of abstraction. The abstraction hierarchy is not rigid: it is constructed on the fly as the structure of the problem demands. Purple nodes are operations at different levels of abstraction, and green nodes are primitive actions. Note that, for instance, the primitive actions in the sub-tree for plan 2 are executed before plan 8 is constructed. This online, inter-leaved planning allows the details of the construction of plan 8 to depend on the concrete state of the world that came about as a result of the recursive planning and execution of plan 2.

Assume we have a PLAN procedure, which takes as arguments *state*, the current world state, *goal*, a description of the goal set and *abs* a description of the level of abstraction at which planning should take place; it additionally depends on a set of operator descriptions that describe the domain dynamics. The PLAN procedure returns a list $((-, g_0), (a_1, g_1), \ldots, (a_n, g_n))$ where the a_i are operator instances, $g_n = goal$, g_i is the pre-image of g_{i+1} under a_i, and $state \in g_0$. The pre-image g_i is the set of world states such that if the world is in some state g_i and action a_{i+1} is executed, then the next world state will be in g_{i+1}; these pre-images are subgoals that serve as the goals for the planning problems at the next level down in the hierarchy.

The HPN process is invoked by HPN (*state, goal, abs, world*), where *state* is a description of the current state of world; *goal* is a set of world states; *abs* is a structure that specifies, for any goal condition, the number of times it has served as a plan step in the HPN call stack above it; and *world* is an actual robot or a simulator in which primitive actions can be executed. In the prototype system described in this paper, *world* is actually a geometric motion planner coupled with a simulated or physical robot.

HPN(*state*, *goal*, *abs*, *world*):
  p = PLAN(*state*, *goal*, *abs*)
  for (a_i, g_i) **in** p
    while *state* $\in g_{i\text{-}1}$ **and not** *state* $\in g_i$
      if ISPRIM(a_i)
        state = *world*.EXECUTE(a_i)
      else
        HPN(*state*, g_i, NEXTLEVEL(*abs*, a_i), *world*)
    if not *state* $\in g_i$ **return**

HPN starts by making a plan p to achieve the top-level goal. Then, it executes the plan steps, starting with action, a_1. Each plan step is executed repeatedly, until either its desired post-condition, g_i, holds in the environment, which means that the execution has been successful, or until its pre-condition, g_{i-1}, ceases to hold in the environment, which means that the suffix of the plan starting with this step can no longer be expected to achieve the goal. If the pre-condition becomes false, then execution of the plan at this level is terminated and control is returned to the level of abstraction above.

This process will re-plan and re-execute actions until the goal is reached, as long as the goal is reachable from every state in the determinized model.

4 Pre-image Backchaining in Belief Space

When there is uncertainty about the current state of the world, we plan in the space of beliefs about the world state, instead of the space of world states itself. In this work, we use probability distributions over world states as belief states. Planning in this space enables actions to be selected because they cause the robot to gain information that will enable appropriate physical actions to be taken later in the plan, for instance.

Planning in belief space is generally quite complex, because it seems to require representing and searching for trajectories in a very high-dimensional continuous space of probability distributions. This is analogous to the problem of finding plans in very high-dimensional continuous space of configurations of a robot and many objects. We take direct advantage of this analogy and use symbolic predicates to specify limited properties of belief states, as our previous approach [10] does for properties of geometric configurations. So, for instance, we might characterize a set of belief states by specifying that "the probability that the cup is in the cupboard is greater than 0.95." Pre-image backchaining allows the construction of high-level plans to achieve goals articulated in terms of those predicates, without explicitly formalizing the complete dynamics on the underlying continuous space.

Traditional belief-space planning approaches either attempt to find entire policies, mapping all possible belief states to actions [9, 17, 19] or perform forward search from a current belief state, using the Bayesian belief-update equation to compute a new belief state from a previous one, an action and an observation [16]. In order to take advantage of the approach outlined above to hierarchical planning and execution, however, we will take a pre-image backchaining approach to planning in belief space.

HPN in belief space The basic execution strategy for HPN need not be changed for planning in belief space. The only amendment, shown below, is the need to perform an update of the belief state based on an observation resulting from executing the action in the world:

```
BHPN(belief, goal, abs, world):
    p = PLAN(belief, goal, abs)
    for (a_i, g_i) in p
        while belief ∈ g_{i-1} and not belief ∈ g_i
            if ISPRIM(a_i)
                obs = world.EXECUTE(a_i)
                belief.UPDATE(a_i, obs)
            else
                BHPN(belief, g_i, NEXTLEVEL(abs, a_i), world)
        if not belief ∈ g_i return
```

After each primitive action is executed, an observation is made in the world and the belief state is updated to reflect both the predicted transition and the information contained in the observation *obs*. It is interesting to note that, given an action and an observation, the belief state update is deterministic. However, the particular observation that will result from taking an action in a state is stochastic; that stochasticity is handled by the *BHPN* structure in the same way that stochasticity of action outcomes in the world was handled in the *HPN* structure.

Hierarchical planning and information gain fit together beautifully: the system makes a high-level plan to gather information and then uses it, and the interleaved hierarchical planning and execution architecture ensures that planning that depends on the information naturally takes place after the information has been gathered.

4.1 Symbolic Representation of Goals and Subgoals

When planning in belief space, goals must be described in belief space. Example goals might be "With probability greater than 0.95, the cup is in the cupboard." or "The probability that more than 1 % of the floor is dirty is less than 0.01." These goals describe *sets* of belief states. The process of planning with pre-image backchaining computes pre-images of goals, which are themselves sets of belief states. Our representational problem is to find a compact yet sufficiently accurate way of describing goals and their pre-images.

In traditional symbolic planning, *fluents* are logical assertions used to represent aspects of the state of the external physical world; conjunctions of fluents are used to describe sets of world states, to specify goals, and to represent pre-images. States in a completely symbolic domain can be represented in complete detail by an assignment of values to all possible fluents in a domain. Real world states in robotics problems, however, are highly complex geometric arrangements of objects and robot configurations which cannot be completely captured in terms of logical fluents. However, logical fluents can be used to characterize the domain at an abstract level for use in the upper levels of hierarchical planning.

We will take a step further and use fluents to characterize aspects of the robot's *belief state*, for specifying goals and pre-images. For example, the condition "With probability greater than 0.95, the cup is in the cupboard,", can be written using a fluent such as *PrIn*(*cup, cupboard,* 0.95), and might serve as a goal for planning. For any fluent, we need to be able to test whether or not it holds in the current belief state, and we must be able to compute the pre-image of a set of belief states described by a conjunction of fluents under each of the robot's actions. Thus, our description of operators will not be in terms of their effect on the state of the external world but in terms of their effect on the fluents that characterize the robot's belief. Our work is informed by related work in partially observed or probabilistic regression (back-chaining) planning [2, 5, 18]. In general, it will be very difficult to characterize the exact pre-image of an operation in belief space; we will strive to provide an approximation that supports the construction of reasonable plans and rely on execution monitoring and replanning to handle errors due to approximation.

In the rest of this paper, we provide examples of representations of belief sets using conjunctions of logical fluents, for both discrete and continuous domains, and illustrate them on example robotics problems.

5 Coarse Object Pose Uncertainty

We have a pilot implementation of the HPN framework on a Willow Garage PR2 robot, demonstrating integration of low-level geometric planning, active sensing, and high-level task planning, including reasoning about knowledge and planning to gain information. Figure 2 shows a planning problem in which the robot must move the blue box to another part of the table. The actual state of the world is shown on the left, and the mode of its initial belief is shown on the right.

Objects in the world are detected by matching known object meshes to point clouds from the narrow stereo sensor on the robot; example detections are shown in Fig. 3. As with any real perceptual system, there is noise in both the reported poses and identities of the objects. Furthermore, there is significant noise in the reported pose of the robot base, due to wheel slip and other odometric error. There is additional error in the calibration of the stereo sensor and robot base. The state

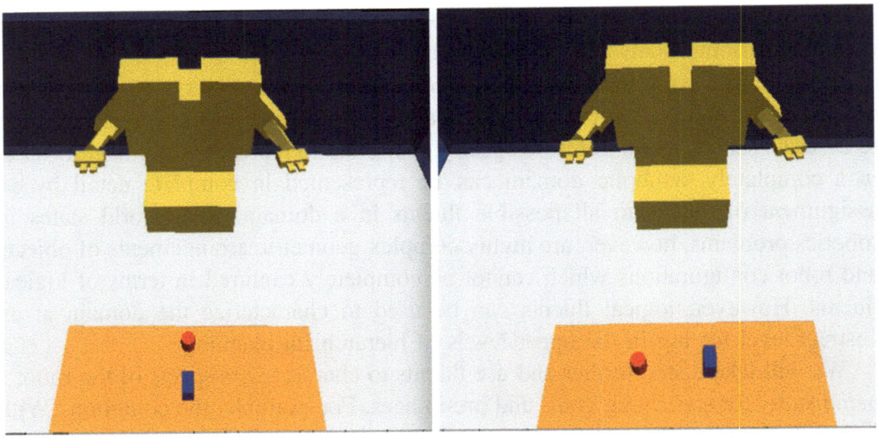

Fig. 2 The situation on the *left* is the real state of the world; the one on the *right* represents the mode of the initial belief

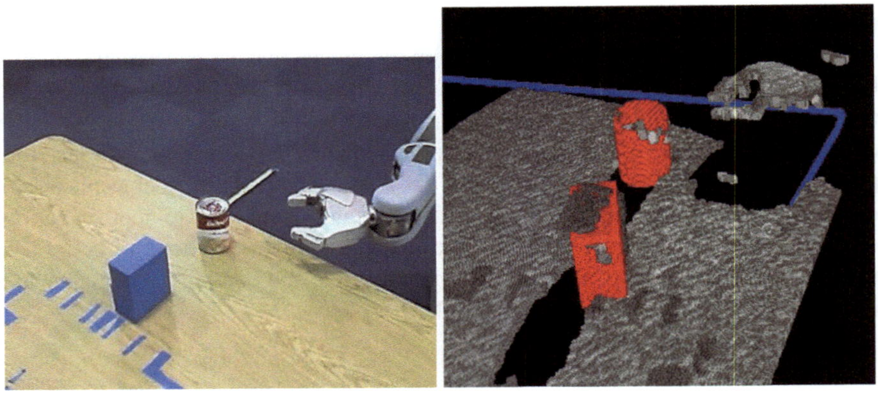

Fig. 3 Point cloud (on *right*) for scene (on *left*); *red* points correspond to the models at their perceived poses

estimation process for the positions of objects is currently very rudimentary: object detections that are significantly far from the current estimate are rejected; those that are accepted are averaged with the current estimate. A rough measure of the accuracy of the estimate is maintained by counting the number of detections made of the object since the last time the robot base moved. A detailed description of the geometric representation and integration of task and motion planning is available [10]. Here, we emphasize uncertainty handling in the real robot. Here are three of the relevant operator descriptions:

$\text{PICK}(O,M) : \mathit{KHolding} = O$:
pre: $\mathit{KClearX}(\mathit{PickSwept}(M),O) = T$,
$\mathit{KCanPickFrom}(O,M) = T, \mathit{KHolding} = \text{NOTHING}$,
$\mathit{KRobotNearLoc}(M) = T, \mathit{LocAccuracy}(O,M,3) = T$

$\text{LOOKAT}(O,M,N) : \mathit{LocAccuracy}(O,M,N) = T$
pre: $\mathit{KRobotNearLoc}(M) = T, \mathit{LocAccuracy}(O,M,N-1) = T$

$\text{MOVEBASE}(M) : \mathit{KRobotNearLoc}(M) = T$
pre: $\mathit{KClearX}(\mathit{MoveSwept}(M),()) = T$
sideEffect: $\mathit{LocAccuracy}(O,M',N) = F$

The PICK operator describes conditions under which the robot can pick up an object O, making the fluent *KHolding* have the value O. The fluent name, *KHolding*, is meant to indicate that it is a condition on belief states, that is, that the robot *know* that it is holding object O. The primitive pick operation consists of (1) calling an RRT motion planner to find a path for the arm to a 'pregrasp' pose, (2) executing that trajectory on the robot, (3) grasping the object, then (4) lifting the object to a 'postgrasp' pose.

The operator description has a free variable M, which describes a trajectory for the robot base and arm, starting from a home pose through to a pose in which the object is grasped; one or more possible values of M are generated by a procedure that takes constraints on grasping and motion into account and uses an efficient approximate visibility-graph motion planner for a simple robot to find a path. This is not the exact path that the motion primitives will ultimately execute, but it serves during the high-level planning to determine which other objects need to be moved out of the way and where the base should be positioned while performing the pick operation. The preconditions to the primitive pick operation are that: the swept volume of the path for the arm be known to be clear, with the exception of the object to be picked up; that the object O be known to be in a pose that allows it be picked up when the robot base pose is the one specified by M; that the robot is known not to be currently holding anything; that the robot base is known to be near the pose specified by M, and that the pose of the object O is known with respect to the robot's base pose in M with accuracy level 3 (that is, it has to have had at least three separate good visual detections of the object since the last time the base was moved).

The LOOKAT operator looks at an object O from the base pose specified in motion M, and can make the fluent *LocAccuracy*(O, M, N) true, which means that the accuracy of the estimated location of object O is at least level N. The primitive operation computes a head pose that will center the most likely pose of object O in its field of view when the robot base is in the specified pose and moves the head to that pose. The image-capture, stereo processing, and object-detection processes are running continuously and asynchronously, so the primitive does not need to explicitly call them. This operation achieves a location accuracy of N if the base is

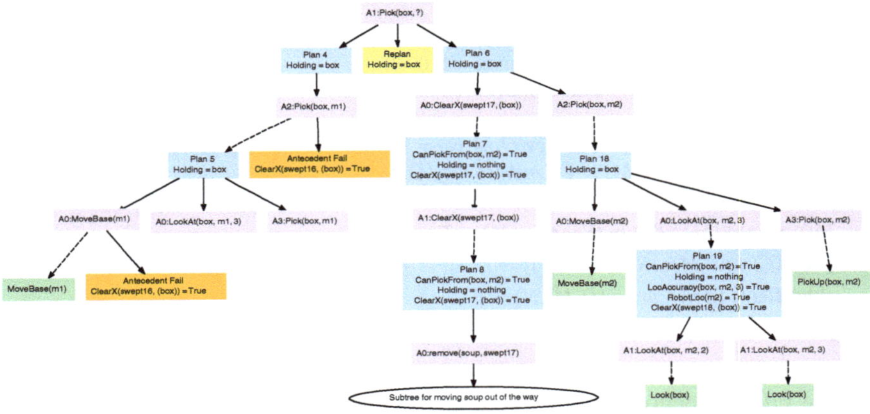

Fig. 4 Partial planning and execution tree for picking up the box, in which the robot notices that the soup can is in the way and removes it

known to be in the appropriate pose and the object had been previously localized with accuracy $N - 1$.

The MOVEBASE operator moves to the base pose specified in motion M. The primitive operation (1) calls an RRT motion planner to find a path (in the full configuration space of the base and one arm—the other arm is fixed) to the configuration specified in M and (2) executes it. It requires, as a precondition, that the swept volume of the suggested path be known to be clear. Importantly, it also declares that it has the effect of invalidating the accuracy of the estimates of all object poses.

Figure 4 shows a fragment of the planning and execution tree resulting from an execution run of the system. At the highest level (not shown) a plan is formulated to pick up the box and then place it in a desired location. This tree shows the picking of the box. First, the system makes an abstract plan (Plan 4) to pick the box, and then refines it (Plan 5) to three operators: moving the base, looking at the box until its location is known with sufficient accuracy, and then executing the pick primitive. The robot base then moves to the desired pose (green box); once it has moved, it observes the objects on the table and updates its estimate of the poses of all the objects. In so doing, it discovers that part of the precondition for executing this plan has been violated (this corresponds to the test in the last line of code for BHPN) and returns to a higher level in the recursive planning and execution process. This is indicated in the planning and execution tree by the orange boxes, showing that it fails up two levels, until the high-level PICK operation is planned again. Now, Plan 6 is constructed with two steps: making the swept volume for the box clear, and then picking up the box. The process of clearing the swept volume requires picking up the soup can and moving it out of the way; this process generates a large planning and execution tree which has been elided from the figure. During this process, the robot had to move the base. So Plan 18 consists of moving the base to an appropriate pose to pick up the box, looking at the box, and then picking it up. Because the robot was able to visually detect the box after moving the base, the LOOKAT operation only needs to gather two additional detections of the object, and then, finally, the robot picks up the box.

6 Fine Object Pose Uncertainty

In the previous example, we had a very coarse representation of uncertainty: the robot's pose was either known sufficiently accurately or it wasn't; the degree of accuracy of a position estimate was described by the number of times it had been observed. For more fine-grained interaction with objects in the world, we will need to reason more quantitatively about the belief states. In this section, we outline a method for characterizing the information-gain effects of operations that observe continuous quantities in the environment. We then illustrate these methods in a robotic grasping example.

6.1 Characterizing Belief of a Continuous Variable

We might wish to describe conditions on continuous belief distributions, by requiring, for instance, that the mean of the distribution be within some value of the target and the variance be below some threshold. Generally, we would like to derive requirements on beliefs from requirements for action in the physical world. So, in order for a robot to move through a door, the estimated position of the door needs to be within a tolerance equal to the difference between the width of the robot and the width of the door. The variance of the robot's estimate of the door position is not the best measure of how likely the robot is to succeed: instead we will use the concept of the *probability near mode* (PNM) of the distribution. It measures the amount of probability mass within some δ of the mode of the distribution. So, the robot's prediction of its success in going through the door would be the PNM with δ equal to half of the robot width minus the door width.

For a planning goal of $PNM(X, \delta) > \theta$, we need to know expressions for the regression of that condition under the a and o in our domain. In the following, we determine such expressions for the case where the underlying belief distribution on state variable X is Gaussian, the dynamics of X are stationary, a is to make an observation, and the observation o is drawn from a Gaussian distribution with mean X and variance σ_o^2.

For a one-dimensional random variable $X \sim \mathcal{N}(\mu, \sigma^2)$,

$$PNM(X, \delta) = \Phi\left(\frac{\delta}{\sigma}\right) - \Phi\left(\frac{-\delta}{\sigma}\right) = \mathrm{erf}\left(\frac{\delta}{\sqrt{2}\sigma}\right),$$

where Φ is the Gaussian CDF. If, at time t the belief is $\mathcal{N}(\mu_t, \sigma_t^2)$, then after an observation o, the belief will be

$$\mathcal{N}\left(\frac{\mu_t\sigma_0^2 + o\sigma_t^2}{\sigma_0^2 + \sigma_t^2}, \frac{\sigma_0^2\sigma_t^2}{\sigma_0^2 + \sigma_t^2}\right)$$

So, if $PNM(X, \delta) = \theta_t = \operatorname{erf}\left(\frac{\delta}{\sqrt{2}\sigma}\right)$ then

$$PNM(X_{t+1}, \delta) = \theta_{t+1} = \operatorname{erf}\left(\frac{\delta}{\sqrt{2}}\sqrt{\frac{\sigma_0^2 + \sigma_t^2}{\sigma_0^2\sigma_t^2}}\right)$$

Substituting in the expression for σ_t^2 in terms of θ_t, and solving for θ_t, we have:

$$\theta_t = PNMregress(\theta_{t+1}, \delta, \sigma_o^2) = \operatorname{erf}\left(\sqrt{\operatorname{erf}^{-1}(\theta_{t+1})^2 - \frac{\delta^2}{2\sigma_o^2}}\right).$$

So, to guarantee that $PNM(X_{t+1}, \delta) > \theta_{t+1})$ holds after taking action a and observing o, we must guarantee that $PNM(X_t, \delta) > PNMregress(\theta_{t+1}, \delta, \sigma_o^2)$ holds on the previous time step.

6.2 Integrating Visual and Tactile Sensing for Grasping

In general, visual sensing alone is insufficient to localize an object sufficiently to guarantee that it be grasped in a desired pose with respect to the robot's hand. Attempts to grasp the object result in tactile feedback that significantly improve localization. The Willow Garage reactive grasping ROS pipeline [8] provides a lightweight approach to using the tactile information, assuming that the desired grasp orientation is fixed relative to the robot (rather than the object). Other work [7] takes a belief-space planning approach to this problem. It constructs a fixed-horizon search tree (typically of depth 2), with branches on possible observations. That approach is effective, but computationally challenging due to (1) re-planning on every step; (2) a large branching factor and (3) the fact that a completely detailed belief-state update must be computed at each node in the forward search tree.

In this section, we outline an approach of intermediate complexity. It makes a complete plan using a deterministic and highly abstract characterization of preimages in belief space, which is very efficient. It can handle objects of more general shapes and different grasping strategies, and automatically trades off the costs and benefits of different sensing modalities.

Operator descriptions We have reformulated the *pick* operator from Sect. 5, with the accuracy requirement for the location of O specified with the fluent $PNMLoc(O,\ \theta,\ \delta_p)$, which is true if and only if the probability that object O's location is within δ_p of its most likely location is greater than θ. The definition of near is now on three-dimensional object poses with threshold δ_p, selected empirically.

$\text{PICK}(O,M) : KHolding = O$:
pre: $KClearX(PickSwept(M),O) = T$,
$KCanPickFrom(O,M) = T, KHolding = \text{NOTHING}$,
$KRobotNearLoc(M) = T, PNMLoc(O,\theta,\delta_p) = T$
cost: $1/\theta$

The threshold θ is a free parameter here. If δ_p is the amount of pose error that can be tolerated and still result in a successful grasp, then this operator says that if we know the object's location to within δ_p with probability θ, then we will succeed in grasping with probability θ. This success probability is reflected in the cost model.

We have two ways of acquiring information. The first is by looking with a camera. The field of view is sufficiently wide that we can assume it always observes the object, but with a relatively large variance, σ^2_{vision}, in the accuracy of its observed pose, and hence a relatively small increase in the PNM.

LOOK$(O,\ \theta,\ \delta)$: $PNMLoc(O,\ \theta,\ \delta)$ = T:
pre: $PNMLoc(O,\ PNMRegress(\theta,\ \delta,\ \sigma^2_{vision}),\ \delta)$ = T
cost: **1**

The second method of gaining information is by attempting to grasp the object. If the probability of the object's pose being within the hand's size of the mode is very low, then an attempt to grasp is likely to miss the object entirely and generate very little information. On the other hand, if the probability of being within a hand's size of the mode is high, so that an attempted grasp is likely to contact the object, then observations from the touch sensors on the hand will provide very good information about the object's position.

$\text{TRYGRASP}(O,\theta,\delta) : PNMDoorLoc(D,\theta,\delta) = \text{T}$:
pre: $KClearX(PickSwept(M),O) = T$,
$KCanPickFrom(O,M) = T, KHolding = \text{NOTHING}$,
$KRobotNearLoc(M) = T$
$PNMLoc(O, PNMRegress(\theta,\delta,\sigma^2_{tactile}),\delta) = \text{T}$,
$PNMLoc(O,\theta_{handSize}, handSize/2) = \text{T}$
cost: $1/\theta_{handSize}$

In this operator description, we use the *PNMRegress* procedure as if the information gained through an attempted grasp were the result of an observation with Gaussian error being combined with a Gaussian prior. That is not the case, and so this is a very loose approximation; in future work, we will estimate this non-Gaussian PNM pre-image function from data.

Both the GRASP and TryGrasp operations are always executed with the mode of the belief distribution as the target pose of the object. This planning framework could support other choices of sensing actions, including other grasps or sweeping the hand through the space; it could also use a different set of fluents to describe the belief state, including making distinctions between uncertainty in position and in orientation of the object.

State estimation In order to support planning in belief space, we must implement a state estimator, which is used to update the belief state after execution of each primitive action in *BHPN*. The space of possible object poses is characterized by x, y, θ, assuming that the object is resting on a known stable face on a table of known height: x and y describe positions of the object on the table and θ describes its orientation in the plane of the table.

The grasping actions are specified by [*targetObjPose*, *grasp*], where *targetObjPose* is a pose of the object and *grasp* specifies the desired placement of the hand with respect to the object. An action consists of moving back to a pregrasp configuration with the hand open, then performing a guarded move to a target hand pose and closing the fingers. The fingers are closed compliantly, in that, as soon as one finger contacts, the wrist moves away from the contacting finger at the same time the free finger closes. The resulting two-finger contact provides a great deal of information about the object's position and orientation. During the guarded move, if any of the tactile sensors or the wrist accelerometer is triggered, the process is immediately halted. In the case of an accelerometer trigger, the fingers are compliantly closed.

The observations obtained from grasping are of the form

$$[handPose, gripperWidth, trigger, contacts],$$

where *handPose* is the pose of the robot's hand (reported based on its proprioception) at the end of the grasping process; *gripperWidth* is the width of the gripper opening; *trigger* is one of {*rightTip, leftTip, accel, None*} indicating whether the motion was stopped due to a contact from one of the fingertip sensors, the accelerometer, or was not stopped; and *contacts* is a list of four Boolean values indicating whether each of four sensors (left finger tip, right finger tip, left internal pad, and right internal pad) is in contact.

In order to perform state estimation, we need to provide an observation model Pr $(o|s, a)$, which specifies the probability (or density) of making a particular observation o if the object is actually in pose s and the robot executes the action specified by a. There is stochasticity in the outcome of commanded action a due to both to error in the robot's proprioception and in its controllers, causing it to follow a trajectory t that differs from the one specified by a. So, we can formulate the observation probabilities as

$$\Pr(o|s,a) = \int_t \Pr(o|s,t)\Pr(t|s,a).$$

Decomposing o into c (contacts and trigger) and h (hand-pose and gripper width), we can write

$$\begin{aligned}\Pr(o|s,a) &= \int_t \Pr(c,h|s,t)\Pr(t|s,a).\\ &= \int_t \Pr(c|s,t)\Pr(h|t)\Pr(t|s,a)\end{aligned}$$

However, the integral is too difficult to evaluate, so we approximate it by considering the trajectory t that maximizes the right hand side

$$\Pr(o|s,a) \approx \max_t \Pr(c|s,t)\Pr(h|t)\Pr(t|s,a).$$

This, too, is difficult to compute; we approximate again by assuming that there is little or no error in the actual contact sensors, and so we consider the set of trajectories that would generate the contacts c if the object were at pose s, $\tau(c,s)$, so

$$\Pr(o|s,a) \approx \max_{\tau(c,s)} \Pr(h|t)\Pr(t|s,a).$$

We consider the class of trajectories that are rigid displacements of the commanded trajectory a that result in contacts c, and such that no earlier point in the trajectory would have encountered an observed contact. We compute the configuration space obstacle for the swept volume of the gripper relative to the object placed at s. Facets of this obstacle correspond to possible observed or unobserved contacts. We first compute a_h, which is the displacement of a that goes through the observed hand pose h, and then seek the smallest displacement of a_h that would move it to $\tau(c, s)$; this is the smallest displacement δ that produces a contact in the relevant facet of the configuration-space obstacle. We assume that the displacements between actual and observed trajectories are Gaussian, so we finally approximate $\Pr(o|s, a)$ as $G(\delta, 0, \sigma^2)$ where G is the Gaussian PDF for value δ in a distribution with mean 0 and variance σ^2.

The relationship between observations and states is highly non-linear, and the resulting state estimates are typically multi-modal and not at all well approximated by a Gaussian. We represent the belief state with a set of samples, using an adaptive resolution strategy to ensure a high density of samples in regions of interest, but also to maintain the coverage necessary to recover from unlikely observations and to maintain a good estimate of the overall distribution (not just the mode).

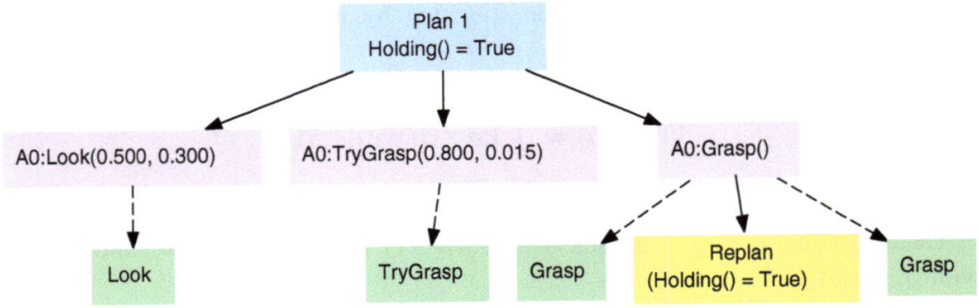

Fig. 5 Planning and execution tree for grasping, showing the initial plan, and a re-execution

Results Using *BHPN* with the operator descriptions and state estimator described in this section, we constructed a system that starts with a very large amount of uncertainty and has the goal of believing with high probability that it is holding an object. Figure 5 shows an example planning and execution tree resulting from this system. It plans to first look, then do an attempted grasp to gain information, and then to do a final grasp. It executes the *Look* and the *TryGrasp* steps with the desired results in belief space, and then it does another grasp action. At that point, the confidence that the robot is actually grasping the object correctly is not high enough. However, the PNM precondition of the *Grasp* action still holds, so it replans locally and decides to try the *Grasp* action again, and it works successfully.

We compared three different strategies in a simulation of the PR2: using only the *TryGrasp* action to gain information, using only the *Look* action to gain information, and using a combination of the *Look* and *TryGrasp* actions, mediated by the planner and requirements on PNM. We found the following average number of actions required by each strategy:

- *TryGrasp* only: 6.80 steps
- *Look* only: 73 steps (estimated)
- *LookandTryGrasp*: 4.93 steps

The *Look* operation has such high variance that, although in the limit the PNM would be sufficiently reduced to actually grasp the object, it would take a very long time. The estimate of 73 is derived using the PNM regression formulas (and in fact the planner constructs that plan). The *TryGrasp*-only strategy works reasonably well because we limited the region of uncertainty to the workspace of the robot arm, and it can reasonably easily rule out large parts of the space. Were the space larger, it would take considerably more actions. Using the *Look* action first causes the PNM to be increased sufficiently so that subsequent *TryGrasp* actions are much more informative. It also sometimes happens that, if the noise in the *TryGrasp* observations is high, the PNM falls sufficiently so that replanning is triggered at the high level and a new *Look* action is performed, which refocuses the grasp attempts on the correct part of the space. These results are preliminary; our next steps are to integrate this planning process with the rest of the motion and manipulation framework.

7 Conclusion

This paper has described a tightly integrated approach, weaving together perception, estimation, geometric reasoning, symbolic task planning, and control to generate behavior in a real robot that robustly achieves tasks in complex, uncertain domains. It is founded on these principles: (1) Planning explicitly in the space of the robot's *beliefs* about the state of the world is necessary for intelligent information-gathering behavior; (2) Planning with simplified domain models is efficient and can be made robust by detecting execution failures and replanning online; (3) Combining logical and geometric reasoning enables effective planning in large state spaces; and (4) Online hierarchical planning interleaved with execution enables effective planning over long time horizons.

References

1. J.L. Barry, L. Pack Kaelbling, T. Lozano-Pérez, DetH*: Approximate hierarchical solution of large markov decision processes, in *IJCAI* (2011)
2. C. Boutilier, Correlated action effects in decision theoretic regression, in *UAI* (1997)
3. S. Cambon, R. Alami, F. Gravot, A hybrid approach to intricate motion, manipulation and task planning. Int. J. Robot. Res. **28** (2009)
4. T. Erez, W. Smart, A scalable method for solving high-dimensional continuous POMDPs using local approximation, in *UAI* (2010)
5. C. Fritz, S.A. McIlraith, Generating optimal plans in highly-dynamic domains, in *UAI* (2009)
6. K. Hauser, Randomized belief-space replanning in partially-observable continuous spaces, in *WAFR* (2010)
7. K. Hsiao, L.P. Kaelbling, T. Lozano-Perez, Task-driven tactile exploration, in *RSS* (2010)
8. K. Hsiao, S. Chitta, M. Ciocarlie, E.G. Jones, Contact-reactive grasping of objects with partial shape information, in *IROS* (2010)
9. L.P. Kaelbling, M.L. Littman, A.R. Cassandra, Planning and acting in partially observable stochastic domains. Artif. Intell. **101** (1998)
10. L.P. Kaelbling, T. Lozano-Pérez, Hierarchical task and motion planning in the now, in *ICRA* (2011)
11. S.M. LaValle, *Planning Algorithms* (Cambridge University Press, Cambridge, 2006)
12. T. Lozano-Pérez, M. Mason, R.H. Taylor, Automatic synthesis of finemotion strategies for robots. Int. J. Robot. Res. **3**(1) (1984)
13. B. Marthi, S. Russell, J. Wolfe, Combined task and motion planning for mobile manipulation, in *ICAPS* (2010)
14. E. Plaku, G. Hager, Sampling-based motion planning with symbolic, geometric, and differential constraints, in *ICRA* (2010)
15. R. Platt, R. Tedrake, L. Kaelbling, T. Lozano-Perez, Belief space planning assuming maximum likelihood observations, in *RSS* (2010)
16. S. Ross, J. Pineau, S. Paquet, B. Chaib-draa, Online planning algorithms for pomdps. J. Artif. Intell. Res. (2008)
17. S. Sanner, K. Kersting, Symbolic dynamic programming for first-order POMDPs, in *AAAI* (2010)
18. R.B. Scherl, T.C. Son, C. Baral, State-based regression with sensing and knowledge. Int. J. Softw. Inf. **3** (2009)

19. R.D. Smallwood, E.J. Sondik, The optimal control of partially observable Markov processes over a finite horizon. Oper. Res. **21**, 1071–1088 (1973)
20. N.E. Du Toit, J.W. Burdick, Robotic motion planning in dynamic, cluttered, uncertain environments, in *ICRA* (2010)
21. R. Waldinger, Achieving several goals simultaneously, in *Machine Intelligence*, vol. 8 (Ellis Horwood Limited, Chichester, 1977) (Reprinted in J. Allen, J. Hendler, A. Tate, eds., *Readings in Planning* (Morgan Kaufmann, 1990))
22. D.S. Weld, Recent advances in AI planning. AI Mag. **20**(2), 93–123 (1999)
23. S.W. Yoon, A. Fern, R. Givan, FF-replan: a baseline for probabilistic planning, in *ICAPS* (2007)

Realtime Informed Path Sampling for Motion Planning Search

Ross A. Knepper and Matthew T. Mason

Abstract Robot motions typically originate from an uninformed path sampling process such as random or low-dispersion sampling. We demonstrate an alternative approach to path sampling that closes the loop on the expensive collision-testing process. Although all necessary information for collision-testing a path is known to the planner, that information is typically stored in a relatively unavailable form in a costmap. By summarizing the most salient data in a more accessible form, our process delivers a denser sampling of the free space per unit time than open-loop sampling techniques. We obtain this result by probabilistically modeling—in real time and with minimal information—the locations of obstacles, based on collision test results. We demonstrate up to a 780 % increase in paths surviving collision test.

1 Introduction

The motion planning problem is to find a path or trajectory that guides the robot from a given start state to a given goal state while obeying constraints and avoiding obstacles. The solution space is high dimensional, so motion planning algorithms typically decompose the problem by searching for a sequence of shorter, local paths, which solve the original motion planning problem when concatenated.

Each local path comprising this concatenated solution must obey motion constraints and avoid obstacles and hazards in the environment. Many alternate local paths may be considered for each component, so planners select a combination of paths that optimizes some objective function. In order to generate such a set of

R.A. Knepper (✉)
Department of Computer Science, Cornell University, Gates Hall,
Ithaca, NY, USA
e-mail: rak@cs.cornell.edu

M.T. Mason
Robotics Institute, Carnegie Mellon University, 5000 Forbes Ave,
Pittsburgh, PA, USA
e-mail: mason@ri.cmu.edu

H.I. Christensen and O. Khatib (eds.), *Robotics Research*,
Springer Tracts in Advanced Robotics 100, DOI 10.1007/978-3-319-29363-9_23

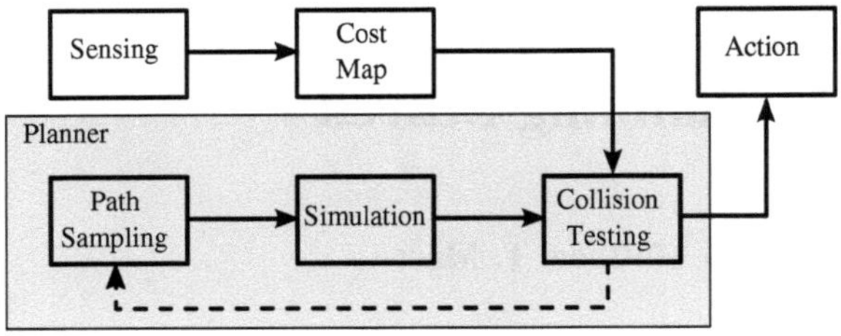

Fig. 1 Typical data flow within a robot closes the loop around the sense-plan-act cycle, but the planner itself runs open-loop. We close the planning loop, informing path sampling with the results of collision-testing earlier paths

candidate paths, the planner must generate many more candidate paths, each of which must be verified against motion constraints and collision-tested prior to consideration. Motion planners generate this large collection of paths by sampling—most often at random or from a low-dispersion sequence.

All the information needed to find collision-free path samples exists within the costmap, but the expensive collision test process prevents that information from being readily available to the planner. A negative collision-test result (i.e. no collision) is retained for future consideration, but a positive collision-test result is typically thrown away because the path is not viable for execution. Such planners may later waste time sampling and testing similar paths that collide with the same obstacle.

This policy of discarding information about colliding paths highlights a major inefficiency, which especially impacts realtime planning. Every detected collision provides a known obstacle location. This observation may not seem significant at first, as all detected collisions represent known obstacles in the costmap. However, not all such obstacles are equally relevant to a given local planning problem, and so we can benefit by storing relevant costmap obstacles in a form more immediately available to the planner. We argue that the planner may derive increased performance by feeding back the set of collision points, known from prior collision tests, to the path sampling process, as in Fig. 1.

1.1 Path Sampling and Path Parametrization

The general path sampling problem is to supply a sequence of distinct paths $\{p1, p2, \ldots\} = \mathscr{P} \subset \mathscr{X}$, the continuum space of paths. Often, these paths are not parametrized directly by their geometry but instead are described by their means of generation. For instance, some planners consider only straight-line paths. Given a current robot state $x_0 \in X$, the configuration space, a straight-line path is uniquely specified by connecting x_0 to an arbitrary sampled state $x_f \in X$. In such planners, it is expected that the robot is able to execute arbitrary paths, and so the boundary value problem is easy to solve because it is under-constrained.

Definition 1 Given start and end states, the **boundary value problem** (BVP) is to find any feasible path from the start to the goal (i.e. the local planning problem). A variant of this problem is to find the shortest such path. □

Some classes of robotic systems possess velocity constraints that limit the direction in which they may move instantaneously. The most well known example of these nonholonomic constraints is the difficulty of parallel parking a car. In such highly constrained, underactuated systems, the set of feasible paths $\mathscr{F}$ is much smaller than the space of all paths, $\mathscr{X}$. Thus, an arbitrary path sample drawn from $\mathscr{X}$ is unlikely to be in the feasible set $\mathscr{F}$. In such cases, the BVP is difficult to solve.

For constrained systems, we may avoid solving the BVP by instead sampling in U, the space of actions. Suppose we have a "black box" model of the robot's response to a control input, which is a mapping $M : U \rightarrow \mathscr{F}$. Sampling in the control space offers several advantages: (1) all sampled paths trivially obey motion constraints; and (2) we may precompute a diverse set of such paths. For a mobile robot, these paths are independent of initial position and heading, depending only on their derivatives (we ignore external interference such as wind or wheel slip). Therefore, a relatively small lookup table suffices to describe an expressive set of robot motions.

1.2 Prior Work

The motion planning community has invested considerable effort in the topic of non-uniform and adaptive sampling. The literature on this topic largely concerns probabilistic roadmaps (PRMs), which sample states rather than paths. Hsu et al. [11] provide a survey of recent work in non-uniform sampling for PRMs. We touch on a few of the broad approaches here.

One approach employs a fixed strategy to bias configuration sampling towards narrow corridors—parts of the C-space that are less likely to be sampled on their own due to their small measure [1, 9, 17]. These works all restrict the sampling consideration to points, whereas we sample directly in the space of paths, using a distribution that varies in reaction to new collision test results.

The non-uniform sampling field has largely moved towards such adaptive strategies. For instance, Jaillet et al. [12] restrict sampling to size-varying balls around nodes in an RRT to avoid testing paths that would go through obstacles. Yu and Gupta [19] perform sensor-based planning in which a PRM is incrementally constructed based on the robot's partial observations of obstacles. Exploratory motions are selected by maximizing information gain. Another recent adaptive approach is to construct a meta-planner with several tools at its disposal; such planners employ multiple sampling strategies [10] or multiple randomized roadmap planners [16], based on a prediction of which approach is most effective in a given setting.

One important feature of our work is the use of information from all collision tests, including positive results, to minimize entropy in a model approximating obstacle locations. The work of Burns and Brock [3–6] bears considerable resemblance to ours in this regard. They describe an adaptive model of obstacle locations in C-space based on previous collision test results. Their model utilizes locally weighted learning to select state [4] or path [5] samples that reduce model uncertainty. We likewise develop a model of obstacle location, although ours inhabits the workspace, and its simplicity is better suited to realtime applications. Burns and Brock subsequently observe, as we do, that model refinement is not an end in itself, but merely a means to the end of finding collision-free paths [6]. We proceed from this observation to consider what level of refinement is appropriate, in the context of constrained paths, based on the maximum width of corridor we are willing to miss discovering.

Separately, Burns and Brock describe an entropy-guided approach to the selection of configuration samples likely to unify two connected components of the PRM graph [3]. In later work [6], they augment this approach with the notion of utility, which combines information gain regarding obstacle locations with the value of a resulting sample for solving the planning problem. Utility-guided sampling selects the configuration expected to solve the eventual planning problem most efficiently.We take a similar approach, in that we sample a combination of paths intended to navigate the space and to refine our obstacle model.

2 Informed Path Sampling Approach

In closing the loop on path sampling, we must feed back knowledge of obstacles reachable by the robot (in the form of collision-test results) into the sample space of paths, be it X or U, so that we can suppress from the sampled path sequence future paths intersecting those regions of the workspace.

A collision can be described as an ordered pair $c = (p, s)$, with $p \in \mathscr{F}$ and $s \in I = [0, 1]$, a time/distance parameter describing where on the path the collision occurred. A path is a mapping $p : I \rightarrow X$. Thus, c maps directly into a state $x \in X$, identifying the location of an obstacle. However, this collision state is special because it is known to be reachable by an action $u \in U$. In fact, x is probably reachable by a continuum of other actions, of which we can easily precompute a sampled subset for each possible collision point.

Our approach to path sampling feedback in highly constrained systems is to keep the feedback entirely within the action space, by which such paths are parametrized. We first collision test some paths drawn from a low-dispersion sequence [7]. After finding the first collision point, its location biases future action samples. We may construct, a priori, a list of correspondences between possible action samples to accelerate this process.

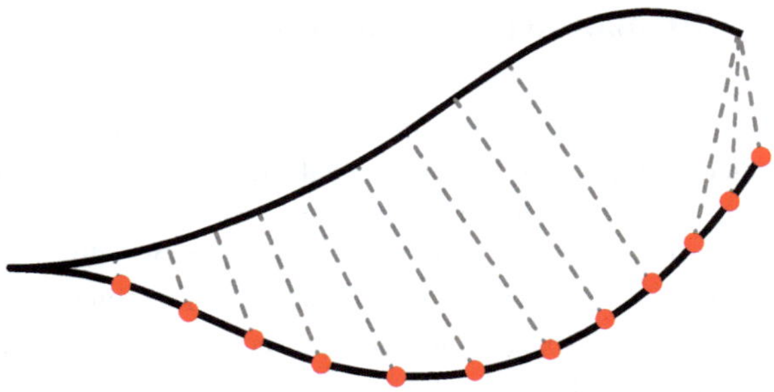

Fig. 2 Given a path set of N paths, each discretized into M points, the proximity look-up table (PLUT) stores for each ordered pair of paths a list of shortest distances to each discrete point on the second path. Thus, there are a total of MN^2 unique PLUT entries

Often, collision-test routines are able to identify the precise location where a collision occurred. Knowing that workspace point w is part of an obstacle, we may eliminate from our sampled sequence all paths passing through the set $cs(w)$, the set of robot configurations $x_i \in X$ in which the robot intersects the workspace point w. To eliminate these paths, we must store the list of actions by which they are parametrized.

In order to identify the set of paths passing unacceptably close to an obstacle point w, we precompute a proximity look-up table (PLUT), as shown in Fig. 2. Suppose our precomputed path set $\mathscr{P}$ contains N paths, each discretized into M points. The PLUT stores, for every ordered pair of paths $(p_i, \ p_j)$ in $\mathscr{P}$, the shortest distance to the kth discretized point on p_j.

$$\mathrm{PLUT}(p_i,p_j,k) = \min_{w \in p_i} d\left(w,p_j\left(\frac{k}{M}\right)\right), \tag{1}$$

where $d(w_1, \ w_2)$ gives the Euclidean distance between two points.

Now, given a collision $c = (p_j, \ s)$, we would like to find out if another path p_i would collide with the same obstacle. We simply query the PLUT as

$$\mathrm{PLUT}(p_i,p_j,sM) < r_r, \tag{2}$$

where r_r is the radius of the robot (or an inscribed circle of the robot). When this condition holds, the collision test would fail. Knowing this, we may eliminate the path without a test, and spend the CPU time considering other paths.

However, we may go beyond short-circuiting the collision-test to estimating the probability distribution on obstacle locations using the *principle of locality*, which states that points inside an obstacle tend to occur near other points inside an obstacle. We propose a series of models of locality and two path sampling problems, which we address using our models. The key to success of this approach is that the final evaluation be less costly than the collision tests it replaces.

3 Probabilistic Foundations

In this paper, we develop a series of probabilistic models that enable us to rapidly select paths for collision test that maximize one of two properties. First, in order to find valid robot motions, we must sample a selection of collision-free paths for execution. Second, we wish to sample broadly within the free space of paths, including in proximity to obstacles. The precision with which we know the obstacle/free-space boundaries directly relates to the size of narrow corridor we expect to find.

The workspace comprises a set of points divided into two categories: obstacle and free. The function

$$\text{obs} : W \rightarrow \beta, \tag{3}$$

where $\beta = \{\text{true, false}\}$ reveals the outcome that a particular workspace point w is either inside (true) or outside of an obstacle. Building on such outcomes, we then describe an event of interest. A path collision-test takes the form

$$\text{pct} : \mathscr{P} \rightarrow \beta, \tag{4}$$

which returns the disjunction of obs(w) for all w within the swept volume (or *swath*) of the path. A result of true indicates that this path intersects an obstacle.

Using these concepts as a basis, we pose two problems:

1. **Exploitation**. We are given a set of workspace points inside obstacles, $C = \{w_1, \ldots w_m\}$ and a set of untested paths $\mathscr{P}_{unknown} = \{p_1, \ldots p_n\}$. Knowing only a finite subset of the continuum of obstacle-points, find the path that minimizes the probability of collision:

$$p_{next} = \underset{P_i \in \mathscr{P}}{\arg\min}\ \Pr(\text{pct}(\text{p}_i, C)). \tag{5}$$

2. **Exploration**. Suppose we have a model of uncertainty U ($\mathscr{P}_{safe}$, C) over the collision status of a set of untested paths, $\mathscr{P}_{unknown}$ in terms of a set of tested paths and known obstacle-points. Find the path $p_{next} \in \mathscr{P}_{unknown}$ giving the greatest reduction in expected uncertainty:

$$\begin{aligned} U_{exp}(p_i) = {} & U\big(\mathscr{P}_{safe} \cup \{p_i\}, C\big) \Pr(\neg\text{pct}(p_i, C)) \\ & + U\big(\mathscr{P}_{safe}, C \cup \{c_i\}\big) \Pr(\text{pct}(p_i, C)) \end{aligned} \tag{6}$$

$$p_{next} = \underset{P_i \in \mathscr{P}_{unknown}}{\arg\max}\ U\big(\mathscr{P}_{safe}, C\big) - U_{\text{exp}}(p_i). \tag{7}$$

These two considerations are essentially the same as those encapsulated in the utility function of Burns and Brock [6] (see Sect. 1.2).

4 Locality

Thus far, we have demonstrated how a single failed collision test may serve to eliminate an entire set of untested paths from consideration because they pass through the same obstacle point. We may extend this approach one step further using the principle of locality.

Definition 2 The **principle of locality** states that if a robot state is in collision with an obstacle, then that state is contained in a neighborhood ball of obstacle points. □

Two contributing factors combine to produce the locality effect. First, the non-zero volume of the robot means that even a point obstacle results in a set of robot states $cs(t)$ in collision with that point. The second factor is that real-world obstacles occupy some volume in space.

Given a known collision point, we employ the principle of locality to define a function expressing the probability that a new path under test is in collision with the same obstacle. A locality model takes the following general form:

$$\mathrm{loc}(p_i \mid \mathrm{C}) = \Pr(\mathrm{pct}(p_i) \mid \mathrm{C}) \tag{8}$$

Here, C may be a single point collision outcome or a set of collisions. If omitted, it is assumed to be the set of all known collisions.

This function depends on many factors, including the size and shape of the robot as well as the distribution on size and shape of obstacles in the environment. The most important parameter, however, is the distance between the new path and the known collision site. Thus, we may establish a rapidly computable first-order locality model in which we abstract away the size and shape of obstacles using a single distribution on radius, as in Fig. 3. We discuss several intermediate locality model formulations before coming to the final form.

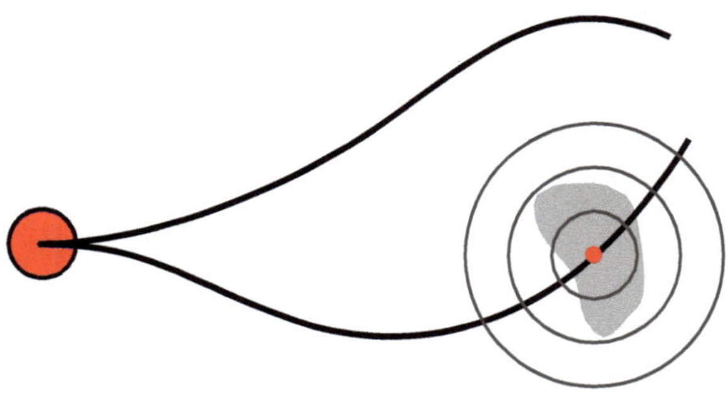

Fig. 3 The robot (*red disc* at *left*) considers two paths. The bottom path fails its collision test. The locality model does not know the full extent of the obstacle (*gray*), but it can approximate the obstacle using a probability distribution (concentric *circles*) and can estimate the likelihood of the top path colliding

4.1 General Locality Model

By explicitly modeling locality, we may reason about which paths are more or less likely to be in collision with any known obstacle, even with only partial information about its location. A path sampling algorithm, when informed by a locality model, provides a path sequence ordered by likelihood of collision, given currently known collision sites. We propose here a general model of locality that can be expected to produce collision-free path samples with high probability.

In constructing a general locality model, we abstract away many parameters; we consider both the robot and obstacles to be balls (in $\mathbb{R}^2$ or $\mathbb{R}^3$), and the obstacles are assumed uniform in radius. We relax some of these assumptions later, in Sect. 4.4. For now, these restrictions permit us to simplify the model by removing bearing from consideration. Thus, the general model's prediction of future collisions is purely a function of *range* from the known collision site to the path. The fixed radii of both the obstacles (r_o) and the robot (r_r) result in the intuitive notion of locality: that its influence is over a limited range only.

The precise formulation of the general locality model, as depicted in Fig. 4, is based on maintaining a probability distribution on possible locations of the obstacle centroid, given a known collision site. In this naive model, the location of the centroid is described by a uniform distribution over $B(r_o)$, a ball of radius r_o centered at the colliding position of the robot. A path *pi* sweeps out a swath $S(p_i)$ of width $2r_o + 2r_r$. Any non-empty set $B(r_o) \cap S(p_i)$ represents some probability of collision. This general model then predicts that the probability of collision is

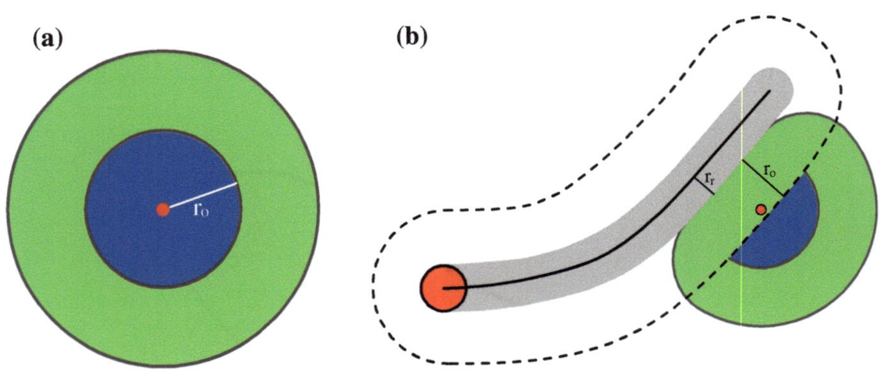

Fig. 4 The general locality model: **a** Given a point known to be in collision with an obstacle (center *red dot*), the blue inner circle of radius r_o represents possible locations of the centroid of the obstacle. The *green* outer circle comprises points possibly occupied by some part of the obstacle. **b** The probability that a new candidate robot path is collision-free equals the fraction of possible obstacle centroids outside a swath of width $2r_o + 2r_r$ The *blue region* represents the set of possible centroids, while the *green* region depicts possible obstacle extents

$$\mathrm{loc}_{general}(p_i \mid c) = \frac{|B(r_o) \cap S(p_i)|}{|B(r_o)|}. \tag{9}$$

If we regard p_i as a straight line, then in 2D the probability of collision is the ratio of the area of a circular segment to the area of the whole circle, which is [2]:

$$f_{segment}(r) = \begin{cases} \frac{1}{\pi r_e^2}\left(r_e^2 \cos^{-1}\frac{r-r_e}{r_e} - (r-r_e)\sqrt{2r_e r - r^2}\right) & \text{if } 0 \le r \le 2r_e \\ 0 & \text{otherwise,} \end{cases} \tag{10}$$

where r is the range between the path and the collision point. We call *re* the range of effect, which we set equal to r_o here.

Definition 3 The **range of effect** of a known collision point describes the radius around that point at which paths are regarded to be at risk of collision with the known obstacle. □

4.2 Simple Locality Model

We now propose an even simpler locality model, which closely approximates (10) but makes use of the existing PLUT. Instead of path area, we consider only the point on the new path most closely approaching the known collision point. This new locality model employs the raised cosine distribution:

$$f_{rcd}(r) = \begin{cases} \frac{1}{2r_e}\left[1 + \cos\left(\pi \frac{r}{2r_e}\right)\right] & \text{if } 0 \le r \le 2r_e \\ 0 & \text{otherwise.} \end{cases} \tag{11}$$

For a straight or gently curving path, this approximation is very good. Then, the probability that a new path p_i will collide with the same obstacle represented by the previous collision $c = (p_j, s)$ is simply

$$\mathrm{loc}_{simple}(p_i \mid c) = f_{rcd}(\mathrm{PLUT}(p_i, p_j, sM) - r_r). \tag{12}$$

Note that here we are no longer maintaining an explicit probability distribution on the location of an obstacle but instead a heuristic estimate of the risk of a single path relative to a single collision site.

4.3 Handling Multiple Collision Sites

Given a known collision site, both (9) and (12) provide a tool for selecting a candidate path to minimize the probability of collision. However, we have not yet

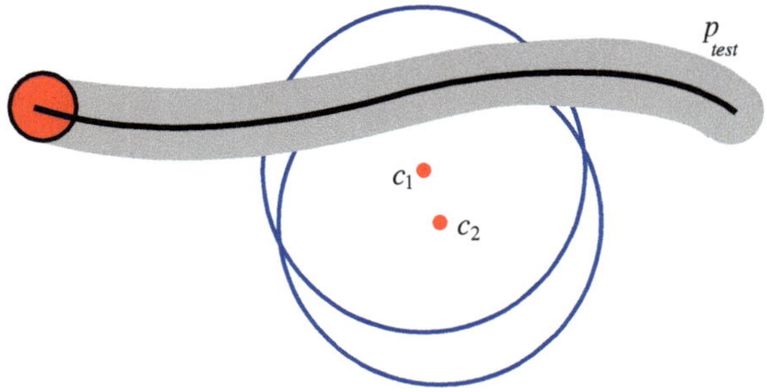

Fig. 5 Two collision sites c_1 and c_2 are located in close proximity. Intuition suggests that c_2 should be ignored when computing the risk of collision of path p_{test}. Either the two sites belong to the same obstacle, or else the obstacle at c_2 is "blocked" by the obstacle at c_1

addressed the issue of multiple known collision sites. The likelihood that two workspace points have the same obstacle outcome correlates strongly with the distance between them, by virtue of describing the same obstacle. The estimate of collision likelihood for an untested path depends on what statistical independence assumptions we make among known collision points.

Figure 5 depicts a situation in which two collision sites appear to be correlated. However, many possible policies for estimating statistical independence among a set of collision points, such as clustering, are complex to compute and implement.

In contrast, we may conservatively assume that all collisions are independent, in which case basic probability theory states that

$$\operatorname{loc}(p) = \operatorname{loc}(p \mid \{c_1,\ldots,c_n\}) = 1 - \prod_{i\in\{1,\ldots,n\}} (1 - \operatorname{loc}(p \mid c_i)). \tag{13}$$

If some collision sites are actually part of the same obstacle, then we are overestimating the likelihood of collision for p. In the absence of any knowledge regarding correlation, however, the most conservative policy is the safest. Thus, we now have the means to address Problem 1, Exploitation:

$$p_{next} = \underset{p_i\in\mathcal{P}}{\arg\min}\ \operatorname{loc}(\mathrm{p}_i). \tag{14}$$

In the next section, we explore an information theoretic approach to safely adjusting this pessimistic model.

4.4 Adaptive Locality Model

The locality models presented in Sects. 4.1–4.2 incorporate only positive collision test outcomes. Those static models conservatively estimate an obstacle distribution

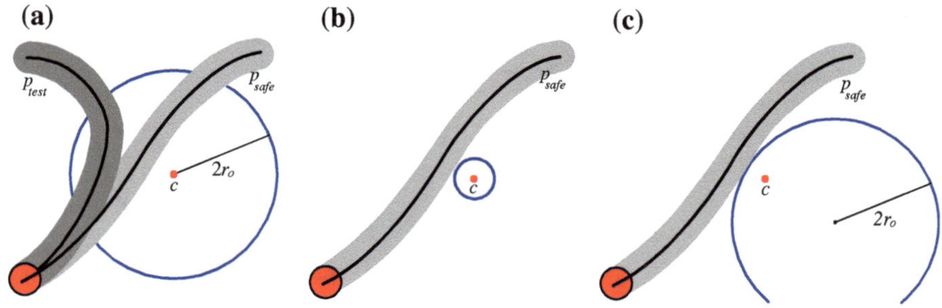

Fig. 6 **a** Given a collision point c and neighboring collision-free path p_{safe}, the blue circle represents a distribution on obstacle locations, some of which are invalidated by p_{safe}. The more distant candidate path p_{test} is not at risk of collision with the obstacle represented by c. **b** and **c** Two simple hypotheses on obstacle scale and position explain these two results. The distribution shown in Fig. 4b is simpler to represent during online path sampling

spread over a large but finite range of effect. We now construct an *adaptive* locality model capable of incorporating both positive and negative collision-test outcomes.

If we should happen to discover a safe path p_{safe} passing within collision site c's range of effect, then we may use this new information to refine the obstacle model of c. In effect, we adjust the locality function to act over a smaller range in the direction of p_{safe}. As Fig. 6a shows, no path p_{test} that is separated from c by p_{safe} can possibly be at risk of collision with this obstacle. This adaptive model effectively relaxes the earlier, rigid assumptions on obstacle size and independence of collision sites. In modeling such geometric relations, we depart from prior work addressing locality.

Following an update to the model, all future probability estimates involving c incorporate this new information. Although the independence assumption may initially make nearby paths like p_{test} appear riskier than they should (Fig. 5), the adaptive model rapidly cancels out this effect after finding a safe path to shrink each collision point's range of effect.

In addressing the problem of how to adaptively adjust obstacle distributions in reaction to a collision-free path, a variety of approaches present themselves. One possible approach is to shrink the range of effect for the obstacle at c, as in Fig. 6b, which supposes that the obstacle is smaller than initially thought. Another approach, to shift the entire distribution away from the safe path as in Fig. 6c, assumes that the obstacle size was correctly estimated, but its position was off.

We adopt a compromise position. We prefer that the collision site remains the center of a distribution in order to keep range checks efficient via look-up table. However, we also prefer to avoid altering the range of effect of the opposing side, about which we have no new data. We therefore split the range of effect into several regions of influence ("sides") centered around each collision site. In 2D, we have left and right sides of the obstacle, as in Fig. 7. In 3D, the division is topologically more arbitrary, although we split the obstacle into four sides.

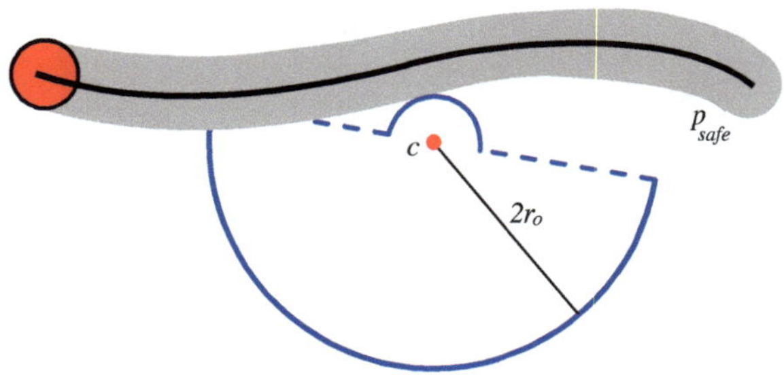

Fig. 7 The range of effect on each side of collision site c is maintained separately. The left range began at $2r_o$, but it was reduced after successfully collision-testing path p_{safe}

In splitting the locality model into several directions, we require a rule to consistently associate each path with a particular side of the collision point. The sides are defined relative to the pose of the robot before executing the path. The sides meet at the line **a**, an axis running through the start pose and the collision point. We assign names to the sides describing their position relative to the robot's frame of reference. Sides are determined by

$$\textit{left} = \operatorname{sgn}(\mathbf{t} \times \mathbf{p} \cdot \mathbf{u}) \tag{15}$$

$$\textit{top} = \operatorname{sgn}(\mathbf{t} \times \mathbf{p} \cdot \mathbf{a} \times \mathbf{u}), \tag{16}$$

where **u** denotes the robot's up vector, **p** the projection of c onto the path, and **t** the tangent vector of the path at this point, as in Fig. 8. These sides may be precomputed for each path. In 2D, it is particularly convenient to augment the PLUT with a sign indicating on which side of the path each possible collision point lies.

Figure 9 shows a family of paths on the left side of an obstacle. We deem each path equally likely to collide with the obstacle because they each approach equally near to the collision point, c. This assignment of paths to a single side of an obstacle places assumptions on the path's shape. We assume here that curvature is bounded and that paths are reasonably short. Previous work by Knepper et al. [15] thoroughly discusses these assumptions.

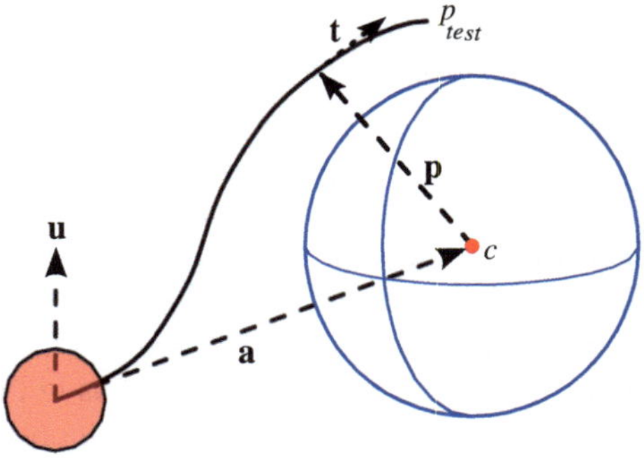

Fig. 8 In three dimensions, the adaptive locality model's range of effect is split into four sides. The robot's up vector, **u**, and the vector pointing toward the collision point, **a**, are used to define which of four sides the path p_{test} is on. As illustrated, the path is on the *top-left side*

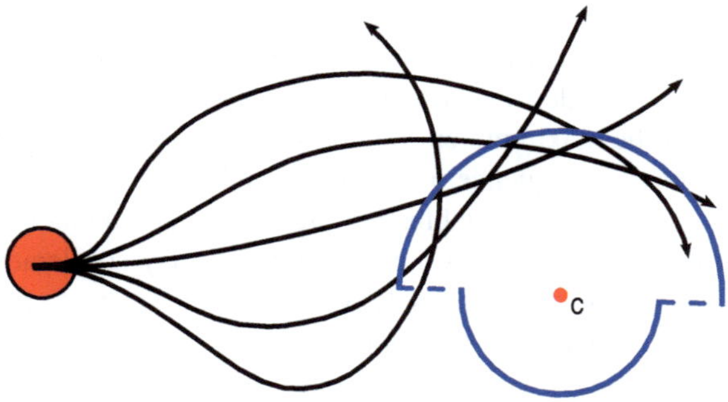

Fig. 9 A family of paths, all of which pass to the *left* of the collision site, c. Despite the variety of shapes, each path intrudes equally into the *left* range of effect of c, and thus they would each reduce its *left* range of effect equally

5 Path Entropy

Having established an adaptive locality model, we now consider a means to reap maximal advantage from its predictive capabilities in order to solve Problem 2, Exploration. It is important to select paths for collision test that cause the model to rapidly converge to an accurate description of obstacles, while simultaneously minimizing failed collision tests. Given a set of collision sites, the best path to collision test is that path with maximum entropy according to the current model parameters.

Definition 4 An untested path's **entropy** (sometimes called Shannon entropy) refers to the expected amount of information about the safety of other untested paths that would be gained from collision-testing it. A path's entropy is

$$H(\text{pct}(p)) = -\Pr(\text{pct}(p)) \log \Pr(\text{pct}(p)) - \Pr(\neg\text{pct}(p)) \log \Pr(\neg\text{pct}(p)).\Box \quad (17)$$

In order to maximize our understanding of the true distribution of obstacles with the fewest possible samples, we choose to sample the maximum entropy path:

$$p_{test} = \arg\max_{p_i \in \mathscr{P}} H(\text{pct}(p_i)). \quad (18)$$

Based on current information, the maximum entropy path has maximal uncertainty with regard to its collision with obstacles; its probability of collision is nearest to 50 %. Testing this particular path will therefore increase total knowledge more than any other. The result will be either a path that significantly reduces the range of effect for some known collision point(s) or a new collision point that is far from known collisions. In either case, model accuracy increases with maximal utility.

The policy of maximizing entropy was proposed by Jaynes [13] for the purpose of estimating an unknown distribution. Maximum entropy has been specifically applied to decision theory [8], as we employ it here. The decision that maximizes entropy is the one that minimizes the worst case possible outcome. In our case, the worst outcome is rediscovering a known result because it wastes computation time for no gain. This outcome takes two forms: testing a path passing through a known collision site, or retesting a known-safe path. This Γ-minimax approach [18] is capable of reasoning simultaneously about an entire family of probability distributions, called Γ—in our case, a range of theories about obstacle extent.

If the maximum entropy policy is pursued repeatedly, path selection proceeds to discover a sequence of safe paths and collision sites that are progressively nearer to each other, thus establishing precisely the boundaries separating the obstacles from free space. Knowing these boundaries may accelerate the process of sampling and testing paths more densely within the free space. In prior work [15], we provide one possible approach to this process.

Refining these boundaries is a process of diminishing returns, however. As we discover safe paths progressively closer to obstacles, the margin of uncertainty becomes so low that additional maximum entropy path samples provide negligible advantage for a variety of reasons: (1) the cost of collision testing a path often increases with greater obstacle proximity; (2) there is a computational cost associated with path sampling that is proportional to the number of known collision sites; and (3) paths close to obstacles are poor choices for execution. These factors could be incorporated into a utility function [6], but this remains as future work. Thus, we should not exclusively pursue the maximum entropy sampling policy, but also select path samples far from obstacles to maximize safety and path diversity.

We utilize several strategies to combine exploration and exploitation in a hybrid approach. In the absence of any uncertainty from our locality model (such as before the first collision site has been discovered), we sample from a low-dispersion sequence. In the presence of uncertainty, we compute the fraction f of the total allowed plan time that has already elapsed (recall that we assume realtime planning). With probability f we pursue an exploitation (obstacle avoidance) sampling strategy, whereas with probability $1 - f$ we instead pursue an exploration (boundary finding) strategy to refine our locality model. Sampled paths very near to known obstacles are set aside without testing for later use, since they make poor candidates for traversal. This threshold distance reflects a willingness to overlook very narrow corridors while other options avail themselves. We search these risky paths if there is time at the end of the plan cycle, after exhausting other paths for test.

6 Experimental Results

We conducted a set of experiments in simulation in order to obtain a sufficient quantity of trials to recognize statistically meaningful trends. Experimental setup is as in our earlier work [14], in which trials comprise sets of one-hundred planning problems; in each one, the curvature-constrained robot attempts to navigate through randomly generated 2D point-obstacle environments, each based on a query comprising start and goal poses. The robot moves continuously while replanning at a fixed rate. A heuristic function selects goal-directed local paths from a set of 2,401 paths. A low-fidelity global planner helps guide the robot to the goal. In order to assess the effect of various path sampling strategies, we varied replan cycle time.

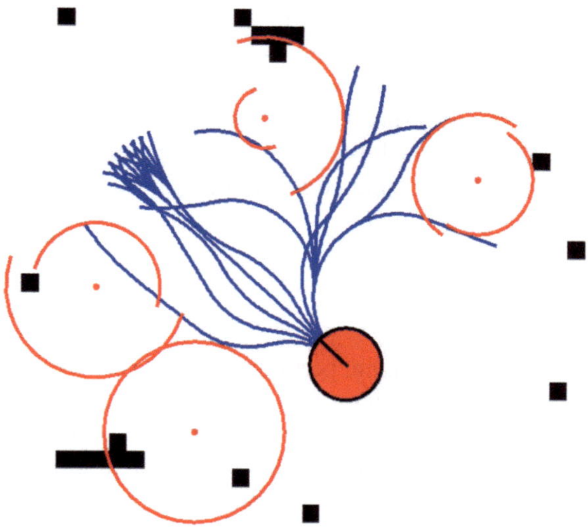

Fig. 10 Simulator depiction of locality model, showing: collision-free paths (*blue*), known collision sites (*red dots*), and their ranges of effect (concentric semicircles). Note that the *dots* correspond to the nearest edge of each C-space obstacle—the point most relevant from the robot's current pose. Not all obstacles (black) are relevant to the current local plan

Figure 10 shows a screen capture of our simulator, indicating known collision sites (in C-space) and ranges of effect. We consider four path sampling strategies:

- Low dispersion—sequence generated by the Green-Kelly algorithm [7]
- Avoid obstacles—pure exploitation; sample as far as possible from obstacles
- Find boundaries—pure exploration; selects maximum entropy path
- Hybrid approach—combines "avoid obstacles" and "find boundaries" strategies.

We present results on three of the strategies in Fig. 11. We see an increase in safe paths produced per unit time for both pure obstacle avoidance (up to 7.8x) and

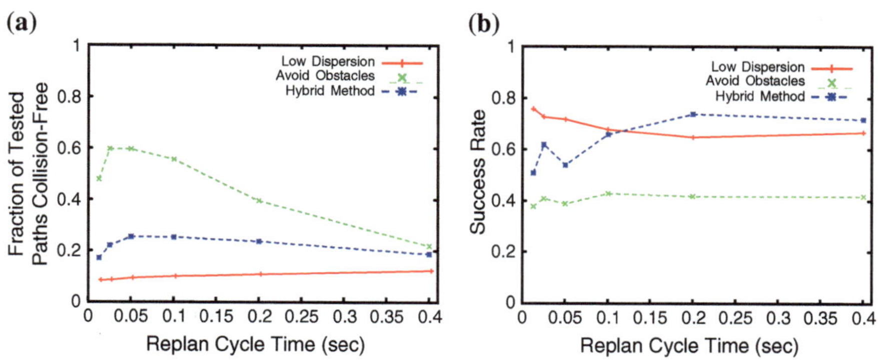

Fig. 11 All tests in these plots were run at an obstacle density of 2 %. **a** The locality model provides up to a 7.8x increase in the fraction of paths collision-free per replan cycle. As replan cycle time increases, we see regression toward the mean, as a larger total fraction of available paths are collision tested. **b** In success rate at solving planning queries, we see that the "avoid obstacles" strategy suffers in performance, whereas the hybrid approach, which strikes a balance between exploration and exploitation of the locality model, performs increasingly well as it has more time to gather information

the hybrid approach (up to 3x), compared to a fixed low dispersion sequence. However, pure obstacle avoidance sampling suffers a drop in performance at solving planning queries, which the hybrid approach overcomes. Note that the low-dispersion sampler actually declines slightly in performance as more samples are allowed, which is consistent with earlier results [14]. The hybrid planner shows a trend of increasing performance as it improves its locality model.

7 Discussion

In real time planning, performance is sensitive to the computational cost associated with collision testing. Alternatives that alleviate some of that computation can be beneficial, provided that such alternatives are computationally efficient themselves. In this paper, we present a strategy for informed path sampling that guides the search away from obstacles and towards safe or unexplored parts of the workspace. Although obstacle information is already available to the planner in costmap form, we obtain a significant increase in performance by representing the most salient subset of those obstacles in a more immediately accessible form.

We utilize a proximity look-up table to accelerate this process. Even so, our statistical model describing nearby obstacles and their relationship is necessarily simple. This model makes use of the principle of locality to search appropriately far from obstacle locations already discovered in prior collision tests to maximally reduce uncertainty. Using our probabilistic locality model, we trade off between exploration and exploitation in order to discover a variety of safe paths while largely avoiding searching colliding paths.

Acknowledgment This work is sponsored by the Defense Advanced Research Projects Agency. This work does not necessarily reflect the position or the policy of the Government. No official endorsement should be inferred.

References

1. N.M. Amato, O.B. Bayazit, L.K. Dale, C. Jones, D. Vallejo, OBPRM: an obstacle-based PRM for 3D workspaces, in *Workshop on the Algorithmic Foundations of Robotics*, Houston, TX, USA, March (1998), pp. 155–168
2. W.H. Beyer (ed.), *CRC Standard Mathematical Tables and Formulae* (CRC Press, Boca Raton, 1991)
3. B. Burns, O. Brock, Information theoretic construction of probabilistic roadmaps, in *International Conference on Intelligent Robots and Systems*, Las Vegas, NV, USA, Oct (2003)
4. B. Burns, O. Brock, Sampling-based motion planning using predictive models, in *International Conference on Robotics and Automation*, Barcelona, Spain, April (2005)
5. B. Burns, O. Brock, Single-query entropy-guided path planning, in *International Conference on Robotics and Automation*, Barcelona, Spain, April (2005)
6. B. Burns, O. Brock, Toward optimal configuration space sampling, in *Robotics: Science and Systems*, Cambridge, MA, USA, June (2005)

7. C. Green, A. Kelly, Toward optimal sampling in the space of paths, in *International Symposium of Robotics Research*, Hiroshima, Japan, Nov (2007)
8. P.D. Grünwald, A.P. Dawid, Game theory, maximum entropy, minimum discrepancy and robust bayesian decision theory. Ann. Stat. **32**(4) (2004)
9. D. Hsu, T. Jiang, J. Reif, Z. Sun, The bridge test for sampling narrow passages with probabilistic roadmap planners, in *International Conference on Robotics and Automation*, Taipei, Taiwnn, Sept (2003)
10. D. Hsu, G. Sánchez-Ante, Z. Sun, Hybrid PRM sampling with a cost-sensitive adaptive strategy, in *Proceedings of the IEEE International Conference on Robotics and Automation*, pp. 3885–3891 (2005)
11. D. Hsu, J.C. Latombe, H. Kurniawati, On the probabilistic foundations of probabilistic roadmap planning. Intl. J. Robot. Res. **25**(7), 627–643 (2006)
12. L. Jaillet, A. Yershova, S.M. LaValle, T. Simeon, Adaptive tuning of the sampling domain for dynamic-domain RRTs, in *International Conference on Intelligent Robots and Systems* (2005)
13. E.T. Jaynes, Information theory and statistical mechanics. Phys. Rev. **106**(4), 620–630 (1957)
14. R.A. Knepper, M.T. Mason, Empirical sampling of path sets for local area motion planning, in *International Symposium of Experimental Robotics*, Athens, Greece, July (2008)
15. R.A. Knepper, S.S. Srinivasa, M.T. Mason, An equivalence relation for local path sets, in *Workshop on the Algorithmic Foundations of Robotics*, Singapore, Dec (2010)
16. M. Morales, L. Tapia, R. Pearce, S. Rodriguez, N.M. Amato, A machine learning approach for feature-sensitive motion planning, in *Workshop on the Algorithmic Foundations of Robotics*, Utrecht/Zeist, The Netherlands, July (2004), pp. 361–376
17. A.F. van der Stappen, V. Boor, M.H. Overmars, The Gaussian sampling strategy for probabilistic roadmap planners, in *International Conference on Robotics and Automation* (1999), pp. 1018–1023
18. B. Vidakovic, *Γ-Minimax: A Paradigm for Conservative Robust Bayesians, Lecture Notes in Statistics*, vol. 152 (Springer, New York, 2000), pp. 241–259
19. Y. Yu, K. Gupta, An information theoretic approach to view point planning for motion planning of eye-in-hand systems, in *International Symposium on Robotics* (2000), pp. 306–311

Asymptotically Near-Optimal Is Good Enough for Motion Planning

James D. Marble and Kostas E. Bekris

Abstract Asymptotically optimal motion planners guarantee that solutions approach optimal as more iterations are performed. There is a recently proposed roadmap-based method that provides this desirable property, the PRM* approach, which minimizes the computational cost of generating the roadmap. Even for this method, however, the roadmap can be slow to construct and quickly grows too large for storage or fast online query resolution. From graph theory, there are many algorithms that produce sparse subgraphs, known as spanners, which can guarantee near optimal paths. In this work, a method for interleaving an incremental graph spanner algorithm with the asymptotically optimal PRM* algorithm is described. The result is an asymptotically *near*-optimal motion planning solution. Theoretical analysis and experiments performed on typical, geometric motion planning instances show that large reductions in construction time, roadmap density, and online query resolution time can be achieved with a small sacrifice of path quality. If smoothing is applied, the results are even more favorable for the near-optimal solution.

1 Introduction

Probabilistic roadmap planners [16] utilize an offline phase to build up knowledge about the configuration space (C-space) and solve many practical motion planning problems. Traditionally, many of these planners focus on feasibility and may return paths of low quality; considerably different from the optimal ones. Path quality can be measured in terms of clearance, or smoothness [34], but this work focuses on path quality measures that are metric functions such as length or traversal time.

J.D. Marble (✉) · K.E. Bekris
Computer Science and Engineering Department, University of Nevada,
1664 N. Virginia St., MS 171, Reno, NV 89557, USA
e-mail: jmarble@cse.unr.edu

K.E. Bekris
e-mail: bekris@cse.unr.edu

H.I. Christensen and O. Khatib (eds.), *Robotics Research*,
Springer Tracts in Advanced Robotics 100, DOI 10.1007/978-3-319-29363-9_24

Smoothing can be used to improve path quality and algorithms exist that produce roadmaps with paths that are deformable to optimal ones [13, 31]. Hybridization graphs [29] combine multiple solutions into a higher quality one that uses the best portions of each input path. These techniques, however, can be expensive for the online resolution of a query, especially when multiple queries must be answered.

Alternatively, it is possible to construct larger, denser roadmaps that better sample the *C*-space by investing more preprocessing time. For instance, a planner that attempts to connect a new sample to every existing node in the roadmap will eventually provide optimal solutions, a property known as asymptotic optimality. While roadmaps with this property are desirable for their high path quality, their large size can be problematic. Large roadmaps impose significant costs during construction, storage, transmission and online query resolution, so they may not be feasible for some applications. The recently proposed `k-PRM*` algorithm [14] minimizes the number of neighbors each new sample has to be connected to while still providing asymptotic optimality. Even so, the density of roadmaps produced by `k-PRM*` can be very high, resulting in slow online query resolution times, as shown in Fig. 1.

This paper argues that a viable alternative is to compute roadmaps with asymptotically near-optimal guarantees. By relaxing the optimality guarantees, it is possible to construct roadmaps that are sparser, faster to build, and can answer queries more quickly while providing solution paths with near-optimal guarantees. Additionally, these roadmaps tend to return a solution in the same homotopic class as the optimum one, so smoothing brings path quality even closer to optimal in practice.

The theoretic foundations for this work lie in graph theory. In particular, graph spanners are sparse subgraphs with guarantees on path quality. Roadmaps with Useful Cycles [25] are, in fact, spanners. Edges that pass the "usefulness" test are added to the roadmap because not doing so would violate the guarantees about path quality. In Sect. 3, this idea is applied to an asymptotically optimal roadmap planner to produce the Incremental Roadmap Spanner (`IRS`) algorithm. This planner can incrementally construct an asymptotically near-optimal roadmap faster

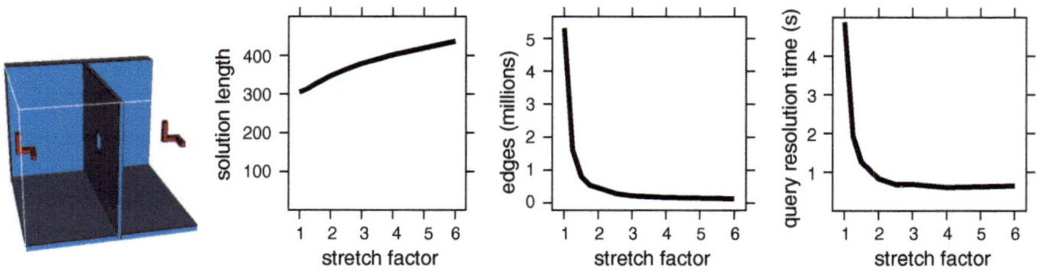

Fig. 1 Relaxing the requirement to find asymptotically optimal solutions, a large decrease in roadmap density and solution query time can be achieved for the problem shown on the *left side*. The larger this relaxation (stretch factor), the sparser the roadmap. These results are averaged over 1000 random queries on 10 runs of 50,000 vertex roadmaps. The `k-PRM*` algorithm corresponds to a stretch factor equal to 1, which results in a considerably higher query resolution time

than an asymptotically optimal one can be constructed. The resulting graph is also sparse, which is beneficial in any application where small roadmaps are preferable.

1.1 Related Work

There has been a plethora of techniques on sampling and connecting configurations to achieve *computational efficiency* in roadmap construction [2, 11, 28, 30, 32, 35]. Certain algorithms, such as the Visibility-based Roadmap [32], Incremental Map Generation [35] and the Reachability Roadmap [8] focus on returning a connected roadmap that covers the entire *C*-space. A reachability analysis suggested that connecting roadmaps is more difficult than covering the *C*-space [10]. A method is available to characterize the contribution of a sample to the exploration of the *C*-space [23], but it does not address how connectivity contributes to path quality.

Work on creating roadmaps that contain high quality paths has been motivated by the objective to efficiently resolve queries without the need for a post-processing optimization step of the returned path. Certain techniques aim to compute all different homotopic solutions [13, 31]. Another approach inspired by Dijkstra's algorithm extracts optimal paths from roadmaps for specific queries [18] but may require very dense roadmaps. The Useful Cycles approach implicitly creates a roadmap spanner with small number of edges [25] and has been combined with the Reachability Roadmap Method to construct connected roadmaps that cover 2D and 3D *C*-spaces and provide high quality paths [9]. A method filters nodes from a roadmap if they do not improve path quality measures [26].

Tree-based algorithms [12, 19] focus on single query planning, but they do not provide the preprocessing properties of roadmaps. A tree is already a sparse graph and it becomes disconnected when removing edges. `Bi-RRT` produces arbitrarily bad paths with high probability and can miss high quality paths [14, 24]. Anytime `RRT` [6] is an approach that incrementally improves paths.

Roadmaps require a solution to the two-point boundary value problem, i.e., computing a path that connects exactly two states of a moving system. On the other hand, tree-based kinodynamic planners can operate on such environments. Such planners can benefit, however, from an approximate roadmap to estimate distances between states and the goal region that take into account *C*-space obstacles. Such distance estimates can be used as a heuristic in tree expansion to bias the selection of states closer to the goal and solve problems with dynamics faster [4, 20].

The `RRG`, `RRT`*, and `PRM`* family of algorithms [14] provide asymptotic optimality for general configuration spaces. `RRG` and `RRT`* are based on `RRT`, a tree-based planner. The Anytime `RRT`* approach [15] extends `RRT`* with anytime planning in dynamic environments and can incrementally improve path quality. `PRM`* is a modification of standard `PRM`. The proposed technique is based on a variation of `PRM`*, i.e., `k-PRM`*, during the offline, roadmap-building step. More details on `k-PRM`* will be provided in Sect. 2.1.

1.2 Contribution

The contributions of this paper are the following:

1. The paper proposes a method for quickly constructing sparse roadmaps that provide high quality solutions to motion planning queries (IRS).
2. It provides an analysis of IRS showing the following:
   a. IRS has an asymptotic time complexity close to that of k-PRM*.
   b. Roadmaps constructed by IRS will provide solutions asymptotically near-optimal in the same spaces that k-PRM* would provide asymptotic optimality.
3. Experimental evidence showing that roadmaps constructed by IRS have the following properties:
   a. faster construction time
   b. sparsity, which results in faster online query resolution
   c. higher path quality than the theoretical lower bounds
   d. even more favorable comparison when smoothing is applied

Specifically, a method similar to Useful Cycles [25] is formalized as a graph spanner algorithm and is interleaved with k-PRM*. It is shown that a constant factor of the asymptotic optimality (asymptotic near-optimality), is maintained. IRS can incrementally construct a roadmap spanner in a continuous space, while most graph spanner algorithms are formulated to operate on an existing roadmap. Because IRS operates on an asymptotically optimal planner, it provides asymptotically near-optimal guarantees that the Useful Cycles approach did not since it employed a constant number of neighbors. IRS produces sparser roadmaps faster than k-PRM*. Previous work by the authors [22] has utilized state-of-the-art graph spanner algorithms with good complexity performance to prune the edges of a roadmap constructed with k-PRM*.

The proposed method balances two extremes in terms of motion planning solutions. On one hand, the connected component heuristic for PRM can connect the space very quickly with a very sparse roadmap, but can produce very poor solutions. On the other hand, the k-PRM* algorithm provides asymptotically optimal roadmaps that may be very dense and slow to construct. With IRS it is possible to tune the solution quality degradation relative to k-PRM* and select a parameter, the stretch factor, that will return solutions arbitrarily close to the optimal ones, while still constructing sparse roadmaps relatively fast.

The theoretical guarantees on path quality that this technique provides are tested empirically in a variety of motion planning problems in $SE(3)$. In all of these environments, the majority of the edges can be removed while increasing mean path length by only a small amount, which can be further reduced by utilizing path smoothing. Path degradation is most pronounced for paths that are very short, while longer paths are less affected. The sparsity of the roadmaps produced is a valuable feature in itself, but a marked decrease in construction time is also measured.

2 Foundations

A robot can be abstracted as a point in a d-dimensional configuration-space (C-space) where the set of collision-free configurations define $C_{\text{free}} \subset C$ [21]. The experiments performed for this paper use the space of rigid body configurations (*SE* (3)) as the C-space, but the proposed method is applicable to any C-space that is also a metric and probability space for reasons described in Sect. 3. Once C_{free} can be calculated for a particular robot, one needs to specify initial and goal configurations to define an instance of the path planning problem:

Definition 1 (*The Path Planning Problem*) Given a set of collision-free configurations $C_{\text{free}} \subset C$, initial and goal configurations $q_{\text{init}}, q_{\text{goal}} \in C_{\text{free}}$, find a continuous curve $\sigma \in \Sigma = \{\rho | \rho : [0,1] \to C_{\text{free}}\}$ where $\sigma(0) = q_{\text{init}}$ and $\sigma(1) = q_{\text{goal}}$.

The `PRM` algorithm [16] can find solutions to the Path Planning Problem by sampling configurations in C_{free} then trying to connect them with local paths. It starts with an empty roadmap and then iterates until some stopping criterion is satisfied. For each iteration, a configuration in C_{free} is sampled and the k-nearest neighbors are found. For each of these neighbors, an attempt is made to connect them to the sampled configuration with a local planner. If such a curve in C_{free} can be found, an edge between the two configurations is added to the roadmap.

Once the stopping criterion is met, the offline phase of the algorithm is complete. The result is a graph $G = (V, E)$ that reflects the connectivity of C_{free} and can be used to answer query paIRS of the form $(q_{\text{init}}, q_{\text{goal}}) \in C_{\text{free}} \times C_{\text{free}}$ where q_{init} is the starting configuration and q_{goal} is the goal configuration. This type of graph is known as a roadmap. The procedure for querying it is to add q_{init} and q_{goal} to the roadmap is similar to the way sampled configurations are added during the offline phase. Then, a discrete graph search is performed to find a path on the roadmap between the two configurations.

2.1 k-PRM*

Asymptotic optimality is the property of an algorithm that, given enough time, solutions to the path planning problem will almost surely converge to the optimum as defined by some cost function.

Definition 2 (*Asymptotic Optimality in Path Planning*) An algorithm is *asymptotically optimal* if, for any path planning problem $(C_{\text{free}}, q_{\text{init}}, q_{\text{goal}})$ and cost function $c : \Sigma \to \mathbb{R}_{\geq 0}$ that admit a robust optimal solution with finite cost c^*, the probability that it will find a solution of cost c^* converges to 1 as the number of iterations approach infinity.

Asymptotic optimality is defined only for *robustly feasible* path planning problems as defined in the presentation of `k-PRM`* [14]. Such problems have a minimum clearance around the optimal solution and cost functions with a continuity property.

The standard PRM, as described above, attempts to connect sampled configurations to a fixed number, k, of nearest neighbors. It has been shown that this prevents the algorithm from providing asymptotic optimality [14]. The k-PRM* algorithm rectifies this deficiency by making k a logarithmic function of the size of the roadmap. Specifically, $k(n) = k_{\mathrm{PRM}} \log n$ where $k_{\mathrm{PRM}} > e(1 + 1/d)$. It has been proven that roadmaps constructed with this variation will almost surely converge to optimal solutions [14]. That is, k-PRM* is asymptotically optimal.

2.2 *Graph Spanners*

A graph spanner, as formalized in [27], is a sparse subgraph. Given a weighted graph $G = (V, E)$, a subgraph $G_S = (V, E_S \subset E)$ is a t-spanner if for all pairs of vertices $(v_1, v_2) \in V$, the shortest path between them in G_S is no longer than t times the shortest path between them in G. Because t specifies the amount of additional length allowed, it is known as the *stretch factor* of the spanner.

A simple method for spanner construction, which is a generalization of Kruskal's algorithm for the minimum spanning tree, is given in Algorithm 1 [1]. Instead of accepting only edges that connect disconnected components, this algorithm accepts edges that provide shortcuts. Kruskal's algorithm is recovered by setting t to a large value.

Algorithm 1 GRAPHSPANNER(V, E, t)

1: sort E by non-decreasing weight
2: $E_S \leftarrow \emptyset, G_S \leftarrow (V, E_S)$
3: **for all** $(v, u) \in E$ **do**
4:    **if** SHORTESTPATH$(V, E_S, v, u) > t \cdot$WEIGHT(v, u) **then**
5:       $E_S \leftarrow E_S \cup \{(v, u)\}$
6: **return** G_S

The inclusion criteria on line 1 ensures that no edges required for maintaining the spanner property are left out. From the global ordering of the edges performed on line 1, it has been shown that the number of edges retained by this algorithm is reduced from a potential $O(n^2)$ to $O(n)$ [1].

The quadratic number of shortest path queries puts the time complexity into $O(n^3 \log n)$, however, much work has been done to find algorithms that reduce this. Most perform clustering in a preprocessing step and one method uses this idea to reduce the time complexity to $O(m)$, where m is the number of edges [3].

A number of spanner algorithms take advantage of the implicit weights of a Euclidean metric space to speed up the process. Most of these operate on *complete*

Euclidean graphs [7, 17]. These algorithms can't be used on roadmaps that operate in C-spaces with obstacles because obstacles remove edges of the complete graph.

A concept central to the proposed technique and graph spanners is that of asymptotic *near*-optimality. This is a relaxation of asymptotic optimality that permits an algorithm to converge to a solution that is within t times the cost of the optimum.

Definition 3 (*Asymptotic Near-Optimality in Path Planning*) An algorithm is *asymptotically near-optimal* if, for any path planning problem $(C_{\text{free}}, q_{\text{init}}, q_{\text{goal}})$ and cost function $c : \Sigma \rightarrow \mathbb{R}_{\geq 0}$ that admit a robust optimal solution with finite cost c^*, the probability that it will find a solution of cost $c \leq tc^*$ for some stretch factor $t \geq 1$ converges to 1 as the number of iterations approach infinity.

3 Approach

The high-level approach is to combine the construction of an asymptotically optimal roadmap with the execution of a spanner algorithm. These two tasks can be performed sequentially (first the roadmap, then the spanner), or incrementally by interleaving the individual steps of each task.

3.1 Sequential Roadmap Spanner

An asymptotically optimal roadmap G generated by k-PRM* can be used to construct an asymptotically *near*-optimal roadmap G_S by selectively removing existing edges. Simply applying an appropriate spanner algorithm to the roadmap accomplishes this and has been studied by the authors in previous work [22].

This approach can be wasteful, however. Every edge in G has been checked for collisions, but many of these will be discarded by the spanner algorithm. An incremental spanner algorithm that accepts or rejects on an edge-by-edge basis, such as Algorithm 1, is a better alternative. If all collision detection is deferred until after the offline phase of k-PRM* is complete (similar to lazy PRM [5]) then edges rejected by the spanner algorithm can skip collision detection all together. Since collision detection is frequently the most expensive operation in motion planning, the time savings could be substantial. Additionally, the graph spanner algorithm used [3] was not able to provide a large reduction in edges for low stretch values. For example, in one environment it was only able to reduce the edge count by 15.3 % for a stretch factor of 3. The incremental method described in the next section was able to achieve a 70.5 % edge reduction for the same experimental parameters.

A potential downside to using an incremental spanner algorithm is higher time complexity. If the entire graph is available at once, then an algorithm that has linear time complexity and guarantees on the number of edges removed can be used [3]. In practice, however, the time saved by avoiding collision detection dominates the additional cost of using an incremental spanner algorithm.

3.2 Incremental Roadmap Spanner

The Incremental Roadmap Spanner (`IRS`, Algorithm 2) takes the idea of a sequential roadmap spanner one step further. Here, roadmap and spanner construction are interleaved. When the roadmap algorithm adds an edge, the spanner algorithm can reject it before collision detection and before it is added to the roadmap. Before discussing the implementation of the algorithm, some subroutines must be defined:

SAMPLEFREE uniformly samples a random configuration in C_{free}.

NEAR(V, v, k) returns the k configurations in set V that are nearest to configuration v with respect to a metric function. This can be implemented as a linear search through the members of V or as something more involved, such as a nearest neighbors search in a *kd*-tree.

COLLISIONFREE(v, u) detects if there is a path between configurations v and u in C_{free}. A local planner plots a curve in C from v to u. Points along this curve are tested for membership in C_{free} and if any fail, the procedure returns false.

STOPPINGCRITERIA determines when to stop looking for a solution. Some potential stopping criteria are:

- a solution of sufficient quality has been found
- allotted time or memory have been exhausted
- enough of the C-space has been covered

or some combination of these or other criteria.

WEIGHT(v, u) returns the positive edge weight of (v, u). In the context of motion planning, the weight of an edge is frequently the cost of moving the robot from configuration v to configuration u along the curve provided by a local planner.

SHORTESTPATH(V, E, v, u) returns the cost of the shortest path between v and u. Note that the actual shortest path cost isn't required, just whether it is larger than t times the weight of the edge (v, u). Instead of naively applying a full graph search, a length variation of Dijkstra's algorithm search can be employed. Additionally, edges that connect two disconnected components can be added without doing any kind of search because the shortest path cost in this case is infinite.

Algorithm 2 INCREMENTALROADMAPSPANNER(t)

1: $V \leftarrow \emptyset, E \leftarrow \emptyset$
2: **while** !STOPPINGCRITERIA() **do**
3: $\quad v \leftarrow$ SAMPLEFREE()
4: $\quad k \leftarrow \lceil e \cdot (1 + \frac{1}{d}) \log(\|V\|) \rceil$
5: $\quad U \leftarrow$ NEAR(V, v, k)
6: $\quad$ sort U by non-decreasing distance from v
7: $\quad$ **for all** $u \in U$ **do**
8: $\quad\quad$ **if** SHORTESTPATH(V, E, v, u) $> t \cdot$WEIGHT(v, u) **then**
9: $\quad\quad\quad$ **if** COLLISIONFREE(v, u) **then**
10: $\quad\quad\quad\quad E \leftarrow E \cup \{(v, u)\}$
11: $\quad V \leftarrow V \cup \{v\}$
12: **return** (V, E)

First, the roadmap $G = (V, E)$ is initialized to empty (line 2). Then, it iterates until StoppingCriteria returns true (line 2). For each iteration, SampleFree is called and returns $v \in C_{\text{free}}$ (line 2). The number of nearest neighbors is calculated (line 2). The k-nearest neighbors of v, U are found by calling Near(V, v, k) (line 2). If Near does not return U ordered by distance from v, then it must be sorted on line 2. For each potential edge connecting v to a neighbor in U, the inclusion criteria must be met before it is added to the roadmap. First, if a path exists between v and u with cost less than t times the weight of (v, u) then that edge can be rejected because it contributed little to path quality (line 2). Second, the edge must be checked for collision (line 2), as it is with other variations of PRM. If the local planner does not succeed in finding a curve in C_{free} the edge is rejected. If the edge passes both inclusion tests, it is added to the roadmap (line 2). Finally, the sampled vertex is added to the roadmap (line 2) and the next iteration is started.

A notable departure that IRS makes from Algorithm 1 is that the edges are not ordered globally. This ordering is not required to preserve solution quality, however, as will be shown in Sect. 3.3. A *local* ordering of potential edges is performed on line 2. This doesn't affect the theoretical bounds on solution quality, but can be seen as a heuristic that improves the sparsity of the final roadmap.

Since the spanner property is tested before the collision check, many expensive collision checks can be avoided. This property greatly improves running time for practically sized roadmaps, as shown in Sect. 4.

3.3 Analysis

Theorem 1 IRS *is asymptotically near-optimal.*

Proof The proof of Theorem 1 relies on the asymptotic optimality of k-PRM* [14], on which IRS is based. The proposed technique has two differences from k-PRM*.

First, on line 2, the potential neighbors of a newly added vertex are ordered by non-decreasing distance from the new vertex. This would have no effect on the asymptotic optimality of k-PRM* because edges are rejected based solely on collisions with obstacles. The order that edges are tested for collisions has no effect on those tests.

The second difference, on line 2, adds an additional acceptance criterion. This has the effect of making the edges in a roadmap output by IRS a subset of those output by k-PRM*.

Consider a pair of such roadmaps, $G = (V, E)$ returned by k-PRM* and $G_S = (V, E_S \subset E)$ returned by IRS. For each rejected edge (v, u) in E/E_S, there was a path from v to u with a cost less than t times the weight of (v, u). This invariant is enforced by line 2.

The shortest path σ in G between any two points $a, b \in V$ has cost c^*. This path may contain edges that are in E but not in E_S. For each of these edges (v, u), there exists an alternate path in G_S, with cost $c_{(v,u)} \leq t \cdot w(v, u)$. Therefore, there is a path σ_S between a and b in G_S with cost $\sum_{(v,u)}^{\sigma_S} c_{(v,u)} \leq t \cdot c^*$. In other words, since each detour is no longer than t times the cost of the portion of the optimal path it replaces, the sum cost of all of the detours will not exceed t times the total cost of the optimal path.

3.4 Time Complexity

- Time complexity breakdown for k-PRM* (ignoring constant time operations):
  - For each of n iterations:

    NEAR (ε-approximate): log n
    For each log n neighbors:

    COLLISIONFREE: $\log^d p$

For each of the n iterations of k-PRM*, a nearest neighbors search and collision checking must be performed. An ε-approximate nearest neighbor search can be done in log n time producing $k(n)$ = log n neighbors. The edge connecting the current iteration's sample to each of these neighbors must be checked for collision at a cost of $\log^d p$ time, where p is the number of obstacles. So, the total running time of k-PRM* is $O(n \cdot (\log n + \log n \cdot \log^d p))$, which simplifies to $O(n \cdot \log n \cdot \log^d p)$.

- Time complexity breakdown for IRS (ignoring constant time operations):
  - For each of n iterations:

    NEAR (ε-approximate): log n
    neighbor ordering: $\log n \cdot \log \log n$

For each $\log n$ neighbors:

CollisionFree: $\log^d p$
ShortestPath > t ·Weight: $t^d \log n \cdot \log(t^d \log n)$ (average case)

For IRS, two additional steps are performed. At every iteration, $k = \log n$ neighbors must be sorted by distance to the sampled vertex. Because many implementations of Near return the list of neighbors in order of distance, the cost of this can be zero. If this is not the case, however, the cost of sorting these k neighbors is $O(k \log k)$, and since $k = \log n$, the final result is: $O(\log n \cdot \log(\log n))$.

A shortest path search must also be performed for each potential edge. If done across the entire roadmap, this has a cost of $O(m \log n) = O(n \log^2 n)$, where m is the number of edges and $m = n \log n$, since each node is connected to at most $\log n$ neighbors. The shortest path algorithm will expand only the vertices with path cost from v that is lower than $t \cdot w(v, u)$. This is due to the fact that the algorithm requires only knowledge about the existence of a path between v and u shorter than that. When $t = 1$, the number of nodes expanded by such a search is, at most, $k(n) \in O(\log n)$. Assuming a uniform sampling distribution, the expected number of nodes that may be expanded when $t > 1$ is proportional to $t^d \log n$ (based on the volume of a d-dimensional hyper-sphere). This brings the *expected* time complexity of IRS to

$$O\left(n \cdot \left(\log n + \log n \cdot \log\log n + \log n \cdot \left(\log^d p + t^d \log n \cdot \log\left(t^d \log n\right)\right)\right)\right)$$

which simplifies to : $O\left(n \cdot t^d \log n \cdot t^d \log n \cdot \log\left(t^d \log n\right)\right)$, or for fixed t and d :

$$O\left(n \cdot \log^2 n \cdot \log\log n\right)$$

which is asymptotically slower than k-PRM*. However, as will be shown experimentally in Sect. 4, the very large constants involved in collision checking that can be skipped by IRS makes this a faster algorithm for roadmaps of practical size.

3.5 Size Complexity

The number of edges in a spanner produced by Algorithm 1 is $O(n)$ with constants related to t and d. This bound does not hold for IRS because, without a global ordering of edges, there is no guarantee that the minimum spanning tree is contained within the spanner. Because of this, the space complexity cannot be bounded lower than that provided by k-PRM*, which is $O(n \log n)$. The experimental results, however, suggest that the dominant term is linear as the number of edges added at each iteration appear to converge to a constant.

4 Evaluation

All experiments were run using the Open Motion Planning Library (OMPL) [33] on 2 GHz processors with 4 GB of memory. Four representative environments were chosen from those distributed with OMPL in addition to the simple environment shown in Fig. 1.

alpha puzzle: The entire free space is highly constrained in this classical motion planning benchmark (Fig. 2a).

apartment: The piano is the robot in this environment with a very expensive collision detection cost (Fig. 2b).

bug trap: A rod shaped robot must orient itself to fit through the narrow passage and escape a three-dimensional version of the classical bug trap (Fig. 2c).

cubicles: Two floors of loosely constraining walls and obstacles must be navigated (Fig. 2d).

For each environment, five roadmaps of 50,000 vertices were generated by `k-PRM`, `k-PRM`*. For `IRS`, 10 different roadmaps were generated for each of 8 stretch factors. On each of these roadmaps, 1000 random start and goal pairs were queried and various qualities of the resulting solutions were measured.

4.1 Construction Time

Although the expected asymptotic time complexity of `IRS` is worse than that of `k-PRM`*, the large constants involved in collision detection dominate the running time in these experiments. Since `IRS` reduces the number of collision checks required, running time is reduced the higher the stretch factor is. This is shown in Fig. 3, where a stretch factor of $t = 2$ allows `IRS` to construct a 50,000 node roadmap in under half the time of `k-PRM`*. The diminishing returns shown for higher stretch factors reflect the larger area of the graph that must be searched for shortest paths.

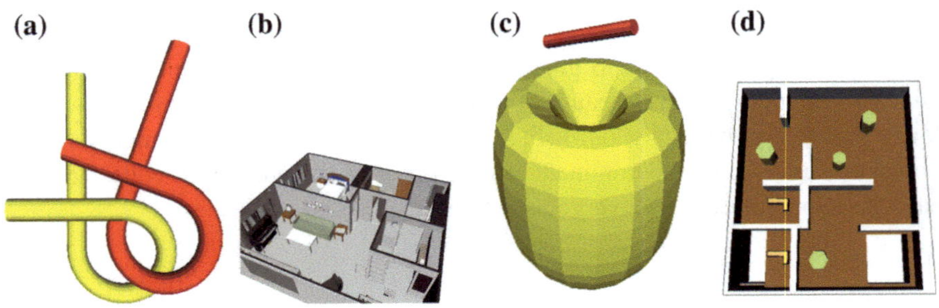

Fig. 2 Environments used in the experiments with example configurations for the robot. **a** Alpha puzzle. **b** Apartment. **c** Bug trap. **d** Cubicles

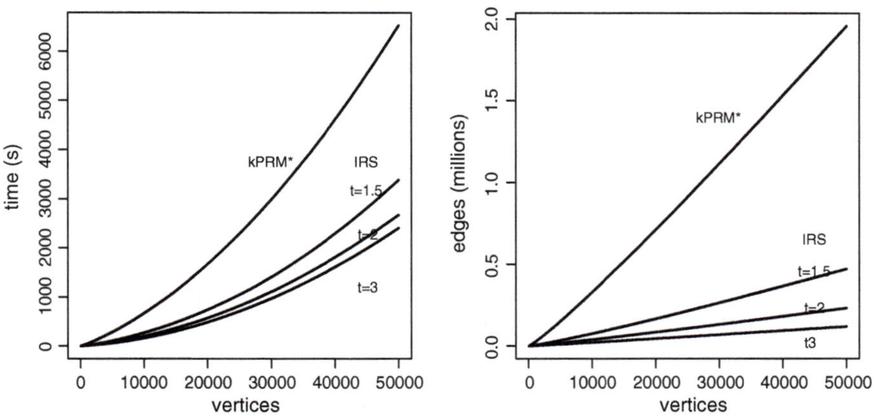

Fig. 3 Construction time and roadmap density for k-PRM* and IRS in the apartment environment averaged over 5 runs. Despite a higher asymptotic complexity, practically sized roadmap spanners can be more quickly constructed because fewer edges must be checked for collision. A dramatic reduction in the number of edges is shown for stretch factor as low as 1.5

4.2 *Space Requirements*

While each roadmap contains the same number of vertices (50,000), the space required for connectivity information is reduced by up to 95 %. Environments with a more connected free space had a larger reduction in the number of edges because they had more edges that could be removed while still maintaining connectivity (Fig. 4).

4.3 *Solution Quality*

In Fig. 5, path quality is measured by querying a roadmap with 1000 random start and goal configurations. The lengths of the resulting paths increase as the number of

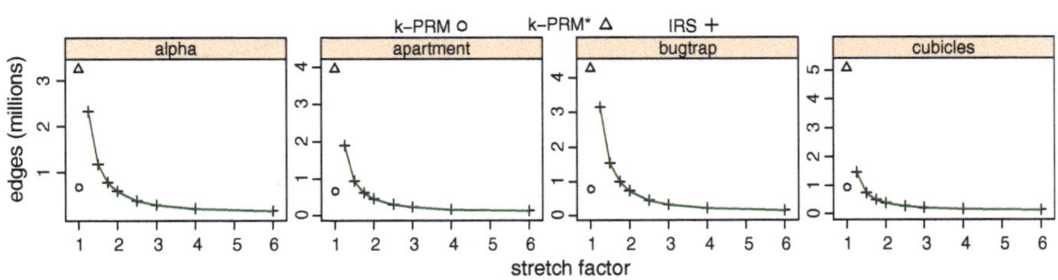

Fig. 4 The roadmap density in 10 different 50,000 node roadmaps for each environment. Environments with denser obstacles have fewer edges than k-PRM* but gain a smaller improvement from IRS

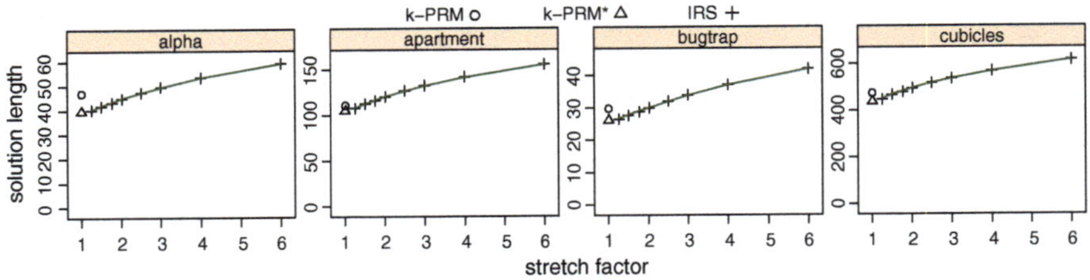

Fig. 5 Mean solution length of 1000 random query pairs on 10 different 50,000 node roadmaps for each environment. Solution length increases when the stretch factor is increased, but the mean increase is much less than the upper bound guaranteed by the algorithm

edges in the spanner is reduced. For these random starting and goal configurations, the average extra cost is much shorter than the worst case guaranteed by the stretch factor.

It is interesting which paths are most affected by this increase in path length. The worst degradation happens for short paths, where taking a detour of even a single vertex can increase the path length by a large factor. Path quality degradation in `IRS` is plotted in Fig. 6 as a function of its length in a roadmap generated by `k-PRM*`. All shortest paths to 10 random vertices are averaged over 5 different roadmaps.

In Fig. 7, the trade-off between roadmap density and path quality is directly compared. Although `k-PRM` can provide either a sparser roadmap or higher quality solutions than `IRS`, it cannot provide a better trade-off except in the apartment environment.

Fig. 6 Mean path length degradation for different stretch factors as a function of the path length in the original roadmap

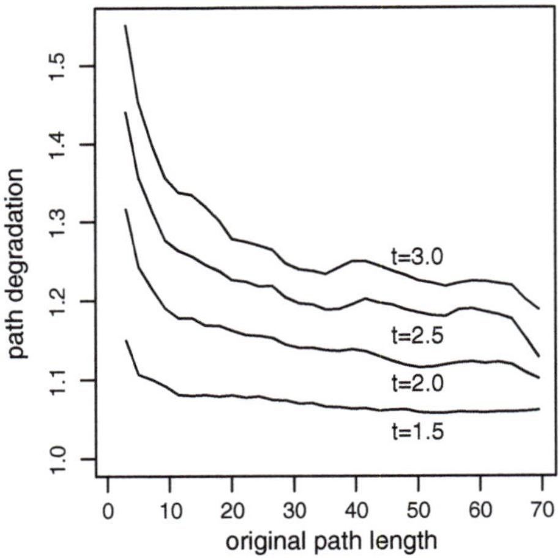

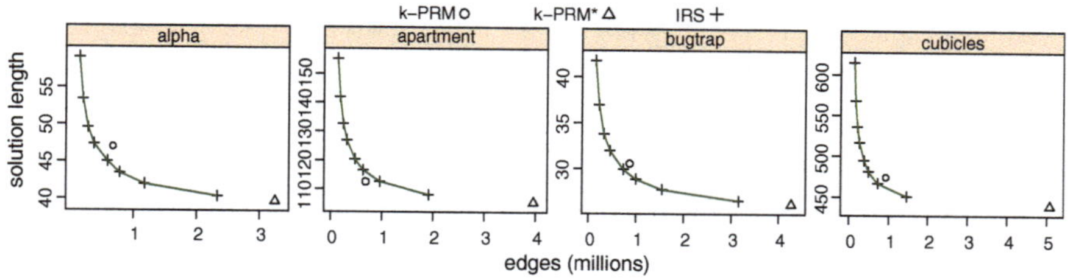

Fig. 7 Trade-off between solution length and roadmap density averaged over 1000 random query pairs on 10 different 50,000 node roadmaps for each environment. Each data point for `IRS` represents a different stretch factor from 1.5 to 6

4.4 *Query Resolution Time*

Query resolution time includes the time it takes to connect the start and goal configurations to the roadmap and perform an A* search. The connection time is not affected by the number of edges in the roadmap, only the number of vertices. Since each roadmap has the same number of vertices, the roadmap connection time is fixed and variations in query resolution time can be attributed to differences in the running time of the A* search. As shown in Fig. 8, this variation is very large. Removing edges from the roadmaps reduces the query resolution time by up to 70 %.

4.5 *Effects of Smoothing*

For each query, a simple approach for path smoothing was tested. Nonconsecutive vertices on the path that were near each other were tested for connectivity. If they could be connected, then the path is shortened by removing intervening vertices. This is a greedy and local method for smoothing, but it executes in less than40 ms,

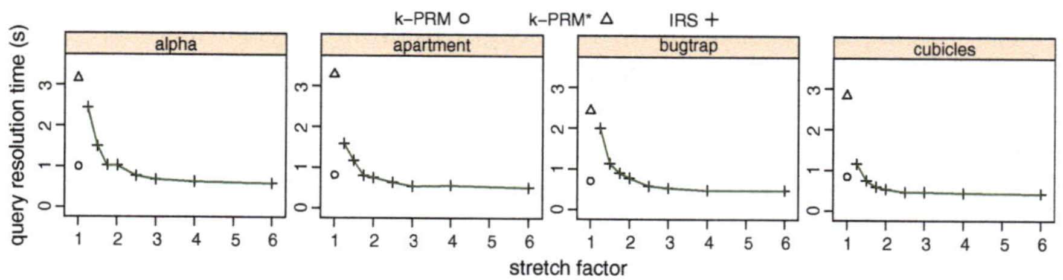

Fig. 8 Mean query resolution time of 1000 random query pairs on 10 different 50,000 node roadmaps for each environment. The online portion of the algorithms is reported by this chart and includes the time it takes to connect the start and goal configurations as well as the A* graph search. Higher stretch factors produce roadmaps with fewer edges that can be searched more quickly

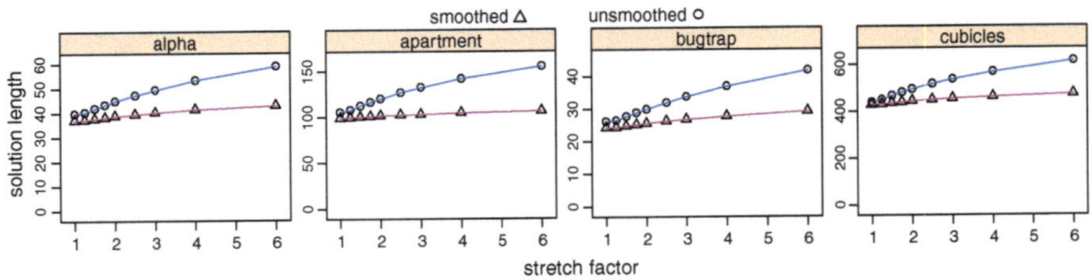

Fig. 9 A comparison between smoothed and unsmoothed solution lengths of 1000 random query pairs on 10 different 50,000 node roadmaps for each environment. The relative solution length increase for higher stretch factors is lower for the smoothed solutions. This reflects the ability of spanners to preserve homotopic path classes. Note that a stretch factor of $t = 1$ corresponds to `k-PRM*`

and produced impressive results (Fig. 9). Note that the gap between the smoothed and unsmoothed solution length indicates that both `k-PRM*` and `IRS` have yet to converge to the optimal solution.

The smoothing time increases as the sparsity of the roadmap increases. This reflects the larger number of vertices in solutions from these roadmaps. However, the time taken to smooth the solutions is two orders of magnitude shorter than the query resolution time in these experiments, although it may become consequential in other environments or for other smoothing techniques.

5 Discussion

This work shows that it is practical to compute sparse roadmaps in *C*-spaces that guarantee asymptotically near-optimal paths. The experimental results suggest that it is possible for these roadmaps to have considerably fewer edges than roadmaps with asymptotically optimal paths, while resulting in much smaller degradation in path quality relative to the almost optimal roadmaps. The stretch factor parameter provides the ability to tune this trade-off. The experiments confirm that roadmaps constructed with lower stretch factors can have higher path quality but are denser.

The existing approach removes only edges from the original roadmap. Since, however, the roadmap is embedded in a continuous *C*-space, it may be that nodes of the roadmap are redundant for the computation of near optimal paths. Future work will investigate how to remove nodes from roadmaps so that the quality of a path answering a query in the continuous space is guaranteed not to get worse than the specified stretch factor.

Finally, it is important to study the relationship of the resulting spanner roadmaps with methods that guarantee the preservation of the homotopic path classes in the *C*-space [13]. Intuitively, homotopic classes tend to be preserved by the spanner because the removal of an important homotopic class will have significant effects in the path quality.

References

1. I. Althöfer, G. Das, D. Dobkin, D. Joseph, J. Soares, On sparse spanners of weighted graphs. Discrete Comput. Geom. **9**(1), 81–100 (1993)
2. N.M. Amato, O.B. Bayazit, L.K. Dale, C. Jones, D. Vallejo, OBPRM: an obstacle-based PRM for 3D workspaces, in *Workshop on the Algorithmic Foundations of Robotics (WAFR)* (1998), pp. 155–168
3. S. Baswana, S. Sen, A simple and linear time randomized algorithm for computing sparse spanners in weighted graphs. Random Struct. Algorithms **30**(4), 532–563 (2007)
4. K. Bekris, L. Kavraki, Informed and probabilistically complete search for motion planning under differential constraints, in *First International Symposium on Search Techniques in Artificial Intelligence and Robotics (STAIR)* (Chicago, IL, 2008)
5. R. Bohlin, L. Kavraki, Path planning using lazy PRM, in *IEEE International Conference on Robotics and Automation (ICRA)*, vol. 1 (San Francisco, CA, 2000), pp. 521–528
6. D. Ferguson, A. Stentz, Anytime RRTs, in *IEEE/RSJ International Conference on Intelligent Robots and Systems (IROS)* (Beijing, China, 2006), pp. 5369–5375
7. J. Gao, L.J. Guibas, A. Nguyen, Deformable spanners and applications. Comput. Geometry Theory Appl. **35**(1–2), 2–19 (2006)
8. R. Geraerts, M.H. Overamrs, Creating small graphs for solving motion planning problems, in *IEEE International Conference on Methods and Models in Automation and Robotics (MMAR)* (2005), pp. 531–536
9. R. Geraerts, M.H. Overmars, Creating high-quality roadmaps for motion planning in virtual environments, in *IEEE/RSJ International Conference on Intelligent Robots and Systems (IROS)* (Beijing, China, 2006), pp. 4355–4361
10. R. Geraerts, M.H. Overmars, Reachability-based analysis for probabilistic roadmap planners. J. Robot. Auton. Syst. (RAS) **55**, 824–836 (2007)
11. L.J. Guibas, C. Holleman, L.E. Kavraki, A probabilistic roadmap planner for flexible objects with a workspace medial-axis-based sampling approach, in *IEEE/RSJ International Conference on Intelligent Robots and Systems (IROS)*, vol. 1 (1999), pp. 254–259
12. D. Hsu, R. Kindel, J.C. Latombe, S. Rock, Randomized kinodynamic motion planning with moving obstacles. Int. J. Robot. Res. **21**(3), 233–255 (2002)
13. L. Jaillet, T. Simeon, Path deformation roadmaps, in *Workshop on the Algorithmic Foundations of Robotics (WAFR)* (New York City, NY, 2006)
14. S. Karaman, E. Frazzoli, Sampling-based algorothms for optimal motion planning. Int. J. Robot. Res. **30**(7), 846–894 (2011)
15. S. Karaman, M. Walter, A. Perez, E. Frazzoli, S. Teller, Anytime motion planning using the RRT, in *IEEE International Conference on Robotics and Automation (ICRA)* (Shanghai, China, 2011)
16. L.E. Kavraki, P. Svestka, J.C. Latombe, M. Overmars, Probabilistic roadmaps for path planning in high-dimensional configuration spaces. IEEE Trans. Robot. Autom. (TRA) **12**(4), 566–580 (1996)
17. J.M. Keil, Approximating the complete Euclidean graph, in *1st Scandinavian Workshop on Algorithm Theory* (Springer, London, UK, 1988), pp. 208–213
18. J. Kim, R.A. Pearce, N.M. Amato, Extracting optimal paths from roadmaps for motion planning, in *IEEE International Conference on Robotics and Automation (ICRA)*, vol. 2 (Taipei, Taiwan, 2003), pp. 2424–2429
19. S.M. LaValle, J.J. Kuffner, Randomized kinodynamic planning. Int. J. Robot. Res. **20**(5), 378–400 (2001)
20. Y. Li, K.E. Bekris, Learning approximate cost-to-go metrics to improve sampling-based motion planning, in *IEEE International Conference on Robotics and Automation (ICRA)* (Shanghai, China, 2011)
21. T. Lozano-Perez, Spatial planning: a configuration space approach. IEEE Trans. Comput. 108–120 (1983)

22. J.D. Marble, K.E. Bekris, Computing spanners of asympotically optimal probabilistic roadmaps, in *IEEE/RSJ International Conference on Intelligent Robots and Systems (IROS)* (San Francisco, CA, 2011)
23. M.A. Morales, R. Pearce, N.M. Amato, Metrics for analyzing the evolution of C-space models, in *IEEE International Conference on Robotics and Automation (ICRA)* (2006), pp. 1268–1273
24. O. Nechushtan, B. Raveh, D. Halperin, Sampling-diagrams automata: a tool for analyzing path quality in tree planners, in *Workshop on the Algorithmic Foundations of Robotics (WAFR)* (Singapore, 2010)
25. D. Nieuwenhuisen, M.H. Overmars, Using cycles in probabilistic roadmap graphs, in *IEEE International Conference on Robotics and Automation (ICRA)* (2004), pp. 446–452
26. R. Pearce, M. Morales, N. Amato, Structural improvement filtering strategy for PRM, in *Robotics: Science and Systems (RSS)* (Zurich, Switzerland, 2008)
27. D. Peleg, A. Scha¨ffer, Graph spanners. J. Graph Theory **13**(1), 99–116 (1989)
28. E. Plaku, K.E. Bekris, B.Y. Chen, A.M. Ladd, L.E. Kavraki, Sampling-based roadmap of trees for parallel motion planning. IEEE Trans. Robot. Autom. (TRA) **21**(4), 587–608 (2005)
29. B. Raveh, A. Enosh, D. Halperin, a little more, a lot better: improving path quality by a path-merging algorithm. IEEE Trans. Rob. **27**(2), 365–370 (2011)
30. G. Sanchez, J.C. Latombe, A single-query, bi-directional probabilistic roadmap planner with lazy collision checking, in *International Symposium on Robotics Research* (2001), pp. 403–418
31. E. Schmitzberger, J.L. Bouchet, M. Dufaut, D. Wolf, R. Husson, Capture of homotopy classes with probabilistic roadmap, in *IEEE/RSJ International Conference on Intelligent Robots and Systems (IROS)* (Switzerland, 2002), pp. 2317–2322
32. T. Simeon, J.P. Laumond, C. Nissoux, Visibility-based probabilistic roadmaps for motion planning. Adv. Robot. J. **41**(6), 477–494 (2000)
33. I.A. Sucan, M. Moll, L.E. Kavraki, The open motion planning library (OMPL) (2010). http://ompl.kavrakilab.org
34. R. Wein, J. van den Berg, D. Halperin, Planning high-quality paths and corridors amidst obstacles. Int. J. Robot. Res. **27**(11–12), 1213–1231 (2008)
35. D. Xie, M. Morales, R. Pearce, S. Thomas, J.L. Lien, N.M. Amato, Incremental map generation (IMG), in *Workshop on Algorithmic Foundations of Robotics (WAFR)* (New York City, NY, 2006)

Robust Adaptive Coverage for Robotic Sensor Networks

Mac Schwager, Michael P. Vitus, Daniela Rus and Claire J. Tomlin

Abstract This paper presents a distributed control algorithm to drive a group of robots to spread out over an environment and provide adaptive sensor coverage of that environment. The robots use an on-line learning mechanism to approximate the areas in the environment which require more concentrated sensor coverage, while simultaneously exploring the environment before moving to final positions to provide this coverage. More precisely, the robots learn a scalar field, called the weighting function, representing the relative importance of different regions in the environment, and use a Traveling Salesperson based exploration method, followed by a Voronoi-based coverage controller to position themselves for sensing over the environment. The algorithm differs from previous approaches in that provable robustness is emphasized in the representation of the weighting function. It is proved that the robots approximate the weighting function with a known bounded error, and that they converge to locations that are locally optimal for sensing with respect to the approximate weighting function. Simulations using empirically measured light intensity data are presented to illustrate the performance of the method.

M. Schwager (✉)
GRASP Lab, University of Pennsylvania, 3330 Walnut St,
Philadelphia, PA 19104, USA
e-mail: schwager@mit.edu

M. Schwager · D. Rus
Computer Science and Artificial Intelligence Lab, MIT,
77 Massachusetts Ave., Cambridge, MA 02139, USA
e-mail: rus@csail.mit.edu

M.P. Vitus
Aeronautics and Astronautics, Stanford University,
Stanford, CA 94305, USA
e-mail: vitus@stanford.edu

C.J. Tomlin
Electrical Engineering and Computer Sciences, UC Berkeley,
Berkeley, CA 94708, USA
e-mail: tomlin@eecs.berkeley.edu

H.I. Christensen and O. Khatib (eds.), *Robotics Research*,
Springer Tracts in Advanced Robotics 100, DOI 10.1007/978-3-319-29363-9_25

1 Introduction

In this paper we present a distributed control algorithm to command a group of robots to explore an unknown environment while providing adaptive sensor coverage of interesting areas within the environment. This algorithm has many applications in controlling teams of robots to perform tasks such as search and rescue missions, environmental monitoring, automatic surveillance of rooms, buildings, or towns, or simulating collaborative predatory behavior. As an example application, Japan was hit by a major earthquake on March 11, 2011 that triggered a devastating tsunami causing catastrophic damage to the nuclear reactors at Fukushima. Due to the risk of radiation exposure, humans could not inspect (or repair) the nuclear reactors, however, a team of robots could be used to monitor the changing levels of radiation. Using the proposed algorithm, the robots would concentrate on areas where the radiation was most dangerous, continually providing updated information on how the radiation was evolving. This information could be used to notify people in eminent danger of radiation exposure due to the changing conditions. Similarly, consider a team of waterborne robots charged with cleaning up an oil spill. Our controller allows the robots to distribute themselves over the spill, learn the areas where the spill is most severe and concentrate their efforts on those areas, without neglecting the areas where the spill is not as severe.

Sensor coverage algorithms have been receiving a great deal of attention in recent years. Cortés et al. [6] considered the problem of findin an optimal sensing configuration for a group of mobile robots. They used concepts from locational optimization [8, 26] to control the robots based upon gradient descent of a weighting function which encodes the sensing quality and coverage of the environment. This weighting function can be viewed as describing the importance of areas in the environment. The control law for each robot is distributed and only depends on the robot's position and the positions of its neighbors'. However, all robots are required to know the weighting function a priori which restricts the algorithm from being deployed in unknown environments. There have been several extensions to this formulation of coverage control. In [7], the robots were assumed to have a limited sensing or communication range. Pimenta et al. [18] incorporated heterogeneous robots, and extended the algorithm to handle nonconvex environments. The work [13] used a distributed interpolation scheme to recursively estimate the weighting function. Similarly, [23] removed the requirement of knowing the weighting function a priori by learning a basis function approximation of the weighting function on-line. This strategy has provable convergence properties, but requires that the weighting function lies in a known set of functions. The purpose of the present work is to remove this restriction, greatly broadening the class of weighting functions that can be approximated.

Similar frameworks have been used for multi-robot problems in a stochastic setting [1]. There are also a number of other notions of multi-robot sensor coverage (e.g. [3, 5, 11, 16]), but we choose to adopt the locational optimization approach for its interesting possibilities for analysis and its compatibility with existing ideas in adaptive control [15, 21, 25].

As noted above, this work extends [23] by removing restrictions on the weighting function, so that a much broader class of weighting functions can be provably approximated. Typically, the form of the weighting function is not known a priori, and if this is not accounted for directly then the learning algorithm could chatter between models or even become unstable. Also, in simulations performed with a realistic weighting function, the original algorithm only explores in a local neighborhood of the robots resulting in a poor approximation of the weighting function. However, the algorithm we propose here explores the entire space, successfully learning the weighting function with provable robustness. The robots first partition the environment and perform a Traveling Sales Person (TSP) based distributed exploration, so that the unknown weighting function can be adequately approximated. They then switch, in an asynchronous and distributed fashion, to a coverage mode in which they deploy over the environment to achieve positions that are advantageous for sensing. The robots use an on-line learning mechanism to approximate the weighting function. Since we do not assume the robots can perfectly approximate the weighting function, the parameter adaptation law for learning this function must be carefully constructed to be robust to function approximation errors.

Without specifically designing for such robustness, it is known that many different types of instability [10] can occur. Several techniques have been proposed in the adaptive control literature to handle this kind of robustness, including using a dead-zone [17, 20], the σ-modification [10], and the e_1-modification [14]. We chose to adapt a dead-zone technique, and prove that the robots learn a function that has bounded difference from the true function, while converging to positions that are locally optimal for sensing with respect to the learned function.

The paper is organized as follows. In Sect. 2 we introduce notation and formulate the problem. In Sect. 3 we describe the function approximation strategy and the control algorithm, and we prove the main convergence result of the paper. Section 4 gives the results of a numerical simulation with a weighting function that was determined from empirical measurements of light intensity in a room. Finally, conclusions and future work are discussed in Sect. 5.

2 Problem Formulation

In this section we build a model of the multi-robot system, the environment, and the weighting function defining areas of importance in the environment. We then formulate the robust adaptive coverage problem with respect to this model.

Let there be n robots with positions $p_i(t)$ in a planar environment $Q \subset \mathbb{R}^2$. The environment is assumed to be compact and convex.[1] We call the tuple of all robot

[1]These assumptions can be relaxed to certain classes of nonconvex environments with obstacles [2, 4 18].

positions $P = (p_1, \ldots, p_n) \in Q^n$ the configuration of the multi-robot system, and we assume that the robots move with integrator dynamics

$$\dot{p}_i = u_i, \tag{1}$$

so that we can control their velocities directly through the control input u_i. We define the Voronoi partition of the environment to be $V(P) = \{V_1(P), \ldots, V_n(P)\}$, where

$$V_i(P) = \{q \in Q \mid \|q - p_i\| \leq \|q - p_j\|, \forall j \neq i\},$$

and $\|\cdot\|$ is the l^2-norm. We think of each robot i as being responsible for sensing in its associated Voronoi cell V_i. Next we define the communication network as an undirected graph in which all robots whose Voronoi cells touch share an edge in the graph. This graph is known as the Delaunay graph. Then the set of neighbors of robot i is defined as $\mathcal{N}_i := \{j \mid V_i \cup V_j \neq \emptyset\}$.

We now define a weighting function over the environment $\phi : Q \mapsto \mathbb{R}_{>0}$ (where $\mathbb{R}_{>0}$ denotes the strictly positive real numbers). This weighting function is *not known* by the robots. Intuitively, we want a high density of robots in areas where $\phi(q)$ is large and a lower density where it is small. Finally, suppose that the robots have sensors with which they can measure the value of the weighting function at their own position, $\phi(p_i)$ with very high precision, but that their quality of sensing at arbitrary points, $\phi(q)$, degrades quadratically in the distance between q and p_i. That is to say the cost of a robot at p_i sensing a point at q is given by $\frac{1}{2}\|q - p_i\|^2$. Since each robot is responsible for sensing in its own Voronoi cell, the cost of all robots sensing over all points in the environment is given by

$$\mathcal{H}(P) = \sum_{i=1}^{n} \int_{V_i(P)} \frac{1}{2} \|q - p_i\|^2 \phi(q) dq. \tag{2}$$

This is the overall objective function that we would like to minimize by controlling the configuration of the multi-robot system.[2]

The gradient of $\mathcal{H}$ can be shown[3] to be given by

$$\frac{\partial \mathcal{H}}{\partial p_i} = - \int_{V_i(P)} (q - p_i)\phi(q) dq = -M_i(P)(C_i(P) - p_i), \tag{3}$$

[2]We have pursued an intuitive development of this cost function, though more rigorous arguments can also be made [24]. This function is known in several fields of study including the placement of retail facilities [8] and data compression [12].

[3]The computation of this gradient is more complex than it may seem, because the Voronoi cells $V_i(P)$ depend on P, which results in extra integral terms. Fortunately, these extra terms all sum to zero, as shown in, e.g. [19].

where we define $M_i(P) := \int_{V_i(P)} \phi(q)dq$ and $C_i(P) := 1/M_i(P) \int_{V_i(P)} q\phi(q)dq$. We call M_i the mass of the Voronoi cell i and C_i its centroid, and for efficiency of notation we will henceforth write these without the dependence on P. We would like to control the robots to move to their Voronoi centroids, $p_i = C_i$ for all i, since from (3), this is a critical point of $\mathscr{H}$, and if we reach such a configuration using gradient descent, we know it will be a local minimum. Global optimization of $\mathscr{H}$ is known to be NP-hard, hence it is standard in the literature to only consider local optimality.

2.1 Approximate Weighting Function

Note that the cost function (2) and its gradient (3) rely on the weighting function $\phi(q)$, which is not known to the robots. In this paper we provide a means by which the robots can approximate $\phi(q)$ online in a distributed way and move to decrease (2) with respect to this approximate $\phi(q)$.

To be more precise, each robot maintains a separate approximation of the weighting function, which we denote $\hat{\phi}(q,t)$. These approximate weighting functions are generated from a linear combination of m static basis functions, $\mathscr{K}(q) = [\mathscr{K}_1(q) \cdots \mathscr{K}_m(q)]^{\mathrm{T}}$, where each basis function is a radially symmetric Gaussian of the form

$$\mathscr{K}_j(q) = \frac{1}{2\pi\sigma} \exp\left\{ \frac{\left\| q - \mu_j \right\|^2}{2\sigma^2} \right\}, \tag{4}$$

with fixed width σ and fixed center μ_j. Furthermore, the centers are arranged in a regular grid over Q. Each robot then forms its approximate weighting function as a weighted sum of these basis functions $\hat{\phi}_i(q,t) = \mathscr{K}(q)^{\mathrm{T}} \hat{a}_i(t)$, where $\hat{a}_i(t)$ is the parameter vector of robot i. Each element in the parameter vector is constrained to lie within some lower and upper bounds $0 < a_{\min} < a_{\max} < \infty$ so that $\hat{a}_i(t) \in [a_{\min}, a_{\max}]^m$. This function approximation scheme is illustrated in Fig. 1. Robot i's approximation of its Voronoi cell mass and centroid can then be defined as $\hat{M}_i(P,t) := \int_{V_i(P)} \hat{\phi}_i(q,t)dq$ and $\hat{C}_i(P,t) := 1/\hat{M}_i(P,t) \int_{V_i(P)} q\hat{\phi}_i(q,t)dq$, respectively. Again, we will drop the dependence of $\hat{M}_i$ and $\hat{C}_i$ on (P, t) for notational simplicity.

We measure the difference between an approximate weighting function and the true weighting function as the L^∞ function norm of their difference, so that the best approximation is given by

$$a := \underset{\hat{a} \in [a_{\min}, a_{\max}]^m}{\arg\min} \; \max_{q \in Q} \left| \mathscr{K}(q)^{\mathrm{T}} \hat{a} - \phi(q) \right|, \tag{5}$$

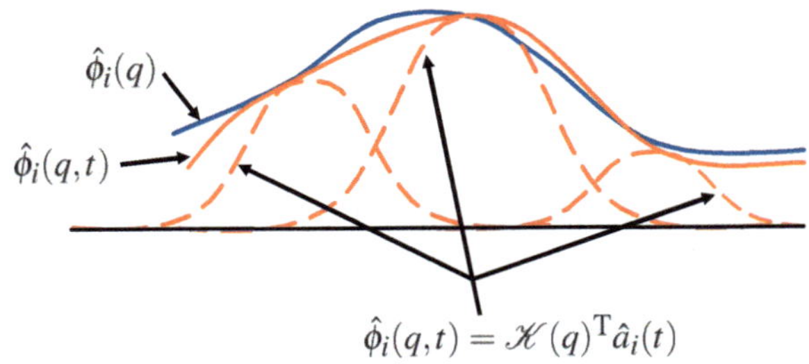

Fig. 1 The weighting function approximation is illustrated in this simplified 2-D schematic. The true weighting function $\phi(q)$ is approximated by robot i to be $\hat{\phi}_i(q,t)$. The basis function vector $\mathcal{K}(q)$ is shown as three Gaussians (*dashed curves*), and the parameter vector $\hat{a}_i(t)$ denotes the weighting of each Gaussian

and the optimal function approximation error is given by

$$\phi_\varepsilon(q) := \mathcal{K}(q)^{\mathrm{T}}a - \phi(q). \tag{6}$$

It will be shown in the proof of Theorem 1 that the L^∞ norm gives the tightest approximation bound with our proof technique. The only restriction that we put on $\phi(q)$ is that it is bounded over the environment, or equivalently, the approximation error is bounded, $|\phi_\varepsilon(q)| \leq \phi_{\varepsilon_{\max}} < \infty$. We assume that the robots have knowledge of this bound, $\phi_{\varepsilon_{\max}}$. The theoretical analysis in the previous work [23] was not robust to function approximation errors in that it required $\phi_\varepsilon(q) \equiv 0$. One of the main contributions here is to formulate an algorithm that is provably robust to function approximation errors. We only require that the robots have a known bound for the function approximation error, $\phi_{\varepsilon_{\max}}$.

Finally, we define the parameter error as $\tilde{a}_i(t) := \hat{a}_i(t) - a$. In what follows we describe an online tuning law by which robot i can tune its parameters, $\hat{a}_i$, to approach a neighborhood of the optimal parameters, a. Our proposed controller then causes the robots to converge to their approximate centroids, $p_i \to \hat{C}_i$ for all i. An overview of the geometrical objects involved in our set-up is shown in Fig. 2.

3 Robust Adaptive Coverage Algorithm

In this section we describe the algorithm that drives the robots to spread out over the environment while simultaneously approximating the weighting function online. The algorithm naturally decomposes into two parts: (1) the parameter adaptation law, by which each robot updates its approximate weighting function, and (2) the control algorithm, which drives the robots to explore the environment before moving to their final positions for coverage. We describe these two parts in separate sections and then prove performance guarantees for the two working together.

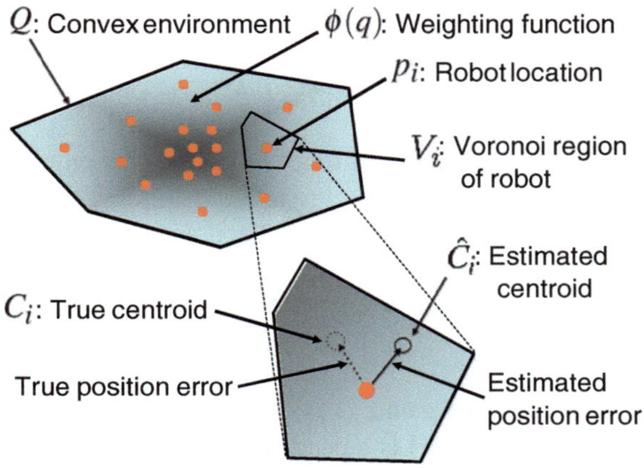

Fig. 2 A graphical overview of the quantities involved in the controller is shown. The robots move to cover a bounded, convex environment Q their positions are p_i, and they each have a Voronoi region V_i with a true centroid C_i and an estimated centroid $\hat{C}_i$. The true centroid is determined using a sensory function $\phi(q)$, which indicates the relative importance of points q in Q. The robots do not know $\phi(q)$, so they calculate an estimated centroid using an approximation $\hat{\phi}_i(q)$ learned from sensor measurements of $\phi(q)$

3.1 Online Function Approximation

The parameters $\hat{a}_i$ used to calculate $\hat{\phi}_i(q,t)$ are adjusted according to a set of adaptation laws which are introduced below. First, we define two quantities,

$$\Lambda_i(t) = \Lambda_0 + \int_{s\in\Omega_i(t)} \mathscr{K}(s)\mathscr{K}(s)^{\mathrm{T}} ds, \text{and } \lambda_i(t) = \int_{s\in\Omega_i(t)} \mathscr{K}(s)\phi(s) ds, \quad (7)$$

where $\Omega_i(t) = \{s |\, s = p_i(\tau) \text{ for some } \tau \in [0,t]\}$ is the set of points in the trajectory of p_i from time 0 to time t, and Λ_0 is a positive definite matrix. The quantities in (7) can be calculated differentially by robot i using $\dot{\Lambda}_i(t) = \mathscr{K}_i(t)\mathscr{K}_i(t)^{\mathrm{T}}|\dot{p}_i(t)|$ with initial condition Λ_0, and $\dot{\lambda}_i(t) = \mathscr{K}_i(t)\phi_i(t)|\dot{p}_i(t)|$ with zero initial conditions, where we introduced the shorthand notation $\mathscr{K}_i(t) := \mathscr{K}(p_i(t))$ and $\phi_i(t) := \phi(p_i(t))$. We require that $\Lambda_0 > 0$, though it can be arbitrarily small. This will ensure that $\Lambda_i(t) > 0$ for all time because $\int_{s\in\Omega_i(t)} \mathscr{K}(s)\mathscr{K}(s)^{\mathrm{T}} ds \geq 0$ and the sum of a positive semi-definite matrix and a positive definite matrix is positive definite. This, in turn, ensures that $\Lambda_i^{-1/2}$ always exists, which will be crucial in the control law and proof of convergence below. As previously stated, robot i can measure $\phi_i(t)$ with its sensors. Now we define another quantity

$$F_i = \frac{\int_{V_i} \mathcal{K}(q)(q-p_i)^{\mathrm{T}} dq \int_{V_i} (q-p_i)\mathcal{K}(q)^{\mathrm{T}} dq}{\int_{V_i} \hat{\phi}_i(q) dq}. \tag{8}$$

Notice that F_i can also be computed by robot i as it does not require any knowledge of the true weighting function, ϕ.

The "pre" adaptation law for $\hat{a}_i$ is now defined as

$$\dot{\hat{a}}_i^{\mathrm{pre}} = -\gamma B_{\mathrm{dz}}(\Lambda_i \hat{a}_i - \lambda_i) - \zeta \sum_{j \in \mathcal{N}_i} l_{ij}(\hat{a}_i - \hat{a}_j) - kF_i \hat{a}_i. \tag{9}$$

where γ, ζ, and k are positive gains, l_{ij} is the length of the shared Voronoi edge between robots i and j, and $B_{\mathrm{dz}}(\cdot)$ is a dead zone function which gives a zero if its argument is below some value. We will give B_{dz} careful attention in what follows as it is the main tool to ensure robustness to function approximation errors. Before describing the dead zone in detail, we note that the three terms in (9) have an intuitive interpretation. The first term is an integral of the function approximation error over the robot's trajectory, so that the parameter $\hat{a}_i$ is tuned to decrease this error. The second term is the difference between the robot's parameters and its neighbors' parameters. This term will be shown to lead to parameter consensus; the parameter vectors for all robots will approach a common vector. The third term compensates for uncertainty in the centroid position estimate, and will be shown to ensure convergence of the robots to their estimated centroids. A more in-depth explanation of each of these terms can be found in [23].

Finally, we give the parameter adaptation law by restricting the "pre" adaptation law so that the parameters remain within their prescribed limits $[a_{\min}, a_{\max}]$ using a projection operator. We introduce a matrix I_{proj_i} defined element-wise as

$$I_{\mathrm{proj}_i} := \begin{cases} 0 \text{ for } a_{\min} < \hat{a}_i(j) < a_{\max} \\ 0 \text{ for } \hat{a}_i(j) = a_{\min} \text{ and } \dot{\hat{a}}_i^{\mathrm{pre}}(j) \geq 0 \\ 0 \text{ for } \hat{a}_i(j) = a_{\max} \text{ and } \dot{\hat{a}}_i^{\mathrm{pre}}(j) \leq 0 \\ 1 \text{ otherwise,} \end{cases} \tag{10}$$

where (j) denotes the jth element for a vector and the jth diagonal element for a matrix. The entries of I_{proj_i} are only nonzero if the parameter is about to exceed its bound. Now the parameters are changed according to the adaptation law

$$\dot{\hat{a}}_i = \Gamma\left(\dot{\hat{a}}_{\mathrm{pre}_i} - I_{\mathrm{proj}_i} \dot{\hat{a}}_{\mathrm{pre}_i}\right), \tag{11}$$

where $\Gamma \in \mathbb{R}^{m \times m}$ is a diagonal, positive definite gain matrix. Although the adaptation law given by (11) and (9) is notationally complicated, it has a straightforward interpretation, it is of low computational complexity, and it is composed entirely of quantities that can be computed by robot i.

As mentioned above, the key innovation in this adaptation law compared with the one in [23] is the dead zone function B_{dz}. We design this function so that the parameters are only changed in response to function errors that could be reduced with different parameters. More specifically, the minimal function error that can be achieved is ϕ_ε, as shown in (6). Therefore if the integrated parameter error $(\Lambda_i\hat{a}_i - \lambda_i)$ is less than ϕ_ε integrated over the robot's path, we have no reason to change the parameters. We will show that the correct form for the dead zone to prevent unnecessary parameter adaptation is

$$B_{\mathrm{dz}}(x) = \begin{cases} 0 & \text{if } C(x) < 0 \\ \frac{x}{\|x\|} C(x) & \text{otherwise,} \end{cases} \tag{12}$$

where $C(x) := \left\|\Lambda_i^{1/2}\right\| \left(\left\|\Lambda_i^{-1/2} x\right\| - \left\|\Lambda_i^{-1/2} \beta_i\right\| \phi_{\varepsilon_{\max}} - \left\|\Lambda_i^{-1/2} \Lambda_0\right\| a_{\max} \right)$ and $\beta_i := \int_{s \in \Omega_i(t)} \mathscr{K}(s) ds$. This condition can be evaluated by robot i since $\beta_i(t)$ can be computed differentially from $\dot{\beta}_i = \mathscr{K}_i |\dot{p}_i|$ with zero initial conditions, we have already seen how to compute Λ_i, and $\phi_{\varepsilon_{\max}}$, $a_{\max}$, and Λ_0 are known.

3.2 Control Algorithm

We propose to use a control algorithm that is composed of a set of control modes, with switching conditions to determine when the robots change from one mode to the next. The robots first move to partition the basis function centers among one another, so that each center is assigned to one robot, then each robot executes a Traveling Salesperson (TSP) tour through all of the basis function centers that have been assigned to it. This tour will provide sufficient information so that the weighting function can be estimated well over all of the environment. Then the robots carry out a centroidal Voronoi controller using the estimated weighting function to drive to final positions. We call the first mode the "partitioning" mode, the second the "exploration" mode, and the third the "coverage" mode. This sequence of control modes is executed asynchronously in a distributed fashion, during which the function approximation parameters are updated continually with (9) and (11).

For each robot we define a mode variable $\mathscr{M}_{ii} \in \{\text{partition, explore, cover}\}$. In order to coordinate their mode switches, each robot also maintains an estimate of the modes of all the other robots, so that $\mathscr{M}_{ij}$ is the estimate by robot i of robot j's mode, and $\mathscr{M}_i := (\mathscr{M}_{i1}, \ldots, \mathscr{M}_{in})$ is an n-tuple of robot i's estimates of all robots' modes. Furthermore, the modes are ordered with respect to one another by partition < explore < cover, so that the max $(\mathscr{M}_i, \mathscr{M}_j)$ function is the maximum between each element of the two mode estimate tuples, $\mathscr{M}_i$ and $\mathscr{M}_j$, according to this ordering. These mode estimates are updated using the flooding communication protocol described below. We first describe the algorithmic structure of the controller, then define the behavior within each mode, and finally prove the convergence of the coupled control algorithm and learning algorithm to a desirable final configuration.

The two algorithms below run concurrently in different threads. Algorithm 1 defines the switching conditions between control modes, and Algorithm 2 describes the flooding protocol that each robot uses to maintain its mode estimates.

Algorithm 1 Switching Control Algorithm (executed by robot i)

Require: Communication with Voronoi neighbors.
Require: Knowledge of position, p_i, in global coordinate frame.
Require: Knowledge of the total number of robots n.
Require: Knowledge of flooding algorithm (Algorithm 2) update period, T.
Require: Access to mode estimates $\mathscr{M}_i$ updated from Algorithm 2.

```
while 𝓜_i ≠ (explore, ..., explore) do
    if 𝓜_ii == explore then
        u_i = [0, 0]^T
    else
        u_i = u_i^partition
    end if
    if Distance to mean of basis function centers < ε^partition and 𝓜_ii == partition then
        𝓜_ii = explore
    end if
end while
Compute TSP tour through basis function centers 𝒩_i^μ
Wait for nT seconds with u_i = [0, 0]^T
Execute TSP tour
𝓜_ii = cover
while 𝓜_ii == cover do
    u_i = u_i^cover
end while
```

Algorithm 2 Mode Estimate Flooding (executed by robot i)

Require: The network is connected.
Require: The robots have synchronized clocks with which they broadcast during a pre-assigned time slot.

```
Initialize 𝓜_i = (partition, ..., partition)
while 1 do
    if Broadcast received from robot j then
        𝓜_i = max(𝓜_i, 𝓜_j) (where partition < explore < cover)
    end if
    if Robot i's turn to broadcast then
        Broadcast 𝓜_i
    end if
end while
```

The control laws within each mode are then defined as follows. In the partition mode, each robot uses the controller

$$u_i^{\text{partition}} = k\left(\frac{1}{\left|\mathcal{N}_i^{\mu}\right|}\sum_{j\in\mathcal{N}_i^{\mu}} \mu_j - p_i\right), \tag{13}$$

where $\mathcal{N}_i^{\mu} := \{\mu_j | \|\mu_j - p_i\| \leq \|\mu_j - p_k\| \forall k \neq i\}$ is the set of the closest basis function centers to robot i, μ_j are the basis function centers from (4), and $\left|\mathcal{N}_i^{\mu}\right|$ is the number of elements in $\mathcal{N}_i^{\mu}$. In the explore mode, each robot drives a tour through each basis function center in its neighborhood,μ_j for $j \in \mathcal{N}_i^{\mu}$. Any tour will do, but a good choice is to use an approximate TSP tour. Finally, for the "cover" mode, each robot moves toward the centroid of its Voronoi cell using

$$u_i^{\text{cover}} = k(\hat{C}_i - p_i), \tag{14}$$

where k is the same positive gain from (9).

Using the above control and function approximation algorithm, we can prove that all robots converge to the estimated centroid of their Voronoi cells, that all robots function approximation parameters converge to the same parameter vector, and that this parameter vector has a bounded error with the optimal parameter vector. This is stated formally in the following theorem.

Theorem 1 (Convergence) *A network of robots with dynamics (1) using Algorithm 1 for control, Algorithm 2 for communication, and (9) and (11) for online function approximation has the following convergence guarantees:*

$$\lim_{t\to\infty} \left\|p_i(t) - \hat{C}_i(p,t)\right\| = 0 \quad \forall i, \tag{15}$$

$$\lim_{t\to\infty} \left\|\hat{a}_i(t) - \hat{a}_j(t)\right\| = 0 \quad \forall i,j, \tag{16}$$

and $$\lim_{t\to\infty} \|\hat{a}_i(t) - a\| \leq \frac{\sum_{j=1}^{n} 2\left\|\Lambda_j(t)^{1/2}\right\|\left(\left\|\Lambda_j(t)^{-1/2}\beta_j(t)\right\|\phi_{\varepsilon_{\max}} + \left\|\Lambda_j(t)^{-1/2}\Lambda_0\right\|a_{\max}\right)}{\text{mineig}\left(\sum_{j=1}^{n}\Lambda_j(t)\right)} \quad \forall i. \tag{17}$$

Proof The proof has two parts. The first part is to show that all robots reach "cover" mode and stay in "cover" mode. The second part uses a Lyapunov type proof technique similar to the one in [23] to show that once all robots are in "cover" mode, the convergence claims of (15)–(17) follow.

Firstly, the "partition" mode simply implements a K-means clustering algorithm [9], in which the basis function centers are the points to be clustered, and the robots move to the cluster means. This algorithm is well-known to converge in the sense that for any $\varepsilon^{\text{partition}}$ there exists a time $T_i^{\text{partition}}$ at which the distance between the robot and the centers' mean is less than $\varepsilon^{\text{partition}}$, therefore all robots will reach

$\mathscr{M}_{ii}$ = explore at some finite time. After this time, according to Algorithm 1, a robot will remain stopped until all of its mode estimates have switched to "explore." Suppose the first robot to achieve $\mathscr{M}_i$ = (explore,..., explore) does so at time T_f. This means that at some time in the past all the other robots, j, have switched to $\mathscr{M}_{jj}$ = explore and stopped moving, but none of them have $\mathscr{M}_j$ = (explore,..., explore) (otherwise they would be the first). Therefore at T_f all robots are stopped. Suppose the last robot to achieve $\mathscr{M}_i$ = (explore,..., explore) does so at T_l. From the properties of Algorithm 2 we know that $T_l - T_f \leq nT$ (the maximum time between the first robot to obtain $\mathscr{M}_i$ = (explore,..., explore) and the last robot to do so is nT).At time T_l, the first robot to have $\mathscr{M}_i$ = (explore,..., explore) will still be stopped, because it waits for nT seconds after achieving $\mathscr{M}_i$ = (explore,..., explore), hence when any robot obtains $\mathscr{M}_i$ = (explore,..., explore), all other robots are stopped. Even though the robots may compute their TSP tours at different times, they are all at the same positions when they do so. Therefore, each basis function center is in at least one robot's TSP tour. Consequently, when all robots have completed their TSP tours, mineig $\left(\sum_{i=1}^n \Lambda_i\right)$ will be similar in size to maxeig $\left(\sum_{i=1}^n \Lambda_i\right)$ making the bound in (17) small.

Each tour is finite length, so it will terminate in finite time, hence each robot will eventually enter the "cover" mode. Furthermore, when some robots are in "cover" mode, and some are still in "explore" mode, the robots in "cover" mode will remain inside the environment Q. This is because Q is convex, and since $\hat{C}_i \in V_i \subset Q$ and $p_i \in Q$, by convexity, the segment connecting the two is in Q. Since the robots have integrator dynamics (1), they will stay within the union of these segments over time, $p_i(t) \in \cup_{\tau > 0}(C_i(\tau) - p_i(\tau))$, and therefore remain in the environment Q. Thus at some finite time, T^{cover}, all robots reach "cover" mode and are at positions inside Q.

Now, define a Lyapunov-like function

$$\mathscr{V} = \mathscr{H} + \frac{1}{2}\sum_i \tilde{a}_i^{\mathrm{T}} \Gamma^{-1} \tilde{a}_i, \tag{18}$$

which incorporates the sensing cost $\mathscr{H}$, and is quadratic in the parameter errors $\tilde{a}_i$. We will use Barbalat's lemma to prove that $\dot{\mathscr{V}} \to 0$ and then show that the claims of the theorem follow. Barbalat's lemma requires that $\mathscr{V}$ is lower bounded, non-increasing, and uniformly continuous. $\mathscr{V}$ is bounded below by zero since $\mathscr{H}$ is a sum of integrals of strictly positive functions, and the quadratic parameter error terms are each bounded below by zero.

Now we will show that $\dot{\mathscr{V}} \leq 0$. Taking the time derivative of $\mathscr{V}$ along the trajectories of the system and simplifying with (3) gives

$$\dot{\mathscr{V}} = \sum_i \left(-\left\|\hat{C}_i - p_i\right\|^2 k\hat{M}_i + \tilde{a}_i^{\mathrm{T}} kF_i\hat{a}_i + \tilde{a}_i^{\mathrm{T}} \Gamma^{-1} \dot{\hat{a}}_i \right).$$

Substituting for $\dot{\hat{a}}_i$ with (11) and (9) gives

$$\dot{\mathscr{V}} = -\sum_i \left(\left\|\hat{C}_i - p_i\right\|^2 k\hat{M}_i + \gamma \tilde{a}_i^{\mathrm{T}} B_{\mathrm{dz}}(\Lambda_i \tilde{a}_i + \lambda_{\varepsilon_i}) + \zeta \tilde{a}_i^{\mathrm{T}} \sum_{j \in N_i} l_{ij}\left(\hat{a}_i - \hat{a}_j\right) + \tilde{a}_i^{\mathrm{T}} I_{\mathrm{proj}_i} \dot{\hat{a}}_{\mathrm{pre}_i} \right),$$

where $\lambda_{\varepsilon_i} := \int_{s \in \Omega_i(t)} \mathscr{K}(s)\phi_\varepsilon(s) ds + \Lambda_0 a$. Rearranging terms we get

$$\dot{\mathscr{V}} = -\sum_i \left(\left\|\hat{C}_i - p_i\right\|^2 k\hat{M}_i + \gamma \tilde{a}_i^{\mathrm{T}} B_{\mathrm{dz}}(\Lambda_i \tilde{a}_i + \lambda_{\varepsilon_i}) + \tilde{a}_i^{\mathrm{T}} I_{\mathrm{proj}_i} \dot{\hat{a}}_{\mathrm{pre}_i} - \zeta \sum_{j=1}^{m} \hat{\alpha}_j^{\mathrm{T}} L(P) \hat{\alpha}_j \right), \tag{19}$$

where $\hat{\alpha}_j := \left[\hat{a}_{1j} \ldots \hat{a}_{mj}\right]^T$ is the jth element in every robot's parameter vector, stacked into a vector, and $L(P) \geq 0$ is the graph Laplacian for the Delaunay graph which defines the robots communication network (please refer to the proof of Theorem 2 in [23] for details). The first term inside the first sum is the square of a norm, and therefore is non-negative. The third term in the first sum is non-negative by design (please see the proof of Theorem 1 from [23] for details). The second sum is nonnegative because $L(P) \geq 0$. Therefore, the term with the dead-zone operator B_{dz} is the only term in question, and this distinguishes the Lyapunov construction here from the one in the proof of Theorem 1 in [23].

We now show that the dead-zone term is also non-negative by design. Suppose the condition $C(\Lambda_i \tilde{a}_i + \lambda_{\varepsilon_i}) < 0$ from (12). Then $B_{\mathrm{dz}}(\Lambda \tilde{a}_i + \lambda_{\varepsilon_i}) = 0$ and the term is zero. Now suppose $C(\Lambda_i \tilde{a}_i + \lambda_{\varepsilon_i}) \geq 0$. In that case we have

$$0 \leq \left\|\Lambda_i^{1/2}\right\| \left(\left\|\Lambda_i^{-1/2}(\Lambda_i \tilde{a}_i + \lambda_{\varepsilon_i})\right\| - \left\|\Lambda_i^{-1/2} \beta_i\right\| \phi_{\varepsilon_{\max}} - \left\|\Lambda_i^{-1/2} \Lambda_0\right\| a_{\max} \right),$$

which implies

$$\begin{aligned} 0 &\leq \left\|\Lambda_i^{-1/2}(\Lambda_i \tilde{a}_i + \lambda_{\varepsilon_i})\right\| - \left\|\Lambda_i^{-1/2} \beta_i\right\| \phi_{\varepsilon_{\max}} - \left\|\Lambda_i^{-1/2} \Lambda_0\right\| a_{\max} \\ &\leq \left\|\Lambda_i^{1/2} \tilde{a}_i + \Lambda^{-1/2} \lambda_{\varepsilon_i}\right\| - \left\|\Lambda_i^{-1/2} \left(\int_{\Omega_i(t)} \mathscr{K}(s)\phi_\varepsilon(s) ds + \Lambda_0 a \right)\right\| \\ &\leq \left\|\Lambda_i^{1/2} \tilde{a}_i + \Lambda_i^{-1/2} \lambda_{\varepsilon_i}\right\|^2 - \left(\Lambda_i^{-1/2} \lambda_{\varepsilon_i}\right)^{\mathrm{T}} \left(\Lambda_i^{1/2} \tilde{a}_i + \Lambda_i^{-1/2} \lambda_{\varepsilon_i}\right) \\ &= \left(\Lambda_i^{1/2} \tilde{a}_i\right)^{\mathrm{T}} \left(\Lambda_i^{1/2} \tilde{a}_i + \Lambda_i^{-1/2} \lambda_{\varepsilon_i}\right) \\ &= \tilde{a}_i^{\mathrm{T}} (\Lambda_i \tilde{a}_i + \lambda_{\varepsilon_i}). \end{aligned} \tag{20}$$

Then from the definition of the dead-zone operator, $B_{\mathrm{dz}}(\cdot)$, we have

$$\tilde{a}^{\mathrm{T}} B_{\mathrm{dz}}(\Lambda_i \tilde{a}_i + \lambda_{\varepsilon_i}) = \frac{\tilde{a}^{\mathrm{T}}(\Lambda_i \tilde{a}_i + \lambda_{\varepsilon_i})}{\|\Lambda_i \tilde{a}_i + \lambda_{\varepsilon_i}\|} C(\Lambda_i \tilde{a}_i + \lambda_{\varepsilon_i}) \geq 0,$$

since the numerator was shown to be non-negative in (20), and $C(\cdot)$ is non-negative by supposition. Therefore the dead-zone term is non-negative in this case as well. We conclude therefore that $\dot{\mathscr{V}} \leq 0$.

The final condition to prove that $\dot{\mathscr{V}} \to 0$ is that $\dot{\mathscr{V}}$ must be uniformly continuous, which is a technical condition that was shown in [23], Lemmas 1 and 2. These lemmas also apply to our case, since our function $\dot{\mathscr{V}}$ is the same as in that case except for the dead-zone term. Following the argument of those lemmas, the dead-zone term has a bounded derivative everywhere, except where it is non-differentiable, and these point of non-differentiability are isolated. Therefore, it is Lipschitz continuous and hence uniformly continuous, and we conclude by Barbalat's lemma that $\dot{\mathscr{V}} \to 0$.

Now we show that $\dot{\mathscr{V}} \to 0$ implies the convergence claims stated in the theorem. Firstly, since all the terms in (19) are non-negative, each one must separately approach zero. The first term approaching zero gives the position convergence (15), the last term approaching zero gives parameter consensus (16) (again, see [23] for more details on these). Finally, we verify the parameter error convergence (17). We know that the dead-zone term approaches zero, therefore either $\lim_{t\to\infty} \tilde{a}_i^{\mathrm{T}}(\Lambda_t \tilde{a}_i + \lambda_{\varepsilon_i}) = 0$, or $\lim_{t\to\infty} C(\Lambda_i \tilde{a}_i + \lambda_{\varepsilon_i}) \leq 0$. We already saw from (20) that $\tilde{a}_i^{\mathrm{T}}(\Lambda_t \tilde{a}_i + \lambda_{\varepsilon_i}) = 0$ implies $C(\Lambda_i \tilde{a}_i + \lambda_{\varepsilon_i}) \leq 0$, thus we only consider this later case. To condense notation at this point, we introduce $d_i := \left\|\Lambda_i^{1/2}\right\| \left(\left\|\Lambda_i^{-1/2} \beta_i\right\| \phi_{\varepsilon_{\max}} - \left\|\Lambda_i^{-1/2} \Lambda_0\right\| a_{\max}\right)$. Then from $\lim_{t\to\infty} C(\Lambda_i \tilde{a}_i + \lambda_{\varepsilon_i}) \leq 0$ we have

$$0 \geq \lim_{t\to\infty} \left(\left\|\Lambda_i^{1/2}\right\| \left\|\Lambda_i^{-1/2}(\Lambda_i \tilde{a}_i + \lambda_{\varepsilon_i})\right\| - d_i \right) \geq \lim_{t\to\infty} (\|\Lambda_i \tilde{a}_i + \lambda_{\varepsilon_i}\| - d_i),$$

and because of parameter consensus (16), $0 \geq \lim_{t\to\infty} \left(\left\|\Lambda_j \tilde{a}_i + \lambda_{\varepsilon_j}\right\| - d_j\right)$ for all j. Then summing over j, we have

$$\begin{aligned} 0 &\geq \lim_{t\to\infty} \left(\sum_{j=1}^{n} \|\Lambda_j \tilde{a}_i + \lambda_{\varepsilon_j}\| - \sum_{j=1}^{n} d_j \right) \\ &\geq \lim_{t\to\infty} \left(\left\| \sum_{j=1}^{n} \Lambda_j \tilde{a}_i + \sum_{j=1}^{n} \lambda_{\varepsilon_j} \right\| - \sum_{j=1}^{n} d_j \right) \\ &\geq \lim_{t\to\infty} \left(\left\| \sum_{j=1}^{n} \Lambda_j \tilde{a}_i \right\| - \left\| \sum_{j=1}^{n} \lambda_{\varepsilon_j} \right\| - \sum_{j=1}^{n} d_j \right). \end{aligned}$$

The last condition has two possibilities; either $\lim_{t\to\infty}\left(\left\|\sum_{j=1}^{n}\Lambda_j\tilde{a}_i\right\| > \left\|\sum_{j=1}^{n}\lambda_{\varepsilon_j}\right\|\right)$, in which case

$$0 \geq \lim_{t\to\infty}\left(\left\|\sum_{j=1}^{n}\Lambda_j\tilde{a}_i\right\| - \left\|\sum_{j=1}^{n}\lambda_{\varepsilon_i}\right\| - \sum_{j=1}^{n}d_j\right) \geq \lim_{t\to\infty}\left(\left\|\sum_{j=1}^{n}\Lambda_j\tilde{a}_i\right\| - 2\sum_{j=1}^{n}d_j\right), \tag{21}$$

where the last inequality uses the fact that $\left\|\sum_{j=1}^{n}\lambda_{\varepsilon_i}\right\| \leq \sum_{j=1}^{n}d_j$. Otherwise, $\lim_{t\to\infty}\left(\left\|\sum_{j=1}^{n}\Lambda_j\tilde{a}_i\right\| < \left\|\sum_{j=1}^{n}\lambda_{\varepsilon_j}\right\|\right)$, which implies $\lim_{t\to\infty}\left(\left\|\sum_{j=1}^{n}\Lambda_j\tilde{a}_i\right\| < \|\sum_{j=1}^{n}d_j\|\right)$ which in turn implies (21), thus we only need to consider (21). This expression then leads to $0 \geq \lim_{t\to\infty}\left(\text{mineig}\left(\sum_{j=1}^{n}\Lambda_j\right)\|\tilde{a}_i\| - 2\sum_{j=1}^{n}d_j\right)$, and dividing both sides by mineig $\left(\sum_{j=1}^{n}\Lambda_j\right)$ (which is strictly positive since $\sum_{j=1}^{n}\Lambda_j > 0$), gives (17). □

4 Simulation Results

The proposed algorithm is tested using the data collected from previous experiments [22] in which two incandescent office lights were placed at the position (0, 0) of the environment, and the robots used on-board light sensors to measure the light intensity. The data collected during these previous experiments was used to generate a realistic weighting function which cannot be reconstructed exactly by the chosen basis functions. In the simulation, there are 10 robots and the basis functions are arranged on a 15×15 grid in the environment.

The proposed robust algorithm was compared against the standard algorithm from [23], which assumes that the weighting function can be matched exactly. The true and optimally reconstructed (according to (5)) weighting functions are shown in Fig. 3a, b, respectively. As shown in Fig. 3c, d, with the robust and standard algorithm respectively, the proposed robust algorithm significantly outperforms the standard algorithm and reconstructs the true weighting function well. The robot trajectories for the robust and standard adaptive coverage algorithm are shown in Fig. 4a, b, respectively. Since the standard algorithm doesn't include the exploration phase, the robots get stuck in a local area around their starting position which causes the robots to be unsuccessful in learning an acceptable model of the weighting function. In contrast, the robots with the robust algorithm explore the entire space and reconstruct the true weighting function well.

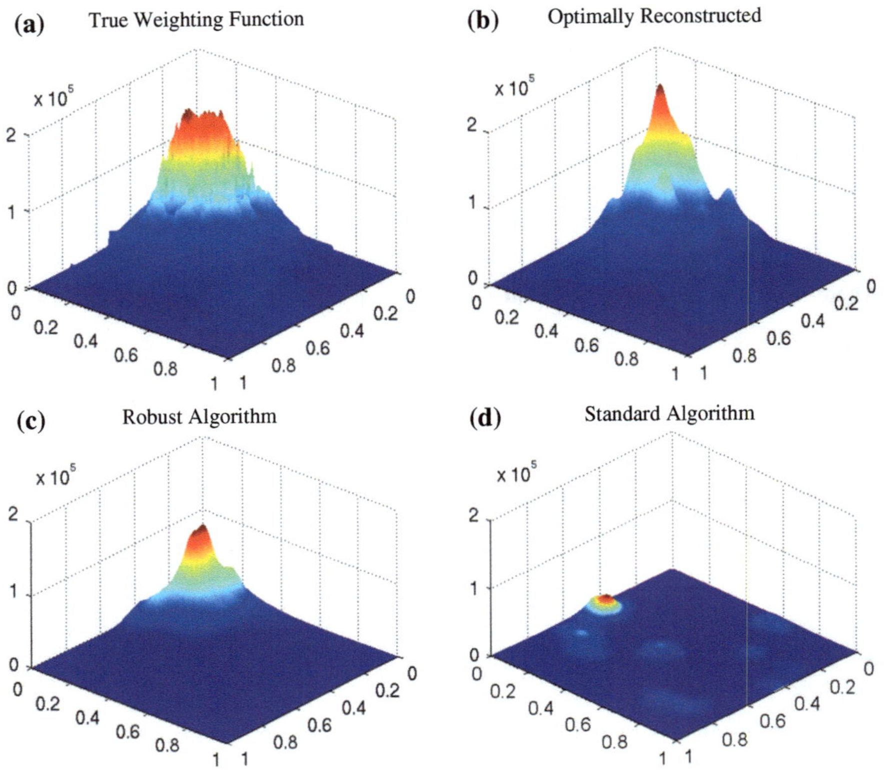

Fig. 3 A comparison of the weighting functions. **a** The true weighting function. **b** The optimally reconstructed weighting function for the chosen basis functions. **c** The weighting function for the proposed algorithm with deadzone and exploration. **d** The previously proposed algorithm without deadzone or exploration

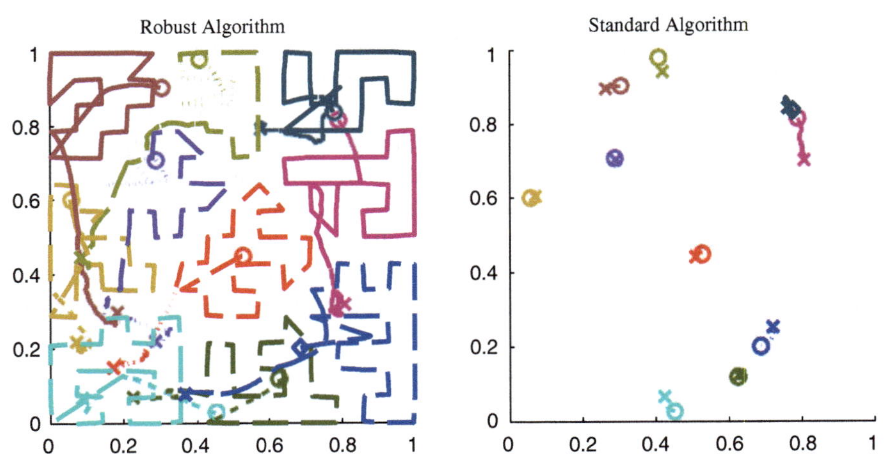

Fig. 4 The vehicle trajectories for the different control strategies. The initial and final vehicle position is marked by a circle and cross, respectively

5 Conclusions

In this paper we formulated a distributed control and function approximation algorithm for deploying a robotic sensor network to adaptively monitor an environment. The robots robustly learn a weighting function over the environment representing where sensing is most needed. The function learning is enabled by the robots exploring the environment in a systematic and distributed way, and is provably robust to function approximation errors. After exploring the environment, the robots drive to positions that locally minimize a cost function representing the sensing quality of the network. The performance of the algorithm is proven in a theorem, and demonstrated in a numerical simulation with an empirical weighting function derived from light intensity measurements in a room. The authors are currently working toward hardware experiments to prove the practicality of the algorithm.

Acknowledgments This work was funded in part by ONR MURI Grants N00014-07-1-0829, N00014-09-1-1051, and N00014-09-1-1031, and the SMART Future Mobility project. We are grateful for this financial support.

References

1. A. Arsie, E. Frazzoli, Efficient routing of multiple vehicles with no explicit communications. Int. J. Robust Nonlinear Control **18**(2), 154–164 (2007)
2. A. Breitenmoser, M. Schwager, J.C. Metzger, R. Siegwart, D. Rus, Voronoi coverage of non-convex environments with a group of networked robots, in *Proceeding of the International Conference on Robotics and Automation (ICRA 10)*, pp. 4982–4989, Anchorage, Alaska, USA (2010)
3. Z.J. Butler, D. Rus, Controlling mobile sensors for monitoring events with coverage constraints, in *Proceedings of the IEEE International Conference of Robotics and Automation*, pp. 1563–1573, New Orleans, LA (2004)
4. C. Caicedo, M. Žefran, Performing coverage on nonconvex domains, in *Proceedings of the 2008 IEEE Multi-Conference on Systems and Control*, pp. 1019–1024 (2008)
5. H. Choset, Coverage for robotics-A survey of recent results. Ann. Math. Artif. Intell. **31**, 113–126 (2001)
6. J. Cortés, S. Martínez, T. Karatas, F. Bullo, Coverage control for mobile sensing networks. IEEE Trans. Robot. Autom. **20**(2), 243–255 (2004)
7. J. Cortés, S. Martínez, F. Bullo, Spatially-distributed coverage optimization and control with limited-range interactions. *ESIAM: Control Optimisation Calc. Var.* **11**, 691–719 (2005)
8. Z. Drezner, *Facility Location: A Survey of Applications and Methods*. Springer Series in Operations Research. (Springer, New York, 1995)
9. R. Duda, P. Hart, D. Stork, *Pattern classification* (Wiley, New York, 2001)
10. P. Ioannou, P.V. Kokotovic, Instability analysis and improvement of robustness of adaptive control. Automatica **20**(5), 583–594 (1984)
11. D.T. Latimer IV, S. Srinivasa, V. Shue, H. Choset, S. Sonne, A. Hurst, Towards sensor based coverage with robot teams, in *Proceedings of the IEEE International Conference on Robotics and Automation*, vol. 1, pp. 961–967 (2002)
12. S.P. Lloyd, Least squares quantization in pcm. IEEE Trans. Inf. Theory **28**(2), 129–137 (1982)

13. S. Martínez, Distributed interpolation schemes for field estimation by mobile sensor networks. IEEE Trans. Control Syst. Technol. **18**(2), 419–500 (2010)
14. K. Narendra, A. Annaswamy, A new adaptive law for robust adaptation without persistent excitation. IEEE Trans. Autom. Control **32**(2), 134–145 (1987)
15. K.S. Narendra, A.M. Annaswamy, *Stable Adaptive Systems* (Prentice-Hall, Englewood Cliffs, 1989)
16. P. Ögren, E. Fiorelli, N.E. Leonard, Cooperative control of mobile sensor networks: adaptive gradient climbing in a distributed environment. IEEE Trans. Autom. Control, **49**(8), 1292–1302 (2004)
17. B. Peterson, K. Narendra, Bounded error adaptive control. IEEE Trans. Autom. Control **27**(6), 1161–1168 (1982)
18. L.C.A. Pimenta, V. Kumar, R.C. Mesquita, G.A.S. Pereira. Sensing and coverage for a network of heterogeneous robots, in *Proceedings of the IEEE Conference on Decision and Control*, Cancun, Mexico (2008)
19. L.C.A. Pimenta, M. Schwager, Q. Lindsey, V. Kumar, D. Rus, R.C. Mesquita, G.A.S. Pereira, Simultaneous coverage and tracking (SCAT) of moving targets with robot networks, in *Proceedings of the Eighth International Workshop on the Algorithmic Foundations of Robotics (WAFR 08)*, Guanajuato, Mexico (2008)
20. C. Samson, Stability analysis of adaptively controlled systems subject to bounded disturbances. Automatica **19**(1), 81–86 (1983)
21. S.S. Sastry, M. Bodson, *Adaptive Control: Stability, Convergence, and Robustness* (Prentice-Hall Inc, Upper Saddle River, 1989)
22. M. Schwager, J. McLurkin, J.J.E. Slotine, D. Rus, From theory to practice: Distributed coverage control experiments with groups of robots, in *Experimental Robotics: The Eleventh International Symposium*, Vol. 54, pp. 127–136. (Springer, 2008)
23. M. Schwager, D. Rus, J.J. Slotine, Decentralized, adaptive coverage control for networked robots. Int. J. Robot. Res. **28**(3), 357–375 (2009)
24. M. Schwager, D. Rus, J.J. Slotine, Unifying geometric, probabilistic, and potential field approaches to multi-robot deployment. Int. J. Robot. Res. **30**(3), 371–383 (2011)
25. J.J.E. Slotine, W. Li, *Applied Nonlinear Control* (Prentice- Hall, Upper Saddle River, 1991)
26. A. Weber, *Theory of the Location of Industries*. (The University of Chicago Press, Chicago, 1929). Translated by Carl. J. Friedrich

A Multi-robot Control Policy for Information Gathering in the Presence of Unknown Hazards

Mac Schwager, Philip Dames, Daniela Rus and Vijay Kumar

Abstract This paper addresses the problem of deploying a network of robots to gather information in an environment, where the environment is hazardous to the robots. This may mean that there are adversarial agents in the environment trying to disable the robots, or that some regions of the environment tend to make the robots fail, for example due to radiation, fire, adverse weather, or caustic chemicals. A probabilistic model of the environment is formulated, under which recursive Bayesian filters are used to estimate the environment events and hazards online. The robots must control their positions both to avoid sensor failures and to provide useful sensor information by following the analytical gradient of mutual information computed using these online estimates. Mutual information is shown to combine the competing incentives of avoiding failure and collecting informative measurements under a common objective. Simulations demonstrate the performance of the algorithm.

M. Schwager (✉) · P. Dames · V. Kumar
GRASP Lab, University of Pennsylvania,
3330 Walnut St, Philadelphia, PA 19104, USA
e-mail: schwager@seas.upenn.edu

P. Dames
e-mail: pdames@seas.upenn.edu

V. Kumar
e-mail: kumar@seas.upenn.edu

M. Schwager · D. Rus
Computer Science and Artificial Intelligence Lab, MIT,
32 Vassar St, Cambridge, MA 02139, USA
e-mail: rus@csail.mit.edu

H.I. Christensen and O. Khatib (eds.), *Robotics Research*,
Springer Tracts in Advanced Robotics 100, DOI 10.1007/978-3-319-29363-9_26

1 Introduction

Networks of robotic sensors have the potential to safely collect data over large scale, unknown environments. They can be especially useful in situations where the environment is unsafe for humans to explore. In many such situations, robots are also susceptible to hazards. It is important to design exploration and mapping algorithms that are hazard-aware, so that the robotic sensor network can effectively carry out its task while minimizing the impact of individual robot failures. In this paper we propose an algorithm, based on an analytic expression for the gradient of mutual information, that enables a robotic sensor network to estimate a map of events in the environment while avoiding failures due to unknown hazards.

Consider, for example, the recent tragic accident at the Fukushima nuclear power plant in Japan, which sustained critical damage from a large earthquake and tsunami in March, 2011. The algorithm we propose here could be used by a team of flying quadrotor robots with cameras to inspect the plant for structural damage, keeping human workers at a safe distance. With our algorithm the robots could build a map of the areas that are likely to have structural damage, while simultaneously building a map of the radiation hazards, to avoid failure due to radiation exposure. This scenario is illustrated in Fig. 1. Both the event map and the hazard map are estimated online using a recursive Bayesian filter, where the event map is estimated from evidence of structural damage seen by the cameras, and the hazard map is estimated by the previous failures of other robots. The robots move along the

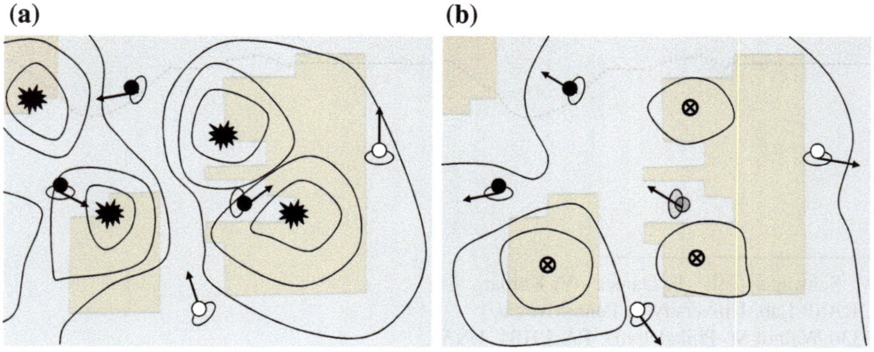

Fig. 1 The tragic accident at the Fukushima nuclear power plant in Japan is a fitting scenario for our algorithm. Hypothetical maps of the events and hazards are shown over an image of the Fukushima plant from http://maps.google.com/. On the *left*, the events of interest are structural defects represented by the explosion symbols, and the contour lines represent the probability of detecting these defects. Sensors move to determine where the structural defects are by increasing the informativeness of their sensor readings. Black circles represent sensors that see a defect while white circles do not see a defect. On the *right*, the hazards are radiation sources represented by the $\otimes$ symbol, and the contours represent the probability of failure due to radiation damage. By moving to increase informativeness, the robots implicitly avoid hazards that may cause failure thereby preventing information from being collected. The grayed-out robot in the center has failed due to the high radiation levels. **a** Event map. **b** Hazard map

gradient of mutual information, which gives the direction of expected maximum information gain given the current maps, thereby driving the exploration of the environment. Our algorithm could also be used, for example, to monitor forest fires while avoiding fire damage, taking ocean measurements while avoiding damage from adverse weather, or mapping a chemical spill site while avoiding failure from caustic chemicals.

In all of these examples, the robots must move to both avoid hazards and provide useful sensor information. Although these two objectives may seem to be in conflict with one another, they are in fact complementary. If we want to map the events as precisely as possible, we implicitly want the robots to avoid hazardous areas, since the failure of a robot makes it unable to contribute to estimating the event map in the future. We use the gradient of mutual information to move the sensors so that their next measurements are as informative as possible. The gradient strategy blends the avoidance of hazards and the seeking of information into one probabilistically consistent objective. We propose a probabilistic model of the environment, the sensors, and the task, and derive the Bayesian filters for updating the event and hazard maps. We then prove a general theorem showing that the analytical gradient of mutual information has a simple form similar to mutual information itself. To our knowledge, this is the first proof of such an expression to appear in the literature. The mutual information gradient is then used to control the robots. We do not consider decentralization of the algorithm in this paper, though that will be a central concern of future work, and several existing methods can be adapted for decentralization.

1.1 Related Work

Mutual information is one of the fundamental quantities in information theory [6, 20] and has been used extensively as a metric for robotic sensor network control and static sensor placement. For example, in [1, 7] mutual information is used as a metric for driving robotic sensor networks in gridded environments for target tracking and exploration tasks. Also, [8] focused on decentralization and scalability using particle filters to approximate mutual information for target tracking. Recently in [10] the gradient of mutual information was used to drive a network of robots for general environment state estimation tasks, and a sampling method was employed to improve computational efficiency. In [5] a mutual information method was used to control robots building maps of radiation intensity in an environment. The property of submodularity of mutual information was used in [11, 12] for placing static sensors at provably near-optimal positions for information gain. The technique was extended to Gaussian processes in [13]. Approximations on information gain for static sensor placement were derived in [3] and an informative trajectory planning algorithm was presented in [2]. In a different but related application, [22] uses mutual information to place static sensors to provide localization information to a mobile robot.

Our method differs in at least two important ways from those described above. Firstly, our work is specifically concerned with estimating and avoiding environmental hazards as well as estimating events. To our knowledge no works combining these objectives using mutual information have appeared in the literature. Secondly, we use an analytically derived expression for the gradient of mutual information for control. All the works above, save one, use grid-based finite difference methods to increase mutual information. The exception is [10], which employs the same analytical gradient of mutual information as we do here, but we have reserved the presentation of the proof of that result for this paper. In [5] the authors show that the gradient of mutual information approaches zero asymptotically in their problem, and thereby avoid computing it in their controller. To the authors knowledge, the only other appearance of a gradient of mutual information in a general form is derived in [18] in the context of channel coding.

Many other methods that do not use mutual information have been proposed for mapping and exploring environments with robotic sensor networks. For example [15] uses the error variance of a distributed Kalman filter to drive robots to estimate environmental fields. In [19] a Voronoi based coverage algorithm from [4] is augmented with online learning and exploration to estimate a scalar field in the environment. Similarly, [16] expanded this coverage algorithm with online interpolation of an environmental field. Artificial potential fields have been used in [9] for multi-robot deployment and exploration, and [14] uses a probabilistic coverage model for multi-robot deployment.

The question we address in this paper is: How do we choose the next positions $x_1, \ldots, x_n$ to make the next Bayesian estimate of the event state as precise as possible? As already described, implicit in this question is the tendency to avoid hazards because a failed robot is an uninformative robot. However, in our scenario, as in real life, all the robots will eventually fail. To counteract the depletion of robots, we let there be a base station located in the environment that deploys new robots to replace ones that have failed. We let the rate of releasing new robots balance the rate of failed ones, so that the total number of robots is constant at all times. However many other interesting possibilities exist.

The rest of this paper is organized as follows. We define notation and formulate the problem in Sect. 2. We derive the Bayesian filters to estimate the hazards and events in Sect. 3. In Sect. 4 we derive the analytical gradient of mutual information and specialize it for our hazard-aware exploration problem. Finally, Sect. 5 presents the results of numerical simulations and conclusions and future work are discussed in Sect. 6..

2 Problem Formulation

Consider a situation in which n robots move in a planar environment $Q \subset \mathbb{R}^2$. The robots have positions $x_i(t) \in Q$ and we want to use them to sense the state of the environment while avoiding hazardous areas that may cause the robots to fail. Let

Table 1 List of symbols

x_i	Position of sensor i
y_i^t	Reading of sensor i at time t
f_i^t	Failure status of sensor i at time t
q_j	Centroid position of grid cell j
s_j	Event state in grid cell j
h_j	Hazard state in grid cell j
x	Stacked vector of sensor positions
$y^{1:t}$	Time history of measurements up to t
$f^{1:t}$	Time history of failures up to t
Q	Environment
s	Full event state of environment
h	Full hazard state of environment

the positions of all the robots be given by the vector $x = \left[x_1^T \cdots x_n^T\right]$. The robots give simple binary sensor measurements $y_i \in \{0, 1\}$ indicating wether or not they have sensed an event of interest near by. They also give a signal to indicate their failure status $f_i \in \{0, 1\}$, where $f_i = 1$ means that the robot has failed. Denote the random vector of all sensor outputs by $y = [y_1 \cdots y_n]^T$ and the vector of all failure statuses as $f = [f_1 \cdots f_n]^T$ (Table 1).

The task of the robot network is to estimate the state of the environment with as little uncertainty as possible. While the robots move in continuous space, we introduce a discretization of the environment to represent the environment state. We let the state of the environment be modeled by a random field $s = \left[s_1 \cdots s_{m_s}\right]^T$, in which each random variable s_j represents the value of the field at a position $q_j \in Q$ and m_s is the number of discrete locations. Each of these random variables takes on a value in a set $s_j \in S$, and the environment state has a value $s \in \mathcal{S} = S^{m_s}$. Similarly, the hazard level, which is related to the probability of failure of the robot, is modeled as a random field $h = \left[h_1^T \ldots h_{m_h}^T\right]$ in which h_k represents the hazard level at position $q_k \in Q$, and takes on a value in a set H. Then the hazard state of the whole environment has a value $h \in \mathcal{H} = H^{m_h}$ In general, S and H may be infinite sets, however one special case of interest is $S = \{0, 1\}$, and $H = \{0, 1\}$, so the state and hazard distributions denote the presence or absence of a target or a hazard, respectively, in a grid cell. Note that the use of the phrase grid cell refers to an element in the discretization of the environment, which need not be a square grid. We will work with the more general framework to include the possibility that some areas may be more important than others, or that there may be multiple events or hazards in a single grid cell. Also note that the discretization of the environment for state and hazard estimation need not be the same, for example we might need more precise localization of events than hazards. Let $\phi^0(s)$ and $\psi^0(h)$ denote the robots' initial guess at the distribution of the state and the hazards, respectively, which can be uniform if we have no prior information about the events or hazards.

Furthermore, in our scenario the robots have some probability of failure due to the hazards in the environment. Let the probability of failure of a robot at x_i due to hazard level h_j at location q_j be given by $\mathbb{P}(f_i = 1 \mid h_j = 1) = \alpha(x_i, q_j)$ assume that the hazards act independently of one another and that the probability of failure when infinitely far away from a hazard is given by $P_{\mathrm{f,far}}$, so that

$$\mathbb{P}(f_i = 0 \mid h) = (1 - P_{\mathrm{f,far}}) \prod_{j \mid h_j=1} \mathbb{P}(f_i = 0 \mid h_j) = (1 - P_{\mathrm{f,far}}) \prod_{j \mid h_j=1} (1 - \alpha(x_i, q_j)). \tag{1}$$

In words, the probability of a robot not failing is the product of the probability of it not failing due to any individual hazard. This gives the probability of a robot failing due to any number of hazards in the environment as

$$\mathbb{P}(f_i = 1 \mid h) = 1 - (1 - P_{\mathrm{f,far}}) \prod_{j \mid h_j=1} (1 - \alpha(x_i, q_j)). \tag{2}$$

When a robot fails, its sensor will output 0 with probability 1, that is, its sensor reading gives no indication of whether or not there is an event of interest near by, giving the conditional probability $\mathbb{P}(y_i = 1 \mid f_i = 1, s) = 0$. In this case, the sensor will naturally provide no further information about event or hazard locations.

If the robot does not fail, the sensor output, y_i, is a Bernoulli random variable with the probability of $y_i = 1$ due to a state value s_j at position q_j given by $\mathbb{P}(y_i = 1 \mid f_i = 0, s_j) = \mu(x_i, q_j)$, and the probability that $y_i = 0$ the complement of this. We again assume that state locations act independently on the robot's sensor and that the probability of a false positive reading is P_{fp} so that

$$\mathbb{P}(y_i = 0 \mid f_i = 0, s) = (1 - P_{\mathrm{fp}}) \prod_{j \mid s_j=1} (1 - \mu(x_i, q_j)). \tag{3}$$

Then the probability that a robot's sensor gives $y_i = 1$ for a given environment state is the complement of this,

$$\mathbb{P}(y_i = 1 \mid f_i = 0, s) = 1 - (1 - P_{\mathrm{fp}}) \prod_{j \mid s_j=1} (1 - \mu(x_i, q_j)). \tag{4}$$

We defer the discussion of specific forms of the functions α and μ to Sect. 5, however potential choices for the case where a hazard or event is present would be a decreasing exponential, a Gaussian function, or, in the simplest case, a constant (e.g. close to 1) inside some distance to q_j and some other constant (e.g. close to zero) outside. We will see that when μ or α have compact support in Q, there are computational benefits.

Now we consider the group of robots together. We derive three quantities that will be used in the Bayesian filters and control law; the likelihood function of the sensor measurements given the failures, the events, and the hazards, $\mathbb{P}(y\,|\,f,s,h)$; the likelihood function of the failures given the events and the hazards, $\mathbb{P}(f\mid s,h)$; and the likelihood function of the sensor measurements given the events and the hazards, $\mathbb{P}(y\mid s,h)$.

For $\mathbb{P}(y\,|\,f,s,h)$, assume that each robot's measurement is conditionally independent of the hazards and the other robots' measurements given its own failure status and the environment state, $\mathbb{P}(y\,|\,f,s,h)=\prod_{i=1}^{n}\mathbb{P}(y_i\,|\,f_i,s)$. Supposing we know the failure status of each robot, we can compute the measurement likelihood to be

$$\mathbb{P}(y\,|\,f,s,h)=\prod_{i\,|\,f_i=0}\mathbb{P}(y_i\,|\,f_i=0,s)\prod_{j\,|\,f_j=1}\mathbb{P}(y_j\,|\,f_j=1,s),$$

but

$$\prod_{j\,|\,f_j=1}\mathbb{P}(y_j\,|\,f_j=1,s)=\prod_{j\,|\,y_j=0,f_j=1}\mathbb{P}(y_j\,|\,f_j=1,s)\prod_{j\,|\,y_j=1,f_j=1}\mathbb{P}(y_j\,|\,f_j=1,s),$$

and the set $\{j\,|\,y=1,f_j=1\}=0'$ and $\mathbb{P}(y_j=0\,|\,f_j=1,s)=1$, therefore this product reduces to 1. Then we have

$$\begin{aligned}\mathbb{P}(y\,|\,f,s,h)&=\prod_{i\,|\,y_i=0,f_i=0}\mathbb{P}(y_i=0\,|\,f_i=0,s)\prod_{j\,|\,y_j=1,f_j=0}\mathbb{P}(y_j=1\,|\,f_j=0,s)\times 1\\&=\prod_{i\,|\,y_i=0,f_i=0}\left(\prod_{k=1}^{m_s}(1-\mu(x_i,q_k))\right)\prod_{j\,|\,y_j=1,f_j=0}\left(1-\prod_{l=1}^{m_s}(1-\mu(x_j,q_l))\right)\end{aligned}\tag{5}$$

where we use (4) and (3) to get the last equality.

Next we derive the failure likelihood, $\mathbb{P}(f\mid s,h)$. Conditioned on knowledge of the hazards, the failures are independent of the events and each other, so $\mathbb{P}(f\mid s,h)=\prod_{i=1}^{n}\mathbb{P}(f_i\mid h)$. Then using (1) and (2) we obtain

$$\mathbb{P}(f\mid s,h)=\prod_{i\,|\,f_i=0}\prod_{k=1}^{m_h}(1-\alpha(x_i,q_k))\prod_{j\,|\,f_j=1}\left(1-\prod_{l=1}^{m_h}(1-\alpha(x_j,q_l))\right).\tag{6}$$

Finally, in the case that we want to predict an information gain for a future measurement, the failures are not yet known, so we will need the quantity $\mathbb{P}(y\mid s,h)$. This leads to

$$\begin{aligned}\mathbb{P}(y \mid s, h) &= \sum_{f \in \{0,1\}^n} \mathbb{P}(y \mid f, s, h)\mathbb{P}(f \mid s, h) = \sum_{f \in \{0,1\}^n} \prod_{i=1}^{n} \mathbb{P}(y_i \mid f_i, s)\mathbb{P}(f_i \mid h) \\ &= \prod_{i=1}^{n} \sum_{f_i \in \{0,1\}} \mathbb{P}(y_i \mid f_i, s)\mathbb{P}(f_i \mid h),\end{aligned} \tag{7}$$

where, as we already saw, $\mathbb{P}(f_i = 0 \mid h) = \prod_{j=1}^{m_h} (1 - \alpha(x_i, q_j))$ and $\mathbb{P}(f_i = 1 \mid h) = 1 - \mathbb{P}(f_i = 0 \mid h)$ and similarly $\mathbb{P}(y_i = 0 \mid f_i = 0, s) = \prod_{j=1}^{m_s} (1 - \mu(x_i, q_j))$ and $\mathbb{P}(y_i = 1 \mid f_i = 0, s) = 1 - \mathbb{P}(y_i = 0 \mid f_i, s)$ and finally $\mathbb{P}(y_i = 0 \mid f_i = 1, s) = 1$ and $\mathbb{P}(y_i = 1 \mid f_i = 1, s) = 0$. Next we use these quantities to derive a recursive Bayesian filters for maintaining a distribution over all possible event and hazard states.

3 Bayesian Estimation

As our robots move about in the environment, we wish to make use of their measurements and failure statuses at each time step in order to recursively estimate the events and hazards in the environment. We will show in this section, surprisingly, that the event and hazard estimates are either statistically independent, or they are deterministically linked (knowledge of either one fully determines the other). An unexpected consequence of the natural formulation in Sect. 2 is that there can be no statistical dependence between the event and hazard estimates except these two extremes. We let the robots collect measurements y^t synchronously at times $t = 1, 2, \ldots$, and we denote the tuple of all measurements up to time t by $y^{1:t} = (y^1, \ldots, y^t)$. We use a similar notation for failures, so that $f^{1:t} = (f^1, \ldots, f^t)$ is the tuple of all failure statuses up to time t. Furthermore, define the event distribution estimate up to time t to be $\phi^t(s) := \mathbb{P}(s \mid y^{1:t}, f^{1:t})$ and the hazard distribution up to time t to be $\psi^t(h) := \mathbb{P}(h \mid y^{1:t}, f^{1:t})$. The main result of this section is stated in the following theorem.

Theorem 1 (Bayesian Filtering) *The distributions for hazards and events given all information up to time t are independent with* $\mathbb{P}(s, h \mid y^{1:t}, f^{1:t}) = \phi^t(s)\psi^t(h)$, *assuming that h and s are not deterministically linked, and that their initial distributions are independent,* $\mathbb{P}(s, h) = \phi^0(s)\psi^0(h)$ *Furthermore,* $\phi^t(s)$ *and* $\psi^t(h)$ *can be computed recursively with the Bayesian filters*

$$\phi^t(s) = \frac{\mathbb{P}(y^t \mid f^t, s)\phi^{t-1}(s)}{\sum_{s \in \mathcal{S}} \mathbb{P}(y^t \mid f^t, s)\phi^{t-1}(s)}, \tag{8}$$

and

$$\psi^t(h) = \frac{\mathbb{P}(f^t \mid h)\psi^{t-1}(h)}{\sum_{h\in\mathscr{H}} \mathbb{P}(f^t \mid h)\psi^{t-1}(h)}. \tag{9}$$

In the case that the events and the hazards are deterministically linked, the Bayesian filter update for the distribution is given by

$$\mathbb{P}(s \mid y^{1:t}, f^{1:t}) = \frac{\mathbb{P}(y^t \mid f^t, s)\mathbb{P}(f^t \mid s)\mathbb{P}(s \mid y^{1:t-1}, f^{1:t-1})}{\sum_{s\in\mathscr{S}} \mathbb{P}(y^t \mid f^t, s)\mathbb{P}(f^t \mid s)\mathbb{P}(s \mid y^{1:t-1}, f^{1:t-1})}. \tag{10}$$

Proof We will argue the existence of these two distinct cases by mathematical induction. We first prove (9) and then use it to prove (8). We will then derive (10) directly from the assumption that s and h are deterministically related.

To obtain an inductive argument, suppose that at $t-1$ the hazard estimate $\psi^{t-1}(h) = \mathbb{P}(h \mid y^{1:t-1}, f^{1:t-1}) = \mathbb{P}(h \mid f^{1:t-1})$ is independent of the sensor measurements $y^{1:t-1}$. Then the recursive Bayesian filter update for time t gives

$$\psi^t(h) = \frac{\mathbb{P}(y^t, f^t \mid h)\psi^{t-1}(h)}{\sum_{h\in\mathscr{H}} \mathbb{P}(y^t, f^t \mid h)\psi^{t-1}(h)}.$$

Now assuming that h and s are not deterministically related, we get $\mathbb{P}(y^t, f^t \mid h) = \mathbb{P}(y^t \mid f^t, h)\mathbb{P}(f^t \mid h) = \mathbb{P}(y^t \mid f^t)\mathbb{P}(f^t \mid h)$, where the last equality is because given the failure, f^t, the measurement, y^t, is independent of the hazards, h, as described in the previous section. This leads to

$$\psi^t(h) = \frac{\mathbb{P}(y^t \mid f^t)\mathbb{P}(f^t \mid h)\psi^{t-1}(h)}{\mathbb{P}(y^t \mid f^t)\sum_{h\in\mathscr{H}} \mathbb{P}(f^t \mid h)\psi^{t-1}(h)},$$

and we can cancel the factor of $\mathbb{P}(y^t \mid f^t)$ from the numerator and denominator to obtain (9). Now notice that $\psi^t(h) = \mathbb{P}(h \mid y^{1:t}, f^{1:t}) = \mathbb{P}(h \mid f^{1:t})$ remains independent of the measurements at time t. The initial distribution, $\psi^0(h)$, must be independent of $y^{1:t}$ (because no measurements have been collected yet), therefore by mathematical induction the hazard estimate distribution conditioned on the failures is always independent of the measurements.

Using a similar mathematical induction argument, suppose that the hazard and event estimates are independent given the measurements and failures up to time $t-1$, so $\mathbb{P}(h, s \mid y^{1:t-1}, f^{1:t-1}) = \phi^{t-1}(s)\psi^{t-1}(h)$. Then the Bayesian update for their joint distribution at time t is given by

$$\mathbb{P}(s,h \mid y^{1:t}, f^{1:t}) = \frac{\mathbb{P}(y^t, f^t \mid s,h)\phi^{t-1}(s)\psi^{t-1}(h)}{\sum_{s\in\mathcal{S}}\sum_{h\in\mathcal{H}}\mathbb{P}(y^t, f^t \mid s,h)\phi^{t-1}(s)\psi^{t-1}(h)}.$$

Factoring the numerator using the conditional independences described in Sect. 2, we get

$$\mathbb{P}(s,h \mid y^{1:t}, f^{1:t}) = \frac{\mathbb{P}(y^t \mid f^t, s)\mathbb{P}(f^t \mid h)\phi^{t-1}(s)\psi^{t-1}(h)}{\sum_{s\in\mathcal{S}}\sum_{h\in\mathcal{H}}\mathbb{P}(y^t \mid f^t, s)\mathbb{P}(f^t \mid h)\phi^{t-1}(s)\psi^{t-1}(h)},$$

and separating terms that depend on s from those that depend on h yields

$$\mathbb{P}(s,h \mid y^{1:t}, f^{1:t}) = \frac{\mathbb{P}(y^t \mid f^t, s)\phi^{t-1}(s)}{\sum_{s\in\mathcal{S}}\mathbb{P}(y^t \mid f^t, s)\phi^{t-1}(s)}\frac{\mathbb{P}(f^t \mid h)\psi^{t-1}(h)}{\sum_{h\in\mathcal{H}}\mathbb{P}(f^t \mid h)\psi^{t-1}(h)},$$

We recognize the right most fraction as the Bayesian update from (9), and the left most expression can be factored as $\mathbb{P}(s,h \mid y^{1:t}, f^{1:t}) = \mathbb{P}(s \mid h, y^{1:t}, f^{1:t})\psi^t(h)$, which gives

$$\mathbb{P}(s \mid h, y^{1:t}, f^{1:t})\psi^t(h) = \frac{\mathbb{P}(y^t \mid f^t, s)\phi^{t-1}(s)}{\sum_{s\in\mathcal{S}}\mathbb{P}(y^t \mid f^t, s)\phi^{t-1}(s)} \mid \psi^t(h).$$

The fraction on the right is independent of h, so we conclude that $\mathbb{P}(s \mid h, y^{1:t}, f^{1:t}) = \mathbb{P}(s \mid y^{1:t}, f^{1:t}) = \phi^t(s)$, and we obtain the Bayesian update in (8). Therefore if the estimate distributions of s and h are independent at time $t-1$ they will also be so at time t, and by induction, if their initial distributions are independent then they will remain so for all time.

Finally, in the case that the hazards and the events are deterministically related, the standard recursive Bayesian filter yields

$$\mathbb{P}(s \mid y^{1:t}, f^{1:t}) = \frac{\mathbb{P}(y^t, f^t \mid s)\mathbb{P}(s \mid y^{1:t-1}, f^{1:t-1})}{\sum_{s\in\mathcal{S}}\mathbb{P}(y^t, f^t \mid s)\mathbb{P}(s \mid y^{1:t-1}, f^{1:t-1})},$$

which factors straightforwardly to the expression in (10). □

For the remainder of the paper, we consider the case where the hazards and the events are not deterministically related, so the filtering equations are given by (8) with (5) for the event update, and (9) with (6) for the hazard update. The robots use these filter equations to maintain estimates of the events and hazards in the environment. Next we consider using these estimates to derive an information seeking controller.

4 Control Using the Mutual Information Gradient

In this section we will derive an analytic expression for the gradient of mutual information in general terms, then specialize it to use the distributions of events and hazards found with the Bayesian filters in the previous section. This will lead to an information seeking controller for our robots.

In information theory [6, 20], the mutual information between two random vectors s and y is defined as

$$I(S;Y) = \int_{y\in\mathcal{Y}} \int_{s\in\mathcal{S}} \mathbb{P}(s,y) \log \frac{\mathbb{P}(s,y)}{\mathbb{P}(s)\mathbb{P}(y)} dsdy.$$

where we let e be the base of the logarithm, and $\mathcal{S}$ and $\mathcal{Y}$ are the range of s and y, respectively. We write s and y as though they were continuous valued for simplicity, though similar expressions can be written for discrete valued and general random variables. Mutual information indicates how much information one random variable gives about the other. In our scenario, we want to position our robots so that their next measurement gives the maximum amount of information about the event distribution.

Consider the situation in which the distribution $\mathbb{P}(s,y)$ depends on a parameter vector $x \in \mathbb{R}^{2n}$. We write $\mathbb{P}_x(s,y)$ to emphasize this dependence. Likewise, let $\mathbb{P}_x(s) := \int_{y\in\mathcal{Y}} \mathbb{P}_x(s,y)dy$, $\mathbb{P}_x(y) := \int_{s\in\mathcal{S}} \mathbb{P}_x(s,y)ds$, and

$$I_x(S;Y) := \int_{y\in\mathcal{Y}} \int_{s\in\mathcal{S}} \mathbb{P}_x(s,y) \log \frac{\mathbb{P}_x(s,y)}{\mathbb{P}_x(s)\mathbb{P}_x(y)} dsdy, \tag{11}$$

In our case x is the positions of the robots, but the following result holds in a general context. We can compute the gradient of the mutual information with respect to the parameters x using the following theorem.

Theorem 2 (Mutual Information Gradient) *Let random vectors s and y be jointly distributed with distribution $\mathbb{P}_x(s,y)$ that is differentiable with respect to the parameter vector $x \in \mathbb{R}^{2n}$ over $Q^n \subset \mathbb{R}^{2n}$. Also, suppose that the support $\mathcal{S} \times \mathcal{Y}$ of $\mathbb{P}_x(s,y)$, does not depend on x. Then the gradient of the mutual information with respect to the parameters x over Q^n is given by*

$$\frac{\partial I_x(S;Y)}{\partial x} = \int_{y\in\mathcal{Y}} \int_{s\in\mathcal{S}} \frac{\partial \mathbb{P}_x(s,y)}{\partial x} \log \frac{\mathbb{P}_x(s,y)}{\mathbb{P}_x(s)\mathbb{P}_x(y)} \,|\, dsdy. \tag{12}$$

Proof The theorem follows straightforwardly by applying the rules of differentiation. Notably, several terms are identically zero, yielding the simple result. Differentiating (11) with respect to x, we can move the differentiation inside the integrals since $\mathcal{S}$ and $\mathcal{Y}$ do not depend on the parameters x. Then applying the chain rule to the integrand results in

$$\frac{\partial I(S;Y)}{\partial x} = \int_{y\in\mathcal{Y}}\int_{s\in\mathcal{S}} \frac{\partial \mathbb{P}(s,y)}{\partial x}\log\frac{\mathbb{P}(s,y)}{\mathbb{P}(s)\mathbb{P}(y)}dsdy + \int_{y\in\mathcal{Y}}\int_{s\in\mathcal{S}} \mathbb{P}(s,y)\frac{\mathbb{P}(s)\mathbb{P}(y)}{\mathbb{P}(s,y)}$$
$$\times\left[\frac{\partial \mathbb{P}(s,y)}{\partial x}\frac{1}{\mathbb{P}(s)\mathbb{P}(y)} - \frac{\partial \mathbb{P}(s)}{\partial x}\frac{\mathbb{P}(y)\mathbb{P}(s,y)}{(\mathbb{P}(s)\mathbb{P}(y))^2} - \frac{\partial \mathbb{P}(y)}{\partial x}\frac{\mathbb{P}(s)\mathbb{P}(s,y)}{(\mathbb{P}(s)\mathbb{P}(y))^2}\right]dsdy,$$

where we have suppressed the dependence on x to simplify notation. Bringing $1/(\mathbb{P}(s)\mathbb{P}(y))$ in front of the brackets gives

$$\frac{\partial I(S;Y)}{\partial x} = \int_{y\in\mathcal{Y}}\int_{s\in\mathcal{S}} \frac{\partial \mathbb{P}(s,y)}{\partial x}\log\frac{\mathbb{P}(s,y)}{\mathbb{P}(s)\mathbb{P}(y)}dsdy$$
$$+\int_{y\in\mathcal{Y}}\int_{s\in\mathcal{S}}\left[\frac{\partial \mathbb{P}(s,y)}{\partial x} - \frac{\partial \mathbb{P}(s)}{\partial x}\mathbb{P}(y\,|\,s) - \frac{\partial \mathbb{P}(y)}{\partial x}\mathbb{P}(s\,|\,y)\right]dsdy.$$

Consider the three terms in the second double integral. We will show that each of these terms is identically zero to yield the result in the theorem. For the first term we have

$$\int_{y\in\mathcal{Y}}\int_{s\in\mathcal{S}} \frac{\partial \mathbb{P}(s,y)}{\partial x}dsdy = \frac{\partial}{\partial x}\int_{y\in\mathcal{Y}}\int_{s\in\mathcal{S}} \mathbb{P}(s,y)dsdy = \frac{\partial}{\partial x}1 = 0.$$

For the second term we have

$$\int_{y\in\mathcal{Y}}\int_{s\in\mathcal{S}} \frac{\partial \mathbb{P}(s)}{\partial x}\mathbb{P}(y\,|\,s)dsdy = \int_{s\in\mathcal{S}} \frac{\partial \mathbb{P}(s)}{\partial x}\left(\int_{y\in\mathcal{Y}} \mathbb{P}(y\,|\,s)dy\right)ds$$
$$= \frac{\partial}{\partial x}\int_{s\in\mathcal{S}} \mathbb{P}(s)ds = 0,$$

and the third term follows similarly if we interchange y and s. □

Remark 1 The result holds for the general definition of mutual information and makes no assumptions as to the distribution of the random variables, or the form of the dependence of $\mathbb{P}_x(s,y)$ on its parameters. The result also holds for generally distributed random variables including discrete valued ones (although we have written the theorem for continuous valued ones).

Remark 2 It is interesting that the gradient of $I_x(S;Y)$ has the same form as $I_x(S;Y)$ itself, except that the first occurrence of $\mathbb{P}_x(s,y)$ is replaced by its gradient with respect to x. To the authors' knowledge, this analytic expression for the gradient of mutual information has not been reported in the literature despite the proliferation of gradient based methods for maximizing mutual information in various applications ranging from channel coding [17, 18], to medical imaging alignment [21], to the control of roboticsensor networks [7]. In [18] the authors derive an expression that can be shown to be equivalent to a special case of Theorem 2 in which $\mathbb{P}(s)$ is not a function of x.

Remark 3 Arbitrary derivatives of mutual information can also be calculated, though the fortuitous cancellations do not necessarily hold for higher order derivatives. In our robotic sensing scenario, for example, it would be interesting to compute the Hessian of mutual information to examine the coupling between the control laws for neighboring robots.

We will use the result in Theorem 2 to design a controller for our robotic sensor network. Writing the result in terms of quantities that we already know, we have

$$\begin{aligned} \frac{\partial I_x(S;Y)}{\partial x} = & \sum_{y \in \{0,1\}^n} \sum_{s \in \mathcal{S}} \sum_{h \in \mathcal{H}} \frac{\partial \mathbb{P}_x(y \mid s, h)}{\partial x} \phi^t(s) \psi^t(h) \\ & \times \log \frac{\sum_{h \in \mathcal{H}} \mathbb{P}_x(y \mid s, h) \psi^t(h)}{\sum_{s \in \mathcal{S}} \sum_{h \in \mathcal{H}} \mathbb{P}_x(y \mid s, h) \phi^t(s) \psi^t(h)}, \end{aligned} \tag{13}$$

where $\phi^t(s)$ and $\psi^t(h)$ come from the Bayesian filter Eqs. (8) and (9), respectively, and $\mathbb{P}_x(y \mid s, h)$ comes from (7). The only remaining necessary term is the gradient of the measurement likelihood, which we can compute straightforwardly from (7) to be

$$\begin{aligned} \frac{\partial \mathbb{P}_x(y_i = 1 \mid s, h)}{\partial x} = & \ \mathbb{P}(f_i = 0 \mid h) \mathbb{P}(y_i = 0 \mid s, h) \sum_{j \mid s_j = 1} \frac{1}{1 - \mu(x_i, q_j)} \frac{\partial \mu(x_i, q_l)}{\partial x_i} \\ & - \mathbb{P}(f_i = 0 \mid h) \mathbb{P}(y_i = 1 \mid s, h) \sum_{k \mid h_k = 1} \frac{1}{1 - \alpha(x_i, q_k)} \frac{\partial \alpha(x_i, q_k)}{\partial x_i}, \end{aligned} \tag{14}$$

and when $y_i = 0$ it is simply the negative of this, $\frac{\partial \mathbb{P}_x(y_i = 0 \mid s, h)}{\partial x} = - \frac{\partial \mathbb{P}_x(y_i = 1 \mid s, h)}{\partial x}$. We propose to use an information seeking controller of the form

$$x_i(t+1) = x(t) + k \frac{\frac{\partial I_x(S;Y)}{\partial x_i}}{\left\| \frac{\partial I_x(S;Y)}{\partial x_i} \right\| + \varepsilon}, \tag{15}$$

where $k > 0$ is a maximum step size, and $\varepsilon > 0$ is a small factor to prevent singularities when a local minimum of mutual information is reached. Although this controller uses a gradient, it is not, strictly speaking, a gradient controller and a formal analysis of its behavior would be expected to be quite difficult. This is because the mutual information changes at each time step due to the integration of new sensor measurements into the Bayesian event estimate and hazard estimate. Intuitively, the controller moves the robots in the direction of the highest immediate expected information gain. An alternative approach would be to use a finite time horizon over which to plan informative paths using the information gradient. Indeed, our controller can be seen as a special case of this with a horizon length of one time step. Such an approach would have an exponential time complexity in the length of the horizon, however, making even a short time horizon computationally impractical.

Empirically, the controller drives the robots to uncertain areas while veering away from suspected hazard sites, learned through the failures of previous robots. The robots eventually come to a stop when they estimate the event state s of the environment with high confidence (i.e. when the entropy of $\phi^t(s)$ approaches zero). While hazard avoidance does not explicitly appear in the control law, it is implicit as an expected robot failure will decrease the amount of information gained at the next time step. This is true even when robots are replaced, since the mutual information gradient is agnostic to the replacement of sensors. It only considers the information gain it expects to achieve with its current robots by moving them in a particular direction. So when deciding in what direction to move, each robot balances the benefit of information received with the possibility of failure by moving in that direction. As future work, it would also be interesting to explicitly account for sensor losses by enforcing a total sensor loss budget, or a maximum sensor loss rate.

4.1 Computational Considerations

Unfortunately, one can see from (13) that the computation of $\partial I_x(S;Y)/\partial x$ is, in general, in $O(n2^n \,|\mathscr{H}|\,|\mathscr{S}|)$ where n is the number of sensors. For example, if the hazard and event states are both binary, we have a complexity of $O(n2^{n+m_s+m_h})$, where m_s is the number of event grid cells and m_h is the number of hazard grid cells. Therefore this algorithm is not computationally practical for even a moderate number of robots or grid cells.

We are currently investigating two methods for decreasing the complexity of the control strategy, which will be detailed in a paper that is soon to follow. One involves a successive grid refinement procedure. In this procedure the robots begin with a coarse grid. Those grid cells that reach a threshold probability of containing an event or hazard are re-partitioned into a finer grid, while those grid cells with small probability are lumped together into a large cell. This procedure has the benefit of delivering arbitrarily fine event and hazard maps. The simulations described in Sect. 5 use a simplified version of this procedure. The second method that we are investigating is to assume a limited correlation structure among the grid cells in the environment. If two grid cells are statistically independent when they are separated by more than a fixed distance, the algorithm complexity reduces significantly. There also exist other methods for overcoming computational limitations, which include using Monte Carlo methods to approximate the sums in (13) in a decentralized way, as in [10], or particle filters as in [8].

5 Simulation Results

We carried out Matlab simulations with the controller (15) over $\mathbb{R}^2$ with the sensor detection probability

$$\mu(x_i, q_j) = P_{\text{fp}} + \frac{1 - P_{\text{fp}} - P_{\text{fn}}}{1 + \exp\left(c\left(\left\|x_i - q_j\right\| - r_{\text{sense}}\right)\right)},$$

with $P_{\text{fp}} = 0.01$, $P_{\text{fn}} = 0.05$, $c = 1$ $r_{\text{sense}} = 2$, and the robot failure probability

$$\alpha(x_i, q_j) = P_{\text{f,far}} + \left(1 - P_{\text{f,far}} - P_{\text{s,near}}\right) \exp\left(-\frac{\left\|x_i - q_j\right\|^2}{2\sigma_{\text{fail}}^2}\right),$$

with $P_{\text{f,far}} = 0$, $P_{\text{s,near}} = 0.1$, $\sigma_{\text{fail}} = 1.25$ and uniform initial event and hazard distributions. The control gains used in (15) were $k = 0.1$ and $\varepsilon = 10^{-10}$. We used a simplified version of the grid refinement procedure described in Sect. 4.1 in which the environment was first divided into a 4×4 grid until the entropy of the event distribution dropped below 0.1, indicating a high degree of certainty in the estimates at that resolution. Then the entire grid was refined to 8×8, and when the entropy again dropped below 0.1, it was refined to 16×16. At any time if the probability of occupancy for an event (or hazard) was less than 10^{-15} for a cell, its occupancy probability was set to zero, the event (or hazard) distribution was renormalized appropriately, and that cell was ignored in all future event (or hazard) distribution updates. This successive grid refinement and pruning was found to dramatically improve computation speed.

The results of simulations with three robots, three events, and one hazard are shown in Fig. 2. The failure of two robots can be seen in Fig. 2a by the discontinuities in the green and black robot paths (new robots are introduced at the lower left corner of the environment to compensate for the failed ones). The event distribution is shown in Fig. 2b where the locations of the three events have been localized down to one grid square. Similarly, Fig. 2c shows the hazard distribution, in which the hazard has been localized to a 4×4 block of grid squares. The robots do not seek to refine the hazard estimate because it is sufficient for them to avoid the hazard an continue mapping the events. The decreasing trend in the entropy (or

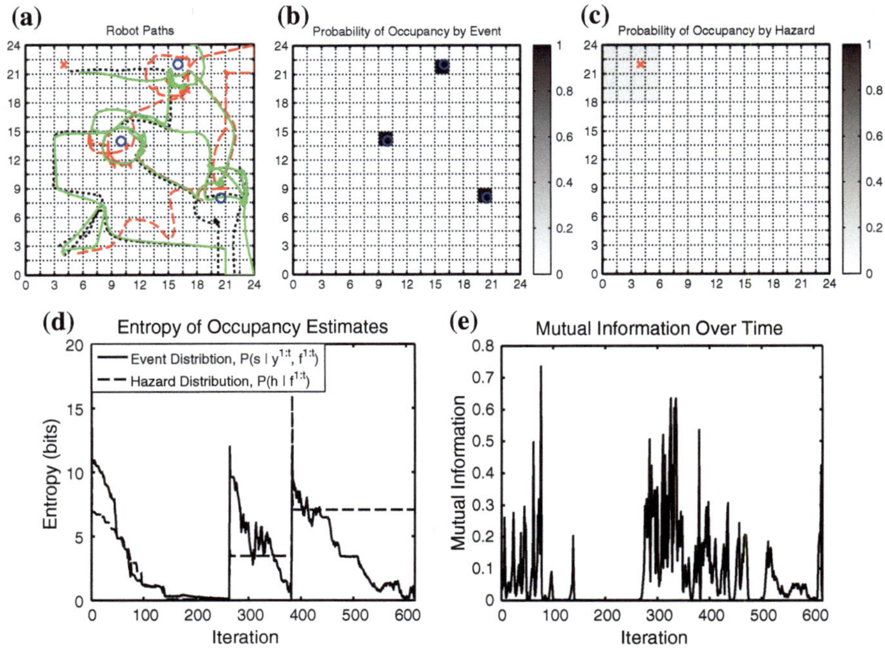

Fig. 2 This figure shows simulation results for three robots in an environment initially divided into a 4 × 4 grid, then refined to an 8 × 8 grid at the 266th iteration, and finally to a 16 × 16 grid at the 383rd iteration. Frame 2(a) shows the paths of the three robots which start from the *lower left* of the environment. The red '×' marks the location of a hazard and the three blue 'O's show the locations of events. The robots with the solid green and dotted black paths both fail once when they come too close to the hazard, after which two replacement robots are introduced at the *lower left*. Frames 2(b) and 2(c) show the final event and hazard distribution estimate, respectively. The events have been localized down to one grid square and the hazard down to a 4 × 4 region. Frame 2(d) shows the entropy, a measure of uncertainty, of the event (*solid*) and hazard (*dotted*) estimates. Both entropies decrease as the robots learn the event and hazard locations. They jump at the 266th and 383rd iteration when the grid is refined, as expected. The mutual information (or the expected drop in entropy) verses time is shown in frame 2(e). **a** Robot paths. **b** Event estimate. **c** Hazard estimate. **d** Event and hazard estimate entropy. **e** Mutual information

uncertainty) of the event and hazard estimates can be seen in Figs. 2d. The mutual information, shown in Fig. 2e, can be interpreted as the *expected* decrease in entropy, hence it tends to be large when there are large entropy drops. The entropy jumps at iteration 266 and 383, when the grid is refined to 8 × 8 and 16 × 16, respectively.

6 Conclusions

In this paper we proposed a multi-robot control policy that utilizes measurements about events of interest to locally increase the mutual information, while also using the history of robot failures to avoid hazardous areas. The central theoretical contribution of this paper is a new analytical expression for the gradient of mutual information presented in Theorem 2, which provides a principled approach to exploration by calculating robot trajectories that lead to the greatest immediate information gain. Despite minimal data from the binary sensors and a binary failure signal, the robot team is able to successfully localize events of interest and avoid hazardous areas. The event state and hazard fields over the environment are estimated using recursive Bayesian filters. The main drawback of the approach is high computational complexity, which makes the controller difficult to compute in realistic environments. We are currently investigating techniques for reducing the complexity and decentralizing the algorithm to run over a multi-robot network.

Acknowledgements This work was funded in part by ONR MURI Grants N00014-07-1-0829, N00014-09-1-1051, and N00014-09-1-1031, and the SMART Future Mobility project. We are grateful for this financial support.

References

1. F. Bourgault, A.A. Makarenko, S.B. Williams, B. Grocholsky, H.F. Durrant-Whyte, Information based adaptive robotic exploration, in *Proceedings of the IEEE International Conference on Intelligent Robots and Systems (IROS 02)* (2002), pp. 540–545
2. H.-L. Choi, J.P. How, Continuous trajectory planning of mobile sensors for informative forecasting. Automatica **46**(8), 1266–1275 (2010)
3. H.-L. Choi, J.P. How, Efficient targeting of sensor networks for large-scale systems. IEEE Trans. Control Syst. Technol. **19**(6), 1569–1577 (2011)
4. J. Cortés, S. Martínez, T. Karatas, F. Bullo, Coverage control for mobile sensing networks. IEEE Trans. Robot. Autom. **20**(2), 243–255 (2004)
5. R.A. Cortez, H.G. Tanner, R. Lumia, C.T. Abdallah, Information surfing for radiation map building. Int. J. Robot. Autom. **26**(1), 4–12 (2011)
6. T. Cover, J. Thomas, *Elements of Information Theory*, 2 edn. (Wiley, 2006)
7. B. Grocholsky, Information-Theoretic Control of Multiple Sensor Platforms. Ph.D. thesis, University of Sydney (2002)
8. G.M. Hoffmann, C.J. Tomlin, Mobile sensor network control using mutual information methods and particle filters. IEEE Trans. Autom. Control **55**(1), 32–47 (2010)
9. A. Howard, M.J. Matarić, G.S. Sukhatme, Mobile sensor network deployment using potential fields: a distributed, scalable solution to the area coverage problem, in *Proceedings of the 6th International Symposium on Distributed Autonomous Robotic Systems (DARS02)*, Fukuoka, Japan (2002)
10. B.J. Julian, M. Angermann, M. Schwager, D. Rus, A scalable information theoretic approach to distributed robot coordination, in *Proceedings of the IEEE/RSJ Conference on Intelligent Robots and Systems (IROS)*, San Francisco, CA, USA (2011)
11. A. Krause, C. Guestrin, Near-optimal observation selection using submodular functions, in *Proceedings of 22nd Conference on Artificial Intelligence (AAAI)*, Vancouver, Canada (2007)

12. A. Krause, C. Guestrin, A. Gupta, J. Kleinberg, Near-optimal sensor placements: maximizing information while minimizing communication cost, in *Proceedings of Information Processing in Sensor Networks (IPSN)*, Nashville, TN (2006)
13. A. Krause, A. Singh, C. Guestrin, Near-optimal sensor placements in gaussian processes: theory, efficient algorithms and empirical studies. J. Mach. Learn. Res. **9**, 235–284 (2008)
14. W. Li, C.G. Cassandras, Distributed cooperative coverage control of sensor networks, in *Proceedings of the IEEE Conference on Decision and Control, and the European Control Conference*, Seville, Spain (2005)
15. K.M. Lynch, I.B. Schwartz, P. Yang, R.A. Freeman, Decentralized environmental modeling by mobile sensor networks. IEEE Trans. Robot. **24**(3), 710–724 (2008)
16. S. Martínez, Distributed interpolation schemes for field estimation by mobile sensor networks. IEEE Trans. Control Syst. Technol. **18**(2), 419–500 (2010)
17. D.P. Palomar, S. Verdú, Gradient of mutual information in linear vector gaussian channels. IEEE Trans. Inf. Theory **52**(1), 141–154 (2006)
18. D.P. Palomar, S. Verdú, Representation of mutual information via input estimates. IEEE Trans. Inf. Theory **53**(2), 453–470 (2007)
19. M. Schwager, D. Rus, J.J. Slotine, Decentralized, adaptive coverage control for networked robots. Int. J. Robot. Res. **28**(3), 357–375 (2009)
20. C.E. Shannon, A mathematical theory of communication. Bell Syst. Tech. J. **27**, 379–423 (1948)
21. P. Viola, W.M. Wells III, Alignment by maximization of mutual information. Int. J. Comput. Vis. **24**(2), 137–154 (1997)
22. M.P. Vitus, C.J. Tomlin, Sensor placement for improved robotic navigation, in *The Proceedings of Robotics: Science and Systems*, Zaragoza, Spain (2010)

Motion Planning Under Uncertainty Using Differential Dynamic Programming in Belief Space

Jur van den Berg, Sachin Patil and Ron Alterovitz

Abstract We present an approach to motion planning under motion and sensing un-certainty, formally described as a continuous partially-observable Markov decision process (POMDP). Our approach is designed for non-linear dynamics and observation models, and follows the general POMDP solution framework in which we represent beliefs by Gaussian distributions, approximate the belief dynamics using an extended Kalman filter (EKF), and represent the value function by a quadratic function that is valid in the vicinity of a nominal trajectory through belief space. Using a variant of differential dynamic programming, our approach iterates with second-order convergence towards a linear control policy over the belief space that is locally-optimal with respect to a user-defined cost function. Unlike previous work, our approach does not assume maximum-likelihood observations, does not assume fixed estimator or control gains, takes into account obstacles in the environment, and does not require discretization of the belief space. The running time of the algorithm is polynomial in the dimension of the state space. We demonstrate the potential of our approach in several continuous partially-observable planning domains with obstacles for robots with non-linear dynamics and observation models.

1 Introduction

Motion planning under uncertainty, or belief-space planning, has received considerable interest in the robotics community over the past decade. The objective is to plan a path (or rather a control policy) for a robot in partially-observable state

J. van den Berg (✉) · S. Patil · R. Alterovitz
Department of Computer Science, University of North Carolina at Chapel Hill, Chapel Hill, USA
e-mail: berg@cs.unc.edu

S. Patil
e-mail: sachin@cs.unc.edu

R. Alterovitz
e-mail: ron@cs.unc.edu

H.I. Christensen and O. Khatib (eds.), *Robotics Research*,
Springer Tracts in Advanced Robotics 100, DOI 10.1007/978-3-319-29363-9_27

spaces with spatially varying degrees of motion and sensing uncertainty, such that the expected cost (as defined by a user-specified cost-function) is minimized. Optimal solutions lead the robot through regions of the state space where the most information on the state is gained through sensing and the least information is lost due to motion uncertainty in order to maximize, for instance, the probability of reaching a specified goal location while avoiding collisions with obstacles. This problem is formally described as a partially-observable Markov decision process (POMDP), on which a large body of work is available in the literature.

Solutions to POMDPs are known to be extremely complex [17], since they attempt to compute a control policy over the *belief space*, which in the most general formulation is an infinite-dimensional space of all possible probability distributions over the (finite-dimensional) state space. Solutions based on discrete or discretized state and action spaces are inherently subject to the "curse of dimensionality", and have only been successfully applied to very small and low-dimensional state spaces.

In this paper, we present a method to approximate a locally optimal solution to the POMDP problem with continuous state and action spaces and non-linear dynamics and observation models, where we assume a belief can be represented by a Gaussian distribution. Our approach uses a variant of differential dynamic programming to perform value iteration, where the value function is approximated using a quadratization around a nominal trajectory, and the belief dynamics is approximated using an extended Kalman filter. The result is a linear control policy over the belief space that is valid in the vicinity of the nominal trajectory. By executing the control policy, a new nominal trajectory is created around which a new control policy is constructed. This process continues with second-order convergence towards a locally-optimal solution to the POMDP problem. Unlike general POMDP solvers that have an exponential running time, our approach does not rely on discretizations and has a running time that is polynomial in the dimension of the state space.

Our approach builds off of and generalizes a series of previous works that have addressed the same problem of creating applicable approximations to the POMDP problem. Our work combines and extends ideas from previous work in order to overcome their key limitations. In particular, our approach (i) does not assume maximum-likelihood observations, (ii) does not assume fixed estimator or control gains, (iii) does not require discretizations of the state and action spaces, (iv) runs in polynomial time, (v) takes into account obstacles, and (vi) converges towards a locally-optimal control policy given an initial nominal trajectory. We do assume that the dynamics and observation models and cost functions are sufficiently smooth, and that the belief about the state of the robot is well described by only its mean and its variance. We show the potential of our approach in several illustrative partially-observable domains containing obstacles for robots with non-linear dynamics and observation models.

2 Previous Work

Partially observable Markov decision processes (POMDPs) [22] provide a principled mathematical framework for planning under uncertainty in partially-observable environments. They are known to be of extreme complexity [17], and can only be directly applied to problems with small and low-dimensional state spaces [14]. Recently, several POMDP algorithms have been developed that use approximate value iteration with point-based updates [1, 15, 16, 18]. These have been shown to scale up to medium-sized domains. However, they rely on discretizing the state space or the action space, making them inevitably subject to the "curse of dimensionality". The methods of [3, 5, 8, 21] handle continuous state and action spaces, but maintain a global (discrete) representation of the value function over the belief space. In contrast, our approach is continuous and approximates the value function in parametric form only in the regions of the belief space that are relevant to solving the problem, allowing for a running time polynomial in the dimension of the state.

Another class of works, to which our method is directly related, assume a linear-quadratic Gaussian (LQG) framework to find approximately locally optimal feedback policies. In the basic LQG derivation [2], motion and sensing uncertainty have no impact on the resulting policy. As shown in [23], the LQG framework can be extended such that it accounts for state and control dependent motion noise, but still implicitly assumes full observation (or an independent estimator) of the state. Several approaches have been proposed to include partial and noisy observations such that the controller will actively choose actions to gain information about the state. Belief roadmaps [20] and icLQG [9] combine an iterative LQG approach with a roadmap, but this approach does not compute a (locally) optimal solution. The approaches of [6, 7, 19] incorporate the variance into an augmented state and use the LQG framework to find a locally-optimal control policy. However, these approaches assume *maximum-likelihood observations* to make the belief propagation deterministic. LQG-MP [24] removes this assumption, but only evaluates the probability of success of a given trajectory, rather than constructing an optimal one. Belief trees [4] overcome this limitation by combining a variant of LQG-MP with RRT* to find an optimal trajectory through belief space. Vitus and Tomlin [25] propose an alternative solution that involves solving a chance constrained optimal control problem. However, these approaches does not solve a POMDP as they assume fixed control gains along each section of the trajectory independent of the context. The work of [13] takes into account state and control dependent motion and observation noise by an interleaved iteration of the estimator and the controller. This approach, however, does not allow for obstacles and converges towards a locally-optimal controller that assumes fixed estimator gains. Our approach combines and generalizes these approaches as it does not assume maximum-likelihood observations, does not assume fixed control or estimator gains, and takes into account the existence of obstacles in the environment to compute locally-optimal policies that maximize the probability of reaching a goal location while avoiding collisions.

3 Preliminaries and Definitions

We begin by defining POMDPs in their most general formulation (following [22]). Then, we specifically state the instance of the problem we discuss in this paper.

3.1 General POMDPs

Let $\mathcal{X} \subset \mathbb{R}^n$ be the space of all possible states $\mathbf{x}$ of the robot, $\mathcal{U} \subset \mathbb{R}^m$ be the space of all possible control inputs $\mathbf{u}$ of the robot, and $\mathcal{Z} \in \mathbb{R}^k$ be the space of all possible sensor measurements $\mathbf{z}$ the robot may receive. General POMDPs take as input a stochastic dynamics and observation model, here given in probabilistic notation:

$$\mathbf{x}_{t+1} \sim p[\mathbf{x}_{t+1}|\mathbf{x}_t, \mathbf{u}_t], \quad \mathbf{z}_t \sim p[\mathbf{z}_t|\mathbf{x}_t], \tag{1}$$

where $\mathbf{x}_t \in \mathcal{X}, \mathbf{u}_t \in \mathcal{U}$, and $\mathbf{z}_t \in \mathcal{Z}$ are the robot's state, control input, and received measurement at stage t, respectively.

The *belief* $b\,[\mathbf{x}_t]$ of the robot is defined as the distribution of the state $\mathbf{x}_t$ given all past control inputs and sensor measurements:

$$b[\mathbf{x}_t] = p[\mathbf{x}_t|\mathbf{u}_0, \ldots, \mathbf{u}_{t-1}, \mathbf{z}_1, \ldots, \mathbf{z}_t]. \tag{2}$$

Given a control input $\mathbf{u}_t$ and a measurement $\mathbf{z}_{t+1}$, the belief is propagated using Bayesian filtering:

$$b[\mathbf{x}_{t+1}] = \eta\, p[\mathbf{z}_{t+1}|\mathbf{x}_{t+1}] \int p[\mathbf{x}_{t+1}|\mathbf{x}_t, \mathbf{u}_t] b[\mathbf{x}_t] d\mathbf{x}_t, \tag{3}$$

where η is a normalizer independent of $\mathbf{x}_{t+1}$. Denoting belief b$[\mathbf{x}_t]$ by $\mathbf{b}_t$, and the space of all possible beliefs by $\mathcal{B} \subset \{\mathcal{X} \to [0, 1]\}$, the belief dynamics defined by Eq. (3) can be written as a function $\beta : \mathcal{B} \times \mathcal{U} \times \mathcal{Z} \to \mathcal{B}$:

$$\mathbf{b}_{t+1} = \beta[\mathbf{b}_t, \mathbf{u}_t, \mathbf{z}_{t+1}]. \tag{4}$$

Now, the challenge of the POMDP problem is to find a control policy $\pi_t : \mathcal{B} \to \mathcal{U}$ for all $0 \leq t < \ell$, where ℓ is the time horizon, such that selecting the controls $\mathbf{u}_t = \pi_t[\mathbf{b}_t]$ minimizes the objective function:

$$\mathop{\mathrm{E}}_{\mathbf{z}_1,\ldots,\mathbf{z}_\ell}\left[c_\ell[\mathbf{b}_\ell] + \sum_{t=0}^{\ell-1} c_t[\mathbf{b}_t, \mathbf{u}_t]\right], \tag{5}$$

for given immediate cost functions c_ℓ and $c_{t.}$ The expectation is taken given the stochastic nature of the measurements.

A general solution approach uses *value iteration* [22], a backward recursion procedure, to find the control policy π_t for each stage t:

$$v_\ell[\mathbf{b}_\ell] = c_\ell[\mathbf{b}_\ell] \tag{6}$$

$$v_t[\mathbf{b}_t] = \min_{\mathbf{u}_t}\left(c_t[\mathbf{b}_t, \mathbf{u}_t] + \mathop{\mathrm{E}}_{\mathbf{z}_{t+1}}[v_{t+1}[\beta[\mathbf{b}_t, \mathbf{u}_t, \mathbf{z}_{t+1}]]]\right) \tag{7}$$

$$\pi_t[\mathbf{b}_t] = \operatorname*{argmin}_{\mathbf{u}_t}\left(c_t[\mathbf{b}_t, \mathbf{u}_t] + \mathop{\mathrm{E}}_{\mathbf{z}_{t+1}}[v_{t+1}[\beta[\mathbf{b}_t, \mathbf{u}_t, \mathbf{z}_{t+1}]]]\right), \tag{8}$$

where $v_t[\mathbf{b}_t] : \mathcal{B} \rightarrow \mathbb{R}$ is called the value function at time t.

3.2 Problem Definition

The complexity of POMDPs stems from the fact that $\mathcal{B}$, the space of all beliefs, is infinite-dimensional, and that in general the value function cannot be expressed in parametric form. We address these challenges in our approach by representing beliefs by Gaussian distributions, approximating the belief dynamics using an extended Kalman filter, and approximating the value function by a quadratization around a nominal trajectory through the belief space.

Specifically, we assume we are given a (non-linear) stochastic dynamics and observation model, here given in state-transition notation:

$$\mathbf{x}_{t+1} = \mathbf{f}[\mathbf{x}_t, \mathbf{u}_t, \mathbf{m}_t], \quad \mathbf{m}_t \sim \mathcal{N}[\mathbf{0}, I], \tag{9}$$

$$\mathbf{z}_t = \mathbf{h}[\mathbf{x}_t, \mathbf{n}_t], \quad \mathbf{n}_t \sim \mathcal{N}[\mathbf{0}, I], \tag{10}$$

where $\mathbf{m}_t$ is the motion noise and $\mathbf{n}_t$ is the measurement noise, each drawn from an independent Gaussian distribution with (without loss of generality) zero mean and unit variance. Note that the motion and sensing uncertainty can be state and control input dependent through manipulations on $\mathbf{m}_t$ and $\mathbf{n}_t$ within the functions $\mathbf{f}$ and $\mathbf{h}$, respectively.

The belief, denoted $\mathbf{b}_t = (\hat{\mathbf{x}}_t, \Sigma_t)$, is assumed to be defined by the mean $\hat{\mathbf{x}}_t$ and variance Σ_t of a Gaussian distribution $\mathcal{N}[\hat{\mathbf{x}}_t, \Sigma_t]$ of the state $\mathbf{x}_t$. Similar to the general case, our objective is to find a control policy $\mathbf{u}_t = \pi_t\,[\mathbf{b}_t]$ that minimizes the cost function $\mathrm{E}\left[c_\ell[\mathbf{b}_\ell] + \sum_{t=0}^{\ell-1} c_t[\mathbf{b}_t, \mathbf{u}_t]\right]$. In our case, we assume in addition that the Hessian matrix $\frac{\partial^2 c_\ell}{\partial \mathbf{b} \partial \mathbf{b}}[\mathbf{b}]$ is positive-semidefinite for all $\mathbf{b}$, and that the Hessian matrix $\frac{\partial^2 c_t}{\partial \mathbf{u} \partial \mathbf{u}}[\mathbf{b}, \mathbf{u}]$ is positive-definite for all $\mathbf{b}$, $\mathbf{u}$, and t. Further, we assume that the initial belief $\mathbf{b}_0 = (\hat{\mathbf{x}}_0, \Sigma_0)$ is given.

4 Approach

To approximate a locally optimal solution to the Gaussian POMDP problem as formulated above, we follow the general solution approach as sketched in Sect. 3.1. First, we approximate the belief dynamics using an extended Kalman filter. Second, we approximate the value function using a quadratic function that is locally valid in the vicinity of a *nominal trajectory* though the belief space. We then use a variant of differential dynamic programming to perform the value iteration, which results in a linear control policy over the belief space that is locally optimal around the nominal trajectory. We then iteratively generate new nominal trajectories by executing the control policy, and repeat the process until convergence to a locally-optimal solution to the POMDP problem. We discuss each of these steps in this section, and analyze the running time of our algorithm.

4.1 Belief Dynamics and the Extended Kalman Filter

Given a current belief $\mathbf{b}_t = (\hat{\mathbf{x}}_t, \Sigma_t)$, a control input $\mathbf{u}_t$, and a measurement $\mathbf{z}_{t+1}$, we let the belief evolve using the *extended Kalman filter* (EKF). The EKF is widely used for state estimation of non-linear systems [26], and uses the first-order approximation that for any vector-valued function $\mathbf{f}[\mathbf{x}]$ of a stochastic variable $\mathbf{x}$ we have:

$$\mathrm{E}[\mathbf{f}[\mathbf{x}]] \approx \mathbf{f}[\mathrm{E}[\mathbf{x}]], \quad \mathrm{Var}[\mathbf{f}[\mathbf{x}]] \approx \frac{\partial \mathbf{f}}{\partial \mathbf{x}}[\mathrm{E}[\mathbf{x}]] \cdot \mathrm{Var}[\mathbf{x}] \cdot \frac{\partial \mathbf{f}}{\partial \mathbf{x}}[\mathrm{E}[\mathbf{x}]]^T. \tag{11}$$

The EKF update equations are then given by:

$$\hat{\mathbf{x}}_{t+1} = \mathbf{f}[\hat{\mathbf{x}}_t, \mathbf{u}_t, \mathbf{0}] + K_t(\mathbf{z}_{t+1} - \mathbf{h}[\mathbf{f}[\hat{\mathbf{x}}_t, \mathbf{u}_t, \mathbf{0}], \mathbf{0}]), \tag{12}$$

$$\Sigma_{t+1} = (I - K_t H_t)\Gamma_t, \tag{13}$$

where

$$\Gamma_t = A_t \Sigma_t A_t^T + M_t M_t^T, \quad A_t = \frac{\partial \mathbf{f}}{\partial \mathbf{x}}[\hat{\mathbf{x}}_t, \mathbf{u}_t, \mathbf{0}] \quad M_t = \frac{\partial \mathbf{f}}{\partial \mathbf{m}}[\hat{\mathbf{x}}_t, \mathbf{u}_t, \mathbf{0}],$$

$$K_t = \Gamma_t H_t^T (H_t \Gamma_t H_t^T + N_t N_t^T)^{-1}, \quad H_t = \frac{\partial \mathbf{h}}{\partial \mathbf{x}}[\mathbf{f}[\hat{\mathbf{x}}_t, \mathbf{u}_t, \mathbf{0}], \mathbf{0}], \quad N_t = [\mathbf{f}[\hat{\mathbf{x}}_t, \mathbf{u}_t, \mathbf{0}], \mathbf{0}].$$

Note that all of these matrices are functions of $\mathbf{b}_t$ and $\mathbf{u}_t$. Equations (12) and (13) define the (non-linear) belief dynamics. The second term of Eq. (12), called the *innovation* term, depends on the measurement $\mathbf{z}_{t+1}$. Since the measurement is unknown in advance, the belief dynamics are *stochastic*. Using Eq. (10) and the assumptions of Eq. (11), the innovation term is distributed according to $\mathcal{N}[\mathbf{0}, K_t H_t \Gamma_t]$.

Defining $\mathbf{b}_t = \begin{bmatrix} \hat{\mathbf{x}}_t \\ \text{vec}[\Sigma_t] \end{bmatrix}$ as a true vector, containing the mean $\hat{\mathbf{x}}_t$ and the columns of the variance Σ_t (obviously, in our implementation we exploit the symmetry of Σ_t to eliminate the redundancy), the belief dynamics are given by:

$$\mathbf{b}_{t+1} = \mathbf{g}[\mathbf{b}_t, \mathbf{u}_t] + \mathbf{w}_t, \quad \mathbf{w}_t \sim \mathcal{N}[\mathbf{0}, W[\mathbf{b}_t, \mathbf{u}_t]], \tag{14}$$

where

$$\mathbf{g}[\mathbf{b}_t, \mathbf{u}_t] = \begin{bmatrix} \mathbf{f}[\hat{\mathbf{x}}_t, \mathbf{u}_t, \mathbf{0}] \\ \text{vec}[(I - K_t H_t)\Gamma_t] \end{bmatrix}, \quad W[\mathbf{b}_t, \mathbf{u}_t] = \begin{bmatrix} K_t H_t)\Gamma_t & 0 \\ 0 & 0 \end{bmatrix}. \tag{15}$$

4.2 Value Iteration

We perform value iteration backward in time to find a locally optimal control policy using a variant of differential dynamic programming [10]. We approximate the value function v_t [$\mathbf{b}$] as a quadratic function of the form

$$v_t[\mathbf{b}] = \frac{1}{2}\mathbf{b}^T S_t \mathbf{b} + \mathbf{b}^T \mathbf{s}_t + s_t, \tag{16}$$

with $S_t \geq 0$, that is approximately valid around a nominal trajectory in belief space $(\bar{\mathbf{b}}_0, \bar{\mathbf{u}}_0, \ldots, \bar{\mathbf{b}}_\ell, \bar{\mathbf{u}}_\ell)$, which we assume is given (we will discuss initialization and iterative convergence of the nominal trajectory to a locally optimal trajectory in the next subsection).

For the final time $t = \ell$, the value function v_ℓ is approximated by setting $S_\ell = \frac{\partial^2 c_\ell}{\partial \mathbf{b} \partial \mathbf{b}}[\bar{\mathbf{b}}_\ell]$, $\mathbf{s}_\ell = \frac{\partial c_\ell}{\partial \mathbf{b}}[\bar{\mathbf{b}}_\ell] - S_\ell \bar{\mathbf{b}}_\ell$, and $s_\ell = c_\ell[\bar{\mathbf{b}}_n] - \bar{\mathbf{b}}_\ell^T \mathbf{s}_\ell - \frac{1}{2}\bar{\mathbf{b}}_\ell^T S_\ell \bar{\mathbf{b}}_\ell$, which amounts to a second-order Taylor expansion of c_ℓ around the point $\bar{\mathbf{b}}_\ell$. The value functions and the control policies for the stages $\ell > t \geq 0$ are computed by backward recursion—following Eq. (7), we get:

$$\begin{aligned} v_t[\mathbf{b}] &= \min_{\mathbf{u}}(c_t[\mathbf{b}, \mathbf{u}] + \mathrm{E}[v_{t+1}[\mathbf{g}[\mathbf{b}, \mathbf{u}] + \mathbf{w}_t]]) \\ &= \min_{\mathbf{u}}\Big(c_t[\mathbf{b}, \mathbf{u}] + \frac{1}{2}\mathbf{g}[\mathbf{b}, \mathbf{u}]^T S_{t+1} \mathbf{g}[\mathbf{b}, \mathbf{u}] + [\mathbf{b}, \mathbf{u}]^T \mathbf{s}_{t+1} + s_{t+1} \\ &\quad + \frac{1}{2}\mathrm{tr}[S_{t+1} W[\mathbf{b}, \mathbf{u}]]\Big) \end{aligned} \tag{17}$$

$$= \min_{\mathbf{u}}(q_t[\mathbf{b}, \mathbf{u}]), \tag{18}$$

where q_t [$\mathbf{b}$, $\mathbf{u}$] groups together the terms in Eq. (17). The trace-term in Eq. (17) follows from the fact that E[$\mathbf{x}^T$ $Q\mathbf{x}$] = E[$\mathbf{x}$]^T Q E[$\mathbf{x}$] + tr[Q Var[$\mathbf{x}$]] for any stochastic variable $\mathbf{x}$. It is this term that ensures that the stochastic nature of the belief dynamics is accounted for in the value iteration. To approximate the optimal value of $\mathbf{u}$ as a function of $\mathbf{b}$ we take the second-order Taylor expansion of q_t [$\mathbf{b}$, $\mathbf{u}$] in $(\bar{\mathbf{b}}_t, \bar{\mathbf{u}}_t)$:

$$v_t[\mathbf{b}] \approx \min_{\tilde{\mathbf{u}}} \left(\frac{1}{2} \begin{bmatrix} \tilde{\mathbf{b}} \\ \tilde{\mathbf{u}} \end{bmatrix}^T \begin{bmatrix} C_t & E_t^T \\ E_t & D_t \end{bmatrix} \begin{bmatrix} \tilde{\mathbf{b}} \\ \tilde{\mathbf{u}} \end{bmatrix} + \begin{bmatrix} \tilde{\mathbf{b}} \\ \tilde{\mathbf{u}} \end{bmatrix}^T \begin{bmatrix} \mathbf{c}_t \\ \mathbf{d}_t \end{bmatrix} + e_t \right), \tag{19}$$

where $\tilde{\mathbf{u}} = \mathbf{u} - \bar{\mathbf{u}}_t$ and $\tilde{\mathbf{b}} = \mathbf{b} - \bar{\mathbf{b}}_t$, and

$$\begin{aligned} C_t &= \frac{\partial^2 q_t}{\partial \mathbf{b} \partial \mathbf{b}} [\bar{\mathbf{b}}_t, \bar{\mathbf{u}}_t], \quad D_t = \frac{\partial^2 q_t}{\partial \mathbf{u} \partial \mathbf{u}} [\bar{\mathbf{b}}_t, \bar{\mathbf{u}}_t], \quad E_t = \frac{\partial^2 q_t}{\partial \mathbf{u} \partial \mathbf{b}} [\bar{\mathbf{b}}_t, \bar{\mathbf{u}}_t], \\ \mathbf{c}_t^T &= \frac{\partial q_t}{\partial \mathbf{b}} [\bar{\mathbf{b}}_t, \bar{\mathbf{u}}_t], \quad \mathbf{d}_t^T = \frac{\partial q_t}{\partial \mathbf{u}} [\bar{\mathbf{b}}_t, \bar{\mathbf{u}}_t], \quad e_t = q_t[\bar{\mathbf{b}}_t, \bar{\mathbf{u}}_t]. \end{aligned} \tag{20}$$

Equation (19) is then solved by expanding the terms, taking the derivative with respect to $\tilde{\mathbf{u}}$ and equating to 0 (for $\tilde{\mathbf{u}}$ to be actually a minimum, D_t must be positive-definite—we will discuss this issue in Sect. 4.4). We then get the solution:

$$\tilde{\mathbf{u}} = -D_t^{-1} E_t \tilde{\mathbf{b}} - D_t^{-1} \mathbf{d}_t. \tag{21}$$

Hence, the control policy for time t is linear and given by:

$$\mathbf{u}_t = \pi_t(\mathbf{b}_t) = L_t \mathbf{b}_t + \mathbf{l}_t, \quad L_t = -D_t^{-1} E_t, \quad \mathbf{l}_t = -D_t^{-1}(\mathbf{d}_t - E_t \bar{\mathbf{b}}_t) + \bar{\mathbf{u}} t. \tag{22}$$

Filling Eq. (21) back into Eq. (19) gives the value function v_t [$\mathbf{b}$] as a function of only $\mathbf{b}$ in the form of Eq. (16). Expanding and collecting terms gives:

$$S_t = C_t - E_t^T D_t^{-1} E_t, \tag{23}$$

$$\mathbf{s}_t = \mathbf{c}_t - E_t^T D_t^{-1} \mathbf{d}_t - S_t \bar{\mathbf{b}}_t, \tag{24}$$

$$s_t = e_t - \frac{1}{2} \mathbf{d}_t^T D_t^{-1} \mathbf{d}_t - \bar{\mathbf{b}}_t^T \mathbf{s}_t - \frac{1}{2} \bar{\mathbf{b}}_t^T S_t \bar{\mathbf{b}}_t. \tag{25}$$

This recursion then continues by computing a control policy for stage $t - 1$.

4.3 Iteration to a Locally-Optimal Control Policy

The above value iteration gives a control policy that is valid in the vicinity of the given nominal trajectory. To let the control policy converge to a local optimum, we iteratively update the nominal trajectory using the most recent control policy, as in differential dynamic programming [10]. Given the initial belief $\mathbf{b}_0 = (\hat{\mathbf{x}}_0, \Sigma_0)$, and an (arbitrary) initial series of control inputs $\bar{\mathbf{u}}_0^{(0)}, \ldots, \bar{\mathbf{u}}_{\ell-1}^{(0)}$, which can be obtained using RRT motion planning [11], for instance, let the initial control policy be given by $L_t^{(0)} = 0$ and $\mathbf{l}_t^{(0)} = \bar{\mathbf{u}}_t^{(0)}$ for all t. We then compute the nominal trajectory $(\bar{\mathbf{b}}_t^{(i)}, \bar{\mathbf{u}}^{(i)})$ of the i'th iteration (starting with $i = 0$) by forward integrating the control policy in the deterministic (zero-noise) belief dynamics:

$$\bar{\mathbf{b}}_0^{(i)} = \mathbf{b}_0, \quad \bar{\mathbf{u}}_t^{(i)} = L_t^{(i)}\bar{\mathbf{b}}_t^{(i)} + \mathbf{l}_t^{(i)}, \quad \bar{\mathbf{b}}_{t+1}^{(i)} = \mathbf{g}\left[\bar{\mathbf{b}}_t^{(i)}, \bar{\mathbf{u}}_t^{(i)}\right], \tag{26}$$

Then, using the value iteration procedure as described above given the nominal trajectory of the i'th iteration, we find the control policy, i.e. the matrices $L^{(i+1)}$ and vectors $\mathbf{l}^{(i+1)}$ for the $i + 1$'th iteration. We then recompute a nominal trajectory using Eq. (26), and reiterate. This lets the control policy converge to a locally optimal trajectory with a second-order convergence rate [12].

4.4 Ensuring Convergence

To ensure that the above algorithm in fact converges to a locally-optimal control policy, the algorithm must be augmented with some subtle but important changes, common to approaches based on differential dynamic programming [10, 12, 27].

First, to make sure that in each step of the value iteration we actually minimize the value function (rather than maximizing it), matrix D_t must be positive definite, which is not the case in general. In addition, to ensure that the trajectory iteration converges to a local optimum, the matrices S as well as the entire matrix $\begin{bmatrix} C_t & E_t^T \\ E_t & D_t \end{bmatrix}$ of Eq. (19) must be positive-semidefinite [12]. To enforce these requirements, while retaining the most amount of second-order information about the value function, we proceed as follows. Let $R_t = \frac{\partial^2 c_t}{\partial \mathbf{u} \partial \mathbf{u}}[\bar{\mathbf{b}}_t, \bar{\mathbf{u}}_t]$, which is by definition positive-definite (see Sect. 3.2). Note that $c_t[\mathbf{b}, \mathbf{u}]$ is one of the terms of $q_t[\mathbf{b}, \mathbf{u}]$, of which D_t is the Hessian with respect to $\mathbf{u}$. Then we decompose the matrix $\tilde{Q}_t = \begin{bmatrix} C_t & E_t^T \\ E_t & D_t - R_t \end{bmatrix}$ into $Q_t = V \Lambda V^T$ such that Λ is a diagonal matrix containing Q_t's eigenvalues. Second, we construct $\tilde{\Lambda}$ by setting all negative elements of Λ to 0, and construct $\tilde{Q}_t = V\tilde{\Lambda}V^T$. Matrix $\tilde{Q}_t$ is now a positive-semidefinite version of Q_t . Subsequently, we let $\begin{bmatrix} \tilde{C}_t & \tilde{E}_t^T \\ \tilde{E}_t & \tilde{D}_t \end{bmatrix} = \tilde{Q}_t + \begin{bmatrix} 0 & 0 \\ 0 & R_t \end{bmatrix}$, such that $\tilde{D}_t$ is positive definite, and is positive-semidefinite. Now, the matrices C_t, D_t, and E_t are replaced by $\tilde{C}_t$, $\tilde{D}_t$, and $\tilde{E}_t$, respectively, in Eqs. (21)–(25). If these changes are made, it is automatically guaranteed that the matrices S_t are positive-semidefinite. Note that s_ℓ is positive-semidefinite too, since $\frac{\partial^2 c_\ell}{\partial \mathbf{b} \partial \mathbf{b}}[\bar{\mathbf{b}}_\ell]$ is positive-definite by definition (see Sect. 3.2).

These changes do not affect the second-order convergence characteristic of the algorithm. However, as with Newton's method, this second order convergence is only achieved if the current nominal trajectory is already close to the locally-optimal trajectory. If the current nominal trajectory is "far away" from the local optimum, using second-order approaches may overshoot local-minima, which significantly slows down convergence, or even results in divergence. To address this issue, we make an additional change to the algorithm, following [27]. We limit the increment to the control policy by adding a parameter ε to Eq. (21): $\tilde{\mathbf{u}} = -\tilde{D}_t^{-1}\tilde{E}_t\tilde{\mathbf{b}} - \varepsilon\tilde{D}_t^{-1}\mathbf{d}_t$. Initially, $\varepsilon = 1$, but each time a new control policy is computed that creates a trajectory with higher cost than the previous nominal trajectory (the cost of a trajectory is evaluated using value iteration as above without updating the control policy), the new trajectory is rejected, and ε is reduced by a factor $\beta < 1$, and the iteration continues. When a new trajectory is accepted, ε is reset to 1. This change is equivalent to using backtracking line search to limit the step size in Newton's method and guarantees convergence to a locally-optimal control policy [27].

4.5 Running Time Analysis

Let us analyze the running time of our algorithm. The dimension of the state is n, and we assume for the sake of analysis that the dimension of the control inputs and the measurements are $O(n)$. As the belief contains the covariance matrix of the state, the dimension of a belief is $O(n^2)$. Hence, the matrix C_t of Eq. (20) contains $O(n^4)$ entries. Each of these entries is computed using numerical differentiation, and requires multiplying matrices of size $n \times n$ within the belief propagation, taking $O(n^3)$ time. Hence, the total running time of a single step of the value iteration is $O(n^7)$. A complete cycle of value iteration takes ℓ steps (ℓ being the time horizon), bringing the complexity to $O(\ell n^7)$. The number of such cycles needed to obtain convergence cannot be expressed in terms of n or ℓ, but as noted before, our algorithm converges with a second-order rate to a local optimum.

5 Environments with Obstacles

We presented our approach above for general immediate cost functions $c_\ell[\mathbf{b}]$ and $c_t[\mathbf{b}, \mathbf{u}]$. In typical LQG-style cost functions, the existence of obstacles in the environment is not incorporated, while we may want to minimize the probability of colliding with them. We incorporate obstacles into the cost functions as follows.

Let $\mathcal{O} \subset \mathcal{X}$ be the region of the state space that is occupied by obstacles. Given a belief $\mathbf{b}_t = (\hat{\mathbf{x}}_t, \Sigma_t)$, the probability of colliding with an obstacle is given by the integral over $\mathcal{O}$ of the probability-density function of $\mathcal{N}[\hat{\mathbf{x}}_t, \Sigma_t]$. As described in [24], this probability can be approximated by using a collision-checker to compute the number $\sigma[\mathbf{b}_t]$ of standard-deviations one may deviate from the mean before an obstacle is hit. A lower-bound on the probability of *not* colliding is then given by $\gamma[n/2, \sigma[\mathbf{b}_t]^2/2]$, where γ is the regularized gamma function, and n the dimension of the state. A lower-bound on the total probability of not colliding along a trajectory is subsequently computed as $\prod_{t=0}^{\ell-1} \gamma[n/2, \sigma[\mathbf{b}_t]^2/2]$, and this number should be *maximized*. To fit this objective within the minimizing and additive nature of the POMDP objective function, we note that maximizing a product is equivalent to minimizing the sum of the negative logarithms of the factors. Hence, we add to $c_t\,[\mathbf{b}, \mathbf{u}]$ the term $f[\sigma[\mathbf{b}]] = -\log\gamma[n/2, \sigma[\mathbf{b}]^2/2]$ to account for the probability of colliding with obstacles (note that $f[\sigma[\mathbf{b}]] \geq 0$), potentially multiplied by a scaling factor to allow trading-off with respect to other costs (such as the magnitude of the control input).

While the above approach works well, it should be noted that in order to compute the Hessian of $q_t\,[\mathbf{b}, \mathbf{u}]$ at $\hat{\mathbf{b}}_t$ (as is done in Eq. (20)), a total of $O(n^4)$ collision-checks with respect to the obstacles need to be performed, since the obstacle term $f\,[\sigma\,[\mathbf{b}]]$ is part of $q_t\,[\mathbf{b}, \mathbf{u}]$. As this can be prohibitively costly, we can instead approximate the Hessian of $f\,[\sigma\,[\mathbf{b}]]$ using linearizations, which involves only $O(n^2)$ collision checks. To this end, let us approximate $f\,[\sigma]$ by a second-order Taylor expansion at $\sigma[\hat{\mathbf{b}}_t]$:

$$f[\sigma[\mathbf{b}]] \approx \frac{1}{2}a(\sigma[\mathbf{b}] - \sigma[\bar{\mathbf{b}}_t])^2 + b(\sigma[\mathbf{b}] - \sigma[\bar{\mathbf{b}}_t]) + f[\sigma[\bar{\mathbf{b}}_t]], \tag{27}$$

where $a = \frac{\partial^2 f}{\partial\sigma\partial\sigma}[\sigma[\bar{\mathbf{b}}_t]]$ and $b = \frac{\partial f}{\partial\sigma}[\sigma[\bar{\mathbf{b}}_t]]$ (note that this requires only one collision-check). Now, we approximate $(\sigma[\mathbf{b}] - \sigma[\bar{\mathbf{b}}_t])$ using a first-order Taylor expansion:

$$\sigma[\mathbf{b}] \approx (\mathbf{b} - \bar{\mathbf{b}}_t)^T\mathbf{a} + \sigma[\bar{\mathbf{b}}_t] \quad \Leftrightarrow \quad \sigma[\mathbf{b}] - \sigma[\mathbf{b}_t] \approx \tilde{\mathbf{b}}^T\mathbf{a}, \tag{28}$$

where $\mathbf{a}^T = \frac{\partial\sigma}{\partial\mathbf{b}}[\bar{\mathbf{b}}_t]$ (note that this requires $O(n^2)$ collision-checks). By substituting Eqs. (28) in (27), we get

$$f[\sigma[\mathbf{b}]] \approx \frac{1}{2}\tilde{\mathbf{b}}^T(a\mathbf{a}\mathbf{a}^T)\tilde{\mathbf{b}} + \tilde{\mathbf{b}}^T(b\mathbf{a}) + f[\sigma[\bar{\mathbf{b}}_t]]. \tag{29}$$

Hence, $a\mathbf{a}\mathbf{a}^T$ is an approximate Hessian of the obstacle term $f\,[\sigma\,[\mathbf{b}]]$ of $q_t\,[\mathbf{b}, \mathbf{u}]$ that requires only $O(n^2)$ collision-checks to compute. Further, $a\mathbf{a}\mathbf{a}^T$ is guaranteed to be positive-semidefinite since $a > 0$.

6 Results

We evaluate our approach in two scenarios with obstacles: a point robot with linear dynamics that is navigating in a 2-D light-dark environment (adapted from Bry and Roy [4]) and a non-holonomic car-like robot with second-order dynamics moving in a partially-observable environment with spatially varying sensing capabilities.

Our method takes as input a collision-free trajectory to the goal. A naïve trajectory computed using an uncertainty-unaware planner might stray very close to the obstacles in the environment and accumulates considerable uncertainty during execution. We show that our method improves the input trajectory to compute a locally-optimal trajectory and a corresponding control policy that safely guides the robot to the goal, even in the presence of large motion and measurement noise.

6.1 Light-Dark Environment

We consider the case of a point robot moving in a 2-D environment with obstacles shown in Fig. 1. The robot localizes itself using measurements from sensors in the environment, the reliability of which varies continuously as a function of the horizontal coordinate of the robot's position. The experiment is set up such that the robot needs to move away from the goal in order to better localize itself before moving through the narrow passage and reaching the goal.

We assume the following linear dynamics model with control-dependent noise:

$$\mathbf{x}_{t+1} = \mathbf{f}[\mathbf{x}_t, \mathbf{u}_t, \mathbf{m}_t] = \mathbf{x}_t + \mathbf{u}_t + M[\mathbf{u}_t] \cdot \mathbf{m}_t, \tag{30}$$

where the $\mathbf{x}_t = (x, y) \in \mathbb{R}^2$ is the robot's position, the control input $\mathbf{u}_t \in \mathbb{R}^2$ is the robot's velocity, and the matrix $M[\mathbf{u}_t]$ scales the motion noise $\mathbf{m}_t$ proportional to the

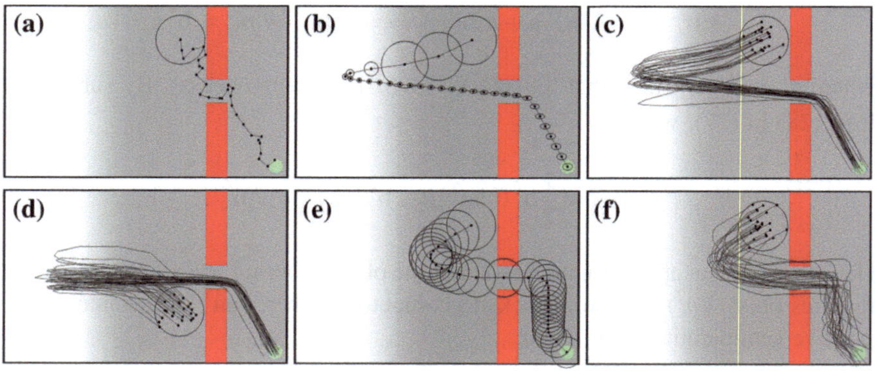

Fig. 1 Point robot moving in a 2-D environment with obstacles. **a** An initial collision-free trajectory is computed using an RRT planner. **b** Nominal trajectory and the associated beliefs of solution computed using our method. The robot moves away from the goal to better localize itself before reaching the goal with significantly reduced uncertainty. Execution traces of the robot's true state starting from the initial belief (**c**) and a different initial belief (**d**), while following the computed control policy. **e** Nominal trajectory computed by ignoring the innovation term in the belief dynamics. The optimization is unable to progress sufficiently to the region of the environment with reliable sensing, resulting in considerable uncertainty in the robot state near the obstacles and at the goal. **f** Execution traces of the robot's true state starting from the initial belief and ignoring the innovation term are much noisier as compared to the execution traces shown in (**c**)

control input $\mathbf{u}_t$. We assume the following linear observation model with state-dependent noise:

$$\mathbf{z}_t = \mathbf{h}[\mathbf{x}_t, \mathbf{n}_t] = \mathbf{x}_t + N[\mathbf{x}_t] \cdot \mathbf{n}_t, \tag{31}$$

where the measurement vector $\mathbf{z}_t \in \mathbb{R}^2$ consists of noisy measurements of the robot's position and the matrix $N[\mathbf{x}_t]$ scales the measurement noise based on a sigmoid function of the horizontal coordinate of the robot's position x (as shown in Fig. 1). The robot is able to obtain reliable measurements in the bright region of the environment, but the measurements become noisier as the robot moves into the dark regions.

We use the following definitions of $c_l[\mathbf{b}_l]$ and $c_t[\mathbf{b}_t, \mathbf{u}_t]$ in the cost function to be minimized (Eq. (5)):

$$c_\ell[\mathbf{b}_\ell] = \hat{\mathbf{x}}_\ell^T Q_\ell \hat{\mathbf{x}}_\ell + \mathrm{tr}[Q_\ell \Sigma_\ell], \quad c_t[\mathbf{b}_t, \mathbf{u}_t] = \mathbf{u}_t^T \mathbf{R}_t \mathbf{u}_t + \mathrm{tr}[Q_t \Sigma_t] + f[\sigma[\mathbf{b}_\ell]], \tag{32}$$

where the term $\hat{\mathbf{x}}_\ell^T Q_\ell \hat{\mathbf{x}}_\ell + \mathrm{tr}[Q_\ell \Sigma_\ell] = \mathrm{E}\left[\mathbf{x}_\ell^T Q_\ell \mathbf{x}_\ell\right]$ encodes the final cost of arriving at the goal, $\mathbf{u}_t^T \mathbf{R}_t \mathbf{u}_t$ penalizes the control effort along the trajectory, $\mathrm{tr}[Q_t \Sigma_t]$ penalizes the uncertainty, and $f[\sigma[\mathbf{b}_t]]$ encodes the obstacle cost term. We use recurring state and control cost matrices of $Q_t = I$ and $R_t = I$ and the final cost matrix, $Q_\ell = 10\ell I$ in our experiments.

Results and discussion: In our experiment, we provide a collision-free initial trajectory computed using an RRT planner [11] (Fig. 1a) as input to our method. The control policy convergence took 2.75 s on a 3.33 Ghz Intel® i7™ PC. Figure 1b shows the nominal trajectory and associated beliefs of the solution computed by our method. The robot's trajectory visits the region of the environment with reliable sensing for better localization before moving through the narrow passage.

We simulated the robot's execution of the computed control policy using the given dynamics and measurement models with synthetic noise. Figure 1c shows the traces of the true state of the robot $\mathbf{x}$ across 25 simulations where the initial state of the robot $\mathbf{x}_0$ is sampled from the initial belief $\mathbf{b}_0$. We also initialized the robot state from a different initial belief to evaluate the robustness of the control policy. The 25 execution traces from these runs are shown in Fig. 1d. Even if the initial belief is far away from the nominal trajectory, the control policy is able to safely guide the robot to the goal. We also evaluated our method quantitatively by computing the percentage of executions in which the robot was able to avoid obstacles across 1000 simulation executions for 10 random initial beliefs. In our experiments, in 93 % (standard deviation: 3 %) of the executions, the robot was able to safely traverse the narrow passage without colliding with obstacles.

Figure 1e shows the nominal trajectory computed by ignoring the innovation term in Eq. (12), i.e. making the assumption that all future observations will obtain their maximum-likelihood measurement estimates. Under this assumption, the optimization is unable to progress sufficiently to the region of the environment with reliable sensing, which results in considerable uncertainty in the robot state near the obstacles and the goal. As expected, the execution traces from 25 simulations (Fig. 1f) are considerably noisier as compared to the execution traces obtained using our method. The expected cost of the solution found using our method is 19.9 units while the expected cost of a solution that ignores the innovation term is 143.0 units for the parameters suggested above. This indicates that it is important to take into account the true belief of the robot while computing the control policy and ignoring the innovation term can lead to sub-optimal policies.

Our solution also agrees with the solution found by Bry and Roy [4] for this experiment. Our method directly optimizes the path rather than relying on RRT*, resulting in an order of magnitude faster computation times.

6.2 Non-holonomic Car-Like Robot

We consider the case of a non-holonomic car-like robot navigating in a 2-D environment with obstacles shown in Fig. 2. We initialize our method with a collision-free trajectory to the goal which is computed using an RRT planner [11].

The state $\mathbf{x} = (x, y, \theta, v) \in \mathbb{R}^4$ of the robot consists of its position $(x,\ y)$, its orientation θ and speed v. The control input vector $\mathbf{u} = (a, \phi)$ consists of an

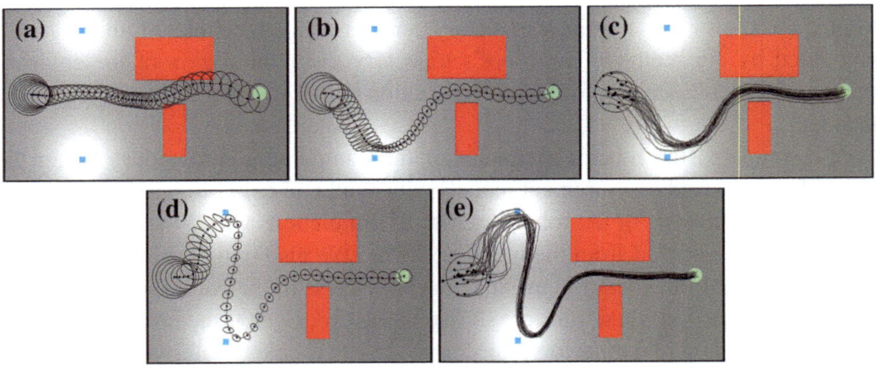

Fig. 2 A car-like robot with second order dynamics moving in a 2-D environment with obstacles. The robot obtains measurements from two beacons (marked by blue squares) and an on-board speedometer. **a** An initial collision-free trajectory is computed using an RRT planner. **b** Nominal trajectory computed using our method. Notice how the car-like robot localizes itself by moving closer to the beacon before reaching the goal. **c** Execution traces of the robot's true state starting from the initial belief for the control policy computed in (**b**). The jaggedness of the paths is due to the large amount of artificial motion and measurement noise introduced in the simulation. The control policy is safely able to guide the robot to the goal, in spite of the large amount of noise. **d** Nominal trajectory computed by varying the cost matrices (Q_t = 10I). The robot tries to reduce the uncertainty in its state by visiting both the beacons. **e** Execution traces of the robot's true state starting from the initial belief for the control policies computed in (**d**)

acceleration a and the steering wheel angle ϕ. The motion uncertainty is scaled by a constant matrix M. This gives the following non-linear dynamics model:

$$\mathbf{x}_{t+1} = \mathbf{f}[\mathbf{x}_t, \mathbf{u}, \mathbf{m}_t] = \begin{bmatrix} x_t + \tau v_t \cos\theta_t \\ y_t + \tau v_t \sin\theta_t \\ \theta_t + \tau v_t \tan(\phi)/d \\ v_t + \tau a \end{bmatrix} + M\mathbf{m}_t \tag{33}$$

where τ is the time step and d is the length of the car-like robot.

The robot localizes itself using signal measurements from two beacons b_1 and b_2 placed in the environment at locations $\check{x}_1$, $\check{y}_1$ and $\check{x}_2$, $\check{y}_2$ respectively. The strength of the signal decays quadratically with the distance to the beacon. The robot also measures its current speed using an on-board speedometer. The measurement uncertainty is scaled by a constant matrix N. This gives us the following non-linear observation model:

$$\mathbf{z}_t = \mathbf{h}[\mathbf{x}_t, \mathbf{n}_t] = \begin{bmatrix} 1/\left((x_t - \check{x}_1)^2 + (y_t - \check{y}_1)^2 + 1\right) \\ 1/\left((x_t - \check{x}_2)^2 + (y_t - \check{y}_2)^2 + 1\right) \\ v_t \end{bmatrix} + N\mathbf{n}_t, \tag{34}$$

where the observation vector $\mathbf{z}_t \in \mathbb{R}^3$ consists of two readings of signal strengths from the beacons and a speed measurement from the speedometer. Figure 2a visually illustrates the quadratic decay in the beacon signal strengths in the environment. The robot is able to obtain very reliable measurements in the bright regions of the environment, but the measurements become noisier as the robot moves into the dark regions due to the decreased signal-to-noise ratio.

We consider a similar cost function as Eq. (32) for this experiment and use recurring state and control input cost matrices of $Q_t = I$ and $R_t = I$ and the final cost matrix, $Q_\ell = 10\ell I$, where ℓ is the number of sections along the initial RRT trajectory.

Results and discussion: The control policy computation took 15.3 s on a 3.33Ghz Intel® i7™ PC. Figure 2b shows the nominal trajectory and associated beliefs of the solution computed by our method. The robot moves closer to the beacon for better localization before reaching the goal. In contrast to the initial trajectory (Fig. 2a), the locally-optimal trajectory also moves away from the obstacles and takes a safer path to the goal.

Figure 2c shows the traces of the true state of the robot $\mathbf{x}$ across 25 simulations where the initial state of the robot $\mathbf{x}_0$ is sampled from the a priori belief $\mathbf{b}_0$. We evaluated our method quantitatively by computing the percentage of executions in which the robot was able to avoid obstacles across 1000 simulation executions where the initial state of the robot $\mathbf{x}_0$ is sampled from the a priori belief $\mathbf{b}_0$. In our experiments, 96 % of the executions were collision-free. The results indicate that the computed control policy is safely able to guide the robot to the goal region in

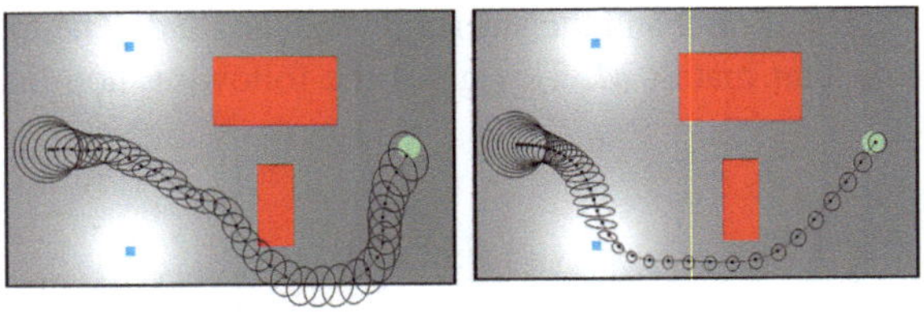

Fig. 3 A different initial trajectory results in a different locally-optimal solution. Our method is able to improve trajectories within a single homotopy class

spite of the large amount of motion and measurement noise encountered during execution.

The cost matrices Q_t and R_t determine the relative weighting between minimizing uncertainty in the robot state and minimizing control effort in the objective function. Figure 2d shows the nominal trajectory of the solution computed by changing the cost matrix $Q_t = 10I$. Notice that the trajectory visits both the beacons for better localization and minimizing uncertainty, at the expense of additional control effort. We simulated 1000 execution runs using the new control policy, of which 98 % were collision-free.

Figure 3 shows the nominal trajectory when a different initial trajectory is provided as input to our method. The presence of obstacles in the environment forces our method to locally optimize trajectories within a single homotopy class. In contrast to considering a large number of initial candidate trajectories for evaluation as in LQG-MP [24], our method would only require trajectory initializations within each homotopy class to compute a globally optimal solution.

7 Conclusion and Future Work

We presented a general approach to motion planning under uncertainty by computing locally-optimal solutions to continuous POMDP problems in environments with obstacles. Our approach generalizes earlier work on Gaussian-based POMDPs by removing several key limiting assumptions, and overcomes the main drawback of approaches based on discretizations of the state space by having a running time that is polynomial ($O(n^7)$) rather than exponential in the dimension of the state.

Our approach has several limitations. First, we represent beliefs using Gaussian distributions. This may not be an acceptable approximation in some applications, for instance ones where multi-modal beliefs are expected to appear. However, the class of problems where Gaussian distributions are applicable is large, as is proven by the widespread use of the extended Kalman filter for state estimation, for instance in mobile robotics. Second, we require the dynamics, observation, and cost functions to be smooth, since our method relies on gradients to iterate towards a locally-optimal solution. Our approach would therefore not work directly in some experimental domains shown in previous work where there are abrupt boundaries between sensing regimes (e.g. inside or outside the field of view of a camera).

Subjects of ongoing and future work include improving the running time of the algorithm. While $O(n^7)$ is polynomial, it may still be too high for robots with complex dynamics and high-dimensional state spaces. The running time can potentially be brought down to $O(n^5)$ if we can avoid computing Hessian matrices. A derivation of our approach based on a quasi-Newton variant of differential dynamic programming [27] may achieve this, and may allow for the direct application of our approach to real-world domains involving complex dynamics such as autonomous quadrotor flight, medical needle steering, or even manipulation of deformable tissue.

Acknowledgments This research was supported in part by the National Science Foundation (NSF) under grant #IIS-0905344 and by the National Institutes of Health (NIH) under grant #R21EB011628.

References

1. H. Bai, D. Hsu, W. Lee, V. Ngo, Monte Carlo value iteration for continuous state POMDPs, in *Workshop on the Algorithmic Foundations of Robotics* (2010)
2. D. Bertsekas, *Dynamic Programming and Optimal Control* (Athena Scientific, 2001)
3. A. Brooks, A. Makarendo, S. Williams, H. Durrant-Whyte, Parametric POMDPs for planning in continuous state spaces. Robot. Auton. Syst. **54**(11), 887–897 (2006)
4. A. Bry, N. Roy, Rapidly-exploring random belief trees for motion planning under uncertainty, in *IEEE International Conference on Robotics and Automation* (2011)
5. S. Candido, S. Hutchinson, Minimum uncertainty robot navigation using information- guided POMDP planning, in *IEEE International Conference on Robotics and Automation* (2011)
6. N. Du Toit, J. Burdick, Robotic motion planning in dynamic, cluttered, uncertain environments, in *IEEE International Conference on Robotics and Automation* (2010)
7. T. Erez, W.D. Smart, A scalable method for solving high-dimensional continuous POMDPs using local approximation, in *Conference on Uncertainty in Artificial Intelligence* (2010)
8. K. Hauser, Randomized belief-space replanning in partially-observable continuous spaces, in *Workshop on the Algorithmic Foundations of Robotics* (2010)
9. V. Huynh, N. Roy, icLQG: combining local and global optimization for control in information space, in *IEEE International Conference on Robotics and Automation* (2009)
10. D. Jacobson, D. Mayne, *Differential Dynamic Programming* (American Elsevier Publishing Company Inc., New York, 1970)
11. S. LaValle, J. Kuffner, Randomized kinodynamic planning. Int. J. Robot. Res. **20**(5), 378–400 (2001)
12. L.-Z. Liao, C. Shoemaker, Convergence in unconstrained discrete-time differential dynamic programming. IEEE Trans. Autom. Control **36**(6), 692–706 (1991)
13. W. Li, E. Todorov, Iterative linearization methods for approximately optimal control and estimation of non-linear stochastic system. Int. J. Control **80**(9), 1439–1453 (2007)
14. L. Kaelbling, M. Littman, A. Cassandra, Planning and acting in partially observable stochastic domains. Artif. Intell. **101**(1–2), 99–134 (1998)
15. H. Kurniawati, D. Hsu, W. Lee, SARSOP: efficient point-based POMDP planning by approximating optimally reachable belief spaces. Robotics: Science and Systems, 2008. configuration spaces. IEEE Trans. Robot. Autom. **12**(4), 566–580 (1996)
16. S. Ong, S. Png, D. Hsu, W. Lee, Planning under uncertainty for robotic tasks with mixed observability. Int. J. Robot. Res. **29**(8), 1053–1068 (2010)

17. C. Papadimitriou, J. Tsisiklis, The complexity of Markov decision processes. Math. Oper. Res. **12**(3), 441–450 (1987)
18. J. Porta, N. Vlassis, M. Spaan, P. Poupart, Point-based value iteration for continuous POMDPs. J. Mach. Learn. Res. **7**, 2329–2367 (2006)
19. R. Platt, R. Tedrake, L. Kaelbling, T. Lozano-Perez, Belief space planning assuming maximum likelihood observations, in *Robotics: Science and Systems* (2010)
20. S. Prentice, N. Roy, The belief roadmap: efficient planning in belief space by factoring the covariance. Int. J. Robot. Res. **28**(1112), 1448–1465 (2009)
21. S. Thrun, Monte Carlo POMDPs. *Advances in Neural Information Processing Systems* (The MIT Press, 2000)
22. S. Thrun, W. Burgard, D. Fox, *Probabilistic Robotics* (MIT Press, 2005)
23. E. Todorov, W. Li, A generalized iterative LQG method for locally-optimal feedback control of constrained nonlinear stochastic systems, in *American Control Conference* (2005)
24. J. van den Berg, P. Abbeel, K. Goldberg, LQG-MP: optimized path planning for robots with motion uncertainty and imperfect state information, in *Robotics: Science and Systems* (2010)
25. M.P. Vitus, C.J. Tomlin, Closed-loop belief space planning for linear, Gaussian systems, in *IEEE International Conference on Robotics and Automation* (2011)
26. G. Welch, G. Bishop, An introduction to the Kalman filter. *Tech. Report TR 95-041*, University of North Carolina at Chapel Hill (2006)
27. S. Yakowitz, Algorithms and computational techniques in differential dynamic programming. Control Dyn. Syst. **31**, 75–91 (1989)

Part IV
Systems and Integration

Rosbridge: ROS for Non-ROS Users

Christopher Crick, Graylin Jay, Sarah Osentoski, Benjamin Pitzer and Odest Chadwicke Jenkins

Abstract We present *rosbridge*, a middleware abstraction layer which provides robotics technology with a standard, minimalist applications development framework accessible to applications programmers who are not themselves roboticists. Rosbridge provides a simple, socket-based programmatic access to robot interfaces and algorithms provided (for now) by ROS, the open-source "Robot Operating System", the current state-of-the-art in robot middleware. In particular, it facilitates the use of web technologies such as Javascript for the purpose of broadening the use and usefulness of robotic technology. We demonstrate potential applications in the interface design, education, human-robot interaction and remote laboratory environments.

1 Introduction

At present, we are at the cusp of a revolution in robotics. For most of the field's history, scientific progress has been hindered by the fact that to have a robot meant investing a great deal in its mechanical engineering and low-level control systems. The result being that every researcher had a different system with different

C. Crick (✉)
Oklahoma State University, Stillwater, OK, USA
e-mail: chriscrick@cs.okstate.edu

G. Jay
Red Hat, Inc., Brisbane, Australia
e-mail: tjay@redhat.com

S. Osentoski
Mayfield Robotics, Redwood City, CA, USA
e-mail: sarah@mayfieldrobotics.com

B. Pitzer
Google, Mountain View, CA, USA
e-mail: pitzer@google.com

O.C. Jenkins
University of Michigan, Ann Arbor, MI, USA
e-mail: ocj@umich.edu

H.I. Christensen and O. Khatib (eds.), *Robotics Research*,
Springer Tracts in Advanced Robotics 100, DOI 10.1007/978-3-319-29363-9_28

capabilities. Furthermore, robots were extremely expensive, both in terms of money and researchers' time. Only very well-funded laboratories could have a robot, and the scope of the robot's activity was constrained by the resources, research focus and imagination of the scientists and engineers that created it.

The emergence of widely-available common robot architectures promises to mitigate the "silo effect" that has heretofore lessened the impact and wider application of research contributions within robotics. Furthermore, developments in robot middleware have begun to create the software engineering infrastructure vital to fostering interoperability and code reuse, a necessary prerequisite to the use of robots on a large scale.

However, the current state of robot middleware is such that users and developers must make a heavy ontological commitment to a particular environment and philosophy in order to use it to its full effect. Furthermore, middleware designers have (perhaps by necessity) assumed that users of their systems would be roboticists themselves, well-versed in the low-level systems programming and complex control and decision algorithms which have always been a part of robotics research. We developed rosbridge to expose these systems to the much wider world of general applications developers, with the hope of unleashing for the first time a "web-scale" revolution in robot availability and accessibility.

2 Background

Several robot middleware system have been proposed to enable code sharing among roboticists. These middleware systems include Player/Stage [8], the Carnegie Mellon Navigation Toolkit (CARMEN) [24], Microsoft Robotics Studio [13], YARP [17], Lightweight Communications and Marshalling (LCM) [12], and ROS [20], as well as other systems [14]. These middleware systems provide common interfaces that allow code sharing and reuse. While middleware systems differ in their design and features, they typically provide a communication mechanism, an API for preferred languages, and a mechanism for sharing code through libraries or drivers. Middleware systems typically require developers to code within the middleware framework, and often within a specified build environment.

At their heart, many of these middleware packages provide a messaging and marshalling protocol between processes running on multiple machines connected in some fashion to robotic hardware. The framework permits, say, a stereo camera to deliver images to a stereo image processor, which in turn can send a depth map to an object recognition routine, which then routes coordinates to an inverse-kinematics driver, which sends motor commands to processes delivering voltages to individual servos. In a complex robot architecture, the number of independent processes and the information that interconnects them quickly becomes massive. Even so, deep down, the system is merely serializing and routing messages, and rosbridge takes advantage of this fact. By way of analogy, web applications have developed huge and complex backends that span continents and perform breathtaking feats of traffic analysis, shaping, routing, data acquisition and conglomeration, but still communicate with browsers and each other over the HTTP protocol. Likewise, robots and their controlling middleware can grow arbitrarily complex on the back end, but with rosbridge they can communicate with an application layer over a single socket and a plain-text protocol.

3 ROS

Rosbridge is designed to work initially within the paradigm established by the ROS middleware system currently maintained by theOpen Source Robotics Foundation. ROS uses a peer-to-peer networking topology; systems running ROS often consist of a number of processes called *nodes*, possibly on different machines, that perform the system's computation. Nodes communicate with each other by passing messages. Under ROS, messages are data structures made up of typed fields. Messages may be made up of standard primitive data types, as well as arrays of primitives. Messages can include arbitrarily nested structures and arrays.

Nodes can use two types of communication to send messages within the ROS framework. The first is synchronous and is called a *service*. Services are much like function calls in traditional programming languages. Services are defined by a string name and a pair of messages: a request and a response. The response returns an object which may be arbitrarily complex, ranging from a simple boolean indicating success or failure to a large point cloud data structure. Only one node can provide a service of a specific name.

The second type of communication is asynchronous and is called a *topic*. Topics are streams of objects that are published by a node. Other nodes, "listeners", may subscribe by registering a handler function that is called whenever a new topic object becomes available. Unlike services, listener nodes are unable to use their subscription to the topic to communicate to the publisher. Multiple nodes may concurrently publish and/or subscribe to the same topic and a single node may publish and/or subscribe to multiple topics.

Unlike many other robot middleware systems, ROS is more than a set of libraries that provide only a communication mechanism and protocol. Instead, nodes are developed within a build system provided by ROS. The intent is that a system running ROS should be comprised of many independent modules. The build system is built on top of CMake [16], which performs modular builds of both nodes and the messages passed between them.

Furthermore, ROS has assimilated a number of tools, algorithms and systems which can serve as a basis for complex robot control. Thus a full suite of ROS packages provides vision processing algorithms [3], 3D point cloud interpretation [21] and simultaneous localization and mapping (SLAM) [10], among many others. This represents the largest effort to date to foster a robotics community that supports code-sharing and building on the prior work of others. This alone serves as reason for applying the rosbridge architecture to ROS initially.

4 Rosbridge

Rosbridge provides an additional level of abstraction on top of ROS, as depicted in Fig. 1. Rosbridge treats all of ROS as a "back end". This shields application developers from needing intimate knowledge of low-level control interfaces, middleware build systems and sophisticated robotic sensing and control algorithms. At a bare minimum they must understand the build and transportation mechanisms of the middleware package. Rosbridge layers a simple socket serialization protocol over all of this complexity, on top of which application developers of all levels of experience can create applications.

ROS abstracts individual robot capabilities, allowing robots to be controlled through messages. It also provides facilities for starting and stopping the individual ROS nodes providing these capabilities. Rosbridge encapsulates these two aspects of ROS, presenting to the user a unified view of a robot and its environment. The Rosbridge protocol allows access to underlying ROS messages and services as serialized JSON objects, and in addition provides control over ROS node execution and environment parameters (Fig. 2).

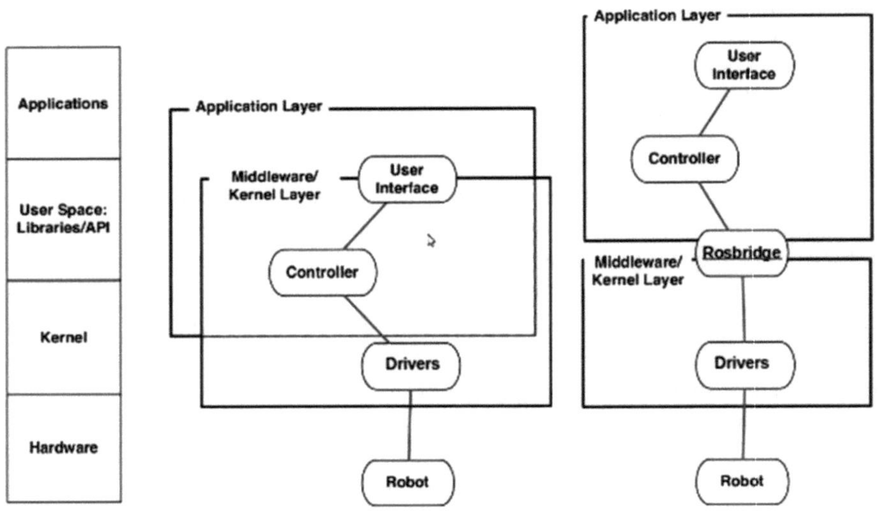

Fig. 1 Recreating traditional abstraction layers in robotics with rosbridge. As depicted at *left*, software development depends on well-established layers of abstraction. Developers and engineers working at each layer possess very different skill sets, but the enterprise succeeds due to well-defined abstractions and interfaces. At present, robotics must deal with all of these layers at once, limited by both their own skills and by the unwieldiness inherent in poorly-abstracted systems (*center*). At *right*, rosbridge attempts to establish a more clear abstraction boundary to address this problem

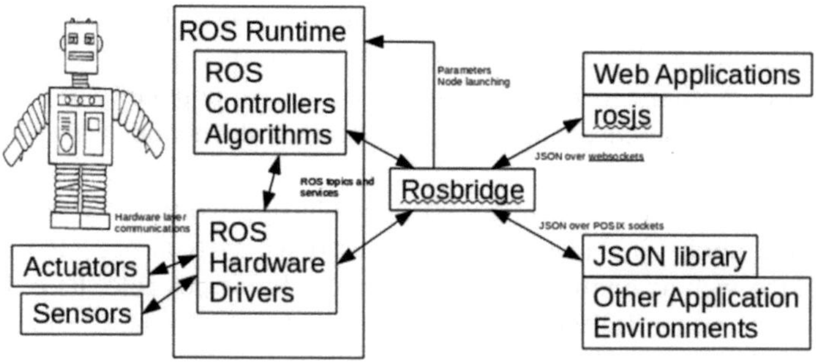

Fig. 2 Rosbridge serializes all applicable ROS topics and services over a single socket interface

Rosbridge allows simple message handling over both HTML5 websockets and standard POSIX IP sockets. For example, a simple Python client which handles data being published on a ROS topic called "/sensorPacket" can be written, simply, as

```
host_sock = socket.socket(socket.AF_INET, socket.SOCK_STREAM)
host_sock.connect((host_address, host_port))
host_sock.send('raw\r\n\r\n')
host_sock.send('\x00{"receiver":"/rosbridge/subscribe","msg":["/sensorPacket",0,]}\xff')
while True:
    incoming = source_socket.recv(1024)
    #handle sensorPacket data
```

This paradigm can be exploited in any language that supports IP sockets, which is to say, all of them. Thus rosbridge enables robot application development in a user's language of choice.

5 ROSJS

Computing paradigms have developed over the years, from batch systems to timeshared mainframes to standalone desktops to client-server architectures to ubiquitous web-based applications. Current technology allows transparent administration, redundant storage, and instantaneous deployment of software running on wildly heterogenous platforms, from smartphones to multicore desktops. This relatively new and extremely ecosystem has spawned a population of users who understand basic web technologies such as HTML and Javascript [7]. Familiarity with basic web technologies extends beyond expert application developers to users who would not necessarily call themselves programmers, but who nevertheless use the web for all manner of creation and communication and are familiar with the basic technologies. One of the goals of rosbridge is to broaden robotics to this vast untapped population of writers, artists, students, and designers. Javascript has become the default language of the web and as such is one of the most popular languages in the world. We hope to leverage a small part of that popularity to open robotics to an entirely new audience and to make working with robotics easier for those who are already familiar.

Because this is one of rosbridge's primary goals, we have provided a large and full-featured rosbridge library in Javascript, known as rosjs. rosjs is designed to integrate ROS with the web as unobtrusively and universally as possible. Its only advanced dependency is on the HTML5 [19] technology of websockets. Currently browsers such as Safari, Opera, and Chrome fully support them, as does the nightly build of Firefox. Universality has been one of the key factors in the success of the web, and accordingly rosjs is implemented as a simple Javascript library, completely agnostic with respect to preferred development frameworks. Rosbridge is built using serialized JSON objects, which are themselves basic Javascript object syntax.

rosjs is now a large library supporting many complex features for visualization and interaction with sophisticated ROS-based manipulation and navigation algorithms. However, it can be used for extremely simple code. The following demonstrates how little Javascript code is required to send navigation commands to a robot.

```
<html><head>
<script>type="text/javascript" src="ros.js"</script>
...
var ros = new connection("ws://10.100.0.100:9090")
...
ros.publish('/cmd_vel','geometry_msgs/Twist',
            '{"linear:{"x":'+x+',"y":0,"z":0},"angular:{"x":0,"y":0,"z":'+z+'}}');
...
```

JSON is simple enough that the serialization can be done by hand, as in the above example. However, many JSON libraries exist to make the construction easier and less error-prone.

rosjs was designed to meet the needs of developers with web programming experience. There are multiple advantages to the ability to develop robot applications in the browser. Web browsers are familiar and widely-used interfaces, even by non-techical users. Allowing users to access robots through the internet may provide insights into new applications for robotics, as well be used as a tool to recruit potential scientists to the field. Javascript allows for rapid and flexible user interface and visualization development. Applications developed within a web browser are also portable across platforms, and updates and new functionality can be easily provided.

6 Rosbridge in Remote Laboratories

While middleware systems allow for code sharing and reuse, many researchers are limited by the overhead (and sometimes pure impossibility) of reproducing results on similar platforms. Large platforms like mobile manipulators are expensive and difficult to obtain for researchers at smaller institutions or companies. It is rare for researchers to have access to common platforms, let alone shared data, especially in fields focused on active learning or those requiring user studies. Additionally, the great difficulty in reproducing experimental results has hindered the robotics field for many years. It is often difficult to assess which proposed approaches per- form best. In fields where online learning and user demonstrations are required, researchers do not perform research on common platforms, let alone on shared data. A remote lab where users can compare results and share experimental data will help provide a more scientific basis for comparison.

A remote robotic laboratory would allow researchers to run experiments and compare against results produced on a common platform. We developed rosbridge and its supporting rosjs libraries in part to support the development of experimental infrastructure for the creation of remote robotic laboratories.

Figure 3 depicts a remote lab interface developed with rosjs to support research into learning from demonstration. Users can access a PR2 robot to demonstrate pick-and-place tasks, specifically setting a table. In addition, they can observe the robot's actions through a variety of sensors and camera streams, all provided through the rosbridge framework. During each session data is logged and stored in a publicly available repository. Custom controllers and learning algorithms, provided in public code repositories, can use the data and provide policies for desired tasks on the robot.

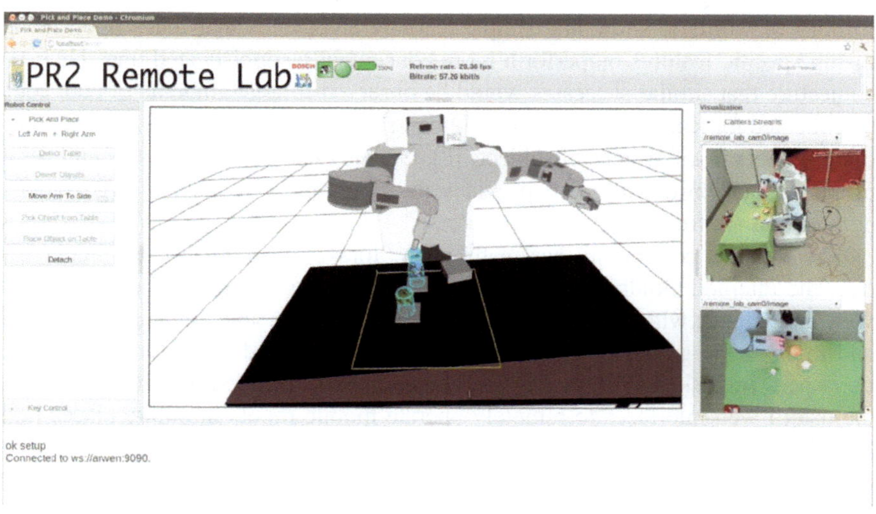

Fig. 3 A complex remote lab interface using rosjs and WebGL

There are many technical challenges to address when creating such a remote lab. The functionality provided by rosjs is instrumental to overcoming them. A web interface is required so that users can work with the robot remotely. The user must have some way of controlling the robot, either with code or through teleoperation. Users must also be able to visualize the result of the control. Security measures are required for the safety of the robot.

7 Rosbridge in Human-Robot Interaction

One of the strengths of rosbridge (and its Javascript application-layer library rosjs) is its support for quickly and easily creating remote user interfaces. Much of the teleoperation work in robotics has traditionally been aimed at tasks where robots operate in environments that are hazardous to human users, such as robotic surgery [18], search and rescue [5], and outer space [1]. In these applications, users are typically experts who have devoted a significant amount of training time to the difficult task of controlling the robot and interacting with its interfaces. Our goal with rosbridge is to allow application developers to create interfaces that are intuitive even for novice users.

Furthermore, even expert user interface designers are not necessarily experts in ROS or robotics generally. The expertise needed for developing rewarding and intuitive interactions over a simple Javascript web interface, however, is widespread and generally available.

Rosbridge has the potential to increase the number of people using, interacting with and programming robots. A recent trend in machine learning has examined the use of truly large data sets for learning rather than attempting to generalize from a small amount of data. Researchers in data mining and machine translation have able to take advantage of Google's index of billions of crowdsourced documents and trillions of words to show that simple learning algorithms that focus upon recognizing specific features outperform more conceptually sophisticated ones [11]. We conjecture that similar successes would be observed if large amounts of data could be collected for learning with robots. Human-robot interaction studies, to date, more often number in the dozens of subjects [2]. Opening up robots to the vast number of users on the world wide web provides the opportunity to gain a large number demonstrations from many different users.

The robotics community has made a few forays into human robot interaction over the internet. Goldberg et al. placed a robot in a garden and allowed users to view and interact with the robot over the web. Users were able to plant seeds, water, and monitor the garden [9]. Taylor and Trevelyan created a remote lab in which users perform tasks involving brightly colored blocks [23]. Schulz et al. examined the use of web interfaces to remotely operate mobile robots in public places [22]. This worked focused on letting remote users interact with humans within the robots' environment and did not examine the effect of the visualizations in a learning task. Burgard and Schulz have explored handling delay in remote operation/teleoperation of mobile robots using predictive simulation for visualization [4].

In previous work, we have used rosbridge to leverage precisely this large network effect [6]. HRI research into the character of interfaces and visualizations which lead to successful human teaching of robot behavior was able to draw on a large pool of participants and develop 276 use cases and eighty thousand points of data.

8 Rosbridge in Education

The simplicity and system independence of rosbridge make it a very powerful tool for programming and robotics education. The ease of hooking into a robot system using simple sockets and text-based JSON messages means that students have a very gentle learning curve. In addition, programming languages and environments that have been expressly designed for educational purposes can easily be extended to communicate with rosbridge.

Figure 4 shows robotics development in the Scratch environment [15], a visual programming system designed for children to learn and understand programming concepts. A very simple extension to Scratch allows students interact with robots programmatically. The system has been used by middle-school students, who were able to program robots to perform basic closed loop behaviors such as line following and bump exploration, without ever being aware of the underlying complexities of ROS itself.

We are currently developing higher-level courses to take advantage of rosbridge, as well. At the college level, robotics classes have traditionally spent a great deal of time just "hacking on the machine", dealing with and learning about the massive infrastructure necessary to get robots to do useful things. Rosbridge short-circuits this process, allowing students to spend more time learning about higher-level

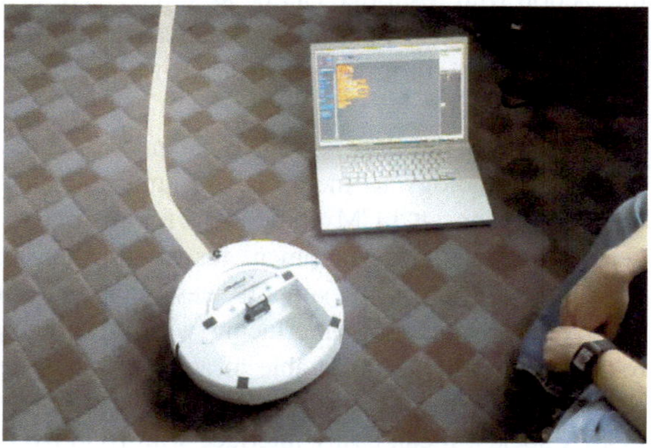

Fig. 4 Robotic control using the Scratch educational programming environment

control and perception and less time wondering how to extract images from a camera stream or compile behaviors in an abstruse and poorly-documented programming environment.

9 Rosbridge Without ROS

In addition to extending ROS, rosbridge can be extended to provide similar functionality for other middleware systems. The messaging protocol at the core of most robot middleware can be translated into JSON objects just as ROS messages can, and passed through the same sockets using the same interface. Our goal is to not only extend ROS to but to also advocate that this additional level of abstraction may be beneficial to other middleware systems.

We are currently developing rosbridge support for the LCM system [12]. This will create a common interface for robots running ROS and LCM to send messages to each other, and for application developers to write software that can support robots running either system.

10 Conclusions and Future Work

In this paper, we described rosbridge, a high-level middleware abstraction layer that exposes robot functionality to developers as a simple interaction over a socket. In addition, we have developed rosjs, a Javascript library on top of rosbridge, that supports extensive interaction and visualization of higher-level ROS constructs. We believe that web-based interaction with robots provides the largest potential pool of new users and developers, and so expanding and enhancing rosjs has consumed the largest share of our development resources. However, the rosjs framework also serves as a model for the development of other libraries for other languages. Interaction with rosbridge can be as simple as desired—no more than sending text strings over sockets—but of course advanced functionality should be developed to support whatever tasks a user wishes. Rosbridge enables that development in purely agnostic fashion.

We plan to develop rosbridge as much as possible into a simple nexus for robotics technology to meet general application development. Already we have begun work on an LCM component for rosbridge, with the hope of supporting general application development for robots using either form of middleware. In addition, we hope to support device manufacturers who need a simple, low-overhead means of interfacing with other computer systems. An embedded robot system might not be able to accommodate the computational demands of a ROS system onboard, but if it can send and receive plain-text messages over a POSIX socket, it can easily interface with ROS over rosbridge. This would greatly expand the technological resources available to the robot.

Rosbridge enables ROS to communicate with the web, applications developers to communicate with robots, and robotics researchers to communicate with each other. All of these are necessary for robots to succeed in the world.

References

1. H. Aldridge, W. Bluethmann, R. Ambrose, M. Diftler, Control architecture for the robonaut space humanoid, in *Proceedings of the First IEEE/RAS Conference on Humanoid Robotics* (2000)
2. C.L. Bethel, R.R. Murphy, Use of large sample sizes and multiple evaluation methods in human-robot interaction experimentation, in *AAAI Spring 2009 Symposium: Experiment Design for Real-World Systems* (2009)
3. G.R. Bradski, V. Pisarevsky, Intel's computer vision library: applications in calibration, stereo segmentation, tracking, gesture, face and object recognition, in *Proceedings of the IEEE Conference on Computer Vision and Pattern Recognition* (2000)
4. W. Burgard, D. Schulz, *Beyond Webcams: An Introduction to Online Robots, Chap, Robust Visualization for Online Control of Mobile Robots* (MIT Press, 2002), pp. 241–258
5. J.L. Casper,R. Robin, A.J. Murphy, Workflow study on human-robot interaction in user, in *Proceedings of the 2002 IEEE International Conference on Robotics and Automation* (2002)
6. C. Crick, S. Osentoski, G. Jay, O.C. Jenkins, Human and robot perception in large-scale learning from demonstration, in *Proceedings of the 6th ACM/IEEE Conference on Human—Robot Interaction* (2011)
7. ECMA-262: ECMAScript language specification, 5th edn. (2009), URL http://www.ecmainternational.org/publications/standards/Ecma-262.htm
8. B. Gerkey, R.T. Vaughan, A. Howard, The player/stage project: tools for multi-robot and distributed sensor systems, in *Proceedings of the 11th International Conference on Advanced Robotics* (2003), pp. 317–323
9. K. Goldberg, H. Dreyfus, A. Goldman, O. Grau, M. Gržinić, B. Hannaford, M. Idinopulos, M. Jay, E. Kac, M. Kusahara, (eds.), *The Robot in the Garden: Telerobotics and Telepistemology in the Age of the Internet* (MIT Press, Cambridge, 2000)
10. G. Grisetti, C. Stachniss, W. Burgard, Improved techniques for grid mapping with raoblackwellized particle filters. IEEE Trans. Robot. **23**(1), 34–46 (2007)
11. A. Halevy, P. Norvig, F. Pereira, The unreasonable effectiveness of data. IEEE Intell. Syst. 8–12 (2009)
12. A. Huang, E. Olson, D. Moore, LCM: lightweight communications and marshalling, in *Proceedings of the IEEE/RSJ International Conference on Intelligent Robots and Systems* (2010)
13. J. Jackson, Microsoft Robotics Studio: A Technical Introduction. IEEE Robot. Autom. Mag. 82–87 (2007)
14. J. Kramer, M. Scheutz, Development environments for autonomous mobile robots: a survey. Auton. Robots 101–132 (2007)
15. J. Maloney, M. Resnick, N. Rusk, B. Silverman, E. Eastmond, The scratch programming language and environment. Trans. Comput. Educ. **10**(4), 1–15 (2010)
16. K. Martin, B. Hoffman, *Mastering CMake: A Cross-platform Build System* (Kitware Inc, 2008)
17. G. Metta, P. Fitzpatrick, L. Natale, YARP: yet another robot platform. Int. J. Adv. Robot. Syst. 43–48 (2006)
18. A.M. Okamura, Methods for haptic feedback in teleoperated robot-assisted surgery. Ind. Robot. **31**(6), 499–508 (2004)
19. M. Pilgrim, *HTML5: Up and Running* (O'Reilly Media, 2010)

20. M. Quigley, B. Gerkey, K. Conley, J. Faust, T. Foote, J. Leibs, E. Berger, R. Wheeler, A. Ng, Ros: an open-source robot operating system, in *Proceedings of the Open-Source Software Workshop of the International Conference on Robotics and Automation* (2009)
21. R.B. Rusu, S. Cousins, 3D is here: point cloud library (PCL), in *Proceedings of the IEEE Conference on Robotics and Automation (ICRA)* (2011)
22. D. Schulz, W. Burgard, D. Fox, S. Thrun, A.B. Cremers, Web interfaces for mobile robots in public places. IEEE Robot. Autom. Mag. **7**, 48–56 (2000)
23. K. Taylor, J. Trevelyan, A telerobot on the world wide web, in *National Conference of the Australian Robot Association* (1995)
24. K. Wyobek, E. Berger, H.V. der Loos, K. Salisbury, Perspectives on standardization in mobile robot programming: the Carnegie Mellon Navigation (CARMEN) toolkit, in *Proceedings of the IEEE/RSJ International Conference on Intelligent Robots and Systems* (2003), pp. 2436–2441

Introduction to the Robot Town Project and 3-D Co-operative Geometrical Modeling Using Multiple Robots

Ryo Kurazume, Yumi Iwashita, Koji Murakami and Tsutomu Hasegawa

Abstract This paper introduces the author's research project called the "Robot Town Project". Service robots, which co-exist with humans and provide various services in daily life, must have sufficient ability to sense changes in the environment and deal with a variety of situations. However, since the daily environment is complex and unpredictable, it is almost impossible with current methods to sense all the necessary information using only a robot and the attached sensors. One promising approach for robots to co-exist with humans is to use IT technology, such as a distributed sensor network and network robotics. As an empirical example of this approach, the authors have started Robot Town Project. The aim of this research project is to develop a distributed sensor network system covering an area of a block in a town in which there are many houses, buildings, and roads, and manage robot services by monitoring events that occur in the town. This paper introduces currently available technologies including an RFID-tag-based localization system, distributed sensor systems for moving object tracking, and object management systems using RFID tags. For the construction of 3-D geometrical models of large-scale environments, a measurement and modeling system using a group of multiple robots and an on-board laser range finder is also introduced.

R. Kurazume (✉)
Kyushu University, 744, Motooka, Nishi-ku, Fukuoka 8190395, Japan
e-mail: kurazume@ait.kyushu-u.ac.jp

Y. Iwashita
Jet Propulsion Laboratory, M/S 198-235 4800 Oak Grove Drive, Pasadena 91109, CA, USA
e-mail: Yumi.Iwashita@jpl.nasa.gov

K. Murakami
Kyushu Sangyo University, 2-3-1, Matsukadai, Higashi-ku, Fukuoka 8138503, Japan
e-mail: mura.k@ip.kyusan-u.ac.jp

T. Hasegawa
Kumamoto National College of Technology, 2659-2, Suya, Koshi-shi 8611102, Kumamoto, Japan
e-mail: hasegawa@kumamoto-nct.ac.jp

H.I. Christensen and O. Khatib (eds.), *Robotics Research*,
Springer Tracts in Advanced Robotics 100, DOI 10.1007/978-3-319-29363-9_29

1 Introduction

The demand for service robots that can co-exist with humans and provide various services in daily life is expected to increase in the next few decades. These robots must have sufficient ability to sense changes in the environment and cope with a variety of situations. However, since the daily environment is complex and unpredictable, it is almost impossible with current methods to sense all the necessary information using only a robot and the attached sensors. One of the promising approaches to develop service robots which co-exist with humans is using IT technology, such as a distributed sensor network and network robotics. The basic idea of this approach is that robots provide a variety of services based on environmental information not only from on-board sensors, but also from sensor networks in the environment. An empirical example of the above approach has been implemented in the research project called the "Robot Town Project" (Fig. 1). The aim of this research project is to develop a distributed sensor network system covering an area of a block in a town in which there are many houses, buildings, and roads, and manage robot services by monitoring events that occur in the town. The sensed events are notified to the "Town Management System, TMS", and each robot receives appropriate information about the surroundings and instructions for proper services. There are several researches for the embedded sensor systems in

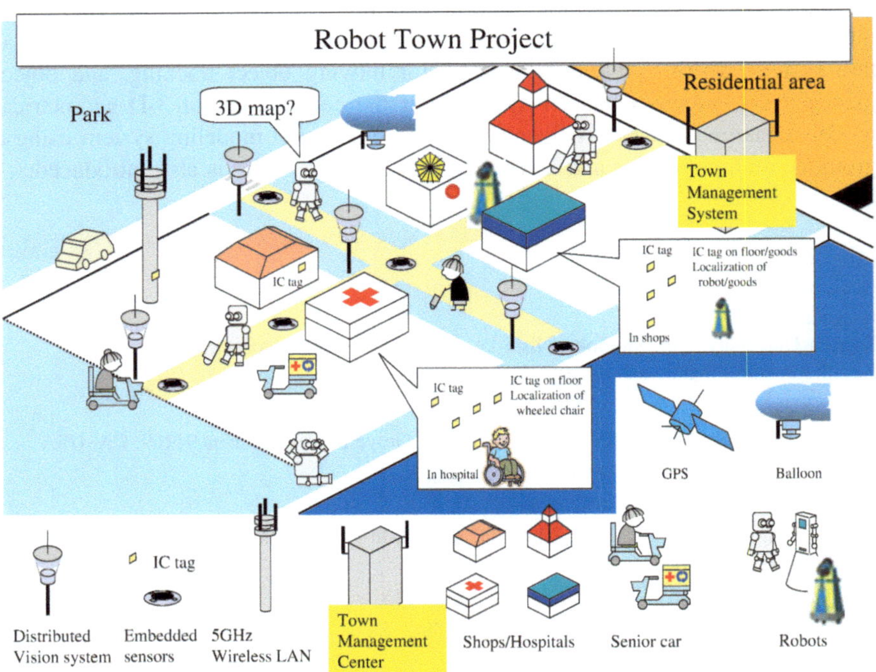

Fig. 1 Concept for robot town

daily human life environment [4, 9, 15, 19, 21, 25, 27]. However, the most of researches so far are limited to area of a single room or a few rooms.

This paper provides a brief introduction to the Robot Town Project and introduces some technologies developed in this project, including an RFID-tag-based localization system, a distributed sensor system for moving object tracking, and an object management system using RFID tags. For the construction of 3-D geometrical models of large-scale environments, a measurement and modeling system using a group of multiple robots and an on-board laser range finder is also introduced, and large-scale geometrical modeling experiments in indoor and outdoor environments are presented.

2 Robot Town Project

This section introduces the core technologies in the Robot Town Project, including the current implementation of TMS and distributed sensor systems using RFID tags, laser range finders, and cameras.

2.1 Town Management System, TMS

TMS is the core technology in this project. A prototype TMS in a home environment has already been developed and tested for human-robot co-operation based on practical scenarios. Figure 2 shows the experimental house for the Robot Town Project.

The TMS consists of a database and API libraries (Fig. 3). The database is developed in MySQL and stores object information acquired using distributed sensors embedded in the environment and the robots: moving object information such as position and velocity of robots and humans; and environmental information

Outward Indoor

Fig. 2 Experimental house

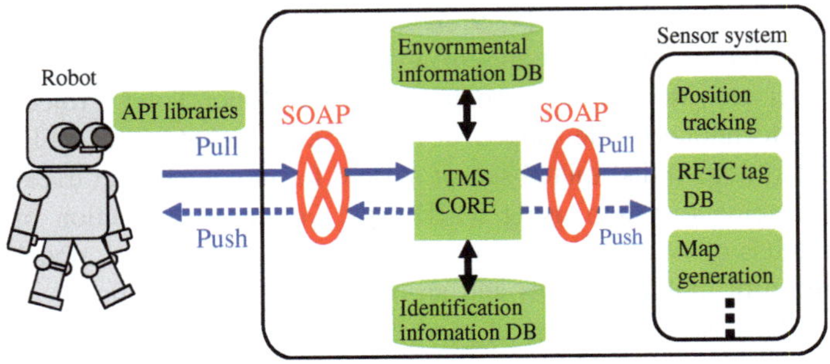

Fig. 3 Town management system, TMS

including semantic and metric maps. SOAP (Simple Object Access Protocol) is adopted for the interface protocol, and a web-based service is provided. Robots (and operators) are able to access the database using APIs and obtain necessary information for providing service to humans.

2.2 *Distributed Sensor Systems*

Information stored in TMS is updated regularly using distributed sensors and robots. In the experimental house of the Robot Town Project shown in Fig. 2, several distributed sensor systems are installed. For example, two types of RFID tags are placed uniformly on the floor. Robots that are equipped with an RFID tag reader can identify their positions by reading the tags and then querying the TMS. To detect the movement of humans, laser range finders and cameras are placed on the floor.

2.2.1 RFID Tag System for Robot Localization

Under the carpet of the floor, 4,000 passive RFID tags are placed as shown in Fig. 4. These tags consist of two types: a high frequency (13.56 MHz) passive RFID tag (Texas Instruments RI-I01-112A, ISO15693) and a low frequency (34.2 kHz) passive RFID tag. The high-frequency RFID tags are placed at an interval of 12.5 cm, while the low-frequency RFID tags are placed at an interval of 25 cm. The yard of the house is also covered by 400 RFID tags.

Robots can identify their positions by reading these tags and querying TMS about their positions based on the tag labels to give a position accurate to within a few centimeters. Detectable areas of the tags are 1 cm for high-frequency tags and 40 cm for low-frequency tags.

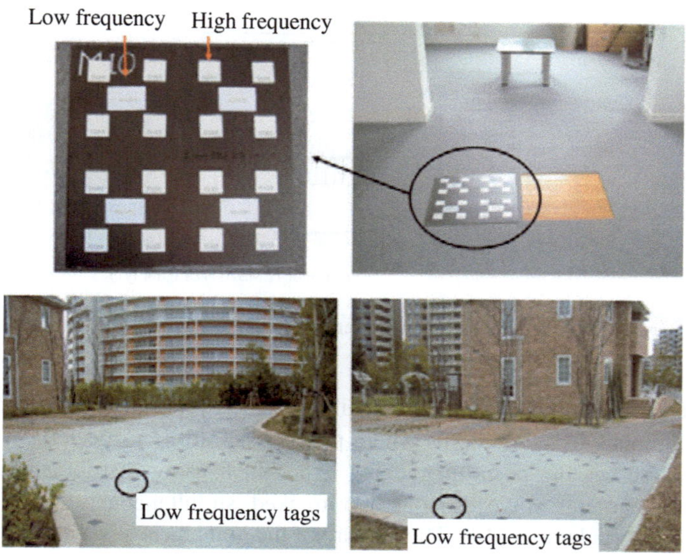

Fig. 4 RFID tag system in and around the experimental house

2.2.2 Laser and Vision Sensor System

As mentioned above, the position of a robot can be detected by the RFID tag systems if the robot is equipped with an RFID tag reader. However, the position of humans and robots that are not equipped with RFID tag readers cannot be identified using this system. Therefore, the experimental house is equipped with several distributed sensors for detecting the motion of humans and robots [7, 14].

For example, 2-D laser range finders (SICK LMS-200) and cameras (Point Grey Dragonfly2) are placed 1 m above the floor, and not only the position but also the posture of the humans is tracked [8]. Figures 5 and 6 show the system configuration

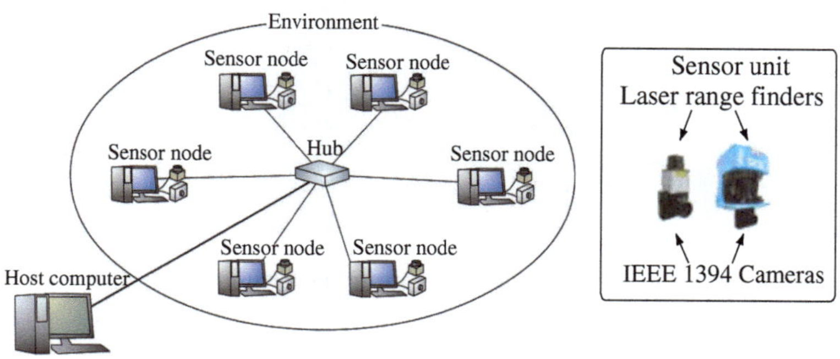

Fig. 5 Tracking system using laser range finders and cameras

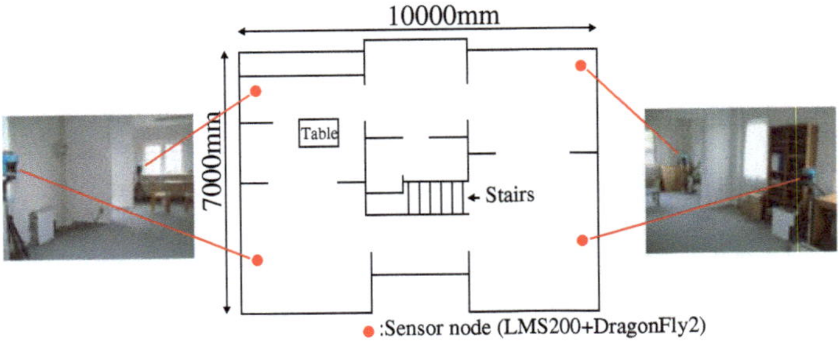

Fig. 6 Sensor positions in the experimental house

and the sensor positions in the house. Ceiling pyroelectric sensors are also used for detecting humans in some rooms where privacy should be respected.

Figure 7 shows the position tracking experiment using distributed laser range finders and cameras. Five persons are tracked simultaneously using a single MCMC/SIR particle filter [14]. The posture of the humans is also detected using the cameras (Fig. 8) [8].

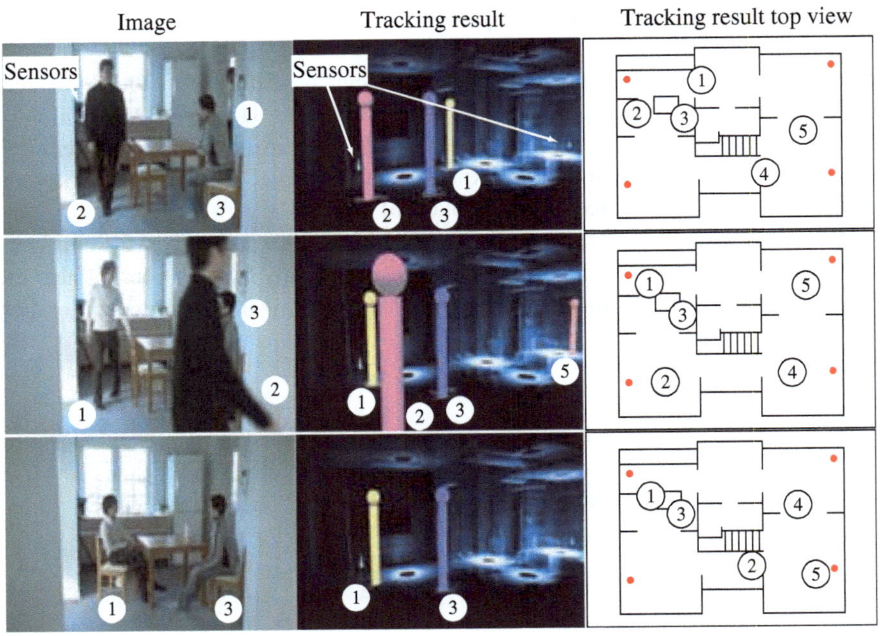

Fig. 7 Tracking experiments of humans in the experimental house

Fig. 8 Posture estimation of humans using captured camera images

2.2.3 Object Management System Using RFID Tags

All of the objects such as drinks, clothes, and shoes in the experimental house are managed by TMS using the attached RFID tags, and robots can identify these objects by reading the tags and querying the identification numbers to TMS. RFID tag readers are placed on the cabinets and refrigerators (Fig. 9) [18] to recognize objects placed in them. On the other hand, robots query the TMS about the necessary object information and can know their positions and current status, such as in-use or empty.

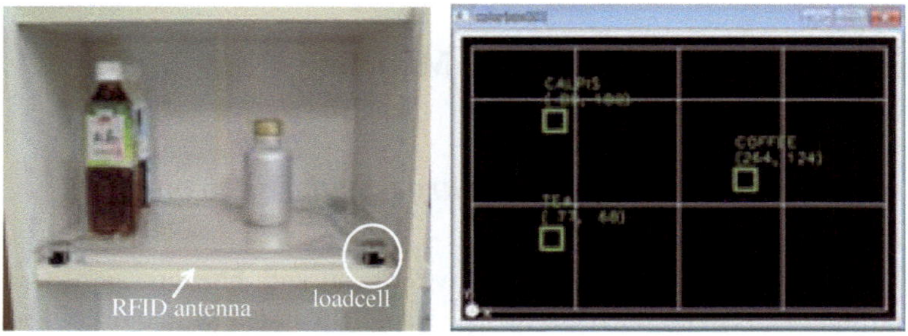

Fig. 9 Intelligent shelf

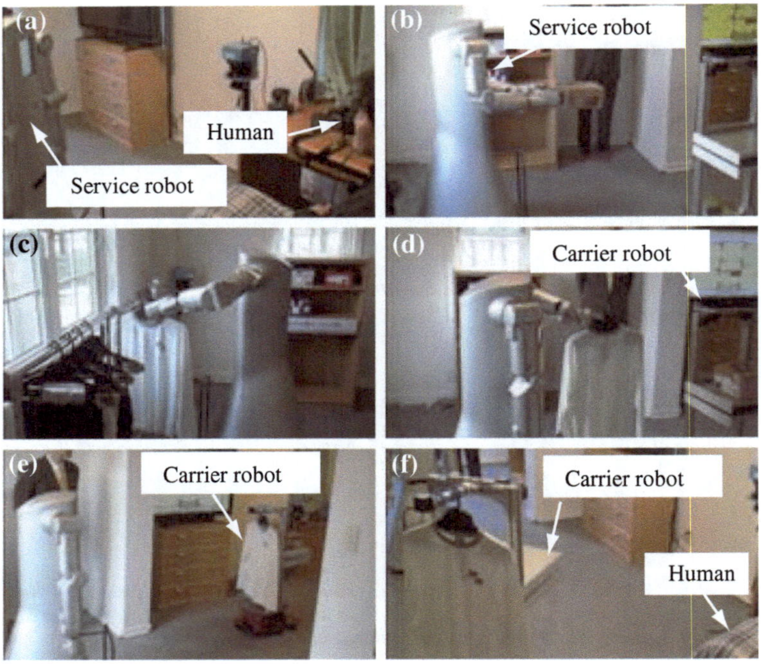

Fig. 10 Robot service experiment

2.2.4 Experiments for Robot Service

Experiments using two types of service robots were conducted as shown in Fig. 10. Using voice, a human operator ordered a service robot to bring shoes and cloth (a), the robot queried the TMS about these positions and retrieved them based on the attached RFID tags (b, c), and then a carrier robot conveyed them to the human operator (d–f). All the information from the RFID tags on a floor, laser and vision sensors, and service robots was managed centrally by the TMS.

3 Environmental Geometrical Modeling Using Mobile Robots

3.1 Geometrical Modeling of Large-Scale Environments

For a service robot working in a daily environment with humans, an accurate map of a surrounding environment is indispensable. In order to avoid collision with environments and provide various services safely, these maps should contain not only 2-D information which is popular in some robotic applications such as path planning, but also 3-D information such as stair step height or 3-D positions of obstacles. However, most of maps used for a service robot are created by hand currently, and this task, especially for creating detailed 3-D maps, is quite laborious and these maps are not re-created very often. Therefore, from the point of view of efficiency and maintainability, these maps should be constructed using robots themselves automatically.

For constructing 3-D models of large-scale environments using a 3-D range sensor such as a laser range finder, a number of partial range images are taken from different viewpoints around the targets. These images are aligned using post-processing procedures, such as the ICP algorithm [1, 2]. However, in general, when sufficiently exact scan pose estimates are not available, a human operator registers the positions before applying the ICP method in order to ensure that the images converge to the proper poses. Also, all of the images must contain sufficient feature shapes and must sufficiently overlap each other, which requires dense scanning from a number of positions, in order to precisely register the range images using the ICP algorithm.

Another approach that requires no post-processing procedures such as the ICP algorithm can also be considered, which involves the precise identification of the position of the range sensor at each measurement. As an example of this approach, several systems that use GPS [23, 33] to determine the position of the range sensor have been proposed. However, special instruments and techniques, such as a Real-time Kinematic (RTK) system or the Virtual Reference Station (VRS) method, are required in order to achieve highly precise position identification using current GPS.

This section introduces a 3-D measurement system for large-scale environments using a group of mobile robots and an on-board laser range finder [12, 13]. The proposed system uses the Co-operative Positioning System (CPS) [10, 11] for multiple robots, which has been proposed as a highly precise position identification technique for mobile robots. This system can construct 3-D models of large-scale environments without any post-processing procedure such as the ICP algorithm or manual intervention. In addition, it is possible to register range images even if the number of measurements is few and the data is sparse. It is also possible to construct a 3-D model in environments where GPS is not available, such as in an indoor environment.

3.2 Related Work

The proposed system is related to the Simultaneous Localization And Mapping (SLAM) method [3, 5, 16, 22, 29–32], which has attracted a great deal of attention in the robotics community. In the proposed system, the obtained 3-D model can be refined by applying ICP as shown in [12]. Obviously, the refined measurement position from ICP can also be fed back to the positioning system. This closed-loop control will increase the accuracy of both the 3-D model and the robot position like

SLAM systems. The following two elements are essential to create a high-accuracy environmental map with 3D measurement robot systems.

(1) High accuracy self-localization system
(2) High accuracy measurement system for surrounding environment

There are several issues to be considered for Item (1). The self-localization method that has been proposed until now, for example, odometry, does not have the ability to measure the self-location of a mobile robot with high accuracy in a bumpy area or in an environment with pitch differences. The SLAM system requires the characteristic features of the environment and has the same problems as odometry, which is the accumulation of measurement error caused by the measurement device. To reduce error accumulation, loop detection and refinement of the obtained models and paths are the principal, critical issues in SLAM.

Besides these methods, co-operative localization using a team of mobile robots also has been attracting much attention as a highly-precise self-localization technique so far [6, 11, 17, 20, 24, 26, 28]. In this method, robots are localized sequentially and alternatively by observing the positions of other robots in a team instead of natural or artificial stable landmarks. The first idea of the co-operative localization was introduced by Kurazume et al. [11]. Each mobile robot in a team repeats to move and stop, acts as a mobile landmark. The basic algorithm of this method is also introduced in Sect. 3.3 in this paper.

Concerning Item (2), systems using a laser range finder are effective and often used from the point of view of cost and accuracy. Originally, laser range finders were large-scale, expensive, and intolerant of vibrations. However, recently, smaller and more inexpensive laser range sensors have been developed and are readily available.

3.3 Co-operative Positioning System (CPS)

Let us consider the system in which a mobile robot equipped with an on-board laser range finder moves around a measurement target and scans the target from several different positions. If all of the measurement positions are identified with high accuracy, the range data acquired at each position can be converted to the global co-ordinate system by a simple co-ordinate transformation calculation.

To achieve highly accurate positioning of mobile robots, Kurazume et al. proposed the Co-operative Positioning System (CPS) [11]. In this system, multiple robots with highly precise measurement devices for their mutual positions are controlled co-operatively. This can lead, compared to conventional positioning techniques, to high positioning accuracy, even in unknown and uneven environments.

The basic principle of CPS is as follows: Divide the robots into group A and group B. Group A remains stationary and acts as a landmark while group B moves. Group B then stops and acts as a landmark for group A. This alternating behavior is

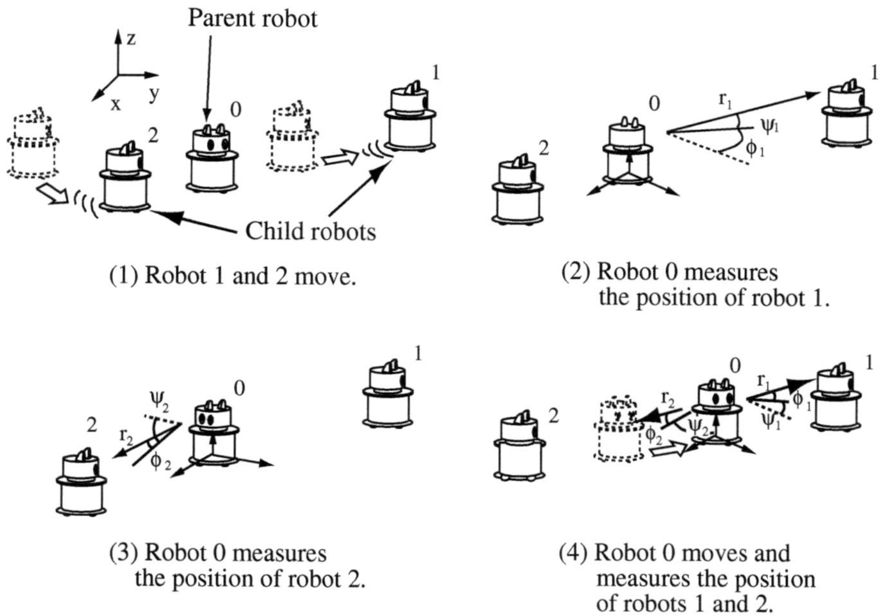

Fig. 11 Co-operative positioning system (CPS)

repeated until the target position is reached. By using the concept of "portable landmarks," CPS has a far lower accumulation of positioning errors than dead reckoning and can work in three dimensions, which is not possible by ordinary dead reckoning. In addition, CPS can be used in an unknown environment, since there is no need to place landmarks beforehand.

An example of CPS is shown in Fig. 11. This example is for a robot system consisting of a parent robot with a sensing device such as a laser range finder and two child robots. The sensing device can measure the relative positions of the child robots from the parent robot. First, assume that the initial position of the parent robot is measured or defined beforehand.

(1) Child robots 1 and 2 are moved and stopped.
(2) The parent robot measures the distance, azimuth, and elevation angles to child robot 1 and identifies the position of child robot 1.
(3) The position of child robot 2 is identified in the same manner as in Step 2.
(4) The parent robot moves and stops. The distances, azimuth, and elevation angles to child robots 1 and 2 are then measured, and the position of the parent robot is calculated using the triangular surveying technique.
(5) Repeat Steps 1 through 4 until the target position is reached.

Though the principle of CPS is simple, a position calculation that suppresses error accumulation is rather complicated [10]. In CPS, although the accuracy is quite high, measurement errors are gradually accumulated with the characteristics of the errors depending on the moving histories of the robots. To minimize error accumulation by taking the moving histories into account, a nonlinear least squared method based on the sequential estimation of error covariance matrices is proposed. In this method, the error accumulation is repeatedly estimated by calculating the error covariance matrices by taking the accuracy of a sensing device and the relative positions between robots into account. The position of each robot is determined using the accumulated error covariance matrices so that the positioning error is minimized at each positioning. In addition, several optimum moving strategies which minimize the error accumulation are proposed. Note that it is possible to refine the error accumulation after closing a loop by the parent robot using techniques developed in SLAM [3, 5, 16, 22, 29–32]. The experimental results after applying ICP was reported in [12].

3.4 Construction of a 3-D Environmental Map Using Multiple Robots

This section introduces a measurement system for the precise construction of a 3-D environmental map by combining CPS for multiple robots and a laser range finder. In this system, mobile robots move around a large-scale target and scan the target by an on-board 3-D laser range finder from several viewpoints. Each measurement position is precisely identified by CPS using a parent robot and two child robots. First, the sixth CPS machine model, called CPS-VI, which is equipped with a 2-D laser range finder and a scanning mechanism, is introduced, and the experimental results for the construction of indoor and outdoor environmental maps by CPS-VI are given.

3.4.1 Sixth CPS Machine Model (CPS-VI)

Figure 12 shows the sixth CPS machine model, CPS-VI. This system consists of a parent robot and two child robots. The parent robot is equipped with an on-board 2-D laser range finder (LMS 151, Sick), a high-precision two-axes altitude sensor (MD900T, Applied Geosystems), an automatic leveling system (Rizumu, AS-21), and a total station for the survey (GPT-9005A, TOPCON Ltd.), which is used to measure the relative positions of the child robots. Even if the body is tilted on a slope, the body inclination is compensated by an automatic leveling system, and precise positioning of the robots is achieved. The 2-D laser range finder can acquire 2-D slit-like range data within the range of 50 m and 270°. The parent robot has a rotating table on the top of its body, and by rotating the table around the vertical axis while scanning using the 2-D laser range finder, 360° 3-D range images as shown in Fig. 13 are acquired in 37.8 s.

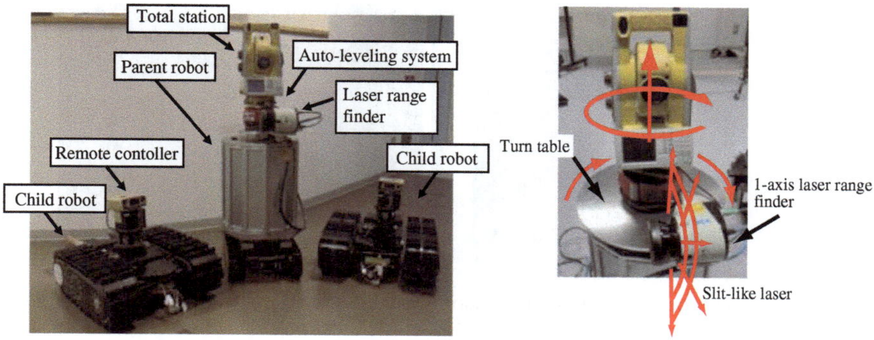

Fig. 12 The developed tunnel shape measurement system

Fig. 13 Range data obtained in one scan

3.4.2 Experiments to Construct the Indoor and the Outdoor Environmental Maps

Experiments for constructing 3-D maps are carried out using CPS-VI in a long corridor environment (Fig. 14). In these experiments, the parent robot and the two child robots moved, stopped, and identified their positions 38, 7, and 8 times respectively, respectively. Total moving distance of the parent robot was 210 m. After identifying their positions, the parent robot captured 3-D range images at each stationary position by rotating the on-board 2-D laser range finder around the vertical axis. The obtained range images are transformed into the global co-ordinate system by simple co-ordinate transformation calculations using the positioning result measured by CPS. No post-processing procedure is applied. In these experiments, the parent robot repeated the laser scanning from 33 positions and obtained about 40,340,000 points. Figures 14 and 15 shows the total view of the resulting 3-D map.

Fig. 14 Photo and measured shape of corridor

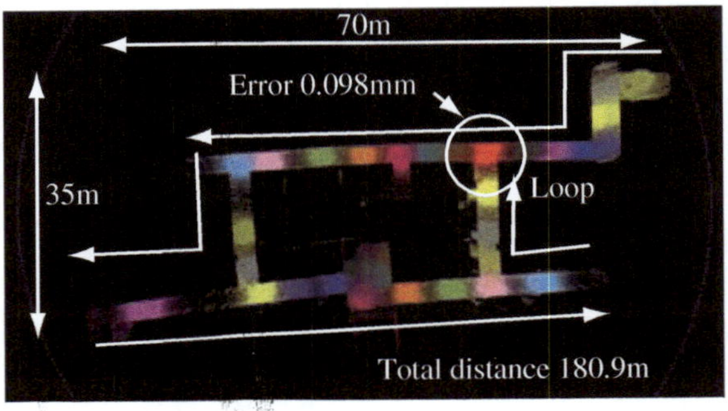

Fig. 15 3D model of corridor

To verify the accuracy of the obtained 3D model, we compared the measured positions of a same point (door corner) before and after the movements along the corridor with a loop and estimated the modeling error. The error between these points is 0.098 m, compared to the moving distance of the parent robot of 180.9 m, that is, the error is only 0.054 % of the total distance traveled.

Next, a construction experiment was done to obtain 3-D environmental maps in the outdoor environments with the difference in height of 5 m. The robots moved around buildings and the outer walls of a building were measured from 20 positions and 3-D models were constructed of the building. The path of the parent robot and the obtained map are shown in Figs. 16 and 17. The modeling error in this experiment is 0.116 m, compared to the total travel distance of the parent robot of 343 m, which is 0.034 % of the distance traveled.

Fig. 16 Measured shapes and errors in outdoor environment

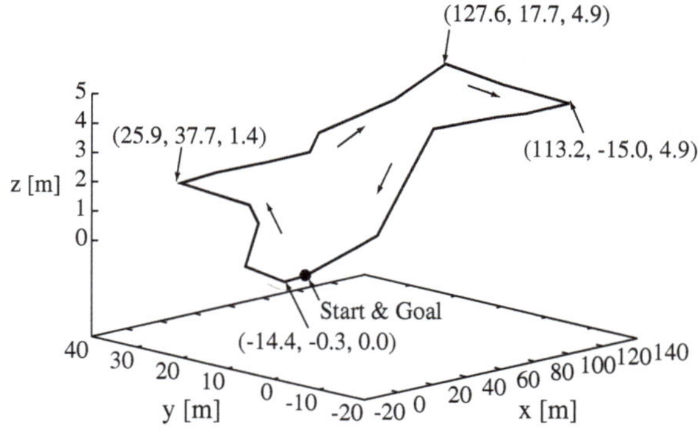

Fig. 17 Path of parent robot

3.4.3 Experiment to Construct a Map of Large-Scale Urban District

Furthermore, using the proposed system, an urban district environmental map of the Fukuoka Island City area, which is located in the Higashi-ku district of Fukuoka City, Japan, was created. The experiment environment consists of a house and a park, where the difference in height is approximately 2 m. Some of the results are shown in Fig. 18. In this experiment, the parent robot measured its surroundings at 45 different viewpoints and moved around up to 115.8 m in x-direction and 131.8 m in the y-direction while traveling a total of 543.4 m. Compared to the 44 movements of the parent robot, the child robots traveled and changed their locations

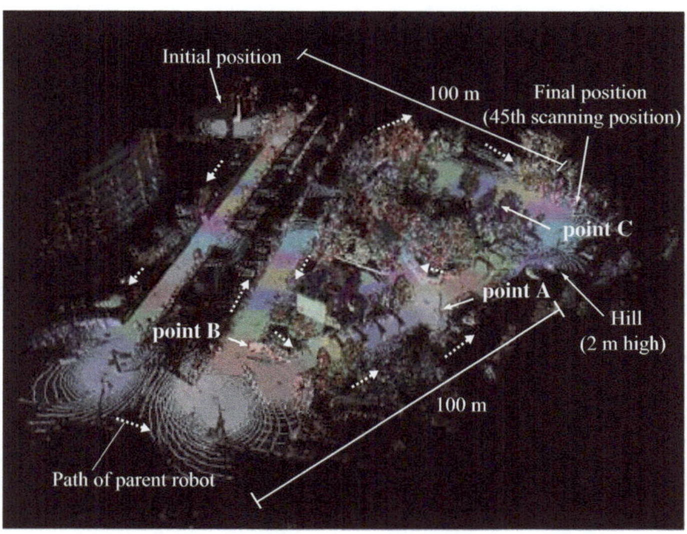

Fig. 18 Obtained 3-D environmental map (overall view)

8 times. We evaluated the errors of the 3D model in three points (Point A, B, and C) indicated in Fig. 18. The errors between the local models measured in the different positions in these points were 57 mm, 372 mm, and 494 mm, respectively. The experiment shows that the proposed system has the ability to create a precise 3-D environmental map, even in a large outdoor environment.

3.5 Application for the Digital Archive of Large-Scale Cultural Heritage Sites

Finally, an example of the application of the proposed system for the digital archive of a large-scale Japanese cultural heritage site is considered.

The Dazaifu Shrine (Dazaifu Tenmangu) (Fig. 19) located in Fukuoka, Japan, was established in 919 in memory of Sugawara no Michizane, a famous Japanese scholar, politician, and poet. The main shrine of the Dazaifu Tenmangu was built in 1591 and is registered as an important cultural property of Japan. The size of the main shrine and the yard are about 250 m × 100 m.

We conducted a 3-D digital archive experiment on the main shrine of Dazaifu Tenmangu and the vast garden by the robot system proposed in this paper. Figure 19 shows the view of the main shrine of Dazaifu Tenmangu. The parent robot moved and measured the Dazaifu Tenmangu from 76 places on the inside and outside of the main shrine and the garden.

Fig. 19 Main shrine of Dazaifu Tenmangu (Dazaifu Shrine)

4 Conclusions

This paper briefly introduces the research project Robot Town Project. The aim of this research project is to develop a distributed sensor network system covering an area of a block in a town in which there are many houses, buildings, and roads, to manage robot services by monitoring events that occur in the town. The sensed events are notified to the TMS, and each robot receives appropriate information regarding the surroundings and instructions for proper service. This paper introduced the developed systems in this project, including RFID-tag-based localization system, distributed sensor systems for moving object tracking, and object management system using RFID tags.

In addition, a 3-D measurement system using multiple mobile robots was introduced for geometrical modeling of large-scale environments. This system is composed of multiple mobile robots and an on-board laser range finder, and a highly precise positioning technique named the Co-operative Positioning System (CPS) is adopted to localize the robots. Measurement experiments in unknown and large in-door/outdoor environments were successfully carried out using the newly developed multiple-robot system, called CPS-VI.

Acknowledgements This work was partially supported by Grant-in-Aid for Scientific Research (B) (23360115).

References

1. P.J. Besl, N.D. McKay, A method for registration of 3-d shapes. IEEE Trans. Pattern Anal. Mach. Intell. **2**(14), 239–256 (1992)
2. Y. Chen, G. Medioni, Object modelling by registration of multiple range images. Image Vis. Comput. **3**(10), 145–155 (1992)
3. D.M. Cole, P.M. Newman, Using laser range data for 3d slam in outdoor environment, in *Proceedings of IEEE International Conference on Robotics and Automation*, 2006, pp. 1556–1563
4. I.A. Essa, Ubiquitous sensing for smart and aware environments: Technologies toward the building of an aware home, in *Position Paper for the DARPA/NSF/NIST workshop on Smart Environment*, 1999
5. A. Howard, Multi-robot simultaneous localization and mapping using particle filters. Int. J. Robot. Res. **25**(125), 1243–1256 (2006)
6. A. Howard, M.J. Matarić, G.S. Sukhatme, Putting the 'i' in 'team': an ego-centric approach to cooperative localization, in *Proceedings of IEEE International Conference on Robotics and Automation (ICRA)*, 2003, pp. 868–892
7. Y. Iwashita, R. Kurazume, T. Mori, M. Saito, T. Hasegawa, Model-based motion tracking system using distributed network camera, in *Proceedings of IEEE International Conference on Robotics and Automation*, 2010, pp. 3020–3025
8. Y. Iwashita, M. Saito, R. Kurazume, T. Hasegawa, Motion tracking in daily environment using distributed image and laser sensors, in *The First International Workshop on Human Behavior Sensing*, 2010

9. C.D. Kidd, R. Orr, G.D. Abowd, C.G. Atkeson, I.A.E. an Blair MacIntyre, E.D. Mynatt, T. Starner, W. Newstetter, The aware home: a living laboratory for ubiquitous computing research, in *Proceedings of the Second International Workshop on Cooperative Buildings, Integrating Information, Organization, and Architecture*, 1999, pp. 191–198
10. R. Kurazume, S. Hirose, An experimental study of a cooperative positioning system. Auton. Robot. **8**(1), 43–52 (2000)
11. R. Kurazume, S. Nagata, S. Hirose, Cooperative positioning with multiple robots, in *Proceedings of IEEE International Conference on Robotics and Automation*, vol. 2, 1994, pp. 1250–1257
12. R. Kurazume, Y. Noda, Y. Tobata, K. Lingemann, Y. Iwashita, T. Hasegawa, Laser-based geometric modeling using cooperative multiple mobile robots, in *Proceedings of IEEE International Conference on Robotics and Automation*, 2009, pp. 3200–3205
13. R. Kurazume, Y. Tobata, Y. Iwashita, T. Hasegawa, 3d laser measurement system for large scale architectures using multiple mobile robots, in *Proceedings of the 6th International Conference on 3-D Digital Imaging and Modeling (3DIM2007)* (2007)
14. R. Kurazume, H. Yamada, K. Murakami, Y. Iwashita, T. Hasegawa, Target tracking using sir and mcmc particle filters by multiple cameras and laser range finders, in *Proceedings of IEEE/RSJ International Conference on Intelligent Robots and Systems*, 2008, pp. 3838–3844
15. J.H. Lee, K. Morioka, A. Ando, H. Hashimoto, Cooperation of distributed intelligent sensors in intelligent environment. IEEE/ASME Trans. Mechatr. **9**(3), 535–543 (2004)
16. M.D. Marco, A. Garulli, A. Giannitrapani, A. Vicino, Simultaneous localization and map building for a team of cooperating robots: a set membership approach. IEEE Trans. Robot. Autom. **19**(2), 1243–1256 (2003)
17. L. Montesano, J. Gaspar, J. Santos-Victor, L. Montano, Cooperative localization by fusing vision-based bearing measurements and motion, in *Proceedings of IEEE/RSJ International Conference on Intelligent Robots and Systems*, 2005, pp. 2333–2338
18. K. Murakami, T. Hasegawa, K. Shigematsu, F. Sueyasu, Y. Nohara, B. Ahn, R. Kurazume, Position tracking system for commodities in a daily life environment, in *Proceedings of IEEE/RSJ International Conference on Intelligent Robots and Systems*, 2010, pp. 3712–3718
19. Y. Nakauchi, T. Fukuda, K. Noguchi, T. Matsubara, Intelligent kitchen: Cooking support by lcd and mobile robot with ic-labeled objects, in *Proceedings of IEEE/RSJ International Conference on Intelligent Robots and Systems*, 2005, pp. 2464–2469
20. E. Nerurkar, S. Roumeliotis, A. Martinelli, Distributed maximum a posteriori estimation for multi-robot cooperative localization, in *Proceedings of IEEE International Conference on Robotics and Automation*, 2009, pp. 1402–1409
21. Y. Nishida, T. Hori, T. Suehiro, S. Hirai, Sensorized environment for self-communication based on observation of daily human behavior, in *Proceedings of IEEE/RSJ International Conference on Intelligent Robots and Systems*, 2000, pp. 1364–1372
22. A. Nüchter, H. Surmann, K. Lingemann, J. Hertzberg, S. Thrun, 6d slam with an application in autonomous mine mapping, in *Proceedings of IEEE International Conference on Robotics and Automation*, 2004, pp. 1998–2003
23. K. Ohno, T. Tsubouchi, S. Yuta, Outdoor map building based on odometory and rtk-gps positioning fusion, in *Proceedings of IEEE International Conference on Robotics and Automation*, 2004, pp. 684–690
24. S. Panzieri, F. Pascucci, R. Setola, Multirobot localization using interlaced extended Kalman filter, in *Proceedings of IEEE/RSJ International Conference on Intelligent Robots and Systems*, 2006, pp. 2816–2821
25. A. Pentland, Smart rooms. Sci. Am. 54–62 (1996)
26. I. Rekleitis, G. Dudek, E. Milios, Multi-robot cooperative localization: a study of trade-offs between efficiency and accuracy, in *Proceedings of IEEE/RSJ International Conference on Intelligent Robots and Systems*, 2002, pp. 2690–2696
27. T. Sato, T. Harada, T. Mori, Environment-type robot system "robotic room" featured by behavior media, behavior contents, and behavior adaptation. IEEE/ASME Trans. Mechatr. **9** (3), 529–534 (2004)

28. J. Spletzer, A. Das, R. Fierro, C. Taylor, V. Kumar, J. Ostrowski, Cooperative localization and control for multi-robot manipulation, in *Proceedings of IEEE/RSJ International Conference on Intelligent Robots and Systems*, vol. 2, 2001, pp. 631–636
29. S. Thrun, A probabilistic online mapping algorithm for teams of mobile robots. Int. J. Robot. Res. **20**(5), 335–363 (2001)
30. S. Thrun, M. Montemerlo, The graph slam algorithm with applications to large-scale mapping of urban structures. Int. J. Robot. Res. **25**(5–6), 403–429 (2006)
31. R. Triebel, P. Pfaff, W. Burgard, Multi-level surface maps for outdoor terrain mapping and loop closing, in *Proceedings of IEEE/RSJ International Conference on Intelligent Robots and Systems*, 2006, pp. 2276–2282
32. J. Weingarten, R. Siegwart: Ekf-based 3d slam for structured environment reconstruction, in *Proceedings of IEEE/RSJ International Conference on Intelligent Robots and Systems*, 2005, pp. 2089–2094
33. H. Zhao, R. Shibasaki, Reconstructing a textured cad model of an urban environment using vehicle-borne lase range scanners and line cameras. Mach. Vis. Appl. **14**, 35–41 (2003)

Soft Mobile Robots with On-Board Chemical Pressure Generation

Cagdas D. Onal, Xin Chen, George M. Whitesides and Daniela Rus

Abstract We wish to develop robot systems that are increasingly more elastic, as a step towards bridging the gap between man-made machines and their biological counterparts. To this end, we develop soft actuators fabricated from elastomer films with embedded fluidic channels. These actuators offer safety and adaptability and may potentially be utilized in robotics, wearable tactile interfaces, and active orthoses or prostheses. The expansion of fluidic channels under pressure creates a bending moment on the actuators and their displacement response follows theoretical predictions. Fluidic actuators require a pressure source, which limits their mobility and mainstream usage. This paper considers instances of mechanisms made from distributed elastomer actuators to generate motion using a chemical means of pressure generation. A mechanical feedback loop controls the chemical decomposition of hydrogen peroxide into oxygen gas in a closed container to self-regulate the actuation pressure. This on-demand pressure generator, called the pneumatic battery, bypasses the need for electrical energy by the direct conversion of chemical to mechanical energy. The portable pump can be operated in any orientation and is used to supply pressure to an elastomeric rolling mobile robot as a representative for a family of soft robots.

Keywords Soft robotics · Fluidic elastomer actuators · Chemical pressure generation · Distributed actuation · Smart materials

---

C.D. Onal (✉) · D. Rus
Computer Science and Artificial Intelligence Laboratory, Massachusetts Institute of Technology, Cambridge, MA, USA
e-mail: cagdas@csail.mit.edu

D. Rus
e-mail: rus@csail.mit.edu

X. Chen · G.M. Whitesides
Department of Chemistry and Chemical Biology, Harvard University, Cambridge, MA, USA
e-mail: xchen@gmwgroup.harvard.edu

G.M. Whitesides
e-mail: gwhitesides@gmwgroup.harvard.edu

H.I. Christensen and O. Khatib (eds.), *Robotics Research*,
Springer Tracts in Advanced Robotics 100, DOI 10.1007/978-3-319-29363-9_30

1 Introduction

There is currently a need to develop robotic devices that rely upon new high-performance soft actuators. Compliance allows conformation, which is desirable for adaptability in the device's negotiation with the world. A low minimum stiffness ensures safety in human interaction [1]. Many application areas will benefit from advances in practical soft actuation mechanisms including medical robotics, artificial muscles, and human interaction devices such as tactile or haptic interfaces.

Many candidate materials are under investigation to realize soft actuators. Many of these studies focus on materials that convert electrical energy to mechanical energy utilizing electroactive polymers [2] including dielectric elastomers [8, 12], electrostrictive polymers [14], and piezoelectric polymers [5]. The electrical operation principle of these actuators constrain their utility, as they require large electric fields or deformable electrodes [8].

In this paper, we describe a new soft mobile robot and demonstrate its locomotion. The robot relies on two novel robotic components: pressure-controlled soft actuators and a portable power source for these actuators.

The pressure-operated low-power soft actuators, called fluidic elastomer actuators (FEAs) [4] comprise synthetic elastomer films operated by the expansion of embedded channels under pressure. Pressure readily creates stresses inside the elastomer, which strains the material for actuation. Once pressurized, the actuator keeps its position with little or no additional energy consumption. The FEAs in this work consume 27.5 mJ of energy for 20.7 kPa pressure input. The stiffness of an FEA increases with applied pressure inside the embedded channels. This makes the material more resistant to disturbances once actuated. In case of a power failure, it can be designed to either go limb or keep its last position, both of which may be useful for safety in different scenarios.

Pressure is a convenient actuation source as it induces local deformation in a soft substrate [19], giving a large actuation range limited only by the mechanical strength of the material. In general, using direct mechanical energy in the form of pressure bypasses the need for electrical energy and its constraints. On the other hand, an important limitation on fluidic actuation is the necessity of a pressure source [7].

Our solution is to utilize a chemical approach to achieve portable and silent pressure generation. This is the equivalent of a battery for fluidic systems as it offloads pressure generation to a controlled gas generating chemical process. Specifically, we report on-demand pressure generation by the mechanical self-regulation of the decomposition of hydrogen peroxide (H_2O_2) into oxygen (O_2) gas in a closed container. The pressure output of the pneumatic battery powers FEAs for practical applications. An aqueous H_2O_2 solution is the fuel in these devices. It provides high power and is easily replaced with a fresh solution when depleted. Pure H_2O_2 has a theoretical energy density of 2.7 kJ/g, one of the highest in common monopropellant fuels. While this optimal value cannot be fully utilized at room temperature operation, it indicates the potential of H_2O_2 as a practical power source. A key feature of our portable pump design is its rotation-invariant usage, which enables the battery to operate in any orientation.

Hydrogen peroxide has been used previously as a monopropellant [21] and recently in robotics applications [6, 18, 20]. A recent work utilized H_2O_2 to build a self-powered microfluidic lab-on-a-chip system [13]. Using H_2O_2 to generate pressure has the benefit of using no conventional power source for operation. Hydrogen peroxide naturally decomposes into oxygen and water with no harmful byproducts at a slow rate. This reaction speeds up in the presence of a catalyst [15]. Once this exothermic reaction starts, it continues until all of the H_2O_2 is consumed or the catalyst is removed. In previous works, relief valves were utilized to periodically vent the gas to keep the pressure build-up under control [18].

The contributions of this work are as follows:

- Modeling, design, fabrication, and evaluation of fluidic elastomer actuators.
- Modeling, design, fabrication, and evaluation of a portable power source for the actuators.
- A robot built using six FEAs and one portable power source, and locomotion experiments.

The organization of this paper is as follows. Section 2 briefly outlines the architecture of a soft robot that utilizes fluidic actuation for mobility and/or manipulation. We identify components that need to be developed to realize such a robot. In Sect. 3 we describe and analyze a bending-type composite fluidic elastomer actuator theoretically and experimentally. In Sect. 4, we present a method of self-regulation of the catalyzed decomposition of hydrogen peroxide using a mechanical feedback loop that controls the reaction based on the pressure inside the pump. We present the theory behind this concept and experimentally verify its operation. Finally, in Sect. 5, we illustrate the application of the pneumatic battery to power a rolling mobile robot made of six bending FEAs. The result is a complete chemically powered fluidic elastomer actuation system.

2 Soft Robot Architecture

The basic architecture of a soft robot made of fluidic actuators is shown in Fig. 1. The soft robot architecture includes all the traditional components of a robot system (i.e. control, perception, actuation). The unique modules are the fluidic actuator system, the pressure source for the actuators, and the interface of the actuators to the robot's control system.

The actuation power source for this application needs to be a pressure generation device that is portable in order to be incorporated in the robotic body. Support hardware comprises miniature valves that are also preferably energy-efficient. The next section describes the actuation modules for the soft robot.

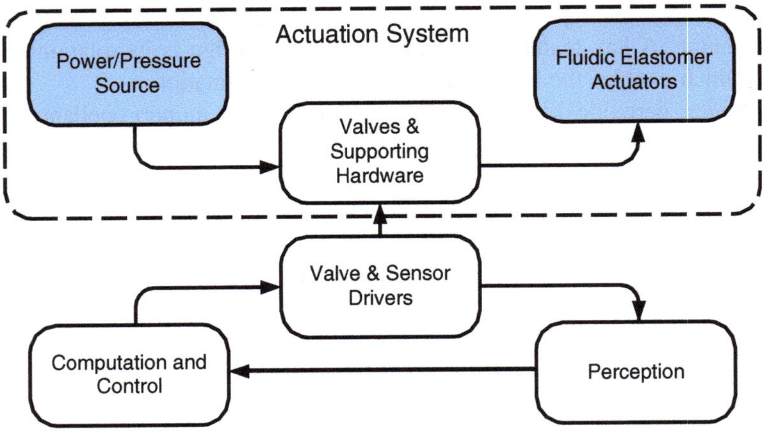

Fig. 1 The architecture of a soft robot using fluidic actuation. Components addressed in this work are shaded in *blue*

3 Fluidic Elastomer Actuators Modeling and Experiments

The operational principle behind our design for FEAs relies on using pressure for actuation. In an elastomeric substrate, expansion of embedded fluidic channels due to pressure input creates overall deformation of the actuator. By the inclusion of appropriate physical constraints, this deformation is forced to follow desired motions.

The design of bending FEAs utilized in this work is shown in Fig. 2. Parallel rectangular fluidic channels reside inside the elastomeric film. The channels are connected on two ends in a meandering arrangement. Without any constraints, the material tends to undergo tensile deformation. With an inextensible thin layer placed on one side, we can convert this axial deflection to an out-of-plane bending motion. Using certain patterns of constraints, we can also induce torsion or localized deformations on the same material.

The FEAs were fabricated by molding in two layers. Molds were created using a Stratasys™ Prodigy Plus™ fused deposition modeling system. The first layer was a 5 mm thick elastomer with open channels on one side. The second layer was a 1.3 mm thick solid elastomer, same length and width as the first piece. The two pieces were attached together in the thickness direction using an uncured thin layer of the same material as glue such that the open channels were sealed off. For a bending actuator, the second layer also embedded a fabric mesh as an inextensible thin sheet to constrain the axial deformation of this layer. The curing time for each step was about 24 h.

Note that FEAs are modular in nature, such that placing these actuators in various arrangements yield a multitude of functionalities. Some of these functionalities

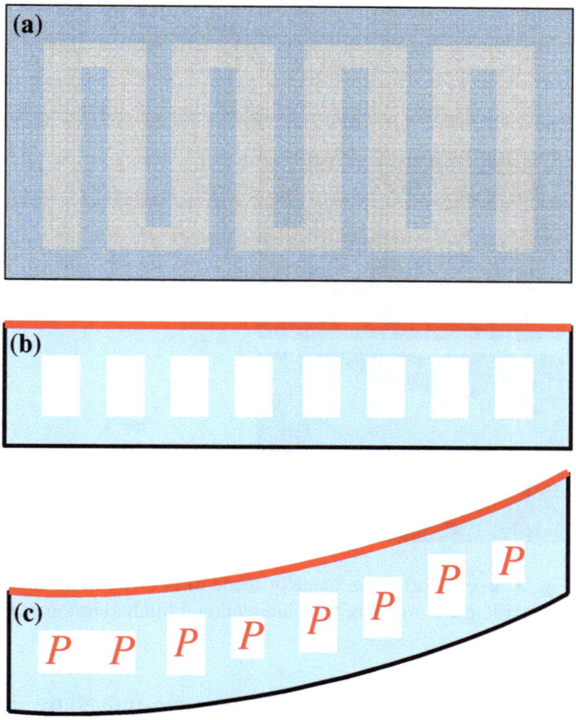

Fig. 2 The design of bending-type fluidic elastomer actuators. Fluidic channels are embedded in the elastomer and an inextensible thin layer is placed on the *top* (depicted in *red*) to induce bending. Top- and side-view sketches of the actuator are shown in (**a**) and (**b**), respectively. The bending motion due to the pressure in the channels is depicted in (**c**)

are demonstrated in Fig. 3. A parallel combination of bending actuators creates a 1-D linear positioner. Placing two bending actuators together in such a way that they sandwich and share a single inextensible layer yields bidirectional bending. A serial arrangement of these bidirectional bending FEAs in alternating bending directions create a soft kinematic chain similar to an octopus arm.

In what follows, we derive a basic static model of displacement based on the geometry and material properties of the actuator. Applied pressure P inside the rectangular channels with height h_c and length l_c creates axial stresses σ_x in the material with height h_t and length l_t given as:

$$\sigma_x = P\frac{h_c}{h_t - h_c}. \tag{1}$$

The resulting strain ε_x is a nonlinear function of the induced stresses. The total axial deformation δ_x of the material is the combination of the individual expansions of n channels resulting in:

$$\delta_x = n l_c \varepsilon_x(\sigma_x). \tag{2}$$

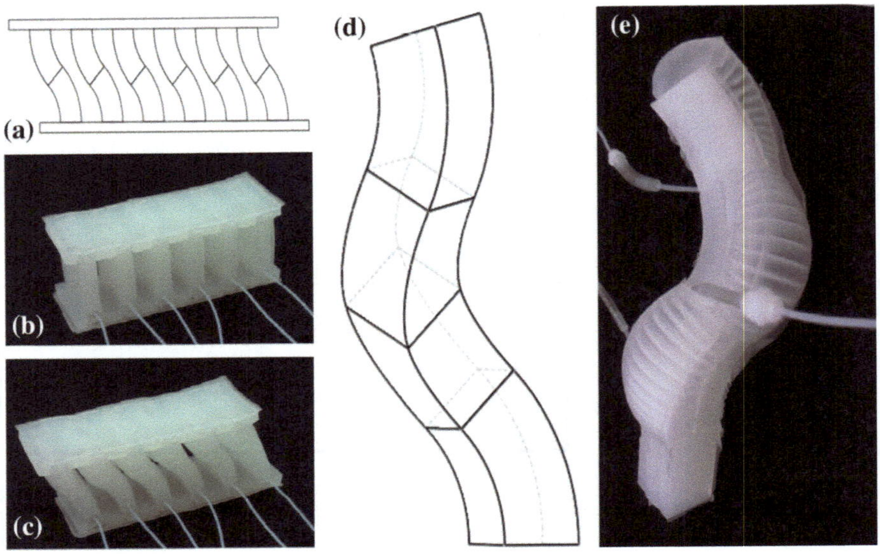

Fig. 3 **a–c** A soft linear actuator made of twelve fluidic elastomer bending actuators. **d–e** A soft kinematic chain made of four bidirectional fluidic elastomer bending actuators arranged in series

In this case, the elastomer is further constrained by an inextensible thin sheet on one side, which causes the actuator to undergo bending deformation. The inextensible sheet constitutes the neutral axis of bending in the composite. Thus, the total bending angle θ can be calculated from the contributions of each channel as:

$$\theta = 2n \arctan \frac{l_c \varepsilon_x(\sigma_x)}{2h_c}. \tag{3}$$

Finally, the total out-of-plane displacement δ_y of the actuator under these conditions is written as:

$$\delta_y = \frac{l_t}{\theta}(1 - \cos\theta). \tag{4}$$

The experimental setup for the investigation of the bending displacement of an FEA under pressure input is shown in Fig. 4. The actuator is clamped on one end and pressure is supplied to the fluidic channels from the side at the base. The out of-plane bending deflection is measured by image processing of a calibrated camera feed.

The bending displacement measurements in Fig. 4 were made by image processing in Matlab, using a Logitech^{TM} Webcam Pro 9000 camera attached to a custom setup, clamping the FEAs on one end and tracking the tip of the actuator

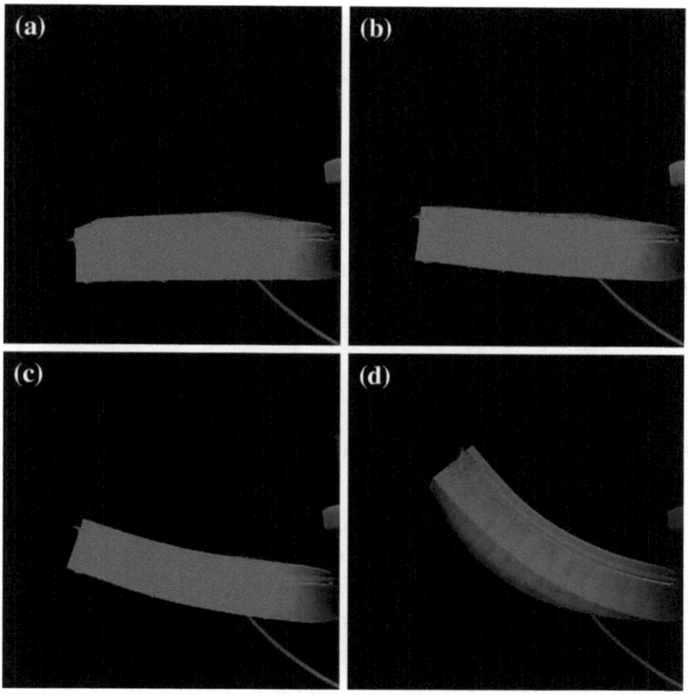

Fig. 4 Photo sequence of a fluidic elastomer bending actuator with large actuation range. As the pressure increases from 0 psi in (**a**) to 3 psi in (**d**), the embedded channels expand laterally and bend the composite due to an inextensible thin sheet on the top layer. Upon removal of the pressure, the initial configuration in (**a**) is restored. Input pressure values are 1, 2, and 3 psi (6.9, 13.8, and 20.7 kPa) from (**b**) to (**d**), respectively. The expansion of fluidic channels is visible in (**d**)

using color segmentation. The vertical position of the actuator tip was tracked for this measurement. FEAs were placed vertically, such that the bending axis coincided with the direction of gravity. Pressure measurements were made in Matlab, using a Honeywell™ ASDX030 gage pressure sensor and a National Instruments™ NI USB-6008 data acquisition system.

The elastomer samples used for the experiments were 38.1 mm long, 38.1 mm wide, and 6.35 mm thick stripes of Smooth-on™ Ecoflex™ Supersoft 0030 silicone rubber. They embedded 13 fluidic channels that were 1 mm long, 33 mm wide, and 3 mm thick.

The experimental deflection response of a silicone rubber FEA for pressure inputs ranging from 1 psi to 3 psi (6.9–20.7 kPa) is depicted in Fig. 5 with the corresponding simulation results according to the given model shown by the dashed curve. The stress-strain relationship of the material is determined by a power function fit to the experimental true stress and strain data.

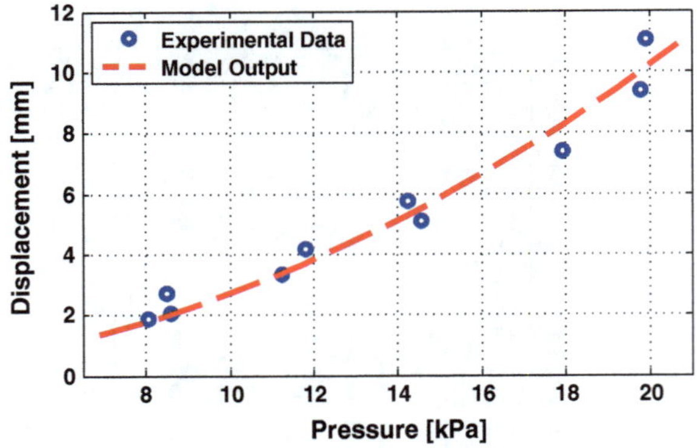

Fig. 5 Out-of-plane displacement of a fluidic elastomer actuator versus the measured pressure input. *Hollow blue circles* are experimental data. The *dashed curve* is the model output. The non-linearity in this curve stems from the stress-strain relationship of the polymer

4 Self-regulated Pressure Generation

The FEAs require a small pressure source of comparable size. In this section, we describe the design, modelling and implementation of a chemically operated portable pressure source we call a pneumatic battery.

The pneumatic battery mechanism that enables self-regulation in pressure generation from an aqueous H_2O_2 solution is depicted in Fig. 6 with a sketch and corresponding prototype. 50 % wt. H_2O_2 solutions were acquired from Sigma-Aldrich™ and diluted with deionized water as needed.

The pump has a cylindrical body, which makes it rotationally invariant. On one side resides an elastomeric deflector that embeds a cylindrical air chamber at atmospheric pressure sealed off from the pump by a thin circular membrane. The deflection of the membrane is dependent upon the pressure in the pump. Self-regulation is achieved by this deflection, creating a mechanical feedback loop.

As catalyst, 0.2 mm thick sheets of 92.5 % pure silver (Ag) are placed on the deflector, around the membrane. The membrane is offset from another disk-shaped elastomeric layer with a boss by a defined distance. With increasing pump pressure, the membrane deflects inside and pulls the soft layer towards the catalyst pack. At a cut-off pressure value, the opposite layer completely conforms to the catalyst surface and stops the reaction. The deflector was molded in two parts from Ecoflex™ 0030 silicone rubber and glued using an uncured thin layer of the same material such that the air chamber was sealed off. The air chamber inside the deflector was sealed at ambient conditions.

An outlet is placed on the other side of the pump to use the generated gas pressure for actuation. The gas is filtered by a polytetrafluoroethylene (PTFE) membrane with sub-micron pores. The filters were Whatman™ 7582-004 WTP Range PTFE membranes with 47 mm diameter and 0.2 μm pore size. The hydrophobic nature of this filter keeps the solution inside the pump while allowing the gas to be removed. The rotational invariance of the mechanism makes it a good

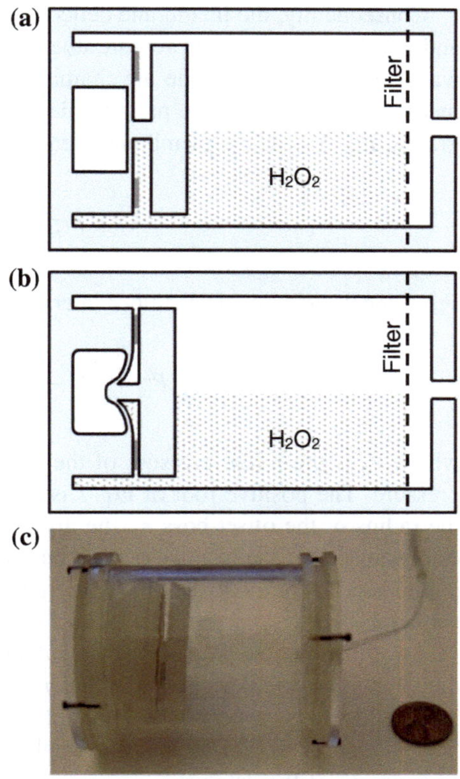

Fig. 6 Side-view sketches (**a**, **b**) and the prototype (**c**) of the self-regulating chemical pneumatic pump mechanism, using hydrogen peroxide as a fuel. A deflector on the *left side* deforms with increasing pressure and completely seals off the catalyst pack (*gray*) from the solution at a tuned critical pressure in (**b**), effectively stopping the reaction. The gas is filtered through a hydrophobic membrane filter on the *right side* before the outlet

candidate for devices that do not necessarily have a defined constant direction of gravity, since it is operational in any orientation.

The prototype shown in Fig. 6 is made from a cylindrical hollow acrylic container 50.8 mm diameter and 3.2 mm thickness, laser machined acrylic lids and custom silicone rubber seals. The deflector is attached to the left lid. The PTFE filter and a pipe fitting are placed on the right lid.

The critical pressure of the pump P_c is tuned based on the following theoretical study. Static plate deflection theory predicts that the deflection w of a clamped circular membrane with radius r_m under a pressure difference $\Delta P = P_c - P_{in}$ is:

$$w(r) = \frac{\Delta P r_m^4}{64K}\left(1 - \left(\frac{r}{r_m}\right)^2\right)^2, \tag{5}$$

where K is the flexural rigidity and ν is the Poisson's ratio of the plate. If the air chamber is connected to the atmosphere, its internal pressure P_{in} remains constant and this equation is enough to engineer the necessary amount of offset or membrane thickness for a given target cut-off pressure. For safety, to reduce the possibility of H_2O_2 leakage, this work uses a closed air chamber.

Consequently, the membrane deflection decreases the volume of the air chamber and increases its internal pressure due to the ideal gas law. With sufficiently thick walls, the volume V of the air chamber with initial height h is solely due to the membrane deflection. The pressure dependant chamber volume is written by integrating Eq. 5 over the membrane area as:

$$V = \pi r_m^2 \left(h - \frac{\Delta P r_m^4}{192K} \right). \tag{6}$$

From ideal gas law, air chamber internal pressure must satisfy:

$$\frac{r_m^4}{192K} P_{in}^2 + \left(h - \frac{P_c r_m^4}{192K} \right) P_{in} - hP_o = 0, \tag{7}$$

where P_o is the initial pressure of the air chamber, typically equal to atmospheric pressure. The positive root of Eq. 7 is the final pressure in the air chamber. Given the radius of the offset boss r_o, the displacement of the opposite layer towards the catalyst pack is calculated as $w(r_o)$ from Eq. 5. We use this theory to tune the design parameters in order to achieve a certain critical pressure suitable for the FEA actuation needs.

An experimental pressure self-regulation data is displayed in Fig. 7. For this demonstration, we designed a deflector such that a silicone rubber membrane with 3 mm thickness deflects for about 2.7 mm under a critical pump pressure of 51.7 kPa (7.5 psi). We used a boss height of 2.5 mm to ensure proper conformation and sealing of the catalyst.

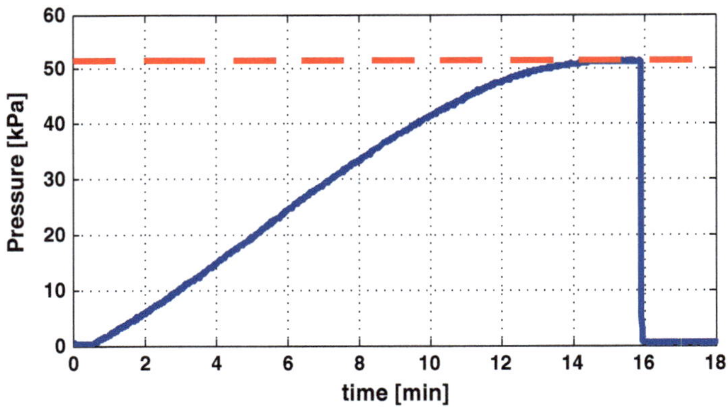

Fig. 7 Experimental results of pressure self-regulation in the pump using an approximately 10 % H_2O_2 solution in water. The reaction stops at a critical pressure value of 51.7 kPa (7.5 psi) due to the deflection of the membrane sealing the catalyst off the solution. The target cut-off pressure is indicated by the dashed line. The pump is vented at 16 min

5 A Soft Mobile Rolling Robot Prototype Powered by the Pneumatic Battery

Using the FEAs in Sect. 3 and the pneumatic battery in Sect. 4, we designed and fabricated a soft mobile robot (see Fig. 8). The robot has a cylindrical body with 80 mm diameter and 63 mm length. It is made of six equally spaced bending FEAs cantilevered on the periphery of the robot, parallel to the cylinder surface. The FEAs have the same dimensions as in Sect. 3 and their weight is 12.5 g. They act as flaps to bend out when actuated and apply torque, pushing the body forward. The actuators are fabricated separately and attached to the robot using silicone epoxy. The cylindrical body is molded from the same elastomeric material and an acrylic cylinder is placed in the center, which becomes the peroxide pump body when assembled. The total weight of the robot without the H_2O_2 solution is 210 g.

The first experiment we conducted aimed to measure the capability of the H_2O_2 pump to supply pressure to a single FEA, which is analyzed in Fig. 9. In this experiment, the pressure in the pump is measured while fluidic channels in the elastomer are pressurized and vented continuously with a 2 s period using the generated gas.This figure depicts that the pressure generation in the pump is canceled by the gas usage of the actuator at around 24.1 kPa (3.5 psi). This data is averaged over several hundred runs.

The integration of the chemical pressure generator to functional devices made of fluidic elastomer actuators is exemplified by the hexagonal rolling mechanism with 6 FEAs in Fig. 10. The hydrogen peroxide pump rests in the center and constitutes the body and the payload of the roller in addition to providing on-board pressure. The internal volume of the pump is approximately 50 ml. A 30 ml (36 g) fresh 50 % H_2O_2 aqueous solution is used for these experiments.

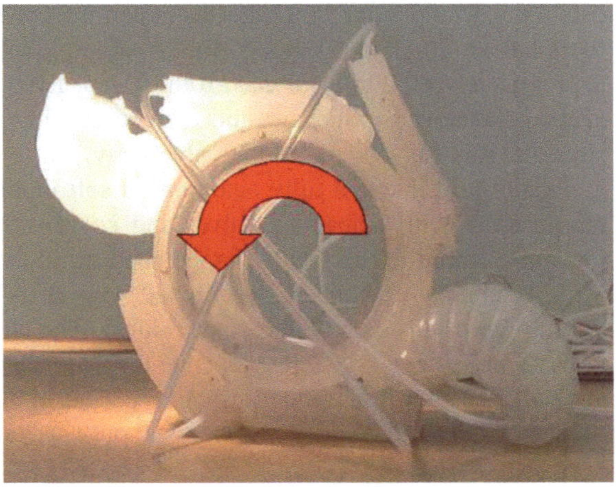

Fig. 8 Prototype of a rolling mobile robot using six bending-type fluidic elastomer actuators to propel itself forward. The *hollow cylinder* in the center is the housing for the pneumatic battery

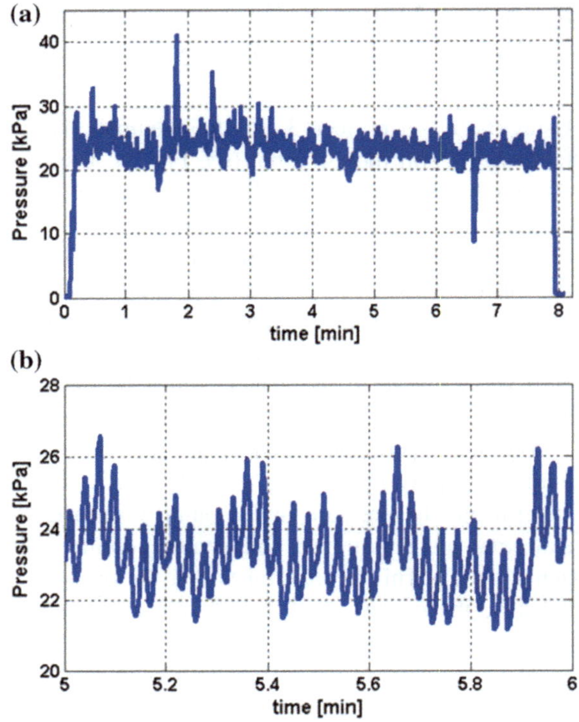

Fig. 9 Pressure inside the pump settles to a finite value while driving an FEA in (**a**). Zoomed-up view of the same experiment shows spikes in the data corresponding to each actuation period in (**b**)

The rolling motion is also selected to underline the utility of the inherent rotational invariance in this chemical pump design. This soft robotic system uses an external valve array, controlled by an operator to demonstrate the proof-of-concept of on-board pressure generation with a controlled chemical reaction.

Pressurized gas taken from the outlet fitting at the center of the pump feeds the external solenoid valve array. In this experiment, we actuated each FEA manually to induce rolling. It takes 7 s for a single rolling step. The body travels at approximately 40.2 mm for each rolling step. Even though the pneumatic battery keeps generating pressure, we did not exceed one full rotation in this experiment since the robot is tethered to the valve array.

Finally, to achieve a fully autonomous soft mobile robot, we fabricated a prototype roller with embedded valves and control as shown in Fig. 11. Six commercial solenoid latching valves are embedded in the elastomeric body under each FEA. A circular custom PCB equipped with an ATtiny13A microcontroller and driving electronics handle the control logic. The electronics and valves have low power requirements, suitable to operate from miniature LiPoly batteries. The valves are activated in a time sequence with a period of 10 s for each rolling step. This robot can roll itself on a flat surface without user intervention.

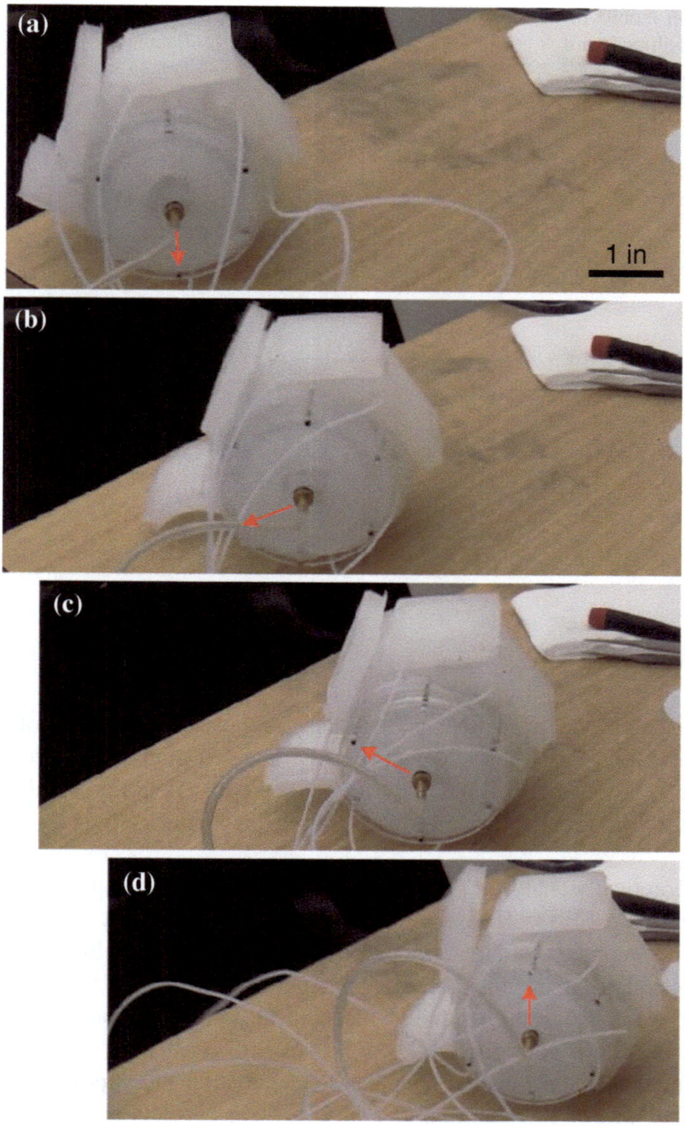

Fig. 10 Photo sequence of a pneumatic rolling mechanism made from six FEAs, rolling on a flat surface using pressure generated by an on-board hydrogen peroxide pump. Three rolling steps are shown from **(a)**–**(d)**. The actuated flaps visible in **(b)**–**(d)** generate the necessary force to take one rolling step. The output tubing from the pump coils with each rolling step. *Red arrows* are augmented to better illustrate the motion

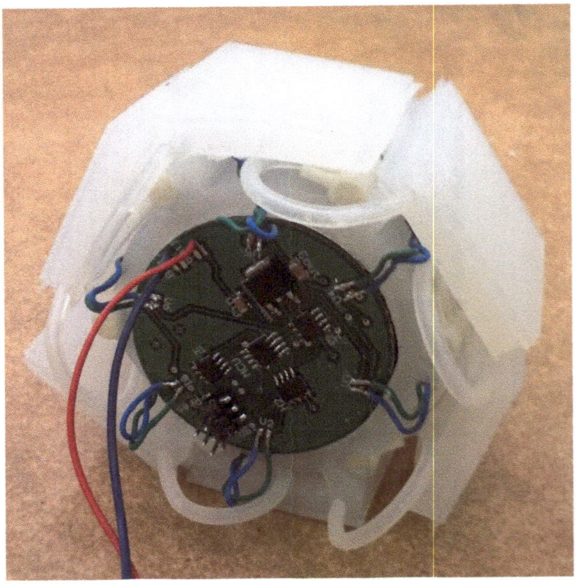

Fig. 11 Soft mobile roller prototype with on-board control electronics and embedded valves

6 Discussion

We discussed a soft robot that uses a class of pressure-operated soft actuators, tagged the fluidic elastomer actuators (FEAs). The soft mobile robot consisted of 6 FEAs that bend out and roll the body forward. We observed that an algorithm that actuates the FEAs in sequence induces rolling locomotion of such a soft robotic design.

We theoretically analyzed and experimentally studied the displacement response of bending-type FEAs. The resulting theory provided insight to FEA operation based on the geometry and material properties of the elastomer. Pressure generation was offloaded to a chemical process, namely the catalyzed decomposition of hydrogen peroxide. With a unique mechanical self-regulation mechanism, this chemical reaction is controlled to keep the pressure constant at a predefined value. This chemomechanical generator, called the pneumatic battery, powered FEAs with no electrical energy consumption as a proof-of-concept demonstration.

Silent and portable operation of this mechanism provides an important step towards the common application of soft fluidic actuators in functional devices. The cheap and fast fabrication of FEAs in addition to their inherent safety makes them useful in human interactions. Potential applications include artificial muscles [17], assistive or rehabilitative devices, haptic or tactile displays and interfaces. Such applications will benefit from a distributed arrangement of these actuators in arbitrary 3D shapes.

Silver oxidizes when exposed to air, which leads to catalyst degradation. Switching to an alternative catalyst such as platinum may be one solution. Also, it has been suggested that a high pH may help this reaction and tin becomes an effective catalyst in a basic solution [18]. A thorough investigation of the pressure build-up rate in the pump for different catalysts and pH values is necessary for optimized operation.

We haven't considered the temperature in the pump. Especially for high H_2O_2 concentrations, the temperature increase becomes large and affects the cut-off pressure value. An open air chamber would circumvent this problem, with a loss of safety from peroxide leakage.

While we demonstrated pressure generation in a practical size, some open questions remain to be answered. For example, can we further miniaturize this chemical pump such that it is embedded inside the elastomer substrate? Microfabrication technologies certainly offer the means to build scaled-down versions of the same design. The catalyst area, however, would be much smaller for a millimeter scale prototype. A distributed structure with many mini-pumps working in parallel may be helpful. On the other hand, a change in the current design to utilize a scaled-down Kipp generator to constitute the self-regulation mechanism in smaller scales seems promising.

Currently, commercial solenoid valves drive the actuators. We are interested in miniature valves that can be embedded in the actuator body itself to achieve a fully embedded actuation system that is directly addressable. Recent studies on micro-fabricated valves [10, 11] and microfluidic multiplexers [16] provide some options. We recently developed custom compact and energy-efficient valves controlled by electropermanent magnets to this end [9].

Similarly, we are interested in soft sensing elements that can be placed inside the FEAs. These sensors will provide feedback on the shape of the actuator and improve controllability [3].

Acknowledgements This work was done in the Distributed Robotics Laboratory at MIT with partial support from the DARPA DSO "Chembots" project (W911NF-08-C-0060). We are grateful to DARPA and to our collaborators.

References

1. A. Albu-Schaffer, O. Eiberger, M. Grebenstein, S. Haddadin, C. Ott, T. Wimbock, S. Wolf, G. Hirzinger, Soft robotics. IEEE Robot. Autom. Mag. **15**, 20–30 (2008)
2. Y. Bar-Cohen, *Electroactive Polymer EAP Actuators as Artificial Muscles: Reality, Potential and Challenges*, 2nd edn. (SPIE Press, 2004)
3. R.C. Chiechi, E.A. Weiss, M.D. Dickey, G.M. Whitesides, Eutectic gallium-indium (egain): a moldable liquid metal for electrical characterization of self-assembled monolayers. Angew. Chem. **47**, 142–144 (2008)
4. N. Correll, C.D. Onal, H. Liang, E. Schoenfeld, D. Rus, Soft autonomous materials using active elasticity and embedded distributed computation, in *12th Internatoinal Symposium on Experimental Robotics* (New Delhi, India, 2010)
5. Y. Fu, E.C. Harvey, M.K. Ghantasala, G.M. Spinks, Design, fabrication and testing of piezo electric polymer pvdf microactuators. Smart Mater. Struct. **15**(1), S141 (2006)

6. M. Goldfarb, E.J. Barth, M.A. Gogola, J.A. Wehrmeyer, Design and energetic characterization of a liquid-propellant-powered actuator for self-powered robots. IEEE/ASME Trans. Mechatron. **8**(2), 254–262 (2003)
7. H. Kazerooni, Design and analysis of pneumatic force generators for mobile robotic systems. IEEE/ASME Trans. Mechatron. **10**(4), 411–418 (2005)
8. C. Keplinger, M. Kaltenbrunner, N. Arnold, S. Bauer, Rntgens electrode-free elastomer actuators without electromechanical pull-in instability. PNAS **107**(10), 4505–4510 (2010)
9. A. Marchese, C.D. Onal, D. Rus, Soft robot actuators using energy-efficient valves controlled by electropermanent magnets, in *IEEE/RSJ International Conference on Intelligent Robots and Systems (IROS)* (2011) (to appear)
10. B. Mosadegh, C.H. Kuo, Y.C. Tung, Y.S. Torisawa, T. Bersano-Begey, H. Tavana, S. Takayama, Integrated elastomeric components for autonomous regulation of sequential and oscillatory flow switching in microfluidic devices. Nat. Phys. **6**(6), 433–437 (2010)
11. K.W. Oh, C.H. Ahn, A review of microvalves. J. Micromech. Microeng. **16**(5), R13 (2006)
12. R. Pelrine, R. Kornbluh, Q. Pei, J. Joseph, High-speed electrically actuated elastomers with strain greater than 100 %. Science **287**(5454), 836–839 (2000)
13. L. Qin, O. Vermesh, Q. Shi, J.R. Heath, Self-powered microfluidic chips for multiplexed protein assays from whole blood. Lab Chip **9**, 2016–2020 (2009)
14. A.W. Richards, G.M. Odegard, Constitutive modeling of electrostrictive polymers using a hyperelasticity-based approach. J. Appl. Mech. **77**(1), 014502 (2010)
15. I.A. Salem, M. El-Maazawi, A.B. Zaki, Kinetics and mechanisms of decomposition reaction of hydrogen peroxide in presence of metal complexes. Int. J. Chem. Kinet. **32**(11), 643–666 (2000)
16. T. Thorsen, S.J. Maerkl, S.R. Quake, Microfluidic large-scale integration. Science **298**(5593), 580–584 (2002)
17. D. Trivedi, C. Rahn, W. Kier, I. Walker, Soft robotics: biological inspiration, state of the art, and future research. Adv. Bionics Biomech. **5**(2), 99–117 (2008)
18. F. Vitale, D. Accoto, L. Turchetti, S. Indini, M.C. Annesini, E. Guglielmelli, Low temperature h2o2-powered actuators for biorobotics: thermodynamic and kinetic analysis, in *Proceedings IEEE International. Conference on Robotics and Automation* (2010), pp. 2197–2202
19. K. Wait, P. Jackson, L. Smoot, Self locomotion of a spherical rolling robot using a novel deformable pneumatic method, in *2010 IEEE International Conference on Robotics and Automation (ICRA)* (2010), pp. 3757–3762
20. Y. Wang, R.M. Hernandez, D.J. Bartlett, J.M. Bingham, T.R. Kline, A. Sen, T.E. Mallouk, Bipolar electrochemical mechanism for the propulsion of catalytic nanomotors in hydrogen peroxide solutions. Langmuir **22**, 10,451–10,456 (2006)
21. J.C. Whitehead, Hydrogen peroxide propulsion for smaller satellites, in *Proceedings of AIAA/USU Conference on Small Satellites* (1998)

Computational Human Model as Robot Technology

Yoshihiko Nakamura

Abstract The study of computational approach to human *understanding* has been the history of artificial intelligence. The robotics developments in algorithms and software have prepared the powerful research tools that were not available when the study of intelligence started from unembodied frameworks. The computational human model is a large field of research. The author and the colleagues have studied by focussing on behavioral modeling and anatomical modeling. The aims of study on human modeling are double faces of a coin. One side is to develop the technological foundation to predict human behaviors including utterance for robots communicating with the humans. The other side is to develop the quantitative methods to estimate the internal states of the humans. The former directly connected to the development of robotic applications in the aging societies. The latter finds fields of application in medicine, rehabilitation, pathology, gerontology, development, and sports science. This paper survey the recent research of the authors group on the anatomical approach to the computational human modeling.

1 Introduction

The study of artificial intelligence arose from a question "*how can one make machines understand things?*" [1]. *Understanding* shall be followed by prediction and then by action. *Understanding* is formally defined only in the relationship between two or more. One may think of a solitary machine on a planet in the outer space analyzing its environments to *understand*. Even in this case *understanding* is defined between the machine and who receives the information. Although the author hardly imagines *understanding* defined only in the relationship among artificial things and not inheriting anything defined in the relationship among the

Y. Nakamura (✉)
Department of Mechano-Informatics, The University of Tokyo,
7-3-1 Hongo, Bunkyo-ku, Tokyo 113-8656, Japan
e-mail: nakamura@ynl.t.u-tokyo.ac.jp

H.I. Christensen and O. Khatib (eds.), *Robotics Research*,
Springer Tracts in Advanced Robotics 100, DOI 10.1007/978-3-319-29363-9_31

human beings, it is still possible to say that it may be emergent among isolated artificial things, which is certainly a scientific matter of discussion.

The author's interests, however, are in *understanding* defined involving the human beings. Since the prediction and action of the others would be a function without which any animal species could have survived, we are talking about *understanding* as a communication function defined in the relationship among the human beings as an inheritance from the biological evolution. The basic principle of *understanding* would be finding the relationship of cause and effect. The animal species have spent a large resource for developing the function for survival. The human beings have developed the brain for the function with special focus on the social relationship. It is an author's scope that the way how we *understand* the nature and the environments is also based on the function to understand social relationship of cause and effect. In this sense, the function to *understanding* a human before oneself, namely predicting the human's future action and utterance is the baseline of the study.

The author and colleagues focus on the study of computational approach to human *understanding*. The aims of study are double faces of a coin. One side is to develop the technological foundation to predict human behaviors including utterance for robots communicating with the humans. The other side is to develop the quantitative methods to estimate the internal states of the humans. The former directly connected to the development of robotic applications in the aging societies. The latter finds fields of application in medicine, rehabilitation, pathology, gerontology, development, and sports science.

The components of research tools are common technologies developed in robotics kinematics, dynamics, and optimization. The anatomical structures of human body and the captured behavioral data provide the information to the phenomenological approach. The approach is based on computer simulation. Simon [2] raised in his book of 1969 an interesting question "*how can a simulation ever tell us anything that we do not already know?*" and answered in a affirmative way in particular on *simulation of poorly understood systems*. The data and the computational technologies of robotics allow the useful computational approach to the poorly understood problem on estimating the internal states of the human.

1.1 Computing from Behavior

The computational model from behavioral data was first being inspired by the coevolution theory of Deacon [3] and the Mimesis theory of Donald [4] where the mimetic acquisition is the model of symbol emergence. The symbolic system and the brain concurrently evolved in the coevolution theory by mutually interacting under the evolutional pressure for survival. The idea that the statistical modeling to allow the bidirectional computation of perception and action was also come to mind when we learned and excited about the Mirror neuron hypothesis [5, 6]. The unified bidirectional computation was so attractive as an architecture for the computational model of behavior system even it was still controversial in biological neuroscience.

Inamura et al. [7, 8] applied the Hidden Markov Model to develop proto-symbol system of the human whole-body motion. The pseud-distance measure was used to introduce topology into the set of proto-symbols. Takano et al. [9] demonstrated the human-robot communication in realtime using the proto-symbol system for both perception and action engines and a motion capturing system. The demonstration was shown at EXPO2005 in the style of a fighting event of a human and a small-size humanoid robot. Unsupervised automatic segmentation and clustering were studied by [10–12].

The motion prediction based on the symbolic and structured database was developed and demonstrated as "*the Crystal Ball*" [13]. The integration of the motion symbol system with the natural language system has been studied by Takano et al. [14–16] based on the common mathematical framework of statistics.

1.2 Computing from Anatomy

Coordination of human motion is a central paradigm of neuroscience [17, 18]. For the computational modeling of human musculoskeletal system there is rich literatures in biomechanics [19–21]. We developed the whole body musculoskeletal model for applying robotic algorithms of kinematics, dynamics, and optimization [22]. The study has been extended by integrating spinal nerves and mathematical model of neural connectivity in the spine. The musculoskeletal computational model has been applied to various professional subjects such as athletes, a dram player, a singer and so on. The neuromuscular model has been focussed to construct the reflex model of such as spinal reflex and cutaneous reflex.

In the following sections the development of the computational model of human from anatomical knowledge is to be explained in details.

2 Musculoskeletal Model with Grouped Muscles [23]

The musculoskeletal human model consists of a *musculo-tendon network* and a *skeleton*. The skeleton is a set of rigid links connected by mechanical joints, while the musculo-tendon network is composed of the elements to drive and/or constrain the bones including muscles, tendons, ligaments, and cartilages. The model used in this paper is basically the same as the one used in [22] but contains the following improvements:

- Many small muscles around the spine are added to reduce the joint torque errors observed in [22]. The number of muscles grew from 366 to 989.
- The muscles are grouped according to their role (for example, extend the left knee), which will be used in the inverse dynamics computation later.

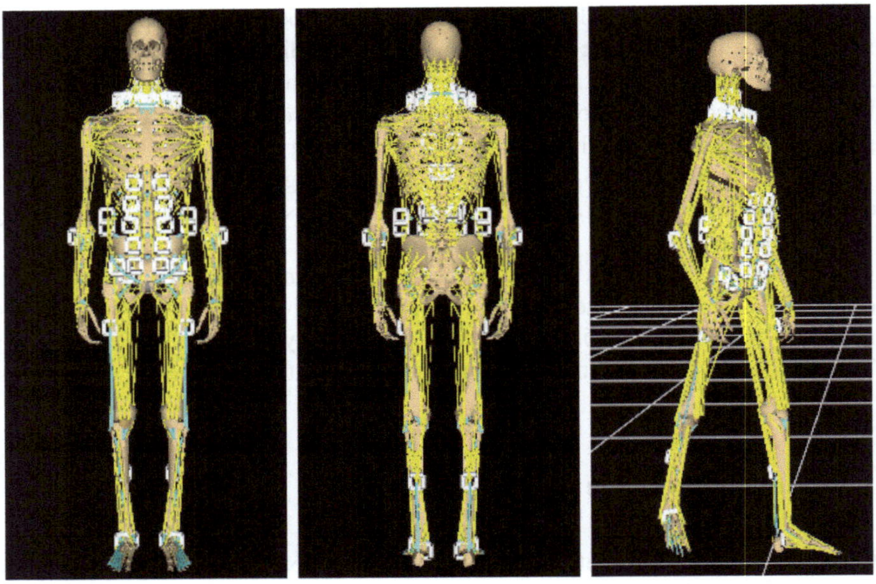

Fig. 1 The musculoskeletal human model used in this paper. *Left* front view, *center* back view, *right* a snapshot from a captured walk motion

Figure 1 shows the new model and Table 1 summarizes the complexity of the model used in this paper.

In [22] we proposed an inverse dynamics algorithm for the musculoskeletal human model, which computes the ground contact forces and muscle tensions by solving the following equation:

$$\tau_G = J^T f + J_C^T \tau_C \tag{1}$$

Table 1 Complexity of the musculoskeletal model

Musculo-tendon network	
Muscles	989
Tendons	50
Ligaments	117
Cartilages	34
Total wires	1190
Muscle groups	78
Virtual links	72
Skeleton	
Bones	200
Bone groups	53
Total DOF	155

Table 2 Neural transmission speed of each nerve fiber [33, 34]

Type of fiber	Transmission speed (m/s)
α motor fiber	100
Ia fiber	75
Ib fiber	75
II fiber	55

where

τ_G generalized forces
J Jacobian matrix of the wire length w.r.t. the generalized coordinates
f wire tensions
J_C Jacobian matrix of the contact points w.r.t. the generalized coordinates
τ_C ground contact forces.

The solution consists of the following two steps:

1. Compute the ground contact forces τ_C by only considering the rows of Eq. (1) corresponding to the 6 DOF of the hip joint. We apply quadratic programming to solve the optimization problem (Table 2).
2. Eliminate τ_C from Eq. (1):

$$\tau'_G = \tau_G - J_C^T \tau_C = J^T f. \quad (2)$$

and compute the muscle tensions by solving another optimization by either linear programming (LP) or quadratic programming (QP).

In this paper, we adopt the same method as [22] for step (1). The rest of this section describes our improvement of the algorithm for step (2).

In [22] we found that LP solves our complex optimization problem much faster than QP. However, the inherent problem of LP is that the resulting muscle tensions can be both temporally and spatially discontinuous. *Spatial discontinuity* implies that the tensions of geometrically close muscles can be completely different, which is not likely to happen in human body.

We solve this problem by considering a measure of variation of tensions of the muscles in each group. The new LP formulation is summarized as follows:

For constant vectors with positive components a_τ, a_f, and a_m, find δ_τ, δ_f, δ_m, and f that minimize

$$Z = a_\tau^T \delta_\tau + a_f^T \delta_f + a_m^T \delta_m \quad (3)$$

subject to

$$-\delta_\tau \leq \tau'_G - J^T f \leq \delta_\tau \quad (4)$$

$$\delta_\tau \geq 0 \quad (5)$$

$$-\delta_f \leq f - f^* \leq \delta_f \tag{6}$$

$$\delta_f \geq 0 \tag{7}$$

$$-f_{max} \leq f \leq 0 \tag{8}$$

$$-\delta_m \leq E_G f \leq \delta_m \tag{9}$$

$$\delta_m \geq 0 \tag{10}$$

where the third term of Eq. (3) has been added to consider the muscle tension variation. The detailed description of the equations will be given in the subsequent paragraphs.

The first term of Eqs. (3), (4) and (5) try to minimize the error of Eq. (2) to assure the physical validity of the result. Instead of including Eq. (2) as an equality condition, we relax the problem because Eq. (2) may not have an exact solution. By including the term $a^T\delta_\tau$ in the objective function Eq. (3), we can obtain the minimum possible values of for elements of δ_τ, which are constrained to be positive by the inequality condition Eq. (5). On the other hand, Eq. (4) ensures that the error of Eq. (3) is smaller than δ_τ. By combining these constraints, we can minimize the error of Eq. (3).

The second term of Eqs. (3), (6) and (7) tries to bias f towards a given desired wire tension vector f^*. The user can, for example, specify the relationship between a pair of flexor and extensor muscles by supplying appropriate values to f^*. For example, this information could be obtained by electromyograph (EMG) measurements for biomechanical applications [24]. Simply setting $f^* = 0$ gives the minimum wire tensions.

Equation (8) represents the upper and lower bounds of the wire tensions where $f_{max} \geq 0$ is the vector of maximum muscle tensions. The elements of f_{max} can be selected independently for each muscle. We may also consider the muscle length and its velocity to compute f_{max} by Hill's muscle model [25].

Finally, the third term of Eqs. (3), (9) and (10) are included to equalize as much as possible the tensions of muscles in the same group. In quadratic programming, this term could be replaced by adding squared sum of wire tensions to the evaluation function. Suppose group m includes n_m muscles and $\mathcal{G}_m$ denote the set of their indices. The average tension of group m is computed by

$$\bar{f}_m = \frac{1}{n_m} \sum_{k \in \mathcal{G}_m} f_k \tag{11}$$

where f_k is the tension of the k-th muscle. The difference between the average tension and the k-th $(k \in \mathcal{G}_m)$ muscle's tension is

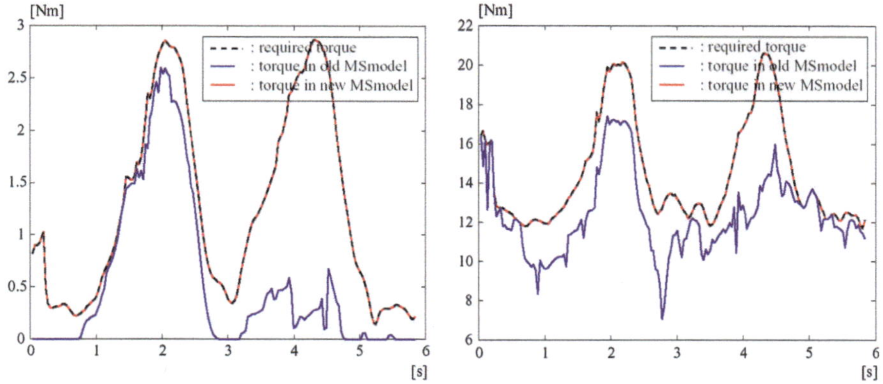

Fig. 2 The joint torques of the third neck vertebra (*left*) and the sixth rib vertebra (*right*). *Black dashed* result of Newton-Euler inverse dynamics computation; *blue* computed from the muscle tensions obtained by the ungrouped model; *red* computed from the muscle tensions obtained by the grouped model

$$\Delta f_{mk} = \bar{f}_m - f_k = E_{Gmk} f \tag{12}$$

where E_{Gmk} is a row vector whose i-th element is $(1 - n_m)/n_m$ for $i = k$, $1/n_m$ for $i \in \mathcal{G}_m$ and $i \neq k$, and 0 otherwise. By collecting $E_{Gmk}(k \in \mathcal{G}_m)$ for all groups and stacking them vertically, we obtain the matrix E_G in Eq. (9).

We verified the effect of the improvements by performing inverse dynamics computation for a motion in which the subject bends the spine back and forth. Figure 2 is the results of inverse dynamics computations using the model in [22] and the improved one. In each graph, the dashed black line represents the required joint torque obtained by Newton-Euler inverse dynamics computation, while the blue and red lines are the joint torques computed from the muscle tensions obtained by the old and new models, respectively. The new model yields much more precise and smooth muscle tensions.

3 Independent Component Analysis of Muscle Tensions [26]

We apply a statistical method called independent component analysis (ICA) [27] to analyze the muscle tension data. This method estimates the original signals underlying multi-dimensional time sequence data, assuming that the data are represented as a linear mixture of nongaussian and mutually independent signals. ICA tries to compute the different signal sources, while Principal Component Analysis (PCA) tries to obtain the common sources.

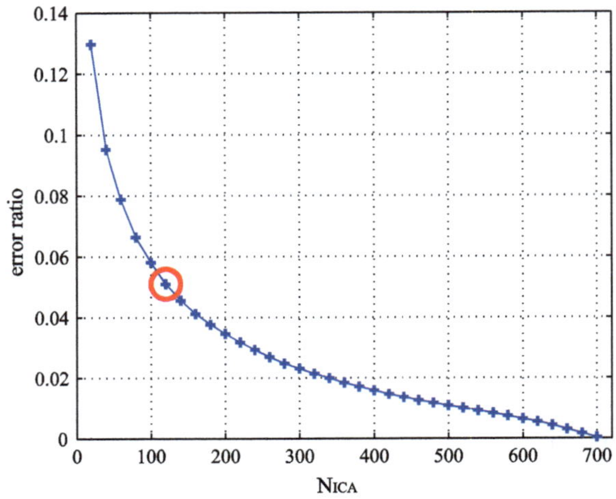

Fig. 3 Variation of error between original and reconstructed muscle tensions with number of independent components

We compute a constant matrix $W_{ICA} \in R^{N_w \times N_{ICA}}$, which maps the vector of independent signals $s \in R^{N_{ICA}}$ to the computed muscle tensions $f \in R^{Nw}$:

$$f = W_{ICA} s \tag{13}$$

where N_w is the number of muscles included in the musculoskeletal model ($N_w = 989$ in our model), N_{ICA} ($N_{ICA} < N_w$) is the manually-selected number of independent component sources.

We apply ICA to the muscle tension data of 989 wires computed for 1000 frames (10 s) of motion capture data. Figure 3 shows the errors between the original and reconstructed muscle tensions for various choices of N_{ICA}.

According to the anatomical knowledge, all the muscles in the human body are controlled by 124 left/right and anterior/posterior rami of the spinal nerve. The 989 muscles in our model are connected to 120 of these rami. We, therefore, verify the validity of applying ICA with $N_{ICA} = 120$. Figure 3 shows that the error slowly increases for smaller N_{ICA} until $N_{ICA} = 120$, and the error is 5.11 % at $N_{ICA} = 120$. The error increases rapidly for $N_{ICA} < 120$. Figure 4 compares the original and reconstructed muscle tensions with $N_{ICA} = 120$. Figure 5 represents the average and variance of the errors between original and reconstructed tensions of all muscles. These results indicate that we can represent the 989 tensions by as few as 120 independent components. Based on these results and facts, we choose $N_{ICA} = 120$ in the rest of the paper.

We next confirm that the independent components and spinal neural signals have one-to-one correspondence to show the possibility of using the independent components s as the spinal neural signals. Let us define a matrix $C \in R^{N_{ICA} \times N_w}$, whose ($i$, j)-th element is 1 if i-th ramus and j-th muscle are connected and 0 otherwise. Then we calculate

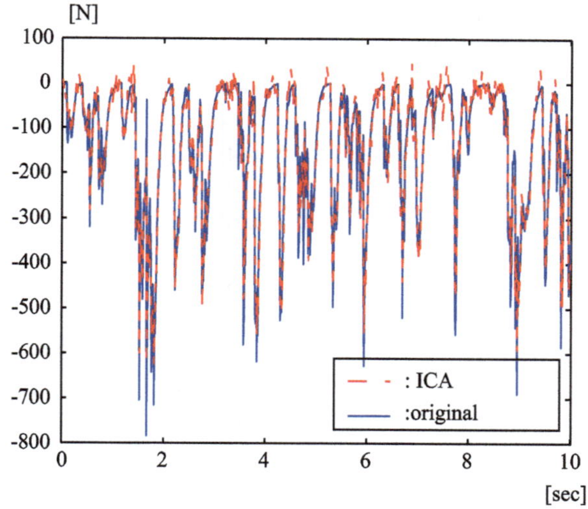

Fig. 4 Original and reconstructed muscle (Soleus) tensions. *Blue line* original muscle tensions; *Dashed red line* muscle tensions reconstructed by ICA

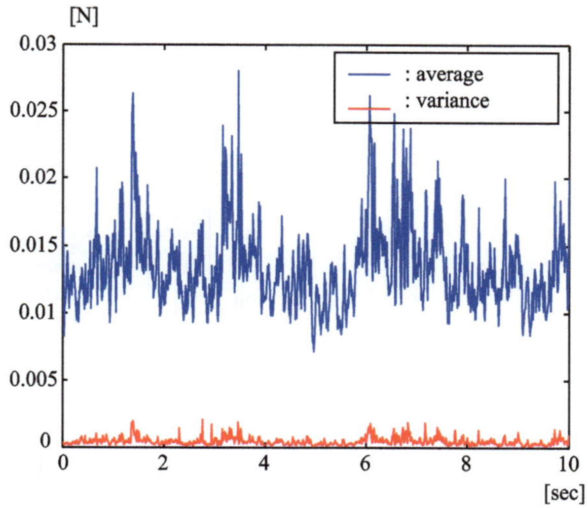

Fig. 5 Average and variance of error between original and reconstructed tensions of all wires

$$P = CW_{ICA} \tag{14}$$

where the (*i*, *j*)-th element of $P \in R^{N_{ICA} \times N_{ICA}}$ represents the sum of elements of W_{ICA} corresponding to the *j*-th independent components and muscles connected to the *i*-th ramus. If there is one-to-one correspondence between the independent components and rami, each row and column of *P* would have only one outstanding peak. Figure 6 shows the value of 5 representative rows of *P*, where the horizontal axis represents the column indices. It is obvious from the graph that most of the lines (the rows of *P*) have only one outstanding peak at different columns. Figure 7 marks the column index of the element where each row of *P* has its peak, in which

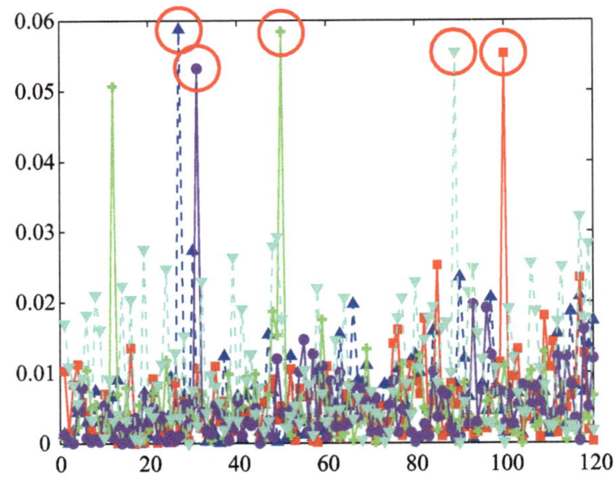

Fig. 6 Values of selected rows of *P* displayed in the order of column

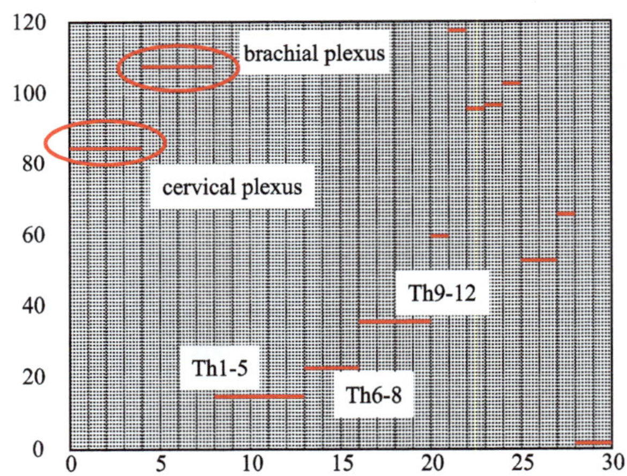

Fig. 7 The location of maximum column in each row of *P*

the horizontal axis represents the indices of the rami and the vertical axis represents the indices of the independent components. In this case, the rami marked by an oval in the figure which correspond to the neuroplexuses share the same independent component. Each group of rami Th1–Th5, Th6–Th8, and Th9–Th12 also shares the same independent component. We suspect that the reason for these duplications of independent components is that the walk motion used in this experiment did not activate the muscles connected to these rami, which are abdominal muscles. Except for these rami, we can observe that almost each ramus corresponds to a different independent component.

4 Reflex Model and Time Delay Identification [28]

The nerve network is composed of the following elements: α motor neurons, γ motor neurons, and interneurons, the efferent pathway, muscle, muscle spindle, the Golgi tendon organ, and the afferent pathway. Figure 8 shows our neuromuscular network model. The model is constructed as a six-layered neural network, which represents:

1. $N_{NJ,i}(i = 1,\ldots, n_m)$ (filled circle): Neuromuscular junctions on the muscles, where n_m represents the total number of muscles. This layer receives and integrates the motion command signal from the α motor neuron in the spinal nerve ramus. The integrated signal activates the muscle and produces tension.
2. $N_{MS,i}(i = 1,\ldots, n_m)$ (filled square): Muscle spindles that measure the muscle length and its velocity. The values are computed by forward or inverse kinematics.
3. $N_{GT,i}(i = 1,\ldots, n_m)$ (filled triangle): The Golgi tendon organs that measure the muscle tensions. The values can be computed from the muscle activity (the motion command signal) using the Hill-Stroeve muscle model [29, 30] or from the inverse kinematics and dynamics.

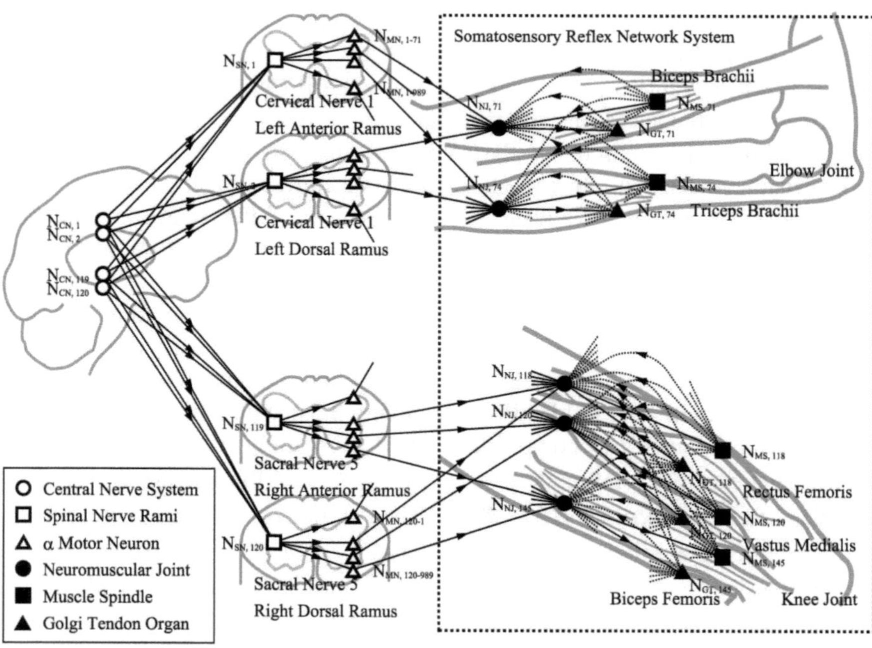

Fig. 8 The neuromuscular network modeled with 6 layered neural network. Each layer represents central nerve system, spinal nerve rami, α motor neuron, neuromuscular joint, muscle spindle and Golgi tendon organ. The part enclosed by *dashed rectangle* represents the somatosensory reflex network model

These layers are connected to each other as follows: *solid lines* from N_{NJ} to N_{MS} and N_{GT}, *dashed lines* from N_{MS} and N_{GT} to N_{MS}.

The weight parameters of the somatosensory reflex model are identified using experimental human motion data. The muscle length, velocity and tension are computed using the inverse kinematics and dynamics computations [31], and the muscle activity is computed using the physiological muscle model [29, 30]. We train this network model so that it outputs the computed muscle activity at N_{NJ} when the somatosensory information is fed back to N_{MS} and N_{GT}.

The reflex arc consists of the proprioceptive sensory receptors, the Ia and II nerve fibers, the interneurons and α motor neuron, and the α motor fiber. The total time delay in somatosensory reflex therefore consists of:

- Signal transmission by the Ia/II fiber and the α motor fiber.
- Time between the beginning of muscle extension to the beginning of muscle spindle actual potential discharge.
- Synaptic transmission from the Ia/II fiber to the α motor neuron or interneuron.
- Signal transmission from the end plate to the muscle fiber.
- Diffusion of action potential along the muscle fiver.
- Induction of muscle contraction by the action potential (excitation-contraction coupling).

The time delay for a particular muscle can be estimated from physiological properties. The delay (1) (δT) can be estimated by dividing the length of the fiber between a spinal nerve ramus and muscle [32] by the neural signal transmission speed shown in Table 3 [33, 34]. The delays (2)–(6) (δt) have been investigated experimentally. In Quadriceps, for example, the time delay caused by (1) is 16 ms,[1] and the time delay caused by (2)–(6) is 9–14 ms. So the total time delay of monosynaptic extension reflex of Quadriceps is therefore 25–30 ms.[2]

We identify and cross validate the somatosensory reflex network with some different time-delay condition $\delta T' = \delta T + \delta t(\delta t = 0, 5, 10, 15, 30, 60\text{ ms})$ to confirm this.

Our hypothesis is that the nature and evolution would have formed human motions that best consistent with the muscular dynamics and the neuro physiological constraints. The hypothesis may allow to find the time delay searching for the most accurate results in identification among those with various time delays.

The inverse kinematics computation based on a n_{DOF} (= 143)-DOF skeleton model calculates the joint angle data $\theta \in \mathscr{R}^{n_{DOF} \times T}$, and the lengths of n_m(= 989) muscles and their velocities l, $\dot{l} \in \mathscr{R}^{n_m \times T}$. Then an inverse dynamics calculation is carried out to obtain the generalized force data $\tau_G \in \mathscr{R}^{n_{DOF} \times T}$ and we estimate the muscle tensions $f \in \mathscr{R}^{n_m \times T}$ using a biological muscle model and mathematical optimization. Finally the muscle activity $a \in \mathscr{R}^{n_m \times T}$ is computed based on the

[1]The distance between Quadriceps and spinal nerve ramus is 800 mm.

[2]This delay is often observed as the latency of knee-jerk reflex.

Table 3 Mapping between EMG electrode and muscle

# of channel	Name of muscle
ch01/09	Right/left rectus femoris
ch02/10	Right/left vastus lateralis
ch03/11	Right/left tibialis anterior
ch04/12	Right/left gluteus maximus os
ch05/13	Right/left biceps femoris caput longum
ch06/14	Right/left biceps femoris caput breve
ch07/15	Right/left gastrocnemius
ch08/16	Right/left soleus

biological muscle model that represents the relationship between muscle tension, activity, length, and its velocity [29, 30].

We estimate the sensor activities from those physical quantities. The somatosensory information of the i-th muscle associated with the somatosensory reflex are m_i, the activity of muscle spindle, and g_i, the activity of Golgi tendon organ. The former feeds back the information of muscle length and its velocity, and the latter feeds back the muscle tension as follows:

$$m_i(t) = 4.3\dot{l}_i(t)^{0.6} + 2l_i(t) + \delta m_i \ (t = 1, \ldots, T) \quad (15)$$

$$g_i(t) = f_i(t). \quad (16)$$

Equation (15) represents muscle spindle model proposed by Prochazka and Gorassini [35] that considers the discharge rate of the Ia nerve fiber. a_i, m_i, $g_i \in \mathscr{R}^T (i = 1, \ldots, n_m)$ are normalized to $[0 - 1]$.

Then we identify the parameters of the somatosensory reflex model. First, the somatosensory information fed back by the proprioceptive sensory receptors, m_{ref}, $g_{ref} \in \mathscr{R}^{n_{SN}n_m^2}$, are computed considering the time delay of nerve signal transmission. The reflex arc between muscles goes through one or more spinal rami, and we consider them separately. If the i-th muscle and j-th muscle are connected via the k-th spinal nerve ramus:

$$\begin{aligned} m_{ref}(n_{SN}n_m(i-1) + n_{SN}(k-1) + j)(t) \\ = m_j(t - \delta T'_{i,j,k}) \end{aligned} \quad (17)$$

$$\begin{aligned} g_{ref}(n_{SN}n_m(i-1) + n_{SN}(k-1) + j)(t) \\ = g_j(t - \delta T'_{i,j,k}) \end{aligned} \quad (18)$$

where $\delta T'_{i,j,k}$ is the time delay caused by the nerve signal transmission from i-th muscle to j-th muscle via k-th spinal nerve ramus computed from the nerve length and neural transmission speed. Then we use the simple back propagation to obtain the weight parameters $W_{ref,m}$ and $W_{ref,g} (\in \mathscr{R}^{n_m \times n_{SN}n_m^2})$ that satisfy:

$$a(t) = \sigma(W_{ref,m}\, m_{ref}(t) + W_{ref,g}\, g_{ref}(t)) \quad (19)$$

where $\sigma(*)$ is the sigmoid function:

$$\sigma(x) = 2\left(\frac{1}{1+e^{-x}} - \frac{1}{2}\right). \quad (20)$$

We consider the anatomical neuronal binding between the spinal nerve ramus and the muscles [36, 37] as the constraints of these weight matrices. If the *i*-th muscle and *j*-th muscle are not anatomically connected via the *k*-th spinal nerve rams, $(i,\ n_{SN} n_m(i-1) + n_{SN}(k-1) + j)$-th elements of $W_{ref,m}$ and $W_{ref,g}$ are constrained to be 0. The convergence calculation continues until the residue at the muscle activity:

$$\delta a(t) = a(t) - \sigma(W_{ref,m}\, m_{ref}(t) + W_{ref,g}\, g_{ref}(t)) \quad (21)$$

becomes sufficiently small. The following three types of motions are measured for the analysis:

1. Step motion in 100 step/min by Subject A ($DATA_{100}$).
2. Step motion in 170 step/min by Subject A ($DATA_{170}$).
3. Step motion with changing its speed from 120 step/min to 150 step/min in 6 s by Subject A ($DATA_{120-150}$).
4. Jump motion by Subject B ($DATA_{jump}$).
5. Squat motion by Subject B ($DATA_{squat}$).

The speed of stepping is controlled with a metronome.

First, we train the model with seven different time delays using the motion data $DATA_{120-150}$. Figure 9 shows the result of identification of the somatosensory reflex network model. The standard back-propagation algorithm [38] is carried out for the training where the learning rate is 0.01, forgetting rate is 0.001, and the number of iteration is 1000. The horizontal axis represents time [sec] and vertical axis represents the normalized activity of right Vastus Lateralis. The top and bottom graph show the first and last 2.5 s of $DATA_{120-150}$ respectively. The black dashed line represents the muscle activity computed using the musculoskeletal model, and the solid lines represent the reconstructed activity using the identified somatosensory reflex networks with different time delays as shown in the figure.

Then we apply these somatosensory reflex network models to the other motion data $DATA_{100}$, $DATA_{170}$, $DATA_{jump}$, and $DATA_{squat}$ for cross validation. The cross validation is performed for each of the seven time delays and the resulting errors are evaluated.

Figure 10 show the results of the cross validation. In each graph, the horizontal axis represents time [sec] and vertical axis represents the normalized activity of right Rectus Femoris. The top graph shows the result of $DATA_{jump}$ and bottom graph shows the result of $DATA_{squat}$ in Fig. 10. The line types and colors are same as in Fig. 9.

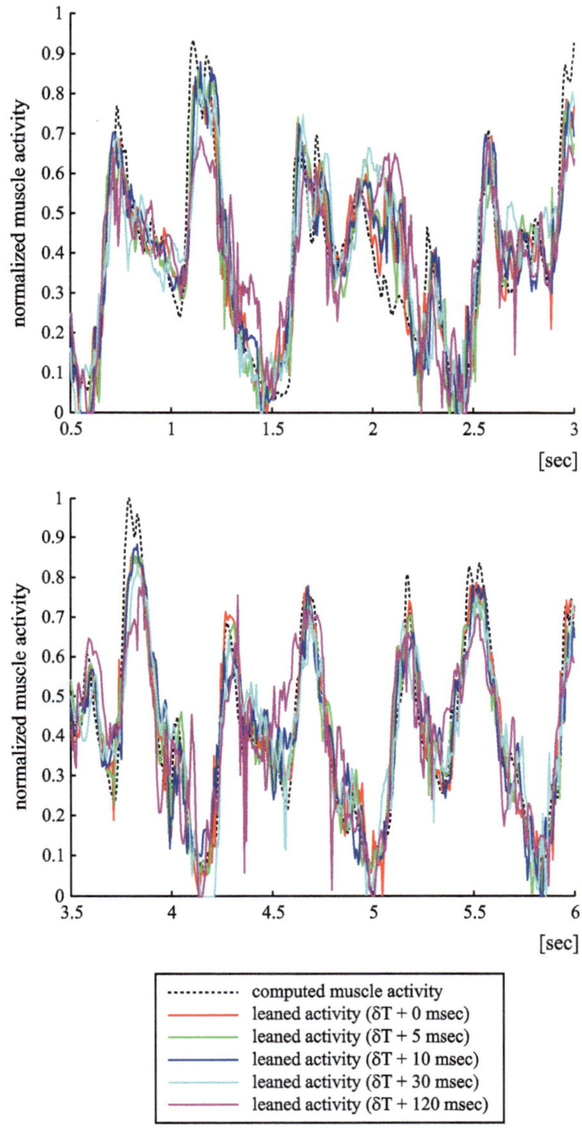

Fig. 9 The computed activity of right Vastus Lateralis and reconstructed data for $DATA_{120-150}$ using the identified neuromusculoskeletal system model. *Black dashed line* computed muscle activity, *red solid line* reconstructed data by somatosensory reflex system whose time delay is $\delta T + 0$ ms, *green solid line* $\delta T + 5$ ms, *blue solid line* $\delta T + 10$ ms, *cyano solid line* $\delta T + 30$ ms, *magenta solid line* $\delta T + 120$ ms. *Top* first part of stepping motion, *bottom* last part of stepping motion

Figure 11 summarizes the results. The horizontal axis represents the offset of time delay, and the vertical axis represents the average error of the lower-body normalize muscle activity and its standard deviation. The black dashed line is the result of $DATA_{120-150}$, the red solid line is the result of $DATA_{100}$, the green solid line is the result of $DATA_{170}$, the blue solid line is the result of $DATA_{jump}$, and the cyano solid line is the result of $DATA_{squat}$.

The result of identification shows that the somatosensory reflex network models can learn the muscle activity pattern with only 2–4 % error. The difference of

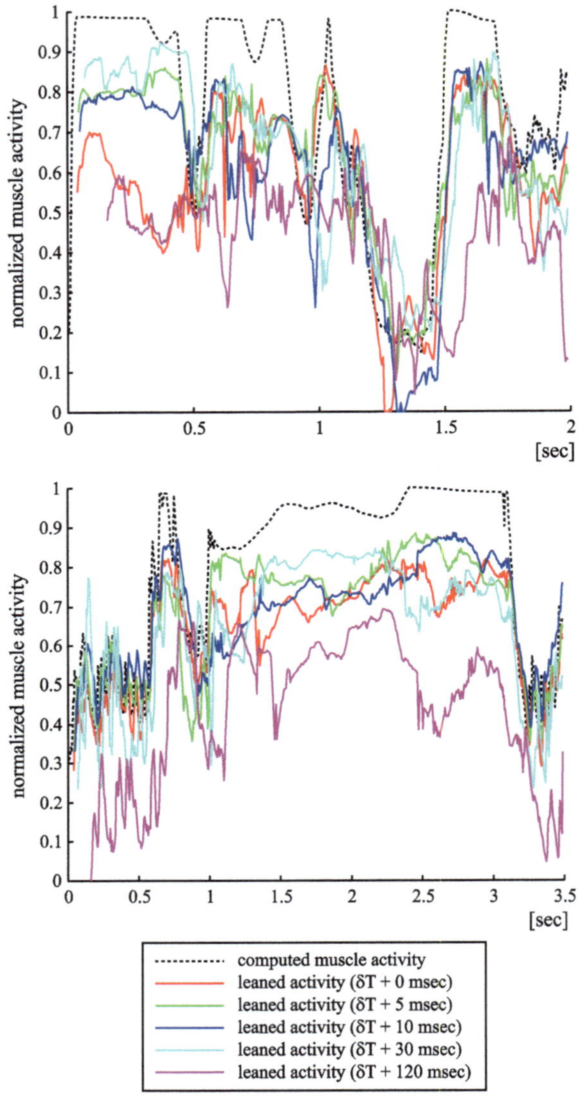

Fig. 10 The computed activity of right Rectus Femoris and reconstructed data for $DATA_{jump}$ and $DATA_{squat}$ using the identified neuromusculoskeletal system model. The parameter of somatosensory reflex is identified using $DATA_{120-150}$. The color of lines are same as in Fig. 9 *Top* $DATA_{jump}$, *bottom* $DATA_{squat}$

time-delay offset has little impact on the result of identification if the offset is less than 30 ms. If the time delay offset is greater than 60 ms, the wave shape of reconstructed muscle activity becomes inaccurate in both timing and amplitude.

The cross validations using $DATA_{100}$ and $DATA_{170}$ show that the identified somatosensory reflex network model can estimate the muscle activity with 2 % error when the time-delay offset is less than 10 ms. The difference due to the time-delay offset is more significant than at the identification. Both $DATA_{100}$ and $DATA_{170}$ show the minimum error when the time-delay offset is 10 ms. In particular, the simulated muscle activity with 10 ms offset is more accurate at the peaks

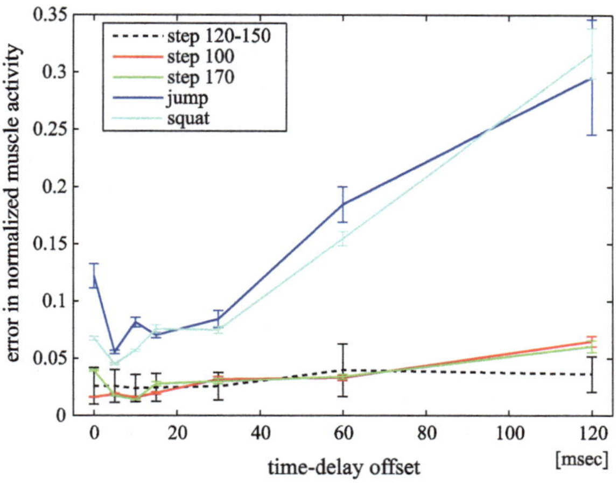

Fig. 11 Errors and their variances between computed and estimated muscle activities. *Black dashed line* $DATA_{120-150}$, *red solid line* $DATA_{100}$, *green solid line* $DATA_{170}$, *blue solid line* $DATA_{jump}$, *cyano solid line* $DATA_{squat}$

of the wave form than the others. This time delay is within the range of measured delay of knee-jerk reflex (between $\delta T + 9$ and $\delta T + 14$ ms). The precision of the reconstruction is somewhat surprising because the novel stepping motions are much slower or faster than the learned motion, and the muscle usage, especially the co-contraction pattern, is likely to change depending on the speed. Our result suggests that humans apply similar control strategies for a wide range of instances of the same behavior.

The result of cross validations using the motion data $DATA_{jump}$, $DATA_{squat}$ show that the identified somatosensory reflex network model can estimate the muscle activity with 5 % error when the time-delay offset is 5 ms. These motions are significantly different from stepping, so the usage of synergist and antagonist muscles and the pattern of co-contraction must be entirely different. For example, the co-contraction of the muscles around knee joint at the preparatory phase of jumping motion is not included in the normal stepping motion, and it can be computed only with the measured EMG. The reconstructed activities of Rectus Femoris and Vastus Intermedius have impulsive shapes that are same as those seen in the activities computed using the dynamics computation and optimization,[3] if the offset of time delay is appropriately selected. The difference due to the time-delay offset is much more significant than the cross validations with $DATA_{100}$ and $DATA_{170}$.

The comparison between the results of cross validation using $DATA_{100}$, $DATA_{170}$, and $DATA_{jump}$, $DATA_{squat}$ shows the interesting results about the offset of time delay. Figure 11 shows that the error changes particularly in $DATA_{jump}$ and $DATA_{squat}$, though it does not change so much in $DATA_{100}$ and $DATA_{170}$. From statistical point of view, the error does not change so much if the patterns of data used for the identification and cross validation are similar (e.g. between $DATA_{120}$

[3]We use the measured EMG to estimate the activity of Rectus Femoris, so its co-contraction during the jump and squat motions appears in the computed activities.

$_{-150}$, $DATA_{100}$, and $DATA_{170}$), or there is no relationship between the somatosensory information and muscle activity. The phenomenon that the relation between the time-delay offset and the error of cross validation is not ever-increasing or -decreasing and has a minimum value suggests that the reflex system is optimized for a particular time delay. Furthermore, this length of time delay roughly matches the value estimated from the anatomical structure and geometry of the human body.

The result that the somatosensory reflex network model identified using the stepping motion can be generalized to jumping and squat motions suggests that there is a possibility that the human body can generate whole-body motions only from simple motion command signals. In this model, muscle activity patterns can be generated by giving the first few frames of muscle length, its velocity and tension. This information will work as the trigger for the somatosensory reflex network model to generate the muscle activity of the rest of motion.

5 Dermatome and Cutaneous Reflex Model [39]

The skin is clearly segmented by the nerves, which is known as dermatome. The most part of body except for the face is segmented by the spinal nerves. One would not be surprised if the segment of skin and the group of muscles which are associated with the same or closer spinal nerve would have more mutual synaptic connections. The reflex by cutaneous sensation can be studied based on the anatomical kinematics and dynamics like the spinal reflex among skeletal muscles as we have seen in the previous section. The dermatome segmented by spinal nerves are illustrated by Fig. 12 as shown in the reference of anatomy [40].

We made preliminary experiments to study the cutaneous reflex by remotely generating cutaneous sensation using small vibrating devices. The devices were attached on each side of legs at the dorsum of foot, heel, shank, posterior surface of knee, and anterior and posterior surfaces of thigh. Table 4 shows the correspondence of the positions of vibrating devices and the spinal nerves.

The vibrating devices of 315 Hz and 3G was made by using the vibrating motors commonly used for mobile phones. The twelve devices were attached. Only one of them vibrated for 0.9 s at a time without letting the subject know of the location and the start timing. The vibrating sensation by the ith device among $n_s(= 12)$ can be represented by a step function of s_i as follows (Fig. 13):

$$s_i = u(t) \tag{22}$$

$$u(t) = \begin{cases} 1 & \text{if stimulation,} \\ 0 & \text{otherwise.} \end{cases} \tag{23}$$

The cutaneous reflex model was constructed by integrating the cutaneous sensation to the spinal reflex model of skeletal muscles discussed in the previous

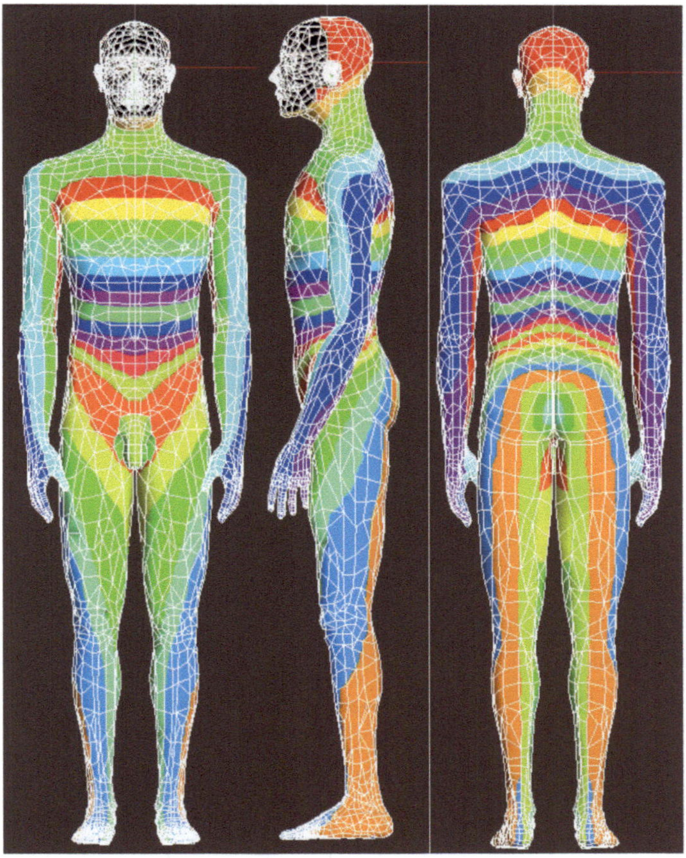

Fig. 12 3D colored dermatome model

Table 4 The connection between skin and spinal nerves

Part of skin	Spinal nerves
Dorsum of foot	L5
Heel	L5, S2
Shank	L4
Posterior surface of knee	L3, S2
Anterior surface of thigh	L3
Posterior surface of thigh	S2

section as seen in Fig. 14. The learnt neural network for the spinal reflex of skeletal muscles were used and the new nodes and arcs of neural network were added to accommodate the cutaneous reflex.

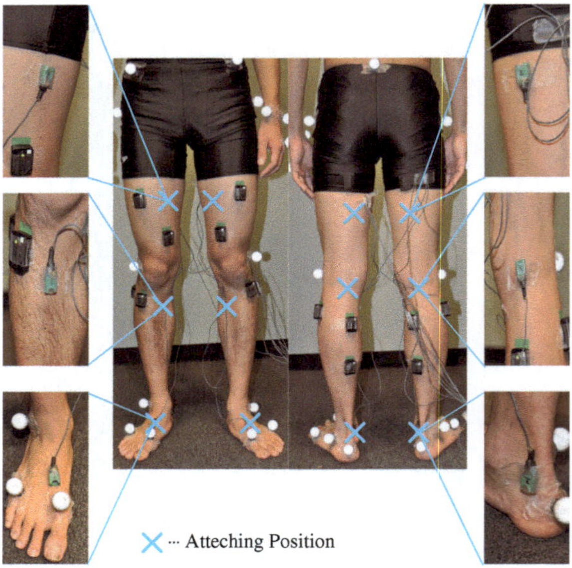

Fig. 13 Atteching position. The vibelation motors are atteched on the dorsum of foot, heel, shank, posterior knee, anterior thigh and posterior thigh

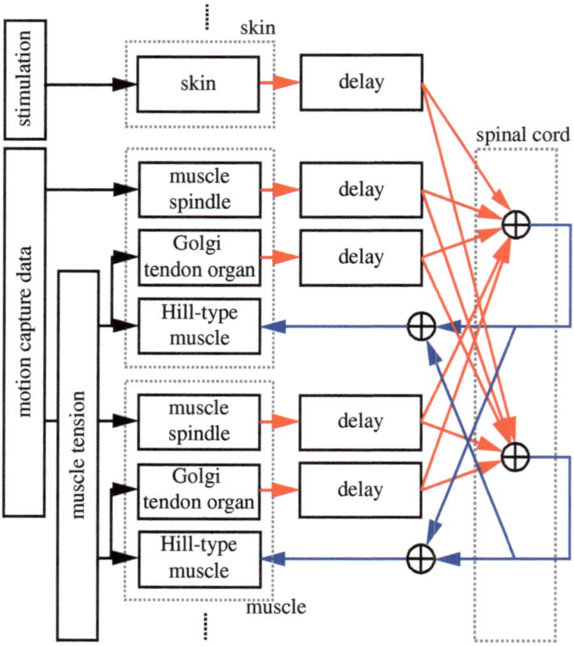

Fig. 14 The block diagram of neural network model

The proprioceptive sensation m_{ref} and g_{ref} and the cutaneous sensation s_{ref} of the ith muscle or skin segment associated with the jth muscle through the kth interneuron in the spine are represented by

$$m_{ref}(t) = m_j(t - \delta t_{i,j,k}) \quad (24)$$

$$g_{ref}(t) = g_j(t - \delta t_{i,j,k}) \quad (25)$$

$$s_{ref}(t) = s_i(t - \delta t_{i,j,k}) \quad (26)$$

where $m_i(t)$ is the activity of muscle spindle of the ith muscle at time t, $g_i(t)$ is the activity of the Golgi tendon organ of the ith muscle at time t, and $s_i(t)$ is the sensation of the ith skin segment at time t. $\delta t_{i,\ j,k}$ implies the time delay when the ith muscle or skin segment is connected to the jth muscle through the kth interneuron in the spine and is determined by length of neural pathway and the transmission rate.

The weight matrices $\mathbf{W}_m$, $\mathbf{W}_g$, $\mathbf{W}_s$ are acquired by the back-propagation learning.

$$\hat{a}(t) = \sigma(\mathbf{W}_m\, m_{ref}(t) + \mathbf{W}_g\, g_{ref}(t) + \mathbf{W}_s\, s_{ref}(t)) \quad (27)$$

where sigmoid function $\sigma(*)$ is represented by

$$\sigma(x) = 2(\frac{1}{1+e^{-x}} - \frac{1}{2}) \quad (28)$$

The iterative computation continues to update and determine the weight matrix $\mathbf{W}_s^*$ by minimizing $\delta a(t)$ given by

$$\delta\hat{a}(t) = a(t) - \sigma(\mathbf{W}_m\, m_{ref}(t) + \mathbf{W}_g\, g_{ref}(t) + \mathbf{W}_s^* s_{ref}(t)) \quad (29)$$

where muscle activity $a(t)$ is the reference signal computed by EMG and the inverse dynamics. Weighting matrices $\mathbf{W}_m$ and $\mathbf{W}_g$ are determined beforehand by training the neural network using the motion data in the absence of the cutaneous sensation.

Using motion capture system (MAC Eagle System), floor force sensors (Kistler), wireless EMGs (Delsys), the data of stationally standing and walking motions are obtained for the cases with and without cutaneous sensation by the vibrating devices. The seven major muscles of lower limbs were used for computation such as Gluteus Maximus, Rectus Femoris, Vastus Medialis, Semimembranosus, Tibialis Anterior, Gastrocnemius, and Soleus.

The results of muscle activity estimation of left Gluteus Maximus and left Semimembranosus are shown in Fig. 15 for the stationally standing motion. Blue dotted lines are the estimation from EMGs and the inverse dynamics. Red solid lines are the output from the neural network trained using the blue dotted lines as the reference signals. The mean error is 1.02×10^{-2}.

Figure 16 shows the results with the learnt neural network with the cutaneous sensation at the left heel using the vibration device. Left Gluteus Maximus and left Semimembranosus are shown for the stationally standing motion. Blue solid lines are the estimated activities of muscles from EMGs and the inverse dynamics. Red solid lines are computed muscle activities using the learnt neural network with the

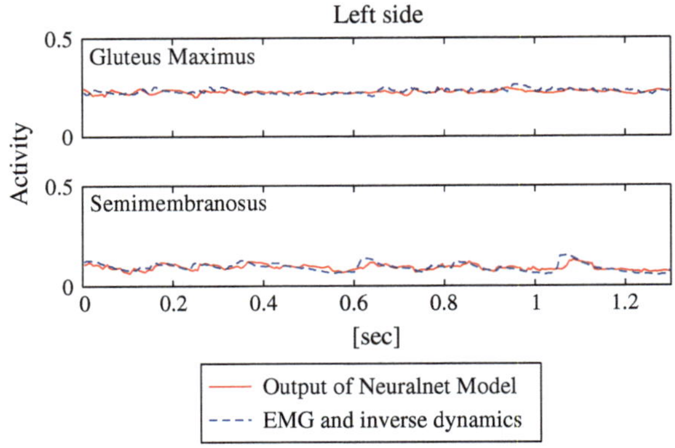

Fig. 15 Computed activities of left Gluteus Maximuis and left Semimembranosus in stationally standing motion without cutaneous sensation

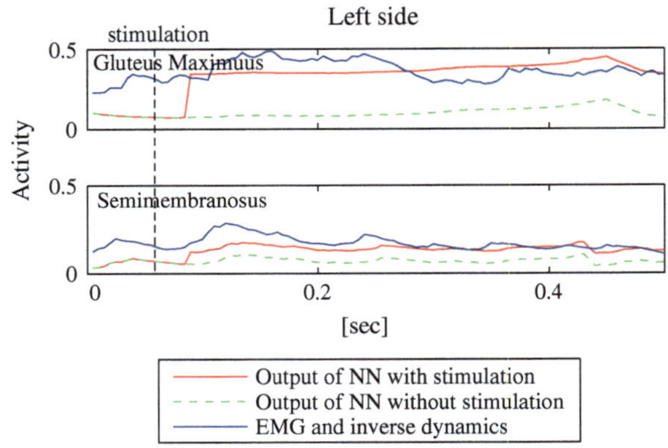

Fig. 16 Computed activities of left Gluteus Maximuis and left Semimembranosus in stationally standing motion with cutaneous sensation at left heel

muscle lengths and their velocities for the stationally standing and the simulated cutaneous sensation of Eq. (23). Green dotted lines are the computed muscle activities without the cutaneous sensation. The cutaneous sensation was given at $t = 0.05$ s and the difference appeared hereafter between the red and green lines. The skin segment of left heel connects to L5 and S2, which have projections to Gluteus Maximus, Semimembranosus. The muscle activities due to the projection compensate the difference between the blue and green lines.

The results of muscle activity estimation of left Rectus Femoris and left Semimembranosus in walking motion without cutaneous sensation is shown in Fig. 17. Blue dotted lines are estimation from EMGs and the inverse dynamics. Red solid lines are the output from the neural network trained using the blue dotted lines as the reference signals. The mean error is 2.67×10^{-2}.

Figure 18 shows the results with the learnt neural network with the cutaneous sensation at the posterior surface of left knee using the vibration device. Left Rectus Femoris and left Semimembranosus are shown for the walking motion. The lines are corresponding to the cases of Fig. 16. The skin segment of posterior surface of

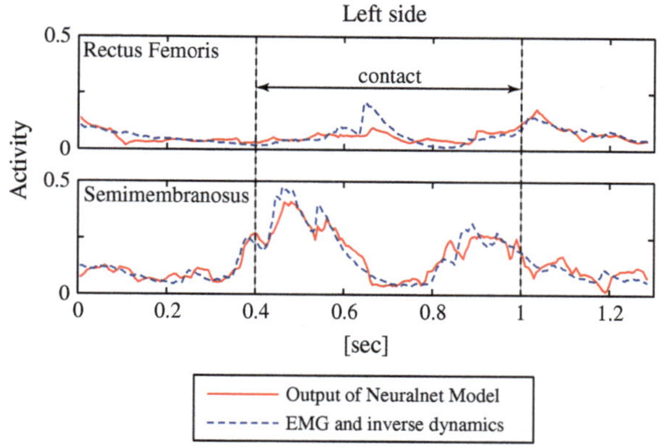

Fig. 17 Computed activities of left Rectus Femoris and left Semimembranosus in walking motion without cutaneous sensation

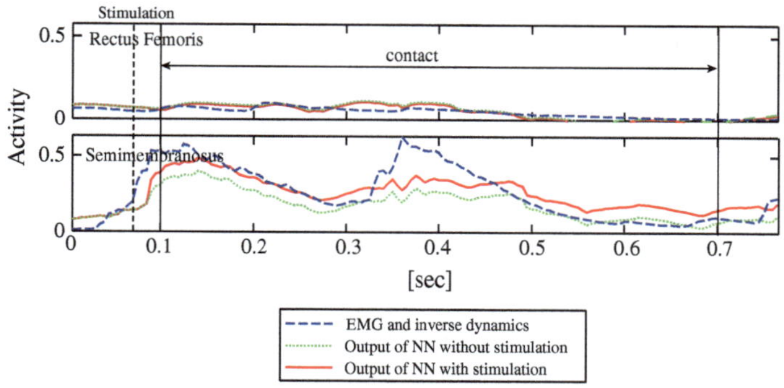

Fig. 18 Computed activities of left Rectus Femoris and left Semimembranosus in walking motion with cutaneous sensation

left knee connects to L3 and S2. S2 has a projection to Semimembranosus, which compensates the difference between the blue and green lines of left Semimembranosus.

The simulation results are shown in the following two figures. Four different cutaneous sensations are applied at stationally standing motion and the data was used for training the neural network. The simulated cutaneous sensation was virtually applied to the left heel and the dosum of left foot respectively in Figs. 19 and 20. The sensation started at $t = 0.05$ s.

Gluteus Maximus, Semimembranosus, and Tibialis Anterior were activated and lifted up the toe with bending the knee in the both cases of sensations. The difference was observed at Soleus, which did not show the reaction against the sensation at the dorsum of foot while it did against that at the heel. The results show that the cutaneous sensation at the heel characterizes Soleus with antagonistic activation.

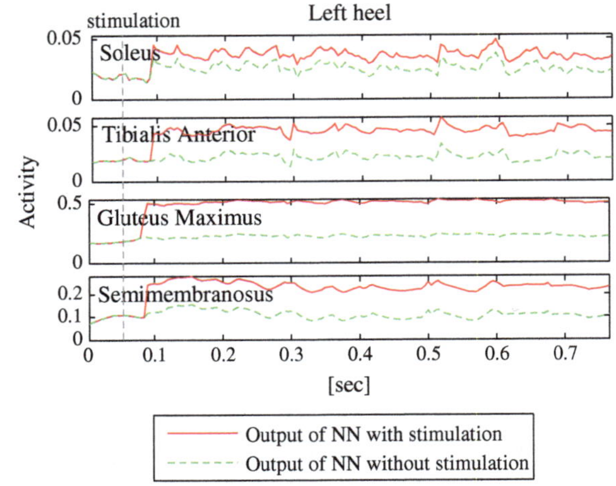

Fig. 19 Simulation with stationally standing motion data with left heel stimulation

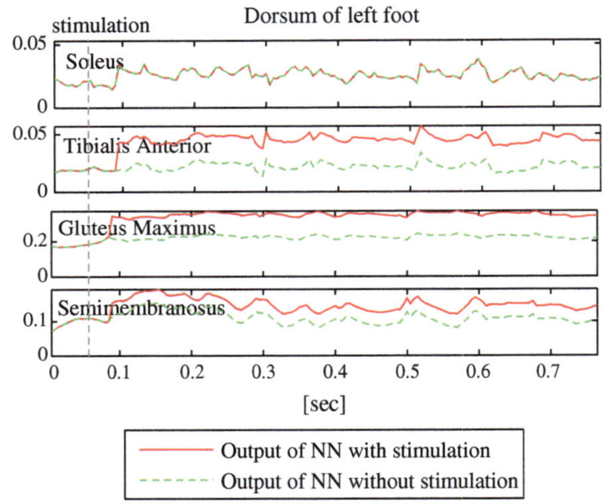

Fig. 20 Simulation with stationally standing motion data with dosum of left foot

6 Summary

It was fortunate that the author could start the series of study with CREST Project (1998–2003) "Development of Brain-Informatics Machines through Dynamical Connection of Autonomous Motion Primitives." The computational foundation of musculoskeletal model of the human whole body was developed. The proto-symbol space and bi-directional computation of human motion data was studied as a mathematical model of the mirror neuron using the Hidden Markov Model.

JSPS Grant-in-Aid-For Scientific Research (S) (2003–2008) "Development of Dynamics-Based Information Processing Model of Intelligence" provided a great support to extend the research. The human-robot communication based on the mathematical model of mirror neuron (Mimesis model) was applied to the realtime human-robot interaction. Computational model of spinal reflex was developed based on the musculoskeletal model of the whole human body.

The current study is under the support of JSPS Grant-in-Aid-For Scientific Research (S) (2008–2013) "Establishing Human-Machine Communication through Kinesiology and Linguistic Integration." The anatomical modeling and behavioral modeling are to be integrated in statistical computation with the natural language model. The communication of humanoid with the human beings are studied.

The development of software for high-definition neuromusculoskeletal model and the large behavioral data base with statistical modeling is currently undergoing being supported by MEXT Next-Generation Supercomputer Strategic Project, Field 1: Predictive Life Science, Medicine, and Pharmaceutical Foundation (2011–2016).

The author would like to acknowledge contributions of many colleagues and students at YNL, Department of Mechano-Informatics, University of Tokyo. In particular, without discussions with Katsu Yamane, Akihiko Murai, and Yuki Ibuka, the coauthors of the papers quoted in this paper, this work would neither have reached the current state and nor have been as enjoyable as it is.

References

1. M. Minsky, *Semantic Information Processing* (The MIT Press, 1969)
2. H.A. Simon, *The Sciences of the Artificial*, 2nd edn. (The MIT Press, 1968, 1981)
3. T. Deacon, *Symbolic Species: The Co-evolution of Language and the Brain* (W.W. Norton and Company Inc, 1997)
4. M. Donald, *Origin of the Modern mind* (Harvard University Press, Cambridge, 1991)
5. V. Gallese, A. Goldman, Mirror neuron and the simulation theory of mind-reading. Trends Cogn. Sci. **2**(12), 493–501 (1998)
6. G. Rizzolatti, L. Fogassi, and V. Gallese, Neurophysiological mechanisms underlying the understanding and imitation of action. Nat. Rev. 661–670 (2001)
7. H. Ezaki, T. Inamura, Y. Nakamura, I. Toshima, Imitation and primitive symbol acquisition of humanoids by the integrated mimesis loop, in *Proceedings of the 18th IEEE International Conference on Robotics and Automation*, pp. 4208–4213 (2001)
8. T. Inamura, I. Toshima, H. Tanie, Y. Nakamura, Embodied symbol emergence based on mimesis theory. Int. J. Robot. Res. **23**(4), 363–377 (2004)
9. W. Takano, K. Yamane, T. Sugihara, K. Yamamoto, Y. Nakamura, Primitive communication based on motion recognition and generation with hierarchical mimesis model, in *Proceedings of the IEEE International Conference on Robotics and Automation*, pp. 3602–2609 (2006)
10. B. Janus, Y. Nakamura, Unsupervised probabilistic segmentation of motion data for mimesis modeling, in *Proceedings of IEEE International Conference on Advanced Robotics*, (2005), pp. 411–417
11. D. Kulic, Y. Nakamura, Scaffolding on-line segmentation of full body human motion patterns, in *Proceedings of the IEEE/RSJ 2008 International Conference on Intelligent Robots and Systems*, pp. 2860–2866 (2008)
12. D. Kulic, H. Imagawa, Y. Nakamura, Online acquisition and visualization of motion primitives for humanoid robots, in *Proceedings of the 18th IEEE International Symposium on Robot and Human Interactive Communication*, pp. 1210–1215 (2009)

13. W. Takano, D. Kulic, H. Imagawa, Y. Nakamura, What do you expect from a robot that tells your future? the crystal ball, in *Proceedings of the IEEE International Conference on Robotics and Automation* (2010)
14. W. Takano, Y. Nakamura, Incremental learning of integrated semiotics based on linguistic and behavioral symbols, in *Proceedings of the IEEE International Conference on Intelligent Robots and Systems*, pp. 2545–2550 (2009)
15. W. Takano, Y. Nakamura, Associative processes between behavioral symbols and a large scale language model, in *Proceedings of the IEEE International Conference on Robotics and Automation*, pp. 2404–2409 (2010)
16. W. Takano, H. Imagawa, Y. Nakamura, Prediction of human behaviors in the future through symbolic inference, in *Proceedings of the IEEE International Conference on Robotics and Automation*, pp. 1970–1975 (2011)
17. T. Flash, N. Hogan, The coordination of arm movements: an experimentally confirmed mathematical model. J. Neurosci. **5**, 1688–1703 (1985)
18. M. Katato, Y. Maeda, Y. Uno, R. Suzuki, Trahectory formation of arm movement by cascade neural network model based on minimum torque-change criterion. Biol. Cybern. **62**(4), 275–288 (1990)
19. S.L. Delp, J.P. Loan, A computational framework for simulating and analyzing human and animal movement. IEEE Comput. Sci. Eng. **2**, 46–55 (2000)
20. T. Komura, P. Prokopow, A. Nagano, Evaluation of the influence of muscle deactivation on other muscles and joints during gait motion. J. Biomech. **37**(4), 425–436 (2004)
21. F.C. Anderson, M.G. Pandy, Static and dynamic optimization solutions for gait are practically equivalent. J. Biomech. **34**, 153–161 (2001)
22. Y. Nakamura, K. Yamane, Y. Fujita, I. Suzuki, Somatosensory computation for man-machine interface from motion capture data and musculoskeletal human model. IEEE Trans. Rob. **21** (1), 58–66 (2005)
23. Y. Nakamura, K. Yamane, A. Murai. "Macroscopic Modeling and Identification of the Human Neuromuscular Network, in *Proceedings of the 28th IEEE EMBS Annual International Conference*, pp. 99–105 (2006)
24. K. Yamane, Y. Fujita, Y. Nakamura, Estimation of physically and physiologically valid somatosensory information, in *Proceedings of IEEE International Conference on Robotics and Automation*, pp. 2635–2641, Barcelona, Spain, April 2005
25. A.V. Hill, The heat of shortening and the dynamic constants of muscle. Proc. Royal Soc. Lond. **B126**, 136–195 (1938)
26. A. Murai, K. Yamane, Y. Nakamura, Modeling and Identification of the Human Neuromusculoskeletal Model's Somatic Refrex Network, in *Proceeding of 2007 JSME Conference on Robotics and Mechatronics (ROBOMEC'07)* (2007) (in Japanese)
27. A. Hyva¨rinen, J. Karhunen, E. Oja, *Independent Component Analysis* (Wiley, 2001)
28. A. Murai, K. Yamane, Y. Nakamura, Effects of Nerve Signal Transmission Delay in Somatosensory Reflex Modeling Based on Inverse Dynamics and Optimization, in *Proceeding of IEEE International Conference on Robotics and Automation, Anchorage, USA* (2010)
29. A. Hill, The heat of shortening and the dynamic constants of muscle. Proc. Royal Soc. Lond. **B126**, 136–195 (1938)
30. S. Stroeve, Impedance characteristics of a neuro-musculoskeletal model of the human arm I: posture control. J. Biol. Cybern. **81**, 475–494 (1999)
31. Y. Nakamura, K. Yamane, Y. Fujita, I. Suzuki, Somatosensory computation for man-machine interface from motion capture data and musculoskeletal human model. IEEE Trans. Rob. **21**, 58–66 (2005)
32. J.H. Warfel, *The Extremities: Muscles and Motor Points* (Lea & Febiger, Philadelphia, 1974)
33. J. Erlanger, H.S. Gasser, *Electrical Signs of Nervous Activity* (University Press, Philadelphia, 1937)
34. D.P.C. Lloyd, C.C. Hunt, A.K. McIntyre, Transmission in fractionated monosynaptic spinak reflex system. J. Gen. Physiol. **38**, 789–799 (1955)

35. A. Prochazka, M. Gorassini, Models of ensemble firing of muscle spindle afferents recorded during normal locomotion in cats. J. Physiol. **507**, 277–291 (1998)
36. C.D. Clemente, *Gray's Anatomy ed 30* (Lea & Febiger, Phyladellphia, 1985)
37. A.M.R. Agur, *Grant's Atlas of Anatomy* (Williams & Wilkins, Baltimore, 1991)
38. A.E. Bryson, Yu-Chi Ho, *Applied Optimal Control* (Blaisdell, New York, 1969)
39. Y. Ibuka, A. Murai, Y. Nakamura, Modeling of Somatic Reflex Network with Cutaneous Sensation,in *Proceeding of JSME Robotics and Mechatronics Conference (ROBOMECH)* (2011)
40. F.H. Netter, *Atlas of Human Anatomy*, 4th edn. (Elsevier, 2006)

Part V
Control

Grasping and Fixturing as Submodular Coverage Problems

John D. Schulman, Ken Goldberg and Pieter Abbeel

Abstract Grasping and fixturing are concerned with immobilizing objects. Most prior work in this area strives to minimize the number of contacts needed. However, for delicate objects or surfaces such as glass or bone (in medical applications), extra contacts can be used to reduce the forces needed at each contact to resist applied wrenches. We focus on the following class of problems. Given a polyhedral object model, set of candidate contacts, and a limit on the sum of applied forces at the contacts or a limit on any individual applied force, compute a set of k contact points that maximize the radius of the ball in wrench space that can be resisted. We present an algorithm, SatGrasp, that is guaranteed to find near-optimal solutions in linear time. At the core of our approach are (i) an alternate formulation of the residual radius objective, and (ii) the insight that the resulting problem is a submodular coverage problem. This allows us to exploit the submodular saturation algorithm, which has recently been derived for applications in sensor placement. Our approach is applicable in situations with or without friction.

1 Introduction

The problem of choosing contact points on an object to securely hold it is relevant to robotics and manufacturing. Robotics is concerned with *grasping*, where a single robot uses a multi-fingered hand to hold an manipulate an object, or multiple robots cooperate to pick up a large object. Manufacturing is concerned with *fixturing*, where a machinist uses locators and clamps to immobilize a part for operations such

J.D. Schulman (✉) · K. Goldberg · P. Abbeel
Department of Electrical Engineering and Computer Science,
University of California, Berkeley, USA
e-mail: joschu@berkeley.edu

K. Goldberg
e-mail: goldberg@berkeley.edu

P. Abbeel
e-mail: pabbeel@cs.berkeley.edu

H.I. Christensen and O. Khatib (eds.), *Robotics Research*,
Springer Tracts in Advanced Robotics 100, DOI 10.1007/978-3-319-29363-9_32

as inspection or machining. In both grasping and fixturing, there are often constraints on the forces that can be applied at the contact points.

Most prior work in this area strives to minimize the number of contacts needed for force closure, known to be four in the plane and seven in three dimensions. However, for delicate objects or surfaces such as glass or bone, extra contacts can be used to reduce the forces needed at each contact to resist applied wrenches.

Researchers have proposed a number of approaches to extend the notions of form and force closure with scalar "quality" measures for grasping or fixturing configurations. We use the wrench-space quality measures proposed by [1, 2]. These elegant measures are based on convex geometry and maximize the disturbance that can be resisted given bounds on the contact forces. See [3] for a review on quality metrics. The metrics we consider are as follows:

- Q_1: The norm of the smallest wrench that can't be resisted, given a constraint on the sum of the normal forces.
- Q_∞: The norm of the smallest wrench that can't be resisted, given a constraint on the maximum normal force.

The notion of norm in wrench space is not well-defined, since the wrench vector has force and torque components, which have different units. Ferrari and Canny [2] address this problem by using the length scale of the object as a dimensional constant that relates forces to torques; we describe a different solution in Sect. 3 based on the minimum-volume ellipsoid containing a set of realistic wrenches.

Our main contributions are as follows:

- Approximate formulas (they are not exact due to discretization) that allow for fast calculation of Q_1 and Q_∞ and their generalizations that incorporate friction.
- A fast algorithm for selection of contact points that maximize Q_1 and Q_∞ out of a set of candidate contacts, which is guaranteed to give a near-optimal solution. The runtime is roughly linear in the number of candidates and the number of contacts.

2 Related Work

The Q_1 metric for grasping and its geometric interpretation was originally proposed by Kirkpatrick et al. [1]. They provided a fast algorithm to find the subset of a set of two-dimensional vectors with maximal *residual radius* (defined in Sect. 3), but suggested that this problem is hard in higher-dimensional spaces. A later paper [4] provides a probabilistic algorithm to optimize the Q_1 metric in the frictionless case. Another article investigates fast methods for calculating the Q_∞ metric but not optimizing it [5].

The problem of contact point selection has also been studied from the perspective of fixture design. Brost and Goldberg [6] provide an algorithm to find viable peg and clamp placements in a planar modular fixturing apparatus, and Brost and Peters [7] extend these results to three dimensions and incorporate quality metrics. These methods require searching through a large number of configurations —exponential in the number of contacts. Wang [8] optimizes a different quality metric, which does not inherently guarantee force closure or force limits, by using the greedy algorithm on a max-det problem.

3 Background: Geometry of Quality Metrics

In this section, we review the geometric interpretation of Ferrari and Canny's quality metrics and introduce some notation. Given k frictionless contact points, suppose that applying unit normal forces at these contacts generates wrenches w_1, $w_2, \ldots, w_k$. Then if we apply normal forces $f_1, f_2, \ldots, f_k$, the total wrench is

$$w = \sum_i f_i w_i. \tag{1}$$

Under the L_1 constraint $\sum_i f_i \leq 1$, the set of attainable wrenches w is ConvexHull(0, $w_1, w_2, \ldots, w_k$). Under the L_∞ constraint $\max_i f_i \leq 1$, the set of attainable wrenches is MinkowskiSum($w_1, w_2, \ldots, w_k$). In both cases, the constraint $f_i \geq 0$ is implied.

The *residual radius* of a compact set C is defined as the distance from the origin to the boundary of C:

$$r_{\text{res}}(C) = \text{dist}(x, \mathbb{R}^n \backslash C). \tag{2}$$

The quality metrics, informally defined in the introduction, are formally defined as follows:

$$Q_1 = r_{\text{res}}(\text{ConvexHull}(w_1, w_2, \ldots, w_k)), \tag{3}$$

$$Q_\infty = r_{\text{res}}(\text{MinkowskiSum}(w_1, w_2, \ldots, w_k)). \tag{4}$$

These quality metrics can be generalized to the case where there is friction. Now the wrench applied through the ith contact is $G_i f_i$, where G_i is a matrix and f_i is a vector describing forces and torques that can be applied at that contact. Let $f^\perp$ be the normal force, and let $f^{||}$ be the vector of frictional components. Then the truncated friction cone for that contact is defined as follows:

$$FC_i = \{G_i f \mid \|Af^{||}\| \leq f^\perp, \quad f^\perp \leq c\}, \tag{5}$$

where the matrix A depends on the friction model and friction coefficients.

If there is an L_1 constraint on the normal forces, $\sum_i f_i^{\perp} = 1$, then the set of attainable wrenches is ConvexHull(FC_1, FC_2,..., FC_k). If there is an L_∞ constraint on the normal forces, $\max_i f^{\perp} = 1$, then the set of attainable wrenches is MinkowskiSum(FC_1, FC_2,..., FC_k). For contacts with friction, the quality functions are defined as

$$Q_1 = r_{\text{res}}(\text{ConvexHull}(FC_1, FC_2, \ldots, FC_k)), \tag{6}$$

$$Q_\infty = r_{\text{res}}(\text{MinkowskiSum}(FC_1, FC_2, \ldots, FC_k)). \tag{7}$$

3.1 Natural Norms in Wrench Space

The quality metrics Q_1 and Q_∞ depend on a norm in wrench space. Since force and torque have different units, it is not obvious how to define the norm of vectors in wrench space, so that we can talk about the size of the ball of wrenches that can be resisted. We propose to define the norm in wrench space in a natural way that depends on the object being immobilized and the disturbances applied to it.

First consider the set of "disturbances"—wrenches that will be applied to the object. This set will typically include the wrench due to gravity, and it might contain all the wrenches arising from forces applied to the surface. Next calculate the minimum-volume enclosing ellipsoid for these points (see [9], 8.4 for a discussion of this problem.) We will change coordinates so that this ellipsoid becomes a unit ball. The most extreme points in our set of expected wrenches will have norm 1.

Specifically, if the ellipsoid is parameterized as

$$\{w \mid \|Aw - x_c\| \leq 1\}. \tag{8}$$

then we transform into normalized coordinates $x = Aw$, so in our new coordinates, the ellipsoid is

$$\{x \| \|x - x_c\| \leq 1\}. \tag{9}$$

For the rest of this paper, we will be working in the space of normalized wrenches, but for clarity, we will refer to these normalized wrenches with the letter w.

Under this metric, $1/Q_\infty$ is the maximum (i.e., worst-case) normal force that needs to be applied at some contact to resist the set of disturbances, and $1/Q_\infty$ is the maximum sum of normal forces. Therefore, we can optimize these quality functions if our goal is to minimize the applied normal forces given a known set of disturbances.

4 Alternate Formulation of Quality Metrics

In this section, we will derive an alternate expression for the residual radius of a convex set. Besides being efficient to evaluate, this alternate form reveals that the radius maximization problem can be solved by methods that exploit submodularity.

For frictionless contact points, we have that

$$Q_1 = r_{\text{res}}(\text{ConvexHull}(w_1, w_2, \ldots, w_k)) = \min_{\|y\|=1} \max_i |y^T w_i|_+ \tag{10}$$

$$Q_\infty = r_{\text{res}}(\text{MinkowskiSum}(w_1, w_2, \ldots, w_k)) = \min_{\|y\|=1} \sum_i |y^T w_i|_+ . \tag{11}$$

Where $|\cdot|_+$ is the "positive part", i.e.,

$$|x|_+ = \begin{cases} x & \text{if } x \geq 0, \\ 0 & \text{otherwise.} \end{cases} \tag{12}$$

For contact points with frictions, where the forces that can be applied at each contact lie in a truncated friction cone FC, we have that

$$Q_1 = r_{\text{res}}(\text{ConvexHull}(w_1, w_2, \ldots, w_k)) = Q_1 = \min_{\|y\|=1} \max_i h_{FC_i}(y), \tag{13}$$

$$Q_\infty = r_{\text{res}}(\text{MinkowskiSum}(w_1, w_2, \ldots, w_k)) = \min_{\|y\|=1} \sum_i h_{FC_i}(y), \tag{14}$$

where $h_{FC_i}(y) = \max_{x \in FC_i} y^T x$.

The key concept for proving these proposition is the *support function*, illustrated in Fig. 1. Given compact, convex set C and point y, the support function $h_C(y)$ is defined as [10]

$$h_C(y) = \max_{x \in C} y^T x \tag{15}$$

For $\|y\| = 1$, the support function tells us the height of C in the direction y. The support function has the following properties:

$$h_{\text{ConvexHull}(C_1, C_2, \ldots, C_n)}(y) = \max_i h_{C_i}(y) \tag{16}$$

$$h_{\text{MinkowskiSum}(C_1, C_2, \ldots, C_n)}(y) = \sum_i h_{C_i}(y) \tag{17}$$

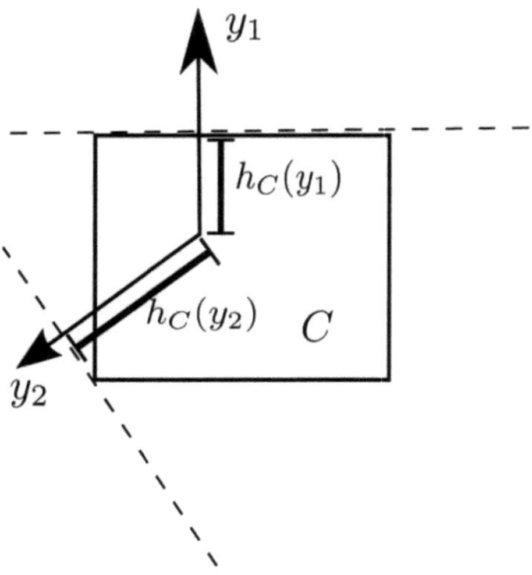

Fig. 1 An illustration of the support function h_C for a rectangular set *C*, evaluated at two unit vectors y_1 and y_2

Proposition 1 *For compact, convex set* C *containing the origin,*

$$r_{res}(C) = \min_{\|y\|=1} h_C(y) \tag{18}$$

$$= \min_{\|y\|=1} \max_{x \in C} y^T x \tag{19}$$

Proof See appendix.

Combining this proposition with Eqs. (16) and (17), we obtain the formulas in Eqs. (13) and (14), which specialize to Eqs. (10) and (11) in the frictionless case.

We can evaluate these formulas by discretizing the sphere (or hypersphere) $\{y \,|\|y\| = 1\}$. As we show in the appendix, if we choose $6n^2$ samples on the 2-sphere or $12n^5$ samples on the 5-sphere, then the error of this approximation of $h_C\,(y)$ is roughly $\frac{\sqrt{p}}{2n}\,|\sin\theta|\,\|x\|$, x and y correspond to the minimum of h_C, and θ is the angle between x and y.

In the frictionless case, the computational cost of evaluating the quality function is just due to calculating the matrix of inner products $y_i \cdot w_j$. In the frictional case, $h_{FCi}\,(y)$ is not much harder to compute. Using the parametrization from Eq. (5), a straightforward calculation shows that for $c = 1$

$$h_{FC_i}(y) = \begin{cases} 1 & \text{if } y \in FC_i \\ \left| y^T G_i^{\|} + \|A^{-1} G_i^{\perp T} y\| \right|_+ & \text{otherwise} \end{cases} \tag{20}$$

where we have written $G_i = \left[G_i^{\perp} | G_i^{\|}\right]$ corresponding to the normal and frictional components.

5 Quality Function Optimization as a Submodular Coverage Problem

Given a set of candidate contact points indexed by $S = \{1, 2,\ldots, k\}$, we would like to find a subset $S' \subseteq S$ to optimize the quality function. In other words, we would like to choose the subset of our truncated friction cones whose convex hull or Minkowski sum has maximal residual radius. So we are solving the following optimization problems:

$$\max_{S' \subset S} Q_1 = \max_{S' \subset S} \min_{\|y\|=1} \max_{i \in S'} h_{FC_i}(y), \tag{21}$$

$$\max_{S' \subset S} Q_\infty = \max_{S' \subset S} \min_{\|y\|=1} \sum_{i \in S'} h_{FC_i}(y), \tag{22}$$

subject to the cardinality constraint $|S'| \leq k$.

Krause et al. [11] introduced the *submodular saturation* algorithm (henceforth called SATURATE), which solves problems where we are trying to optimize the minimum of a collection of objectives:

$$\max_{S' \subset S} \min_i F_i(s'), \quad \text{subject to } |S'| \leq k. \tag{23}$$

When the functions F_i are submodular, there are theoretical performance guarantees on this algorithm. Namely, if we relax the cardinality constraint to $|S'| \leq \alpha k$ (where α is a parameter depending on the problem) and run the algorithm, then the αk-element solution found by SATURATE is guaranteed to be better than the optimal k-element solution. SATURATE is very fast, and involves performing the greedy algorithm several times with a transformed objective function. See the appendix for a review of SATURATE and the value of α.

The expresions $\max_{i \in S'} h_{FC_i}(y)$ and $\sum_{i \in S'} h_{FC_i}(y)$ are both submodular functions of the set S', since max and sum are submodular. Each direction in wrench space y indexes a different objective function that we are trying to optimize. Therefore, optimization of the quality functions over S' has exactly the form given in Eq. (23), so SATURATE can be applied to this problem. To apply the algorithm, we need to calculate $h_{FC_i}(y_j)$ for every pair consisting of a contact point i and a direction in wrench space j. In the frictionless case we just need to calculate $y_j^T w_i$ for each pair. Then to evaluate the objective functions when we perform SATURATE, we merely need to take the max and sum over the columns of a table—an extremely cheap operation.

6 Analysis of Solutions Found by SatGrasp

Figure 2 illustrates the solutions found by SatGrasp, which involves using SATURATE to optimize the right-hand side of Eqs. (10) or (13) on a grid of y. This example illustrates that SATURATE has some look-ahead ability unlike greedy algorithms—it makes a different sequence of choices depending on the number of contacts requested.

Figures 3 and 4 illustrate that SatGrasp can efficiently and sensibly select a large number of contact points. Running a combinatorial search to optimize the objective is impractical, even for merely seven contact points, corresponding to the minimum

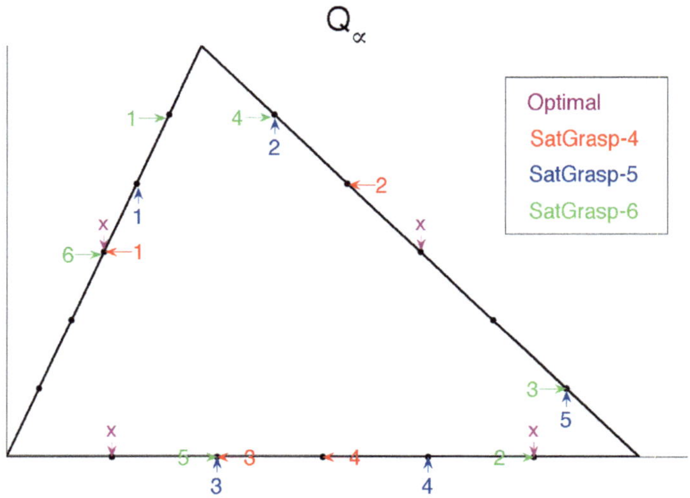

Fig. 2 We selected a subset of 4, 5 or 6 contact points out of 15 candidates, using either the saturate algorithm or an exhaustive search to optimize the Q_∞ metric. *Arrows* indicate which contacts (*black dots*) were chosen, not force direction

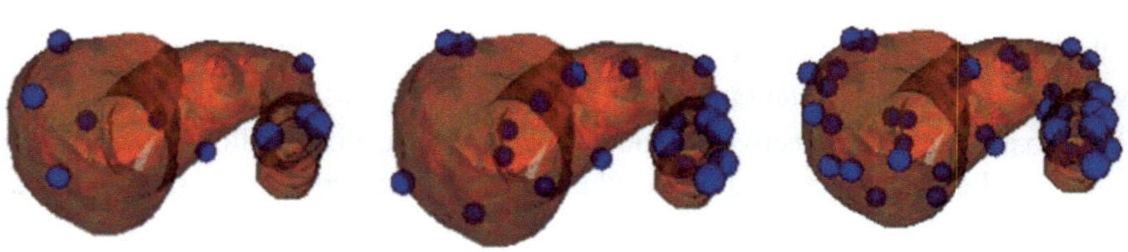

Fig. 3 Fixture locations chosen on a three-dimensional object. From *left* to *right* 10, 20, 40 contacts

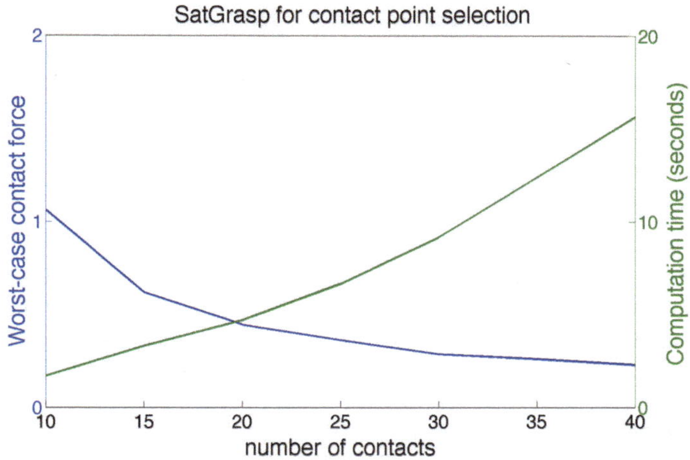

Fig. 4 Maximum contact forces needed to resist a given set of wrenches as the number of contacts is increased. The computational time of SatGrasp procedure remains extremely modest. In contrast, a combinatorial search takes a prohibitive amount of time even to find the minimal number of contacts needed for force closure (namely seven)

for force closure. The larger number of contact points is redundant from the perspective of force closure but useful when there are force limits at each contact. We generated a set of disturbance wrenches that will be applied to the object (unit forces applied at the contacts) and used SATURATE to find a set of contacts that maximize the residual radius of the resulting polytope in wrench space. Because the disturbances have norm of at most 1 in the transformed coordinates (see Sect. 3), $1/r_{\text{res}}$ is the maximum contact force needed to resist all of the disturbances.

7 Conclusions

We used some basic tools from convex geometry to find alternate formulas for the classic quality metrics. Since they are cheap to evaluate, these formulas may be useful in their own right for fast quality computation, especially in the frictional case. However, more interestingly, they reveal that the contact point selection problem can be viewed as a multi-objective submodular coverage problem, and SATURATE provides an efficient, near-optimal solution. We found that our algorithm can select a large number of contact points out of a larger number of candidates in an extremely modest amount of time.

As far as we know, the SATURATE algorithm has exclusively been applied to problems of observation selection and information maximization. Our work hints at an approach that may have other applications in engineering design. We must build our system with a small set of supports so that our system can survive a set of disturbances. Each disturbance has its own cost function, which is submodular in the set of supports. In our work, the supports were contact points and the disturbances are the wrenches that might be applied to the object. Then, one uses SATURATE to find the best set of supports to handle the worst-case disturbance.

8 Extensions and Future Work

These methods could be generalized to select from a set of surface or edge contacts, rather than point contacts. Highly redundant sets of contacts are useful for fixturing of deformable objects and objects that have limits on structural stress. We plan to investigate quality functions for such objects and extend SatGrasp to these settings.

Acknowledgements We would like to thank Mark Meckes for providing the proof to Proposition 1 (see [12]) and James O'Brien for providing us with 3D models.

Appendix

Proof of Proposition 1 We use the fact that for two convex sets A and B, $A \subset B$ if and only if $h_A \preceq h_B$ (i.e., $h_A(x) \leq h_B(x)$ for all x). Let B_r be the ball of radius r around the origin. If $\hat{r} = r_{\text{res}}(C)$, then it follows that

$$B_{\hat{r}} \subseteq C \tag{24}$$

$$h_{B_{\hat{r}}}(y) \leq h_C(y) \text{for all } \|y\| = 1 \tag{25}$$

$$\hat{r} = r_{\text{res}}(C) \leq \min_{\|y\|=1} h_C(y) \tag{26}$$

On the other hand, let $\tilde{r} = \min_{\|y\|=1} h_C(y)$. Then

$$h_{B_{\tilde{r}}}(y) \leq h_C(y) \text{for all } \|y\| = 1 \tag{27}$$

$$h_{B_{\tilde{r}}} \preceq h_C \text{ since } h \text{ is homogeneous} \tag{28}$$

$$B_{\tilde{r}} \subseteq C \tag{29}$$

$$\tilde{r} = \min_{\|y\|=1} h_C(y) \leq r_{\text{res}}(C) \tag{30}$$

Thus $\min_{\|y\|=1} h_C(y) = r_{\text{res}}(C)$ □

Discretization Error in Formula for Residual Radius

We will consider a particular scheme for deterministically sampling the p-dimensional sphere (a subset of $\mathbb{R}^{p+1}$) and bound the error that results when one evaluates the support function $h_C(y)$ at only the sampled points y to approximate its minimum. We take an $n \times n \times \cdots \times n$ grid on every p-dimensional facet of the $p + 1$-dimensional hypercube. This requires $(2p + 2)n^p$ points.

Let y_n = argmin $h_C(y_i)$ be the optimal sampled point, and let y_* = argmin $h_c(y)$ be the exact optimal point. Let x_n = argmax $y_n^T x$.

$$y^T x_n = \|x_n\| \cos \theta_{x_n,y} \tag{31}$$

$$\frac{d}{d\theta} y^T x_n = -\|x_n\| \sin \theta_{x_n,y} \tag{32}$$

$$y_n^T x_n - y_*^T x_n = -\|x_n\| \sin \tilde{\theta} \Delta\theta \tag{33}$$

where $\theta_{y^n,x_n} \leq \tilde{\theta} \leq \theta_{y_*,x_n}$, and $\Delta\theta = \theta_{y_n,x_n} - \theta_{y_*,x_n}$. Next, note that $h(y_*) \geq y_*^T x_n$ so

$$h(y_n) - h(y_*) \leq \|x_n\| |\sin \tilde{\theta} \Delta\theta| \tag{34}$$

The largest possible $\Delta\theta$ occurs at the nearby part of each face, between the vector (1, 0,..., 0) and $\left(1, \frac{1}{2n}, \ldots, \frac{1}{2n}\right)$, where $\cos(\Delta\theta) = \sqrt{1 + p/4n^2}$. It follows that

$$\Delta\theta \leq \frac{\sqrt{p}}{2n} \tag{35}$$

Thus

$$h(y_n) - h(y_*) \leq \|x_n\| |\sin \tilde{\theta} \Delta\theta| \tag{36}$$

$$\leq \|x_n\| | \sin \theta_{x_n,y_n} | \left(\frac{\sqrt{p}}{2n} + \frac{p}{4n^2}\right) \tag{37}$$

Submodular Saturation Algorithm

SATURATE finds solutions to problems where we are simultaneously trying to optimize a collection of submodular objectives.

$$\max_{S' \subset S} \min_i F_i(S'), \quad \text{subject to } |S'| \leq k \tag{38}$$

If we run saturate and request a αk-element solution, it will give a solution that's better than the optimal k-element solution, where

$$\alpha = 1 + \log(\max_{s \in S} \sum_i F_i(s)) \tag{39}$$

This bound applies when the functions F_i take integer values. Thus to apply this result to a general problem of the form in Eq. (38), we must typically rescale and then round the objective functions. In the present problem, we can rescale and round the values $h_{FC_i}(y_j)$, or, in the frictionless case $y_j^T w_i$. Now the αk-element solution found by SATURATE is no longer guaranteed to be better than the optimal k-element solution, however, the difference is small and due to rounding error.

Here we describe the SATURATE algorithm for self-containedness. The key idea is as follows: if the optimal value is c, then there is no benefit if an objective function exceeds c. Thus we define the truncated objective functions

$$\widehat{F}_{i,c}(S) = \max\{F_i(S), c\} \tag{40}$$

If the original objectives F_i are submodular, then the truncated versions are also submodular. Now our goal is to *saturate* all of the objective functions $\widehat{F}_{i,c}(S)$, i.e., achieve the value c. The mean of the truncated functions, $\overline{F}_c$, is also submodular, and it describes progress towards this goal.

$$\overline{F}_c(S) = \sum_i \widehat{F}_{i,c}(S) \tag{41}$$

To optimize $\overline{F}_c$, we can use the greedy algorithm, which is guaranteed to give good solutions since it is submodular. (The greedy k-element solution is bested by the optimal k-element solution by at most a factor of $1 - 1/e$).

We initially don't the largest value of c that our greedy algorithm will successfully achieve (i.e., saturate all of the objective functions), so we use a binary search over the range of possible values. For each c, we greedily optimize $\overline{F}_c$. If we saturate all of the $\widehat{F}_{i,c}$, then we next try a larger c. If we fail to saturate them all, we next use a lower value of c.

SATURATE runs the greedy algorithm about 10 times with the transformed objective function. Thus the running time is roughly linear in $|S|$ and k. It is actually somewhat faster than $O(k)$ because we can use a "lazy greedy" algorithm that does not test elements that are guaranteed to give less improvement than some element that we've already tested.

References

1. D. Kirkpatrick, B. Mishra, C.-K. Yap, Quantitative Steinitz's theorems with applications to multifingered grasping. Discrete Comput. Geom. **7**(1) (1992)
2. C. Ferrari, J. Canny, Planning optimal grasps, in *Proceedings 1992 IEEE International Conference on Robotics and Automation* (1992), pp. 2290–2295
3. B. Mishra, Grasp metrics: optimality and complexity, in *Algorithmic Foundations of Robotics* (1995)
4. M. Teichmann, B. Mishra, Probabilistic algorithms for efficient grasping and fixturing. Algorithmica **26**(3–4), 345–363 (2000)
5. Ch. Borst, M. Fischer, G. Hirzinger, A fast and robust grasp planner for arbitrary 3D objects, in *Proceedings 1999 IEEE International Conference on Robotics and Automation (Cat. No.99CH36288C)*, 3(May) (1999), pp. 1890–1896
6. R.C. Brost, K.Y. Goldberg, A complete algorithm for synthesizing modular fixtures for polygonal parts, in *Proceedings of the 1994 IEEE International Conference on Robotics and Automation* (1994), pp. 535–542
7. R.C. Brost, R.R. Peters, Automatic design of 3-D fixtures and assembly pallets. Int. J. Robot. Res. **17**(12), 1243–1281 (1998)
8. M.Y. Wang, An optimum design for 3-D fixture synthesis in a point set domain. IEEE Trans. Robot. Autom. **16**(6), 839–846 (2000)
9. S.P. Boyd, L. Vandenberghe, *Convex optimization* (Cambridge University Press, 2004)
10. B Gruünbaum, V. Klee, *Convex Polytopes* (Springer, 2003)
11. A. Krause, H. Brendan Mcmahan, Robust submodular observation selection. J. Mach. Learn. Res. **9**, 2761–2801 (2008)
12. Mark Meckes (mathoverflow.net/users/1044). A min-max formula for depth of the origin in a convex set. MathOverflow

A Unified Perturbative Dynamics Approach to Online Vehicle Model Identification

Neal Seegmiller, Forrest Rogers-Marcovitz, Greg Miller and Alonzo Kelly

Abstract The motions of wheeled mobile robots are governed by non-contact gravity forces and contact forces between the wheels and the terrain. Inasmuch as future wheel-terrain interactions are unpredictable and unobservable, high performance autonomous vehicles must ultimately learn the terrain by feel and extrapolate, just as humans do. We present an approach to the automatic calibration of dynamic models of arbitrary wheeled mobile robots on arbitrary terrain. Inputs beyond our control (disturbances) are assumed to be responsible for observed differences between what the vehicle was initially predicted to do and what it was subsequently observed to do. In departure from much previous work, and in order to directly support adaptive and predictive controllers, we concentrate on the problem of predicting candidate trajectories rather than measuring the current slip. The approach linearizes the nominal vehicle model and then calibrates the perturbative dynamics to explain the observed prediction residuals. Both systematic and stochastic disturbances are used, and we model these disturbances as functions over the terrain, the velocities, and the applied inertial and gravitational forces. In this way, we produce a model which can be used to predict behavior across all of state space for arbitrary terrain geometry. Results demonstrate that the approach converges quickly and produces marked improvements in the prediction of trajectories for multiple vehicle classes throughout the performance envelope of the platform, including during aggressive maneuvering.

N. Seegmiller (✉) · F. Rogers-Marcovitz · G. Miller · A. Kelly
Robotics Institute, Carnegie Mellon University, Pittsburgh, USA
e-mail: nseegmil@rec.ri.cmu.edu

F. Rogers-Marcovitz
e-mail: forrest@rec.ri.cmu.edu

G. Miller
e-mail: gmiller@rec.ri.cmu.edu

A. Kelly
e-mail: alonzo@rec.ri.cmu.edu

H.I. Christensen and O. Khatib (eds.), *Robotics Research*,
Springer Tracts in Advanced Robotics 100, DOI 10.1007/978-3-319-29363-9_33

1 Introduction

We concentrate in this paper on the problem of calibrating the faster-than-real-time models that are used in mobile robot predictive control and motion planning. In obstacle avoidance, lane change maneuvers, and path following the predicted motion of the vehicle is the basis for the precise specification of the inputs which will generate the desired behavior. The mapping between inputs and resultant behavior depends critically on terrain conditions which vary significantly over time and space so it cannot be pre-programmed. The mapping must be either perceived from non-contact sensing or learned from experience. We take the latter approach in this paper.

Motion models of ground robots have many uses, but the aspects of wheel-terrain interaction that are needed for accurate models are neither well known nor easily measurable in realistic situations. Published methods are mostly designed to use measurements to estimate present state for feedback controllers. Model-based approaches have been applied to estimate longitudinal wheel slip and to detect immobilization of mobile robots [11]. Analytical models also exist for steering maneuvers of planetary rovers on loose soil [4]. Some published methods lump all of the unknown soil parameters into slip ratios and a slip angle and use velocity measurements to aid in estimation. An Extended Kalman Filter (EKF) and a Sliding Mode Observer have been developed to estimate the slip ratio parameters [12, 7]. An EKF has also been used to estimate the slip angles and longitudinal slippage for a wheeled mobile robot [8]. Visual terrain recognition has also been employed for terrain-dependent slip estimation [2].

Other researchers have addressed the problem of model identification for ground robots. For example, algorithms have been developed to learn soil parameters given wheel-terrain dynamic models [10]. However, there is little precedent in the literature for the calibration of *predictive* models despite the fact that they are fundamental to virtually every decision that a mobile robot makes. The only precedent known to us is [3] where our colleagues constructed an artificial neural network that was trained offline. Our method learns a predictive model by capturing the underlying dynamics as a function of all of input space and it is calibrated on-line based on whatever trajectories the vehicle is executing.

The literature on identifying stochastic differential equations is even less developed, at least in robotics applications. One of the authors [5] presented methods for calibration of odometry error models which are similar to the methods used here. By contrast, [1] presents off-line coordinate ascent methods for tuning Kalman filters automatically. Other than these two references, we can find nothing that even slightly anticipates our efforts here to calibrate stochastic dynamics on-line. In this paper we present a kind of meta Kalman filter, running in real time, which calibrates the uncertainty in the system model used in the pose estimation filter.

Unlike all previous work, our method exploits the excellent short term accuracy of pose sensing that is available on mobile robots (inertial navigation, real-time kinematic GPS, or visual odometry) by using measurements of relative pose rather than velocity. In effect, we integrate the model rather than differentiate the measurements. In our recent initial work, we have developed on-line calibration techniques for learning vehicle slip rates [9]. In this paper, we extend those techniques to a more elegant formulation of the perturbative dynamics that incorporates all of initial condition errors, 3D terrain, and stochastic disturbances, all using the same underlying model. We first develop a somewhat general 3D land kinematic vehicle model in Sect. 2. This model, along with the pose residual observations, is integrated into an EKF in Sect. 3 to calibrate deterministic slip and in Sect. 4 to calibrate residual random disturbance covariance. The results are presented in Sect. 5 along with the experimental set-up which is followed by brief conclusions in Sect. 6.

2 System Modeling

Our fundamental approach is to linearize the system dynamics in order to capture, in a dynamical model, the first order evolution of pose prediction error caused by input disturbances. The disturbances are inputs to deterministic and stochastic differential equations and their values depend on the state of the environment and of the vehicle. Once the model is linearized, the perturbative dynamics provide a derived system model describing the mean behavior of deterministic error. Likewise, stochastic calculus provides the first order evolution of the state covariance, so the same linearization can be used to estimate the remaining random error in a probabilistic sense. In both cases, the calibration process is performed on-line using a Kalman filter and, in this way, the system can adapt rapidly to changes in the terrain. The state vectors in the estimation systems are the parameters which characterize the disturbances as functions over the terrain, the inputs, and the applied forces.

For any vehicle moving over a terrain surface, ignoring the suspension deflections, there are three instantaneous degrees of freedom of motion as long as the vehicle remains in contact with the terrain, (Fig. 1).

The true inputs to a vehicle typically have dimensions of power, force, curvature, linear velocity, or angular velocity. However, we are interested in predictive models of the platform under the influence of its control system, so the system boundary encloses the controller as well. We find a velocity driven model to be most appropriate. This choice also leads to some simplification of the perturbative dynamics because the dynamics are driftless and the transition matrix is available in closed form.

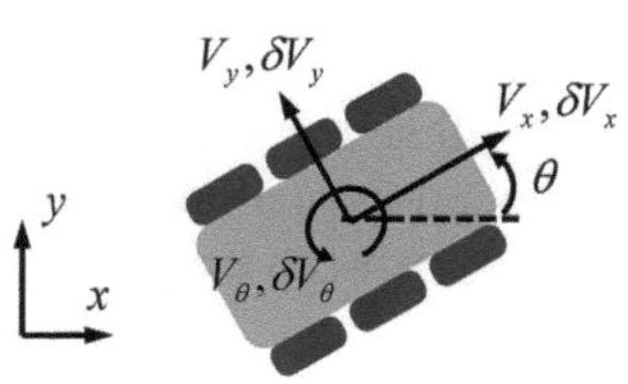

Fig. 1 Vehicle Dynamics. Three degrees of freedom remain in the general case after terrain contact is enforced. Disturbances and inputs are expressed in the body frame

Given the vehicle's commanded linear and angular velocities, we have the following unconstrained kinematic differential equation for the time derivatives of the position and yaw with respect to a ground fixed frame of reference.

$$\begin{bmatrix} \dot{x} \\ \dot{y} \\ \dot{\theta} \end{bmatrix} = \begin{bmatrix} c\theta c\beta & c\theta s\beta s\gamma - s\theta c\gamma & 0 \\ s\theta c\beta & s\theta s\beta s\gamma + c\theta c\gamma & 0 \\ 0 & 0 & \frac{c\gamma}{c\beta} \end{bmatrix} \begin{bmatrix} V_x \\ V_y \\ V_\theta \end{bmatrix} \tag{1}$$

$$c = \cos(), s = \sin(), \gamma = roll, \beta = pitch, \theta = yaw$$

A more precise model would be a 6 degree of freedom unconstrained differential equation subject to 3 constraints requiring the suspension and the roll γ, pitch β, and altitude z to adjust the wheel contact patches to remain in contact with the terrain. This level of precision would complicate the formulation and require a numerical solution for a kind of constrained transition matrix. Such precision is also unwarranted here because the above model is the nominal model only. The context is already one of characterizing deviations from this nominal model.

The terrain is assumed to be rigid and the predicted attitude angles above are computed by a perception system. If the terrain is not rigid, attitude and altitude disturbances can be added to the model, but note that such disturbances do not have the dimensions of velocities so their effects do not compound with time as do errors in dead reckoning. In plainer terms, the attitude error at the end of a 3 s prediction depends, to first order, on the terrain model at that instant rather than the history of attitude errors to that point. Conversely, wheel slip is a velocity disturbance which must be integrated to ascertain its first order effect on terminal position error. For this reason, we omit attitude and altitude errors from the perturbative model and thereby avoid the need to linearize (or even represent) the constraints at all.

Subject to the above caveats, this model is the general case for a vehicle moving on an arbitrary rigid surface with the (maximum possible) three degrees of velocity freedom. The model is relevant to rough terrain motion prediction because it is the inputs, rather than the state, which are confined to 3 degrees of freedom. Our general motion model, with pose vector $\underline{\rho} = [x\ y\ \theta]^\top$ and input vector $\underline{u} = [V_x\ V_y\ V_\theta]^\top$, therefore has the form:

$$\dot{\underline{\rho}} = \underline{f}(\underline{\rho}, \underline{u}) \tag{2}$$

2.1 Linearized Dynamics

We will now develop the linearized perturbation dynamics of this system. The first and most crucial step is to compute the linear relationship, known as the *transition matrix*, between the pose at any point in time and that at any later point in time. We can estimate the effects of the disturbances $\delta\underline{u}(t)$ on the pose errors $\delta\underline{\rho}(t)$ by writing the linear perturbation dynamics:

$$\delta\dot{\underline{\rho}} = F(t)\delta\underline{\rho}(t) + G(t)\delta\underline{u}(t) \tag{3}$$

The Jacobians $F(t)$ and $G(t)$ depend on the reference trajectory and they are taken with respect to the variables defined in Eq. (1). The attitude angles are treated as known inputs based on earlier comments so the system dynamics are really of the form $\dot{\underline{\rho}} = \underline{f}(\underline{\rho}, \underline{u}, \underline{\Omega})$ for attitude angles $\underline{\Omega}$. After a little manipulation (see [6] for detail), the Jacobians reduce to:

$$F(t) = \partial\underline{f}/\partial\underline{\rho} = \begin{bmatrix} 0 & 0 & -\dot{y} \\ 0 & 0 & \dot{x} \\ 0 & 0 & 0 \end{bmatrix} \tag{4}$$

$$G(t) = \partial\underline{f}/\partial\underline{u} = \begin{bmatrix} c\theta c\beta & c\theta s\beta s\gamma - s\theta c\gamma & 0 \\ s\theta c\beta & s\theta s\beta s\gamma + c\theta c\gamma & 0 \\ 0 & 0 & \frac{c\gamma}{c\beta} \end{bmatrix} \tag{5}$$

The transition matrix of the system is easy to derive in this case because all powers of the system Jacobian F vanish. Consider the matrix integrating factor:

$$\Psi(t, \tau) = \int_{\tau}^{t} F(\zeta)\, d\zeta = \begin{bmatrix} 0 & 0 & -\Delta y \\ 0 & 0 & \Delta x \\ 0 & 0 & 0 \end{bmatrix} \tag{6}$$

The predicted *history point displacements* are defined as $\Delta x(t, \tau) = x(t) - x(\tau)$ and $\Delta y(t, \tau) = y(t) - y(\tau)$. It is well-known in linear systems theory that when the above matrix commutes with itself, as it does in this case, its matrix exponential is the transition matrix:

$$\Phi(t, \tau) = exp[\Psi(t, \tau)] = I + \Psi(t, \tau) = \begin{bmatrix} 1 & 0 & -\Delta y \\ 0 & 1 & \Delta x \\ 0 & 0 & 1 \end{bmatrix} \tag{7}$$

It will also be convenient to define the *input transition matrix*:

$$\Gamma(t, \tau) = \Phi(t, \tau)G(\tau) \tag{8}$$

Now the main result for this section is given by what we will call the *vector superposition integral*:

$$\delta\underline{\rho}(t) = \Phi(t, t_0)\delta\underline{\rho}(t_0) + \int_{t_0}^{t} \Gamma(t, \tau)\delta\underline{u}(\tau)\, d\tau \tag{9}$$

This result expresses how the effects of errors in initial conditions $\delta\underline{\rho}(t_0)$ and the input (systematic) disturbances $\delta\underline{u}(\tau)$ are projected forward and integrated over time to produce the errors in the pose.

3 Systematic Model Identification

This section formulates a solution to the problem of identifying the system equations in order to enable better predictions of the effects of disturbances. Because the system state is always the integral of its velocity, producing the correct velocity will produce the correct state, whether the velocity errors are caused by slip or some other phenomena. We will therefore, without loss of generality, represent errors in motion prediction in terms of instantaneous values of forward slip rate δV_x, lateral slip rate δV_y, and angular slip rate δV_θ. Such input perturbations are additive to the inputs so they are expressed in the coordinates of the body frame where they are likely to be constant under steady state conditions.

Of course, the slip rates will certainly depend on the terrain and the trajectory so they will not be constant under non steady-state conditions. Therefore, the general relationship between time varying slip rates and pose errors is not a function; it is the functional given by Eq. (9) that depends on the entire error history. In this case, the relevant theory for finding an unknown function is variational optimal control. The real-time solution of the resulting Euler-Lagrange (partial differential) equations seems ill-advised, so we will use parameterization to convert to a more conventional estimation problem. The small disturbance inputs will be assumed to depend on an unknown set of parameters $\underline{\alpha}$, so that $\delta\underline{u}(\underline{\alpha}) = \left[\delta V_x \;\; \delta V_y \;\; \delta V_\theta\right]^{\mathrm{T}}$. Substituting into Eq. (9):

$$\delta\underline{\rho}(\underline{\alpha},t) = \Phi(t, t_0)\delta\underline{\rho}(t_0) + \int_{t_0}^{t} \Gamma(t, \tau)\delta\underline{u}(\underline{\alpha}, \tau)\, d\tau \tag{10}$$

Once the reference trajectory is specified, this is a vector-valued function of a vector of parameters, and time, which depends on the terrain and the inputs.

3.1 *Linearizing the Perturbation Integral*

For the identification system, we will need the linear algebraic relationship between the error inputs $\delta\underline{u}$ and the error pose $\delta\underline{\rho}(t)$ that is predicted to occur at the end of the prediction interval $(t - t_0)$ on some trajectory $\underline{\rho}(t)$. For our parameterized pose errors, the derivative of Eq. (10) is a Jacobian matrix. Since differentiation can be moved inside the integral sign by Leibniz rule and inside the matrix product by the rules of matrix differentiation, the derivative of Eq. (10) is:

$$J_{\alpha} = \frac{\partial \delta \underline{\rho}(t)}{\partial \underline{\alpha}} = \int_{t_0}^{t} \Gamma(t,\tau) \frac{\partial \delta \underline{u}}{\partial \delta \underline{\alpha}} d\tau = \int_{t_0}^{t} \Gamma(t,\tau) U_{\alpha}(\tau)\, d\tau \tag{11}$$

3.2 *Formulating the Kalman Filter*

This section formulates a Kalman filter that calibrates the predictive model on-line based on experience. We will reinterpret Eq. (1) (with slip velocities added) as a measurement process where the change in pose $\underline{\rho}(t)$ is reinterpreted as an observation of an unknown $\delta \underline{u}$.

$$\underline{h}(\underline{x}) = \Delta \underline{\rho}_{pred} = \int_{t_0}^{t} \begin{bmatrix} c\theta c\beta & c\theta s\beta s\gamma - s\theta c\gamma & 0 \\ s\theta c\beta & s\theta s\beta s\gamma + c\theta c\gamma & 0 \\ 0 & 0 & \frac{c\gamma}{c\beta} \end{bmatrix} \begin{bmatrix} V_x(\tau) + \delta V_x(\tau) \\ V_y(\tau) + \delta V_y(\tau) \\ V_{\theta}(\tau) + \delta V_{\theta}(\tau) \end{bmatrix} \delta\tau \tag{12}$$

Note that the prediction is a suitably nonlinear function of the angular slip $\delta V_{\theta}(\tau)$ because it affects the yaw angle in the rotation matrix inside the integral and the observation is nonlinear in that angle. It is for this reason that we chose to use the full nonlinear prediction in an extended Kalman filter rather than use the linearized dynamics in Eq. (9). We used the linearized dynamics, rather, to compute the Jacobian. We have used the notation $\underline{\rho}(t)$ for the system motion state, and called it a *pose*, in order to distinguish it from the state vector $\underline{x}$ of the identification Kalman filter.

The overall approach is to predict, at regular intervals, given the input trajectory $\underline{u}(t)$ for the last few seconds, a prediction of the change in pose $\underline{\rho}(t) - \underline{\rho}(t_0)$. This prediction is then compared to an actual measurement of the pose change to form an innovation that the filter must explain in terms of errors in the slip parameters. It would be possible to use recently produced terrain models in order to calibrate perception errors as well, but we have chosen to provide the attitude angles from the historical state as known inputs.

Note that our integrated dynamics formulation introduces the three gauge freedoms of motion in the plane due to the introduction of the initial conditions. If all measurements were transformed by a rigid planar transform, the innovations would be unaffected because they must use the initial pose measurement as initial conditions. One approach is to define error states that absorb the error of the initial measurement relative to the others, but we have chosen instead to form the innovation in the coordinates of the initial body frame to eliminate the initial conditions altogether.

3.3 Systematic Model

Our models are predictive, so they must be formulated in terms of predictable quantities regardless of whatever measurements of recent motions may be available. Measurements of applied forces at the wheels are not typically available, but the inertial forces caused by wheel reactions can be measured directly or computed from velocities.

Slip velocities are represented as functions of linear and angular velocity, their products (representing lateral acceleration) and applied gravitational force (computed from attitude angles). This representation includes the *slip angle* of automotive engineering as the coefficient relating lateral slip to longitudinal velocity, but it permits other linear relationships to be learned as well. The approach allows us to learn a model for how slip depends on arbitrary terrain and inputs even as they vary over time. Such a general model is a prerequisite for identifying the system on-line based on whatever trajectories it is executing. None of our vehicles accept a lateral velocity command, so the slip velocity is expressed over commanded speed V_x, and angular velocity V_θ. The result of these formulation decisions is a *slip surface* defined over this input space, expressed in coordinates fixed to the body. Of course, this model can be modified arbitrarily to suit different situations.

$$\begin{aligned} \delta V_x &= \alpha_1 V_x + \alpha_2 |V_\theta| + \alpha_3 V_x |V_\theta| + \alpha_4 g_x \\ \delta V_y &= \alpha_5 V_x + \alpha_6 V_\theta + \alpha_7 V_x V_\theta + \alpha_8 g_y \\ \delta V_\theta &= \alpha_9 V_x + \alpha_{10} V_\theta + \alpha_{11} V_x V_\theta + \alpha_{12} g_x + \alpha_{13} g_y \end{aligned} \tag{13}$$

The quantity $|V_\theta|$ appears in order to force longitudinal slip to be an even function of turn direction. The components of gravity g_x and g_y are computed from the known magnitude of gravity and the attitude angles.

We will be interested in a filter state vector formed from the parameters, $\underline{\alpha}$:

$$\underline{x} = [\alpha_1\ \alpha_2 \ldots \alpha_N]^\top \tag{14}$$

The elimination of the gauge freedoms comes at the cost of introducing a measurement transformation. The transformed measurement is the difference between the measured terminal and initial pose ($\underline{\rho}_{f,meas}$ and $\underline{\rho}_{i,meas}$ respectively) as measured by the pose estimation system, converted to initial pose coordinates.

$$\underline{z} = \underline{\rho}^i_{f,meas} = (R^w_i)^{-1}\left(\underline{\rho}_{f,meas} - \underline{\rho}_{i,meas}\right) \tag{15}$$

The Jacobian is $H = J_\alpha$, a $3 \times N$ matrix derived above in Sect. 3.1. The measurement uncertainty is derived from the uncertainty in the pose estimation system, being careful to express any needed correlations, including the correlation introduced in Eq. (15) by the conversion of coordinates.

4 Stochastic Model Identification

The random error behavior of the system can be "calibrated" in the sense that the covariance of pose predictions can be required to agree with the observed scatter of earlier predictions. Such an approach can be confusing because we are calibrating the equation that normally serves as the covariance dynamics in a Kalman filter with another Kalman filter.

The underlying system dynamics are not linear as described above. It was possible to compute a nonlinear predictive measurement $\underline{h}(\underline{x})$ in the systematic case, but here we must be content with a linear approximation or attempt nonlinear covariance propagation on-line. We chose a linear approximation because the computation must be fast and the trajectories are always relatively short. Once the decision of a linear filter is made, the random error dynamics are easy to derive from the systematic. Recall that the relationship between input noise covariance $\Xi(t)$ and pose covariance $\Pi(t)$ in continuous time is given by what we will call the *matrix superposition integral*:

$$\Pi(t) = \Phi(t, t_0)\Pi(t_0)\Phi(t, t_0)^\top + \int_{t_0}^{t} \Gamma(t, \tau)\Xi(\tau)\Gamma(t, \tau)^\top d\tau \tag{16}$$

All of the (trajectory dependent) matrices above are known except for the input random disturbance covariance $\Xi(t) = Exp[\delta\underline{u}(t)\delta\underline{u}(t)^\top]$. Once it is known, the pose prediction covariance $\Pi(t) = Exp[\delta\underline{\rho}(t)\delta\underline{\rho}(t)^\top]$ can be computed. The notation is chosen to distinguish these matrices from analogous ones called P and Q in the Kalman filter used to estimate them.

The linearization of the prediction integral proceeds analogously to the deterministic case. To save space, we will present only the highlights. The Jacobian of the pose covariance Π taken with respect to the input covariance Ξ is the derivative of a matrix with respect to a matrix—a 4th order tensor. In the simplest case, the noise sources are constants, assumed not to depend on the trajectory, and $\Xi(t)$ is a 3×3 symmetric positive definite matrix with 6 independent elements. The Jacobian can also be regarded as a set of six matrices. The derivative of $\Pi(t)$ with respect to the (i,j) element of Ξ is:

$$J_{\xi ij} = \frac{\partial \Pi(t)}{\partial \xi_{ij}} = \int_{t_0}^{t} \underline{\gamma}_i(t, \tau)\underline{\gamma}_j(t, \tau)^\top d\tau \tag{17}$$

where $\underline{\gamma}_i(t, \tau)$ is the ith column of $\Gamma(t, \tau)$.

While a Kalman filter with a "state matrix" can be defined, it is conceptually simpler to collect the 6 independent elements of Π into a state vector and the independent elements of Ξ into a measurement vector and reorganize the Jacobian as appropriate. The Kalman filter is then analogous to the deterministic case where the measurement is the sample covariance matrix S computed from the pose innovations *after* the systematic component of error has been removed.

$$S(t_k) = \frac{1}{(n-1)} \sum_{k=1}^{k=n} \delta \underline{\rho}^i_{f,lmeas}(t_k) \delta \underline{\rho}^i_{f,meas}(t_k)^\top \tag{18}$$

When calibrating online, the luxury of repeating the same path multiple times to observe scatter is unavailable. It takes just a little effort to define the predicted variance of n samples taken from n different distributions. It can be shown from the total probability theorem [5] that the average of all of the predicted covariance matrices for each of the trajectories is the predicted covariance for the sample. However, our experiments have produced very consistent estimates of covariance based on presenting innovations to the stochastic identification filter one single predicted covariance $\Pi(t)$ at a time.

5 Results

Experiments were conducted on three vehicles. Crusher is a six-wheeled skid-steered vehicle with an advanced active suspension. It is capable of autonomously driving through deserts, mountains, forests, wetlands, and other extreme environments. In order to show applicability to other platforms, tests were also conducted on the LandTamer (skid-steered, hydraulic) and RecBot (Ackerman-steered, electric). In all cases, a high-end IMU and differential GPS unit were used for ground truth motion measurement. Our method should work just as well with visual odometry or any other system that measures motion, on the scale of a few seconds, with error significantly less than the prediction errors being resolved (Fig. 2).

The first results presented are for the Crusher vehicle driving on rough, grassy terrain at Camp Roberts in California. In this dataset Crusher traverses steep slopes of up to 29° (see Fig. 3) which enables the identifier to predict the dependence of slip on the x and y components of the gravity vector, as well as commanded forward and angular velocities. The vehicle is commanded to drive at speeds up to 6 m/s and angular velocities up to 4 rad/s. The improvement in predicted change in pose at the end of 2-s path segments is shown in Fig. 4. The standard deviation of along track error and cross track error are reduced by 71 % and 81 % respectively. The standard deviation of heading prediction error is reduced by 90 %. Note that the mean error is also reduced from 1.9 m to near zero. In these and all other scatter plots, all of the data is processed but only 1000 data points are plotted. The data points are equally spaced in time and span the entire experiment.

The time to converge depends primarily on the initial parameter estimates and the diversity of input path segments. Figure 5a shows the pose prediction performance on holdout Camp Roberts data after calibrating for limited periods ranging

Fig. 2 From *left* to *right* Crusher, LandTamer, RecBot

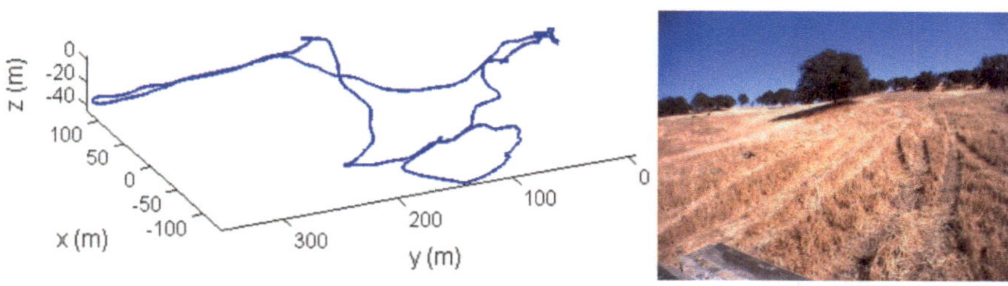

Fig. 3 *Left* The path of the Crusher vehicle in the Camp Roberts test. During the test Crusher traverses steep slopes; roll angles range from –28 to 29° and pitch ranges from –22 to 17°. *Right* Image captured by one of Crusher's cameras; Crusher traverses both a dirt road and tall dry grass

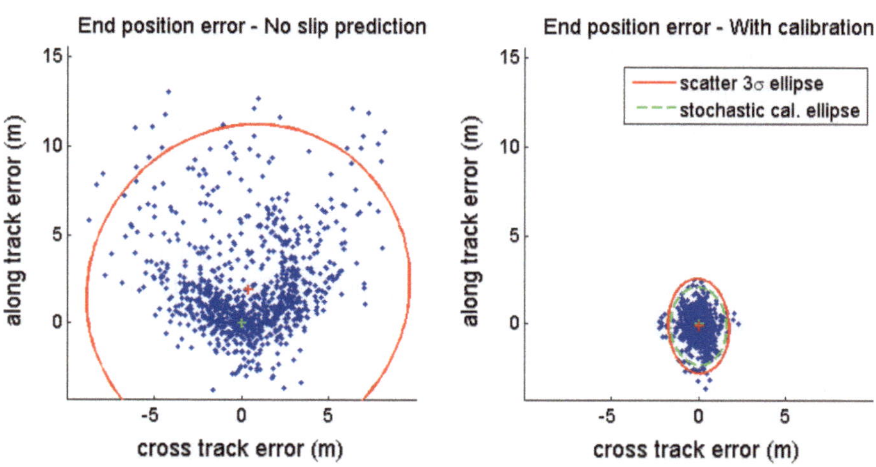

Fig. 4 Scatter plots of along track and cross track error for the Crusher vehicle at Camp Roberts. Each point represents predicted pose error at the end of a 2-s path segment. The *left* figure shows predicted pose error with no slip calibration, the *right* figure shows *online* prediction error during calibration. The three standard deviation error ellipse of the points is shown by the *solid red line*. The *dashed green line* ellipse (just inside the *red* ellipse) is the average pose uncertainty predicted by the stochastic calibration filter

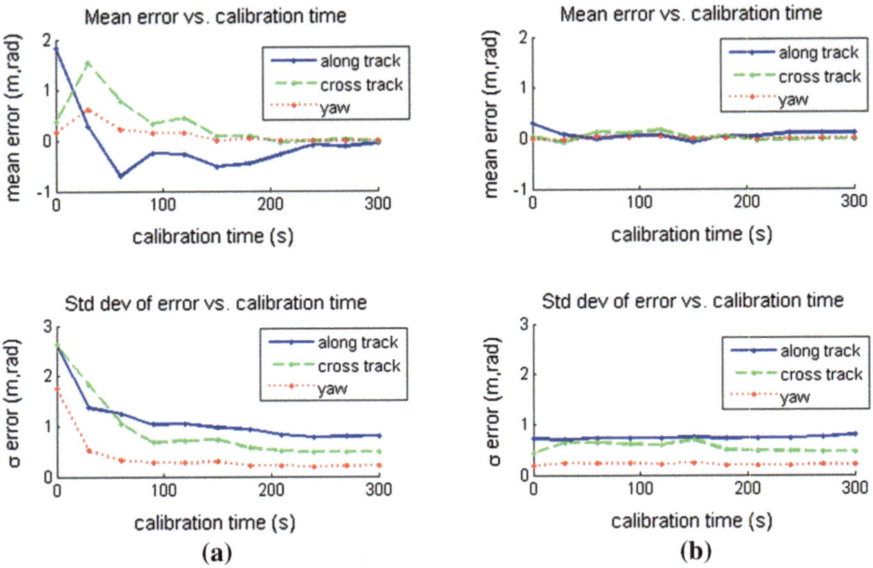

Fig. 5 Plots of the mean and standard deviation of pose prediction error versus calibration time on the Camp Roberts dataset. The model is calibrated for 0–300 s of driving then evaluated on holdout data (the remainder of the 13 min data log). **a** shows results when starting from a completely uncalibrated model (all slip parameters initialized to zero), and **b** shows results when starting from the average calibration on other Crusher datasets

from 0 to 300 s. Based on pose prediction error, the filter converges within 4 min of driving when starting from the uncalibrated case (i.e. all slip parameters initialized to zero). When starting from the mean calibration for the Crusher vehicle on other datasets, the filter converges in only seconds, Fig. 5b.

The slip surfaces learned by the filter on the Camp Roberts dataset are shown in Fig. 6. Note that forward slip is negatively correlated with the absolute value of the commanded angular velocity, $|V_\theta|$. Lateral slip depends primarily on centripetal acceleration ($V_x V_\theta$) as expected. Angular slip is predominantly a linear function of the commanded angular velocity. As expected, the filter learned a positive correlation between forward slip and the x component of the gravity vector ($\delta V_x = \cdots + 0.256 g_x$) and between lateral slip and the y component ($\delta V_y = \cdots + 0.043 g_y$).

Figure 7 provides a visual summary of the extreme cases that are being predicted well in the Crusher datasets. First, it is important to note that Crusher's effective turn rate is only a third of the commanded rate. This is partly due to the fact that four of six wheels are slipping sideways and resisting motion in a tight turn. There is evidently no yaw rate feedback used to compensate these gross errors. Furthermore, when turning on a hill, the wheels on the high side carry no load and are therefore unable to generate traction. The result is that turning becomes impossible and turn commands cause more or less translational motion perpendicular to the terrain

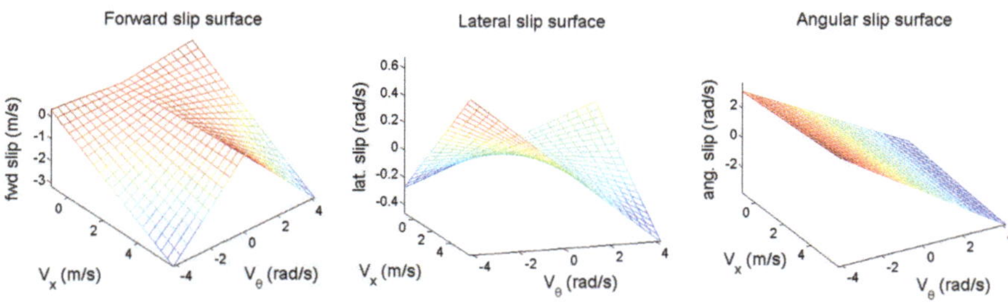

Fig. 6 Plots of the slip surfaces for the Crusher Camp Roberts dataset. These surfaces show the *predicted* forward, lateral, and angular slip as a function of the commanded forward and angular velocity (V_x and V_θ), according to (13). These surfaces correspond to zero x and y components of the gravity vector (i.e. driving on flat ground)

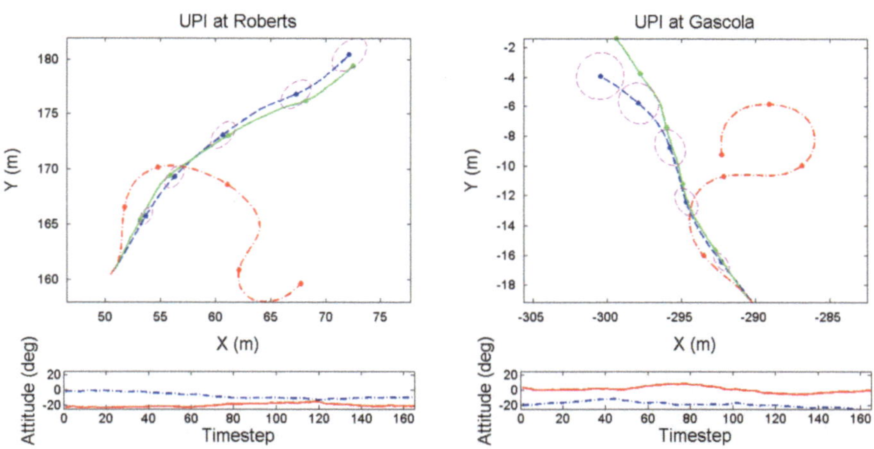

Fig. 7 Predicted and measured paths under different conditions for Crusher. The plan views show predicted paths without slip modeling (*dash-dot red*) with slip modeling (*dashed blue*) and measured path (*solid green*). Attitude signals roll (*solid red*) and pitch (*dash-dot blue*) are shown underneath. The paths are 10 s long. The ellipses are estimates of uncertainty at 2 s intervals. *Left* Crusher diving at Camp Roberts with a –20° roll. Commanded turns in either direction are correctly predicted to be impossible. *Right* Crusher driving uphill at Gascola on 15° to 20° slope. 25 % longitudinal wheel slip is correctly predicted

gradient. These effects and others depend on the orientation of the terrain gradient in the body frame, just as our model is formulated to learn them.

Crusher was also driven on muddy slopes during a rainstorm at Gascola, Pennsylvania (see Fig. 8). Large reductions in motion prediction error were observed similar to the Camp Roberts results despite the difficult weather conditions, Fig. 9. The standard deviation of along track, cross track, and heading error were reduced by 71 %, 82 %, and 87 % respectively.

In another experiment, data was collected on the Land Tamer vehicle in a muddy gravel lot after a heavy rain. Data was collected as the vehicle was commanded to

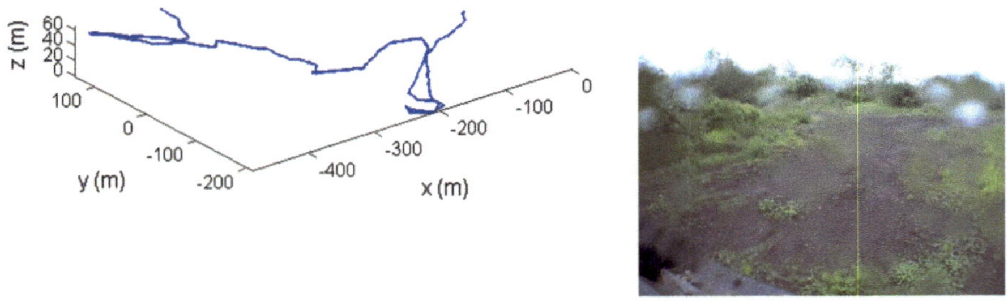

Fig. 8 *Left* Path of the Crusher vehicle in the Gascola test. During the test, roll angles range from –25 to 25° and pitch ranges from –26 to 16°. *Right* Image captured by one of Crusher's cameras; Crusher traverses muddy terrain and vegetation. Note the raindrops on the lens in the image

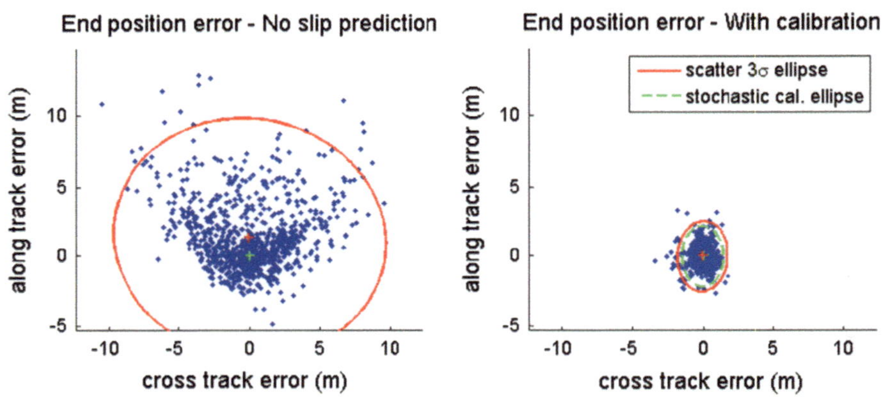

Fig. 9 Scatter plots of along track and cross track error for the Crusher vehicle at Gascola. Each *dot* represents predicted pose error 2 s in the future

drive in circles at various speeds and curvatures (0.25–1.0 m/s and 0.4–0.6 m^{-1}). As seen in the error scatter plot, Fig. 10, the large cross track error bias was almost completely removed by the calibrated wheel slip model.

Finally, tests were also conducted on the RecBot vehicle, a medium size drive-by wire UGV. The vehicle is Ackerman-steered in contrast to the previous skid-steered data sets. Data was collected as the RecBot was driven randomly on a grass lawn for just over five and half minutes at speeds up to 4.8 m/s. The grass was mostly level and flat, except for large tractor treads that added variance to the slip rates. Results for the RecBot vehicle are presented in Fig. 11.

In addition to learning models of wheel slip, our formulation can be used to learn models of the vehicle powertrain. The powertrain model maps nominal velocity commands (V_x, V_θ) to actual wheel angular velocities (as measured by encoders) and is learned online in parallel with slip model. The powertrain modeling is not the dominant effect in our experiments, so we omit the details due to space limitations.

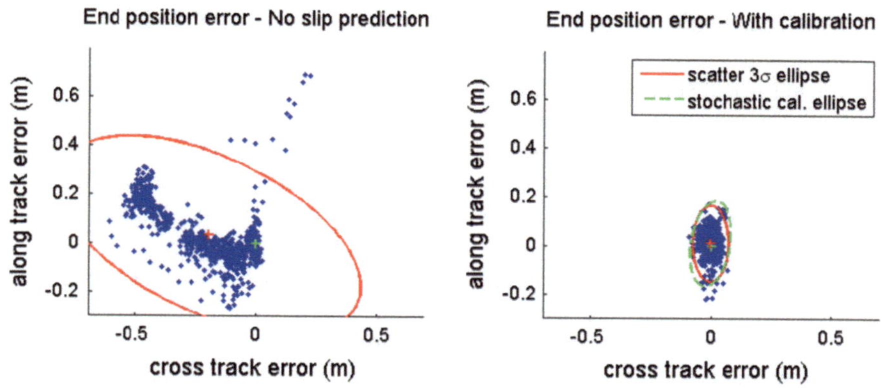

Fig. 10 Scatter plots of along track and cross track error for the LandTamer vehicle driving circles on wet gravel. Each dot represents predicted pose error 2 s in the future

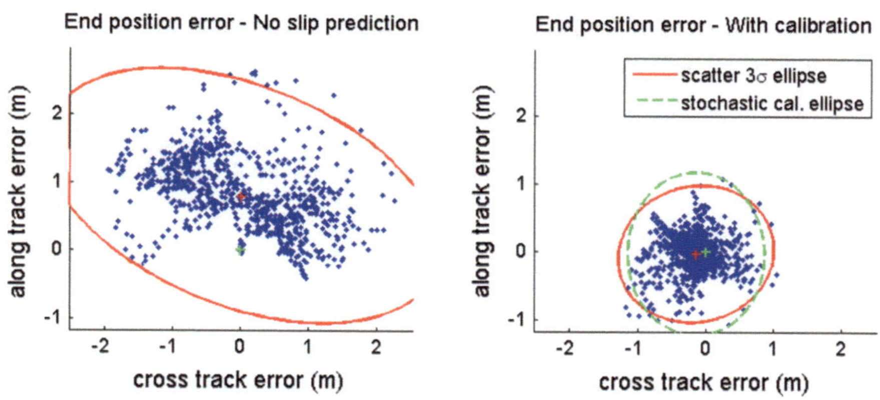

Fig. 11 Scatter plots of along track and cross track error for the RecBot vehicle driving in grassy yard. Each *dot* represents predicted pose error 4 s in the future. Improvements are not as pronounced here because the terrain and the trajectories are less extreme

6 Conclusion

The capacity to adapt to changes in slip and traction while predicting aggressive maneuvers accurately is a fundamental requirement of self-sufficient, high performance robots. Because the relevant mechanical properties of the terrain are not directly observable, predicting motion accurately has always been regarded as a difficult problem. We undertook it anyway because we felt that even a poor model was better than the present state of the art which ignores such issues completely. We have developed a capacity to enable mobile robots to be much more informed about their own mobility, both in terms of the mapping from their inputs to outputs and in terms of the residual random error in that mapping.

While our black box approach teaches us little about the underlying terramechanics, it has been quite successful in predicting motion under conditions of significant wheel slip, on significant slopes, during aggressive maneuvers. It applies without modification across multiple vehicle classes including Crusher which represents the possible worst case of high speed skid steer platforms. Because our vehicle model is velocity-based, and not force-based, skidding with locked brakes and certain other slip scenarios are not predicted, but will be investigated in future work.

The implementation is straightforward and lightweight from a processing point of view. Path integrals are the same computations already performed in model predictive control, but we need only the path followed and only one path segment every few seconds. The systematic and stochastic Kalman filters are of 13 and 6 states respectively and can run at frequencies in excess of 100 Hz. In effect, the computations are negligible compared to the processing load of motion control and pose estimation. We hope to show in future work how our identification system can coexist with a model predictive control system and a pose estimation system so that both benefit from continuously calibrated, more accurate models of the platform motion during aggressive maneuvering.

Acknowledgments This research was made with U.S. Government support under and awarded by the Army Research Office (W911NF-09-1-0557), the Army Research Laboratory (W911NF-10-2-0016), and by the DoD, Air Force Office of Scientific Research, National Defense Science and Engineering Graduate (NDSEG) Fellowship, 32 CFR 168a.

References

1. P. Abbeel, A. Coates, M. Montemerlo, A.Y. Ng, S. Thrun, Discriminative training of Kalman filters, in *Proceedings of the Robotics: Science and Systems* (2005)
2. A. Angelova, L. Matthies, D. Helmick, G. Sibley, P. Perona, Learning to predict slip for ground robots, in *Proceedings of the IEEE International Conference on Robotics and Automation* (2006)
3. M. Bode, Learning the Forward Predictive Model for an Off-Road Skid-Steer Vehicle, Technical report CMU-RI-TR-07-32, Robotics Institute, Carnegie Mellon University, Sept 2007
4. G. Ishigami, A. Miwa, K. Nagatani, K. Yoshida, Terramechanics-based model for steering maneuver of planetary exploration rovers on loose soil. J. Field Robot. **24**(3), 233–250 (2007)
5. A. Kelly, Fast and easy systematic and stochastic odometry calibration, in *Proceedings of the International Conference on Robots and Systems* (2004)
6. A. Kelly, Linearized error propagation in odometry. Int. J. Robot. Res. **23**(2), 179–218 (2004)
7. E. Lucet, C. Grand, D. Sall, P. Bidaud, Dynamic sliding mode control of a four-wheel skid-steering vehicle in presence of sliding, Romansy, Tokyo, Japan, July 2008
8. C.B. Low, D. Wang, Integrated estimation for wheeled mobile robot posture, velocities, and wheel skidding perturbations, in *Proceedings of the IEEE International Conference on Robotics and Automation*. Rome, Italy, Aug 2007
9. F. Rogers-Marcovitz, A. Kelly, On-line mobile robot model identification using integrated perturbative dynamics, in *Proceedings of the International Symposium on Experimental Robotics*, Delhi, India, Dec 2010

10. L. Seneviratne, Y. Zweiri, S. Hutangkabodee, Z. Song, X. Song, S. Chhaniyara, S. Al-Milli, K. Althoefer, The modeling and estimation of driving forces for unmanned ground vehicles in outdoor terrain. Int. J. Model. Ident. Control **6**(1) (2009)
11. C. Ward, K. Iagnemma, A dynamic model-based wheel slip detector for mobile robots on outdoor terrain. IEEE Trans. Robot. **24**(4), 821–831 (2008)
12. J. Yi, H. Wang, J. Zhang, D. Song, S. Jayasuriya, J. Liu, Kinematic modeling and analysis of skid-steered mobile robots with applications to low-cost inertial-measurement-unit-based motion estimation. IEEE Trans. Robot. **25**(6) (2009)

Prediction and Planning Methods of Bipedal Dynamic Locomotion Over Very Rough Terrains

Luis Sentis, Benito R. Fernandez and Michael Slovich

Abstract Although the problem of dynamic locomotion in very rough terrain is critical to the advancement of various areas in robotics and health devices, little progress has been made on generalizing gait behavior with arbitrary paths. Here, we report that perturbation theory, a set of approximation schemes that has roots in celestial mechanics and non-linear dynamical systems, can be adapted to predict the behavior of non closed-form integrable state-space trajectories of a robot's center of mass, given its arbitrary contact state and center of mass (CoM) geometric path. Given an arbitrary geometric path of the CoM and known step locations, we use perturbation theory to determine phase curves of CoM behavior. We determine step transitions as the points of intersection between adjacent phase curves. To discover intersection points, we fit polynomials to the phase curves of neighboring steps and solve their differential roots. The resulting multi-step phase diagram is the locomotion plan suited to drive the behavior of a robot or device maneuvering in the rough terrain. We provide two main contributions to legged locomotion: (1) predicting CoM state-space behavior for arbitrary paths by means of numerical integration, and (2) finding step transitions by locating common intersection points between neighboring phase curves. Because these points are continuous in phase they correspond to the desired contact switching policy. We validate our results on a human-size avatar navigating in a very rough environment and compare its behavior to a human subject maneuvering through the same terrain.

L. Sentis (✉) · B.R. Fernandez · M. Slovich
University of Texas, Austin, TX, USA
e-mail: lsentis@austin.utexas.edu

B.R. Fernandez
e-mail: benito@mail.utexas.edu

M. Slovich
e-mail: slovich@mail.utexas.edu

H.I. Christensen and O. Khatib (eds.), *Robotics Research*,
Springer Tracts in Advanced Robotics 100, DOI 10.1007/978-3-319-29363-9_34

1 State of the Art

In dynamic walking we can classify techniques in various categories: (1) trajectory-based techniques, (2) limit cycle-based techniques, (3) prediction of contact, and (4) hybrids of the previous three.

Trajectory-based techniques are techniques that track a time-based joint or task space trajectory according to some locomotion model such as the *Zero Moment Point* (ZMP). The state of the art of these methods includes generalized multi-contact locomotion behaviors, developed in [1] and more recently, a time delay extension to the ZMP method for locomotion in moderately uneven terrain, developed by [2].

Prediction of contact placement are techniques that use dynamics to estimate suitable contact transitions to produce locomotion or regain balance. In [3], simple dynamic models are used to predict the placement of next contacts to achieve desire gait patterns. Finding feasible CoM static placements given frictional constraints was tackled in [4, 5]. In [6], *stable locomotion*, in the wide sense of not falling down, is studied by providing velocity based stability margins. This work is used to regain stability when the robot's is pushed out, and lead to the concept of Capture Point.

Limit cycle based techniques were pioneered by McGeer [7] through the field of *passive dynamic walking*. In [8] the authors study orbital stability, and the effect of feedback control to achieve asymptotic stability. Optimization of open-loop stability is investigated in [9]. In [10], the authors analyze the energetic cost of bipedal walking and running as well as the role of leg sequencing. In [11], the authors developed a dynamic walker using artificial muscles and principles of stability of passive walkers. In [12], a methodology for the analysis of state-space behavior and feedback control are presented for various physical robots. *Step recovery* in response to perturbations is studied in [13] supported by a linear bipedal model in combination with an orbital energy controller. In [14], the selection of gait patterns based on studying the interplay between robustness against perturbations and leg compliance is investigated.

Hybrid methods include [15], where the stability of passive walkers is studied and a controller obeying the rule, "in order to prevent falling backward the next step, the swing leg shouldn't be too far in front", in the words of the author, is suggested. Stochastic models of stability and its application for walking on moderately rough unmodeled terrain are studied in [16]. The design of non-periodic locomotion for uneven terrain is investigated in [17]. In [18], the authors explore the design of positivity-based controllers to achieve walking on different ground slopes. Optimization-based techniques for locomotion in rough terrains are presented in [19]. Locomotion in very rough terrain is presented in [20], where the authors exploit optimization and static models as a means to plan locomotion. More recently, the authors of [21] have proposed a very efficient planner that can generate a discrete sequence of multi-contact stances using static criteria. Also recently, we made a theoretical contribution in the form of an extended abstract [22] to enable walking at fast speeds in very difficult variable terrain.

2 Summary of Our Approach

We present here a new contribution that tackles rough terrain locomotion by exploring *CoM state-space manifolds* and *transitional contact states*.

Our approach, can be explained algorithmically in terms of various phases, namely (1) geometric planning, (2) perturbation-based CoM phase generation, and (3) dynamic step planning based on locating common intersection points between neighboring CoM phase curves. The geometric planning phase consists of applying standard kinematic planning techniques to obtain initial guesses of feet contact locations and CoM geometric path. Perturbation-based CoM phase generation is our first contribution and consists on: (1) formulating CoM accelerations based on the contact state, (2) incorporating the dependencies between Sagittal and vertical accelerations due to the given CoM geometric path, and (3) using perturbation theory to predict phase curves of the CoM in the vicinity of the step contacts and given initial and final conditions of the step. The step solver is our second contribution and consists on finding step transitions by locating common intersection points between neighboring CoM phase curves. Because these points are continuous in phase they correspond to the desired contact switching policy.

3 Mathematic Derivations

3.1 Dynamic Behavior from Single Contact Point

Dynamic equilibrium (a principle derived from Newton's Laws of Motion and Lagrange-d'Alembert Formalism) states that the sum of acting moments on a moving system equals the net inertial moment. Given a contact scenario, such as the one shown in Fig. 1, this principle translates into the following moment balance expression

$$\sum_{i=1}^{n_s} p_{\text{cop}_i} \times f_{r_i} + \sum_{i=1}^{n_s} m_{r_i} = p_{\text{com}} \times (f_{\text{com}} + M\, g) + m_{\text{com}}, \tag{1}$$

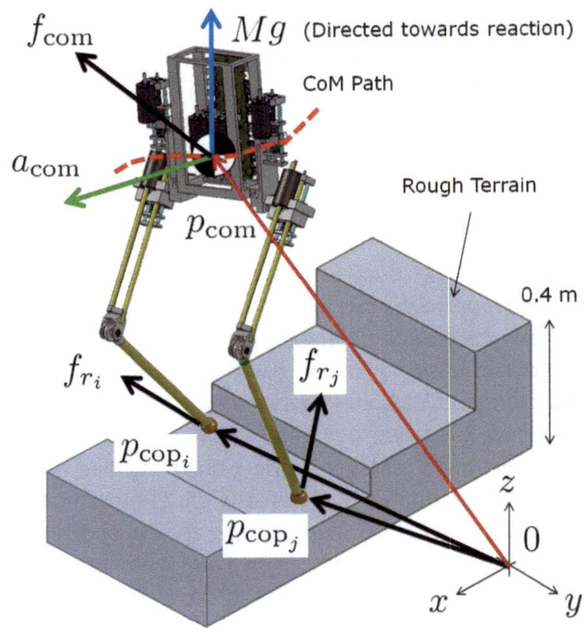

Fig. 1 Definition of coordinates of center of mass (CoM), center of pressures (CoP), and their position coordinates, $p_{com}, p_{cop_i}, p_{cop_j}$. Also shown are reaction forces, f_{r_i}, f_{r_j} and the CoM's acceleration, a_{com}

where p_{cop_i} is the i-th foot contact pressure point (with respect to the coordinate origin), $i = 1, \ldots, n_{s,}$ the number of supporting limbs; f_{r_i} and m_{r_i} are the reaction force and moment at the pressure point; p_{com} is the vector from the origin (of coordinates) to the CoM; f_{com} and m_{com} are the net force and moment acting on the CoM; M is the robot's mass and g is gravitational constant expressed upwards in the direction of the reaction forces.

Due to the complexity of the algorithms, in this paper we will first address locomotion as transitions involving one support limb. Therefore, the above equation becomes

$$p_{cop_k} \times f_{r_k} + m_{r_k} = p_{com} \times (f_{com} + M\,g) + m_{com}, \tag{2}$$

where, k is the limb in contact with the terrain, p_{cop_k} is the limb's Center of Pressure (CoP) point. The above equation is vectorial and represents three orthogonal moments. Because we aim first at controlling planar robots in the Sagittal direction, we consider only solutions that produce accelerations in that direction, i.e.

$$\left[p_{cop_k} \times f_{r_k} = p_{com} \times (f_{com} + M\,g) + m_{com}\right]^{\mathcal{Y}}. \tag{3}$$

where the $\mathcal{Y}$ symbolizes the Sagittal plane (x-coordinate for frontal direction and z-coordinate for vertical).

Considering dynamic equilibrium in forces we obtain

$$f_{r_k} = f_{\text{com}} + M\,g, \tag{4}$$

and therefore we can rearrange Eq. (3) as

$$\left[\left(p_{\text{com}} - p_{\text{cop}_k}\right) \times f_{rk}\right]^{\mathcal{Y}} = m_{\text{com}}^{\mathcal{Y}}. \tag{5}$$

Solving this equation for the CoP in the Sagittal direction leads to the solution

$$p_{\text{cop}_k[x]} = p_{\text{com}[x]} - \frac{f_{r_{[kx]}}}{f_{r_{[kz]}}}\left(p_{\text{com}[z]} - p_{\text{cop}_k[z]}\right) - \frac{m_{\text{com}[y]}}{f_{r[kz]}}. \tag{6}$$

Considering that $f_{r_{[kx]}} = M a_{\text{com}[x]}$, and $f_{r_{[kz]}} = M\left(a_{\text{com}[z]} + g\right)$ we rewrite the above equation as

$$a_{\text{com}[x]} = \frac{\left(p_{\text{com}[x]} - p_{\text{cop}_k[x]}\right)\left(a_{\text{com}[z]} + g\right)}{p_{\text{com}[z]} - p_{\text{cop}_k[z]}} \tag{7}$$

Here, we have assumed a point mass model of the robot, with all of its weight located at its center of mass. As such, there are no inertial moments generated about the center of mass. Also, note that a similar equation could be derived for accelerations in the lateral direction, but for the sake of simplicity we do not consider them in this first study.

3.2 *Integration of Geometric Path*

Considering that $a_{\text{com}[x]} \triangleq \ddot{p}_{\text{com}[x]}$, the above equation is dynamic and nonlinear. As such, the major challenge that it poses is that it does not have a closed form solution, specially if $p_{\text{com}[z]}$ and $a_{\text{com}[z]}$ are time varying. This difficulty corresponds to the case of our study.

Almost all previous work that has addressed Eq. (7) has tackled the solution by simplifying it, constraining CoM trajectories to a fixed height, i.e. $p_{\text{com}[z]} =$ constant. These type of solutions have led to the concept of the Zero Moment Point (ZMP). However, in doing so, locomotion trajectories cannot be considered for arbitrary terrains nor natural motion involving vertical changes of the hip can be predicted. Therefore, our first contribution is on predicting the behavior corresponding to the general case of Eq. (7). Because there are two variables that need to be solved, i.e. the trajectories of the center of mass on the Sagittal and vertical directions, we need to first seed geometric dependencies based on an initial guess.

There are many options to determined these dependencies, ranging from ensuring kinematic constraints, generating biomimetic patterns, or minimizing electric and mechanical power. For the time being, let us pick the option of ensuring kinematic constraints.

In such case, one simple dependency that fulfills the needs is to draw a piecewise linear geometric path of the humanoid's CoM behavior that changes slope with the terrain while complying with kinematic constraints. In Fig. 2 we depict two hypothetical paths, one linear and one sinusoidal. Let us consider the linear case first with a static contact and use it to predict the CoM dynamic behavior. More generally, if the CoM geometric path is piecewise linear, it can be specified through equations of two or more intersecting lines, i.e.

$$p_{\text{com}[z]} = \begin{cases} a_1\, p_{\text{com}[x]} + b_1, & p_{\text{com}} \in \mathbb{P}_1 \\ a_2\, p_{\text{com}[x]} + b_2, & p_{\text{com}} \in \mathbb{P}_2 \\ \quad\vdots & \\ a_N\, p_{\text{com}[x]} + b_N, & p_{\text{com}} \in \mathbb{P}_N \end{cases} \tag{8}$$

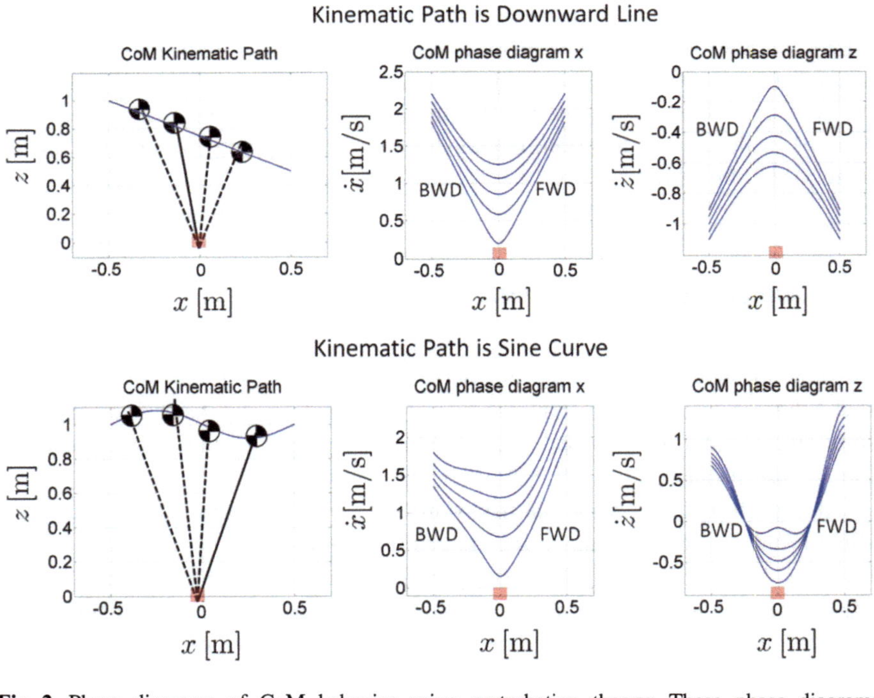

Fig. 2 Phase diagrams of CoM behavior using perturbation theory: These phase diagrams correspond to Matlab simulations of CoM behavior given a foot contact point, a desired CoM kinematic path, and varying boundary conditions given at the apex of the step (i.e. when the CoM is directly above the foot contact point)

where, $\mathbb{P}_k$ represents the path of the CoM over step k, a_i represents the slope of the piecewise lines, and b_i represents the corresponding vertical crossing points. Moreover, the acceleration profile can be extracted by differentiating twice the above piecewise equation, i.e.

$$p_{\text{com}[z]} = a_i\, p_{\text{com}[x]} + b_i \Rightarrow a_{\text{com}[z]} = a_i a_{\text{com}[x]}. \tag{9}$$

Plugging the above acceleration in (7) we get

$$a_{\text{com}[x]} = \frac{\left(p_{\text{com}[x]} - p_{\text{cop}_k[x]}\right)\left(a_i a_{\text{com}[x]} + g\right)}{a_i\, p_{\text{com}[x]} + b_i - p_{\text{cop}_k[z]}}, \tag{10}$$

and since $a_{\text{com}[x]}$ appears both on the left and right hand sides, we can rewrite the equation as

$$a_{\text{com}[x]} = \frac{\left(p_{\text{com}[x]} - p_{\text{cop}_k[x]}\right) \cdot g}{\left(b_i + a_i\, p_{\text{cop}_k[x]} - p_{\text{cop}_k[z]}\right)}. \tag{11}$$

Notice that the denominator and the second term in the numerator above are constants, so the above equation is of the form $\ddot{x} = \beta(x - \alpha)$, which is linear and as such has an exact solution.

However, in the more general case, kinematic paths do not necessarily map to piecewise linear functions, but instead should be based on more sophisticated mappings. For instance, an efficient gait can be produced by following circular arcs, i.e. $p_{\text{com}[z]} = \left(r^2 - p^2_{\text{com}[x]}\right)^{0.5}$. In that case path accelerations for a given step can be expressed by differentiating the arc, i.e.

$$\begin{aligned} a_{\text{com}[z]} = &-\left(r^2 - p^2_{\text{com}[x]}\right)^{-1.5} p^2_{\text{com}[x]} v^2_{\text{com}[x]} \\ &-\left(r^2 - p^2_{\text{com}[x]}\right)^{-0.5} v^2_{\text{com}[x]} \\ &-\left(r^2 - p^2_{\text{com}[x]}\right)^{-0.5} p_{\text{com}[x]}\, a_{\text{com}[x]} \end{aligned} \tag{12}$$

where, r is the radius of the arc. Plugging the above acceleration dependency in (7) we get

$$a_{\text{com}[x]} = \left(p_{\text{com}[x]} - p_{\text{cop}[kx]}\right) \frac{N\left(p_{\text{com}[x]}, v_{\text{com}[x]}, p_{\text{cop}[kx]}\right)}{D\left(p_{\text{com}[x]}, p_{\text{cop}_k[x]}, p_{\text{cop}_k[z]}\right)}, \tag{13}$$

with

$$N \triangleq g - \left(r^2 - p_{\mathrm{com}[x]}^2\right)^{-1.5} p_{\mathrm{com}[x]}^2 v_{\mathrm{com}[x]}^2 - \left(r^2 - p_{\mathrm{com}[x]}^2\right)^{-0.5} v_{\mathrm{com}[x]}^2 \tag{14}$$

$$D \triangleq \left(r^2 - p_{\mathrm{com}[x]}^2\right)^{0.5} - p_{\mathrm{cop}_k[z]} + \left(p_{\mathrm{com}[x]} - p_{\mathrm{cop}_k[x]}\right)\left(r^2 - p_{\mathrm{com}[x]}^2\right)^{-0.5} p_{\mathrm{com}[x]}. \tag{15}$$

The acceleration of Eq. (13) is non-linear and therefore there is no closed-form solution anymore.

If the CoM geometric paths are generated by a more sophisticated planner with more complex kinematic dependencies, the acceleration profile will be non-linear with general expression

$$a_{\mathrm{com}[x]} = \left(p_{\mathrm{com}[x]} - p_{\mathrm{cop}[kx]}\right) \cdot \Phi\left(p_{\mathrm{com}[x]}, v_{\mathrm{com}[x]}, p_{\mathrm{cop}_k[x]}, p_{\mathrm{cop}_k[z]}\right), \tag{16}$$

where, $\Phi(\cdot, \cdot, \cdot, \cdot)$ is a non-linear function, and as such does not have a closed form solution.

3.3 State-Space Behavior Prediction from Perturbation Theory

Our objective is to extract state-space trajectories for arbitrary kinematic CoM paths, $\mathbb{P}_k$. We refer to numerical integration to address the difficulty of solving non-linear differential equations such as Eq. (16). In particular, perturbation theory, has been widely used to solve the trajectory of celestial bodies and complex physical phenomena. Perturbation theory, is a set of methods that enable to approximate solutions from problems that do not have exact solutions, by looking into the solution of an exact related problem. In our case, we have the exact solution of accelerations given positions and contact points and we seek to approximate the solution of the CoM trajectory versus its velocity, i.e. the state-space trajectory.

Let us study our case. For simplicity, we call $x \triangleq p_{com[x]}$ and therefore we can write Eq. (16) as

$$\ddot{x} = f(x, \dot{x}), \tag{17}$$

where $f(x, \dot{x})$ is the RHS of Eq. (16). We assume that $\ddot{x}$ is approximately constant for small perturbations of x. By integrating over a small time period, ϵ (the perturbation), and for boundary conditions $(x_k, \dot{x}_k)$ we approximate the behavior of neighboring points as

$$\dot{x}_{k+1} \approx \dot{x}_k + \ddot{x}_k \epsilon, \tag{18}$$

$$x_{k+1} \approx x_k + \dot{x}_k \epsilon + 0.5\, \ddot{x}_k \epsilon^2. \tag{19}$$

From Eq. (18) we find an expression of the perturbation in terms of the velocities and acceleration, $\epsilon \approx (\dot{x}_{k+1} - \dot{x}_k)/\ddot{x}_k$, and substituting in Eq. (19), with $\ddot{x}_k = f(x_k)$, we get

$$x_{k+1} \approx \frac{\left(\dot{x}_{k+1}^2 - \dot{x}_k^2\right)}{2f(x_k)} + x_k, \tag{20}$$

which is the state-space approximate solution that we were looking for.

The pipeline for finding state-space trajectories goes as follows: (1) choose a very small time perturbations ϵ, (2) given known velocities $\dot{x}_k$ and accelerations $\ddot{x}_k$ and using Eq. (18), we get the next velocity $\dot{x}_{k+1}$, (3) using Eq. (20) we get the next position x_{k+1}, (4) plot the points $(x_{k+1}, \dot{x}_{k+1})$ in the phase plane. We also notice, that we can iterate this recursion both forward and backward. If we iterate backward, we need to choose a negative perturbation ϵ.

Let us apply this method to the case of complex CoM paths as characterized by the general acceleration of Eq. (16). We apply it to two different trajectories, one where the CoM follows a downward linear path, Fig. 2a–c and another one where the CoM follows a sinusoidal wave, Fig. 2d–f. The results of these two studies are shown in Fig. 2. In both cases the contact foot is located at point $\left(p_{\text{cop}[x]}, p_{\text{cop}[z]}\right) = (0,0)[m]$. For both studies, we provide various initial conditions at the apex (i.e. when the CoM is on top of the contact point), corresponding to the initial position and velocity, and using the proposed perturbation method obtain the phase diagram using forward and backward propagation. The reason why the Sagittal phase diagram of the linear CoM path is symmetrical is because Sagittal CoM accelerations are independent of vertical variations. This is not the case when the path is sinusoidal.

4 Motion Planning

Equipped with the perturbation method, which has allowed us to predict phase diagrams given arbitrary CoM paths and contact locations, we are now in the position to use it to plan dynamic walk in a very rough terrain.

4.1 Cascading Multiple CoM Phases

We have built a rough terrain set-up (see Fig. 4) in the Human Centered Robotics Lab at UT which consists of several steps of a variety of heights and widths. Figure 3, shows the resulting data of dynamically walking over this terrain, for both

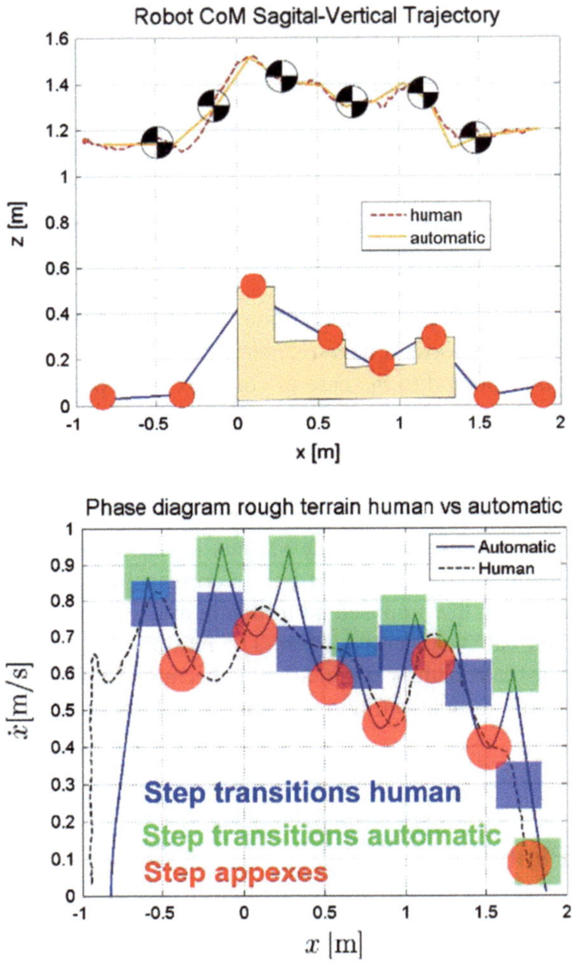

Fig. 3 Concatenation of steps: The *top* graph shows the kinematic trajectory of the human CoM (derived using motion capture) versus a piecewise linear approximation that we use to generate the automatic walking simulation. The *red dots* correspond to the position of the foot contacts. The *bottom* figure shows Matlab plots of Sagittal phase curves for the human and the automatic simulation. The *red circles* correspond to apexes of the steps. The *green squares* correspond to contact transitions of the automatic walk. The *purple squares* correspond to contact transitions of the human walk. Notice, that during the climbing of the first step of the stairs results in a smooth CoM pattern for the human walk. This is due to the smoothening effect of dual contact during the stance phase. This is not the case during the automatic walk because we have neglected the dual contact phase and therefore the transitions between contacts are instantaneous. Besides this difference, the rest of the walk correlates well

our human subject and the automatic planner presented throughout this paper. As we will see in the results section, the automatic planner approximates feet locations and CoM kinematics from the human, and automatically derives the dynamic walk. In particular, the CoM path has been approximated with piecewise linear segments, which are shown laid over the human CoM path extracted from a motion capture process.

Traversing the terrain of Fig. 4 involves making several steps, 7 in our example which are marked with red circles in Fig. 4. We are interested in displaying the phase diagram of the CoM for all steps for both the human and automatic walks. By plotting phase behavior for each step we can determine the intersections between neighboring steps (before and after), and therefore, derive the precise phase points to switch between steps. Because we have derived phase behavior for arbitrary CoM kinematics and feet locations, finding step intersections yields the motion and contact plan needed to dynamically walk over the rough terrain.

Let us focus on the automatic walk of the results section. We have used the perturbation method of Eq. (20) to derive phase curves for every step in forward and backward modes with respect to the contact point. Boundary conditions corresponding to CoM position and velocity are provided at the apex of each step. For this example we have used similar values than the human. However, mimicking the human is not needed in the general case. We do it here to compare results between the planner and the human. The valleys of the bottom graph of Fig. 3 correspond to the deceleration/acceleration pattern of single steps. They are in fact the same type of curves than those predicted in Fig. 2, this time derived for every step of the desired sequence given the desired boundary conditions. As such, the green squares shown in Fig. 3 correspond to the points where two curves from neighboring steps

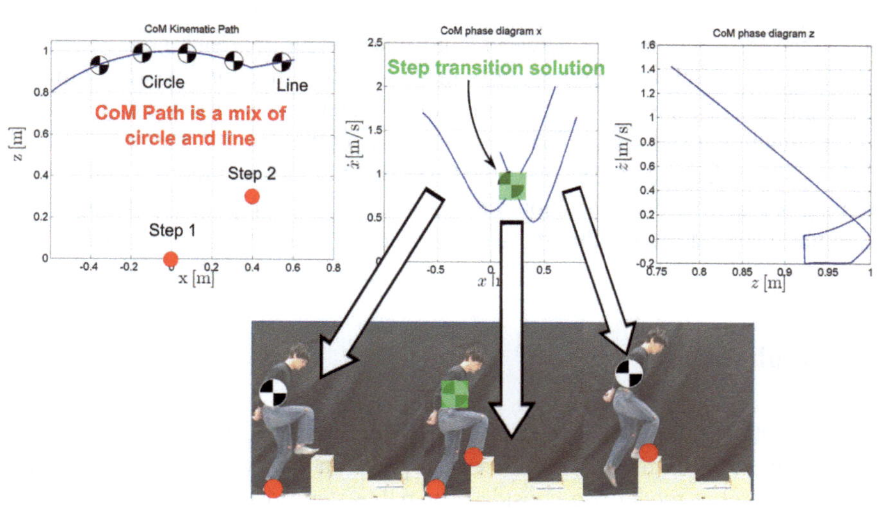

Fig. 4 Step solver: The center graph depicts phase curves for the two steps given the CoM path shown on the *left*. We fit polynomials and find the differential root between the adjacent curves to find the point of intersection

have the same position and velocity and therefore correspond to the necessary contact transitions to switch to the next step.

The pipeline for automatically planning the walk in the rough terrain is therefore as follows: (1) develop CoM geometric path to overcome the terrain, (2) choose boundary conditions, i.e. position and velocity, of the CoM at the apex of each step, (3) using the perturbation method of Eq. (20), predict phase curves for each step, (4) find the phase intersections between neighboring steps which represent the phase point where the transition between steps need to occur, (5) the resulting multi-step phase diagram is the locomotion plan that will be fed to a control stage.

4.2 Phase Intersections to Determine Step Transitions

From Fig. 3, it becomes clear that we derive the locomotion curves by finding the phase intersection between steps. We illustrate this procedure by studying the step to step transition on a particular example. Let us focus on the graphs shown in Fig. 4. The left graph shows our test example, with a CoM path consisting of a circular path first, continued by a line path. The motivation to use different curves is to illustrate the versatility of our method on working with any CoM path. The positions of the first and second step are also shown as red circles. The center graph depicts the phase curves for the first and second steps. We have used boundary conditions equal to $(x_0, \dot{x}_0) = (0, 0.6)$ and $(x_1, \dot{x}_1) = (0.4, 0.45)$ at the apexes of steps 1 and 2 respectively. The pictures showing the human are only to illustrate the switching strategy between steps but they have not been used to derive CoM geometric paths for this particular example.

Because the perturbation method of Eq. (20) is numerical, it is not obvious to derive the intersection point between CoM phases. Our approach goes as follows, (1) fit a polynomial of order 5 using Matlab's polyfit() function, to each of the two CoM phases, (2) subtract the two polynomials and find its roots using Matlab's roots() function, (3) discard imaginary roots, (4) get the point of intersection within CoM position range, and (5) extract the CoM velocity intersection by evaluating the polynomial at the CoM position intersection. If we apply this pipeline to the example of Fig. 4 we get that the step intersection is at $(x_s, \dot{x}_s) = (0.3, 0.7)$.

5 Results

Based on the methods described in the previous sections, we have conducted a study of locomotion in the Sagittal/vertical plane on a very rough terrain. Using a human-size robot model, we consider a variable stepped terrain with height variations between 0 and 40 cm and width variations between 20 and 40 cm. The goal of the planner is to maneuver the robot throughout the total length of the terrain. The speed specifications are determined to cruise the terrain at an average speed of

0.6 m/s, although this choice could be arbitrary. We also assume that the robot starts and finishes with zero velocities and it increases velocity according to a trapezoidal profile similar to that of our human subject. Once more, our planner does not need human specifications, but we use them for comparison (see Fig. 3). Velocity specifications are given only at each new step, corresponding to the moment when the center of mass Sagittal position crosses the corresponding supporting foot, namely the apex of the step. We consider *steps* to be spanned from apex to apex. Also for simplicity, we consider only single-support phases, with instantaneous transition between feet. The contact locations and CoM geometrical path are given by the human subject and we assume a point mass model of the robot, with all of the weight located at its waist. The human subject traversing the terrain is shown in Fig. 6. His height is 184 cm and his weight is 80.5 kg at the time of the experiment.

An analysis of the experiments is shown in the caption text of Figs. 5 and 6.

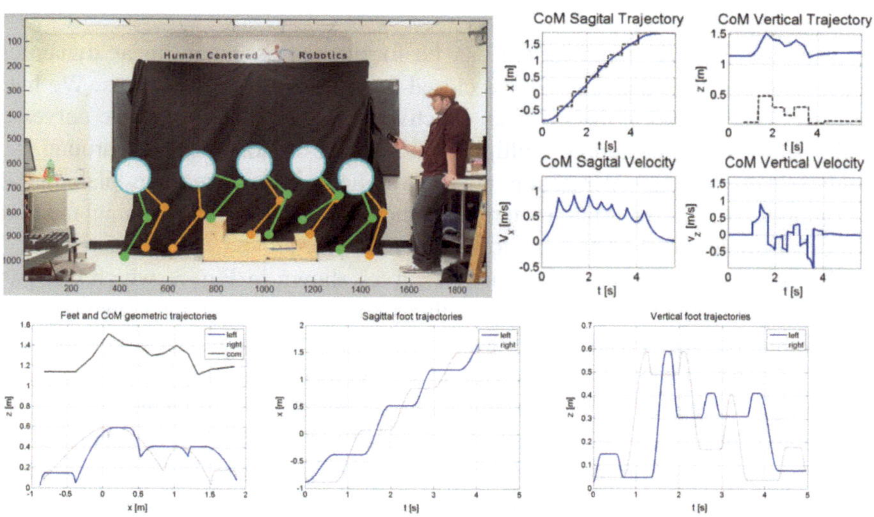

Fig. 5 Automatic locomotion planner: Using the proposed locomotion planner and based on human kinematic data, we create artificial CoM trajectories and determine contact transitions to achieve the desired design specifications of the walk. The snapshots on the *upper left* show a mix reality sequence derive from our planner. Time trajectories of CoM Sagittal and vertical behavior are shown to the *right* and are derived from the phase curves. A separate planner computes feet trajectories to synchronize with CoM behavior and switch step at the desired contact intersections

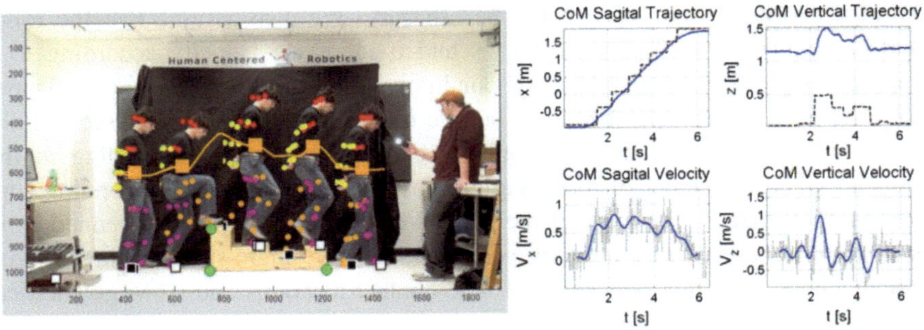

Fig. 6 Data extraction from human walk: A human subject walks over a rough terrain. Marker tracking is implemented and used to extract approximate CoM paths as well as Sagittal and vertical CoM trajectories and velocities

6 Conclusions and Outlook

Locomotion in very rough terrain can be formulated as a non-linear dynamical process. As such, it does not have a closed-form solution in most cases. We have resorted to perturbation theory as an effective tool to predict state space curves of CoM behavior. By cascading multiple phase curves of CoM behavior around step contacts and finding intersection points, we have generalized the planning of locomotion curves for arbitrary terrains. These prediction and planning methods represent important contributions to locomotion.

The strong correlation of locomotion curves shown in Fig. 3, which compare artificial and human walks, demonstrate the validity of our methods. However, to be deployable, our method needs to further include multicontact stages such as when the two feet are in contact with the ground for some period of time. In such case, we will need to derive new dynamic models involving the effect of multicontact. We anticipate, that in such cases the effect of internal forces will play an important role of the acceleration profile. The Multicontact/Grasp matrix of [23] presents a powerful method to derive dynamic behavior given frictional constraints and tension forces between feet. Moreover, during multicontact phases, there will be multiple phase curves that will fulfill frictional constraints. This fact will enable to consider solutions that minimize some criterion such as effort.

Constraining the locomotion paths to the Sagittal-vertical plane has allowed us to tackle rough terrain locomotion effectively. However, practical locomotion needs to include the 3 dimensions of space. In such case, CoM geometric paths need to be planned in the full 3D space and a lateral dynamic model similar to Eq. (7) needs to be consider. Although this work has explore modeling and planning issues, an important component for locomotion is the choice of a controller. Operating in state space opens opportunities to implement robust controllers. We plan to tackle this problem in the context of whole-body compliant control [23]. The proposed methods can be used to tackle a wide variety of issues such as rough terrain locomotion, disturbance robustness, parameter uncertainty, internal force behavior, optimization of performance parameters, and feasibility conditions for planners.

References

1. K. Harada, S. Kajita, K. Kaneko, H. Hirukawa, Zmp analysis for arm/leg coordination, in *Proceedings of the IEEE/RSJ International Conference on Intelligent Robots and Systems*, Las Vegas, USA, October 2003, pp. 75–81
2. S. Kajita, M. Morisawa, K. Harada, K. Kaneko, F. Kanehiro, K. Fujiwara, H. Hirukawa, Biped walking pattern generator allowing auxiliary zmp control, in *Intelligent Robots and Systems, 2006 IEEE/RSJ International Conference on*, October 2006, pp. 2993–2999
3. M. Raibert, *Legged Robots that Balance* (MIT Press, Cambridge, 1986)
4. T. Bretl, S. Lall, Testing static equilibrium for legged robots. IEEE Trans. Rob. **24**(4), 794–807 (2008)
5. C. Collette, A. Micaelli, C. Andriot, P. Lemerle, Robust balance optimization control of humanoid robots with multiple non coplanar grasps and frictional contacts, in *Proceedings of the IEEE International Conference on Robotics and Automation*, Pasadena, USA, May 2008
6. J. Pratt, R. Tedrake, Velocity-based stability margins for fast bipedal walking, in *Fast Motions in Biomechanics and Robotics*, ed. by M. Diehl, K. Mombaur, vol. 340, pp. 299–324 (2006)
7. T. McGeer, Passive dynamic walking. Int. J. Robot. Res. **9**(2), 62–68 (1990)
8. A. Goswami, B. Espiau, A. Kermane, Limit cycles and their stability in a passive bipedal gait, in *Proceedings of the IEEE International Conference on Robotics and Automation*, April 1996, pp. 246–251
9. K. Mombaur, H. Bock, J. Schloder, R. Longman, Self-stabilizing somersaults. IEEE Trans. Robot. **21**(6), 1148–1157 (2005)
10. A. Ruina, J. Bertram, M. Srinivasan, A collisional model of the energetic cost of support work qualitatively explains leg sequencing in walking and galloping, pseudo elastic leg behavior in running and the walk-to-run transition. J. Theor. Biol. **237**, 170–192 (2005)
11. T. Takuma, K. Hosoda, Controlling the walking period of a pneumatic muscle walker. Int. J. Robot. Res. **25**(9), 861–866 (2006)
12. E. Westervelt, J. Grizzle, C. Chevallereau, J. Choi, B. Morris, *Feebdack Control of Dynamic Bipedal Robot Locomotion* (CRC Press, 2007)
13. B. Stephens, C. Atkeson, Modeling and control of periodic humanoid balance using the linear biped model, in *9th IEEE RAS International Conference on Humanoid Robots, 2009. Humanoids 2009*, December 2009, pp. 379–384
14. J. Rummel, Y. Blum, A. Seyfarth, Robust and efficient walking with spring-like legs. Bioinspir. Biomimet. **5**(4), 046004 (2010)
15. M. Wisse, A. Schwab, R. van der Linde, F. van der Helm, How to keep from falling forward: elementary swing leg action for passive dynamic walkers. IEEE Trans. Robot. **21**(3), 393–401 (2005)
16. K. Byl, R. Tedrake, Metastable walking machines. Int. J. Robot. Res. **28**(8), 1040–1064 (2009)
17. I.R. Manchester, U. Mettin, F. Iida, R. Tedrake, Stable dynamic walking over uneven terrain. Int. J. Robot. Res. **30**(3), 265–279 (2011)
18. M. Spong, J. Holm, D. Lee, Passivity-based control of bipedal locomotion. IEEE Robot. Autom. Mag. **14**(2), 30–40 (2007)
19. M. Zucker, J. Bagnell, C. Atkeson, J. Kuffner, An optimization approach to rough terrain locomotion, in *2010 IEEE International Conference on Robotics and Automation (ICRA)*, May 2010, pp. 3589–3595

20. K. Hauser, T. Bretl, K. Harada, J. Latombe, Using motion primitives in probabilistic sample-based planning for humanoid robots, in *Workshop on Algorithmic Foundations of Robotic (WAFR)*, New York, USA, July 2006
21. K. Bouyarmane, A. Kheddar, Multi-contact stances planning for multiple agents, in *Proceedings of the IEEE International Conference on Robotics and Automation*, Shanghai, May 2011
22. L. Sentis, B. Fernandez, Com state space cascading manifolds for planning dynamic locomotion in very rough terrain, in *Proceedings of Dynamic Walking 2011*, Jena, Germany, July 2011
23. L. Sentis, J. Park, O. Khatib, Compliant control of multi-contact and center of mass behaviors in humanoid robots. IEEE Trans. Rob. **26**(3), 483–501 (2010)

Autonomous Navigation of a Humanoid Robot Over Unknown Rough Terrain

Koichi Nishiwaki, Joel Chestnutt and Satoshi Kagami

Abstract The present paper describes the integration of laser-based perception, footstep planning, and walking control of a humanoid robot for navigation over previously unknown rough terrain. A perception system that obtains the shape of the surrounding environment to an accuracy of a few centimeters is realized based on input obtained using a scanning laser range sensor. A footstep planner decides the sequence of stepping positions using the obtained terrain shape. A walking controller that can cope with a few centimeters error in terrain shape measurement is achieved by combining 40 ms cycle online walking pattern generation and a sensor feedback ground reaction force controller. An operation interface that was developed to send commands to the robot is also presented. A mixed-reality display is adopted in order to realize intuitive interfaces. The navigation system is implemented on the HRP-2, a full-size humanoid robot. The performance of the proposed system for navigation over unknown rough terrain is investigated through several experiments.

K. Nishiwaki (✉) · S. Kagami
Digital Human Research Center,
National Institute of Advanced Industrial Science and Technology (AIST),
2-3-26, Aomi, Koto-Ku, Tokyo 135-0064, Japan
e-mail: k.nishiwaki@aist.go.jp

S. Kagami
e-mail: s.kagami@aist.go.jp

K. Nishiwaki · S. Kagami
Japan Science and Technology Agency, CREST, Tokyo, Japan

J. Chestnutt
Boston Dynamics, Inc., 78 Fourth Avenue, Waltham, MA 02451, USA
e-mail: jchestnutt@bostondynamics.com

H.I. Christensen and O. Khatib (eds.), *Robotics Research*,
Springer Tracts in Advanced Robotics 100, DOI 10.1007/978-3-319-29363-9_35

1 Introduction

Biped robots are considered to be suitable for use on terrain having obstacles, discontinuous height changes, and roughness. On the other hand, since biped robots are naturally unstable, enabling a biped robot to walk over such terrain is a challenging problem. The solution to this problem requires an accurate perception system, a path planner that decides step placement, and a robust walking controller.

We developed a walking controller for a humanoid robot capable of handling unknown roughness, such as an unknown height change of up to a few centimeters and an unknown inclination of up to 10° [1]. Recently, small scanning-type laser range sensors that can be mounted on a humanoid robot have become available, and an accuracy of a few centimeters is achieved when observing an area a few meters away from the robot. We use this performance matching to realize the navigation of a humanoid robot over unknown rough terrain.

Successful integrations of on-board sensing, footstep planning, and walking control were reported using a stereo camera system [2], a laser range sensor [3], and a monocular camera [4]. In order to achieve sufficient measurement accuracy and realize walking control, the stereo camera system and laser range sensor are operated under the assumption that the environment consists of horizontal plane segments. The monocular camera requires advance knowledge of the shapes of the objects in the environment. Assumptions or previous knowledge of the shapes of the objects in the environment are not used in the present study.

Although the shape of the surrounding environment can be obtained with sufficient accuracy online, it is difficult to implement knowledge on the environment, such as identifying objects that should not be stepped on. In addition, we do not assume the existence of a global map. We herein present an interface for assigning commands to the robot using the high-level knowledge of a human operator. An outline of the path or movement direction may be the commands given to the robot, and the system plans the locations of steps locally using the obtained terrain shape. The planned footsteps can be checked through a mixed-reality interface.

Figure 1 shows the overview of the autonomous navigation system. We described each component technology in Sect. 2 through Sect. 5. Then, experiments on the full-size humanoid HRP-2 robot are shown in Sect. 6, followed by a discussion and conclusions.

2 Laser-Based Perception System

Terrain shape map generation using a laser range sensor is described in this section. The terrain map for the footstep planning is represented as a grid of cells. Each cell has a height value and an information value that indicates whether the height of the cell is observed. A cell size of 0.02 m × 0.02 m is used for the system presented herein.

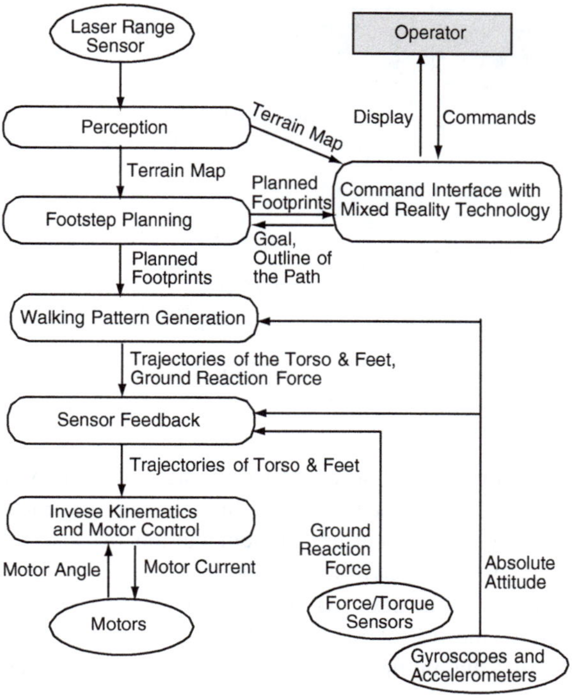

Fig. 1 Overview of the autonomous navigation system

We adopted the UTM-X002S (Hokuyo Automatic Co. LTD.) as a scanning-type laser range sensor. The scanning frequency is 100 Hz, and the angular resolution is 0.625°. The UTM-X002S can perform measurement from a distance of up to 30 m. The sensor is attached to a swinging mechanism, which is mounted to the torso of a humanoid robot as shown in Fig. 2.

Obtained distance data are converted into absolute 3D positions using the position of the robot and the angle of the swinging mechanism. An optical motion capture system (Motion Analysis Corp. Eagle Digital System) is used to localize the

Fig. 2 Scanning-type laser range sensor with a swinging mechanism mounted to the torso of a robot

Fig. 3 Evaluation of the terrain shape measurement (*left* photograph of the experimental setup, *right* obtained terrain map)

robot in an absolute coordinate system. The map is then updated, and the newer observation replaces the height value at the corresponding cell. We used this approach in the current implementation because this approach can better handle changes in the environment and the expected accuracy of the measurement relative to the robot position will be better when the point is measured from nearby. In the future, the use of SLAM technologies for localizing the robot position and integrating the multiple measurement for realizing better accuracy will be investigated.

An experiment to obtain a terrain map on a flat office floor while standing still was carried out to evaluate the accuracy of the map generation. Figure 3 (left) shows the experimental setup. The map region is limited to the light blue area of the floor (5 m × 5 m rectangle). The sensor swung between 15° and 90° from the horizontal plane. Each upward and downward scan required 2 s. The obtained map is shown in Fig. 3 (right). Two sets of arrows show the positions of the left and right feet, and the yellow area indicates cells whose height has not been obtained. Figure 4 shows the height error of the map with respect to the distance from the laser range sensor.

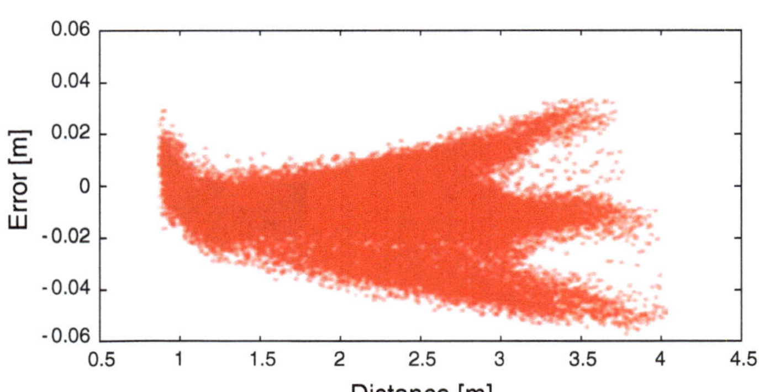

Fig. 4 Distribution of height error for terrain map acquisition

Fig. 5 An example of acquisition of a terrain map for a complex environment

Figure 5 shows an example of a map obtained for complex terrain. The terrain map obtained after walking though the terrain is superimposed on an image captured by a ceiling camera in the right image.

3 Footstep Planning

The footstep planner uses an A* search to generate a sequence of footstep locations to reach a given goal state. Possible foot transitions are restricted by limiting the landing positions of the swing foot relative to the stance foot to a finite number. An example of a limited transition set is shown in Fig. 6. The figure shows the possible landing positions of the right foot as white rectangles and the left stance foot as a gray rectangle. The terrain shape of the stepping position is evaluated. Then, whether the robot can step on this positions judged and the quality of the position is calculated as the cost of the location. The inclination, roughness, stability, largest

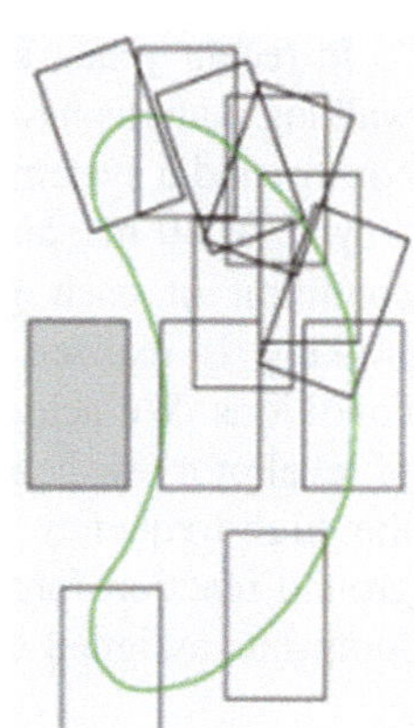

Fig. 6 Example of transition sets for footstep planning

bump, and safety of the terrain are used as metrics for both the abovementioned judgment and the cost calculation [5].

Using fewer possible transitions in the candidate set will facilitate making the planner available online. On the other hand, if the number of possible transitions is too small, a sequence of footprints that traverses rough terrain will not be found because the possible landing position is also limited by the shape of the terrain. We proposed an adaptive action model by which to address this problem [6]. The basic idea is to try to maintain the number of possible transitions even over rough terrain. If a possible transition is not valid because of the terrain shape, an appropriate transition near the original position is searched as an alternative position. This method generates suitable landing position candidates that fit the given terrain shape.

In some cases, we prefer to give not only the goal but also an outline of the path, so that the knowledge and intention of the operator can be transferred to the navigation system more clearly. We built a path planner that takes the guide curve into consideration [7]. A heuristic for the A* search is generated from the guide curve so that the search will be carried out along the curve.

Free leg trajectories that will not hit the terrain while moving are planned after deciding the sequence of the footprints.

4 Walking Control

The requirements for walking control as a part of the autonomous navigation system are the realization of footprints and free leg trajectory assigned online by the footstep planner and managing the error of the terrain shape that is generated by the perception system.

Humanoid walking was commonly realized by constructing a dynamically stable trajectory in advance using dynamics parameters and executing this trajectory with sensor feedback, if needed. This procedure was adopted because bipedal humanoids have a complicated dynamic model.

In recent years, several studies on the online generation of dynamically stable walking patterns have been published (e.g., [8–18]). We extended this approach and constructed a system that generates walking patterns at a very high frequency, e.g., a cycle of 40 ms, and which considers the actual motion of the robot as the initial condition of each generation. Maintenance of the dynamic stability of actual walking is realized by this high-frequency generation from the actual motion conditions. We achieved a walking controller that can handle an unknown change of level of a few centimeters or an unknown inclination of up to 10° by combining the high-frequency walking pattern generation system with a sensor feedback ground reaction force controller [1]. This controller is adopted in order to realize footprints assigned online and to manage the perception error.

4.1 Dynamically Stable Pattern Generation

Figure 7 shows an overview of the dynamically stable trajectory-generation system. The role of this system is to generate dynamically stable walking trajectories that start from the estimated actual motion.

Since trajectory generation takes some time (We currently start each generation at 36 ms before the starting time of 40 ms-trajectory length), the initial conditions at some point in the future should be estimated using the current motion status. Let t_n^{ini} and $t_n^{est} = t_n^{ini} - T_{est}$ be the start time of the nth motion and the decision time for the nth initial condition, respectively ($T_{est} = 0.036$ s in the current implementation). The initial position of the torso in the transformation matrix representation of the nth trajectory is decided by using the estimated absolute position of the torso at t_n^{est} (${}_wT_{est}^t(t_n^{est})$), as follows:

$$ {}_wT_n^t(t_n^{ini}) = {}_wT_{est}^t(t_n^{est})\left({}_wT_{n-1}^t(t_n^{est})\right)^{-1}{}_wT_{n-1}^t(t_n^{ini}) \quad (1) $$

The initial positions of the left and right feet are decided in the same manner.

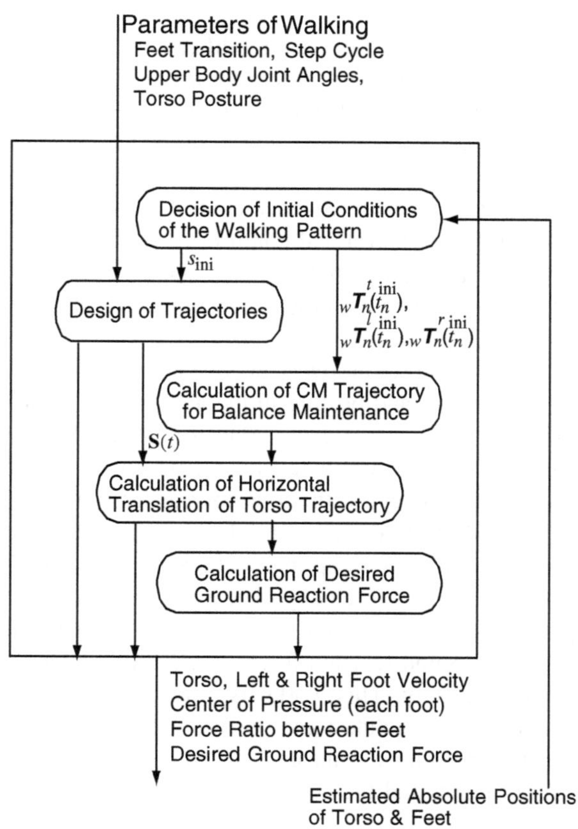

Fig. 7 Overview of dynamically stable trajectory generation

The trajectories for the horizontal torso position are used for maintaining balance. Foot position and posture trajectories and torso height and posture trajectories ($\mathbf{s}(t)$ in Fig. 7) are designed to follow the given command smoothly from the estimated initial conditions.

The initial position of the center of mass is calculated from the estimated initial torso and foot positions $({}_wT_n^t(t_n^{ini}), {}_wT_n^l(t_n^{ini}), {}_wT_n^r(t_n^{ini}))$. Then, the desired trajectory of the center of mass (CM) is generated from the reference ZMP trajectory by applying preview control theory [19]. The desired CM trajectory is converted to a horizontal torso position trajectory by applying a slightly modified version of the resolved momentum control method [20]. Here, the trajectories of the feet and other components of the torso are used in deciding the horizontal torso position.

Finally, inverse dynamics calculations are carried out in order to decide the desired ground reaction force necessary to execute the generated motion. The calculated ground reaction force will be the control reference of the sensor feedback part of the system.

4.2 Sensor Feedback Modification

The goal of sensor feedback modification is to realize the specified torso motion in the absolute coordinate system for a short-term timescale, even if the terrain shape is different than expected. Gradual divergence from the given trajectories will not be a significant problem because the repetitive trajectory generation compensates for the divergence by using the actual motion for the initial conditions. We basically adopted two types of feedback control in the sensor feedback modification: ground reaction force control and control of the absolute posture of the torso. In order to ensure that the torso motion is not sensitive to differences in the terrain shape, we attempt to control the ground reaction force at the feet rather than the actual foot positions.

Figure 8 shows an overview of the sensor feedback control system. Estimation of the position and posture in the absolute coordinate system is carried out in this loop using a cycle of 1 ms. The estimated information is used to decide the orientation of the coordinate system, which is used for the ground reaction force control, and is sent to the trajectory generation system.

Ground reaction force control is implemented as a impedance control of the foot positions with a zero spring coefficient. The inputs of the control are the desired reaction force and velocity. The desired velocity of the feet assigned by the trajectory generation system is modified according to the ground contact phase. The assigned ground reaction force is then modified according to the error of the torso angular velocity. This works as a torso posture damping control and attempts to maintain the absolute torso posture specified by the trajectory generation system. Finally, impedance control is carried out using the desired center of pressure as the control point, and the desired foot positions are decided.

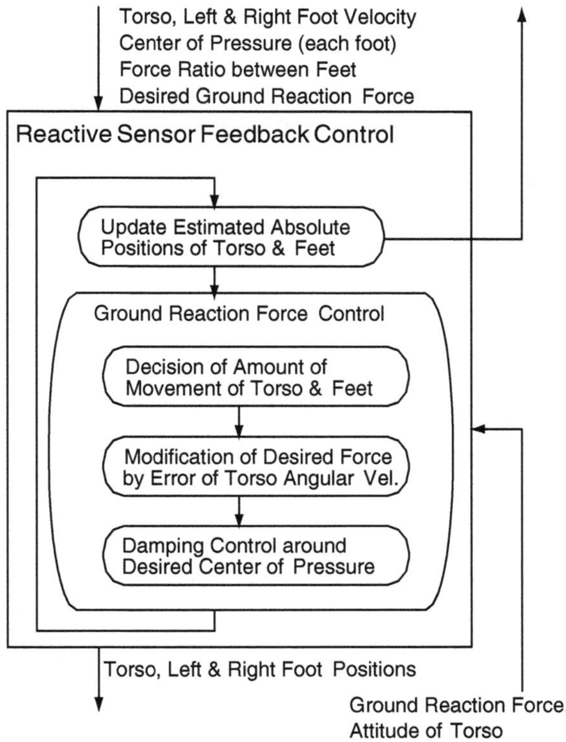

Fig. 8 Overview of the sensor feedback control system

5 Operation Interface

Three different operation interfaces for giving commands to the autonomous navigation system are presented in this section.

5.1 Graphic User Interface

The most basic interface specifies the goal on the obtained terrain map using a GUI (Fig. 9c). The corresponding experimental setup is shown in Fig. 9a. An outline of the path can also be specified using this interface (Fig. 9d). The corresponding experimental setup is shown in Fig. 9b. The knowledge of the operator of the terrain and the preferred route can be transferred to the navigation system by providing an outline of the path. A problem with this interface is that the goal may be in an area in which the terrain shape has not been obtained, as shown in Fig. 9. The operator needs to determine the target environment by some other means and its correspondence to the map shown in the GUI system.

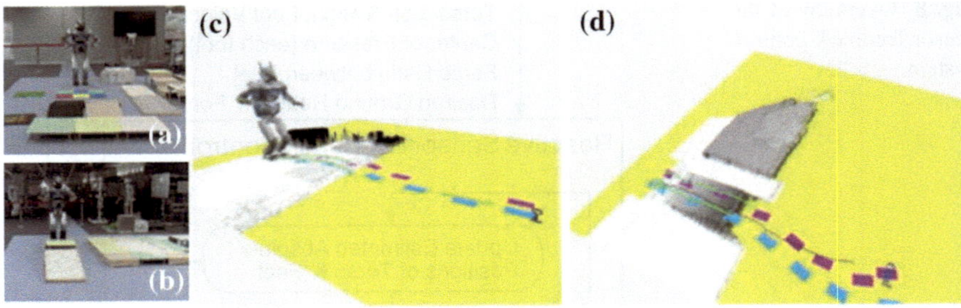

Fig. 9 Graphic user interface for navigation control

5.2 *MR Interactive Interface*

The second interface is a mixed-reality interface. The operator wears a head-mounted display (HMD) and uses a game controller (controller). The positions of both the HMD and the controller are localized by the optical motion capture system in the same absolute coordinate system as that in which the robot is localized. The operator can draw an outline of a path or specify a goal using the controller, and the drawn path and the goal is superimposed on the HMD view, as well as the planned path (Fig. 10). Commands to the robot, such as start walking, can be assigned using the same interface, as shown in Fig. 10 (right). The operator can specify the goal or the outline path intuitively even if the target position has not yet been observed by the perception system. A drawback of this interface is that the operator must be in the operation area in order to specify the positions.

5.3 *Joystick Interface*

In order to overcome the drawback of the previous interface, another mixed-reality interface is introduced. The camera view of the robot is displayed on a monitor, and the desired direction and rotation of locomotion is assigned by manipulating the triangle shape in the view using a joystick (Fig. 11). The footstep planner is used

Fig. 10 Mixed-reality interface for navigation control

Fig. 11 Joystick interface for navigation control

for global path planning rather than local path planning. The footstep planner plans a sequence of a few steps that realizes the desired movement as well as adaptation to the local terrain shape. The interface realized remote control of the autonomous locomotion. However, there is a drawback in that the operator has to become more involved with the locomotion control.

6 Experiments

We implemented the proposed control system on the HRP-2, a full-size 38-DOF humanoid robot. Terrain map building and footstep planning were carried out on a computer outside of the robot, and laser range sensor data and the planned result were sent from/to the robot via Ethernet. At each step the robot takes, the planner begins to compute a new plan based on the position of the robot and the step that is being taken. By computing a new plan at each step, the robot can use the recent perception results and adjust slips, or other deviations from the plan. Footstep planning is repeated more frequently with the joystick interface in order to achieve a better response to the change of the joystick input.

We installed a Core 2 3.06 GHz CPU board in the robot. Dynamically stable walking pattern generation and sensor feedback control were implemented on the CPU together with sensor processing and motor servo control. The dynamically stable trajectory generation runs at 40 ms cycles and requires approximately 30 ms to generate a pattern of 40 ms in length. At the same time, the sensor feedback control runs at a 1 ms cycle for executing the generated trajectories.

Photographs of experiments to investigate autonomous navigation over previously unknown rough terrains are shown in Figs. 12, 13 and 14. Each experiment uses a different interface system described in the previous session.

In the experiment shown in Fig. 12, pebbles were placed on the slope. The GUI was used to assign the commands. The GUI view is superimposed in the top left corner of each photograph. The yellow area indicates that the height is unknown. Since the operator knows the environment, the operator drew an outline of the path, including the area in which the height was unknown. The terrain map was incrementally built during walking, and the obtained map was used online for planning

Fig. 12 Photographs of an experiment on autonomous navigation over unknown terrain (GUI)

the footprints that fit the terrain. Online incremental map building and replanning worked well, and the robot successfully reached the goal by approximately following the provided outline.

Figure 12 shows the case in which the mixed-reality interface was used. The view of the operator through the HMD is shown in the first four photographs and is superimposed at top right corner of each of the other photographs. Starting from the fifth photograph, the view superimposed at the bottom right corner shows the obtained terrain map superimposed on the image captured by the ceiling camera. The robot successfully walked through the previously unknown rough terrain by approximately following the provided outline.

An autonomous navigation experiment involving the joystick interface is conducted as shown in Fig. 14. The view shown to the operator is superimposed at the top left corners of the photographs. The green triangle indicates the commanded

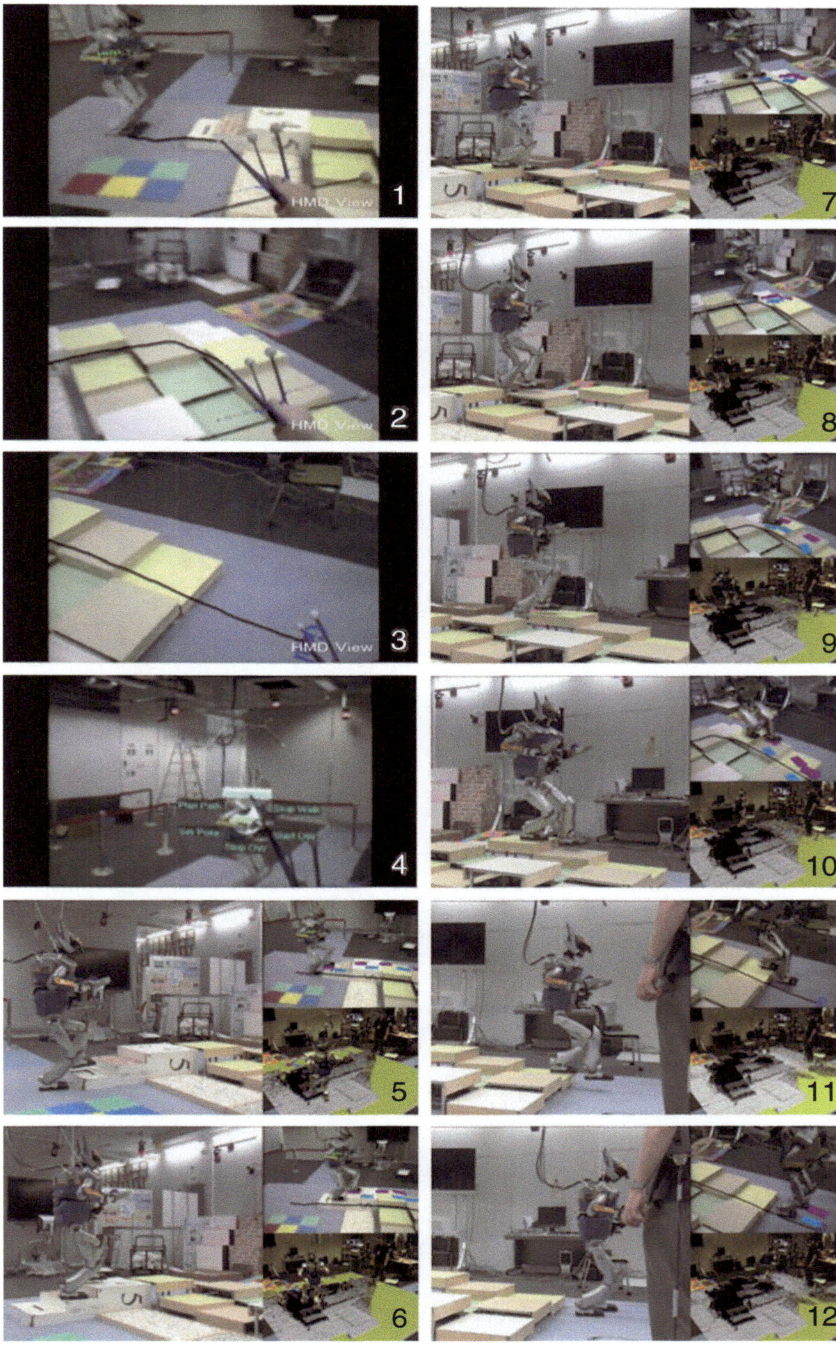

Fig. 13 Photographs of an experiment on autonomous navigation over unknown terrain (mixed-reality interface)

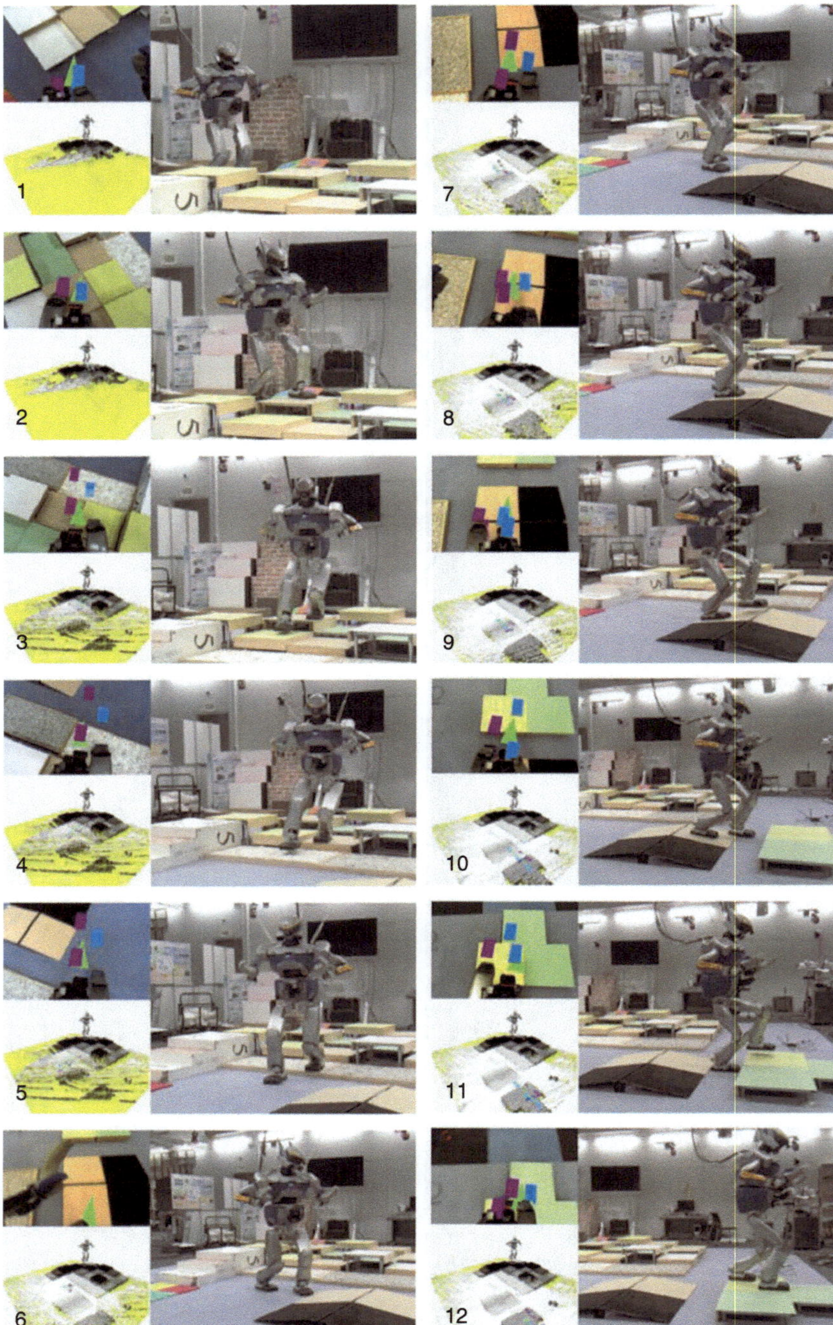

Fig. 14 Photographs of an experiment on autonomous navigation over unknown terrain (joystick interface)

direction and rotation. The obtained map and planned footprints projected onto the map are shown in the bottom left corner of the photographs. The operator successfully guided the robot using the joystick interface by watching only the view shown at top left.

7 Conclusion

An integrated system of online perception, footstep planning, and walking control of a humanoid robot for navigating on previously unknown rough terrains was proposed. A laser-based perception system realized online measurement of the surrounding terrain shape with an accuracy of a few centimeters. The robust walking control enabled a humanoid robot to walk over previously unknown rough terrain with an unknown roughness of a few centimeters. Operation interfaces using a mixed-reality display enabled commands to be fed to the system intuitively.

Localizing the robot using odometry and laser sensor data as well as the construction and use of a larger-scale map will be investigated in future research.

References

1. K. Nishiwaki, S. Kagami, Frequent walking pattern generation that directly uses estimated actual posture for robust walking control, in *Proceedings of International Conference on Humanoid Robots*, 2009, pp. 535–541
2. J.-S. Gutmann, M. Fukuchi, M. Fujita, Real-time path planning for humanoid robot navigation, in *Proceedings of International Joint Conferences on Artificial Intelligence*, 2005, pp. 1232–1237
3. J. Chestnutt, Y. Takaoka, K. Suga, K. Nishiwaki, J. Kuffner, S. Kagami, Biped navigation in rough environments using on-board sensing, in *Proceedings of International Conference on Intelligent Robots and Systems*, 2009, pp. 3543–3548
4. P. Michel, J. Chestnutt, S. Kagami, K. Nishiwaki, J. Kuffner, T. Kanade, GPU-accelerated real-time 3d tracking for humanoid locomotion and stair climbing, in *Proceedings of International Conference on Intelligent Robots and Systems*, 2007, pp. 463–469
5. J. Chestnut, J.J. Kuffner, K. Nishiwaki, S. Kagami, Planning biped navigation strategies in complex environments, in *Proceedings of International Conference on Humanoid Robots*, 2003
6. J. Chestnutt, K. Nishiwaki, J. Kuffner, S. Kagami, An adaptive action model for legged navigation planning, in *Proceedings of International Conference on Humanoid Robots*, 2007, pp. 196–202
7. J. Chestnutt, K. Nishiwaki, J. Kuffner, S. Kagami, Interactive control of humanoid navigation, in *Proceedings of International Conference on Intelligent Robots and Systems*, 2009, pp. 3519–3524
8. S. A. Setiawan, S. H. Hyon, J. Yamaguchi, A. Takanishi, Physical interaction between human and a bipedal humanoid robot—realization of human-follow walking, in *Proceedings of International Conference on Robotics and Automation*, 2009, pp. 361–367
9. K. Nishiwaki, T. Sugihara, S. Kagami, M. Inaba, H. Inoue, Realtime generation of humanoid walking trajectory by mixture and connection of pre-designed motions—online control by

footprint specification, in *Proceedings of International Conference on Robotics and Automation*, 2001, pp. 4110–4115
10. K. Yokoi, F. Kanehiro, K. Kaneko, K. Fujiwara, S. Kajita, H. Hirukawa, A honda humanoid robot controlled by aist software, in *Proceedings of International Conference on Humanoid Robots*, 2001, pp. 259–264
11. Y. Kaneshima, Y. Sugahara, S. Ando, M. Sato, H. ok Lim, A. Takanishi, Quasi real-time motion pattern generation for bipedal humanoid robots, in *Proceedings of Annual Conference on Robotics Society of Japan*, 2001, pp. 987–988 (In Japanese)
12. K. Nishiwaki, S. Kagami, Y. Kuniyoshi, M. Inaba, H. Inoue, Online generation of humanoid walking motion based on a fast generation method of motion pattern that follows desired zmp, in *Proceedings of International Conference on Intelligent Robots and Systems*, 2002, pp. 2684–2689
13. K. Löffler, M. Gienger, F. Pfeiffer, Sensor and control design of a dynamically stable biped robot, in *Proceedings of International Conference on Robotics and Automation*, 2003, pp. 484–490
14. K. Harada, S. Kajita, K. Kaneko, H. Hirukawa, An analytical method on real-time gait planning for a humanoid robot, in *Proceedings of International Conference on Humanoid Robots*, No. 60, 2004
15. T. Sugihara, Y. Nakamura, A fast online gait planning with boundary condition relaxation for humanoid robots, in *Proceedings of International Conference on Robotics and Automation*, 2005, pp. 306–311
16. I.-W. Park, J.-Y. Kim, J. Lee, J.-H. Oh, Online free walking trajectory generation for biped humanoid robot KHR-3 (HUBO), in *Proceedings of International Conference on Robotics and Automation*, 2006, pp. 2667–2672
17. T. Buschmann, S. Lohmeier, M. Bachmayer, H. Ulbrich, F. Pfeiffer, A collocation method for real-time walking pattern generation, in *Proceedings of International Conference on Humanoid Robots*, 2007
18. D. Dimitrov, J. Ferreau, P.-B. Wieber, M. Diehl, On the implementation of model predictive control for on-line walking pattern generation, in *Proceedings of International Conference on Robotics and Automation*, 2008
19. S. Kajita, F. Kanehiro, K. Kaneko, K. Fujiwara, K. Harada, K. Yokoi, H. Hirukawa, Biped walking pattern generation by using preview control of zero-moment point, in *Proceedings of International Conference on Robotics and Automation*, 2003, pp. 1620–1626
20. S. Kajita, F. Kanehiro, K. Kaneko, K. Fujiwara, K. Harada, K. Yokoi, H. Hirukawa, Resolved momentum control: humanoid motion planning based on the linear and angular momentum, in *Proceedings of International Conference on Intelligent Robots and Systems*, 2003, pp. 1644–1650

Hybrid System Identification via Switched System Optimal Control for Bipedal Robotic Walking

Ram Vasudevan

Abstract While the goal of robotic bipedal walking to date has been the development of anthropomorphic gait, the community as a whole has been unable to agree upon an appropriate model to generate such gait. In this paper, we describe a method to segment human walking data in order to generate a robotic model capable of human-like walking. Generating the model requires the determination of the sequence of contact point enforcements which requires solving a combinatorial scheduling problem. We resolve this problem by transforming the detection of contact point enforcements into a constrained switched system optimal control problem for which we develop a provably convergent algorithm. We conclude the paper by illustrating the performance of the algorithm on identifying a model for robotic bipedal walking.

1 Introduction

Bipedal robotic walking has been extensively studied with the principal ambition of achieving anthropomorphic gait. The generation of human-like gait would have a dramatic impact on the design of robotic assistive devices and prosthetics [1, 3, 23]. Unfortunately this goal that had originally motivated many roboticists to study bipedal rather than quadrupedal or hexapedal robotics has been primarily ignored in favor of studying the generation of any type of stable or energy minimizing gait. This transformation of purpose is best reflected in the disparate number of models considered in the literature. These bipedal models are differentiated by not only considering the number of degrees of freedom i.e. the inclusion of knees or feet, but also by considering the temporal ordering of events or discrete phases during walking termed a *domain breakdown*.

R. Vasudevan (✉)
Department of Mechanical Engineering, University of Michigan,
Ann Arbor, MI 48109, USA
e-mail: ramv@umich.edu

H.I. Christensen and O. Khatib (eds.), *Robotics Research*,
Springer Tracts in Advanced Robotics 100, DOI 10.1007/978-3-319-29363-9_36

Traditionally most models of bipedal robots employed a single domain [11, 24, 29], which assumes an instantaneous double support phase and usually excludes the presence of feet (models of this form began with the so called compass gait biped, which did not have knees or feet). Adding feet to the bipedal robot results in the need to extend the domain breakdown beyond a single discrete phase, which is typically done by either adding a phase where the heel is off the ground or a double support phase where both feet are on the ground, or any combination thereof [6, 25]. Adding knees that lock further complicates the picture, forcing the need to have phases in which the knee is locked and unlocked [5, 7, 14, 22]. When one considers a biped with knees and feet, the possible number and type of domain breakdowns becomes daunting. At this point, depending on the researcher and the objective, different temporal orderings have been chosen ranging from one discrete phase to five, e.g., [12] considers one, [6, 25] considers two, [19] considers three, [5, 13] considers four and [22] considers five.

This lack of consistency among models in the literature motivates the desire to determine if there does in fact exist a "universal" model dictated by empirical observation of human gait that should be used by bipedal robots in order to generate human-like gait. Biomechanists have worked diligently employing such observation to detail how the musculoskeletal system behaves during walking [4, 18, 20, 30] and have segmented human walking gait by considering shock absorption, initial limb stability and preservation of progression into seven distinct qualitatively described functional phases [16]; however, since a robotic biped and human employ fundamentally distinct constructs in order to generate gait, these empirical observations alone are insufficient in guiding a bipedal roboticist to a mathematical model of a robotic biped. The idea of employing such observations is not novel in the robotics community either as some roboticists have fixed a robotic model of a biped and tracked the human trajectories in order to generate controllers for this fixed model robotic biped [23]. We instead seek a method to extract a mathematical model of a biped from empirical observation of human gait.

Recently, we devised a means to view these empirical observations in a fashion that allowed for the separation of a mathematical model into a piece that was determined by the construction of the biped, i.e. an intrinsic component corresponding to the Lagrangian, and a piece that reflected a choice made by the roboticist, i.e. an extrinsic component corresponding to the sequence of contact point enforcements [2]. This abstraction of human walking which we detail in this paper, allowed us to segment flat-ground human walking by determining the sequence of constraint enforcements via a function fitting approach [2] and a persistent homology approach [26] which both produced an identical "universal" model of flat-ground human walking. This insight allowed for the construction of a control law via feedback linearization that produced human-like gait [21]. Though this abstract view of human walking allowed for a means to segment human walking, the ad hoc method used to perform this segmentation limited the utility of our approach to flat-ground walking.

The goal of this paper is to resolve the makeshift nature of our system identification scheme. We accomplish this task by translating our determination of the sequence of contact point enforcements into a constrained nonlinear switched system optimal control problem. A switched system is a system whose state is governed by a finite number of differential equations. The control parameter for such systems has a discrete component, the sequence of modes, and two continuous components, the duration of each mode and the continuous input. The difficulty with the optimal control of such systems is that it requires the resolution of a mode scheduling problem. We recently resolved this shortcoming by developing a bilevel hierarchical algorithm that divided the problem into two nonlinear constrained optimization problems [9, 10]. The result, which we describe in this paper, is an algorithm that provides a sufficient condition to guarantee the local optimality of the mode duration and continuous input while decreasing the overall cost via mode insertion. Importantly this provides a way to automatically generate a model of a bipedal robot capable of generating anthropomorphic gait from human data.

This paper is organized as follows: Sect. 2 describes how the Lagrangian of a biped together with a temporal ordering of constraints completely determines a mathematical model of a biped; Sect. 3 describes how to recast the problem of contact point enforcement into a constrained switched system optimal control problem; Sect. 4 describes an algorithm to solve this problem; Sect. 5 describes the performance of the algorithm; and Sect. 6 concludes the paper.

2 From Constraints to Robotic Models

In this section, we begin by describing the tool employed in the literature to describe bipedal walking robots: *hybrid systems* and then show how a Lagrangian for the biped that is intrinsic to it along with the sequence of active constraints (chosen by a robotocist) allows one to explicitly construct a hybrid system model of a biped. Hybrid systems have been studied in a variety of contexts and have been used successfully to model bipeds since they naturally display both discrete and continuous behavior. We begin by defining a class of hybrid systems able to mathematically formalize bipedal walking.

Definition 1 A *hybrid dynamical system in a cycle* is a tuple

$$\mathscr{H}\mathscr{C} = (\Gamma, \mathcal{D}, U, S, \Delta, F),$$

where

- $\Gamma = (V, E)$ is a *directed cycle* where V is the set of vertices with each vertex corresponding to a mode of the system and E is the set of edges,
- $\mathcal{D}$ is the *domain* where $\mathcal{D} \equiv M \times U \subset \mathbb{R}^n \times \mathbb{R}^m$ and M is the state space and is an embedded smooth submanifold of $\mathbb{R}^n$,
- $U \subset \mathbb{R}^m$ is the set of controls where U is a compact connected set containing the origin,

- $S = \{S_e\}_{e \in E}$ is a set of *guards*, where $S_e \subseteq \mathcal{D}$,
- $\Delta = \{\Delta_e\}_{e \in E}$ is a set of *reset maps*, where $\Delta_e : S_e|_M \to M$ is a smooth map,
- $F = \{f_v\}_{v \in V}$, where f_v is a *control vector field* on M, i.e., $f_v(x, u) \in T_x M$.

2.1 Modeling an Unconstrained Biped

To understand how to model a robotic biped, consider a configuration space for the biped Q, i.e., a choice of (body or shape) coordinates for the robot where typically $q : [0, \infty) \to Q$ is a function describing the evolution of a collection of (relative) angles between each successive link of the robot. The Lagrangian of a bipedal robot, $L : TQ \to \mathbb{R}$, can be stated in terms of kinetic and potential energies as:

$$L(q(t), \dot{q}(t)) = K(q(t), \dot{q}(t)) - V(q(t)), \tag{1}$$

where $(q(t), \dot{q}(t)) \in TQ$. The Euler-Lagrange equations yield the equations of motion, which for robotic systems are stated as:

$$D(q(t))\ddot{q}(t) + H(q(t), \dot{q}(t)) = B(q(t))u(t), \tag{2}$$

where $D(q(t))$ is the inertia matrix, $B(q(t))$ is the torque distribution matrix, $B(q(t))u(t)$ is the vector of actuator torques and $H(q(t), \dot{q}(t))$ contains the Coriolis, gravity terms and non-conservative forces grouped into a single vector [15].

The continuous dynamics of the system depend on which constraints are enforced at any given time, while the discrete dynamics depend only on the temporal ordering of constraints. Constraints and their enforcement are dictated by the number of contact points of the system. Explicitly, let the *set of contact points* be $\mathcal{C} = \{c_1, c_2, \ldots, c_k\}$, where each c_i is a specific type of contact possible in the biped, either with the ground or in the biped itself (such as the knee locking). Contact points introduce *holonomic constraints* on the system, which are vector valued functions $\eta_c : Q \to \mathbb{R}^{n_c}$, that must be held constant for a contact point to be maintained, i.e., $\eta_c(q(t)) = \text{constant} \in \mathbb{R}^{n_c}$ fixes the contact point but allows rotation about this point if feasible. It is useful to express the collection of all holonomic constraints in a single diagonal matrix $\eta(q(t)) \in \mathbb{R}^{n_c \times |\mathcal{C}|}$. Another class of constraints that are important are *unilateral constraints*, h_c for $c \in \mathcal{C}$, which are scalar valued functions, $h_c : Q \to \mathbb{R}$, that dictate the set of admissible configurations of the system; that is $h_c(q(t)) \leq 0$ implies that the configuration of the system is admissible for the contact point c. These constraints can also be put in the form of a diagonal matrix $h(q(t)) \in \mathbb{R}^{|Q| \times |Q|}$. A domain breakdown is a directed cycle together with a specific choice of contact points on every vertex of that graph. That is if $\mathcal{C} = \{c_1, c_2, \ldots, c_k\}$ is a set of contact points and $\Gamma = (V, E)$ is a directed cycle then a *domain breakdown* is a function $\mathcal{B} : \Gamma \to \mathbb{Z}_2^k$ such that $B(v)_i = 1$ if c_i is in contact on v and $B(v)_i = 0$ otherwise.

2.2 A Hybrid Dynamical Model of a Biped

We now demonstrate that given a Lagrangian, a directed cycle, and a domain breakdown, a hybrid system can be explicitly constructed. Since the Lagrangian is intrinsic to a robot, this result proves that a domain breakdown, which is determined by the enforced contact points, alone dictates the mathematical model of a biped.

The domain of the hybrid system is the direct product of the tangent space of the configuration space of the robot and the input space U which is dictated by the set of available actuators on the robot. The vector field in each mode is constructed by imposing the constraints as specified by the domain breakdown. For the mode $v \in V$, the holonomic constraints that are imposed are given by:

$$\eta_v(q(t)) = \eta(q(t))\mathcal{B}(v), \tag{3}$$

where the domain breakdown dictates which constraints are enforced. Differentiating the holonomic constraint yields a *kinematic constraint*:

$$J_v(q(t))\dot{q}(t) = 0, \tag{4}$$

where $J_v(q(t)) = \text{RowBasis}\left(\frac{\partial \eta_v(q(t))}{\partial q}\right)$ is a basis for the row space of the Jacobian (this removes any redundant constraints so that J_v has full row rank). The kinematic constraint yields the *constrained dynamics* in that mode:

$$D(q(t))\ddot{q}(t) + H(q(t), \dot{q}(t)) = B(q(t))u(t) + Jv(q(t))F_v(q(t), \dot{q}(t), u(t)) \tag{5}$$

which enforces the holonomic constraint; here D, H and B are as in Eq. (2) and $F_v(q(t), \dot{q}(t), u(t))$ is a *wrench* [15]. To determine the wrench $F_v(q(t), \dot{q}(t), u(t))$, we differentiate the kinematic constraint and combine this result with Eq. (5):

$$F_v(q, \dot{q}, u) = J_v^{-1}(q)\left[D(q)J_v^{-1}(q)\left(\dot{J}_v(q)\dot{q}\right) + H(q, \dot{q}) - B(q)u\right], \tag{6}$$

where we have suppressed the dependence on t in $q, \dot{q}$, and u. Therefore, for $x(t) = (q(t), \dot{q}(t))$, substituting Eq. (6) into Eq. (5) yields the control vector field $f_v(x(t), u(t))$. Importantly, notice that only the holonomic constraints but not the unilateral constraints appear in the control vector field and that the actual position to be maintained by the contact point as dictated by the holonomic constraints never appears inside of the control vector field. We now construct the guards and reset maps for a hybrid system using the domain breakdown. From the wrench $F_v(q(t), \dot{q}(t), u(t))$, one can ensure that the contact point is enforced by considering inequalities on the friction which can be stated in the form: $\mu_v(q(t))^T F_v(q(t), \dot{q}(t), u(t)) \leq 0$, with $\mu_v(q(t))$ a matrix of friction parameters and constants defining the geometry of the contact point (see [13] for more details). These are coupled with the unilateral constraint in each mode, $h_v(q(t)) = h(q(t))\mathcal{B}(v)$, to yield the set of admissible configurations:

$$A_v(q(t)),\dot{q}(t),u(t)) = \begin{bmatrix} \mu_v(q(t))^T F_v(q(t),\dot{q}(t),u(t)) \\ h_v(q(t)) \end{bmatrix} \leq 0. \tag{7}$$

The guard is just the boundary of the domain with the additional assumption that the set of admissible configurations is decreasing, i.e., the vector field is pointed outside of the domain, or for an edge $e = (v, v') \in E$,

$$G_e = \left\{(q,\dot{q},u) \in TQ \times U : A_v(q,\dot{q},u) = 0 \,\text{and}\, \dot{A}_v(q,\dot{q},u) \leq 0\right\}, \tag{8}$$

where we have suppressed the dependence on t in $q, \dot{q}$, and u. The reset map is derived by considering the impact equations which are given by considering the constraints enforced on the subsequent mode. For an edge $e = (v, v') \in E$, the post-impact velocity $\dot{q}(t^+)$ is given in terms of the pre-impact velocity $\dot{q}(t^-)$ and determines the reset map:

$$R_e(q(t^-),\, \dot{q}(t^-)) = \begin{bmatrix} q(t^-) \\ I - D^{-1}J_{v'}^T(J_{v'}D^{-1}J_{v'}^T)^{-1}J_{v'})\dot{q}(t^-) \end{bmatrix}, \tag{9}$$

where I is the identity matrix, $J_{v'}$ is the Jacobian matrix in mode v', D is the inertia matrix, and we have suppressed the dependence on $q(t)$ in the Jacobian and inertial matrices. The result of this analysis is that given a domain breakdown and a bipedal robot (which determines just the unconstrained Lagrangian), the hybrid model for the biped is completely determined. If we then segment human walking by considering the sequence of contact point enforcements, we can extract a hybrid model for a biped that is capable of generating human-like gait. Unfortunately the determination of this sequence from human data requires resolving a combinatorial problem.

3 Switched Optimal Control Formulation

To address this combinatorial problem, remember that given a biped and a set of contact points one can immediately write down a finite set of constrained control vector fields indexed by V since only the holonomic constraints affect the vector field and the value that must be maintained by the contact point as dictated by the holonomic constraint never appears in the control vector field. The set of constraints do not immediately translate into a set of guards since the value of the unilateral constraint is required. In order to determine the sequence of contact point enforcements, we can track the human data by choosing the optimal sequence of vector fields wherein we allow arbitrary switching between vector fields. In this section, we describe how to address this problem.

Any trajectory of a switched system is encoded by a sequence of discrete modes, a corresponding sequence of times spent in each mode, and the continuous input over time. To formalize the optimal control problem, we define four spaces: Σ is the *discrete mode sequence space*, $\mathcal{S}$ is the *transition time sequence space*, $\mathcal{U}$ is the *continuous input space*, and $\mathbb{N}^W$ is the *objective mapping space*, where $W \in \mathbb{N}$. We begin by describing each of these spaces in detail. For notational convenience we define an additional vector field, $f_0(\cdot,\cdot) = 0$, in which the trajectories stop evolving. The discrete mode sequence space is most readily thought of as an infinite dimensional space with elements that contain only a finite number of non-zero vector fields:

$$\Sigma = \bigcup_{N=1}^{\infty} \left\{ \sigma \in (V \cup \{0\})^{\mathbb{N}} | \sigma(j) \in v_j \leq N, \sigma(j) = 0\, j > N \right\}. \tag{10}$$

We define $\#\sigma = \max\{j \in \mathbb{N} | \sigma(j) \neq 0\}$, i.e. $\#\sigma$ is the number of modes in the sequence. Second, let an element of the transition time sequence space be a sequence whose elements correspond to the amount of time spent in each discrete mode:

$$\mathcal{S} = \bigcup_{N=1}^{\infty} \{ s \in l^1 | s(j) \geq 0\, \forall_j \leq N,\ s(j) = 0\, \forall_j > N \}, \tag{11}$$

where l^1 denotes the space of absolutely summable sequences. Third, we define the continuous input space, $\mathcal{U}$:

$$\mathcal{U} = \{ u \in L^2([0,\infty), \mathbb{R}^m) | u(t) \in U,\ \forall t \in [0,\infty) \}. \tag{12}$$

Finally, let the objective mapping space, $\mathbb{N}^W$, be the set of W–tuples with elements in the natural numbers, where W is equal to the number of objectives in our problem, which we soon define. Elements in this space define a mapping between each of our objectives and an element of our discrete mode sequence space. These objectives as we show below are useful in ensuring that our algorithm "sees" the changes in contact point enforcement. We combine these four spaces together to define our optimization space, $\mathcal{X}$, as follows:

$$\mathcal{X} = \{ (\sigma,\ s,\ u,\ w) \in \Sigma \times \mathcal{S} \times \mathcal{U} \times \mathbb{N}^W | s(k) = 0\, \forall k > \#\sigma, \text{ and } \omega(i) \leq \#\sigma\, \forall i \},$$

and we denote $\xi \in \mathcal{X}$ by a 4–tuple $\xi(\sigma,\ s,\ u,\ \omega)$. Maintaining the notion of absolute times, which we call the *jump time sequence* $.\mu : \mathbb{N} \times \mathcal{X} \to [0,\infty)$. is also useful:

$$\mu(i;\xi) = \begin{cases} 0 & \text{if } i = 0 \\ \sum_{k=1}^{i} s(k) & \text{if } i \neq 0. \end{cases} \tag{13}$$

Let $\mu_f(\xi) = \|s\|_{l^1} = \sum_{k=1}^{\infty} s(k)$. We also define, $\pi : [0,\infty) \times \mathcal{X} \to (V \cup \{0\})$ to return the mode corresponding to an absolute time t:

$$\pi(t;\xi) = \begin{cases} \sigma(\min\{i|\mu(i;\xi) > t\}) & \text{if } t < \mu_f(\xi) \\ \sigma(\#\sigma) & \text{if } t = \mu_f(\xi) \\ 0 & \text{if } t > \mu_f(\xi) \end{cases} \tag{14}$$

We suppress the dependence on ξ in μ, π, and μ_f whenever the choice of ξ is clear in context. For notational convenience, we write $\mu(i)$ for μ_i.

Given $\xi \in \mathcal{X}$ and $x_0 \in M$, the corresponding trajectory, $x^{(\xi)}$, is the solution to:

$$\dot{x}(t) = f_{\pi}(t)(\Delta_{(\pi(t-),\pi(t+))}(x(t)), u(t)), \quad \forall t \geq 0,\ x(0) = x_0, \tag{15}$$

where Δ is the reset map (when $\pi(t^-) = \pi(t^+)$ Δ is assumed to be the identity) and where we have suppressed the dependence on x_0 in $x^{(\xi)}$. Let the cost function $J : \mathcal{X} \to \mathbb{R}$ for our optimization problem be defined as:

$$J(\xi) = \int_0^{u_f} L(x^{(\xi)}(t), u(t), t)dt + \sum_{i=1}^{W} \phi_i(x^{(\xi)}(\mu_{\omega(i)}, \mu_{\omega(i)})). \tag{16}$$

Given $\xi \in \mathcal{X}$ and a finite set of functions, $r_j : M \to \mathbb{R}, j \in \mathcal{J}$ we constrain the state by demanding the state satisfy $r_j(x^{(\xi)}(t)) \leq 0$ for each $t \in [0, \mu_f]$ and each $j \in \mathcal{J}$. Moreover given $\xi \in \mathcal{X}$ and a finite set of functions, $g_k : \mathcal{S} \to \mathbb{R}$, $k \in: \mathcal{K}$ we constrain the transition times by requiring they satisfy $g_k(s) \leq 0$ for each $k \in \mathcal{K}$. We compactly describe all the constraints by defining a new function ψ:

$$\psi(\xi) = \max\left(\max_{j\in\mathcal{J}} \max_{t\in[0,\mu_f]} r_j(x^{(\xi)}(t)), \max_{k\in\mathcal{K}} g_k(s) \right). \tag{17}$$

With these definitions, we can state our problem.

Multiple Objective Switched System Optimal Control Problem

$$\begin{aligned} &\min_{\xi\in\mathcal{X}} J(\xi) \\ \text{s.t.}\quad &\psi(\xi) \leq 0 \end{aligned} \tag{18}$$

Since we are interested in the construction of an algorithm that provably converges to minima of our problem, we make the following assumptions on the dynamics, cost, and constraints:

Assumption 1 The functions L and f_v are Lipschitz and differentiable in x and u for all $v \in V$. In addition, the derivatives of these functions with respect to x and u are also Lipschitz.

Assumption 2 The functions ϕ_i, r_j, *and* g_k are Lipschitz and differentiable in their argument for all $i \in \{1, \ldots, W\}, j \in \mathcal{J}$ and $k \in \mathcal{K}$. In addition, the derivatives of these functions with respect to their arguments are also Lipschitz.

Assumption 1, together with the controls being measurable and uniformly bounded functions, is sufficient to ensure the existence, uniqueness, and boundedness of the solution to our differential Eq. (15). Assumption 2 is a standard assumption on the objectives and constraints and is used to prove the convergence properties of the algorithm defined in the next section.

4 Algorithm Design

In this section, we present an optimization algorithm to determine a numerical solution to our problem. Since the focus of this paper is the application to bipedal walking, we focus on presenting a high-level description of the algorithm, but a more explicit description of the various pieces of the algorithm and a proof of its convergence can be found in [9, 10].

Given $\xi \in \mathcal{X}$, the algorithm works by employing a *variation*, ρ, that inserts a mode $\hat{\alpha}$ and control $\hat{u}$ at time $\hat{t}$ into ξ for a length of time determined by the argument to the variation. The algorithm stops when this variation does not produce either a reduction in the cost or infeasibility. Since the initialization of an optimization algorithm can be non-trivial, we construct an algorithm known in the optimization literature as a *Phase I/Phase II* algorithm [17]. A *Phase I* algorithm takes any point in the optimization space and finds a feasible point, and a *Phase II* algorithm finds a local minimum given a feasible point as an initial condition. Our algorithm divides the problem into two nonlinear constrained optimization problems:

Bi-Level Optimization Scheme

Stage 1: Given a fixed discrete mode sequence, employ a *Phase I/Phase II* algorithm to find either a locally optimal transition time sequence and continuous control or a locally optimal infeasible transition time sequence and continuous control.

Stage 2: Given a transition time sequence and continuous control, employ the variation, ρ, to modify the discrete mode sequence to find either a lower cost discrete mode sequence if the initialization point is feasible, or find a less infeasible discrete mode sequence if the initialization point is infeasible. Repeat Stage 1 using the modified discrete mode sequence.

In order to formalize this description, notice first that Stage 1 can be transformed into a classical optimal control problem over the switching instances and continuous control by employing the time-free transformation (Sect. 5 of [10] describes this transformation). Let $\hat{a} : \mathcal{S} \times \mathcal{U} \to \mathcal{S} \times \mathcal{U}$ be a function that solves Stage 1. To formalize Stage 2, we begin by defining the variation, ρ. Given $\xi \in \mathcal{X}$, consider an insertion of a mode, $\hat{a}$ and control, $\hat{u}$, at time $\hat{t}$. This insertion is characterized by $\eta = (\hat{a}, \hat{t}, \hat{u}) \in \mathcal{H}_\xi \equiv \mathcal{Q} \times \mathcal{T}_\xi \times U$ where $\mathcal{T}_\xi = [0, \mu_f]$. Given $\xi \in \mathcal{X}$, and $\eta \in \mathcal{H}_\xi$, we let $\rho : [0, \infty) \to \mathcal{X}$ denoted $\rho^{(\eta)}(\lambda)$ describe this type of insertion (ρ is defined

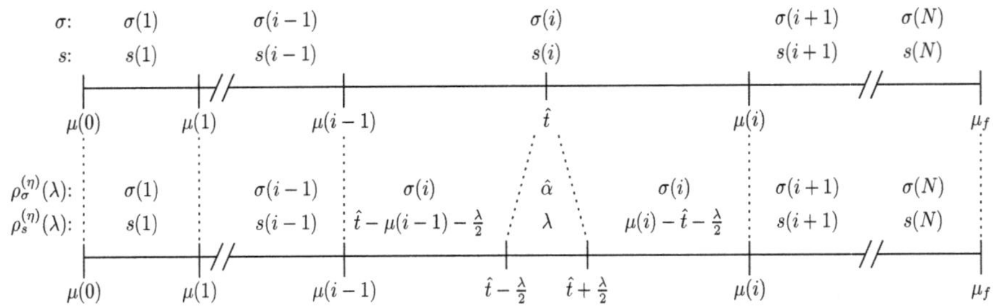

Fig. 1 Diagram illustrating the transition from σ to $\rho_\sigma^{(\eta)}(\lambda)$ and s to $\rho_\sigma^{(\eta)}(\lambda)$. The *top line* is the original σ and s, and the *bottom line* shows the result for $\lambda > 0$

explicitly in Definition 2 in [10], its argument denotes the duration of the insertion, and we have suppressed the dependence on ξ). Figure 1 illustrates a pair (σ, s) after they are modified by the function $\rho^{(\eta)}$.

We employ the variation, ρ, to characterize when an optimal point has been reached. Observe that after Stage 1 the only way to locally reduce the cost or infeasibility is by modifying the discrete mode sequence. Given this procedure, a point $\xi = (\sigma, s, u, w) \in \mathcal{X}$ satisfies our *optimality condition* if (s, u) is a locally optimal solution to Stage 1 and if the best modification of the discrete mode sequence via the variation, ρ, does not produce a decrease in the cost, J, whenever the point ξ is feasible, or does not produce a decrease in the constraint, ψ, whenever the point ξ is infeasible.

Given any function $F : \mathcal{X} \to \mathbb{R}^k$ for some $k \in \mathbb{N}$, let us define the directional derivative of F composed with ρ for $\lambda > 0$ by:

$$D^{(\xi,\eta)}F = \lim_{\lambda \downarrow 0} \frac{1}{\lambda}\left[F\left(\rho^{(\eta)}(\lambda)\right) - F(\xi)\right] \tag{19}$$

In order to check if we have arrived at a point that satisfies the optimality condition, we first study the effect of the variation, ρ, on the cost, J, using the first order approximation of its change due to the insertion, $D^{(\xi,\eta)}J$. If this derivative is negative, then one can argue that it is possible to decrease the cost via the variation. Second, consider the first order approximation of our constraint, ψ, with respect to ρ, denoted by $D^{(\xi,\eta)}\psi$, to determine if the infeasibility decreases due to the insertion. Again if this derivative is negative, then it is possible to decrease the infeasibility via the variation. Using these results, we define an *optimality function*, $\theta : \mathcal{X} \to (-\infty, 0]$:

$$\theta(\xi) = \min_{\eta \in \mathcal{H}_\xi} \varsigma(\xi, \eta), \varsigma(\xi, \eta) = \begin{cases} \max\left\{D^{(\xi,\eta)}J, D^{(\xi,\eta)}\psi + \gamma_1\psi(\xi)\right\} & \text{if } \psi(\xi) \geq 0, \\ \max\left\{D^{(\xi,\eta)}J - \gamma_2\psi(\xi), D^{(\xi,\eta)}\psi\right\} & \text{if } \psi(\xi) > 0, \end{cases} \tag{20}$$

where $\gamma_1, \gamma_2 > 0$ are design parameters. Note that $\theta(\xi) \leq 0$ for each $\xi \in \mathcal{X}$, since given a value ξ we can always perform an insertion that leaves the trajectory unmodified, e.g. given $\xi \in \mathcal{X}$ and $\hat{t} \in \mathcal{T}_\xi$ choose $\eta = (\pi(\hat{t}), \hat{t}, u(\hat{t}))$, hence in that case $D^{(\xi,\eta)}J = D^{(\xi,\eta)}\psi = 0$.

To appreciate the utility of our optimality function consider three cases. First, if at a feasible point, $\psi(\xi) \leq 0$, and if $\theta(\xi) < 0$, then a mode insertion which reduces the cost while remaining feasible is possible. Second, if at an infeasible point, $\psi(\xi) > 0$, and if $\theta(\xi) < 0$, then a mode insertion which reduces the infeasibility without resulting in too large an increase in the cost is possible. Third, if we are at a feasible point and the cost cannot be decreased using the variation ρ, or if we are at an infeasible point and the infeasibility cannot be decreased using the variation ρ, then $\theta(\xi) = 0$. Therefore, the optimality function can serve as a stopping criterion for the Bi-Level Optimization Scheme since its zeros encode points that satisfy our optimality condition. Note that the γ_1, γ_2 terms in the optimality function capture the possibility that the reduction in cost or constraint may result in too large an increase in the infeasibility or cost, and therefore maybe undesirable (Sect. 2.5 of [17] describes this utility).

An insertion that reduces the cost or infeasibility tells us nothing about the length of time to actually perform an insertion for. Comparing our algorithm to a simple finite dimensional optimization algorithm, knowing that an insertion is plausible would compare to knowing that the gradient of the cost function was negative. During finite dimensional optimization, a step size in the direction of the gradient is chosen by performing a linear search. We can apply an identical algorithm in this instance. Given $\alpha, \beta \in (0, 1)$, let the maximum insertion length be:

$$\lambda^{(\xi,\eta)} = \max_{k \in \mathbb{N}} \left\{ \beta^k \middle| \psi(\rho^{(\eta)}(\beta^k)) \leq 0, J(\rho^{(\eta)}(\beta^k)) - J(\xi) \leq \alpha \beta^k \varsigma(\xi, \eta) \right\} \tag{21}$$

whenever $\psi(\xi) \leq 0$, and otherwise let:

Algorithm 1 Optimal Control Algorithm

Data: $\xi_0 = (\sigma_0, s_0, u_0, w_0) \in \mathcal{X}$, $\alpha, \beta \in (0, 1)$, $\gamma_1, \gamma_2 > 0$.
Step 0. Let $(s_1, u_1) = \hat{a}(s_0, u_0)$, $\xi_1 = (\sigma_0, s_1, u_1, w_0)$, and set $j = 1$.
Step 1. If $\theta(\xi_j) = 0$ then stop and return ξ_j.
Step 2. $\xi_{j+1} = a(\xi_j)$, where a is defined as follows:

a. Let $\hat{\eta} = (\hat{\alpha}, \hat{t}, \hat{u}) \in \mathcal{H}_{\xi_j}$ be any point s.t. $\zeta(\xi_j, \hat{\eta}) < 0$, and let $(\tilde{\sigma}_j, \tilde{s}_j, \tilde{u}_j, \tilde{w}_j) = \rho(\lambda^{(\xi,\hat{\eta})}; \xi, \hat{\eta})$.
b. Given $\tilde{\sigma}_j$, let $(s_{j+1}, u_{j+1}) = \hat{a}(\tilde{s}_j, \tilde{u}_j)$.
c. Let $\xi_{j+1} = a(\xi_j) = (\tilde{\sigma}_j, s_{j+1}, u_{j+1}, \tilde{w}_j)$.

Step 3. Replace j by $j + 1$ and go to Step 1.

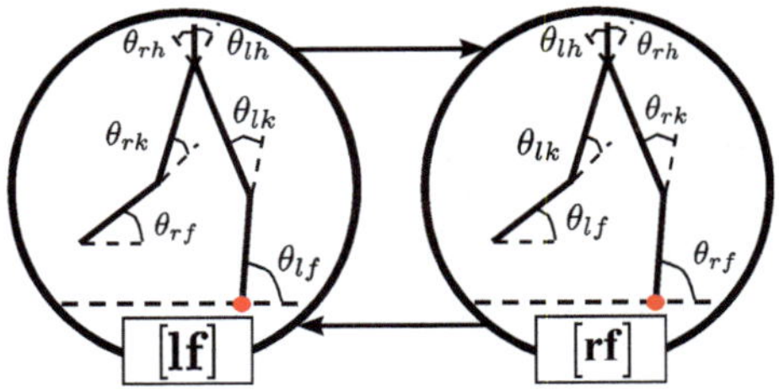

Fig. 2 The 2-mode hybrid system employed in this section. The *lf* (or *rf*) mode has a control vector field where the *lf* (or *rf*) is constrained

$$\lambda^{(\xi,\eta)} = \max_{k \in \mathbb{N}} \left\{ \beta^k \middle| \psi(\rho^{(\eta)}(\beta^k)) - \psi(\xi) \leq \alpha \beta^k \varsigma(\xi,\eta) \right\}. \tag{22}$$

Algorithm 1 describes our numerical method to solve the Multiple Objective Hybrid Optimal Control Problem. Comparing steps of Algorithm 1 with our Bi-Level Optimization Scheme notice that Step (2b) encodes Stage 1 and Step (2a) encodes Stage 2. Due to limited space and since the focus of this paper is the application of this algorithm to bipedal walking, we do not include a proof of the convergence, a derivation of the pieces, or a description of the implementation of our algorithm but these can all be found in [9, 10].

5 Application

In this section, we illustrate the application of our optimal control scheme to the identification of a robotic biped model. We focus our attention on the identification of a model considered in the literature rather than human data in order to gauge the performance of our algorithm.

The model used to generate this ground truth data is a 2D biped with knees and a torso for a total of six links. Figure 2 illustrates this model. Observe that it is a two-mode hybrid system. The two modes denoted *rf* and *lf* arise due to the enforcement of a foot contact point and the model presumes that only one contact point is enforced at any instant in time. The configuration space has coordinates $\Theta = (\theta_{rf}, \theta_{rk}, \theta_{rh}, \theta_{lf}, \theta_{lk}, \theta_{lh})$ (we index these coordinates by $i = 1, \ldots, 6$ for convenience), which means the domain of each mode has coordinates $(\Theta, \dot{\Theta})$ and is a 12-dimensional manifold. Due to space constraints, we do not include the hybrid system description here explicitly, but this information can be found online. We generate a sample stable walking trajectory from this model by employing numerical integration and the techniques described in [21]. The ground truth mode sequence and amount of time spent in each mode can be found in the second row of Table 1. Figure 3 illustrates the sample trajectory that we employ for identification.

To apply our algorithm, we assume that there are two contact points and that only one is enforced at any instant in time. In doing so we arrive at a switched system with two vector fields: *rf* and *lf*. In practice we know the initial condition,

Table 1 The results of our switched optimal control scheme using the model described in Fig. 2 and the sample trajectories drawn in Fig. 3

σ_0	ω_0	s_0	σ_f	ω_f	s_f	Computation time
(*lf*, *rf*)	N/A	(0.445.445)	N/A	N/A	N/A	N/A
(*lf*, *rf*)	1	(0.445, 0.445)	(*lf*, *rf*)	1	(0.445, 0.445)	201 (s)
(*lf*, *rf*)	1	(0, 0.889)	(*lf*, *rf*)	1	(0.445, 0.445)	222 (s)
(*lf*, *rf*)	1	(0.889, 0)	(*lf*, *rf*)	1	(0.445, 0.445)	239 (s)
(*rf*, *rf*)	1	(0.445, 0.445)	(*rf*, *lf*, *rf*, *rf*)	3	(0.01, 0.431, 0, 0.448)	537 (s)
(*lf*, *lf*)	1	(0.445, 0.445)	(*lf*, *lf*, *rf*)	1	(0.424, 0.02, 0.445)	394 (s)
(*rf*, *lf*)	1	(0.445, 0.445)	(*rf*, *lf*, *rf*, *lf*)	3	(0.04, 0.463, 0, 0.406)	568 (s)

The ground truth data is in the second row, the 0 subscript is the initialization, the *f* subscript is the result of our algorithm, and the computation time was determined on an AMD Opteron, 8 core, 2.2 GHz, 16 GB RAM machine

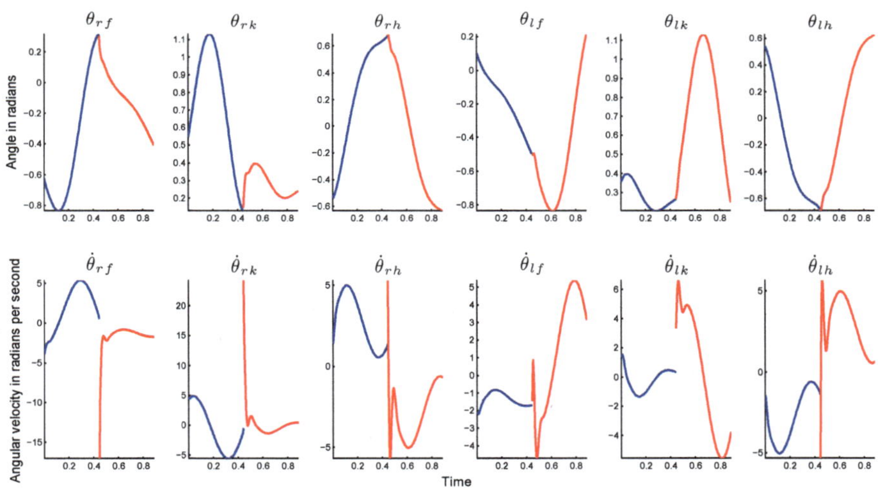

Fig. 3 An example stable trajectory for the hybrid system in Fig. 2 generated using the techniques described in [21]. When the *lf* (or *rf*) mode is active the trajectory is drawn in *blue* (or *red*)

i.e. x_0, of the human data that we wish to track; therefore, we presume our optimal control scheme is also aware of this initial condition.

Denoting the observed trajectory by $(\widehat{\Theta}(t), \dot{\widehat{\Theta}}(t))$, we set the running cost equal to:

$$L(\Theta(t),\dot{\Theta}(t),u(t),t) = 0.001 \int_0^{u_f} \sum_{i=1}^{6} \left(\left\| \dot{\theta}_i(t) - \dot{\hat{\theta}}_i(t) \right\|_2^2 \right) \mathrm{d}t. \tag{23}$$

Observe from the sample trajectories that the transition between different modes is detectable due to the change in velocity due to impact. Since these transition points are particularly important, we include objectives at these transition times $\hat{t}$ equal to:

$$\phi(\Theta(t),\dot{\Theta}(t),t) = 0.05\sum_{i=1}^{6}\left(\left\|\dot{\theta}_i(t) - \dot{\hat{\theta}}_i(\hat{t})\right\|_2^2 + 5(t-\hat{t})^2\right). \tag{24}$$

We constrain $\Theta \in [-\pi, \pi]^6$, $\dot{\Theta} \in [-30, 30]^6$, and the total time to be equal to the length of the sample trajectory. Notice that our dynamics, running cost, objectives, and constraints satisfy the assumptions required in order to ensure the convergence of our optimal control scheme.

To evaluate the robustness of our approach, we consider the behavior of our scheme under a variety of modal sequence initializations. Since it is clear from the data that there are two steps, we focus on two element mode sequence initializations. The result of our algorithm under these different initializations are described in Table 1. Notice that regardless of the modal and transition time sequences chosen for initialization the algorithm nearly converges to the ground truth modal and transition time sequences.

6 Conclusion

This paper presents the first steps toward automatically identifying a hybrid dynamical model for robotic walking from human data. To achieve this goal, the first half of the paper presented a means to view human walking data in a way that allowed for the immediate extraction of a robotic bipedal walking model capable of human-like gait. Unfortunately extracting this model required solving a combinatorial problem. The second half of this paper addresses this shortcoming by reformulating the combinatorial problem into a switched optimal control problem and presenting an algorithm to solve this problem. To illustrate the validity of this approach, our paper presented an example where in place of the human data, simulated data was employed.

While the application of this paper was limited to a simple bipedal model, our result provides a means to automatically extract a model capable of generating task dependent anthropomorphic periodic motion. That is, if we are given data of a person walking upstairs or a person running we are able to construct a mathematical model of this periodic motion that is employable by a robotic biped. In fact, our method in this instance generates a task-driven mathematical model of periodic motion comparable to the notion of a central pattern generator [8]. As such, this result could have important ramifications to our understanding of both robotic and human periodic motion particularly for prosthetic design and rehabilitation. Moreover, several recent papers describe more efficient implementations of switched system optimal control which have consistently found better quality minimizers across a wide variety of benchmarks [27, 28].

Acknowledgements I would like to thank Sam Burden for the many conversations on system identification, Humberto Gonzalez and Maryam Kamgarpour for their help with the development of the switched system optimal control algorithm, Ryan Sinnet for his help with robotic model considered in the Application Section. I am also grateful to Aaron Ames for the many insightful discussions on bipedal walking and system identification and Ruzena Bajcsy for her willingness to try to solve hard problems. This research is supported in part by NSF Awards CCR-0325274, CNS-0953823, ECCS-0931437, IIS-0703787, IIS-0724681, IIS-0840399 and NHARP Award 000512-0184-2009.

References

1. R. Ambrose, H. Aldridge, R. Askew, R. Burridge, W. Bluethmann, M. Diftler, C. Lovchik, D. Magruder, F. Rehnmark, Robonaut: Nasa's space humanoid. IEEE Intell. Syst. Appl. **15**(4), 57–63 (2000)
2. A. Ames, R. Vasudevan, R. Bajcsy, Human-data based cost of bipedal robotic walking, in *Proceedings of the 14th International Conference on Hybrid Systems: Computation and Control* (ACM, 2011), pp. 153–162
3. S. Au, P. Dilworth, H. Herr, An ankle-foot emulation system for the study of human walking biomechanics, in *Proceedings 2006 IEEE International Conference on Robotics and Automation, 2006, ICRA 2006* (IEEE, 2006), pp. 2939–2945
4. G. Bergmann, G. Deuretzbacher, M. Heller, F. Graichen, A. Rohlmann, J. Strauss, G. Duda, Hip contact forces and gait patterns from routine activities. J. Biomech. **34**(7), 859–871 (2001)
5. D.J. Braun, M. Goldfarb, A control approach for actuated dynamic walking in bipedal robots. IEEE Trans. Rob. **25**, 1–12 (2009)
6. J.H. Choi, J.W. Grizzle, Planar bipedal walking with foot rotation. pp. 4909–4916, Portland, OR (2005)
7. S.H. Collins, M. Wisse, A. Ruina, A 3-d passive dynamic walking robot with two legs and knees. Int. J. Robot. Res. **20**, 607–615 (2001)
8. J. Duysens, H. Van de Crommert, Neural control of locomotion; part 1: the central pattern generator from cats to humans. Gait & Posture **7**(2), 131–141 (1998)
9. H. Gonzalez, R. Vasudevan, M. Kamgarpour, S. Sastry, R. Bajcsy, C. Tomlin, A numerical method for the optimal control of switched systems, in *2010 49th IEEE Conference on Decision and Control (CDC)* (IEEE, 2010), pp. 7519–7526
10. H. Gonzalez, R. Vasudevan, M. Kamgarpour, S. Sastry, R. Bajcsy, C. Tomlin, A descent algorithm for the optimal control of constrained nonlinear switched dynamical systems, in *Proceedings of the 13th ACM International Conference on Hybrid Systems: Computation and Control* (ACM, 2010), pp. 51–60
11. A. Goswami, B. Thuilot, B. Espiau, Compass-like biped robot part I : Stability and bifurcation of passive gaits. Rapport de recherche de l'INRIA (1996)
12. J.W. Grizzle, G. Abba, F. Plestan, Asymptotically stable walking for biped robots: analysis via systems with impulse effects. IEEE Autom Control **46**, 51–64 (2001)
13. J.W. Grizzle, C. Chevallereau, A.D. Ames, R.W. Sinnet, 3d bipedal robotic walking: models, feedback control, and open problems, in *NOLCOS*, Bologna, Italy (2010)
14. T. McGeer, Passive walking with knees, in *IEEE International Conference on Robotics and Automation*, Cincinnati, OH (1990)
15. R.M. Murray, Z. Li, S.S. Sastry, *A Mathematical Introduction to Robotic Manipulation*, Boca Raton, FL (1993)

16. J. Perry, J. Burnfield, *Gait Analysis: Normal and Pathological Function*, SLACK (2010)
17. E. Polak, *Optimization: Algorithms and Consistent Approximations* (Springer, 1997)
18. M. Rodgers, Dynamic biomechanics of the normal foot and ankle during walking and running. Phys. Ther. **68**(12), 1822 (1988)
19. T. Schaub, M. Scheint, M. Sobotka, W. Seiberl, M. Buss, Effects of compliant ankles on bipedal locomotion, in *IROS*, St. Louis, Missouri, USA (2009)
20. A. Seireg, R. Arvikar, The prediction of muscular load sharing and joint forces in the lower extremities during walking. J. Biomech. **8**(2), 89–102 (1975)
21. R. Sinnet, M. Powell, R. Shah, A. Ames, A human-inspired hybrid control approach to bipedal robotic walking, in *International Federation on Automatic Control* (2011)
22. R.W. Sinnet, A.D. Ames, 2D bipedal walking with knees and feet: a hybrid control approach, in *48th IEEE Conference on Decision and Control*, Shanghai, P.R. China (2009)
23. S. Srinivasan, E. Westervelt, A. Hansen, A low-dimensional sagittal-plane forward dynamic model for asymmetric gait and its application to study the gait of transtibial prosthesis users. J. Biomech. Eng. **131**, 031003 (2009)
24. R. Tedrake, T. Zhang, H. Seung, Learning to walk in 20 minutes, in *Proceedings of the Fourteenth Yale Workshop on Adaptive and Learning Systems*, New Haven, Connecticut, USA (2005)
25. D. Tlalolini, C. Chevallereau, Y. Aoustin, Comparison of different gaits with rotation of the feet for planar biped. Robot. Auton. Syst. **57**, 371–383 (2009)
26. R. Vasudevan, A. Ames, R. Bajcsy, Persistent homology for automatic determination of human-data based cost of bipedal walking, in *Nonlinear Analysis: Hybrid Systems* (2013), pp. 101–115
27. R. Vasudevan, H. Gonzalez, R. Bajcsy, S. S. Sastry, Consistent approximations for the optimal control of constrained switched systems—part 1: A conceptual algorithm. SIAM J. Control Optim. **51**(6), 4463–4483 (2013)
28. R. Vasudevan, H. Gonzalez, R. Bajcsy, S. S. Sastry, Consistent approximations for the optimal control of constrained switched systems—Part 2: An implementable algorithm. SIAM J. Control Optim. **51**(6), 4484–4503 (2013)
29. E.R. Westervelt, J.W. Grizzle, C. Chevallereau, J. Choi, B. Morris, *Feedback Control of Dynamic Bipedal Robot Locomotion*. Control and Automation. Boca Raton, FL (2007)
30. J. Yang, D. Winter, R. Wells, Postural dynamics of walking in humans. Biol. Cybern. **62**(4), 321–330 (1990)

MIX
Papier aus verantwortungsvollen Quellen
Paper from responsible sources
FSC® C105338

If you have any concerns about our products,
you can contact us on
ProductSafety@springernature.com

In case Publisher is established outside the EU,
the EU authorized representative is:
Springer Nature Customer Service Center GmbH
Europaplatz 3, 69115 Heidelberg, Germany

Printed by Libri Plureos GmbH
in Hamburg, Germany